Advances in Cryogenic Engineering - Materials: Proceedings of the International Cryogenic Materials Conference (ICMC 2021)

IOP Conference Series: Materials Science and Engineering
Volume 1241

Online
19 - 23 July 2021

Editors:

U. (Balu) Balachandran
David Cardwell
Pashupati Dhakal
David Evans
Timothy Haugan
Eric Hellstrom
Jianyi Jiang
Peter Lee
Sastry Pamidi
Tatiana Prikhna
Wolfgang Stautner
Chiara Tarantini
Robert Walsh

ISBN: 978-1-7138-5809-6
ISSN: 1757-8981

Printed from e-media with permission by:

Curran Associates, Inc.
57 Morehouse Lane
Red Hook, NY 12571

Some format issues inherent in the e-media version may also appear in this print version.

Additional copies of this publication are available from:

Curran Associates, Inc.
57 Morehouse Lane
Red Hook, NY 12571 USA
Phone: 845-758-0400
Fax: 845-758-2633
Email: curran@proceedings.com
Web: www.proceedings.com

TABLE OF CONTENTS

IOP Conf. Series: Materials Science and Engineering 1241 (2022) 011001 doi:10.1088/1757-899X/1241/1/011001

Preface

The 2021 Joint Cryogenic Engineering Conference (CEC) and International Cryogenic Materials Conference (ICMC) events were held from July 19 through July 23 virtually, using the *Whova* platform. As at past conferences, the international scope of these meetings was strongly maintained with 23 countries being represented by 641 attendees who gathered virtually to enjoy the joint technical programs and industrial exhibits. In total, 224 papers were submitted for publication of which 215 are published in these conference proceedings.

The program for the joint conferences included a total of 429 presentations organized into 69 sessions - plenary, oral, poster and awards. Four plenary talks gave interesting in-depth updates and overviews on exciting topics with titles: i) *"The Promise of Superconducting Quantum Information Processing"*, ii) *"The Mars Oxygen In-Situ Resource Utilization Experiment (MOXIE)"*, iii) Airbus SAS talk, *"ASCEND - A First Step Towards Cryogenic Electric Propulsion for Aircraft?"*, and iv)) *"ITER Cryogenic Systems – Scale, Complexity, and Innovation"*. Virtual attendance for each of the plenary talks was excellent from ~ 140 to 280. The attendees also convened for three special Joint CEC-ICMC Symposia organized: *i) Hydrogen Technologies for Transportation, ii) Superconducting Quantum Systems,* and *iii) SRF Materials and Systems*. Twelve additional curated Focus Sessions were included in the programming, emphasizing high-quality invited talks. And this included 4 topics new to the conference: i) *Topological Materials for Electronics*, ii) *3D Printing Materials,* iii) *Mechanical Properties of HTS Wires and Cables*, and iv) *LTS and HTS Cables for Fusion*. In addition to the Joint Session *Hydrogen Technologies for Transportation* with 4 Sessions and 9 talks, an extra Focus Session on *Transportation* was held again this year consisting of 6 Sessions and 35 talks. Contributed papers covered a wide range of topics including many aspects and advances in cryogenics and superconductors, along with their applications.

Both CEC and ICMC boards are encouraging student participation, education, and career growth. It was very exciting and rewarding to have 170 students register for the conference this year, which is ~ 2x higher than any other conference held in the last 20 years. And 32 of the students received sponsorship of the registration fees. Virtual attendance of sessions also was very strong, with median average ~ 90 for the 69 sessions held, including ~ 200 for the 4 plenary sessions (*Whova* Statistics). Presenters had the option of presenting live or using pre-recorded talks. Attendees also had the option of watching live or watching prerecorded talks online at the global time of their choosing, and for 3 months after the conference. So, the attendance of talks gradually increased as time progressed, even after the conference dates. The strong session attendance was exciting and very positive and made the conference a more rewarding experience for the presenters and attendees.

The CEC/ICMC'21 virtual exhibit displayed products and services of 20 organizations, all of which are intimately involved in the support and advancement of cryogenics. *Spatial Chat* (a networking platform) was added to provide an interesting and interactive environment for the Cryo Expo attendees. This proved to be a significant value-add for the participating companies. Exhibitors also had the ability to add pictures, videos and conduct live product showcases to interact with attendees. The exhibit booths averaged 87 views each by attendees. Overall, the Cryo Expo experience received strong positive reviews from the exhibiting companies.

The Conference is very grateful for the support it received from its two main Conference Sponsors, *Scientific Instruments, Inc.* and *Sumitomo (SHI) Cryogenics of America, Inc.*, as well as its Conference Supporters, *Cryomech, Inc., Fermi National Accelerator Laboratory, FormFactor Inc., Linde Engineering*, and the *National High Magnetic Field Laboratory*.

Conference Chairs for 2021 were Peter Bradley from NIST, for CEC and Timothy Haugan from U.S. Air Force Research Laboratory for ICMC. Program Chairs were Sastry Pamidi from the Center for Advanced Power Systems for CEC and Shreyas Balachandran from the National High Magnetic Field Laboratory at Florida State University and Tengming Shen from Lawrence Berkeley National Laboratory for ICMC, assisted by the CEC Program Vice Chairs, Chul Kim from Florida State University and Peter Kittel, Consultant. The Exhibit Chair was Austin Capers of Scientific Instruments Inc. who also served as Publicity Chair. Finally, Awards Chairs were Peter Kittel, Consultant, for CEC and Eric Hellstrom from the National High Magnetic Field Laboratory at Florida State University for ICMC. Mike Sumption, ICMC Board president and Tom Nicol, CEC Board president greatly participated in the conference development as part of the leadership team.

ICMC held three short courses virtually on three separate days of July 11, July 17 and July 18 to enable better attendance, and were well attended. On July 11, Peter J. Lee and Tengming Shen presented the course *High-field Superconducting Materials & Conductors*, which was well attended with 27 scientists and students and was well received and appreciated. On July 17, Zlatko Minev and Matthew Hollister presented the course *Quantum Information Science*, which was fully attended with 44 scientists and students. On July 18, Robert Walsh, Ignacio Aviles Santillana and Timothy Horn presented the course *Properties of Structural Materials and Introduction to Additive Manufacturing for Cryogenic Applications*, which was fully attended with 46 scientists and students. The ICMC organizers were very pleased by the strong attendance, which catered to both seasoned professionals and newcomers to cryogenics in engineering and materials.

Also on July 18, the Cryogenic Society of America presented three short courses: *Theory, Modeling and Design of Regenerative Cryocoolers*, a full day course given by R. Radebaugh; *Aspects of Cryostat Design*, a half-day course given by J. Weisend II; *Getting Started with Cool Fuels: Liquefied hydrogen and natural gas*, a half-day course given by J. Leachman. The courses were well attended and catered to both seasoned professionals and newcomers to cryogenics in engineering and materials.

The able leadership of Paula Pair, Annett Cady, Carrie Lian and all the hardworking, enthusiastic, and efficient staff from Centennial Conferences provided outstanding conference management and operations. Thank you, Centennial Conferences, for a truly outstanding and memorable conference experience!

The conference and proceedings were the result of significant volunteer efforts from many people, especially the conference attendees, Centennial Conferences, Focus session chairs, session chairs, reviewers, and editors, who all contributed greatly to the success of the conference. In the six-month run-up leadership from CEC, ICMC, and Centennial met weekly (virtually) to navigate the changing landscape for if-and-how to hold an in-person meeting in Hawai'i or pivot to a virtual on-line meeting. The ever-changing world landscape brought about from Covid and desire to pursue holding the meeting resulted in a compressed timeline for organizing the conference. Encouraged by evolving virtual technology, the leadership pivoted to an on-line virtual format, one that the conference had not done before, that in turn brought numerous challenges. One of the more pressing was how to give attendees the best opportunity to participate from their respective time zones. Productive effort led by Centennial staff and CEC, ICMC board members to carefully evaluate and down-select virtual software systems, Whova and Spatial Chat enabled a very successful conference that encouraged collegial interaction amongst attendees to the greatest extent possible. We extend a special thanks to all who contributed to the successful conference and the proceedings.

We very much look forward to the next CEC/ICMC in Honolulu, Hawai'i, July 9-13, 2023. Hawai'i will offer attendees one of the most beautiful settings in the U.S., with paradise everywhere to view and experience. The conference will be held at the Hawai'i Convention Center, which is close to

IOP Conf. Series: Materials Science and Engineering 1241 (2022) 011001 doi:10.1088/1757-899X/1241/1/011001

the Waikiki Beach strip and hotels, and only 5-10 minutes walking distance to ocean swimming, beachside parks and jogging, and the Ala Moana conference hotel. Many fast service restaurants and food courts are within a 5–10-minute walk, as well as the Ala Moana Mall which is the largest open-air shopping center in the world with almost 350 stores. The local food cuisine reflects the mix of many cultures in Hawai'i, including Hawai'i, Polynesia, Japan, Korea, Philippines, Puerto Rico, Portugal, China, American, and Fusion mixes. The Convention Center will provide comfortably large exhibit/poster and oral presentation rooms.

In addition to the rich spectrum of R&D and industrial cryogenic topics making up the conference, we hope you will enjoy the comforts and attractions of Hawai'i when you are attending. We hope you are able and plan to join us, and we extend you a very warm welcome and Aloha to Honolulu!

Peter Bradley
2021 CEC Conference Chairman

Timothy Haugan
2021 ICMC Conference Chairman

IOP Conf. Series: Materials Science and Engineering 1241 (2022) 011002 doi:10.1088/1757-899X/1241/1/011002

ICMC Awards

ICMC LIFETIME ACHIEVEMENT AWARD 2021

Professor Shi Xue Dou

University of Wollongong
Australia

ICMC has presented its Award for Lifetime Achievement, typically biannually, at the CEC/ICMC conference since its 30th anniversary conference in Keystone in 2005 to an individual to "recognize a lifetime's achievement in advancing the knowledge of cryogenic materials".

The tenth recipient of the ICMC Lifetime Achievement Award is Shi Xue Dou, who follows Edward W. Collings in 2005, David C. Larbalestier in 2007, Kyoji Tachikawa in 2009, René Flükiger in 2011, Harold Weber in 2013, Herbert Freyhardt in 2015, Kozo Osamura in 2017, Archie Campbell in 2018 and Teruo Matsushita in 2019.

Shi Xue Dou is a Distingiushed Professor at University of Wollongong and the founding director of Institute for Superconducting & Electronic Materials at UOW. He received his PhD at Dalhousie University, Canada in 1984, DSc at the University of New South Wales in 1998 and was elected as a Fellow of the Australian Academy of Technological Science and Engineering in 1994. He was awarded the Australian Government's Centenary Medal in 2003 and Medal of Australian Order of Member in 2019 for his contribution to materials science and engineering by Governer General, multiple Australian Professorial Fellowships from 1993 to 2011, the Vice-Chancellor's Senior Excellence Award in 2008, Outstanding Partnership Award in 2012, the ASTS Lifetime Achievement Award in 2018, one of top five Lifetime Achevers in physical science in Australia in 2020, and was named a global leader in electrochemistry, nanotechnology and materials engineering in 2021 by Australian Research Magazine. He is cited as a highly cited researcher in materials science by Thomson Reuters from 2018 to 2020 and double highly cited resaercher in chemistry and materials science in 2021, with citations of 68,000 and an *h-index* of 121 (Scopus). His research focusses on energy and electronic materials. He has supervised and co-supervised 100 PhD students, more than 60 postdoctoral and visiting fellows.

IOP Conf. Series: Materials Science and Engineering 1241 (2022) 011002 doi:10.1088/1757-899X/1241/1/011002

Dou's contributions to HTS include development of manufacture technologies for HTS Bi-2223 wires and coils which lay the foundation for the establishment of a domestic HTS industry in Australian Technology Park and Wollongong -- Australian Superconductors Ltd. Through systematic efforts Dou's group has achieved excellent materials performance characteristics in several HTS superconductors systems, in particular developing innovative approaches in Bi-based HTS tape production. His group also achieved the highest pinning potential in potassium-doped $BaFe_2As_2$ system among all the HTS in which the K doping resulted in the coexistence of static magnetism and superconductivity on a lattice parameter length scale over a wide range of dopant concentrations, leading to an especially good high field performance of the superconductor for applications.

Magnesium diboride (MgB_2) was found to be superconductive at 39 K by Akimitsu's group in 2001. The great challenge is how one can introduce effective pinning centres into MgB_2 to make this material usable for applications. Dou has made a significant breakthrough in the fabrication of MgB_2 superconductor by using nanoscale silicon carbide doping, achieving a record high upper critical field of 46 Tesla and an order of magnitude increase in critical current density in magnetic field. To understand why SiC doping is so special, Dou proposed a dual reaction model explaining the mechanism of nano-scale SiC doping in MgB_2. The model not only can explain the mechanism of nano-scale SiC doping, but also allows the classification of the effect of about 100 different dopants, as well predicting what are the best dopants for achieving optimal superconducting properties. The understanding of the dual reaction model has led to the discovery of the advantages of a range dopants in MgB_2, resulting in a significant enhancement in electromagnetic properties. The power of the dual reaction model for prediction is well demonstrated by the following works after the SiC doping work.

Dou was a principal investigator, leading a number of large consortium projects on the "Technology and Applications of HTS" and a main contributor for the foundation of the spin-off company, the Australian Superconductors Ltd, which is one of few in the world that have full capacity and facilities to produce long length Bi-2223 tapes for both fundamental research and commercial applications. Australian Superconductors Ltd was incorporated into Zenergy Power in 2001, which has designed, built and installed the commercial FCL in the power grid at the Edison Substation of Southern California based on the UOW inductive model. Dou's group has demonastrated a number of application for MgB_2 superconductors including prototype FCL and MRI.

Citation for Dou's award: For his contribution to the significant improvement and understanding of the mechanism for critical currenet and flux pinning via nano-doping to magnesium diboride super-condcutors.

IOP Conf. Series: Materials Science and Engineering 1241 (2022) 011002 doi:10.1088/1757-899X/1241/1/011002

ICMC CRYOGENIC MATERIALS AWARD FOR EXCELLENCE 2021

Dr. Mark Ainslie

University of Cambridge
United Kingdom

The ICMC Cryogenic Materials Award for Excellence is awarded annually at the ICMC Conference, subject to the nomination of an appropriate candidate, to an individual, who is under 40 years of age by the application deadline, to "recognize excellence in advancing the knowledge of cryogenic materials over recent years". The 2021 award was given to Mark Ainslie. Previous awardees are Fumitake Kametani in 2014, Kazumasa Iida in 2016, Tengming Shen in 2017, Anna Kario and Akiyasu Yamamoto in 2018, Kohei Higashikawa in 2019 and Dongliang Wang in 2020.

Mark Ainslie received his doctorate in Electrical Engineering from the University of Cambridge, UK, in 2012. From 2012-2017, he was a Royal Academy of Engineering Research Fellow in the Department of Engineering, University of Cambridge, investigating various aspects of superconducting electric machine design utilizing both wire- and bulk-forms of high-temperature superconducting (HTS) materials. Since 2017, he has been an Engineering and Physical Sciences Research Council (EPSRC) Early Career Fellow in the same institution, investigating portable, high magnetic field charging of bulk superconductors for practical engineering applications.

He has published over 100 peer-reviewed research papers and given over 40 invited talks and short courses at international conferences and summer schools across a broad range of topics in applied superconductivity, many of which include numerical modelling and simulation as a key component. He has developed world-leading and versatile techniques to simulate the multiphysical behaviour of different families of superconducting materials from electromagnetic, thermal and mechanical points of view, in order to use them practically and effectively in a range of applications.

In 2011, he was awarded the European Society for Applied Superconductivity (ESAS) Young Researcher's Award in Large Scale Applications for his PhD work on transport AC loss in HTS coils. His topical review paper, "Modelling of bulk superconductor magnetization," published in Superconductor Science and Technology (SUST) in 2015, was selected in SUST's Highlights of 2015 as well as the SUST 30th Anniversary Collection as the Most Cited Paper of 2015. Leading on from this, he recently published the first textbook *Numerical Modelling of Bulk Superconductor Magnetisation* in the Institute of Physics eBooks™ series.

IOP Conf. Series: Materials Science and Engineering 1241 (2022) 011002 doi:10.1088/1757-899X/1241/1/011002

ICMC BEST PAPER AWARD

The ICMC Best Paper Awards provide an incentive for the production and presentation of high-quality papers at the International Cryogenic Materials Conferences and recognition of authors who, in the judgment of the ICMC Board of Directors, presented the best paper at the preceding conference. Papers are nominated by the reviewers.

In 2021, two awards for the best papers delivered at the 2019 ICMC in Hartford, Connecticut, USA, and published in the *IOP Conference Series: Materials Science and Engineering, Vol. 756, 2020*, were presented at the 2021 virtual conference to the following:

Best Superconducting Materials Paper

A. Nishimura

for his paper

"Neutron irradiation effect on critical current of Nb_3Sn wire under 8 T to 15.5 T"

IOP Conference Series: Materials Science and Engineering, Vol. 756, 2020; 012013

Best Structural Materials Paper

D. Evans

for his paper

"Resins for Superconducting Magnet Construction - An Overview of Requirements, Processing and Properties"

IOP Conference Series: Materials Science and Engineering, Vol. 756, 2020; 012003

ICMC STUDENT MERITORIOUS PAPER AWARD

The ICMC Board of Directors also recognizes students who write high quality papers. All students who applied for and participated in the ICMC Early Registration Fee Waiver program during the 2019 ICMC Hartford Conference were automatically entered for consideration of the ICMC Student Meritorious Paper Award. The papers are ranked on the basis of research merit and quality of writing.

The recipients of the 2021 ICMC Student Meritorious Paper Award are:

S. Telikapalli, M. Kvitkovicova, P. Cheetham, C. Park, C.H. Kim, L. Graber, S. Pamidi

for their paper

"Effect of Non-Uniformity of Electric Field on Breakdown Strength of Cryogenic Gaseous Insulation Media for Superconducting Power Applications"

IOP Conference Series: Materials Science and Engineering, Vol. 756, 2020; 012006

And

S I Hossain, J Jiang, U P Trociewitz, J Lu, E Bosque, Y Kim, D C Larbalestier and E Hellstrom

for their paper

"A study on the extent of Ag protrusions in different TiO_2-coated Bi-2212 wires"

IOP Conference Series: Materials Science and Engineering, Vol. 756, 2020; 012017

IOP Conf. Series: Materials Science and Engineering 1241 (2022) 011003 doi:10.1088/1757-899X/1241/1/011003

2021 International Cryogenic Materials Conference Board of Directors

CONFERENCE AND PROGRAM

Michael Sumption, *President* – The Ohio State University
Timothy Haugan, *Conference Chair* – U.S. Air Force Research Laboratory
Shreyas Balachandran, *Program Chair* – ASC / NHMFL / FSU
Tengming Shen, *Program Chair* – Lawrence Berkeley National Laboratory

BOARD OF DIRECTORS

Shreyas Balachandran – ASC / NHMFL / FSU
U. (Balu) Balachandran, *Chief Technical Editor* – Argonne National Laboratory
David A. Cardwell – University of Cambridge, United Kingdom
David Evans – Advanced Cryogenic Materials Ltd, United Kingdom
Herbert C. Freyhardt – TcSUH / University of Göttingen (retired) and ZFW gGmbH Göttingen, Germany
K. Theodore Hartwig, Jr. – Texas A&M University
Timothy J. Haugan – U.S. Air Force Research Laboratory
Eric E. Hellstrom – NHMFL, Florida State University
Takanobu Kiss – Kyushu University
Juan Knaster – Fusion for Energy at ITER Organisation, France
Yanwei Ma – Institute of Electrical Engineering, Chinese Academy of Science, China
Thomas H. Nicol – Fermi National Accelerator Laboratory
Richard P. Reed – Cryogenic Materials, Inc.
Kenichi Sato – Japan Science and Technology Agency, Japan
Sonja Schlachter – Karlsruhe Institute of Technology
Jun-ichi Shimoyama – Aoyama Gakuin University, Japan
Hyung-Seop Shin – Andong National University, South Korea
Michael D. Sumption – The Ohio State University
Herman H. J. ten Kate – CERN, Switzerland
Robert P. Walsh – NHMFL, Florida State University
Klaus-Peter Weiss – Karlsruhe Institute of Technology

AWARDS COMMITTEE

David A. Cardwell – University of Cambridge, United Kingdom
Lance D. Cooley – NHMFL, Florida State University
K. Theodore Hartwig, Jr. – Texas A&M University
Eric E. Hellstrom, *Chair* – NHMFL, Florida State University
Michael D. Sumption – The Ohio State University
Robert P. Walsh – NHMFL, Florida State University

ICMC Technical Editors

Acknowledgments

The CEC and ICMC Boards wish to thank our supporters, sponsors, and exhibitors who have contributed to the 2021 virtual CEC/ICMC Conference. These organizations chose to step up and support our conference in a time when we quite possibly needed it most. Their contributions helped ensure the success of the 2021 Conference and we are eternally grateful for their generosity.

CONFERENCE SPONSORS

- **Scientific Instruments, Inc**
- **Sumitomo (SHI) Cryogenics of America, Inc.**

CONFERENCE SUPPORTERS

- **Cryomech, Inc.**
- **Fermi National Accelerator Laboratory**
- **FormFactor Inc.**
- **Linde Engineering**
- **National High Magnetic Field Laboratory**

EXHIBITORS

- **Ability Engineering Technology, Inc.**
- **Air Liquide**
- **Cryomagnetics, Inc.**
- **Cryomech, Inc.**
- **Demaco Holland bv.**
- **FormFactor Inc.**
- **Fujikura Ltd**
- **Hyper Tech Research**
- **Lake Shore Cryotronics**
- **Linde Engineering**
- **PBS Velka Bites**
- **Rol-Vac, LP**
- **RUAG Space GmbH**
- **Scientific Instruments**
- **Stirling Cryogenics**
- **Stöhr Armaturen GmbH & Co.KG**
- **Sumitomo (SHI) Cryogenics of America, Inc.**
- **Sunpower, Inc.**
- **Taiyo Nippon Sanso Corporation**
- **WEKA AG**

Producing the **Open Access *IOP Conference Series: Materials Science and Engineering (MSE)*** requires the dedicated effort of many individuals. The quality of the publication relies on the dedication of those who took time from their busy schedules to review and edit papers. The Boards wish to specifically acknowledge the efforts of Centennial Conferences for their editorial support, and the chief technical editors, John Weisend (CEC) and Balu Balachandran (ICMC). Their contributions to ensure a quality publication in a timely manner are greatly appreciated. The chief technical editors are indebted to the individual technical sub-editors for their time spent with reviewers and authors to get the final manuscripts ready for publication. The Boards also wish to thank all students and post-doctoral researchers from a number of universities for their service with the initial format checking of the papers submitted for peer review.

CEC/ICMC 2021 was managed by Paula Pair and Centennial Conferences. Recognizing this is a tremendous team effort, the Boards would especially acknowledge, in addition to Paula, all her staff – Annett Cady, Brion Jacobs, Carrie Lian, Steve Pair, Sydney Pair, Christina Tomhave, and Letitia Xu – for the fantastic job they did wonderfully.

We also would like to recognize the effort and dedication of the attendees in preparing their presentations and papers, presenting virtually, and showing the professionalism and enthusiasm that makes the CEC/ICMC such a great Conference time after time.

IOP Conf. Series: Materials Science and Engineering 1241 (2022) 011006 doi:10.1088/1757-899X/1241/1/011006

Materials Index

IOP Conf. Series: Materials Science and Engineering 1241 (2022) 011006 doi:10.1088/1757-899X/1241/1/011006

IOP Conf. Series: Materials Science and Engineering 1241 (2022) 011006 doi:10.1088/1757-899X/1241/1/011006

Subject Index

IOP Conf. Series: Materials Science and Engineering 1241 (2022) 011006 doi:10.1088/1757-899X/1241/1/011006

IOP Conf. Series: Materials Science and Engineering 1241 (2022) 011006 doi:10.1088/1757-899X/1241/1/011006

IOP Conf. Series: Materials Science and Engineering 1241 (2022) 011007 doi:10.1088/1757-899X/1241/1/011007

INDEX TO VOLUME 1241

K

L

M

N

O

P

R

S

T

U

W

X

Y

Z

Peer review declaration

All papers published in this volume have been peer reviewed through processes administered by the Editors. Reviews (2 required per paper) were conducted by expert referees to the professional and scientific standards expected of a proceedings journal published by IOP Publishing.

- **Type of peer review: Single anonymous**
- **Conference submission management system: Morressier**
- **Number of submissions received: 57**
- **Number of submissions sent for review: 57**
- **Number of submissions accepted: 55**
- **Acceptance Rate (Number of Submissions Accepted / Number of Submissions Received X 100): 96.5%**
- **Average number of reviews per paper: 2.25**
- **Total number of reviewers involved: 59**
- **Format quality control: In addition to the technical reviews, all papers went through an initial quality control review process handled via the peer review system.**
- **Contact person for queries:**
  Name: Annett Cady
  Affiliation: Centennial Conferences
  Email: cecicmc@centennialconferences.com

IOP Conf. Series: Materials Science and Engineering 1241 (2022) 012001 doi:10.1088/1757-899X/1241/1/012001

Tensile Properties of 22Cr-12Ni Austenitic Stainless Steel Thick Plates and Bars at Cryogenic Temperatures

T Kato[1], Y Kato[2], O Umezawa[3], Y Ono[4], T Yuri[4], A Nishimura[5], S Kumagai[6]

[1] Graduate School of Engineering, Yokohama National University, 79-5 Tokiwadai, Hodogaya, Yokohama, 240-8501, Japan

[2] School of Engineering Science, Yokohama National University, 79-5 Tokiwadai, Hodogaya, Yokohama, 240-8501, Japan

[3] Faculty of Engineering, Yokohama National University, 79-5 Tokiwadai, Hodogaya, Yokohama, 240-8501, Japan

[4] National Institute for Materials Science, 1-2-1 Sengen, Tsukuba, 305-0047, Japan

[5] National Institute for Fusion Science, Oroshi, Toki, Gifu 509-5292, Japan

[6] National Institute of Technology, Sendai College, Medeshima, Natori, Miyagi 981-1239, Japan

E-mail: umezawa-osamu-fv@ynu.ac.jp

Abstract. The tensile properties of XM-19 austenitic stainless steels with thick plates (30 mm thickness), extra-thick plates (100 mm thickness), and rectangular bars (square of 14.3 mm) were examined at cryogenic temperatures. Influence of thermomechanical treatments on the steels was demonstrated. Even though their 0.2 % proof stresses were over 1200 MPa at 4.2 K, there were variations in strength depending on location. Namely, the 1/2t specimen (the position of the half in the thickness) of extra-thick plate showed lower 0.2% proof stress, elongation, and reduction of area at 4.2 K than the 1/4t (the position of the quarter in the thickness) and the thick plates. The effect of grain size on the 0.2% proof stress was clearly described by the Hall-Petch relationship for the materials.

1. Introduction

Austenitic steels are the structural materials of choice for high-field superconducting magnets. Especially at 4.2 K, nitrogen strengthened austenitic stainless type 316LN is commonly used for its high strength and thermodynamic stability. In addition, cryogenic steels called "Japanese Cryogenic Steels" have been developed for the superconducting magnets of the Fusion Experimental Reactor in 1980's [1,2]. In recent years, there has been some interests in investigating and developing stronger and tougher weldable alloys for cryogenic structural components, particularly for magnet systems such as fusion demonstration reactor (DEMO) [3], where 0.2% proof stress of the structural material should be over 1200 MPa at 4.2 K.

A proper balance between 0.2% proof stress and fracture toughness at 4.2 K is derived from the grain coarsening. ASME standard XM-19 (22Cr-12Ni-5Mn-2Mo-0.2V-0.2Nb-0.3N, in mass%) stainless steel is strengthened by nitrogen solid-solution and grain refinement with (Nb,V) precipitates and shows the advantages of high strength and excellent weldability [4]. However, thickness dependence of strength and toughness should be taken into account [4]. Limited data are available for the mechanical properties of extra-thick plates of XM-19. In this study, tensile properties of extra-thick plates (100 mm thickness), thick plates (30 mm thickness) and grooved rolled bars for XM-19 steels were evaluated at cryogenic temperatures.

2. Experimental procedure

2.1. Materials

The chemical compositions of the test steels are listed in table 1. The 180 kg steel ingots with ϕ210~ϕ250 × 525 mm were hot-forged into rectangular blocks in a square of 150 mm and hot-rolled. The 100-mm-extra-thick plates (ETPs) solution-treated at 1373 K, 1423 K or 1473 K for 21.6 ks and 30-mm-thick plates (TPs) solution-treated at 1373 K for 21.6 ks were provided. The test pieces were taken from the positions of the half (1/2t) and the quarter (1/4t) in the thickness and parallel to the transverse direction (TD). A 30-mm-thick plate in a square of 30 mm was cold-groove-rolled to a rectangular bar in a square of 14.3 mm. The rectangular bars (RBs) were annealed at 1073, 1173, 1273, 1373 or 1473 K for 3.6 ks followed by water quenched. The test pieces were taken from the bar parallel to the rolling direction (RD).

Table 1. The chemical compositions of test steels.

Steels	Concentration (mass%)										
	C	Si	Mn	P	S	Ni	Cr	Mo	V	Nb	N
ETP	0.031	0.390	4.53	0.015	0.002	12.26	21.95	2.11	0.20	0.200	0.330
TP	0.029	0.410	4.56	0.016	0.002	12.31	21.91	2.11	0.21	0.190	0.329
RB	0.015	0.430	5.47	0.016	0.002	13.32	21.82	1.82	0.11	0.100	0.291

2.2. Tensile tests

Round bar tensile test specimens with 31.25 mm in gauge length and a nominal diameter of 6.25 mm (the length of reduced section is 37.5 mm) were cut from the materials. The tensile test was carried out at 293 K (in air), 77 K (immersed in liquid nitrogen) and 4.2 K (immersed in liquid helium) with the initial strain rate of 2.38×10^{-4} s^{-1}. Tensile properties were evaluated as the average of three tests data for each condition in the ETPs and TPs, and of two in the RBs. The geometries on the specimen surface after the tensile tests were evaluated by laser microscopy to characterize the local necking deformation. Fracture surfaces were analyzed by scanning electron microscopy (SEM) with energy dispersive X-ray spectroscopy (EDS).

3. Results and discussion

3.1. Tensile properties

Figure 1 shows the tensile properties of the TP and ETPs such as 0.2% proof stress, ultimate tensile strength, elongation, and reduction of area. As the test temperature decreased from 293 K to 4.2 K, their strength increased but elongation decreased. The ETPs showed lower ductility at 4.2 K than the TP. Especially, the ETP 1/2t solution-treated at 1373 K showed a poor elongation (figure 1(b)). Their fracture surfaces of the ETPs specimens were flat consisted of facets. Nb segregation was detected on the fracture surface by EDS as shown in figure 2. In the ETPs, the diffusion of Nb during hot-rolling and solution treatment may be insufficient owing to the central segregation in the cast. Thus, Nb compound precipitates may cause the early fracture as described in below.

Figure 3 summarized the tensile properties of the RBs with the annealed at between 1073 K and 1473 K. The dependence of their tensile properties on the test temperatures exhibited almost the same trend to that of ETPs and TP. Both the ultimate tensile strength and 0.2% proof stress were higher as the decrease of annealing temperature, although the elongation and reduction of area were almost the same at 4.2 K for recrystallized materials. The RBs annealed at 1073 and 1173 K showed quite poor ductility.

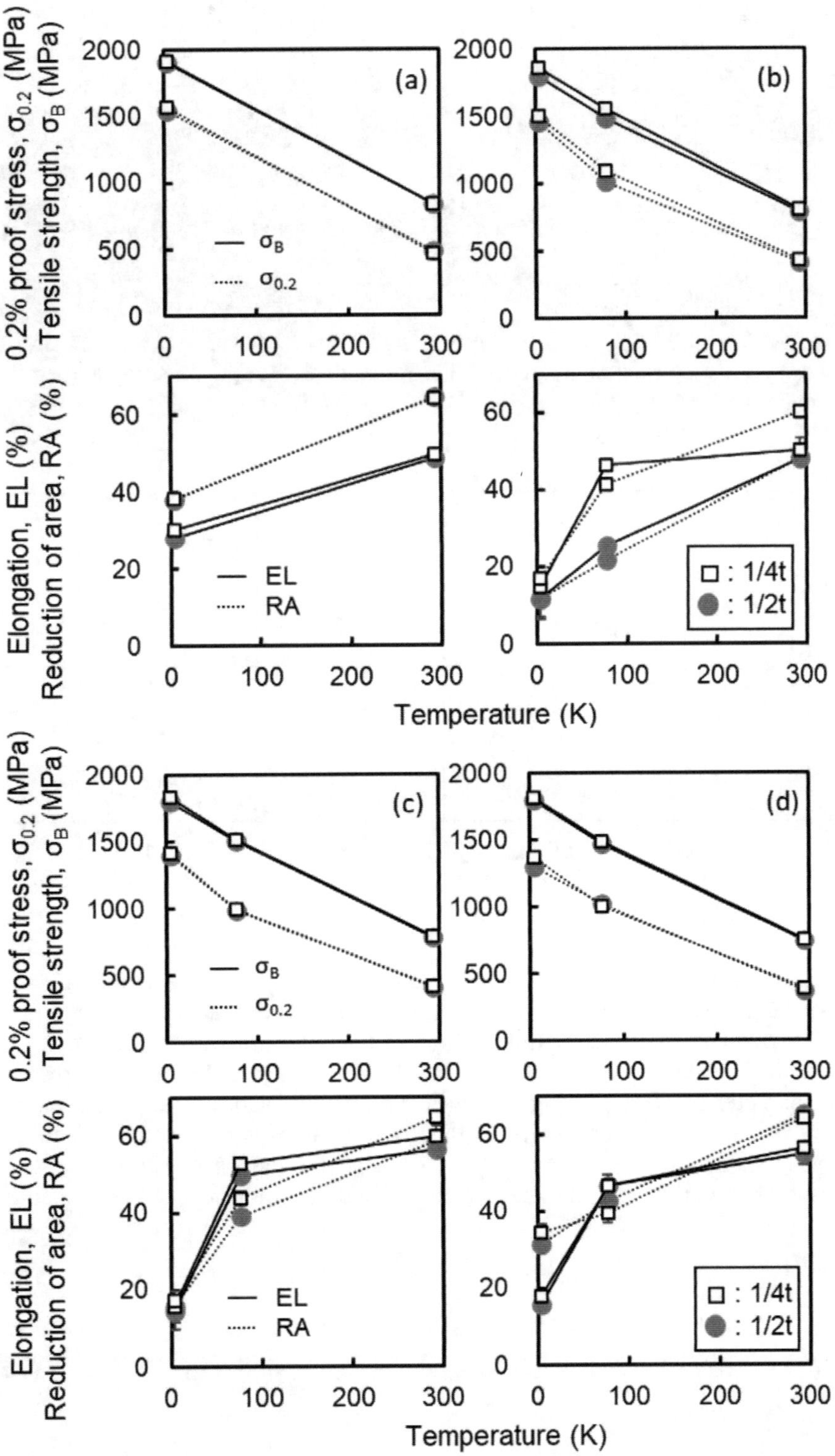

Figure 1. Tensile properties of the plates with the test temperatures: (a) TP, (b) ETP annealed at 1373 K, (c) ETP annealed at 1423 K, and (d) ETP annealed at 1473 K.

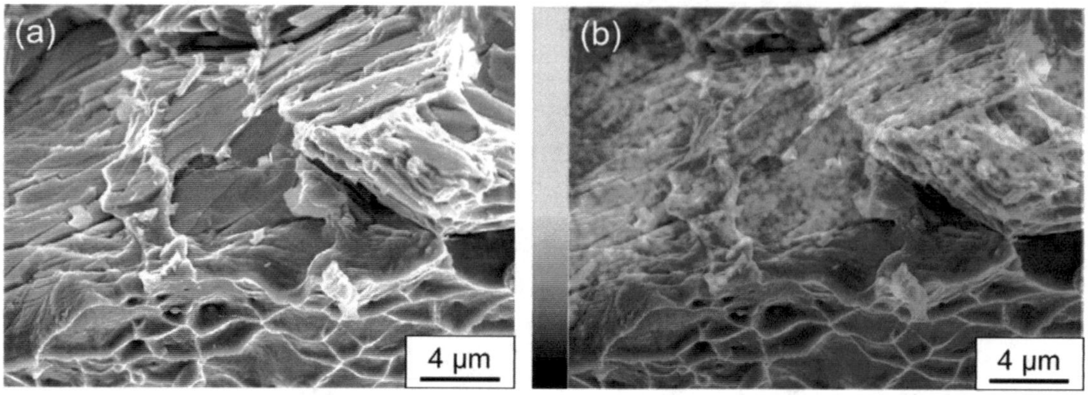

Figure 2. A fracture surface at 77 K of ETP 1/2 annealed at 1373 K (a) and its Nb element mapping image (b).

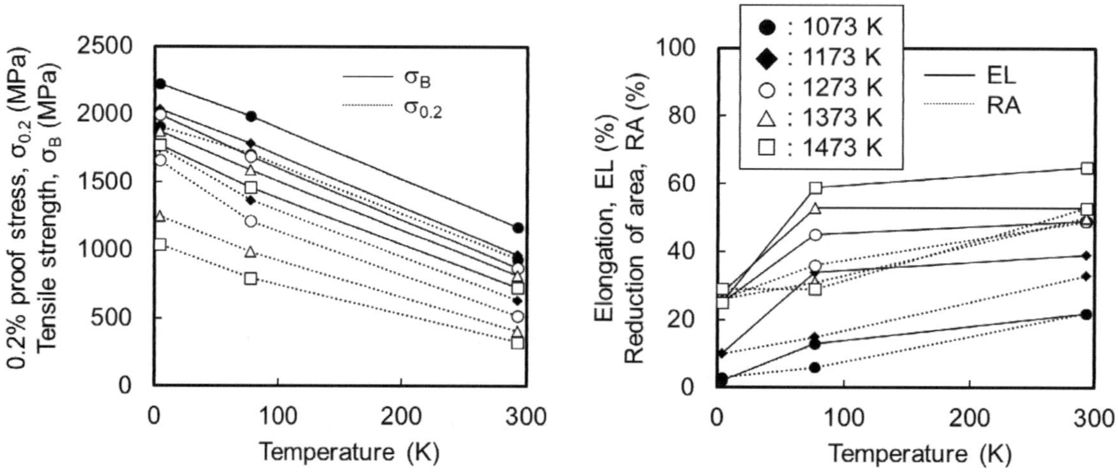

Figure 3. Tensile properties of RBs with the test temperatures.

3.2. Deformation behavior

Figure 4 shows the true stress-true strain curves and work hardening rates at 4.2 K for the EPTs and RB annealed at 1373 K. The EPT 1/2t sample exhibited the early fracture, while the EPT 1/4t and RB exhibited serrations, and higher strains and work hardening rates than the EPT 1/2t, respectively. Figure 5 shows their diameter profiles of gauge length after tensile fracture. The EPTs showed a localized deformation, in which the work hardening in the necked part was not enough. Therefore, the further deformation was localized in the first necked part or caused in the early failure. The RBs, on the other hand, showed a higher work hardening rate, which resulted in a multiple necking and higher elongation.

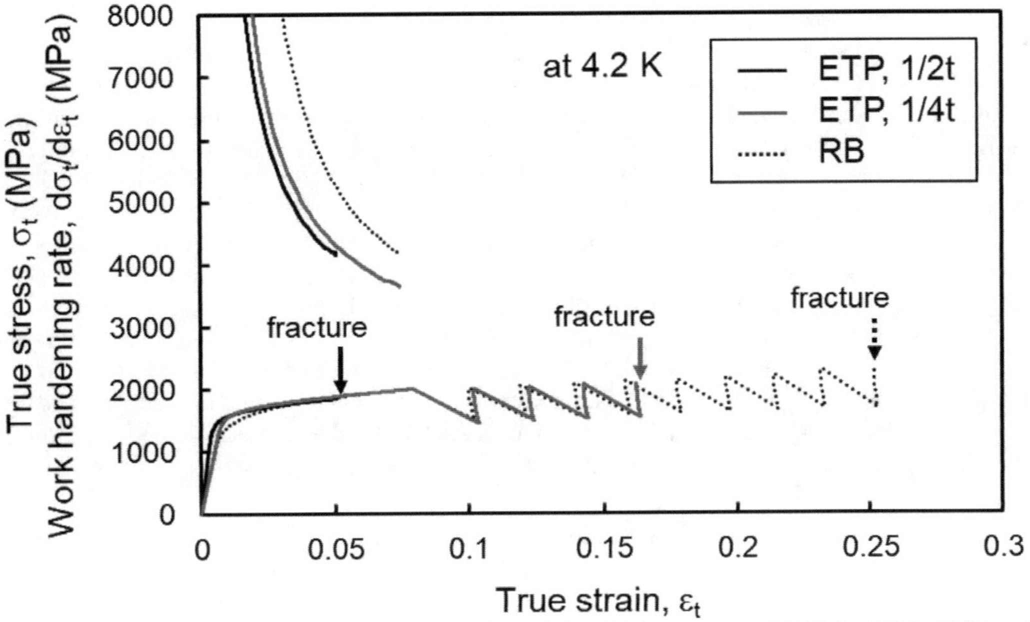

Figure 4. True stress - true strain curves and work hardening rates at 4.2 K for ETPs (1/2t and 1/4t) and RB annealed at 1373 K.

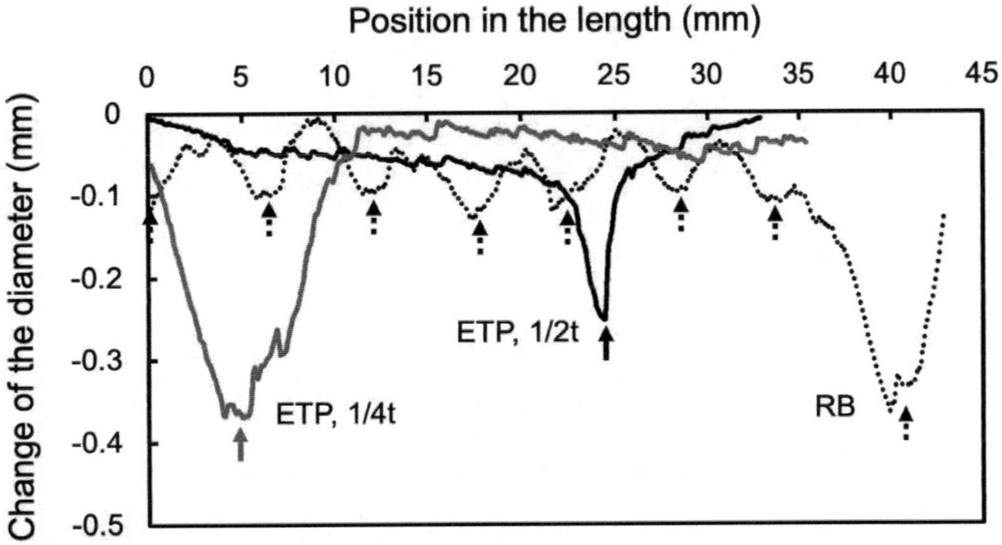

Figure 5. Diameter profiles of gauge length after tensile fracture for ETPs (1/2t and 1/4t) and RB annealed at 1373 K. Arrows indicate fracture points and necking parts, respectively. 0 position is fitted to an end of gauge length.

3.3. Microstructure dependence on 0.2% proof stress

Table 2 represents the average diameter of austenite grains for the materials. While the RBs annealed at 1273, 1373 and 1473 K were recrystallized austenite, the RBs annealed at 1073 and 1173 K were partially recrystallized austenite with σ phase precipitation which was determined by EDS chemical composition analysis.

The effect of grain size on the 0.2 % proof stress of the recrystallized RBs is clearly described by the Hall-Petch relationship as shown in figure 6. The Hall-Petch equation is given as follows:

$$\sigma_y = \sigma_0 + k_y \cdot d^{-1/2}$$

where σ_y is the yield stress, σ_0 is the corresponding stress for large single crystals, k_y is the Hall-Petch coefficient, and d is the average grain size. The Hall-Petch coefficient at 4.2 K is higher than that at 293 K, because the nitrogen strengthening of austenitic stainless steels is remarkable at lower temperature. Furthermore, lower content of N in the RB resulted in the lower 0.2 % proof stress than the ETPs and TP at 4.2 K. The 0.2 % proof stresses of the partially recrystallized RBs at 4.2 K and 77 K were lower than the expected values from the Hall-Petch equations indicated in figure 6, respectively. The dispersion of movable dislocations in the recovered austenitic grains may lower the 0.2 % proof stress. The Hall-Petch relationship was also applicable to the ETPs and TP, in which the trend was the same as the previous work [4]. Movable dislocations dispersed in the matrix during hot-rolling as well as rather inhomogeneous grain structure with Nb segregation for the ETP 1/2t may provide lower 0.2 % proof stress than for the ETP 1/4t and TP at 4.2 K. Therefore, grain refinement is very effective to increase the strength for the XM-19 at cryogenic temperature. From the viewpoint of stronger and tougher weldable alloys for cryogenic structural components, the evaluated materials in the present study exhibited the balance between 0.2% proof stress and fracture toughness as shown in figure 7 [2,5].

Table 2. Average grain diameter of test materials.

Materials	Diameter (μm)					
	1073 K	1173 K	1273 K	1373 K	1423 K	1473 K
ETP, 1/2t	-	-	-	26.0	41.2	50.3
ETP, 1/4t	-	-	-	24.3	45.8	59.0
TP	-	-	-	20.0	-	-
RB	1.5	3.3	8.9	42.5	-	147.5

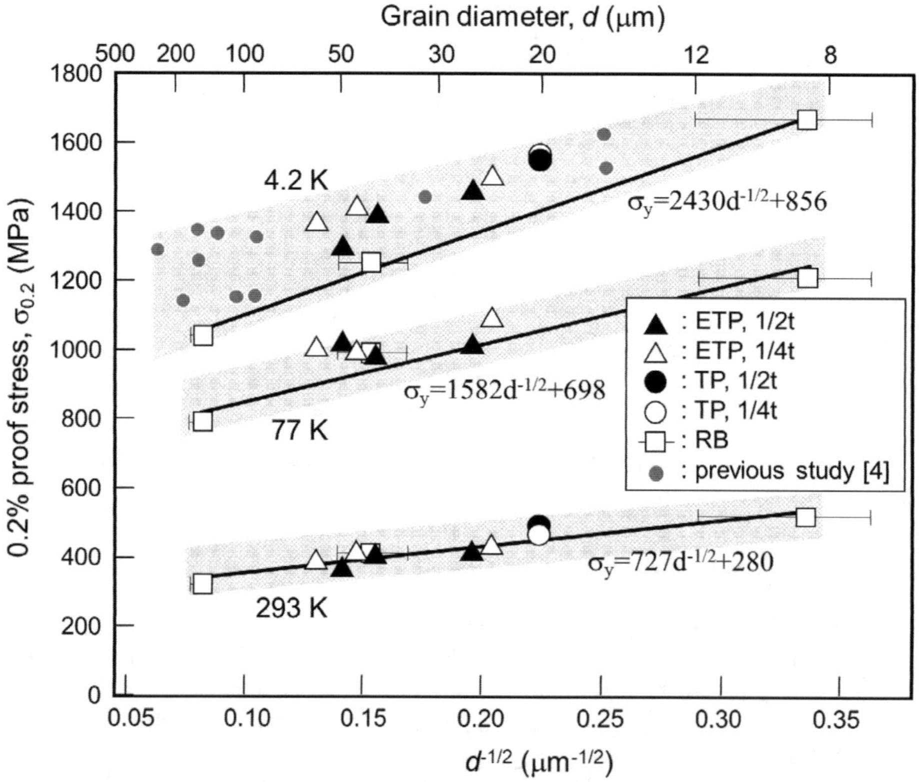

Figure 6. Temperature dependence on Hall-Petch relationship for austenitic stainless steels.

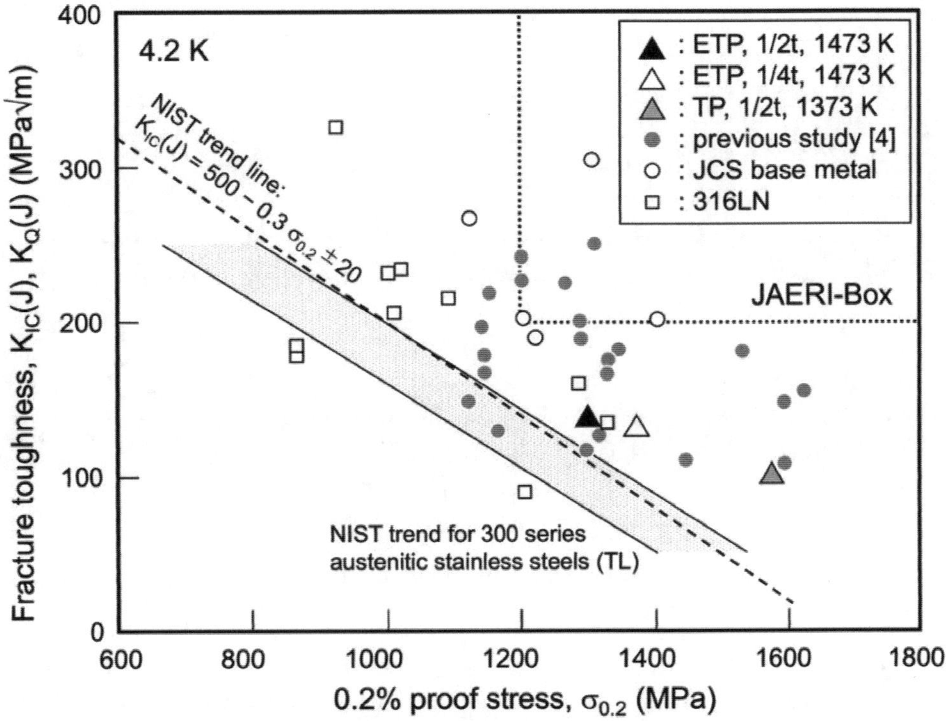

Figure 7. Relationship between fracture toughness and 0.2% proof stress for high nitrogen austenitic stainless steels. JAERI (Japan Atomic Energy Research Institute) -Box is a development target of strength and toughness for magnet systems at 4.2 K (0.2% proof stress, $\sigma_{0.2} > 1200$ MPa, fracture toughness, $K_{IC} > 200$ MPa√m). NIST (National Institute of Standards and Technology) trend exhibits a trade-off relationship between strength and toughness.

4. Conclusions

The tensile properties of XM-19 austenitic steels in the thick plates (30 mm thickness), extra-thick plates (100 mm thickness) and rectangular bars (square of 14.3 mm) were evaluated at cryogenic temperatures. Their 0.2 % proof stresses at 4.2 K were over 1200 MPa and exhibited the temperature dependence by Hall-Petch relationship. The extra-thick plates showed a poor elongation at 4.2 K, especially for solution-treated at lower temperature. It revealed localized deformation and lower work hardening rate.

5. References

[1] Nakajima H, Yoshida K and Shimamoto S 1990 *ISIJ International* **30** 567–578
[2] Umezawa O 2021 *Materials Performance and Characterization* **10(2)** 3–15
[3] Tobita K, Asakura N, Hiwatari R, Someya Y, Utoh H, Katayama K, Nishimura N, Sakamoto Y, Honma Y, Kudo H, Miyoshi Y, Nakamura M, Tokunaga S, Aoki A and the Joint Special Design Team for Fusion DEMO 2017 *Fusion Science and Technology* **72(3)** 537–545
[4] McRae D M, Balachandran S and Walsh R P 2017 *Materials Science and Engineering* **279** 012001
[5] Ono Y, Umezawa O, Nishimura A, Kumagai S and Yuri T, unpublished data

Acknowledgments

Authors acknowledge the financial support of the framework of the Joint Special Design Team for Fusion DEMO contract research program of National Institutes for Quantum and Radiological Science and Technology, Japan in the 2017 and 2018 fiscal years.

IOP Conf. Series: Materials Science and Engineering 1241 (2022) 012002 doi:10.1088/1757-899X/1241/1/012002

Effects of grain size and strain rate on the low-temperature tensile properties of ferrite-austenite duplex stainless steel

N Koga[1], T Nameki[2] and O Umezawa[3,4]

[1] Faculty of Mechanical Engineering, Institute of Science and Engineering, Kanazawa University, Kakuma-machi, Kanazawa, Ishikawa, 920-1192, Japan

[2] Graduate school of Engineering, Yokohama National University, 79-5 Tokiwadai, Hodogaya, Yokohama, 240-8501, Japan

[3] Faculty of Engineering, Yokohama National University, 79-5 Tokiwadai, Hodogaya, Yokohama, 240-8501, Japan

[4] Center of Advanced Innovation Technologies, Vysoká Škola Báňská - Technical University of Ostrava, 17. listopadu 15, 708 33 Ostrava-Poruba, Czech Republic

E-mail: koga-norimitsu@se.kanazawa-u.ac.jp

Abstract. The effects of grain size and strain rate on the low-temperature tensile properties of ferrite-austenite duplex stainless steel were investigated. The coarse- and fine-grain specimens exhibited a grain size of approximately 20 and 8 μm, respectively. At 77 K, the fine-grain specimen exhibited a higher strength and elongation than the coarse-grain specimen. The work-hardening rate at 77 K in both specimens leveled off or increased slightly at late stage of deformation although the work-hardening rate was lower than the true stress. This characteristic work-hardening behavior is caused by the deformation-induced martensitic transformation of metastable austenite. Grain refinement stabilized austenite phase and enhanced the elongation of the material, resulting in the better low-temperature tensile properties of fine grain. The elongation decreased remarkably with an increasing strain rate at 77 K independent of grain size. The characteristic work-hardening was not detected at a high strain rate, indicating that the deformation-induced martensitic transformation did not occur. The strain rate affects the frequency of deformation-induced martensitic transformation, resulting in a change in elongation. Grain refinement effectively improves the low-temperature tensile properties of duplex stainless steels; however, these properties are strongly influenced by the strain rate, where a high strain rate causes a low elongation.

1. Introduction

The temperature dependence of the mechanical properties of metal materials is dominated by their crystal structure [1]. The mechanical properties of body-centered-cubic (bcc) materials, such as Fe and W, exhibit a strong temperature dependence, where the strength of the material increases and elongation drastically decreases at low temperature. In contrast, the mechanical properties of face-centered-cubic (fcc) materials, such as Cu and Ni, exhibit a weak temperature dependence, and show high elongation even at cryogenic temperatures. Recently, Koga et al. [2-3] revealed that the Fe (bcc) and Cu (fcc) dual-phase structure exhibits exceptional strength-elongation balance at cryogenic temperatures from 8 to 77 K, where the strength of the material increases while elongation remains high with a lowering temperature. They further demonstrated that a ferrite (bcc)-austenite (fcc) duplex stainless steel exhibits an exceptional temperature dependence of mechanical properties [4]. The ferrite-austenite duplex stainless steel has a considerably higher strength than the Cu-Fe dual-phase alloy. The simultaneous deformation of the ferrite-austenite phase, deformation-induced martensitic transformation of austenite phase, and ductile fracture of the ferrite phase owing to its chemical composition were key for the exceptional strength-elongation balance at cryogenic temperatures (77 K). On the other hand, it is generally known that grain refinement improves both strength [5] and ductility [6] in metal materials, and the deformation-induced martensitic transformation is strongly influenced by strain rate [7].

In this study, the low-temperature tensile properties of coarse- and fine-grain ferrite-austenite duplex stainless steel were measured at various strain rates, and the effects of the grain size and strain rate on the low-temperature tensile properties were discussed.

2. Experimental procedure

A commercial cold-rolled ferrite-austenite duplex stainless steel (SUS329J4L: Fe-0.02%C-7%Ni-25%Cr-3%Mo-0.1%N, in mass%) was used. The coarse-grain specimen was fabricated by heat-treatment at 1473 K for 600 s and subsequently 1273 K for 30 s followed by air cooling. The heat-treatment for fine-grain specimen was 1273 K for 600 s followed by air cooling. The microstructure was observed by optical microscopy (OM) and scanning electron microscopy (SEM). The phase and crystal orientation were measured by using electron backscattered diffraction method and the resulting data were analyzed with the OIM^{TM} analysis 7.0.1 software. Tensile test specimens with a gage length of 16 mm and width of 4 mm which are identical ratio of gage length to the width with that in Japanese Industrial Standards (JIS)-15B were cut along the rolling direction (RD) from plates. Tensile testing was conducted at an initial strain rate of 5.2×10^{-4}, 5.2×10^{-3}, and 5.2×10^{-2} s^{-1} at 293 K (in ambient air), and 77 K (immersed in liquid nitrogen).

3. Results

3.1. Microstructure

Figure 1 presents the OM images in the RD plane of the specimen annealed at (a) 1473 K and (b)1273 K. Here, ND and TD were normal direction and transverse direction of the plate. The gray and white regions corresponded to ferrite and austenite, respectively. The grain sizes of the specimen annealed at 1473 K and 1273 K, as measured by the intercept method without distinguishing the ferrite and austenite phases following the JISG0551, were approximately 20 μm and 8 μm, respectively. The specimens annealed at 1473 K and 1273 K are hereafter referred to as the "coarse-grain specimen" and "fine-grain specimen", respectively. Figure 2 shows the inverse pole figure of the (a) ferrite and (b) austenite along the RD, i.e., tensile direction, in the fine-grain specimen. The sample exhibited significant textures, such as $\{110\}_{ferrite}$//RD, $\{001\}_{austenite}$//RD, and $\{111\}_{austenite}$//RD. Since these are typical textures developed by cold rolling, the recrystallized grains should inherit these deformation textures. Similar textures were observed in coarse-grain specimen in our previous study [4].

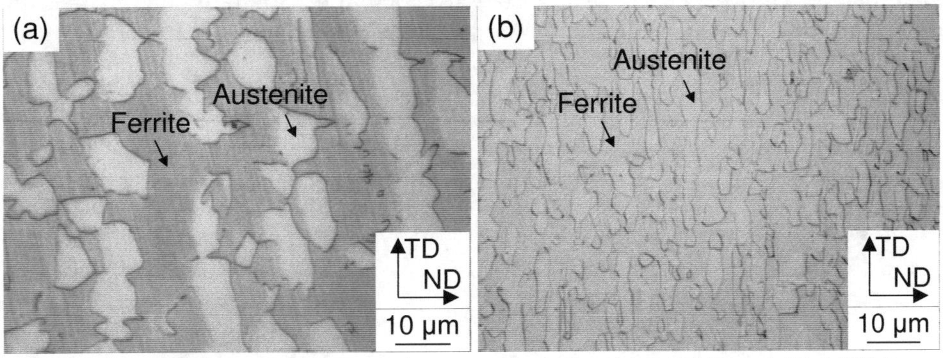

Figure 1. OM images of the specimens annealed at (a) 1473 K and (b) 1273 K.

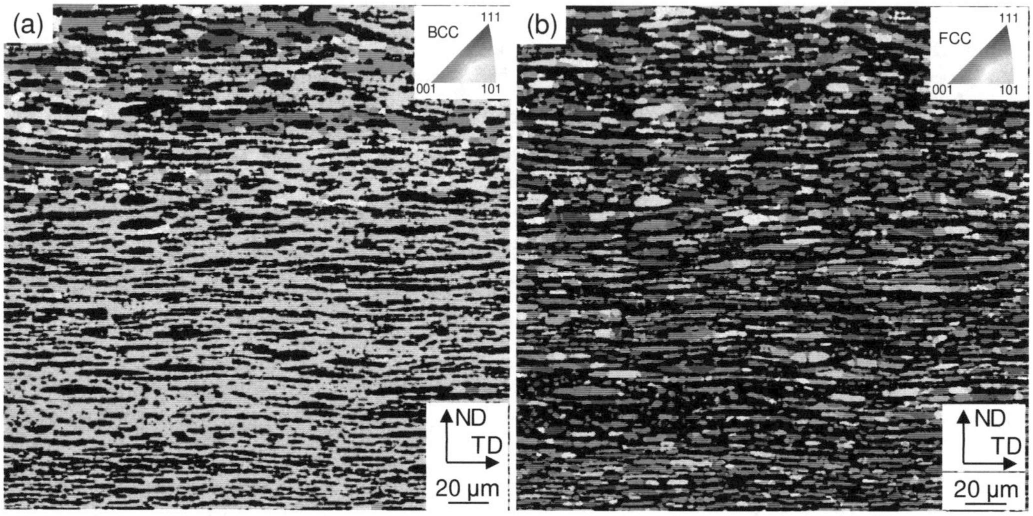

Figure 2. Orientation maps of the (a) ferrite and (b) austenite phases along the RD in the fine-grain specimen.

3.2. Grain size and strain rate dependence of tensile properties

Figure 3 shows nominal stress and nominal strain curves of the fine- and coarse-grain specimens at 293 K and 77 K and at strain rates of 5.2×10^{-4}, 5.2×10^{-3} and 5.2×10^{-2} s^{-1}. Table 1 summarized yield stresses, tensile stresses, total elongations, and work-hardening coefficients at 77 K measured from Fig. 3. At strain rate of 5.2×10^{-4} s^{-1}, the fine-grain specimen exhibited slightly higher strength and elongation than the coarse-grain specimen at 293 K, whereas both the strength and elongation increased significantly with grain refinement at 77 K. In both specimens, at 77 K, the elongation decreased significantly with an increasing strain rate and the strength slightly increased. The work-hardening coefficients slightly decreased with an increasing strain rate in both specimens.

Figure 4 shows true stress and work-hardening rate as functions of true strain (a) at 293 K and 77 K at strain rate of 5.2×10^{-4} and (b) at 77 K at strain rates of 5.2×10^{-3} and 5.2×10^{-2} s^{-1} in the fine- and coarse-grain specimens. At strain rate of 5.2×10^{-4}, the characteristic work-hardening behavior was measured at only 77 K in both specimens (Fig. 4(a)); the work-hardening rate stabilized or slightly increased although it was lower than the true stress. A previous study [4] demonstrated that the characteristic work-hardening behavior was due to the deformation-induced martensitic transformation of the metastable austenite phase, i.e., transformation induced plasticity effect [8]. Therefore, the necking region was strengthened by deformation induced martensite which has much higher strength than ferrite and austenite, resulting in characteristic work-hardening behavior and suppression of necking deformation even after the plastic instability condition. At strain rate of 5.2×10^{-3} s^{-1}, both specimens also exhibited a characteristic work-hardening behavior, similar to that at 5.2×10^{-4}. However, at strain rate of 5.2×10^{-2} s^{-1}, the characteristic work-hardening behavior disappeared (Fig. 4(b)), indicating that the deformation-induced martensitic transformation did not occur.

IOP Conf. Series: Materials Science and Engineering 1241 (2022) 012002 doi:10.1088/1757-899X/1241/1/012002

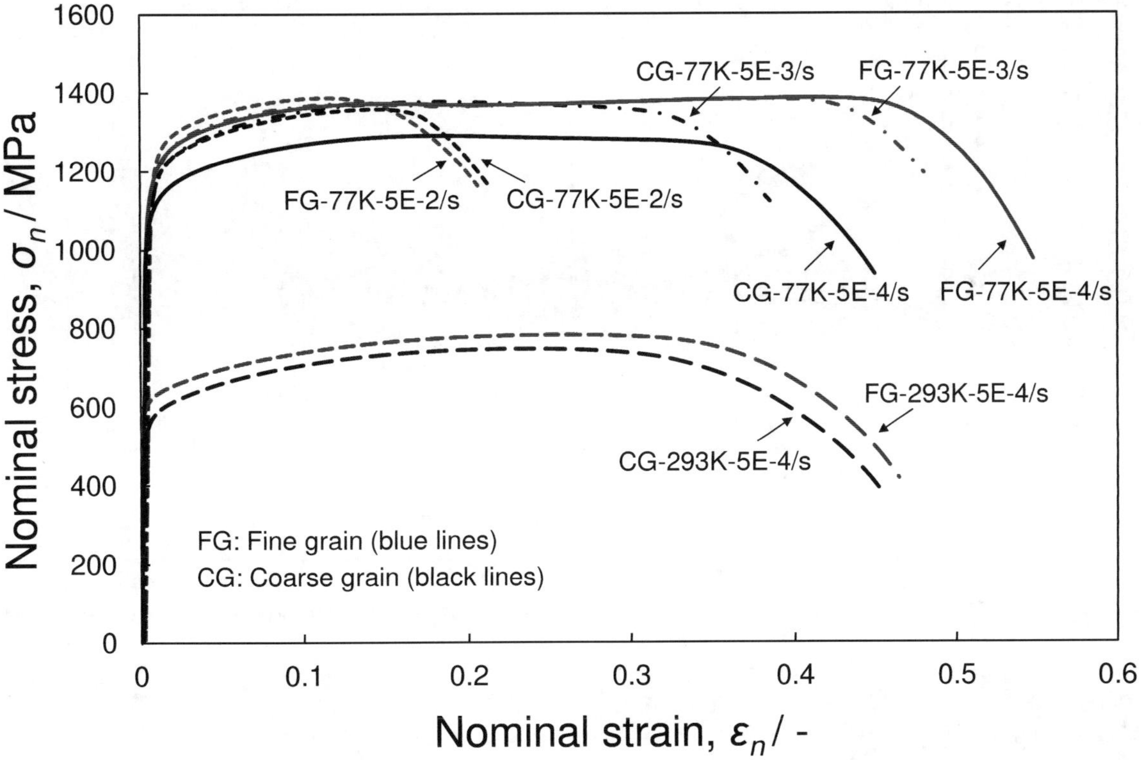

Figure 3. Nominal stress and nominal strain curves of the fine- and coarse-grain specimens at 293 K and 77 K and at strain rates of 5.2×10^{-4}, 5.2×10^{-3} and 5.2×10^{-2} s^{-1}.

Table 1. Yield stresses, tensile stresses, total elongations, and work-hardening coefficients measured from Fig. 3 in the fine- and coarse-grain specimens at strain rates of 5.2×10^{-4}, 5.2×10^{-3} and 5.2×10^{-2} s^{-1}.

	Strain rate (s^{-1})	Yield stress (MPa)	Tensile stress (MPa)	Total elongation (-)	Work-hardening coefficient (-)
Fine grain	5.2×10^{-4}	1120	1355	0.56	0.11
	5.2×10^{-3}	1140	1380	0.48	0.09
	5.2×10^{-2}	1200	1380	0.20	0.08
Coarse grain	5.2×10^{-4}	1060	1290	0.45	0.11
	5.2×10^{-3}	1130	1380	0.38	0.10
	5.2×10^{-2}	1130	1360	0.22	0.09

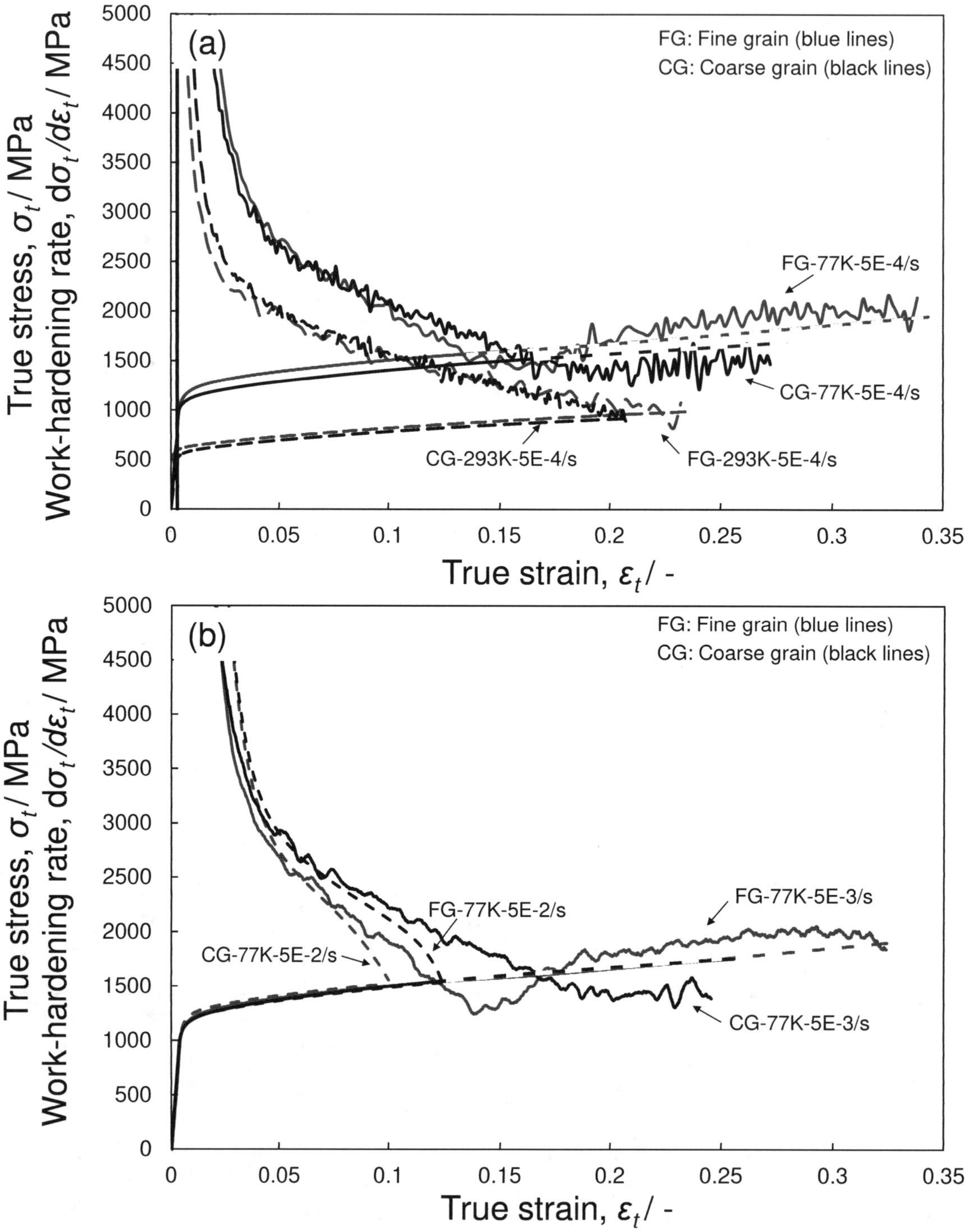

Figure 4. True stress and work-hardening rate as functions of true strain (a) at 293 and 77 K at strain rate of 5.2×10^{-4} and (b) at 77 K at strain rates of 5.2×10^{-3} and 5.2×10^{-2} s^{-1} in the coarse- and fine-grain specimens.

3.3. *Fractography*

Figure 5 shows the fracture surfaces of the (a) coarse- and (b) fine-grain specimens subjected to tensile testing at 77 K at a strain rate of 5.2×10^{-4} s^{-1}. Dimples were observed in both specimens with the fine-grain specimen exhibiting finer dimples than the coarse-grain specimen. This indicates that the void coalescence became more difficult with grain refinement, resulting in high total and necking elongations in the fine-grain specimen.

Figure 6 shows the fracture surfaces of the coarse-grain specimens subjected to tensile testing at 77 K at a strain rate of (a) 5.2×10^{-3} s^{-1} and (b) 5.2×10^{-2} s^{-1}. Dimples were observed at both strain rates, indicating that ductile fracture occurred. However, the dimple size was larger in the specimen subjected to tensile testing at 5.2×10^{-2} s^{-1}. As discussed above, this indicates that void coalescence easily occurred at a strain rate of 5.2×10^{-2} s^{-1}, resulting in low elongation.

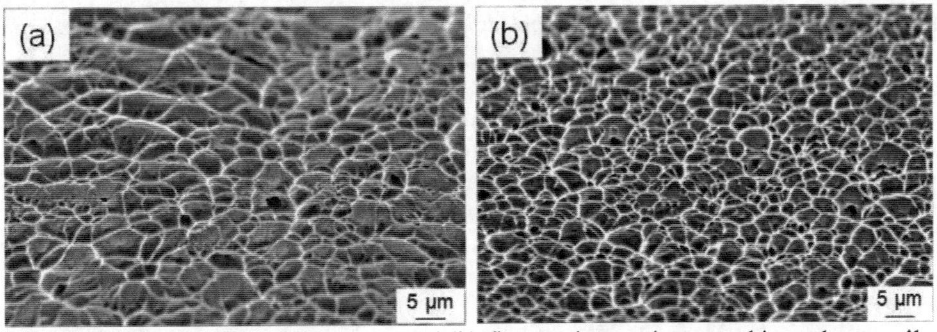

Figure 5. Fracture surfaces of the (a) coarse- and (b) fine-grain specimens subjected to tensile testing at 77 K at a strain rate of 5.2×10^{-3}.

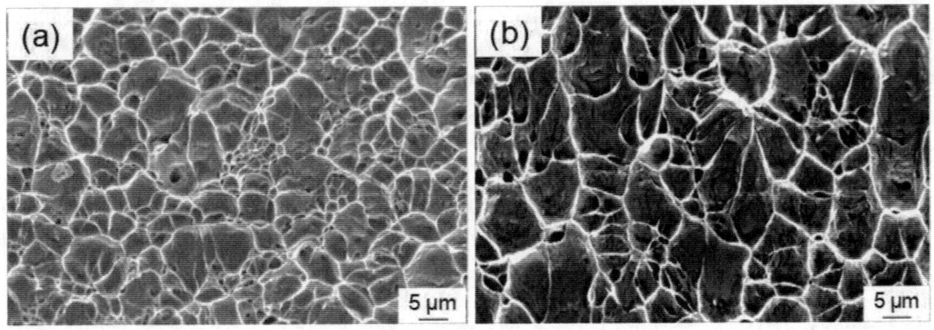

Figure 6. Fracture surfaces of the coarse-grain specimen subjected to tensile testing at 77 K at a strain rate of (a) 5.2×10^{-3} and (b) 5.2×10^{-2} s^{-1}.

4. Discussion

The grain refinement improved both strength and elongation at 77 K. The high strength in the fine-grain specimen was attributed to the grain refinement strengthening [9-10]. While, the high elongation in the fine grain specimen should relate to the TRIP effect. The grain size affects the mechanical stability of the austenite phase and fine grains impart high stability [11]. The stability of austenite influences the mechanical properties, particularly elongation, and the highly stable austenite provides high elongation, that is, high TRIP effect, because the deformation-induced martensitic transformation occurs in the late stage of deformation [12]. Therefore, grain refinement effectively stabilized the metastable austenite in duplex stainless steel, resulting in high elongation in the fine-grain specimen. Furthermore, the grain refinement suppressed the void coalescence as mentioned in Fig. 5, which also enhanced the ductility. Consequently, these positive effects of grain refinement on tensile properties led to the improvement of strength and elongation balance, and it can be concluded that the grain refinement is an effective method to improve the low-temperature tensile properties of the duplex stainless steel.

The elongation at 77 K significantly decreased with the increasing strain rate (Fig. 3) and the deformation induced martensite did not occur at high strain rate. The strain rate and test temperature were correlated, and the *R* value was proposed as follows [13]:

$$R = T \ln (A/\dot{\varepsilon}),$$

where *T* is the test temperature, *A* is a constant and can be assumed to be 10^8, $\dot{\varepsilon}$ is the strain rate. It was demonstrated that the yield stress is determined by the *R* values even though the strain rate and test temperature differ. The calculated *R* values at 77 K and at a strain rate of 5.2×10^{-4} s^{-1} corresponds to that in approximately 60 K at a strain rate of 5.2×10^{-2} s^{-1}. In a previous study [4], the elongation decreased below 77 K because the volume fractions of deformation-induced martensite decreased below that temperature. Therefore, the high strain rate decreased the volume fraction of deformation-induced martensite, resulting in low TRIP effect and elongation.

5. Summary and Conclusions

The low-temperature tensile properties were measured at various strain rates in a coarse- and fine-grain ferrite-austenite duplex stainless steel and the effect of the strain rates and grain sizes on the low-temperature tensile properties was discussed. The main results were summarized as follows.

The fine-grain specimen exhibited higher strength and elongation at 77 K than the coarse-grain specimen. The work-hardening rate stabilized or slightly increased, although the work-hardening rate was lower than the true stress, due to the deformation-induced martensitic transformation of the metastable austenite phase. The fine grain imparts high strength through grain refinement strengthening and the high stability of austenite, results in high strength and elongation at 77 K. The elongation at 77 K decreased with an increasing strain rate independent of the grain size. The characteristic work-hardening behavior, i.e., deformation-induced martensitic transformation, did not occur at high strain rate, results in low elongation. In the ferrite-austenite duplex stainless steel, grain refinement effectively improved the low-temperature tensile properties; however, its elongation was strongly influenced by the strain rate. Therefore, the effect of strain rate on the mechanical properties should be considered when duplex stainless steels are applied to structural materials at cryogenic temperatures.

6. References

[1] Reed R P and Clark A F 1983 *Materials at low temperature* (Ohio: American Society for Metals)
[2] Koga N, Zhang W, Umezawa O, Tschan V, Sas J and Weiss K P 2018 *IOP Conference Series: Mater. Sci. Eng.* **279** 012004
[3] Koga N, Zhang W, Tomono S and Umezawa O 2021 *Mater. Trans.* **62** 57-61
[4] Koga N, Nameki T, Umezawa O, Tschan V and Weiss K P 2021 *Mater. Sci. Eng. A* **801** 140442
[5] Takaki S 1996 *Sanyo Technical Report* **3** 1-7
[6] Torizuka S 2005 *Bull. Iron Steel Inst. Jpn.* **10** 188-195
[7] Livitsanos C P and Thomson P F 1977 *Mater. Sci. Eng.* **30** 93-98
[8] Tamura I 1982 *Met. Sci.* **16** 245–253
[9] E.O.Hall 1951 *Proc. Phys. Soc. A* **64** 747
[10] N.J.Petch 1953 *J. Iron Steel Inst.* **174** 25
[11] Sun B, Ma Y, Vanderesse N, Varanasi R S, Song W, Bocher P, Ponge D and Raabe D 2019 *Acta Mater.* **178** 10-25
[12] Sugimoto K, Usui N, Kobayashi M and Hashimoto S 1992 *ISIJ Inter.* **32** 1311-1318
[13] Bennet P E and Smclar G M 1966 *J. Basic Eng. Jun.* **88** 518-524

Acknowledgement
The authors acknowledge the financial support of the Grant-in-Aid for Scientific Research (KAKENHI) Grant No. 20K1460 and Iketani Science and Technology Foundation (0331027-A).

Friction stir welding of AISI 316LN high strength austenitic stainless steel for cryogenic application

I Aviles Santillana[1], J Debeux[1], M Meyer[1], M Muddamallappa[2], A Polo Picazo[1], H Robe[3], E Rodriguez Castro[1,4], S Sgobba[1], KP Weiss[5]

[1]*CERN, CH-1211 Genève – Switzerland*
[2]*Texas A&M University. 400 Bizzell Street. College Station, 77843, Texas, United States*
[3] *Institut de Soudure, 4 rue Pilâtre de Rozier, 57420 Goin – France*
[4] *University Carlos III of Madrid, Campus Leganes, Av. Universidad 30, 28911 Madrid – Spain*
[5] *KIT, Institute for Technical Physics, 76344 Eggenstein-Leopoldshafen / Germany*

E-mail: Ignacio.Aviles.Santillana@cern.ch

Abstract. Friction stir welding (FSW) is a solid state joining process that uses the heat generated by the friction of a rotating tool and the base material to join materials together. Due to the fact that the material is never melted, and that extensive plastic deformation is introduced in the weld seam, a unique set of properties is achieved. The technique has been extensively used to join aluminium and aluminium alloys, but very few developments are reported on high strength austenitic stainless steel, which is the material of choice for many high energy physics and fusion magnets. This paper contains a comprehensive microstructural and mechanical characterization, including at cryogenic temperature, of an 8 mm thick high strength austenitic stainless steel plates. The steel grade is the high alloy version of AISI 316LN (identified as 1.4429 or - X2CrNiMoN17-13-3 according to European standards). Special attention was given to cryogenic elastic – plastic J – integral testing of the weld seam. To the authors' knowledge, this is the first time fracture toughness at cryogenic temperature on friction stir welded 1.4429 has been measured.

1. Introduction

Austenitic stainless steel is an excellent material of choice for structural cryogenic applications, thanks to a unique combination of strength, ductility, and toughness at service temperature. Furthermore, it is also a very appealing material for superconducting high field magnets thanks to the very low magnetic permeability of the austenitic phase.

At the European Organization for Nuclear Research (CERN), for its flagship project, High Luminosity LHC (HL-LHC), a series of high field magnets will be integrated in the machine [1]. For the new high field dipoles and quadrupoles, an important component is the cold mass vessel, composed of the shrinking cylinder and the end covers. The vessel is fabricated by welded nitrogen strengthened austenitic stainless steel components, consisting of half shells and covers to be joined together by longitudinal and circular butt welds [2] in order to guarantee an excellent mechanical performance at cryogenic temperature.

This work is intended for the circular welds of the vessel; however, the obtained results show a great potential for any nitrogen strengthened austenitic stainless steel joint, intended for use in cryogenic applications.

1.1. *Friction stir welding of austenitic stainless steel*

Friction Stir Welding (FSW) is a solid-state welding process where a non-consumable rotating tool inserted into the material to be welded creates friction and large plastic deformations to generate heating and softening of the material. During tool movement along the welding line, complex material flow occurs at a temperature below the melting point. FSW involves low heat input and process temperature, thus avoiding classical fusion weld imperfections such as hot cracking, lack of fusion and porosities, resulting in in a high performance and low distortion weld. Compared to light alloys (aluminium, magnesium…), FSW of hard material or high melting temperature like steels, stainless steels and nickel-based alloys is more difficult due to the temperature to be reached to perform the operations. For these materials, welding temperature is often considered above 1000°C [3] and FSW tool lifetime (i.e., tool wear) is a major issue. Tool must have high abrasion resistance, toughness, and creep resistance at elevated temperature and under high load [4].

In the last two decades, FSW of austenitic stainless steel (mainly AISI 304) has been studied [5 - 9], but only a few very recent publications are found for high N grades [9, 10]. The technique has been used for crack repair [5], grain refinement of fusion welds and more rarely for real part applications. The welding parameters for austenitic stainless steel are quite different than for light and high conducting materials such as aluminium alloys, and imply a low rotation speed, a low welding speed and a high axial force [6].

2. Materials and methods

2.1. *Raw – material specification*

The material which was chosen for the development of friction stir welding is the material which is currently used for the vessels of the 11 T dipoles: the high alloy version of AISI 316LN (1.4429 or X2CrNiMoN17-13-3). This alloy guarantees a fully austenitic microstructure and is very stable against undesired martensitic transformation.

Moreover, the material has been purchased according to CERN's technical specification for ultra – high vacuum (UHV) components [10]. The stringent requirements of this material specification for products intended for UHV purposes, impose to apply an adapted metallurgy and manufacturing process, aimed at meeting the structure and inclusion limits specified. The process shall include a mandatory ElectroSlag Remelting (ESR) step in order to guarantee very stringent requirements of inclusion content according to method D of ASTM E45 (severity level number shall be at most 1).

Additional requirements on the chemical composition are imposed with respect to EN 10028-7 for Ni, Cr, Mo, Co and N, and especially low P and S content are requested. The product analysis can be found in Table 1:

Table 1. Product analysis of the 8 mm plates used as base material. R stands for remainder.

Element	C	S	N	Cr	Ni	Mn	Si	Mo	P	Co	B	Fe
Weight %	0.010	< 0.002	0.19	17.32	13.05	1.34	0.26	2.60	0.024	0.10	0.002	R

The relative magnetic permeability at room temperature after solution annealing shall be lower than or equal to 1.005 for fields of over 80 000 A/m.
The internal soundness and microstructural homogeneity of the 8 mm thick plates used as base material is guaranteed by the ultrasonic examination of 100% of the volume.

2.2. *FSW welding*

With this base material, the FSW specimens were produced using a specific FSW machine (MTS I-STIR 10). The process parameters of this study were determined after several preliminary trials, and they are the following: rotation speed: 125 rev/min; Welding speed: 50 mm/min; Tilt angle: 1.5 degrees; Axial force: 62 kN. The welding operations were conducted in force control only to ensure the stability and quality along the weld. Pure Argon was used as shielding gas to protect both the welding area and the tool from oxidation. The process was performed using a polycrystalline cubic boron nitride (PcBN) tool and a cooled head. The tool was composed of an 8 mm pin length of and a convex 25 mm diameter shoulder, as shown in Figure 1.

Figure 1. Friction stir welding tool (70% PcBN/30% W-Re), with a 8 mm long conical pin and 25 mm diameter convex shoulder.

2.3. *Metallographic examination*
Detailed metallographic observations were performed at several cross sections of the FSW. An example of a macroscopic observation is shown in **Figure** ***2***, together with a schematic description of the different regions of a FSW, containing the naming convention that will be used to refer to the different regions.

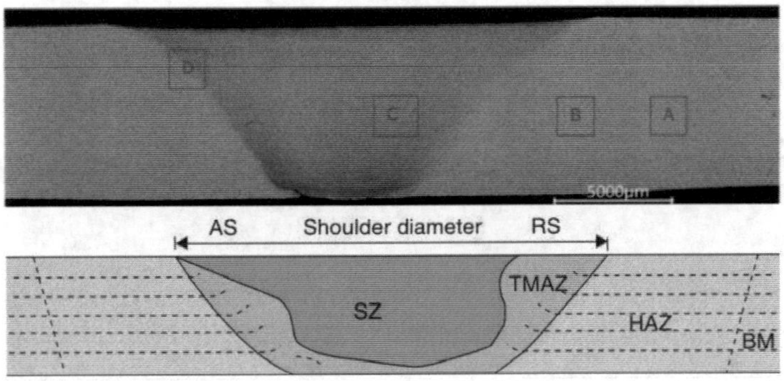

Figure 2. Top) cross section of FSW. Positions marked with letters A – D are the regions where the metallographic observations were carried out. Bottom) schematic of the different regions of a FSW: Advancing side (AS), Retreating side (RS), Stirred zone (SZ), Thermo mechanically affected zone (TMAZ), Heat affected zone (HAZ), Base metal (BM).

Full cross sections were mounted in cold resin, ground and polished down to 1 µm diamond paste. The samples were observed under optical microscope, after chemical etching (Nital® 5% and 10% oxalic acid (etchants #74 and #13b according to the ASTM E407 respectively)). Grain size assessment was done according to ASTM E112, using the linear intercept procedure.

2.4. *Hardness*
A matrix of 7 x 148 Vickers hardness indents (HV2) was performed at a cross section of the FSW. In order to respect the guidelines of ISO 6807 – 1, the distance between indents is 1 mm in thickness and 0.5 mm in the horizontal direction

2.5. *Tensile testing*
Transverse tensile tests were performed in 5 specimens. The geometry and the test parameters are according to ISO 6892-1. The thickness is 6.6 mm (1 mm is removed from the root and 0.4 mm from the face) in order to include as much as possible, the whole cross section of the weld.

Additionally, six flat tensile specimens were extracted according to ISO 6892-4 (gage length: 25 mm; thickness: 3 mm) and have been tested at 4.2 K following the test procedure described in the same standard. They have all been welded via FSW, with the welded zone localized in the middle of the specimen's calibrated length. Thus, transverse tensile tests were performed. Two specimens were extracted from the top (H1 and H2), two from the bottom (B1 and B2) and two at mid-thickness (M1 and M2). 1 mm was removed from the top and the bottom surfaces of the welded coupon before sample extraction to avoid variations in the cross section as well as the contamination observed in 3.1.1.

2.6. *Fracture toughness testing*

Fracture toughness tests were performed on six stainless steel 316LN 6 mm thick compact specimens (T-L orientation) according to ASTM-E1820 at 4 K. As for the tensile specimens, 1 mm was removed from the top and the bottom surfaces of the welded coupon before sample extraction to avoid variations in the cross section as well as the contamination observed in 3.1.1. Three specimens (BM1 – BM3) were extracted from the base material (to be used as a reference), and three specimens (S1 – S3) were extracted with the tip of the notch centred in the FSW's axis (thus, crack growth takes place at the weld bead, as it is shown in Figure 3).

Figure 3. Schematic showing the position of the 'S' specimens with respect to the FSW plates.

An example of a tested specimen as well as the crack opening displacement (COD) gage used to estimate crack growth can be seen in Figure 4. Due to the very high values of toughness expected, at least for the base material [11], an elastic – plastic fracture mechanics' test protocol was put in place: J – tests, single specimen method with the unloading compliance technique for crack estimation. Following the guidelines given by ASTM E1820, after precracking, side grooves were introduced in order to minimize crack deviation and tunnelling.

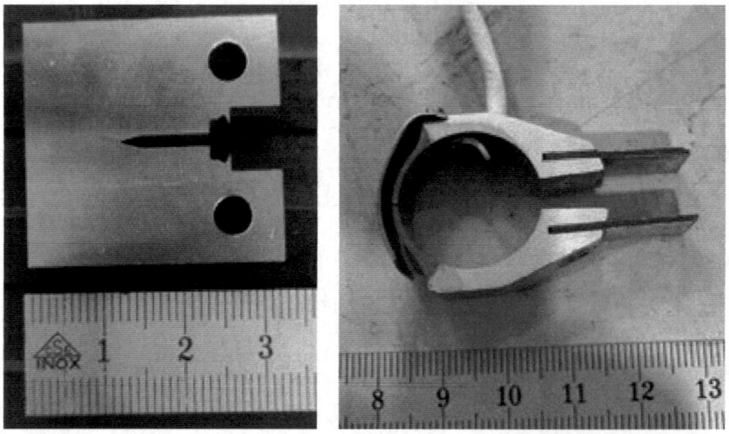

Figure 4. Left) CT specimen after pre cracking. Right) COD gage, courtesy of NHMFL.

IOP Conf. Series: Materials Science and Engineering 1241 (2022) 012003 doi:10.1088/1757-899X/1241/1/012003

3. Results and discussion

3.1. *Weld characterization*

A comprehensive characterization of the weld was performed via metallographic observations and hardness profiles of its cross section:

3.1.1. *Metallographic observations.* As it can be seen in the images below, Figure 5A and Figure 5B, taken at the base metal and heat affected zone respectively, grain size numbers are identical (G = 5) and grains are quite equiaxial. At the SZ, very fine (G = 13) and equiaxial grains are observed (Figure 5C), showing a significant grain refinement due to the FSW process. This effect is observed in Figure 5D, with the transition from the TMAZ and the SZ. The TMAZ corresponds to the area where the deformation and temperature are too low to induce a full recrystallization, thus it is characterized by reoriented and slightly elongated grains due to the shearing efforts introduced by the rotating tool.

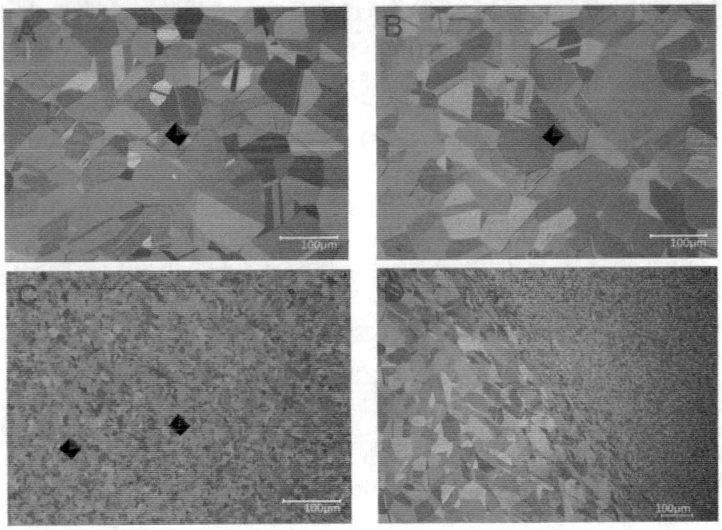

Figure 5. Metallographies selected positions of the FSW. Original magnification: 500x. Some of them contain Vickers indentations. A) Base material; B) HAZ; C)SZ; D) transition between TMAZ and SZ.

Additionally, when looking closer at the weld's root left side, there seems to be a heterogeneity as well as a banded structure. These features were analysed as polished (Figure 6 left) and after Nital® etching (Figure 6 right).

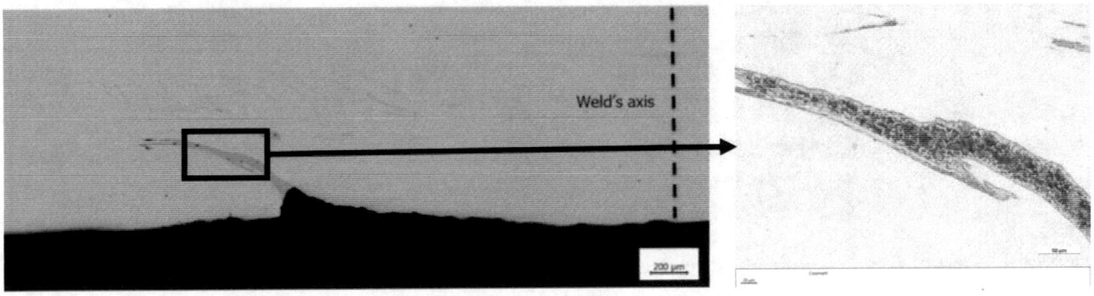

Figure 6. Left) as polished root of the SZ (AS) of FSW. Right) Feature in the orange rectangle. After Nital®. Perlitico – ferritic structure.

Some of them reveal a perlitic – ferritic microstructure, corresponding with inclusions transferred to the weld bead from the anvil used for the FSW. Other secondary phases that did not show any reaction to the etching are most likely a small quantity of δ – ferrite and / or σ – phase. Park et al. already observed (for AISI 304) the formation of these two secondary phases in the advancing side, at the root of the SZ, from introduction of high strain and dynamic recrystallization during FSW [11].

3.1.2. *Hardness profiles*. In order to construct the hardness mapping shown in Figure 7 based on the hardness matrix that was performed, a bilinear interpolation was applied.

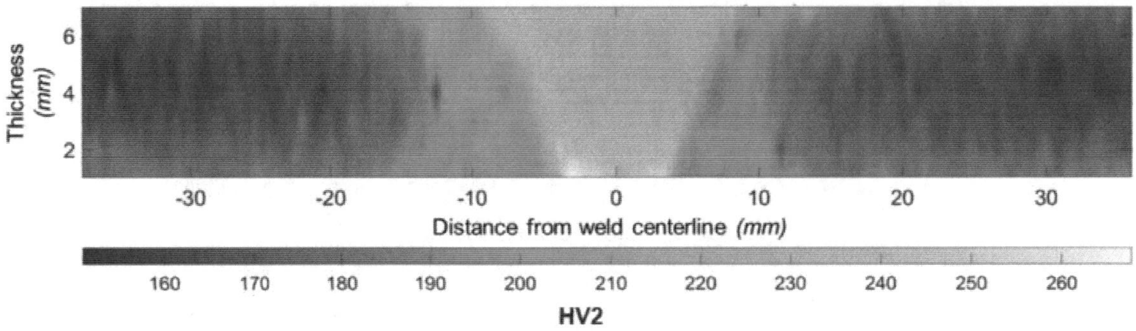

Figure 7. Hardness (HV2) mapping of FSW cross section.

From it, we can observe that all the stirred zones exhibit higher hardness than the base material, induced mainly by a Hall – Petch effect. The highest value of hardness is reached in the centre of the weld's axis, at the nugget. The material hardness changes in the thickness, with the highest values reached at the top and bottom surfaces, where the cooling rate is expected to be higher than at other regions and where extreme grain refinement was observed (see Figure 8). The hardness then decreases in function of the distance from the weld centre line, until it reaches the asymptotic value of the base material at +/- 15mm from the weld's axis. A relatively narrow HAZ is observed, with a similar width on both side of the weld creating a smooth transition to the base metal.

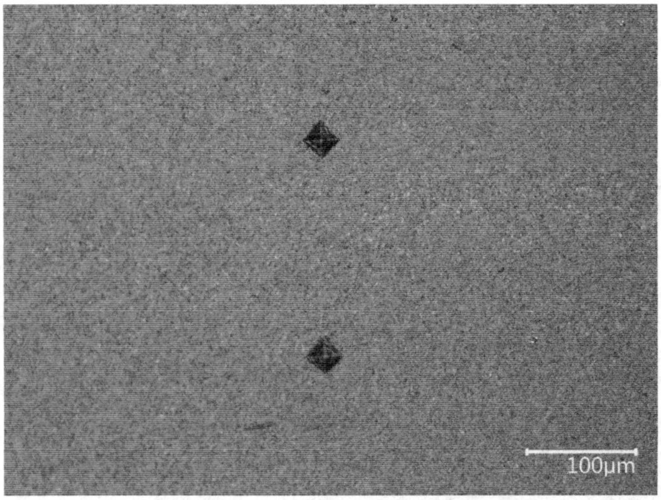

Figure 8. Root of SZ. Original magnification 500x. Vickers indentations are observed. Ultra – fine grain (G=16) is measured.

3.2. *Mechanical testing*
A comprehensive characterization of the FSW was performed, both at room and cryogenic temperatures. At room temperature, tests were performed at the Institut de Soudure (FR) and at cryogenic temperature, tests were performed at CERN (CH).

3.2.1. *Room temperature.* The results of the transverse tensile tests show a joint efficiency > 1 for all the tested specimens (tensile strength is well above the minimum specified value of the parent material; moreover, yield strength is also above the minimum specified value), with the position of the failure always found outside the weld bead and heat affected zone.

3.2.2. *Cryogenic temperature (4.2 K).* Transverse tensile tests and fracture toughness (J – tests) were performed. A detailed analysis of the obtained results is performed here:

Tensile tests results for each specimen are shown in Table 2. Elongation at break is shown for information (the calibrated length is obviously heterogeneous) to highlight the elevated ductility of the specimens at low temperature. For specimen B1, only the tensile strength is shown due to the sliding of the extensometer.

Table 2. Transverse tensile tests results (0.2% proof strength (Rp02); Tensile strength (Rm)and elongation after fracture (A)) of FSW specimens at 4 K

Specimen	Rp_{02} [MPa]	Rm [MPa]	A [%]
H1	1166.3	1483.7	14.6
H2	1190.0	1522.8	16.5
M1	1170.2	1518.1	13.0
M2	1170.2	1530.3	14.5
B1	-	1481.4	-
B2	1153.1	1479.6	16.7

As it can be seen in Table 2, the tensile results show a limited spread of cryogenic tensile properties. Since all the specimens broke well outside the heat affected zone of the weld the joint's strength shows to be higher than the one of the base material.

Figure 9 shows an example of two resistance J - R curves: one obtained for the base material (BM2, left) and one obtained for the FSW (S2, right). They contain all the construction, exclusion and auxiliary lines required for the assessment of J_Q.

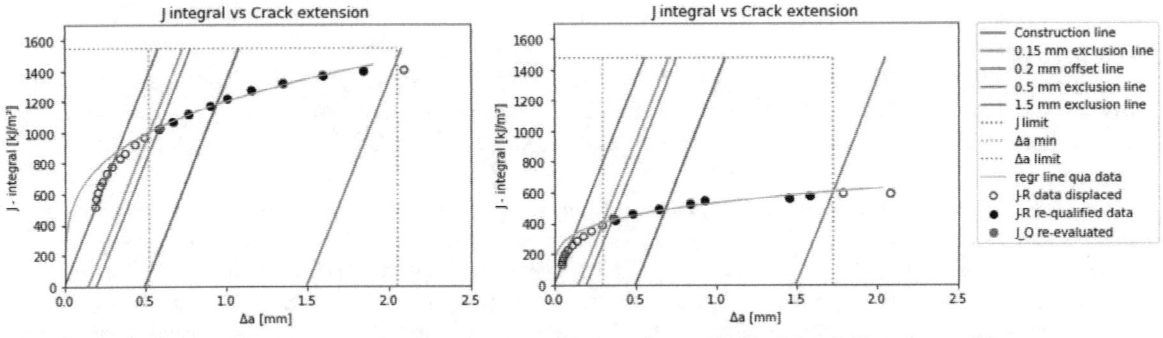

Figure 9. Resistance (J-R) curves. Left) Specimen BM2. Right) Specimen S2.

After crack opening and assessment of the geometrical criteria, only specimen S1 did not fulfil the crack extension requirements of ASTM E1820 due to crack deviation. Thus, the results for this specimen are not exploitable and are not included. For the other 5 specimens, post processing of the data was performed according to [12], even if there is practically no negative crack extension for any of the tested specimens. J values were recalculated using the corrected crack extension values, and K_{JIC} values were calculated from the J_{IC} values. The results are gathered in Table 3:

Table 3. Fracture toughness at cryogenic temperature results

Specimen	BM1	BM2	BM3	S2	S3
J_{IC} [kJ/m^2]	883	1030	732	442	417
K_{JIC} [MPa√m]	430	465	392	304	296

For each group of specimens, the spread is small, showing a high reproducibility of the results. When comparing the average values for each group of specimens (430 ± 13 MPa√m for the base material, 300 ± 5 MPa√m for the FSW) we see a significant drop of fracture toughness, which can be considered as normal for two reasons: one is the typically higher content of inclusions that can act as crack initiators / stress concentrators as suggested by Morris [13]. The second one is the fact that, based on the obtained results of hardness (see 3.2.2) and tensile tests at cryogenic temperature, higher mechanical properties would be expected at the weld seam (including yield strength). It is known that the relationship between fracture toughness and yield strength for a given material in inversely proportional [14], thus explaining the drop of fracture toughness in the weld seam.

However, based on the obtained results, FSW shows higher fracture toughness values at 4 K than similar joints (same base material) performed by fusion welding (Tungsten Inert Gas) as it can be seen in [15], where values of 262 MPa√m were obtained using 1.4453 as filler material and 266 MPa√m when the filler was JK2LB. The reasons behind can be linked to the inherent nature of the FSW explained above: a solid solution joining technique with low heat input. This will be translated in absence of solidification microcracking and other classical fusion welding imperfections, and less propensity to high temperature secondary phases, detrimental for fracture toughness at cryogenic temperature due to its brittleness [15].

4. Summary and Conclusions

Two 8 mm thick high nitrogen AISI 316LN (1.4429) plates have been successfully joined using FSW. Tensile tests at room temperature already showed a welding efficiency larger than 1. Additionally, a comprehensive metallurgical and mechanical cryogenic assessment of the weld coupon was successfully carried out.

The metallurgical assessment and hardness profiles show that the maximum hardness is achieved at the weld's axis, in the top and bottom surface. At these positions, the most pronounced grain refinement is observed, showing a clear Hall – Petch contribution to the hardness. Hardness values at the HAZ seem to have a thermo – mechanical origin. In the TMAZ, there is a combined effect of both contributions.

Transverse tensile tests at 4 K show high yield and tensile strengths. The rupture occurring always outside the welded zone and far from the heat affected zone confirms what was already observed in the hardness profiles: a significant increase of tensile strength of the FSW compared to the base material.

Fracture toughness tests at 4 K of the FSW were successfully performed for the first time in a high nitrogen AISI 316LN (1.4429). Higher values with respect to more conventional fusion welding techniques have been measured, confirming the excellent mechanical performance of the welded joint. Attention should be taken with the contamination observed at the welds' root by transfer from the anvil and from suspected secondary phases, which can have an impact both in the mechanical and the magnetic performance of the FSW.

IOP Conf. Series: Materials Science and Engineering 1241 (2022) 012003 doi:10.1088/1757-899X/1241/1/012003

5. References

[1] Brüning, O., & Rossi, L. (2019). The high-luminosity large hadron collider. Nature Reviews Physics, 1(4), 241-243.
[2] Brüning, O., et al. High-Luminosity Large Hadron Collider (HL-LHC). 2020.
[3] C. D. Sorensen et T. W. Nelson, Friction Stir Welding of Ferrous and Nickel Alloys, Friction Stir Welding and Processing, p. 10.
[4] M. Almoussawi, A. J. Smith, et M. Faraji, Wear of Polycrystalline Boron Nitride Tool During the Friction Stir Welding of Steel, Metallography, Microstructure, and Analysis, vol. 7, no 3, p. 252-267, juin 2018, doi: 10.1007/s13632-018-0439-0.
[5] C. Gunter, M. P. Miles, F. C. Liu, et T. W. Nelson, Solid state crack repair by friction stir processing in 304L stainless steel, Journal of Materials Science & Technology, vol. 34, no 1, p. 140-147, janv. 2018, doi: 10.1016/j.jmst.2017.10.023.
[6] A. P. Reynolds, W. Tang, T. Gnaupel-Herold, et H. Prask, Structure, properties, and residual stress of 304L stainless steel friction stir welds, Scripta Materialia, vol. 48, no 9, p. 1289-1294, mai 2003, doi: 10.1016/S1359-6462(03)00024-1.
[7] S. H. C. Park, *et al.*, Microstructures and properties of friction stir welded 304 austenitic stainless steel, Welding International, vol. 19, no 11, p. 877-881, nov. 2005, doi: 10.1533/wint.2005.3518.
[8] Sunilkumar, D., Muthukumaran, S., Vasudevan, M., & Reddy, M. G. (2020). Effect of friction stir and activated-GTA welding processes on the 9Cr–1Mo steel to 316LN stainless steel dissimilar weld joints. Science and Technology of Welding and Joining, 25(4), 311-319.
[9] Sunilkumar, D., Mathew, J., Muthukumaran, S., & Vasudevan, M. (2020). Friction Stir Welding of 2.25 Cr–1Mo Steel to AISI 316LN Stainless Steel. Transactions of the Indian Institute of Metals, 73, 1689-1693.
[10] CERN 1002. Technical requirements for 1.4429 (X2CrNiMo17-13-3, AISI 316LN) stainless steel sheets / plates for ultra - high vacuum applications at CERN.
[11] S. H. C. Park, Y. S. Sato, H. Kokawa, K. Okamoto, S. Hirano, et M. Inagaki, « Rapid formation of the sigma phase in 304 stainless steel during friction stir welding », Scripta Materialia, vol. 49, no 12, p. 1175-1180, déc. 2003, doi: 10.1016/j.scriptamat.2003.08.022.
[12] J. W. Ekin, “Experimental techniques for low – temperature measurements”, New York, Oxford University press, 2006
[13] Weiss, K., & Nyilas, A. (2006). Specific aspects on crack advance during J-test method for structural materials at cryogenic temperatures. Fatigue & Fracture of Engineering Materials & Structures, 29(2), 83-92.
[14] Morris, J. W. (1993). Steels: for low temperature applications. Encyclopedia of Advanced materials.
[15] Ritchie, R. O. (2011). The conflicts between strength and toughness. Nature materials, 10(11), 817-822.
[16] Aviles Santillana, I., et al., (2018). A comparative study of fracture toughness at cryogenic temperature of austenitic stainless steel welds. Journal of Materials Engineering and Performance, 27(4), 1995-2002.

IOP Conf. Series: Materials Science and Engineering 1241 (2022) 012004 doi:10.1088/1757-899X/1241/1/012004

Study on Irradiation Effect of Insulating Materials for Fusion Superconducting Magnet: Effect of Low-temperature Irradiation

Y Kunitoku, Y Akiyama, Y Manabe, and F Sato

Graduate School of Engineering, Osaka University
2-1 Yamadaoka, Suita, Osaka 565-0871, Japan

yoko-ak@see.eng.osaka-u.ac.jp

Abstract. The purpose of this study is to investigate the change in dielectric performance in the direction of parallel to layers due to the difference in γ-ray irradiation temperature, irradiation atmosphere, and absorbed dose. γ-ray was irradiated using a ^{60}Co source at room temperature, in liquid nitrogen, and in liquid nitrogen temperature gas atmosphere in the absorbed doses of 0, 5, and 10 MGy, respectively, and the dielectric breakdown tests were conducted at room temperature using a high-voltage breakdown tester. To investigate the mechanism of the change in dielectric strength, the density and electron spin resonance (ESR) measurements of the specimens were carried out. After irradiation at room temperature, the dielectric strength increased slightly with the increase in the absorbed dose. The reason for this may be that the molecular density increased due to the promotion of cross-linking reaction by recombination of molecules and reaction of unreacted epoxy groups by γ-ray irradiation. After irradiation in liquid nitrogen, the dielectric strength decreased significantly at 10 MGy, while after irradiation in liquid nitrogen temperature gas atmosphere, the dielectric strength slightly decreased with the increase in absorbed dose. These results suggest that, at low temperature, the freezing of molecular motion limits the movement of radicals generated by irradiation, and that the decomposition reaction is dominant against the crosslinking reaction.

1. Introduction

The materials used in ITER (International Thermonuclear Experimental Reactor) are exposed not only to large electromagnetic forces and cryogenic temperatures but also to radiations of neutron and its secondary radiations such as γ-rays [1][2]. It is necessary to develop materials that can maintain their insulation performance even under these severe conditions. However, the organic polymer materials used for insulating materials are highly radiation-sensitive, so there is a possibility of material degradation and loss of insulation performance due to radiation exposure.

Epoxy resins, which are widely used as electrical insulating materials, are highly radiation-sensitive and tend to deteriorate easily. Therefore, the matrix consists of epoxy resin and cyanate ester with triazine rings that has relatively high radiation resistance [3], will be used as the insulating material in the ITER toroidal field (TF) coils [4]. Glass cloth and polyimide film are used in the composite materials [5]. The use of polyimide film, which has high insulation performance and durability, will provide high insulation performance in the direction vertical to layers. But the insulation performance in the direction parallel to layers is lower than that in the direction vertical to layers because the polyimide film cannot provide insulation in the direction parallel to layers. Since the actual materials are irradiated with neutrons and gamma rays at cryogenic temperatures, it is necessary to study the irradiation effect on the insulating performance of the composite materials considering the irradiation temperature and atmosphere for actual use.

In this study, we focused on the glass cloth and the epoxy resin among the components of composite materials. The glass fiber reinforced plastic (GFRP) was fabricated using these materials, to examine the degradation mechanism of materials by radiation. The purpose of this study is to investigate the change in insulating performance due to γ-ray irradiation by measuring the change in dielectric strength in the direction of parallel to layers, by changing the irradiation temperature, irradiation atmosphere, and absorbed dose.

Firstly, GFRP was prepared by impregnating and curing glass cloth laminated with epoxy resin, and then processed into the shape for the dielectric breakdown test. These specimens were irradiated with γ-ray at room temperature, in liquid nitrogen, and in a liquid nitrogen temperature gas atmosphere using a ^{60}Co source to obtain absorbed doses of 0, 5, and 10 MGy in the same dose rate. Then the specimens were tested for dielectric breakdown at room temperature using a high-voltage breakdown tester. The dielectric strength of the specimens was calculated by dividing the breakdown voltage by the specimen thickness. Based on these results, the change in insulating performance due to irradiation temperature and atmosphere was discussed.

2. Experimental Methods

2.1. Materials
Glass fiber reinforced plastic (GFRP), a composite material, was prepared using epoxy resin and glass cloth. Bisphenol A epoxy resin (diglycidyl ether bisphenol A, DGEBA) [6] was used as the main agent and polyether amine as the hardener. The structural formula is shown in Figure 1, and the reaction of the epoxy group with the amino group is shown in Figure 2. Since ITER is mainly irradiated with neutrons and γ-rays, boron-free S-glass cloth is used as the reinforcing material, which does not cause material degradation due to (n,α) reaction [7][8]. In this study, S-glass cloth which is the same as the actual operating environment was used, although only γ-ray was irradiated.

2.2. Preparation of Test Specimen
A plate of GFRP was prepared and was processed into the specimens for the dielectric breakdown test. 45 sheets of silane-treated S-glass cloth (Arisawa Manufacturing Co., Ltd., Japan) were cut at a size of 65 mm×105 mm, laminated with the fiber direction aligned on the aluminum tray coated with a mold release agent, and fixed by placing a metal plate on the top of it. After that, the glass cloths were dried in the electric furnace at 100°C for 24 h in a vacuum [9][10].

Next, the epoxy resin as the matrix was prepared. Bisphenol A epoxy resin (JER-828, Mitsubishi Chemical, Japan) was used as the main agent, and polyether amine (Baxxodur EC 301, Mitsui Chemicals Fine Chemicals, Japan) was used as the hardener. 100.0 g of the main agent and 32.2 g of the hardener were mixed to achieve the epoxy equivalent weight, and the mixture was stirred in a vacuum for about 3 hours until the air bubbles in the resin disappeared and were completely removed. After that, the mixed resin was filled into a tray that was sufficiently cooled under vacuum, and the glass cloth was impregnated with resin under vacuum for about 2.5 hours. The grass cloth impregnated with resin was then placed in an electric furnace and heat-treated at a rate of 40 °C/h at 70 °C for 3 hours and then at 110 °C for 2 hours to cure the resin. After heat treatment, the trays were cooled to room temperature.

The obtained GFRP plates were processed into the shape shown in Figure 3 using a milling machine and PCB cutter to fabricate the test specimens. In the breakdown test, if the specimen has a short creepage distance, creepage discharge that extends along the surface of the solid will occur before breakdown occurs. Thus, we cut a trench in the specimen, which can increase the creepage distance in the short side direction building walls on both sides, and in the long side direction by building a wall on one side. The reason for the open shape of one of the long sides was to make it easy to measure the thickness of the specimen with a micrometer. In addition, the specimen was shaped so that the dielectric strength could be measured not in the direction of vertical to layers but in the direction of parallel to layers, where the dielectric strength is relatively low.

A total of 45 test specimens were prepared by this method, and the dielectric strength was measured so that the number of samples per a condition was 5.

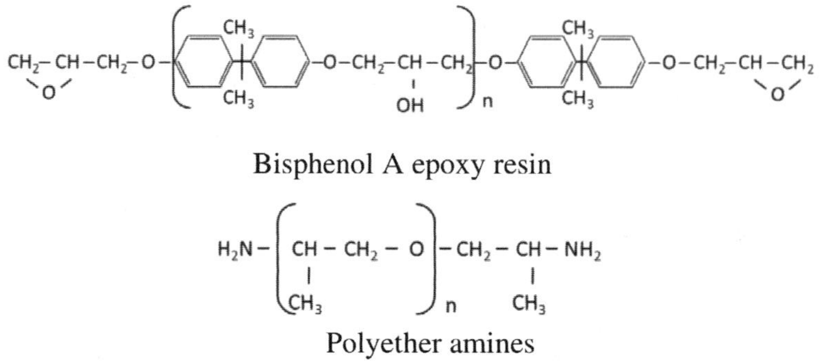

Figure 1. The chemical structures of bisphenol A epoxy resin as a main agent and polyether amine as a hardener [4].

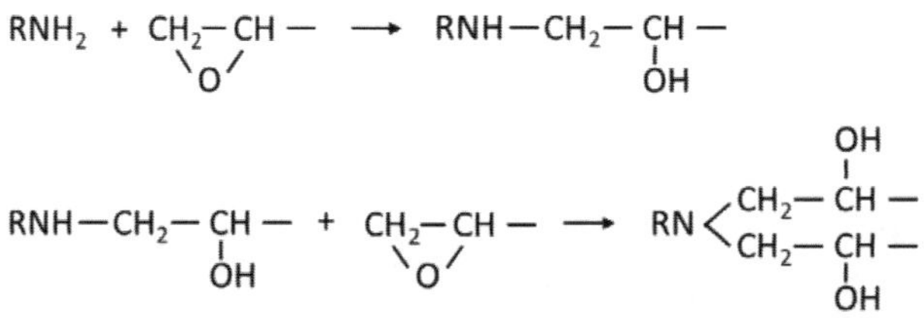

Figure 2. Curing reaction of amino and epoxy groups [4].

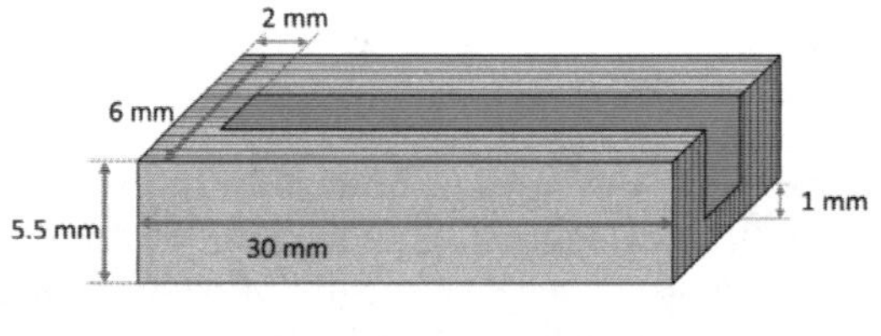

Figure 3. GFRP test specimen used for dielectric breakdown test. The layers show the direction of grass fibers.

2.3. Gamma-ray irradiation

The specimens processed above were irradiated with γ-ray at room temperature, in liquid nitrogen, and in a liquid nitrogen temperature gas atmosphere in the ^{60}Co γ-ray irradiation facility at the Institute of Scientific and Industrial Research (ISIR), Osaka University.

In the irradiation at room temperature (RT), the specimens were placed side by side in polyethylene bags and fixed to the irradiation table. In the irradiation in liquid nitrogen (LNT), the specimen was fixed on the glass rods and were directly immersed in a Dewar bottle (made of stainless steel) filled with liquid nitrogen. In the irradiation in liquid nitrogen temperature gas atmosphere (LNTA), nitrogen-purged freezing vessels were fixed on the glass rods and irradiated in Dewar bottles filled with liquid nitrogen. The irradiation time was 8 h/day including intermittent irradiation. The dose rate was 84 kGy/h, and the absorbed doses were 5 MGy and 10 MGy after 60 and 120 hours of irradiation, respectively.

2.4. Dielectric breakdown test

Dielectric breakdown tests were conducted on the irradiated and the unirradiated specimens using high-voltage dielectric strength tester (7474, Keisoku Giken Co., Ltd. Japan). A needle electrode with a diameter of 1.8 mm and a disc electrode with a diameter of 25 mm made of SUS304 were used. The voltage was increased at a rate of 0.5 kV/s, and the application was stopped when a current of 1 mA or more flowed, and the voltage at that time was used as the breakdown voltage of the specimen. After measuring the breakdown voltage of each specimen, the thickness of each specimen was measured with a micrometer (MDC-25MJ, Mitutoyo Corp. Japan). The dielectric strength per unit thickness of the specimen was calculated by dividing the breakdown voltage by the thickness of the specimen, and the average value was used as the dielectric strength [kV/mm].

2.5. Density measurement

The weight of the specimens in water and air was measured using the specific gravity measurement mode of an electronic balance (Analytical Balance, AUW-D series, Shimadzu Corp. Japan), and the density was calculated using the Archimedes method with equation (1),

$$\rho = \frac{W_a}{W_a - W_l}\rho_l \quad (1)$$

where ρ is density of test piece, W_a is weight in the air, W_l is weight in liquid, and ρ_l is density of distilled water.

2.6. ESR measurement

A different specimen from the one used for the dielectric breakdown test was prepared using the same procedure. The size of the specimen was 2.5×2.5×40 mm to fit into the glass tube for ESR measurement. The ESR measurement of the specimens was carried out using an ESR instrument (JES-RE2X, JEOL, Japan) at room temperature in the air. A central magnetic field was 336.0 mT, and the frequency and power of the microwave was respectively 9.733 GHz and 1.0 mW.

3. Results and discussions

3.1. Dielectric breakdown test

The results of the dielectric breakdown tests of the samples irradiated under each irradiation condition are shown in Figure 4. The dielectric strength of the samples irradiated at room temperature (RT) increased slightly with the increase in the absorbed dose. This may be due to the recombination of molecules or the promotion of cross-linking reaction of unreacted epoxy groups by γ-ray irradiation. On the other hand, in the case of irradiation in liquid nitrogen (LNT), the dielectric strength slightly decreased at the absorbed dose of 5 MGy and significantly decreased at 10 MGy. This may be due to the acceleration of oxidative degradation caused by the ozone generated by irradiation on the oxygen in the air dissolved in the liquid nitrogen.

It is known that when liquid nitrogen is irradiated with γ-rays, the oxygen in the air dissolves into the liquid nitrogen and changes to ozone, which can explode in the presence of ignition sources such as electric sparks or organic materials [11] [12]. Ozone (O_3) oxidizes other substances and decomposes itself into oxygen (O_2). It is also known to react selectively with double bonds, which are one of the unsaturated bonds, as shown in Figure 5. First, ozone is added to the double bond to form the initial ozonide, molozonide. Since molozonide is unstable, it is quickly decomposed into ozonide, which is further transformed into carbonyl compounds. It is thought that the above ozone decomposition decreased the molecular density and the dielectric strength. We have also confirmed by Fourier transform infrared spectroscopy (FT-IR) that the peak of the carbonyl group at 1384 cm^{-1} is significantly increased at 10 MGy irradiation.

In view of the actual operating environment, it is necessary to eliminate the effect of ozone and investigate the simple effect of cryogenic temperature. To reduce the effect of ozone, liquid nitrogen temperature gas atmosphere irradiation (LNTA) was conducted. A gradual decrease in the dielectric strength was observed as the absorbed dose increased. The reason for this is considered to be the freezing of molecular motion caused due to low temperature. It is thought that the molecular chain scission became dominant by freezing of molecular motion, while the cross-linking reaction and recombination of molecular chains were prevented, resulting in a decrease in molecular density and dielectric strength.

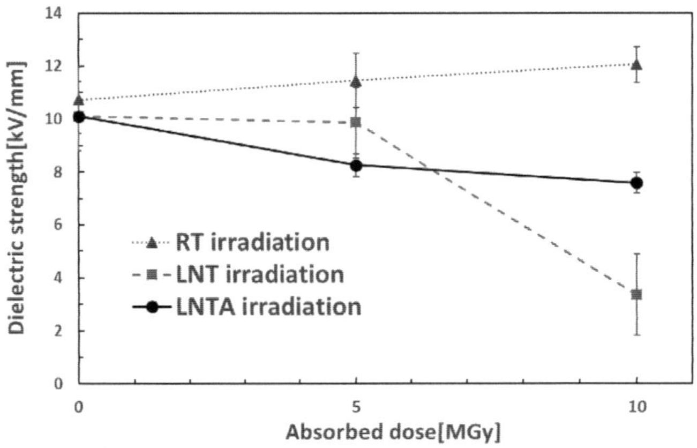

Figure 4. Relationship between absorbed dose and dielectric strength in RT, LNT, and LNTA irradiation.

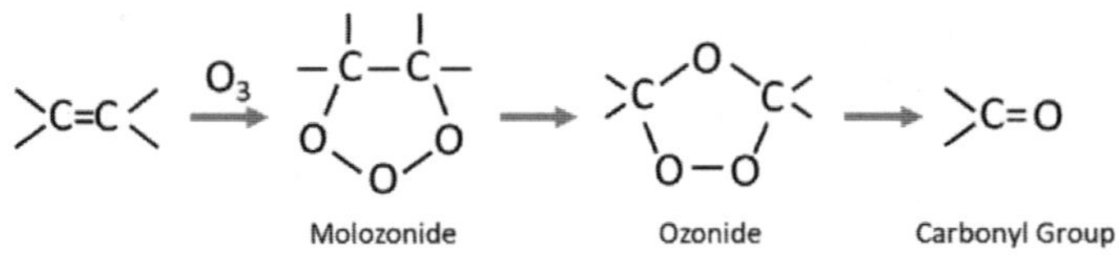

Figure 5. The reaction of ozone with carbon-carbon double bond [11].

3.2. Density measurement

The increase or decrease in dielectric strength could be due to change in the density caused by molecular cross-linking or scission, so we measured the density of these specimens. When a high electric field is applied to a solid material, a current flows rapidly through the solid at a certain voltage, resulting in the dielectric breakdown, in other words, an electron avalanche occurs in the material. The electron avalanche is a phenomenon in which electrons in a material molecule accelerated by an electric field collide with another molecule, ionize the molecule to generate a new accelerated electron. The higher the molecular density, the shorter the time it takes for the electron to collide with another molecule. When the molecular density is high, the electrons collide with the molecule before they receive enough energy to ionize the molecule, and the dielectric breakdown is less likely to occur. On the other hand, when the molecular density is low, the electrons receive enough energy to ionize the molecule before colliding with it. Based on this, the relationship between dielectric strength and molecular density was examined.

The relationship between dielectric strength and density in RT, LNT, and LNTA irradiation is shown in Figure 6. In the room temperature (RT) irradiation, the density decreased at 5 MGy and then increased again at 10 MGy. This may be due to the scission of side chains such as C-O bonds, which have low binding energy, at 5 MGy. On the other hand, at 10 MGy, the radicals generated by the molecular chain scission may have recombined.

The dielectric strength and density of the samples irradiated at liquid nitrogen temperature (LNT and LNTA) were almost proportional to each other. The molecular motion is almost frozen at the liquid nitrogen temperature, and the radicals generated by irradiation cannot move, making it difficult for reactions of cross-linking and recombination to occur. Therefore, the cross-linking and recombination by radicals hardly occurred, and the molecular chain scission reaction is considered to be dominant. Therefore, it is thought that the molecular chain scission produced more sparse parts in the molecular structure than that before irradiation, resulting in lower molecular density and dielectric strength.

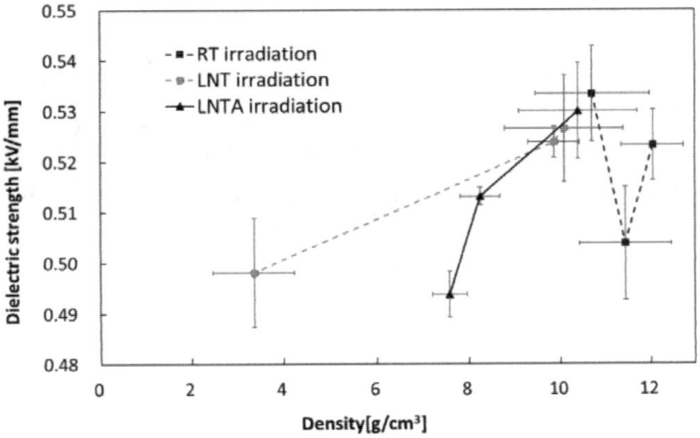

Figure 6. Relationship between dielectric strength and density in RT, LNT, and LNTA irradiation.

3.3. ESR measurement

In the relationship between dielectric strength and density in the room temperature irradiation, the dielectric strength monotonically increased while the density decreased once at 5 MGy and increased again at 10 MGy. The reason was thought to be that the molecular chains with low binding energy were broken at 5 MGy, resulting in a temporary decrease in molecular density, and the molecular chains were cross-linked by the radicals generated by the scission at 10 MGy, resulting in an increase in molecular density. To confirm the chemical reaction at room temperature, ESR measurements were conducted on the samples irradiated at room temperature.

The results of ESR measurements are shown in Figure 7, and the absorbed dose dependence of the radical peaks [13] [14] at 335 mT is shown in Figure 8. The peak intensity of the generated radicals increased up to 7 MGy and then decreased at higher absorbed dose. This trend was similar to the results that the molecular density decreased at 5 MGy and then increased again at 10 MGy. In other words, in the absorbed dose of 5 MGy, the molecular chains with low binding energy were broken and generated the long lifetime radicals, and the radicals promoted recombination and cross-linking reactions at higher absorbed doses. Here, the side chains of the molecules were temporarily broken, whereas the main chains were not, and it is thought that the dielectric strength did not decrease at 5 MGy. The g-value of ESR spectra was 2.3928±0.0004 and did not vary significantly by change in the adsorbed dose, indicating that the same type of radical species once increased and then decreased.

These results revealed the mechanism of molecular cleavage and crosslinking at room temperature, but the dielectric strength increases slightly and monotonically with increasing absorbed dose, regardless of the increase or decrease in density or radicals for the samples irradiated at RT as shown in Figure 4. There are two possible reasons for this; one is that the decrease in density at room temperature is due to side-chain scission, while the decrease in density at liquid nitrogen temperature may be due to main chain scission, because the decrease in density by irradiation is more significant in LNT and LNTA irradiation than in RT as shown in Figure 6. The second reason is that the dielectric strength does not necessarily depend only on the density, but also on the molecular structure and the radicals present in the polymers.

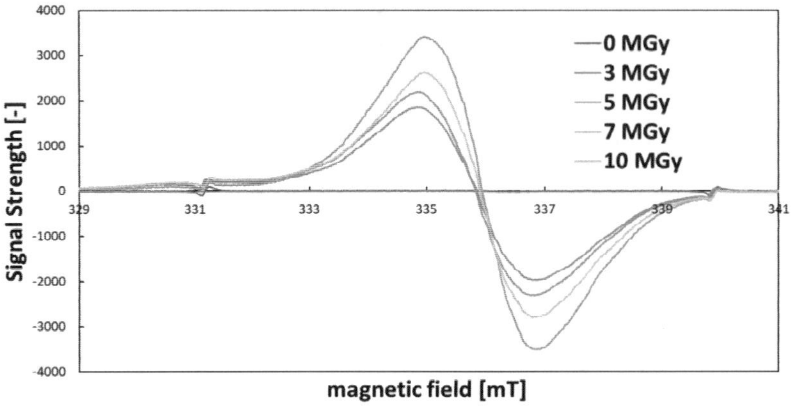

Figure 7. ESR spectra of γ-ray irradiated samples at room temperature.

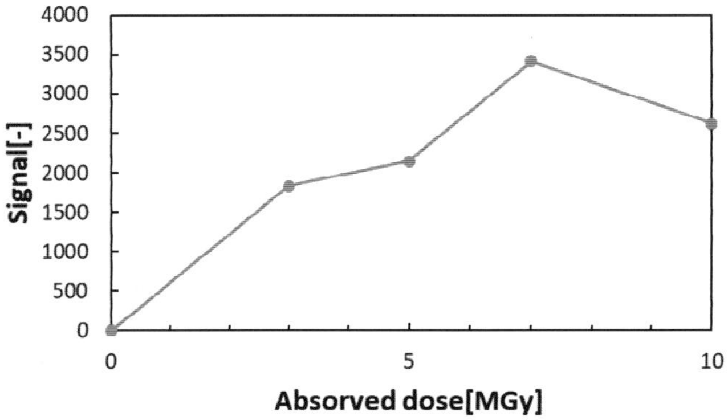

Figure 8. Absorbed dose dependence of radical peaks in ESR spectra at 335 mT.

4. Conclusion

In this study, we investigated the change in dielectric strength of GFRP by γ-ray irradiation in three different environments: at room temperature (RT), in liquid nitrogen (LNT), and in liquid nitrogen temperature gas atmosphere (LNTA).

After irradiation at room temperature (RT), the dielectric strength increased slightly with increase in the absorbed dose. The results of density and ESR measurement suggested that the crosslinking reaction may be caused by the radicals generated by the side chain scission. In the case of irradiation in liquid nitrogen (LNT), oxidative degradation was occurred due to the change of oxygen dissolved in the liquid nitrogen into ozone by γ-ray irradiation, which results in a significant decrease at 10MGy. In the irradiation in liquid nitrogen temperature gas atmosphere (LNTA), where the effect of ozone was reduced, dielectric strength decreased slightly with increase in the absorbed dose. It is considered that the molecular motion was frozen due to the low temperature, which suppress recombination and crosslinking reactions to occur and the decomposition reaction becomes dominant.

These results indicate that the dielectric strength of the specimens irradiated at low temperature may be decreased, whereas that irradiated at room temperature increased. This suggests that the restriction of molecular motion at cryogenic temperature makes the decomposition reaction dominant against the crosslinking reaction. However, the matrix of the samples used in this study is only epoxy resin to simplify the system, so we plan to conduct the research using a mixed resin of cyanate ester and epoxy which may be adopted in actual use.

5. References

[1] ITER Design Description Document 2006 DDD11 ITER_D_22HV5L v2.2
[2] Aymar R 2001 ITER R&D: executive summary: design overview, *Fusion Engineering and Design*, 55 pp.107-118
[3] Mitchella N and Salpietro E 2001 ITER R&D: Magnets: Toroidal Field Model Coil, *Fusion Engineering and Design* 55 pp.171 – 190.
[4] Bittner-Rohrhofer K, Humer K, Fillunger H, Maix R K, Weber H W, Fabian P E and Munshi N A 2004 Radiation effects on the mechanical integrity of novel organic insulators for the ITER magnet coils *Journal of Nuclear Materials* 329-333 (1-3 PART B) pp. 1083-1087
[5] Hemmi T, Nishimura A, Matsui K,. Koizumi N, Nishijima S and Shikama T 2014 Evaluation of Inter-laminar Shear Strength of GFRP Composed of Bonded Glass/Polyimide Tapes and Cyanate-Ester/Epoxy Blended Resin for ITER TF Coils *AIP Conference Proceedings* 1574 pp.154-161
[6] Bittner-Rohrhofer K, Humer K and Weber H W 2002 Low-temperature tensile strength of the ITER-TF model coil insulation system after reactor irradiation, *Cryogenics* Volume 42, Issue 5, pp. 265 – 272
[7] Humer K, Prokopec R, Maix R K, Fillunger H and Weber H W 2009 Insulation systems for superconducting fusion magnets based on cyanate ester blends, *Fusion Engineering and Design* Volume 84 Issue 2-6 pp.960 - 963
[8] Humer K, Prokopec R, Weber H W, Fillunger H and Maix R K 2013 Characterization and qualification of advanced insulators for fusion magnets, *Fusion Engineering and Design* Volume 88 Issue 5 pp.350 – 360
[9] Kobayashi K, Akiyama Y and Nishijima S 2017 Irradiation effect of the insulating materials for fusion superconducting magnets at cryogenic temperature *Journal of Physics: Conference Series* 897(1) 012007
[10] Kito S, Akiyama Y and Nishijima S 2019 Study on irradiation effect of insulating materials for fusion superconducting magnet: Change in electric insulation performance by irradiation, *IOP Conference Series: Materials Science and Engineering*, 502(1) 012195
[11] Brereton S J 1988 Explosion hazard in liquid nitrogen cooled fusion systems *Fusion Technology* Volume 15, Issue 2 pt 2B, pp.833 – 838
[12] Lattimer R P, Kuczkowski R L and Gillies C W 1974 Mechanism of ozonolysis. (a)Microwave spectra, structures, and dipole moments of propylene and trans-2-butene ozonides. (b)Orbital symmetry analysis, *J. Amer. Chem. Soc.*, 96[2] pp. 348-358
[13] Veksli Z, Andreis M and Rakvin B 2000 ESR spectroscopy for the study of polymer heterogeneity *Progress in Polymer Science* Volume 25, Issue 7 , pp.949-986
[14] Diao F, Zhang Y, Liu Y, Fang J and Luan W 2016 γ-Ray irradiation stability and damage mechanism of glycidyl amine epoxy resin, *Nuclear Instruments and Methods in Physics Research Section B: Beam Interactions with Materials and Atoms*, 383, pp.227-233

Acknowledgments

This research was partly supported by National Institute for Fusion Science (NIFS), Japan (NIFS16KERA010)

Epoxy Electret: A Remedy for Partial Discharge at Cryogenic Temperature

Farhina Haque and Chanyeop Park

Department of Electrical and Computer Engineering, Mississippi State University, Mississippi State, MS, 39762, USA

fh317@msstate.edu, chanyeop.park@ece.msstate.edu

Abstract. Epoxy-based composites exhibit mechanical compatibility at cryogenic temperatures. Owing to these properties, composites based on epoxy are used as electrical insulators in high temperature superconducting (HTS) power applications. However, the inevitable presence of voids in solid insulators, triple points, and airgaps at high-voltage conductor-insulator interfaces increase electric fields locally. The intensified electric field around these defects and interfaces is the main cause of partial discharge (PD), which is a dielectric challenge for numerous power applications including HTS cables. Recently, electret has been introduced as a promising solution to mitigate PD activities caused by voids and triple points. In this work, we intend to report the PD mitigation performance of electrets fabricated from epoxy resin, which is suitable for cryogenic power applications.

1. Introduction

In recent years, high temperature superconducting (HTS) power cables have become preferable over conventional cables in serving the increasing electric demand [1]. High power density provided by the HTS power cables substantially reduces the weight and volume of electric power system. HTS power technology is useful in all-electric shipboard and electric aircraft power system, where high-power ratings with reduced weight and size are demanded [2]. Gaseous helium (GHe) is used as a coolant in HTS systems owing to its extended operating temperature range [3]. In the gas-cooled HTS power cable, lapped tape is commonly used for electrical insulation [1], [2], [4]. Lapped tape insulation is helically wrapped around the cable and butt gaps are introduced to avoid mechanical stress. Butt gaps are filled with the coolant that has comparatively low permittivity than the insulation tape. This causes local electric field enhancement; the main cause of partial discharge (PD) [4] that occurs actively at sharp edges, triple points, and cavities leading to dielectric material ageing and the increased risk of device failure.

Numerous studies have been proposed based on geometry [5] and material [6]–[8] to address the challenge of electric field enhancement in the field of high voltage and dielectrics. Recently, an electret-based approach was proposed in [9], in which it was numerally shown that the local field enhancements at sharp edges, triple points, and cavities can be substantially reduced by incorporating electret layers. In [10], the same research group showed experimentally the PD mitigation performance of electrets. Electrets are the electrostatic analogy of permanent magnets holding charge with significantly high lifetime. It has been reported in the literature that the charge decay rate of electret made from Teflon is 1 percent per year [11]. These materials are fabricated from either silicone dioxide (SiO_2)-based inorganic materials or polymer-based organic materials. Based on the triode-corona charging method, we charge polyvinylidene fluoride (PVDF) under various charging conditions including charging voltage, duration, and temperature. The charge stored in the electret surface are utilized to neutralize the intense local electric field and mitigate PD activities [11], [12]. Epoxy-based composites show exceptionally high electrical and mechanical strength at cryogenic temperature [13]. For these reasons, epoxy spacers are widely used in HTS power cables [15]. However, butt gaps filled with cryogens are the main source of PD in HTS power cables. To address this problem, we fabricated electrets by charging epoxy-based composites and utilized them in mitigating PD activities caused by cavities, which is equivalent to butt gaps from a physical standpoint. The PD mitigation performance of uncharged epoxy and charged epoxy (electret) are compared while they are both stressed under same power electronics switching voltage.

2. Electric field neutralization by electret

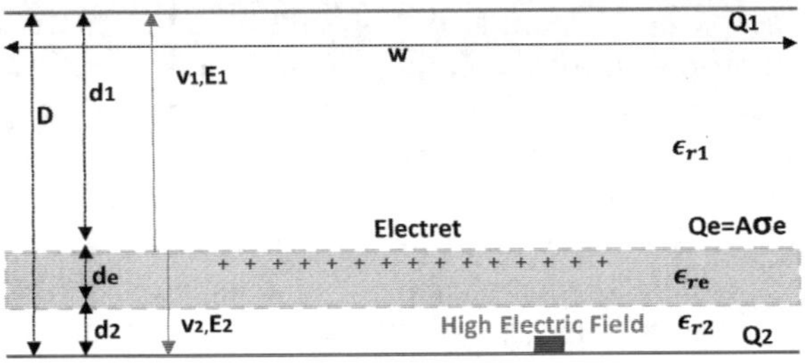

Figure.1. Model used to derive the surface charge density of electret required to neutralize enhanced local electric field.

To demonstrate the electric field neutralization of electret, we assumed a model shown in figure 1. Here, two surfaces with arbitrary surface charge Q_1 and Q_2 is assumed. An electret layer with surface charge density σ_e and thickness d_e is inserted between them. The potential difference between the top surface and the top surface of the electret layer is v_1 and the bottom surface of the electret layer and the bottom surface is v_2. We define two electric fields E_1 and E_2 such that E_1 pointing toward the top surface and E_2 pointing towards the bottom surface. If there is a point of high electric filed on the bottom surface, then the required surface charge density to neutralize the electric field is defined as below [9], [10].

$$\sigma_e = \frac{\epsilon_o \epsilon_{r1} v}{d} \tag{1}$$

3. Fabrication of epoxy-based electret

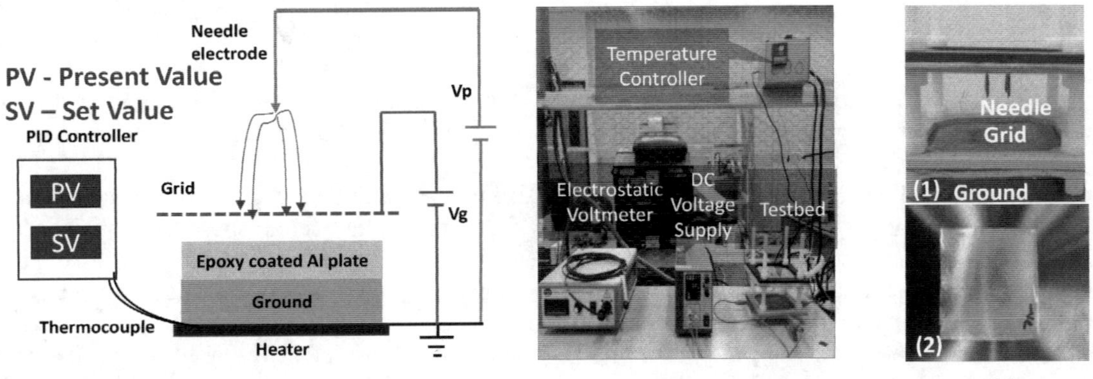

(a) (b) (c)

Figure. 2. Fabrication of electret from epoxy-based composites. (a) Schematic diagram of triode-corona charging method used for charging epoxy coated aluminium plate at elevated temperature, (b) experimental setup used to fabricate electret based on triode-corona charging method, (c) a close view of the testbed (c-1) and epoxy electret (c-2).

An aluminium (Al) plate with thickness 0.25 mm was coated with a thin layer of epoxy. The epoxy used for this purpose was 820 resin that is a very low viscosity resin, and 824 hardener of minimum cure schedule 24 hours was added to the resin in 5:1 volumetric ratio for the curing purpose. The thickness of the epoxy coating was 0.2 mm. Once the coating was done, the Al-plate was placed on an induction heater controlled by a PID controller to accelerate the curing process of the epoxy in air. The epoxy layer was charged based on the triode-corona charging method during the whole curing process. In figure 2(a), the schematic diagram of the triode-corona charging method along with the temperature control system is shown. The needle electrode was supplied with high voltage (17 kV DC) to generate corona discharge that ionizes air surrounding the needle electrode. These ions were forced downwards due to the presence of electric field. The grid electrode, a uniformly spaced copper mesh, was placed 5mm above the epoxy coated Al-plate. The main reason for placing the grid electrode is to create a uniformly distributed electric field such that the electret sample is charged uniformly. Temperature was kept at 60 °C to accelerate the curing process and to increase the stability of the surface charge by allowing charged particles to penetrate into deep traps. Figure 2(b) shows the experimental setup used in the lab to fabricate epoxy electret using the triode corona charging method. For the grid voltage, a DC source was used to supply 600 Vdc to the grid electrode. Needle electrode was supplied with 17 kV DC voltage generated and amplified by a function generator and high voltage amplifier, respectively. After the curing process, we stopped the charging process and measured the surface potential with an electrostatic voltmeter. The surface potential of the epoxy electret achieved by this process was 164 V, which translates 26.17 μC/m^2 surface charge density. To compare the effectiveness of epoxy electret in mitigating PD activities, an uncharged epoxy coated Al-coated sample was prepared and cured under the same temperature.

4. PD mitigation by epoxy electret

4.1. Experimental Description

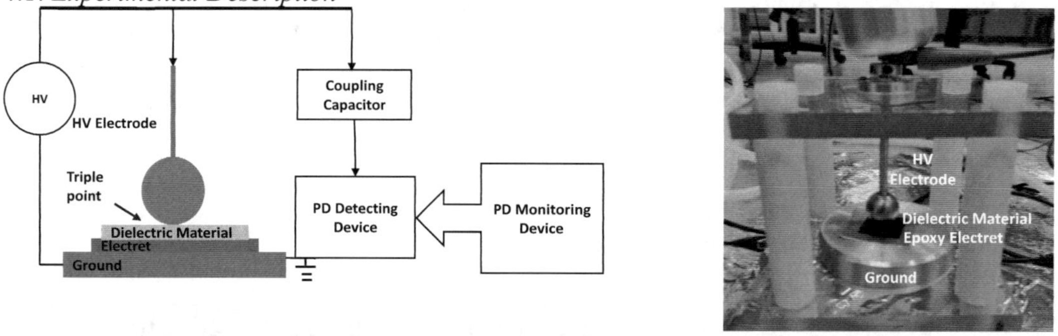

(a) (b)

Figure. 3. Partial Discharge measurement with epoxy-based electret. (a) Schematic diagram of PD measurement circuit designed based on IEC standard 60270 [16], (b) PD measurement testbed used.

With the uncharged epoxy layer and charged epoxy layer, we conducted a series of PD experiments under same square voltage waveform and compared PD signals. PD occurs actively on sharp edges, triple points, and cavities. In this study, a spherical electrode shown in figure 3 was used as the high voltage electrode to avoid the effect of sharp edges causing PD. However, the presence of triple points at the interface of metal electrode, dielectric material, and surrounding medium (air) cause local field enhancements and promote PD occurrence. A 3D printed solid insulator was used as the dielectric material as shown figure 3(a). The micro-scale cavities inherently created in the 3D printed solid insulator are the main sources of PD activities. Below the 3D printed solid insulator, the uncharged epoxy layer was placed. A function generator was used to generate the square voltage, which is amplified with a high voltage amplifier and supplied to the high voltage electrode. PD activities caused by the presence of triple points and cavities are detected with the coupling capacitor and PD

IOP Conf. Series: Materials Science and Engineering 1241 (2022) 012005 doi:10.1088/1757-899X/1241/1/012005

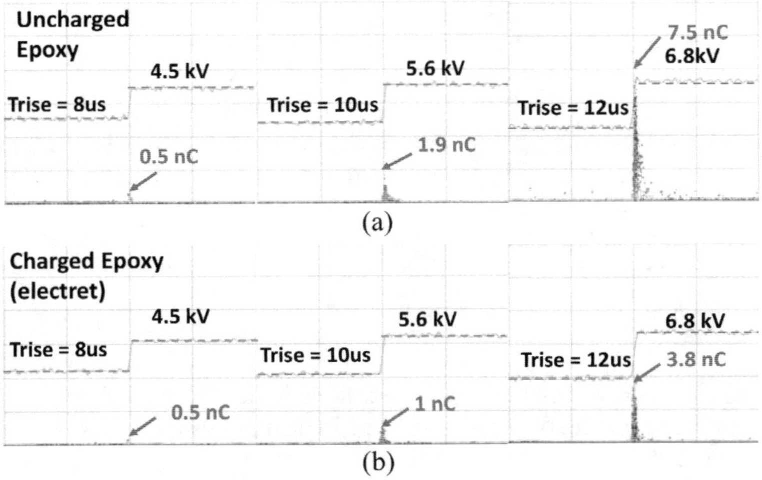

Figure. 4. The PD mitigation performance of uncharged epoxy and charged epoxy stressed under three different square voltage stimuli. In each case, the *dv/dt* was maintained at a constant value. (a) PD signals with uncharged epoxy, (b) PD signals with charged epoxy (electret).

detecting device (Omicron MPD 600). PD signals achieved with the uncharged epoxy were recorded with a computer-based software provided by Omicron. Later, uncharged epoxy was replaced with the charged epoxy (electret) and PD signals under same square voltage were recorded. Figure 3(b) shows the testbed used to conduct the PD measurements.

4.2. Comparison of PD mitigation performance

PD measurements were conducted with uncharged and charged epoxy (electret) layers under same unipolar square voltage waveforms. We applied three different square voltage levels on the high voltage electrode and maintained a constant *dv/dt* of 560 V/μs for all voltage levels. To this end the square voltage magnitude and rise time were both increased systematically. The PD signals achieved at each voltage level with uncharged and charged epoxy are presented in figure 4. Figure 4(a) is the case with uncharged epoxy and figure 4(b) is the case with charged epoxy electret. It is shown that when square voltage magnitude varies between 0 to 4.5 kV with a rise time 8 μs, both uncharged and charged epoxy caused same PD magnitudes of 0.5 nC. Subsequently, the voltage was increased to 5.6 kV and to maintain same *dv/dt* ratio, rise time was set to 10 μs. It is observed that, PD magnitude with uncharged epoxy (1.9 nC) is 90 % higher than that of the charged electret (1 nC). Once the voltage level is further increased and set to 6.8 kV with the rise time of 12 μs, there is a drastic rise in PD magnitudes in the case of the uncharged epoxy (7.5 nC). However, PD magnitudes achieved with the charged epoxy is 3.8 nC which only 50% of the PD magnitude resulted with the uncharged epoxy. This significant reduction in the PD magnitudes achieved with the electrets fabricated from epoxy suggests that the charge stored in the electret surface neutralizes the high electric field caused by triple points and cavities. In order to confirm the results achieved with the epoxy-based electret in PD mitigation, we prepared one more electret sample by charging epoxy coated Al-plate under identical conditions and the results shows significant reduction in the electric field.

5. Discussion

In this paper, epoxy- based electrets are fabricated using the triode corona charging method while curing. We did not use degassing prior to casting. The objective of this paper is to prepare epoxy-based electret and observe if they are able to mitigate PD. Several samples of charged and uncharged epoxy sheets were cured under identical condition that did not involve a degassing process. Despite the possibility of having bubbles, the results show that epoxy-based electrets can mitigate PD. The next step is to fabricate electrets that does not have bubbles by degassing the resin while curing. The charge retention characteristics of epoxy-based electrets have yet to be investigated. Charge retention both at room temperature and at cryogenic conditions will be characterized in our future work.

6. Conclusion

In this study, we experimentally demonstrated the performance of epoxy-based electrets in mitigating PD activities caused by the presence of triple points and cavities. The results show that electrets with a surface charge density of 26.17 $\mu C/m^2$ can reduce PD magnitude by 50%. Further improvements in PD mitigation are expected as improved epoxy-based electrets are developed. With higher surface charge density and increased stability, the epoxy-based electrets will enable PD-free conditions in cryogenic and HTS power applications.

7. References

[1] P. Cheetham, J. Viquez, L. Graber, C. H. Kim, H. Rodrigo, and S. Pamidi, "Novel Design Concept and Demonstration of a Superconducting Gas-Insulated Transmission Line," *IEEE Trans. Appl. Supercond.*, pp. 1–1, 2016, doi: 10.1109/TASC.2016.2641905.

[2] T. Stamm, P. Cheetham, C. Park, C. H. Kim, L. Graber, and S. Pamidi, "Novel gases as electrical insulation and a new design for gas-cooled superconducting power cables," *IEEE Electr. Insul. Mag.*, vol. 36, no. 5, pp. 32–42, Sep. 2020, doi: 10.1109/MEI.2020.9165697.

[3] C. Park, S. Pamidi, and L. Graber, "Boltzmann Analysis of Cryogenic He–H_2 Gas Mixtures as Dielectric Media for High-Temperature Superconducting Power Devices," *IEEE Trans. Appl. Supercond.*, vol. 27, no. 4, pp. 1–6, Jun. 2017, doi: 10.1109/TASC.2016.2637319.

[4] P. Cheetham, W. Kim, C. H. Kim, S. V. Pamidi, L. Graber, and H. Rodrigo, "Use of partial discharge inception voltage measurements to design a gaseous helium cooled high temperature superconducting power cable," *IEEE Trans. Dielect. Electr. Insul.*, vol. 24, no. 1, pp. 191–199, Feb. 2017, doi: 10.1109/TDEI.2016.005909.

[5] H. Reynes, C. Buttay, and H. Morel, "Protruding ceramic substrates for high voltage packaging of wide bandgap semiconductors," in *2017 IEEE 5th Workshop on Wide Bandgap Power Devices and Applications (WiPDA)*, Albuquerque, NM, Oct. 2017, pp. 404–410. doi: 10.1109/WiPDA.2017.8170581.

[6] L. Donzel and J. Schuderer, "Nonlinear resistive electric field control for power electronic modules," *IEEE Trans. Dielect. Electr. Insul.*, vol. 19, no. 3, pp. 955–959, Jun. 2012, doi: 10.1109/TDEI.2012.6215099.

[7] T. Christen, L. Donzel, and F. Greuter, "Nonlinear resistive electric field grading part 1: Theory and simulation," *IEEE Electr. Insul. Mag.*, vol. 26, no. 6, pp. 47–59, Nov. 2010, doi: 10.1109/MEI.2010.5599979.

[8] L. Donzel, F. Greuter, and T. Christen, "Nonlinear resistive electric field grading Part 2: Materials and applications," *IEEE Electr. Insul. Mag.*, vol. 27, no. 2, pp. 18–29, Mar. 2011, doi: 10.1109/MEI.2011.5739419.

[9] C. Park, "Electret: An Entirely New Approach of Solving Partial Discharge Caused by Triple Points, Sharp Edges, Bubbles, and Airgaps," *IEEE Access*, vol. 8, pp. 78354–78366, 2020, doi: 10.1109/ACCESS.2020.2990310.

[10] C. Park, "Electrets: A Remedy for Partial Discharge Caused by Power Electronics Switching," *IEEE Trans. Ind. Electron.*, pp. 1–1, 2020, doi: 10.1109/TIE.2020.3045707.

[11] P. Kotrappa, "Long term stability of electrets used in electret ion chambers," *Journal of Electrostatics*, vol. 66, no. 7–8, pp. 407–409, Jul. 2008, doi: 10.1016/j.elstat.2008.04.008.

[12] F. Haque, C. Park, "Electret Fabrication Under Various Discharge Conditions of Triode Corona Charging and the Partial Discharge Mitigation Performance,", presented at the IEEE Electrical Insulation Conference 2021, Virtual Conference, June 7-28, 2021.

[13] F. Haque, O. Faruqe, and C. Park, "Electret Fabrication Under Various Temperatures and Partial Discharge Mitigation Performance," submitted at the IEEE CEIDP, Vancouver, BC, Canada, Dec. 12-15, 2021.

[14] T. Ueki, S. Nishijima, and Y. Izumi, "Designing of epoxy resin systems for cryogenic use," *Cryogenics*, vol. 45, no. 2, pp. 141–148, Feb. 2005, doi: 10.1016/j.cryogenics.2004.07.002.
[15] W.-J. Kim, H.-J. Kim, J.-W. Cho, Y. S. Choi, and S.-H. Kim, "Electrical and Mechanical Characteristics of Insulating Materials for HTS DC Cable and Cable Joint," *IEEE Trans. Appl. Supercond.*, vol. 25, no. 3, pp. 1–4, Jun. 2015, doi: 10.1109/TASC.2014.2385958.
[16] "High-voltage test techniques – Partial discharge measurements," IEC Standard 60270, p. 61, Dec. 2000 .

IOP Conf. Series: Materials Science and Engineering 1241 (2022) 012006 doi:10.1088/1757-899X/1241/1/012006

Interlaminar shear strength of vacuum pressure impregnated coils with epoxy resin for high field magnet at cryogenic temperature

Chundong Wang[1,2], Yalin Zhao[1,2], Huiming Liu[1], Zhichun Fang[3], Jijun Xin[3], Zhixiong Wu[1], Chuanjun Huang[1], Rongjin Huang[1,2*], Laifeng Li[1,2]

1. CAS Key Laboratory of Cryogenics, Technical Institute of Physics and Chemistry, Beijing 100089, China
2. University of Chinese Academy of Sciences, Beijing 100049, China
3. Songshan Lake Materials Laboratory, Dongguan 523000, Guangdong Province, China

* Corresponding author at: CAS Key Laboratory of Cryogenics, Technical Institute of Physics and Chemistry, Beijing 100049, China.

E-mail: huangrongjin@mail.ipc.ac.cn (R. Huang)

Abstract. The interlaminar shear properties of epoxy impregnated coils are critical parameter to the design of the high field superconducting magnets. The existing test methods of the interlaminar shear strength of the composite insulation systems for the superconducting magnet coils in cryogenic temperature were discussed. The insulation system of the niobium titanium superconductor coated with woven glass fiber reinforced polymer coil were fabricated by vacuum pressure impregnation with IR-3 and CTD-101K resins. Short beam shear tests of the impregnated coil were performed at room temperature and liquid nitrogen temperature (77K). The results show that the interlaminar shear strength of the samples used IR-3 resin are better than CTD-101K both at room and cryogenic temperature. A detailed observation of the failed specimens was also investigated to verify the failure mechanisms by optical microscopy.

1. Introduction

Superconducting magnets are the core components of Magnetic Resonance Imaging (MRI) [1-3], Nuclear Magnetic Resonance (NMR)[4, 5], fusion energy devices[6] and any other high energy physics and nuclear devices. For example, high field MRI plays an important role in biomedical research. The electromagnetic force would cause higher stress and deformation in the coils during the operation. In order to react the electromagnetic load, the insulation system which provide structural support to the coils would be designed during fabrication. Epoxy resin is used as a filler to prevent the movement of superconducting wires caused by the Lorentz force and improve the insulation performance [7]. The mechanical properties of the superconducting coil impregnated with epoxy must be considered in the design of high field superconducting magnets. In particular, the interlaminar shear strength (ILSS) of superconducting wires impregnated with epoxy at cryogenic temperature is often an important design parameter [8]. Designers must be in control of a good understanding of the interlaminar shear behavior of superconducting coil impregnated with epoxy resin at cryogenic temperature. Thus, a lot of testing methods have been developed to evaluate this parameter.

The purpose of this research is to compare the suitability of the existing test methods for interlaminar shear strength of the composite insulation systems in a superconducting magnet coil configuration at cryogenic temperature. Then, the ILSS of superconducting coils used IR-3 and CTD-101K resin at cryogenic temperature were examined and compared. The superconducting coils reinforced epoxy are fabricated by vacuum pressure impregnated (VPI) technology. Short beam shear tests were conducted at RT and 77K and then the failed specimens were observed by optical microscopy.

2. Interlaminar shear strength testing methods

The Iosipescu test, the double-notch test, the short beam shear test, and the four-point shear test are commonly used for determining the ILSS of a Composite material. The Iosipescu testing is considered as a reliable method [9], But the samples are difficult to prepared for impregnated superconducting coils and the structure of the fixture is complicated. Digital image correlation might be used to resolve the strains of the sample, which is difficult to organize at cryogenic temperature. The double-notch shear method is compatible with testing at cryogenic temperature because the fixture is easily adapted to cryostats and the interlaminar shear failure occurs consistently in well-designed samples [10]. However, there are some disadvantages mentioned about the double-notch method. The notch depth in the notched test is difficult to control and the results are highly sensitive and quite scattered. The short beam method is commonly used to measure the ILSS of materials at cryogenic temperature for its advantages of easy operation and the specimen are easy to prepare [11]. X. Ming et al. [12] showed that both the three- and four-point shear test methods should be used cautiously when measuring the ILSS of composite materials because of the high dependence of ILSS on span-to-thickness ratio and high stress concentration affects the real value of ILSS. Considering the characteristics that the fixtures are simple, the samples are easy to prepare and the test is easy to organize at cryogenic temperature, the short beam shear method is selected as the test method.

Many researchers have studied the mechanical properties of superconducting insulating composite material, which can be used for reference in the ILSS test of impregnated superconducting coil. The interlaminar shear properties of woven glass fiber reinforced polymer (GFRP) composite laminates were investigated by Shindo et al [13]. The average value for samples containing diglycidyl ether of bisphenol-A (DGEBA) epoxy and S-glass fabric is 40.8 MPa. Short beam shear test was used for laminate structure insulation manufactured by the glass/polyimide reinforced epoxy resin at 77K by X. Huang et a, and the average value of the ILSS of the Kapton tape with S2-glass samples is 53.9 MPa [14]. Sun et al. [15] prepared an experimental superconducting coil for MRI by modified bisphenol-A type epoxy resin and the vacuum pressure impregnation (VPI) technology. The mechanical properties of superconducting coil and epoxy resin was measured at room temperature (RT) and at liquid nitrogen temperature (77K). It is found that mechanical strength of the epoxy resin reinforced by fiber cloth at 77 K is 78.5 MPa, which is higher than 40.3 MPa at RT. The elastic properties of specimens consisting NbTi superconductor, DGEBA resin with amine hardeners and E-glass cloth were measured by Iain R. Dixon et al [16]. Yu Liu et al. [17] investigated the role of matrix modification by multi-walled carbon nanotubes (MWCNTs) and n-butyl glycidyl ether (BGE) on the interlaminar shear strength (ILSS) of glass fibre reinforced epoxy composites. The ILSS of modified epoxy/glass fibre composites was 41.46 MPa at room temperature, which was higher than that of the unmodified epoxy/glass fibre composite.

As is mentioned above, mechanical properties about superconducting magnets have mainly focused on the properties of fiber-reinforced epoxy resin or superconducting coil itself. The ILSS between superconducting coils and epoxy resin or the shear strength of a single superconducting wire at cryogenic temperature are rarely reported. In order to provide database for the design of superconducting magnets, the ILSS test to evaluate the properties of the resins impregnated coil was performed. Two resin systems were selected for the vacuum pressure impregnated process of the coil and the short beam shear tests of the impregnated coil were performed at room temperature and liquid nitrogen temperature (77K).

IOP Conf. Series: Materials Science and Engineering 1241 (2022) 012006 doi:10.1088/1757-899X/1241/1/012006

3. Experimental and results

3.1. Materials and Specimen preparation

CTD-101K, used extensively in the magnet industry, is a product of Composite Technology Development Inc. [18]. IR-3 was formulated at the Technical Institute of Physics and Chemistry for the purpose of impregnating high-field magnets.

Wire-In-Channel (WIC) superconducting wire with 2.24×1.425 mm was manufactured by Western Superconducting Technologies Co., ltd. As shown in Fig. 1, WIC superconducting wire is plated by chrome and wrapped with 0.2 mm thick braided polyester fiber electrical insulation layer.

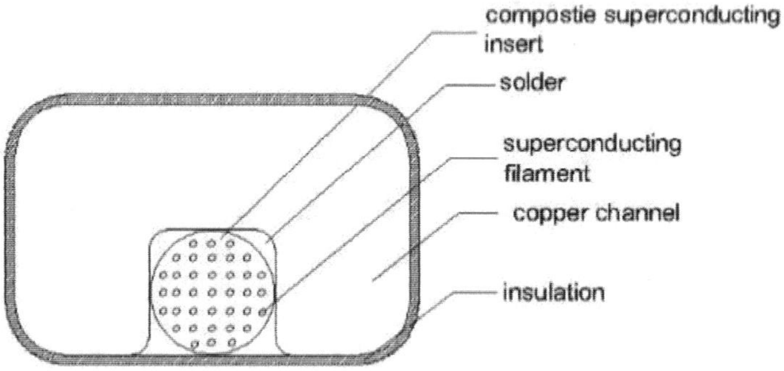

Figure 1. Schematic cross-section of WIC composite superconductor.

The test specimens consist of four layers of four WIC wires, and the coil was fabricated by VPI technique as shown in Fig. 2. The VPI technique was described as follows: The parts of epoxy resins were mixed and then degassed under vacuum at 60℃ about 2 hours to remove the bubbles in the mixture. In the meantime, the mold was coated with release agent and the superconducting coil was placed into the mold. Then the mold was sealed and degassed at 60℃ by a vacuum pump. After the bubble in the mixture was eliminated, the resin valve was opened and the coil was impregnated. Finally, the mold was heated and the coil was cured. The curing process of CTD-101K and IR-3 are same and and the parameters for the VPI are listed in Table 1.

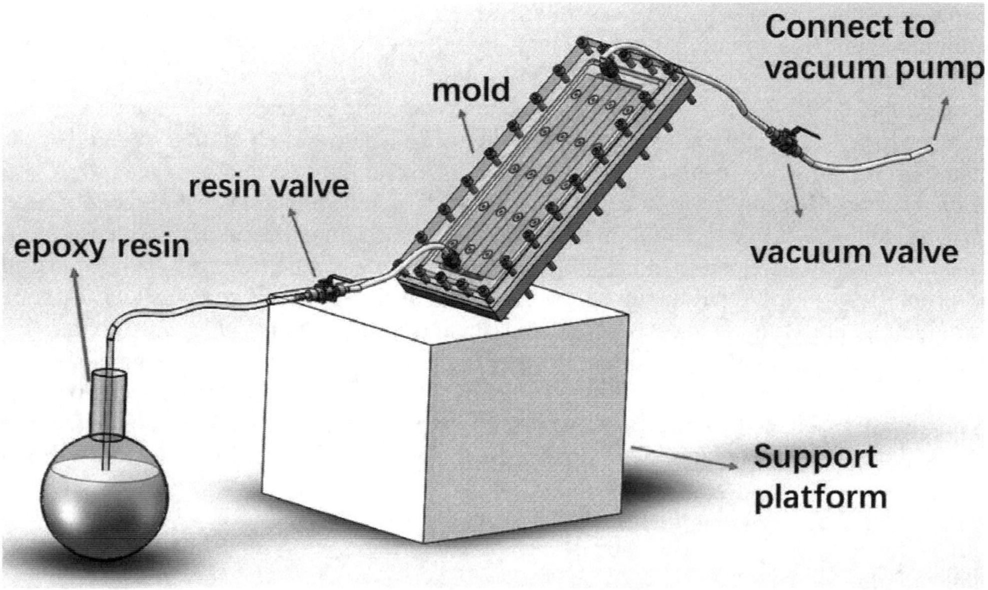

Figure 2. The diagram of VPI

Table 1. Parameters for the VPI and curing procedures for the coil.

Parameter	Methods and value
Resin degassing	Heated by oil bath at 60℃while pumping vacuum for 2h.
Impregnating temperature	The mold was heated to 60℃ and then impregnated while the resin was heated by oil bath at 60℃.
Curing temperature	Heated to 110℃ at a rate of 1.5℃/s and kept 5h, then heated to 125℃ at the same rate and kept 16h.

3.2. Short beam shear test and microscopic examination

As shown in Fig. 3, the specimens for the short beam shear tests were cut by a diamond cutting machine to the following dimensions length l = 40 mm, width b = 9mm and thickness t = 6mm. The short beam shear test setup is shown in Fig. 4. The loading nose diameter d_1 = 6mm and the support roller diameter d_2 = 6mm. A span to thickness ratio s/t of 5 was chosen and the span length s was 30 mm.

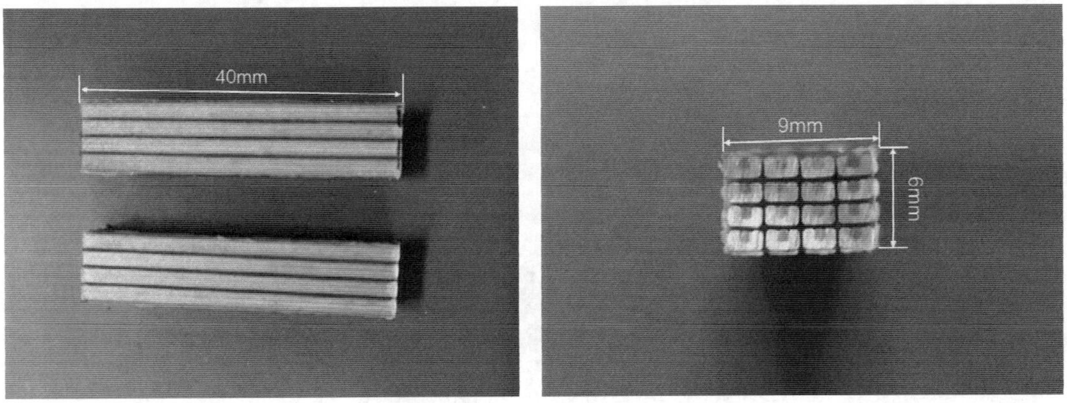

Figure 3. Prepared samples

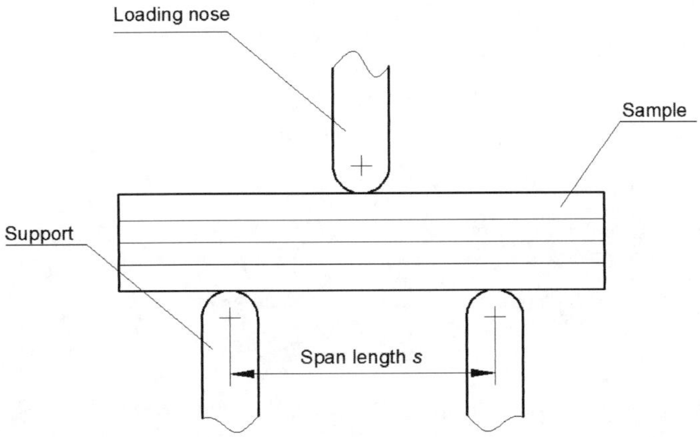

Figure 4. Sketch of short beam shear test

The short beam shear tests were conducted at room temperature and 77 K. Testing at 77 K was accomplished by immersing the test fixture and specimen in liquid nitrogen. The specimens were loaded at a displacement rate of 0.2 mm/min. The load point displacement (deflection) and the load were measured by the transducer and testing machine load cell, respectively. The short beam shear strength is given by:

$$F^{sbs} = \frac{3P_m}{4bt} \quad (1)$$

Where the F^{sbs} is the apparent interlaminar shear strength (MPa) and P_m is the failure load. After the short beam shear tests, the fracture surfaces of the tested samples were examined by optical microscopy.

3.3. Interlaminar shear strength

The typical load-displacement curves were obtained from the ILSS tests of the coils as shown in Fig. 5. It can be seen that curves show different properties at room temperature and cryogenic temperature. Curves show a nearly linear elastic trend during the early stage of loading and rise to the peak load. Then they fall down suddenly which reflects a failure developed in the samples due to interlaminar shear at 77K. However, the curves showing a drop suddenly don't obviously at room temperature, which reflects that the failure mode of the specimens contains interlaminar shear and yield.

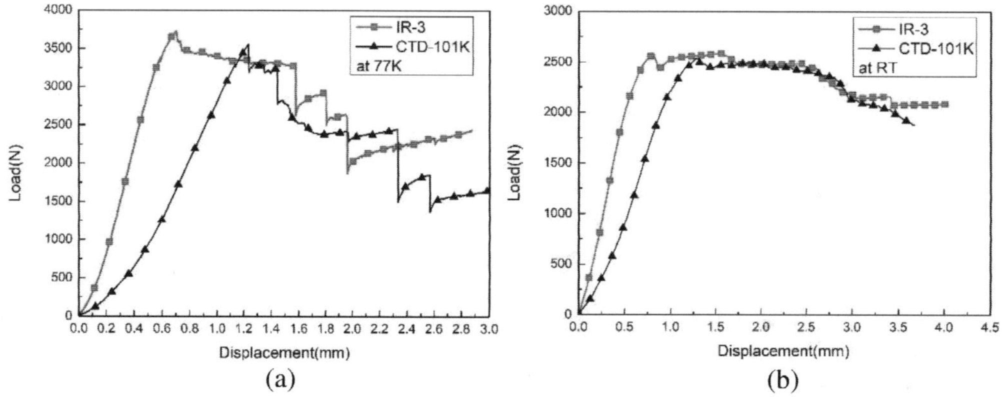

Figure 5. Load versus displacement curves of the coils. (a) coils were tested at 77K, (b) coils were tested at room temperature.

Typical optical micrographs of the fracture surfaces of the failed specimens are shown in Fig. 6. It can be seen that the crack propagates through the interface of epoxy resin and polyester after SBS test at 77K. In order to assess the ILSS value precisely, the obtained results are quoted as the average values with deviation from all samples with interlaminar shear. As shown in Fig. 7, the ILSS of specimens used IR-3 at 77K is 49.8MPa, which is better than that of CTD-101K (45.6MPa). And the ILSS of specimens used IR-3 at room temperature is 39.1MPa, which is better than that of CTD-101K (38.2MPa). The ILSS of the two groups of specimens are very similar at room temperature, but the ILSS of IR-3 are significantly better than that of CTD-101K at low temperature, which suggests that IR-3 is more suitable for superconducting coils impregnated at cryogenic application.

Figure 6. Optical microscope image of fracture surface of the failed specimen

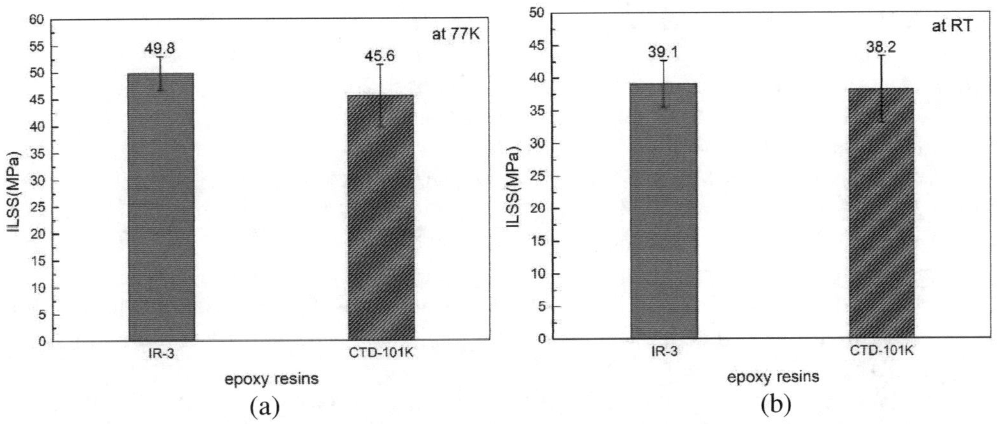

Figure 7. ILSS of specimens. (a) measured at 77K, (b)measured at room temperature

4. Conclusions

In this study, the advantages and disadvantages of four main methods for determining the ILSS of a material are elaborated. The ILSS of WIC superconducting wires used IR-3 and CTD-101K epoxy resins at room temperature and cryogenic temperature were investigated by short beam shear test. The ILSS of specimens used IR-3 are better than CTD-101K both at 77K and room temperature, and the effect was more obvious at cryogenic temperature. Optical microscopy examinations showed that the crack propagates mainly through the interface of epoxy resin and polyester.

5. References

[1] Li, D., Ren, Y., and Wang, H. (2020) Electromagnetic design of actively shielded MRI magnetfor dedicated mice imaging, *Physica C: Superconductivity and its Applications* Vol 575, p.1353657.

[2] Wang, Y., Wang, Q., Wang, L., Qu, H., Liu, Y., and Liu, F. (2020) Electromagnetic design of a1.5T open MRI superconducting magnet, *Physica C: Superconductivity and its Applications* Vol. 570, p. 1353602.

[3] Li, L., Cheng, J., Cui, C., Li, Y., Dai, Y., Hu, X., Liu, J., Wang, L., and Wang, Q. (2017) Numerical Analysis of Mechanical Behavior for a 9.4-T Whole-Body MRI Magnet, *IEEE Transactions on Applied Superconductivity* Vol. 27, pp. 1-5.

[4] Nakamura, T., Tamada, D., Yanagi, Y., Itoh, Y., Nemoto, T., Utumi, H., and Kose, K. (2015) Development of a superconducting bulk magnet for NMR and MRI, *Journal of Magnetic Resonance* Vol. 259, pp. 68-75.

[5] Liu, J., Cheng, J., and Wang, Q. (2013) Evaluation of NbTi Superconducting Joints for 400 MHz NMR Magnet, *IEEE Transactions on Applied Superconductivity* Vol. 23, pp.34-39.

[6] Wang, D., Ren, Y., Zhang, M., and Gao, X. (2021) Design and analysis of the 16T superconducting magnet for CFETR strand test facilities, *Fusion Engineering and Design* Vol. 162, p. 112097.

[7] Brennan, A. B., Miller, T. M., Arnold, J. J., Huang, K. V., Gephart, N. L., and Markewicz, W. D.(1995) Thermomechanical properties of a toughened epoxy for impregnating superconducting magnets, *Cryogenics* Vol. 35, pp. 783-785.

[8] Elawadly, K. M. (2003) On the Interlaminar Shear Stress Response for E-Glass/Epoxy Composite, *Journal of Composite Materials* Vol. 37, pp. 2149-2158.

[9] Bru, T., Olsson, R., Gutkin, R., and Vyas, G. M. (2017) Use of the Iosipescu test for the identification of shear damage evolution laws of an orthotropic composite, *Composite Structures* Vol. 174, pp. 319-328.

[10] Shindo, Y., Wang, R., Horiguchi, K., and Ueda, S. (1999) Theoretical and Experimental Evaluation of Double-Notch Shear Strength of G-10CR Glass-Cloth/Epoxy Laminates at Cryogenic Temperatures, *Journal of Engineering Materials and Technology* Vol. 121, pp. 367-373.

[11] Shen, X. J., Meng, L.-X., Yan, Z.-Y., Sun, C.-J., Ji, Y.-H., Xiao, H.-M., and Fu, S.-Y. (2015) Improved cryogenic interlaminar shear strength of glass fabric/epoxy composites by graphene oxide, *Composites Part B: Engineering* Vol. 73, pp. 126-131.

[12] Xie, M., and Adams, D. F. (1995) Study of three- and four-point shear testing of unidirectional composite materials, *Composites* Vol. 26, pp. 653-659.

[13] Shindo, Y., Takeda, T., Narita, F., Miura, M., Watanabe, S., Koizumi, N., Idesaki, A., and Okuno, K. (2010) Interlaminar shear properties of composite insulation systems for fusion magnets at cryogenic temperatures, *Cryogenics* Vol. 50, pp. 36-42.

[14] Huang, X., Yu, X., Wang, C., Li, G., Song, Y., Lu, K., Niu, E., Gung, C., Clayton, N., and Su, M. (2014) Interlaminar Shear Strength Property of the Glass Fiber/Polyimide Reinforced Epoxy Resin for ITER Feeder Mock-ups, *IEEE Transactions on Applied Superconductivity* Vol. 24, pp. 1-4.

[15] Sun, W., Wu, Z., Huang, C., Zhang, H., Shen, F., Huang, R., Gong, L., Zhou, Y., and Li, L. (2021) Study on cryogenic mechanical properties between superconducting wires and resins for MRI superconducting magnet, *Cryogenics* Vol. 115, p. 103259.

[16] Dixon, I. R., Markiewicz, W. D., and Marshall, W. S. (2000) Axial mechanical properties of epoxy impregnated superconducting solenoids at 4.2 K, *IEEE Transactions on Applied Superconductivity* Vol. 10, pp. 1313-1316.

[17] Liu, Y., Yang, J.-P., Xiao, H.-M., Qu, C.-B., Feng, Q.-P., Fu, S.-Y., and Shindo, Y. (2012) Role of matrix modification on interlaminar shear strength of glass fibre/epoxy composites, *Composites Part B: Engineering* Vol. 43, pp. 95-98.

[18] Fabian, P. E., Munshi, N. A., and Denis, R. J. (2002) Highly radiation-resistant vacuum impregnation resin systems for fusion magnet insulation, *AIP Conference Proceedings* Vol. 614, pp. 295-304.

Acknowledgments

This work was supported by National Key Research and Development Program of China (Nos. 2020YFF01014701 and 2019YFA0704904), Key Research Program of Frontier Sciences of Chinese Academy of Sciences (No. QYZDB-SSW-JSC042), Strategic Priority Research Program of Chinese Academy of Sciences (No. XDB25040300), Youth Innovation Promotion Association of Chinese Academy of Sciences (No. 2013020), and President's International Fellowship Initiative (No.2019VEA0017).

IOP Conf. Series: Materials Science and Engineering 1241 (2022) 012007 doi:10.1088/1757-899X/1241/1/012007

Improved breakdown strength at cryogenic temperature in epoxy nanocomposites by a small amount of SiC nanoparticles

Z C Miao[1,2], P Jia[1,2], Y G Wang[1,2], R J Huang[*1,2] and L F Li [1,2]

1.State Key Laboratory of Technologies in Space Cryogenic Propellants, Technical Institute of Physics and Chemistry, Chinese Academy of Sciences
2. University of Chinese Academy of Sciences, Beijing, China

huangrongjin@mail.ipc.ac.cn

Abstract. The study of the electrical properties of materials at cryogenic temperature is significant for the design of insulation structures of superconducting equipment. Here, epoxy resin was combined with a small amount of KH650 modified SiC nanoparticles. When the dosage of SiC nanoparticles increased from 0.1% to 0.3%, the DC breakdown strength of the composite increase from 83.93 kV/mm to 97.51 kV/mm at room temperature and from 90 kV/mm to 120 kV/mm at 77 K. Furthermore, we used differential scanning calorimeter (DSC) and dynamic thermomechanical analysis (DMA) to analyze the effect of low-content SiC nanoparticles on the thermal behavior of epoxy resin.

1. Introduction

Nuclear fusion is an efficient and clean technology that can solve the future energy crisis. Currently, superconducting magnetic confinement nuclear fusion is recognized as a feasible solution to achieve controllable nuclear fusion [1, 2]. The insulating structure is used for electrical insulation, protection and support of superconducting magnets, plays an important role in system heat dissipation, chemical isolation, protection of superconducting materials, and prevention of conductor quenches [3-5]. The insulating structure of the superconducting device usually selects epoxy resin as the matrix material [6]. The device needs to operate at extremely low temperatures (77k) and will generate high voltage, high field strength (>10T), and strong radiation. This extremely harsh environment puts forward higher requirements on the performance of insulating materials. Therefore, superconducting devices have very high requirements for the selection and design of materials in the insulation system. Under the punishment of high voltage and high magnetic field, insulating materials will undergo severe electrical aging, which will eventually lead to the failure of equipment [7, 8]. Traditional epoxy resins are difficult to meet performance requirements at low temperatures, so it is necessary to study the thermal and electrical behaviors of different composite systems.

Silicon carbide (SiC) is an excellent semiconductor material with high thermal conductivity and strong thermal stability [9]. These characteristics make it suitable for high frequency, high power, and harsh environments. Compared with traditional silicon carbide, silicon carbide nanoparticles have a large specific surface area, excellent mechanical properties and photoelectric properties. It has broad application prospects in the field of reinforced composite materials, nano-electronic devices and sensors [10, 11]. Here, we modified the silicon carbide by using a coupling agent glycidoxypropyltrimethoxysilane(KH560) and then dispersed the silicon nitride in the epoxy resin matrix to obtain a SiC/EP composite material. Through the test of its thermal performance and insulation performance at low temperatures, it can guide the design of the composite material system in an extreme environment.

2. Experiment

2.1. Materials

The resin used in this study was a commercially available diglycidyl ether of bisphenol A epoxy resin (DGEBA, 282) supplied by Nanya Epoxy Co., Ltd, China. The silicon carbide nanoparticles were purchased from Shanghai Macleans Biochemical Technology Co., Ltd, China. The curing agent was a liquid aromatic amine, diethyltoluenediamine(DETDA), supplied by Shanghai Macklin Biochemical Co., Ltd, China. The coupling agent KH560 was purchased from Dongda Fine Chemical Materials Co., Ltd, China.

First, 0.06g silicon carbide nanoparticles were dispersed in 10g acetone, added 2-3 drops of KH560, and then sonicated the dispersion in an ultrasonic cleaner for two hours. Second, poured the dispersion into 60g epoxy resin, rapidly stirred for ten minutes, and ultrasonically treat again for 1 hour. Then, 14.4g DETDA was added, stirred at 60°C for two hours, degassed under a vacuum pump for about 1 hour. Finally, poured the mixture into a mold, and put the mold into a blast oven to cure to obtain a sample. The curing process is 80℃ for 10 hours and 125℃ for 10 hours.

2.2. Characterizations and measurements

The thermal behavior of the material was analyzed by Dynamic Thermomechanical Analysis (DMA,) and Differential Scanning Calorimeter (DSC, German, NETZSCH). The sample of DMA was a 2*12*50mm cube. The sample of DSC was about 6mg powder. The electrical breakdown of the composite was investigated by our self-designed full-temperature electrical breakdown equipment. The sample of it was a 1um thick wafer. The dispersion of nanoparticles in the epoxy matrix was observed by scanning electron microscopy (SEM) using a HITACHI S-4800 microscope (Japan).

3. Result and discussion

3.1. Scanning Electron Microscope (SEM) Characterization

Nanoparticles have a large specific surface area and large surface energy, so they are easier to agglomerate at the nanometer scale. Better dispersion can bring excellent performance to the composite, and the dispersion of nanoparticles has always been a problem to be solved. Figure 1 shows the dispersion of SiC nanoparticles in the epoxy matrix. In Figure a, it can be seen that the dispersion of silicon carbide nanoparticles at very low levels is relatively sparse and there is less agglomeration in the epoxy resin matrix. When the content of nanoparticles increases to 0.5%, it begins to agglomerate and become denser in the epoxy resin matrix. But overall, due to the modification of KH560, the dispersion is decent.

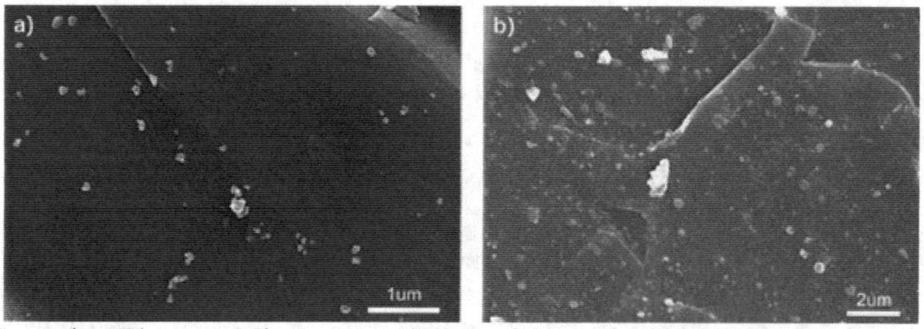

Figure 1. Scanning Electron Microscope (SEM) of (a) 0.1% SiC/EP composite and (b) 0.3% SiC/EP composite.

3.2. *DC Breakdown strength*

Superconducting equipment will generate a higher electric field when operating at extremely low temperatures, so it is necessary to learn the breakdown strength of insulating structures at low temperatures. Figure 2a shows the Weibull distribution after testing the DC breakdown strength of SiC/EP composites of different components. After fitting, it can be concluded that the breakdown strength of the pure EP at 300 K and 77 K are 63.65 and 70.00 kV/mm, the 0.1%SiC/EP composite at 300 K and 77 K are 83.93 and 90.00 kV/mm, the 0.3%SiC/EP composite at 300 K and 77 K are 97.51 and 120 kV/mm, respectively. First of all, we can conclude that the breakdown strength of the composite is higher at low temperatures, indicating better insulation properties at 77 K. The main reason is that the segmental movement of the material is severely hindered at low temperatures, and the free radical reaction caused by high voltage is more difficult to proceed, results in the slow internal degradation reaction of the material. Besides, we found that the breakdown strength is higher as the SiC content increase. Here, we believe that the extremely low content of SiC nanoparticles is the key to the increase of breakdown strength. Although SiC is a semiconductor, in the case of extremely low content, the distance between the nanoparticles is relatively long, and no conductive path is formed. When the electric field reaches the vicinity of the nanoparticles, it can homogenize the electric field and reduce the probability of occurrence of electric field distortion inside the material. Finally, the breakdown strength of the composite improved a lot.

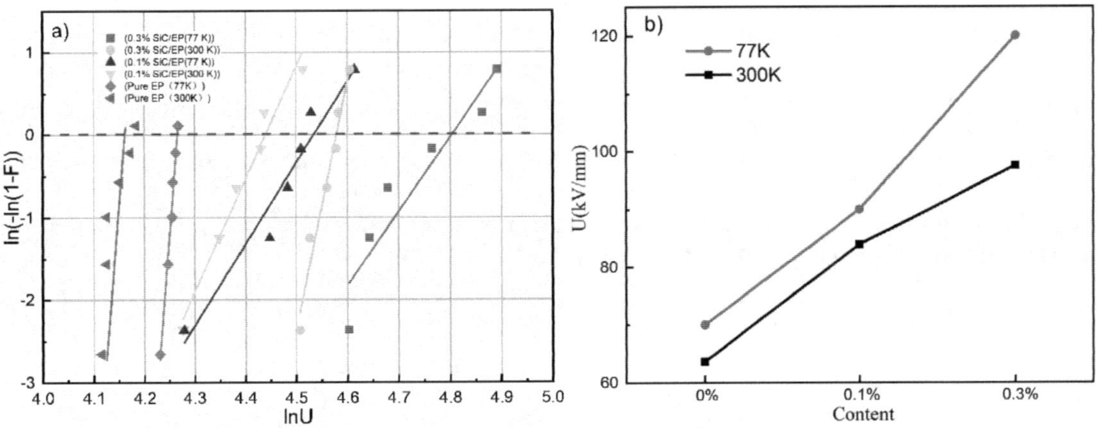

Figure 2. Weibull distribution(a) and Breakdown Strength(b) of different SiC/EP composites

3.3. *DSC & DMA*

Although superconducting device operates at extremely low temperatures, the local overheating of materials can also occur under high magnetic fields and high voltages. Therefore, analyzing the thermal behavior of insulating structures is an important thing. Here we tested different SiC/EP composite materials through DSC. It can be seen from Figure 3 that the glass transition temperatures (Tg) of 0.1% SiC/EP, 0.3% SiC/EP and 0.5% SiC/EP are 124.6℃, 126.3℃ and 128.4℃, respectively, compared with 128.6℃ of pure epoxy resin, there is little change. Therefore, the extremely low content of SiC nanoparticles has less effect on the glass transition temperature of the epoxy resin matrix, which can reflect that the effect on the curing degree of the epoxy resin is negligible. Meanwhlie, Tg also reflects the influence of nanofillers on the segment movement of the resin matrix. As the Tg increases, the resistance to movement at the end of the chain segment increases, which leads to the weakening of the electron multiplication process, and finally increases the breakdown field strength. Even at low temperatures, local high temperatures will be generated during breakdown, so the study of Tg is also meaningful

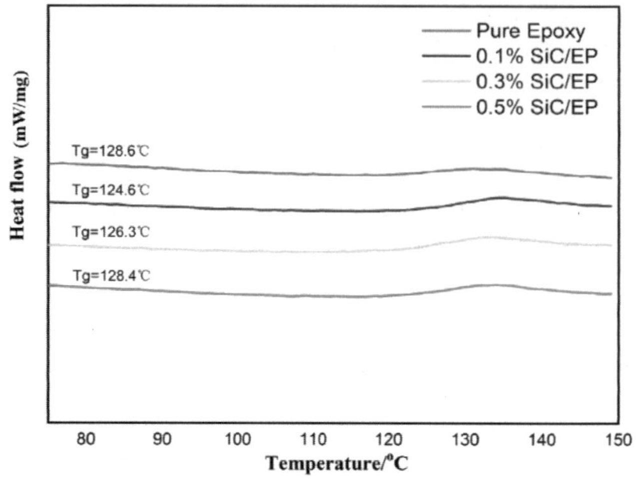

Figure 3. DSC of different SiC/EP composites

Furthermore, we studied the thermomechanical behavior of the material at -100°C to 150°C by DMA. From Figure 4, the composite has a higher modulus at about -80°C, the glass state of the composite hinders the movement of the molecular chain, and energy can only be stored in the form of potential energy. The higher storage modulus indicates the higher the rigidity of the material. It can be seen from the figure that the storage modulus of 0.1% and 0.3% SiC/EP is greatly reduced compared with the pure epoxy, demonstrating that the low content of SiC nanoparticles reduces the rigidity of the material. However, when the content of SiC nanoparticles reaches 0.5%, the rigidity of the material is restored. Tanδ reflects the damping performance of the material. The peak value of tanδ of 0.1% SiC/EP composite is significantly higher than that of other components, indicating that 0.1% SiC nanoparticles can improve the damping performance of the composite.

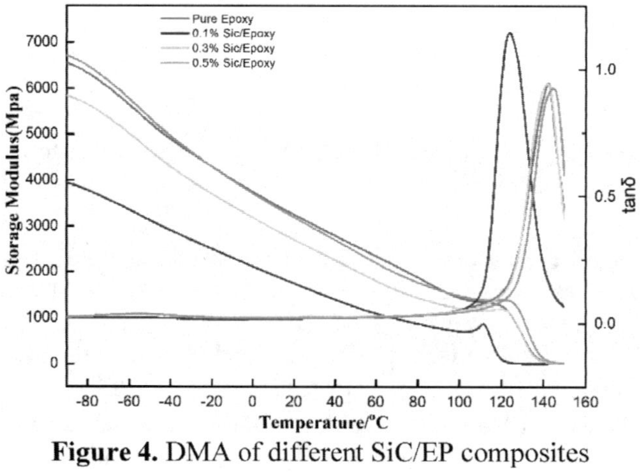

Figure 4. DMA of different SiC/EP composites

4. Conclusion

In this work, we prepared SiC/EP composites by combining a small amount of KH560 modified SiC nanoparticles with epoxy. The thermal behavior and the insulation performance at the low temperature of the composite are studied. The result shows that the low content of SiC nanoparticles can increase the breakdown voltage of epoxy resin, and the voltage increase as the content of nanoparticles increases. At the same time, the low content of nanoparticles has almost no effect on the glass transition of epoxy resin, but reduces the rigidity of the material. These results guide the design and selection of insulating materials for superconducting devices.

5. References

[1] Nishimura A, Muroga T, Takeuchi T, Nishitani T, and Morioka A 2006 *Fusion. Eng. Des.* **81** 1675-81
[2] Iwasa Y and Marken K R 1995 *Phys. Today* **48** 68
[3] J. Schutz 1998 *Cryogenics* **38** 3-12
[4] Hemmi T, Koizumi N, Matsui K, Okuno K, Nishimura A, Sakai M and Asano S 2009 *Fusion. Eng. Des.* **84** 923-927
[5] Evans D 2001 *Physica C* **354** 136-142
[6] Ueki T, Nishijima S and Izumi Y 2005 *Cryogenics* **45** 141-8
[7] Guo X L, Xu F Y, Wang L, Green M A and Amm K 2009 *IEEE T. Appl. Supercon.* **19** 1360-63
[8] Guo X L, Wang L and Green M A 2012 *Cryogenics* **52** 420-7
[9] Shen G, Bando Y, Ye C, Liu B and Golberg D 2006 *Nanotechnology* **17** 3468-72
[10] Kassiba A, Makowska-Janusik M, Boucle J, Bardeau J F, Bulou A, Herlin N, Mayne M and Armand X 2002 *Diam. Relat. Mater.* **11** 1243-47
[11] Rodgers R M, Mahfuz H, Rangari V K, Chisholm N and Jeelani S 2005 *Macromol. Mater. Eng.* **290** 423-9

Acknowledgments
This work was supported by Strategic Priority Research Program of Chinese Academy of Sciences (No. XDB25040300), National Key Research and Development Program of China (No. 2019YFA0704904), Key Research Program of Frontier Sciences of Chinese Academy of Sciences (No. QYZDB-SSW-JSC042), President's International Fellowship Initiative (No. 2019VEA0017), Youth Innovation Promotion Association of Chinese Academy of Sciences (No. 2013020).

Fiber bragg grating sensors to monitor strain response of epoxy resin during curing and cryogenic temperature

H Y Dong[1,2], H M Liu[1], C S Fang[1,2], C J Huang[1*], L F Li[1,2*]

[1]Chinese Academy of Sciences Key Laboratory of Cryogenics, Technical Institute of Physics and Chemistry, Beijing 100190, China
[2]University of Chinese Academy of Sciences, Beijing 100049, China

E-mail: cjhuang@mail.ipc.ac.cn; laifengli@mail.ipc.ac.cn

Abstract. The epoxy resin is one of the critical components of large-scale superconducting magnets which find wide applications in magnetic confinement fusion, high energy accelerator and magnetic resonance imaging. In general, magnetic confinement fusion, high energy accelerator and magnetic resonance will be used at cryogenic temperatures. An approach with respect to real-time monitor of epoxy curing and cryogenic process is practically significant for magnetic reliability, maintainability as well as future mass production. Fiber Bragg Grating (FBG) sensors are excellent candidates of this measurement for their anti-electromagnetic interference characteristics and high sensitivity. The aim of this work is to use the embedded Fiber Bragg Gratings (FBG) to obtain real-time strain response of curing epoxy, and the strain change of the epoxy at cryogenic temperature from 4.2K to 300K. An FBG pre-tensioning device was designed to keep the FBG embedded in the epoxy in a stretched state during the epoxy curing process. Moreover, strain gauge sensors were also embedded to monitor the curing process at the same time. There are subtle differences in strain values obtained by the two sensors, which illustrate higher sensitivity of FBGs. This work verifies the reliability of FBG monitoring the strain response of epoxy during curing process as well as the cryogenic service.

Key words: Fiber Bragg Gratings, epoxy resin, curing process, cryogenic

1. Introduction

Epoxy resin is a key component of the insulation system of large superconducting magnets. In the field of aviation and aerospace, FBGs have been widely used in structure health detection technologies. In the aviation industry, Airbus has developed a series of health detection methods for key components of large passenger aircraft such as A380, such as structural deformation detection technology based on optical fiber sensor network, ultrasonic-based damage detection technology, active multiple damage assessment technology, and Crack detection technology based on electromechanical impedance, etc. [1-4]. In the aerospace field, NASA took the lead in adopting structural health detection technology based on fiber grating sensors since the 1990s, and applied it to real-time stress field detection of composite fuel tanks in spacecraft. [5,6] NASA has also developed a series of methods for real-time structure response informationdetection of space vehicles[7]. These structural health detection technologies have different characteristics. For example, structure corresponding detection based on FBG network requires pre-embedded optical fibers in the material, and the use of external equipment can detect the strain and temperature field distribution of composite materials in time[8]. While large-scale superconducting magnet insulation system is used in a more demanding environment, such as ultra-low temperature, high electric field, high magnetic field, and complex stress. Current structural health detection technology in the aerospace field cannot be directly used in the large-scale superconducting magnet insulation system. In this work, a special was designed to embed FBGs and strain gauges in epoxy resin to monitor real-time strain response of epoxy resin during curing process. A cryogenic experimental setup was established for FBGs to measure thermal expansion coefficient of epoxy resin cooling down to cryogenic temperature.

2. Theory

Fiber Bragg gratings are formed by changing the refractive index of the fiber core to produce small periodic modulation. Due to the perturbation of the periodic refractive index, only a narrow section of the spectrum will be affected. Therefore, when broadband light waves are transmitted in the grating, the incident light will be reflected at the corresponding frequency, and the rest of the transmitted light waves will be almost unaffected. In this way, the fiber grating plays the role of a light wave selective mirror. According to the theory of fiber coupling mode, when broadband light is transmitted in FBG, mode coupling will occur, and the light that meets the reflection condition (2-1) is reflected:

$$\lambda_B = 2n\Lambda \tag{2-1}$$

Where n is the effective refractive index of the fiber core, Λ is the period of the grating plane. The reflection wavelength at the center of the strain sensor is mainly affected by the elastic-optical effect and the thermo-optical effect, and the reflection wavelength change fraction is

$$\frac{\delta\lambda}{\lambda} = (1-P_e)\frac{\delta l}{l} + \frac{\delta n}{n} \tag{2-2}$$

Where $\delta\lambda$ is wavelength shift, δl is axial length change of FBG, δn is effective refractive index change, P_e is effective elasticity coefficient. $\frac{\delta l}{l}$ 和 $\frac{\delta n}{n}$ are proportional to the amount of temperature change at room temperature. $\alpha = \frac{1}{l}\frac{dl}{dT}$, $\xi = \frac{1}{n}\frac{dn}{dT}$, α is thermal expansion coefficient of base material, ξ is thermo-optical coefficient. Substituting α, ξ into the formula(2-2):

$$\frac{\delta\lambda}{\lambda} = [(1-P_e)\alpha + \xi]\Delta T \tag{2-3}$$

At room temperature, the first term in equation (2-3) is much smaller than the second term and can be ignored. Equation (2-3) can be written as:

$$\frac{\delta\lambda}{\lambda} = \xi\Delta T \tag{2-4}$$

The calculation model of the fiber grating temperature sensor can be obtained:

$$\Delta T = \frac{\delta\lambda}{\lambda\xi} \tag{2-5}$$

For fiber grating strain sensor, its fiber grating length change rate $\frac{\delta l}{l} = \varepsilon$, $\xi = \frac{1}{n}\frac{dn}{dT}$, formula (2-2) can be written as:

IOP Conf. Series: Materials Science and Engineering 1241 (2022) 012008 doi:10.1088/1757-899X/1241/1/012008

$$\frac{\delta\lambda}{\lambda} = [(1\text{-}P_e)\varepsilon + \xi]\Delta T \tag{2-6}$$

The calculation model of the fiber grating strain sensor can be obtained:

$$\varepsilon = \frac{1}{1-P_e}\left(\frac{\delta\lambda}{\lambda\Delta T} - \xi\right) \tag{2-7}$$

3. Materials and methods

3.1 Cryogenic-Grade Epoxy Resin

The cryogenic-grade epoxy resin used in this research consists of di-glycidyl ether bisphenol F(DGEBF) and diethyltoluenediamine (DETDA) as curing agent. The weight ratio of the DGEBF and DETDA is 100:24.

3.2 Fiber Bragg Gratings

Fiber Bragg Gratings (FBGs) have been provided by Femtofibertec GmbH Company (Germany) with the following specifications: grating length 6.5mm, reflectivity >75%, surface lattice resonances (SLRs) <25dB. The wide applicable temperature of the FBGs is reported to be 4.2K to 1273.15K.

To compare strain response measured by FBG and strain gauge, an aluminum sample with a FBG a strain gauge pasted on was tested on tensile testing machine. Strain comparison is shown in Figure 1. The value measured by FBG is highly coincident with the theoretical value. While the value measured by the strain gauge lags slightly. High sensitivity and accuracy of FBG have been verified by this experiment.

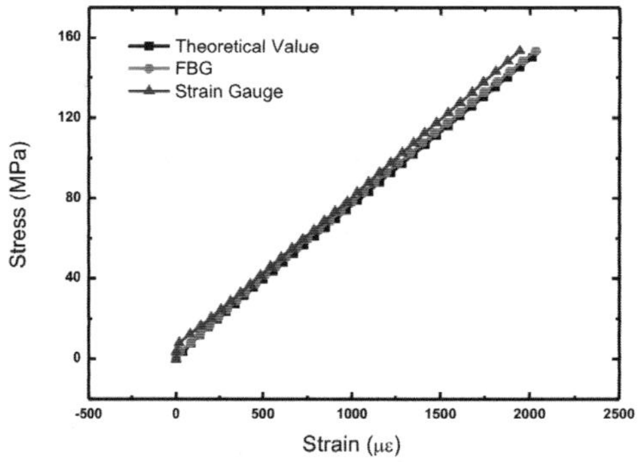

Figure 1. Strain Comparison of aluminum measured by FBG and strain gauge

3.3 Experimental Method

Differential Scanning Calorimetry (DSC, NETZSCH model DSC 404 F3, Erich NETZSCH GmbH & Co. Holding KG, Selb, Germany) was used to decide the curing process of the epoxy resin. Based on the DSC result, 353K/24h+403K/12h curing process for epoxy resin was used in this work.[9]

The strain response of epoxy resin during the curing process was measured using a strain gauge, a quartz with a strain gauge bonded on its surface, a strain gauge meter, two FBGs, a Fiber Bragg Grating interrogator (Micron Optics, Atlanta, GA, USA, SI 155), a temperature monitor, an oven with the maximum heating temperature of 573K, a data collection system, as shown in Figure 2. Strain gauges were used to compare the strain response of epoxy resin measured by FBGs. The real-time strain response and the temperature of the epoxy resin during curing process can be simultaneously recorded by this system. According to Standard ISO 8130-6: 1992[10], the initial solid point of epoxy in this work was determined by rheological behavior of epoxy resin during curing. Initial solid point of epoxy resin was reported at about 80% curing degree, and the time is around 10hours for the curing process in this work.[9] After being completely mixed and evacuated, epoxy resin was poured into silicone mold. The soft boundary of silicone mold enabled epoxy resin to expand freely. FBG passed through the two opposite tiny holes of the silicone mold, thus can embed in epoxy resin. To record the real-time strain response of epoxy during curing and temperature change inside the epoxy, an optimized copper lid was realized, as illustrated in Figure 2. There are four parts attached to the mold, including one channel for strain gauge to embed into epoxy, and three containers to hold FBG, quartz, and temperature sensor. The orientation of sensors does not affect measurement results for epoxy resin is isotropic. To keep FBGs straight during experimental process, a FBG pre-tensioning structure was designed. Two ends of FBG were fixed to FBG fixtures. And pulling tension bars can make FBG straight.

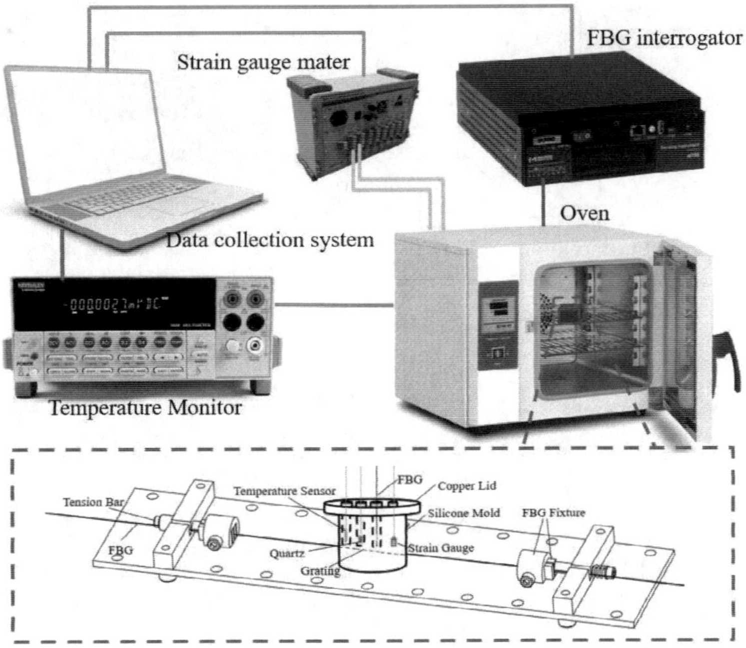

Figure 2. Experimental setup for Fiber Bragg Grating monitoring epoxy resin curing process

To investigate the thermal shrinkage of epoxy at cryogenic temperature with FBGs, a device cooled down by cryocooler was designed, as shown in Figure 3. During this period, the pre-tensioning fixture has been removed. The system consists of one vacuum tank, one adiabatic shield, one test cell, FBG feedthroughs. Aluminum foil was applied to reduce the thermal radiation. The sample can be cooled down to 4.2K. The thermal shrinkage of epoxy resin was also measured by a physical property measurement system (PPMS, Quantum Design model PPMS-14, Quantum Design, Inc. (QD), San Diego, CA, USA).

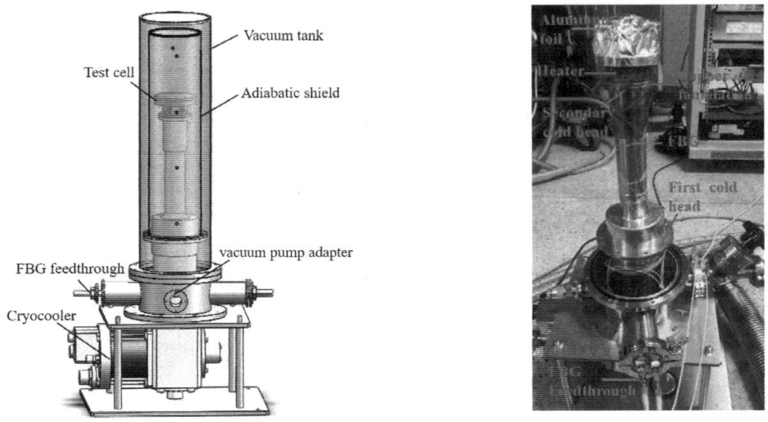

Figure 3. Cryogenic chamber to measure the thermal expansion coefficient of epoxy resin with FBGs

4. Results and discussion

The real-time strain responses of epoxy resin measured by FBGs and strain gauges are shown in Figure 4a. The results are obtained from five experiments. Initial data is the apparent output of real-time strain responses of FBGs and gauges. Expansion and shrinkage of FBG itself will cause wavelength shift. A free FBG was used as temperature compensation. The expansion and contraction of the metal wire of the strain gauge and the strain change of the epoxy resin simultaneously constitute the apparent data. Quartz has been used as temperature compensation sample of strain gauge in this experiment. The strain response data (i.e., temperature compensation) after subtraction of the compensated strain data is the compensation data shown in the figure 4b-d.

As shown in Figure 4b, point a to point b is the first stage of the curing process, when the temperature rises from room temperature (300 K) to the first curing temperature (353 K). At this stage, the epoxy resin system is very soft and has a very low viscosity. FBGs and strain gauges can expand freely. From point b to point c, since the temperature does not change, the strain remains constant during this period. Micro-curing occurred during this period without global changes. Before the point c, the curing has proceeded, and the percentage of curing has increased. The micro-cured molecules float in the uncured soft resin. At the point c, global change is observed. The micro-cured molecules come together to produce larger molecules, which is the starting point for the epoxy resin to become solid. From the point c to the point d, the entire curing process is carried out during this period. The volume is reduced and the strain is reduced. From point d to point f, the curing ratio line becomes almost flat, which means that curing is about to end. As the temperature rises, the epoxy resin and the pressure gauge thermally expand. Then it is cured at a constant temperature, and the epoxy resin shrinks. In this way, the tension is reduced. The last step is from the f point to the g point. During this time the temperature will drop. The cooling shrinkage of the strain gauge wire and the cooling shrinkage of the epoxy resin simultaneously cause the strain to decrease. Both FBGs and strain gauges cannot test the real-time strain response of epoxy before initial solid point (point c). For epoxy resin is liquid during this period, thus FBGs and strain gauges are free in epoxy. Strain responses of epoxy resin measured by FBGs and strain gauges are different, as shown in Figure 4c. This shows that the temperature sensitivity of strain gauge is lower than FBG. A fluctuation is observed from point d to point e, as illustrated in Figure 4d. A large number of small molecules are cross-linked together to become large molecules at this stage.

The thermal shrinkage of the epoxy resin measured by FBG and PPMS during the cooling process is shown in Figure 5. The results show that the real-time strain response of FBG monitoring during cooling down to cryogenic temperature is effective and reliable. However, subtle differences can be observed from the embedded line and the combined line. The reason is that the surface measurement will cool or heat faster than the inside. Naturally, the internal cooling or heating rate will be slower than the surface.

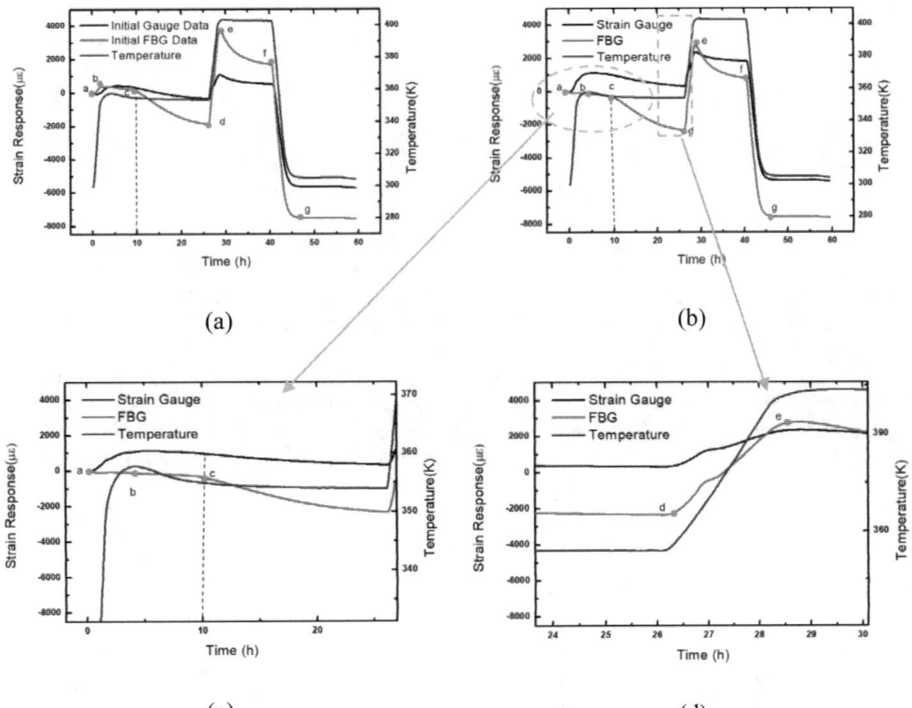

Figure 4. Strain response comparison of epoxy resin measured by FBGs and strain gauges: (a) initial data; (b) compensated data; (c) partial zoom of circle in Figure 4(b); (d) partial zoom of square in Figure 4(b).

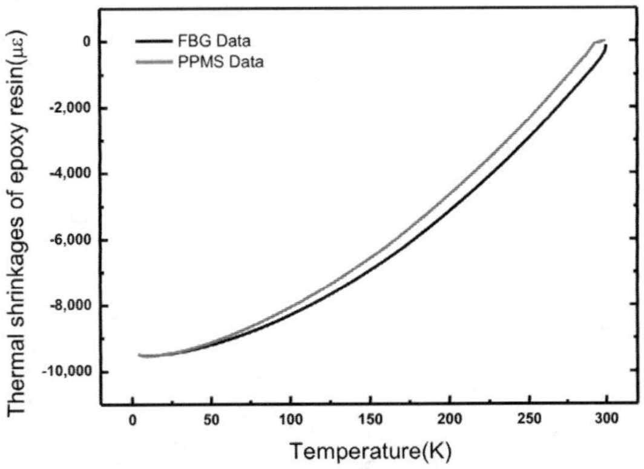

Figure 5. Thermal shrinkage of epoxy resin during cooling down to cryogenic temperature measure by FBG and PPMS

5. Conclusions

In this work, embedded FBGs were successfully applied to monitor the strain response of epoxy resin from the initial solid point during curing process and cooling down to cryogenic temperature. A strain response test system of embedded FBGs was designed to monitor the curing process of epoxy resin. A test platform for measuring the thermal expansion coefficient

IOP Conf. Series: Materials Science and Engineering 1241 (2022) 012008 doi:10.1088/1757-899X/1241/1/012008

of epoxy resin with FBGs was established. Compared to measurement results of PPMS, the reliability of real-time strain response of epoxy resin measured by embedded FBGs was verified. Through experiments, FBGs are found to have better sensitivity and accuracy than strain gauges when monitoring strain response of epoxy resin. Results obtained in this work lay the foundation of the real-time strain monitoring of epoxy-based composite materials with FBGs.

6. References

[1] Ihn J.B., Chang F. K., Sturct. Health Monit., 7(2008): 5-19.
[2] Lau K. T., Mater . Sci. Tech-lond., 30(2014): 1642-1654.
[3] Xing K. J., Fritzen C. P ., Damage Assessment of Structures VII, 347(2007), p153.
[4] Sun Z., Rocha B., Wu K. T., Mrad Z., Int. J. Aerospace Eng., 2013(2013):1-22.
[5] X-33/RLV system health management/vehicle health management, NASA, AIAA-98-1928
[6] X-33/RLV reusable cryogenic tank VHM using fiber optic distributed sensing technology,NASA, AIAA-98-1929.
[7] Cramer K. E., Research developments in nondestructive evaluation and structural health monitoring for the sustainment of composite aerospace structures at NASA, 30th Congress of the International Council of the Aeronautical Science, 2016: 1-7.
[8] Wu M. C., Winfree W. P ., Allison S. G., NASA, Langley Research Center, 2007.
[9] Dong H, Liu H, Nishimura A, Wu Z, Zhang H, Han Y, et al. Sensors. 2020;21:172.
[10] ISO. ISO 8130–6:1992: Coating Powders, Part 6: Determination of Gel Time of Thermosetting Coating Powders at a Given Temperature; ISO: Geneva, Switzerland, 1992

Acknowledgements

This work was supported by the National Natural Science Foundation of China (Grant No.: 51877209) and the National Key Research and Development Program of China (Grant Nos: 2017YFE0301405 and 2017YFE0301403).

Thermal conductivity measurements of nano-particle-filled epoxies

J L Adams[1], A Haight[2], P C Michael[1], L Bromberg[1], and J G Brisson[1]

[1]Massachusetts Institute of Technology, Cambridge MA
[2]Composite Technology Development Inc.

Email: jacoba@mit.edu

Abstract. For conduction-cooled superconducting magnets to be truly effective, it is necessary to have good thermal conduction from the cooling source through the insulating epoxy resin to the magnet winding. We have developed epoxy resins filled with thermally conductive nanoparticles that have significantly increased thermal conductivities while retaining most of their dielectric and mechanical strengths. Base resins, including CTD101K, were filled with alumina, aluminum nitride, or diamond nanoparticles. A wire-based 3ω method was developed to measure the thermal conductivity of the nano-filled epoxies over the temperature range from 4 K to 300 K. A unique aspect of this experimental setup is the 10 μm diameter Invar wire heating element, which serves as a sensitive thermometer down to 4 K. Measurements show the filled resins have improved thermal conductivity, especially at temperatures of 15 K and higher.

1. Introduction

Closed-cycle cryocoolers offer an alternative means for cooling superconducting magnets compared to immersion in a liquid helium bath. However, the design of the cooling pathways in conduction-cooled magnets required significant attention to ensure effective heat removal, especially during transient operation. Effective conduction cooling may be inhibited by the low thermal conductivity epoxy resin that electrically insulates the magnet's electrical power circuit from the cooling circuit. To increase thermal conductance of the epoxy, nano-particle fillers may be added to the resin without sacrificing the epoxies dielectric or mechanical properties. The increase in thermal conductance is beneficial for effective conduction cooling.

A wire based 3ω technique was developed to measure the thermal conductivity of filled epoxy resins in the temperature range from 4.2 K to 300 K. The transient 3ω measurement technique has many advantages over steady-state techniques including in-variance to thermal contact resistance, small sample size, and short equilibration time [1,2]. A unique aspect of this work is the 10 μm Invar wire heating element, as opposed to a photolithographically patterned or sputtered heating element. The 3ω method was first utilized by Cahill in 1990 [2]. An overview of the 3ω method, related techniques, and applications is discussed by Dames [1].

2. 3-ω Theory

Figure 1a) depicts the 3ω measurement geometry of a straight wire embedded in a sample. This wire acts as both a heating element and thermometer. Figure 1b) graphically depicts the electro-thermal transfer functions by which a driving current at frequency 1ω results in an underlying voltage across the wire at frequency 3ω. The 3ω voltage contains information about the thermal properties of the wire-sample system [1].

In more detail, a sinusoidal current at frequency 1ω drives the heater/thermometer (heater for short). Power is dissipated by the heater at a frequency of 2ω due to Joule-heating. The temperature of the heater oscillates at a frequency of 2ω, corresponding to the power dissipated. Since the heater's resistance is a function of temperature, the temperature oscillations at 2ω correspond with resistance oscillations at 2ω. The convolution between the 1ω current times the 2ω resistance oscillations results in an underlying voltage at 3ω. Therefore, the total voltage across the heater contains the large, primary voltage signal at 1ω in addition to the smaller signal at 3ω. The amplitude and phase shift of the voltage signal at 3ω contains information about the thermal properties of the heater/sample system [1,2].

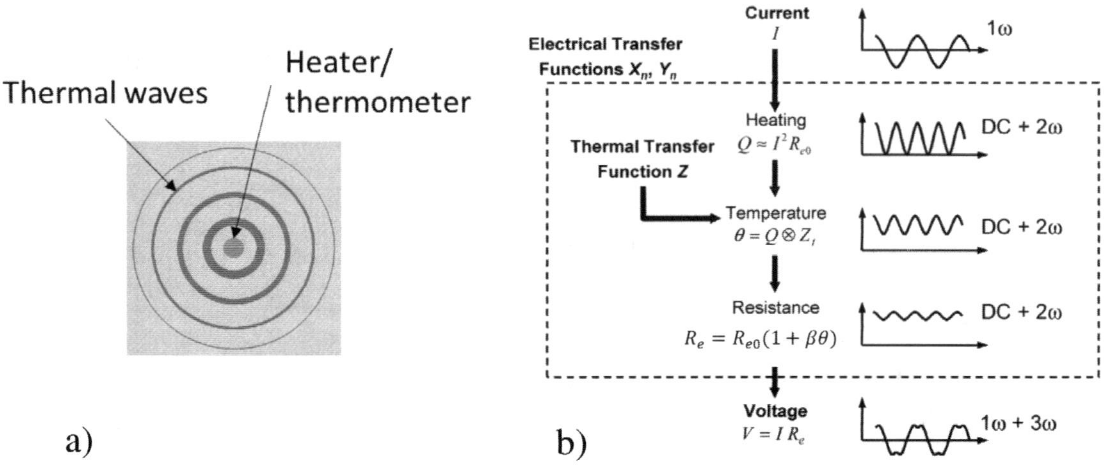

Figure 1. a) Cross-sectional view of 3-ω geometry. A line heater embedded in a semi-infinite solid. b) Graphical relationship of electro-thermal transfer functions where a driving current at frequency 1ω results in a superimposed voltage signal at frequency 3ω [1].

The geometrical configuration displayed in Figure 1a) is well described as a one-dimensional line heater embedded in a semi-infinite solid. The solution to the heat diffusion equation for this geometry with sinusoidal temperature oscillations at the wire is given by Carslaw and Jaeger [3]. Cahill shows that the resulting voltage at 3ω, $V_{3\omega}$, is related to the thermal properties of the surrounding material by the following expression [2].

$$V_{3\omega} = -\frac{V_0^3\beta}{16\pi lkR_h}\left(\ln(2\omega) + \ln\left(\frac{b_h^2}{\alpha}\right) - 1.756\right) - i\frac{V_0^3\beta}{8lkR_h} \tag{1}$$

where V_0 is the applied driving voltage to the heater, β is the temperature coefficient of resistance of the heater, l is the length of the heater, k is the thermal conductivity of the sample, R_h is the resistance of the heater, ω is the driving frequency, b_h is the radius of the heater, and α is the thermal diffusivity of the sample. Voltages are in RMS. The imaginary component in equation (1) represents the out-of-phase temperature oscillations with respect to the driving frequency.

In theory, it is possible to calculate the thermal conductivity of the sample, k, from equation 1 after measuring the magnitude and phase of $V_{3\omega}$ if all other variables are known. In practice however the thermal diffusivity, α, is usually unknown and the heater radius, b_h, is small (~10 μm) and potentially difficult to measure. For this reason, the thermal conductivity is more accurately measured by taking the slope, ζ, of the in-phase 3ω voltage, $V_{3\omega,in}$, with respect to the logarithm of frequency [2].

$$k = \frac{(V_0^3\beta)}{16\pi R_h l * \zeta} \tag{2}$$

where

$$\zeta = \frac{\mathrm{d}(\mathrm{V}_{3\omega,\mathrm{in}})}{d(\ln(2\omega))} \tag{3}$$

The slope method described by equations (2) and (3), requires that $V_{3\omega,in}$ be measured at various 1ω driving frequencies. The driving frequency dictates how far the evanescent thermal wave will penetrate the sample. The characteristic length that the thermal wave penetrates the sample will be referred to as the thermal penetration depth, λ.

The penetration depth, λ, is related to the thermal diffusivity of the sample and the driving frequency by [2]

$$\lambda = \sqrt{\frac{\alpha}{2\omega}} \tag{4}$$

Adjusting the driving frequency, ω, results in a change to the thermal penetration depth. To satisfy the theoretical assumption of a semi-infinite solid, the thermal penetration depth must be less than the thickness of the sample, t_s. If λ is greater than the thickness of the sample, then the thermal properties being measured include materials outside the sample. Simultaneously, the thermal penetration depth must be larger than the radius of the heater to satisfy the assumption of a one-dimensional line heater. If λ is less than the radius of the heater, then the thermal properties of the heater are being measured. To keep errors under 1%, the practical constraints on thermal penetration depth are: $5b_h < \lambda < \frac{t_s}{5}$ [1].

Error associated with interfacial thermal contact resistance between the heater wire and the sample are nulled with the slope method because an interfacial contact resistance has no thickness and no heat capacity. Therefore, an interfacial thermal contact resistance results in a constant offset to $V_{3\omega}$ that is independent of driving frequency. The slope method for evaluating thermal conductivity only relies on the variation of $V_{3\omega}$with respect to driving frequency.

Advantages of the 3ω method include:

- Small samples can be tested since the thermal penetration depth is only ~50-250 μm into the material.
- Short equilibration time of ~30 seconds at each temperature point.
- Thermal contact resistance between the wire and sample does not affect the measured thermal conductivity (thanks to the slope method).
- The wire is embedded in the sample, so parasitic heat losses to the outside environment are negligible.

These advantages allow for simple integration of the wire-based 3ω thermal conductivity measurement into various test cryostats.

3. Sample Preparation

The 3ω method requires that a small diameter (~10 μm) heating element be in intimate thermal contact with the sample. In many 3ω experiments, the heater is deposited onto the surface of the sample via sputtering through a shadow mask, or photolithography [1]. These deposition processes are viable, but in some applications undesirable due to cost, convection effects at the surface, and difficulty adhering to some surfaces. In this work, a commercially available (Goodfellow Corporation) 10 μm diameter Invar wire is used as the heating element. The 12 mm long Invar wire is mechanically clamped between two epoxy sample bars to satisfy the theoretical assumption of a line-heater embedded in a semi-infinite solid.

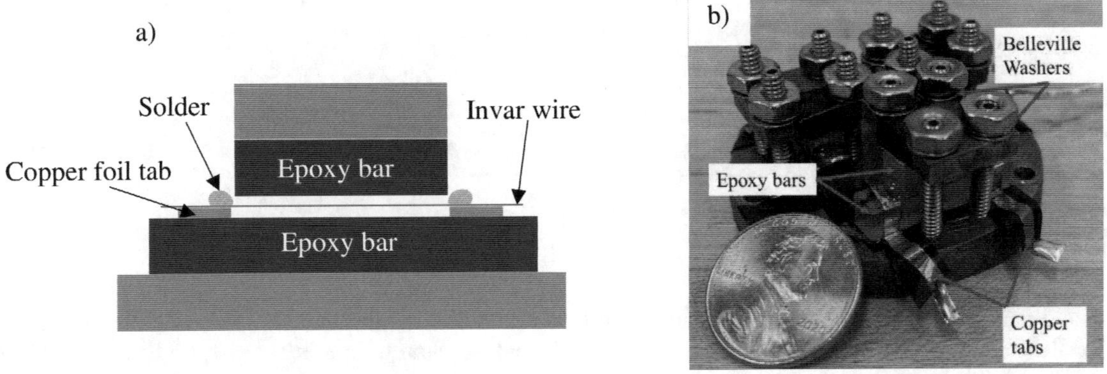

Figure 2. a) Drawing of Invar wire clamped between epoxy bars. b) Image of sample holder which holds three separate samples.

The 3ω method necessitates a heating element with a temperature coefficient of resistance so that the temperature oscillations correspond with resistance oscillations. Invar has a measurable temperature coefficient of resistance down to ~4.2 K allowing for thermal measurements in that temperature region (as opposed to many pure metals which are not sensitive below ~30 K).

Surface-treated nanoparticles with sizes ranging from 20 – 80nm were added into epoxy resins and blended using a FlackTek Speedmixer. Proprietary surface treatments were selected based on the particle type. Mixing speed and times varied depending on the specific resin and particles used. After mixing, the resins were poured into a mold, cured, and subsequently machined into bars with approximate dimensions of 2mm x 3mm x 25mm for testing. The mating surfaces of the epoxy bars are sanded flat and polished to 3000 grit. The epoxy bars are aligned in the clamping fixture and temporarily held in place with two-sided tape. Pre-tinned copper foil tape (~50 μm thick) is placed on the bottom epoxy bar and serves as the electrical contact pad to the Invar wire.

After the epoxy bars are aligned and the copper foil tabs are in place, the Invar wire is stretched between the copper foil tabs and held in place with tape. The Invar wire is inspected under microscope to ensure it is straight and free of impurities or other defects. After inspection, the Invar wire is clamped between the epoxy bars and the nuts are tightened until the stacked Belleville washers are fully compressed. The Belleville washers ensure compression during thermal contraction from 300 K to 4 K.

The Invar wire is soldered to the pre-tinned copper tabs using 96/4 Tin-Silver solder with Superior No. 71 stainless-steel soldering flux. This solder and flux resulted in appropriate wetting of the Invar wire and reliable electrical connections. Any residual flux is cleaned from the electrical contact, and the contact is inspected under microscope for signs of damage or remaining flux. Finally, the electrical contact is potted in epoxy for corrosion protection.

4. Instrumentation and Procedure

4.1. Wheatstone Bridge

The voltage at 1ω is typically 1000 times larger than the voltage at 3ω ($V_{3\omega}$~5-50 μV), so a Wheatstone bridge is used to attenuate the large, 1ω component. A lock-in amplifier (Signal Recovery 7270) drives the Wheatstone bridge and measures the 3ω voltage between points A-B on figure 3. A three-wire bridge configuration eliminates errors associated with non-zero lead wire resistance (R_L). Resistors R_1, R_2, and R_V are high precision with temperature coefficients less than 10 ppm to limit stray 3ω components in the bridge.

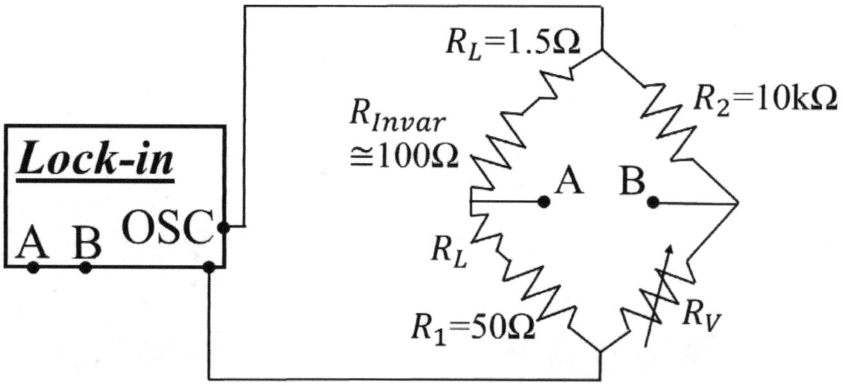

Figure 3. Electrical schematic of Wheatstone bridge with three-wire configuration. R_L is lead wire.

4.2. Cryostat

The sample holder is mounted in a liquid helium cryostat (CryoIndustries DSTAT). The samples are cooled via direct contact with cold, flowing helium vapor. The temperature of both the copper sample holder and the temperature of the flowing helium vapor are controlled with a Lakeshore Model 335 Temperature Controller. The temperature at the sample holder is measured with a calibrated silicon diode thermometer.

4.3. Procedure

The sample holder is wired and mounted in the cryostat. The temperature is set between 4.2 K to 300 K. At each temperature point the Invar wire resistance is measured twice. First with a low power (~1 μW) excitation which is used for calibrating the resistance as a function of temperature (with negligible self-heating of the Invar wire). Then, the resistance is measured again with the high power (~0.5 mW - 1 mW) excitation used for the 3ω measurement. The resistance during the high power 3ω measurement is used to find the local temperature of the Invar wire which is typically ~1 K higher than the temperature of the sample holder. To measure resistance, the variable resistor (R_V in figure 3) is adjusted until the bridge is balanced and then the resistance of R_V, R_1, and R_2 is used to calculate R_{Invar}.

For each temperature point, the 3ω voltage is measured over a range of driving frequencies (usually 0.5-5 Hz) where the magnitude and phase shift are recorded. The thermal conductivity is then evaluated using the "slope method" described by equations (2) and (3).

5. Results and Error Estimation

5.1. Temperature Coefficient

Measuring the temperature coefficient, β, is required for determination of thermal conductivity (equation (2)). The temperature coefficient of the Invar wire varies over the range 4.2 K to 300 K as shown in figure 4. Figure 4 labels the relative error between various Invar samples. Note that the variance among samples is largest near 4 K where the magnitude of the temperature coefficient is smallest and most difficult to measure. Since all the Invar wires plotted in figure 4 are the same material from the same spool, the temperature coefficients should be approximately the same. The variance in measured temperature coefficient is identified as a source of error.

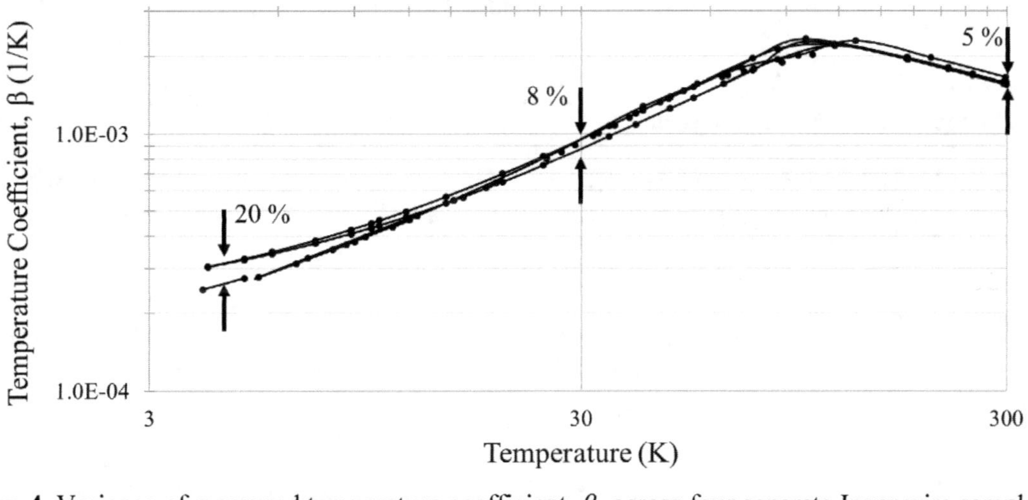

Figure 4. Variance of measured temperature coefficient, β, across four separate Invar wire samples all from the same spool of Invar. Percent difference labeled at various temperatures.

The resistance of Invar is single-valued and monotonically decreases with temperature. Between ~100 K to 300K the resistance of Invar is linear with temperature. Below ~100 K, polynomial functions are curve fit. Figure 4 shows an inflection in β near 100 K, where the trend goes from linear to polynomial.

5.2. Thermal Conductivity Results

Figure 5 shows that the filled epoxies have improved thermal conductivity over the neat samples particularly in the temperature range ~15 K to 300 K. The improvement to thermal conductivity is largest near 300 K, where up to five times increase in conductivity was observed.

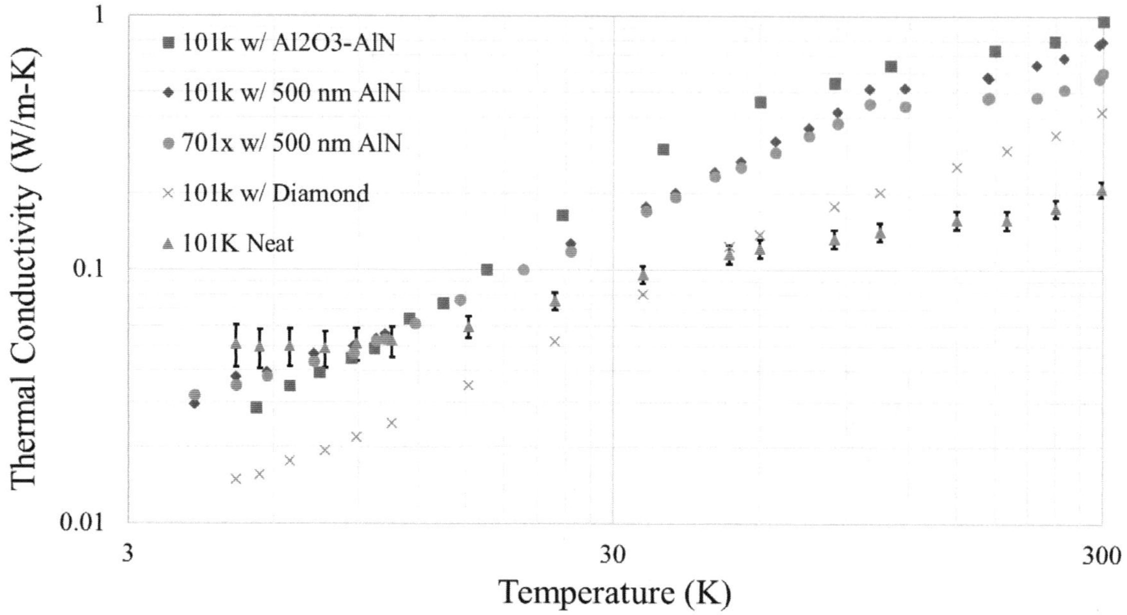

Figure 5. Measured thermal conductivity for nano-particle filled epoxies. Representative error bars shown on the CTD 101K neat datapoints.

In the temperature range of 5 K – 9 K the unfilled, neat sample thermal conductivity plateaus, while the conductivity of the filled samples continues to fall. This "plateau" behavior was also observed in [4]. Reference [4] utilized a steady-state type thermal conductivity measurement technique and is used as a point of comparison with this work.

Table 1. Comparison of measured thermal conductivity for neat CTD101K resin from this work versus Ref. [4]. Values from [4] are approximate since they are read from a graph.

Temperature (K)	This work ($\frac{W}{m-K}$)	Ref. [4] ($\frac{W}{m-K}$)
5	0.05 ± 0.01	~0.048 ± 0.005
15	0.06 ± 0.006	~0.062 ± 0.005
30	0.09 ± 0.007	~0.089 ± 0.005

Table 1 shows good agreement between the thermal conductivities measured in this work versus those in [4], despite the difference in measurement techniques.

5.3. Error Analysis

An error propagation analysis of equation (2) is used to quantify the relative uncertainty from the input variables $(V_0, \beta, R_h, L, \zeta)$. To first order, the error in the temperature coefficient, β, dominates. Dames agrees that β is commonly a large source of uncertainty in 3ω measurements [1].

The temperature coefficient of the Invar wire decreases with temperature (see figure 4). As a result, it becomes increasingly more difficult to measure the miniscule changes in resistance as a function of temperature. For example, near 5 K a change in temperature of 1 K results in a change in resistance of ~10 mΩ on the Invar wire. It is determined that the maximum error in β is ~20% at 4.2 K, with this error decreasing to ~8% by 20 K (see figure 4). The representative error bars for the CTD101K neat sample in figure 5 reflects increasing error at lower temperatures.

6. Conclusion

A 3ω thermal conductivity measurement technique was implemented using a commercially available Invar wire heating element. The Invar wire is a sensitive thermometer down to 4.2 K, as opposed to many pure metals which are insensitive to temperature change below ~30 K. The wire-based 3ω technique allows for inexpensive, relatively fast thermal conductivity characterization of small samples over a wide temperature range of 4.2 K to 300 K.

The Invar wire 3ω technique is validated on nano-particle filled epoxies. The measurements on CTD101K neat samples from this work agree with the values measured in [4] (Table 1). The filled epoxies show improved thermal conductivity over the neat samples, particularly above 15 K. For conduction cooled magnet design, filled epoxies may be beneficial over the un-filled, neat epoxies. Further system level analysis is required to quantify performance improvements from filled epoxies.

7. References

[1] C. Dames, (2013) "Measuring the Thermal Conductivity of Thin Films: 3 Omega and related Electrothermal Methods" Annual review of Heat transfer

[2] D. Cahill, (1990), "Thermal conductivity measurement from 30 to 750 K: the 3ω method" Review of Scientific Instruments 61, 802

[3] H. S. Carslaw and J. C. Jaeger (1959), "*Conduction of Heat in Solids*" (Oxford University Press, Oxford, p. 193.

[4] T Koettig, et. al. (2017), "Thermal conductivity measurements of impregnated Nb3Sn coil samples in the temperature range of 3.5 K to 100 K" IOP Conf. Mater. Sci. Eng.

Acknowledgments

This work was funded in part by the U. S. Department of Energy under Grant Number DE-SC0020828. The equipment and methods were developed under Air Force Contract No. FA8702-15-D-0001.

IOP Conf. Series: Materials Science and Engineering 1241 (2022) 012010 doi:10.1088/1757-899X/1241/1/012010

Characterization of Candidate Insulation Resins for Training Reduction in High Energy Physics Magnets

A E Haight[1], M M Haynes[1], S T Krave[2], J L Rudeiros-Fernandez[3] and T Shen[3]

[1] Composite Technology Development, Inc., 2600 Campus Drive, Suite D, Lafayette, CO 80026

[2] Magnet Systems Department, Mail Stop 315, Fermilab, PO Box 500, Batavia, IL 60510-5011

[3] Lawrence Berkeley National Laboratory, 1 Cyclotron Road Mailstop 47R0112, Berkeley, CA 94720

Email: andrea.haight@ctd-materials.com

Abstract. Superconducting magnets are critical components in particle accelerators and are used to generate and sustain the large magnetic fields needed for High Energy Physics programs. One significant issue with current epoxy insulated Nb_3Sn magnets is the long training process required before stable magnet performance can be realized. It is believed that training can be significantly reduced by addressing magnet quenching through improvements in the epoxy electrical insulation. In this work, two approaches for insulation modification have been undertaken: (1) addition of thermally conductive fillers to help with quench management and (2) development of insulation resins with high strain capability at cryogenic temperatures. This paper will discuss the characterization of these insulation systems to verify their performance prior to evaluation in subscale Nb_3Sn canted cosine theta accelerator dipole magnets.

1. Introduction

Superconducting magnets are critical components in particle accelerators and are used to generate and sustain the large magnetic fields needed high energy physics experiments. Nb_3Sn magnets are the current focus for upgrades to the Large Hadron Collider and are being considered for future, larger machines such as the 100 TeV class Future Circular Collider (FCC). One significant issue with current epoxy insulated Nb_3Sn magnets is the long training process required before stable magnet performance can be realized. It is believed that training can be significantly reduced by addressing magnet quenching through improvements in the epoxy electrical insulation. Modification of epoxy insulation for improved magnet performance has been reported extensively in the literature [1]-[7]. The potential of alternative systems has also been discussed [8].

The main objective of this work is to identify the new resin systems with the most potential for reducing the number of training cycles required for Nb_3Sn superconducting magnets. Two approaches are being considered: (1) addition of thermally conductive fillers to CTD-101K type insulation to better manage heat dissipation during quench and (2) development of toughened resin systems that will exhibit reduced cracking during cooling and energization of the magnet. This paper discusses the characterization of these resin systems in preparation for their evaluation in subscale Canted Cosine Theta (CCT) [9] magnets.

2. Experimental Methods

2.1. Resin Viscosity

Resin viscosity was determined using a Brookfield DV2T viscometer with automated data acquisition. In order to be classified as suitable for vacuum pressure impregnation (VPI), a viscosity below 300 mPa-s and working times of at least 4 hours were sought.

2.2. Mechanical Testing

2.2.1. Neat Resin Tensile Test

Neat resin tensile tests of high strain resin candidates were conducted in accordance with ASTM D638 using a Type IV strain-gaged specimen with a gage length of 35.6 mm. Specimens were tested at both 77K and 293K.

2.2.2. Compression and Short Beam Shear Tests

Laminates were fabricated using Style 6781 S2 glass fabric and the resins developed during the course of the program. Resins included: CTD-101K (reference), CTD-155, CTD-155GN, and CTD-701X. All laminates were fabricated using a vacuum pressure impregnation (VPI) process using a shimmed closed mold and the number of plies and mass of resin necessary to produce laminates with a nominal 50% fiber volume fraction with a laminate thickness of approximately 3.2 mm. Resin transfer was along the length of the laminate. Impregnation pressure was adjusted as needed to accommodate the higher volatility of the CTD-701X resin system. Fiber volume fraction was determined using weights of glass and resin in the laminate.

Short beam shear testing was conducted in accordance with ASTM D2344, nominal specimen dimensions were 28 mm x 6.35 mm x 3.2 mm with a span of 15 mm. The dominant failure mode in all cases was interlaminar shear.

Compressive properties were evaluated using methods established in evaluating ITER resin candidates [10]. Nominal specimen dimensions were 6.35 mm x 6.35 mm x 3.2 mm. Laminates were fixtured using 25 mm platens and tested in the through-thickness direction at both 77 K and 293 K. This configuration was chosen in part because it more closely represents conditions observed in magnet applications. Through-thickness deflections were determined using extensometers. Modulus is reported as "chord modulus," which we define as the slope of the chord drawn between any two specified points on the stress-strain curve; chord modulus calculations are recommended for non-linear materials [11].

2.3. Thermal Expansion

Thermal expansion tests were performed on two or three axes of at least three specimens of each type. Uni-axial strain gages were mounted on the test specimens, one in each direction to be evaluated. An identical strain gage was mounted on a Copper NIST reference standard. A single silicon diode temperature sensor is mounted between one test specimen and the reference material, and all specimens reside within the same vertical plane during testing. The specimens and standards were cooled to cryogenic temperature using either liquid helium or liquid nitrogen and held at cryogenic temperature for a minimum of 5 minutes prior to initiation of the warming cycle. Thermal expansion is measured via strain variance relative to the NIST standard. Thermal expansion data is normalized relative to room temperature and averaged across the three data files.

Shear compression testing was conducted at 77K using the method described in the work of Simon, *et al* [12]. Specimens were fabricated by placing plies of S2 glass cloth between two grit-blasted steel buttons in the impregnation fixture, followed by impregnation with the resin of choice. Steel was chosen as the substrate to facilitate comparison to prior data. Specimens were then placed at an angle in the test fixture for testing. Several angles were tested for each resin type to establish the "envelope" for the system.

3. Results and Discussion

3.1. Evaluation of Thermally Conductive Resin Systems

Thermally conductive resin evaluation focused on a system utilizing the CTD-155 resin system with varying levels of nanoscale filler. All early testing was conducted at the highest possible volume loadings of fillers (35 volume percent) but lower filler contents were evaluated due to the lower resin viscosities that can be achieved. Filled resins were formulated at 15, 20, and 35 volume percent (vol %) of nanofiller and were evaluated for viscosity/pot life and thermal conductivity.

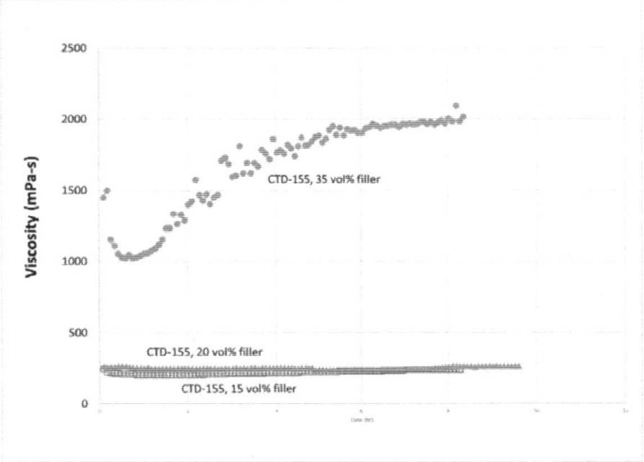

Figure 1. Pot life at 60°C for CTD-155 with various levels of nano-alumina filler.

Viscosity curves obtained at 60°C are presented in Figure 1. The viscosity of the 35 vol % formulation is not at all suitable for the commonly employed VPI processes, whereas both the 20 vol % and 15 vol % systems are suitable.

The thermal conductivity of these resin systems was also verified by Thermtest, Inc. (Fredericton, NB Canada) using the Physical Properties Measurement System (PPMS) by Quantum Design to ensure that significant thermal conductivity was not lost relative to the 35 vol% system. The thermal conductivity of the 20 vol % specimen was 0.4 W/m-K at 90 K (the lowest temperature available at the time of testing), corresponding to that of the 35 vol % system measured during an earlier phase of the program. It is recognized however, that this temperature is not necessarily relevant to a magnet application and that thermal conductivity enhancement at 4K is extremely difficult. Additional efforts on characterizing the thermal conductivity of these materials are described in the paper by Adams, *et. al* in these proceedings [13].

3.1.1. Laminate Evaluation. The 20 vol % CTD-155 resin, hereafter referred to as CTD-155GN, was then used to manufacture a laminate reinforced with S2 glass (50% fiber volume fraction) using a standard VPI process. Generally, the concern using filled resin systems is that the particles will get filtered out by the reinforcement; nanometer-sized particles were used for this work in the hope that the filtering issue can be avoided. This does indeed appear to be the case; inspection of the laminates using optical microscopy shows a uniformly impregnated laminate with no evidence of particle filtering, as shown in Figure 2.

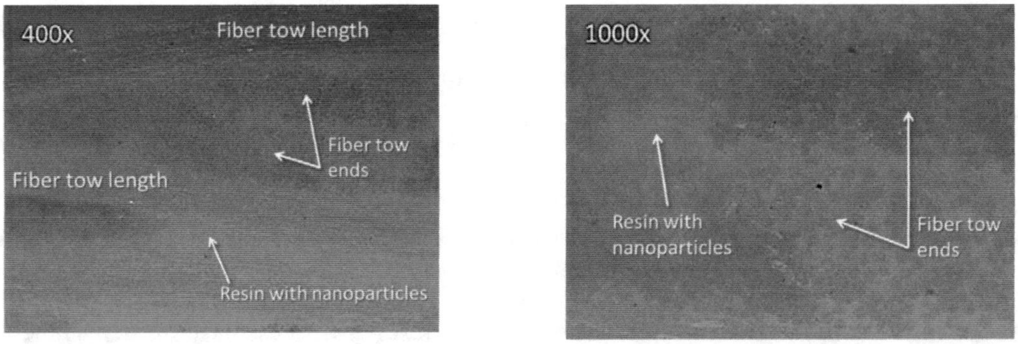

Figure 2. Optical microscopy of S2 glass/CTD-155GN laminate. Notice the uniformity of the laminate; no particle separation visible, resin area is a uniform milky white color with no "clear" spots or voids.

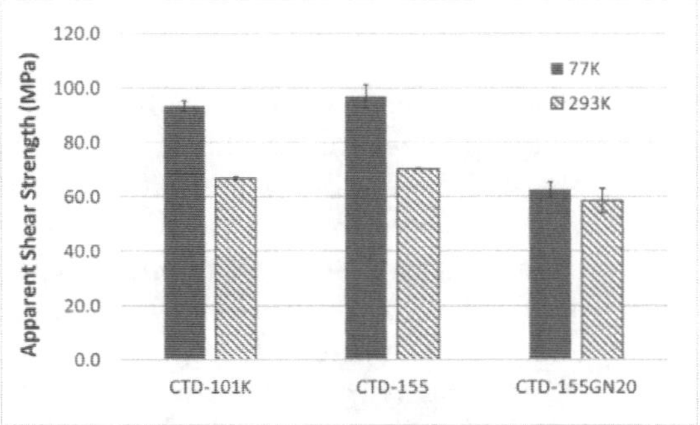

Figure 3. Comparison of shear performance of S2 glass laminates using CTD-155 resin variants. Data for CTD-101K laminates is included for comparison.

Mechanical properties of the laminates were then evaluated at 77 K and at 293 K; results of the short beam shear testing are presented in Figure 3. While the apparent shear strength of the nanoparticle-filled laminate is somewhat reduced relative to laminates manufactured with the unfilled resins, the resulting strength of 60 MPa is more than sufficient for the intended magnet application; modeling indicated that a minimum shear strength of 35 MPa would be required in the insulation to withstand magnet shear stresses [14].

3.1.2. Shear Compression Testing. Since magnet stresses are not strictly limited to shear or compressive stresses, we also conducted shear-compression testing for new resins under development in this work. This test is designed to more accurately mimic the performance of the insulation system in a magnet configuration. Typically, several angles are evaluated to define the shear-compression "envelope" for the insulation system. Results for the CTD-155GN system compared to the often-used CTD-101K insulation system are shown in Figure 4.

The shear compression performance for the CTD-155GN laminate is a bit lower than that of the CTD-101K laminates. We believe that this is due to the larger fraction of adhesive failures in the CTD-155GN specimens, where it appears that the particle loading may have an influence on adhesion to the steel substrates. In contrast, most of the CTD-101K failures were cohesive in nature. The photograph inset of Figure 4 shows both cohesive (A1, A4) and adhesive (A3, A6, A8) failures of the CTD-155GN specimens.

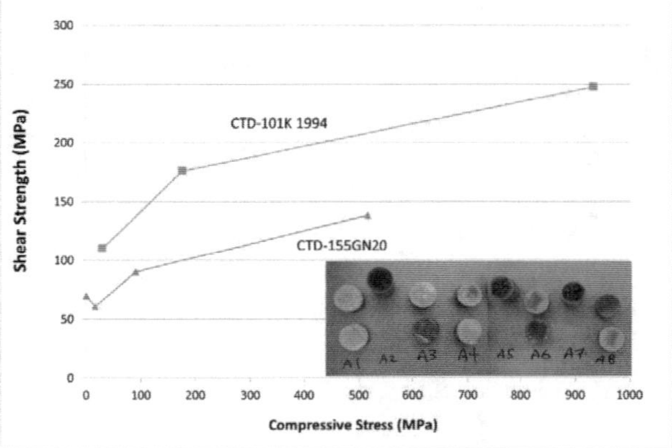

Figure 4. Shear compression test results for S2 glass/CTD-155GN as compared to results for S2 glass/CTD-101K. Inset photograph shows failure surfaces in the test specimens.

3.1.3. Thermal Expansion. Thermal expansion of the CTD-155GN was evaluated over the temperature range 77-293 K. Thermal expansions of the unfilled CTD-155 resin (neat resin and laminate) were included for comparison. The results for both S2 fabric reinforced laminates and neat resin are provided in Figure 5.

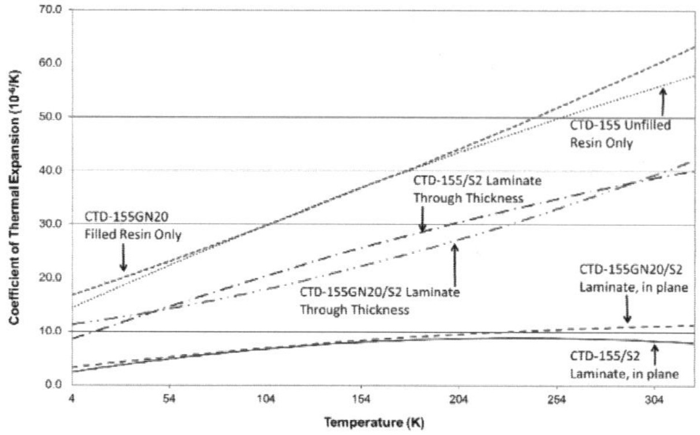

Figure 5. Comparison of thermal expansion for CTD-155 and CTD-155GN20 laminates and unreinforced resins.

As expected, the in-plane thermal expansion of the laminates is lower than the through-thickness thermal expansion of the laminate and both are lower than that of the neat resins. Interestingly, the thermal expansion of the unfilled resin and the nanoparticle-filled resin are not significantly different; one might expect that the filled resin would show lower thermal expansion. It may be that the filler content is lower than is normally used to manage thermal expansion, or that nanofillers do not have as large an effect on thermal expansion as micron-scale fillers, or a combination of the two. Additional studies are underway to better understand these results.

3.2. Evaluation of High Strain Resin Candidates

Several higher strain resin systems were developed during this program in an effort to reduce long training cycles in magnet systems. The systems evaluated to date and a few basic characteristics are listed in Table 1.

Table 1. High strain resin candidates for magnet applications.

Name	Description	T_g, °C (tan δ)	Viscosity at 60°C (mPa·s)	
			Initial	7 hours
CTD-101K	Baseline anhydride-cured epoxy resin	150 (avg QA)	70	100
CTD-103K2	Long pot life 2-part formulation similar to CTD-101K	141	61	61.5 (4.5 hr)
CTD-103LT	Lightly toughened version of CTD-103K2	126	24	71
CTD-155	Anhydride-cured epoxy with reactive rubber toughener	147	125	127
CTD-701X	Extremely low viscosity, tough [15] polyolefin resin system	131	25 (@ 25°C)	70 (90 min, 25°C)
CTD-7.1E	Low viscosity, toughened amine-cured epoxy resin system	72	225	Approx. 2hr

Of these, the most interesting system is CTD-701X, which showed neat resin elongation at failure as high as 5% at 77 K (Figure 6), whereas most of the other systems (e.g., CTD-101K) exhibited elongation at failure on the order of 2% at 77 K.

The following sections discuss a more comprehensive characterization of the CTD-701X system to support the production and testing of a subscale CCT magnet.

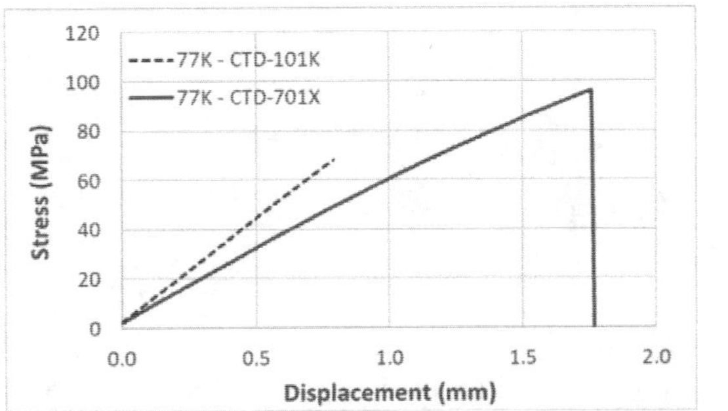

Figure 6. Neat resin stress-deflection curves at 77 K for CTD-101K and CTD-701X

3.2.1. Viscosity and Pot Life

The viscosity and pot life of the CTD-701X were evaluated at 25 °C and the results are shown in Figure 7. The initial formulation showed a significant acceleration in the viscosity increase beginning at approximately 3 hours. An alternative catalyst, that requires an elevated temperature cure (1 hour at 40°C followed by a 1 hour post cure at 120°C), was then evaluated at 25 °C. In this case, the viscosity remained stable over more than 8 hours.

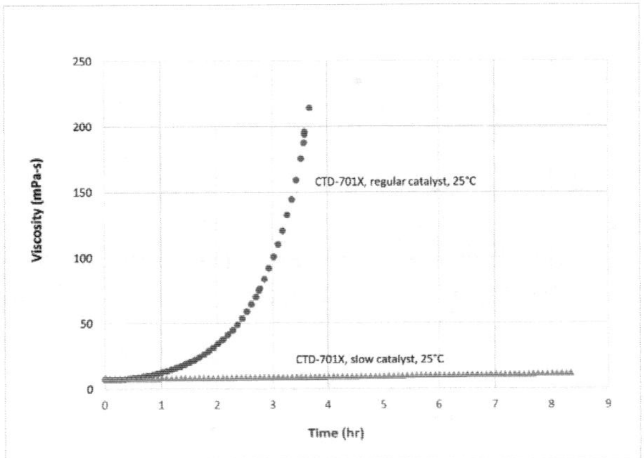

Figure 7. Pot life of CTD-701X at 25°C with two different catalyst systems.

3.2.2. Laminate Mechanical Properties. Shear and compressive properties of laminates fabricated with the high strain candidate systems were also evaluated. Compressive properties are shown in Figure 8 and shear properties are shown in Figure 9.

The compressive properties show very little variation regardless of the resin system. However, there were some differences noted in the shear properties. The epoxy systems (CTD-101K, CTD-101LT, and CTD-155) are largely equivalent, but the CTD-701X has significantly lower shear strength, likely due to the different chemistry (polyolefin vs epoxy) and potentially reduced compatibility with the fiber sizing in the glass laminate. However, the shear strength does appear to be sufficient for the subscale CCT magnet application based on modeling results as described above.

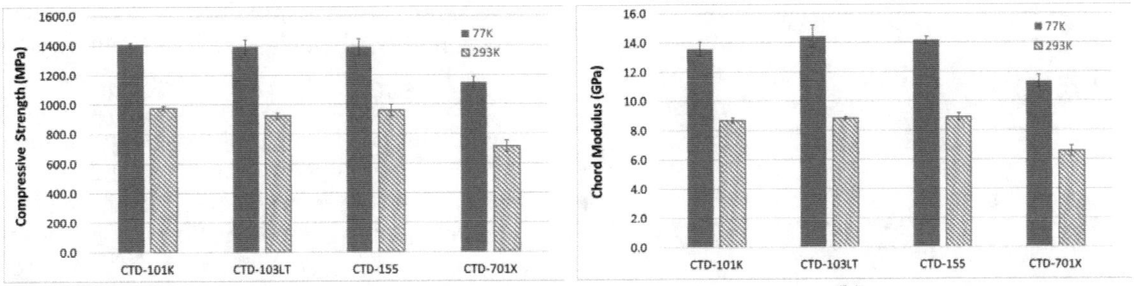

(a) (b)

Figure 8. Compressive properties of S2 glass-reinforced laminates fabricated with high strain resin candidates (a) compressive strength and (b) chord modulus. Data for CTD-101K laminates are included for comparison.

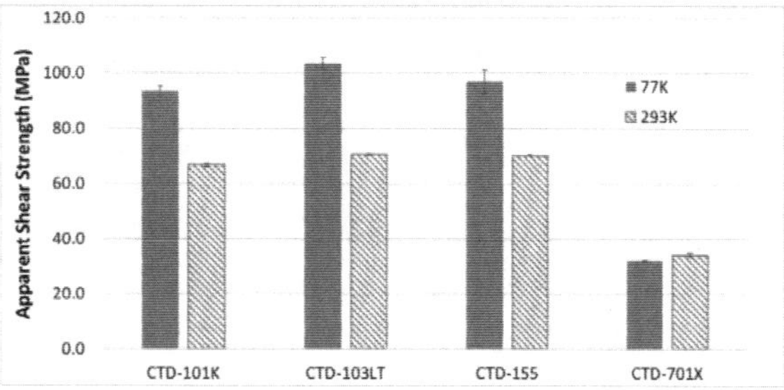

Figure 9. Apparent shear strength of S2 glass-reinforced laminates fabricated with high strain resin candidates. Data for CTD-101K laminates are included for comparison.

3.2.4. Thermal Expansion. Thermal expansion of the CTD-701X was evaluated between 4 K and 293 K. The results for both S2 fabric reinforced laminates and neat resin are provided in Figure 11.

The thermal expansion of this resin system, both with and without reinforcement, is somewhat higher than that observed for the commonly used epoxy resins, including CTD-101K. This trend is also expected in the through-thickness direction, which was not evaluated in this work. The presence of the S2 glass reinforcement does significantly curtail thermal expansion of the resin, however, so there is no reason to expect this material to be problematic in a magnet application provided that care is taken to avoid significant pockets of neat resin in the finished magnet.

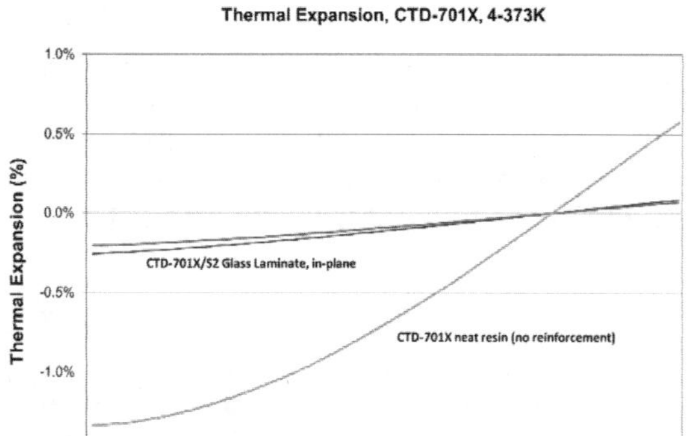

Figure 11. Thermal expansion of CTD-701X with and without reinforcement.

3.2.3. Shear Compression Testing. Shear compression testing as described in Section 3.1.2 was also conducted for laminates using the CTD-701X resin; the results are shown in Figure 10. The CTD-701X results are slightly lower than those observed for the CTD-101K system but are higher than those discussed earlier for the CTD-155GN system. All CTD-701X specimens showed cohesive failures.

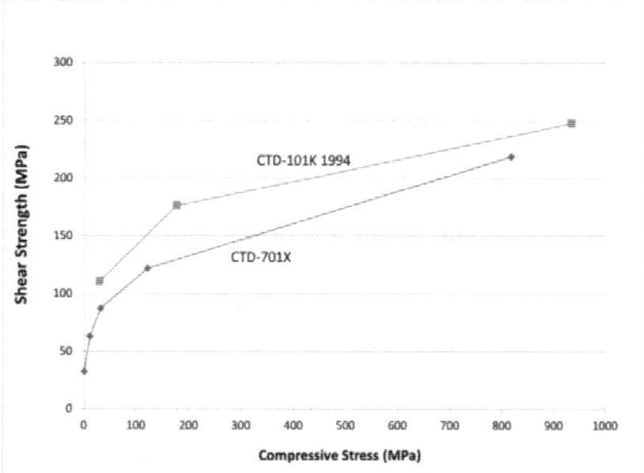

Figure 10. Shear compression test results for S2 glass/CTD-701X as compared to results for S2 glass/CTD-101K.

3.3. Results of Stack Testing [15]

Additional work evaluating these materials in thermal cycling as well as in four and ten-stack configurations also supported recommending materials for subscale CCT evaluation. Two particular results stood out from that work: (1) thermal cycling resulted in only one sample of six exhibiting any cracking over 10 thermal cycles and (2) the CTD-701X resin system in a cable-type configuration (4-stack) showed a higher short beam strength (160% at 77 K) relative to CTD-101K in a comparable configuration. Note that this is different from the laminate-only result presented here and was a large contributor to the decision to pursue subscale CCT magnets with this resin system.

4. Conclusions

This paper has presented results of characterization activities for two different resin systems developed as candidates for reducing training in high energy physics magnets. This characterization effort focused on understanding the resin systems as well understanding the properties of S2 glass laminates fabricated with these resins so to better understand how to apply them in a subscale CCT magnet system.

Based on the testing described herein, the following conclusions can be made for the thermally conductive system CTD-155GN: (1) it is possible to make a high quality composite using a resin system with 20 vol% nanoscale filler (2) thermally conductive formulations are unlikely to offer any improvements in heat dissipation when operating at 4 K, but are valuable for high-temperature superconducting magnets working above 20 K, and (3) addition of nanoscale fillers does not offer any significant improvement in thermal contraction performance for the CTD-155GN system. Thus, testing of this material in a Nb-Ti or Nb_3Sn magnet application is not justified at this time. However, the application of this insulation system for magnets operating at higher temperatures (e.g., magnets based on high temperature superconductors) might be worth considering.

For the high strain system CTD-701X, lower mechanical properties were noted. The compression modulus and strength are lower but are sufficient for superconducting accelerator magnets, for which the design transverse pressure to Rutherford cable is now no larger than 200 MPa. The interlaminar shear strength and the shear strength determined by shear/compression testing are also lower and in the territory of potentially causing more interface failures, and perhaps quench, in superconducting accelerator magnets. However, it is felt that this system provides a critical and useful evaluation of the importance of this parameter on the quench training of superconducting accelerator magnets. The work on four and ten-stack configurations described by Krave *et al.* [15] and summarized in Section 3.3 also supports this evaluation. The extreme resistance to cracking and improved short beam strength in stacks described by Krave *et al.*, coupled with favorable processing (low viscosity and long pot life, resin transfer at ambient temperature) as shown in this work, led to a decision to evaluate this resin system in a subscale CCT magnet. The results of the CCT magnet performance will be reported elsewhere.

5. References

[1] Yin, S, Arbelaez, D, Swanson, J. and Shen, T, 2019 *IEEE Trans. Appl. Supercond.*, **29(5)**, 7800205, doi: 10.1109/TASC.2019.2898124.

[2] Evans, D. and Morgan, J. "Epoxide resins for use at low temperatures," in *Nonmetallic Materials and Composites at Low Temperatures*. Boston, MA, USA: Springer, 1982, pp. 73–87.

[3] Fabian, P, Munshi, N and Denis, R. 2002, *AIP Conf. Proc.,* **614**, 295–304.

[4] Reed, R, Evans, D and Fabian, P, "Development of a new resin system for the US ITER central solenoid model coil," in Advances in Cryogenic Engineering Materials. New York, NY, USA: Springer, 2000, pp. 227– 234.

[5] Reed, R P and Evans, D, 2004, *AIP Conf. Proc.*, **711(1)**, 209– 216.

[6] Munshi, N A *et al.*, 2013 *IEEE Trans. Appl. Supercond.,* **23(3)**, 7700104.

[7] Zhang, Z and Evans, D, "Low-temperature deformation and fracture behaviour of epoxide resins," in Advances in Cryogenic Engineering Materials. New York, NY, USA: Springer, 2000, pp. 251–257.

[8] Toplosky, V J and Walsh, R P, 2006, *AIP Conference Proceedings* **824**, 219.

[9] Arbelaez,D, Bogdanof, T, Brouwer, L, Caspi, S, Dietderich, D, Rudeiros Fernandez, J L, Ferracin, P, Gourlay, S, Hafalia, R, Krutulis, M, Marchevsky, M, Maruszewski, M, Prestemon, S, Reynolds, M, Shen, T, Swanson, J, Teyber, R, Turqueti, M, Vallone, G and Wang, X, "Status of the Nb_3Sn Canted-Cosine-Theta Dipole Magnet Program at Lawrence Berkeley National Laboratory", submitted for publication in *IEEE Trans. Appl. Supercond.*

[10] Fabian, P E, Schutz, J B, Hazelton, C S, and Reed, R P 1994 *Adv. Cryo Eng.* **40**, 1007.

[11] https://www.admet.com/youngs-modulus-tangent-chord-modulus-testing-tips/#:~:text=Chord%20modulus%2D%20Slope%20of%20the,recommended%20for%20non%2Dlinear%20materials.

[12] Simon, N, Reed, R, and Walsh, R, 1992 *Adv. Cryo. Eng. (Mater)* **38,** 363.
[13] Adams, J, Haight, A E, Michael, P C, Bromberg, L, and Brisson, J G, *IOP Conference Series: Materials Science and Engineering, Advances in Cryogenic Engineering*, in press.
[14] Tengming Shen, private communication, August 20, 2020.
[15] Krave, S, Shen T, and Haight A. 2021, *IEEE Trans. Appl. Supercond.* **31(5)**, 400190.

Acknowledgments
The authors would like to acknowledge funding of this work through U. S Department of Energy Small Business Innovation Research (SBIR) Program, Grant No. DE-SC0018701.

IOP Conf. Series: Materials Science and Engineering 1241 (2022) 012011 doi:10.1088/1757-899X/1241/1/012011

Experimental Evaluation of Dielectric Losses of PPLP for Single Phase HTS Cable at Subcooled LN_2 Temperatures

Maalika Sarkar, I de Souza, A Sarkar, A S Gour and V V Rao

Cryogenic Engineering Centre, Indian Institute of Technology, Kharagpur, India, 721302
E-mail: maalika.sarkar123@gmail.com

Abstract. For the operation of high-temperature superconducting (HTS) power cables in liquid nitrogen (LN_2) at high voltage levels, there is a need for reliable and cost-effective insulating materials. This article investigates the dielectric losses in poly-propylene laminated paper (PPLP- Sumitomo) which is used as a cold dielectric for HTS cables. To calculate the dielectric losses, an experimental setup was developed to measure the relative permittivity of PPLP at different temperatures from 65 to 300 K at various frequencies. Using an in-house developed experimental setup, the dielectric loss for different layers of PPLP, at various temperatures and operating frequencies was estimated to evaluate one of the criteria of cold dielectric material and the same was compared with the data at room temperature. The temperature was reduced to 65 K using a vacuum pump-assisted sub-cooling system. The effective relative permittivity increases marginally for subcooled LN_2 impregnated PPLP, however there is reduction in tan delta loss which in turn reduces the dielectric loss and is useful for high voltage HTS based AC cables. This paper provides a comparative study on PPLP performance variation with respect to operating temperature and frequency.

1. Introduction

With the growing demand for electrical energy, high temperature superconducting (HTS) power cables are considered as a highly efficient solution. Towards the practical development of high-voltage high temperature superconducting (HTS) cables, liquid nitrogen (as cryogen) impregnated polypropylene laminated paper (PPLP) is widely used as a cold dielectric material [1–4]. PPLP in the form of tape is often used as insulation for oil-filled power cables because of its low dielectric loss and high dielectric strength [5–7]. PPLP-HTS-125 supplied by Sumitomo Electric Industries Ltd has a total thickness of (120 μm). Each PPLP sheet consists of a PolyPropylene sheet (70 μm) laminated on either side by a layer of kraft paper (25 μm). PPLP strips are helically wound in layers around the core of the HTS power cable and is also the main insulating material for cables. The kraft paper has good impregnating properties due to which these PPLP sheets easily absorb the cryogen (LN_2) which enhances the dielectric properties required for HTS cables [8][9].

AC losses occur in superconducting cables which in turn act as a source of heat to the cryogen. The AC losses are composed of core losses and dielectric losses. The core loss occurs in the HTS tapes and the former, whereas dielectric loss occurs in the cryogen impregnated PPLP wrapped around the HTS tapes and the former. The core loss has two components: hysteresis loss which is related to the magnetic field response of the material, and eddy current loss which is dependent on induced magnetic field and operating frequency. On the other hand, dielectric loss is the energy dissipated by heating of the dielectric material due to the varying electric field [10][11]. In HTS cables, the LN_2 impregnated PPLP is the insulating material. The dielectric losses in high voltage HTS cables are dependent on the dielectric properties, i.e. the relative permittivity and dissipation factor which are dependent on frequency as well as on operating temperature [2][12].

In this paper, the dielectric losses of subcooled liquid nitrogen (below 77 K) impregnated PPLP is experimentally determined for different samples and compared with room temperature (300 K) PPLP and liquid nitrogen (77 K) impregnated PPLP. For each sample, capacitance and dissipation factor were measured using an LCR meter which measures the impedance for the flow of AC current; and the relative permittivity at different temperatures was computed. The experimental setup and procedure are discussed in Section-II, whereas in Section-III the relative permittivity and the dielectric losses are computed. Section-IV summarises the results and discussions.

2. Experimental setup

In order to determine the relative permittivity and dielectric losses of PPLP, a capacitance measurement for various samples was carried out at subcooled temperatures and room temperature. The thermodynamic and electrical properties of nitrogen at atmospheric pressure are shown in Table 1. The complete experimentation involves two major steps- (a) sample preparation (b) measurement of capacitance and dissipation factor ($tan\ \delta$).

Table 1. Thermodynamic and electrical properties of nitrogen at 1×10^5 Pa (1 atm.) [13]

Property	Value
Dielectric permittivity of liquid nitrogen	$\epsilon_l = 1.43\epsilon_0$
Dielectric permittivity of gaseous nitrogen	$\epsilon_g = 1.00\epsilon_0$
Dielectric permittivity of vacuum	$\epsilon_0 = 8.854 \times 10^{-12}\ F \cdot m^{-1}$
Density of liquid nitrogen	$\rho_l = 804.2\ kg \cdot m^{-3}$
Density of gaseous nitrogen	$\rho_g = 4.59\ kg \cdot m^{-3}$
Surface tension of liquid nitrogen	$\sigma_{st} = 8.8 \times 10^{-3}\ J \cdot m^{-2}$
Viscosity of liquid nitrogen	$\mu = 1.65 \times 10^{-4}\ kg \cdot m^{-1} \cdot s^{-1}$
Boiling temperature of liquid nitrogen	$T_s = 77.348\ K$

2.1. Sample preparation

PPLP samples (PPLP-HTS-125) manufactured by Sumitomo Electric Industries Ltd. were used for these experiments. The samples were cut into a circular shape with a diameter of 45 mm and having an area of 1590 mm^2. The various samples used for the experiment and their measured thickness are shown in Table 2. The measured thickness of compressed PPLP is less than the estimated thickness, which is due to tightening of M5 SS nuts and bolts on the sample holder with 10 N-m torque uniformly. In order to remove moisture from samples, they were baked in a vacuum furnace at a temperature of 65^oC, for 45 mins at a pressure of 10^{-2} mbar. An in-house sample holder set up was fabricated; an exploded view of the CAD modelled sample holder is shown in Figure 1. In the CAD modelled exploded view of the sample holder, the label names are shown in Table 3. These are described as below:

(i) Two copper shielding rings were used and placed on the top and bottom of the sample stack. These were grounded to reduce stray capacitance and electromagnetic interference

Table 2. Thickness of test samples.

No. of PPLP layers	Sample names	Measured thickness (mm)	Estimated thickness (mm)
1	L1	0.12	0.12
2	L2	0.22	0.24
3	L3	0.35	0.36
4	L4	0.46	0.48
5	L5	0.57	0.6

during measurement.

(ii) Teflon spacers/bushings was used to hold the copper electrodes together in position and to insulate the copper shielding rings from the copper electrodes.

(iii) The setup consists of two cylindrical copper electrodes of diameter and thickness 37 mm and 5 mm respectively. These electrode diameters are smaller than the PPLP sample to reduce fringing electric field which in turn affects the capacitance measurement.

(iv) The PPLP samples were stacked together and placed in between the copper electrodes.

Teflon coaxial cables were used for the electrical connection with the sample holder. The centre core of the cable was soldered to the copper electrodes and the outer shielding mesh of the cable was connected to the shielding rings for grounding.

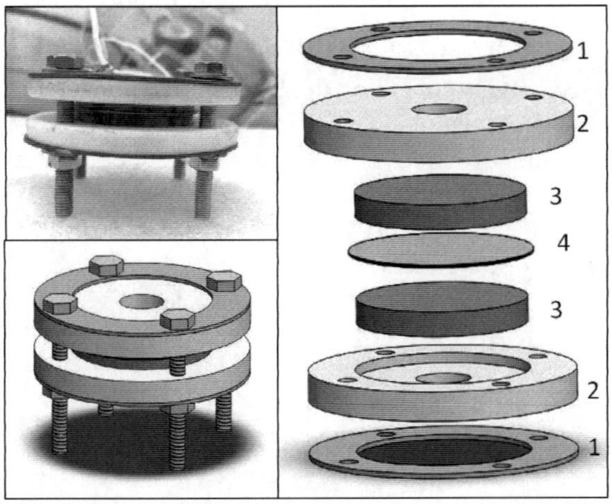

Figure 1. Photograph and 3D CAD model of sample holder.

Table 3. Parts of sample holder.

Label	Part names
1	Copper shielding rings
2	Teflon spacers
3	Copper electrodes
4	PPLP samples

2.2. Measurement of capacitance and dissipation factor (tan δ)

An in-house test setup was developed for measuring the capacitance and dissipation factor (*tan δ*) of the PPLP samples at room, liquid nitrogen and subcooled LN_2 temperatures. An LCR precision meter APLAB 4300R was used to measure capacitance and dissipation factor for various frequencies at 300 K, 77 K, 72 K and 68 K. Prior to capacitance measurement, the LCR meter was calibrated for the short circuit and open circuit tests after connecting the setup. The measurements were carried out using a 4-probe method with the help of shielded teflon wires which were compatible with room and LN_2 temperatures. To perform the experiment, feedthrough pins were connected to the vacuum desiccator box. For the capacitance measurement, two pairs of feedthrough pins were placed 180° apart, to avoid parasitic capacitance. A Lakeshore 325 temperature controller, connected to the PT-100 RTD in the 4-probe configuration was used to record the temperature. The LCR meter and temperature controller were interfaced with a LabVIEW program and the data logged using USB and GPIB interfaces respectively. A schematic and a photograph of the experimental setup are shown in Figure 2 and Figure 3 respectively. For the measurements at liquid nitrogen temperature

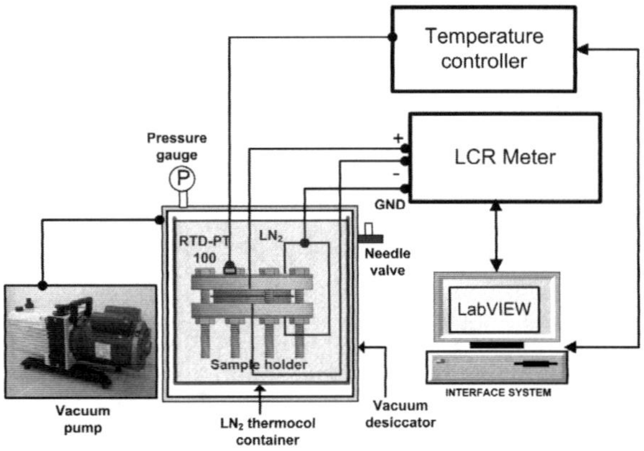

Figure 2. Schematic of the experimental setup.

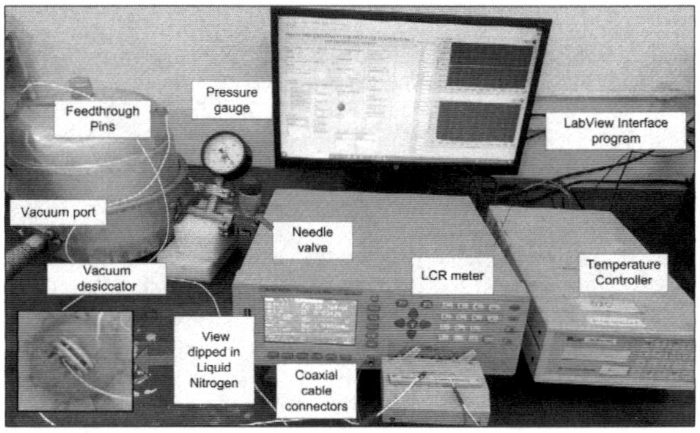

Figure 3. Photograph of the experimental setup.

(77 K) and subcooled liquid nitrogen temperatures (68 K and 72 K), the PPLP samples were initially positioned in the test setup and then placed inside a cylindrical thermocol box which was filled with liquid nitrogen. The measurements were taken after placing the set up completely submerged in liquid nitrogen for a considerable amount of time (20-25 minutes). Once the set-up reaches thermal equilibrium, which was determined when bubbles stop emanating from the set-up, the subcooling unit was switched on. When the temperature reached to 65 K, the pump was switched off and the readings recorded with frequency sweeps.

3. Computation of relative permittivity (ϵ_r), dissipation factor ($tan\ \delta$) and dielectric loss for samples

The relative permittivity (i.e. the dielectric constant, ϵ_r) is obtained by the following formulas:

$$C_{ef} = \frac{\epsilon_r \epsilon_o A}{d} \tag{1}$$

$$\frac{1}{C_{ef}} = \frac{1}{\epsilon_0 A} \sum_{i=1}^{n} \frac{d_i}{\epsilon_{ri}} \tag{2}$$

where, C_{ef} is the capacitance measured by the LCR meter, d is the thickness of the PPLP sample, $A = 1.075 \times 10^{-3}\ m^2$, is the cross sectional area of the copper electrodes, $\epsilon_o = 8.854 \times 10^{-12}\ F/m$, is the permittivity of free space, and ϵ_r is the relative permittivity which is to be determined.

$$\frac{1}{C_{ef}} = \frac{1}{C_1} + \frac{1}{C_2} + \frac{1}{C_3} + ... + \frac{1}{C_n} = \sum_{i=1}^{n} \frac{1}{C_i} \tag{3}$$

Equation (3) shows that the relation reduces to a configuration of capacitors in series where insertion of different layers of PPLP are symbolized by i.

The dielectric power loss (W_d in W) and its value per unit volume (W_v in W/m^3) are given by Equation (4) and Equation (5) respectively.

$$W_d = \omega C_{ef} V_0^2 \tan\delta \tag{4}$$

$$W_v = \omega \epsilon_0 \epsilon_r E^2 \tan\delta \tag{5}$$

where, ω is the operating angular frequency, V_0 is operating voltage (V), δ is dielectric loss angle (rad), $\tan\delta$ is the dissipation factor and E is the electric field strength (V/m).

The equivalent resistance for each measurement is given by Equation (6). It is to be noted that the sample is connected with the LCR meter in a closed circuit [14][15]. The equivalent parallel capacitance C_P and equivalent parallel resistance R_P obtained from the LCR meter have to be used for computation of dissipation factor $\tan\delta$ given by Equation (7). R and $\tan\delta$ are also written in terms of Equation (1) and Equation (6), which shows that the dissipation factor is dependent on the conductivity and equivalent permittivity of the sample.

$$R = \frac{d}{\sigma A} \tag{6}$$

$$\tan\delta = \frac{1}{\omega C_P R_P} = \frac{\sigma}{\omega\epsilon} \tag{7}$$

where, σ is the conductivity (S/m) and R is the resistance (Ω).

4. Results and Discussions

The variation of capacitance for different layers of PPLP samples for the frequency range from 60 Hz to 1 kHz at room temperature and LN_2 temperature are shown in Figure 4 and Figure 5 respectively and the frequency range from 60 Hz to 100 Hz for the subcooled temperatures 72 K and 68 K are shown in Figure 7 and Figure 6 respectively. The result shows that the capacitance for all five layers at room temperature decrease with frequency due to a decrease in polarization whereas at 77 K, 72 K and 68 K, the variation due to frequency is negligible since the polarization decreases and becomes stable [10]. With a drop of capacitance from 300 K to 77 K, the permittivity decreases. The change in permittivity is caused by restricted dipole movement with lower lattice energy at lower temperatures [14]. It is observed that the capacitance value increases with a temperature drop in the sub-cooled state from 77 K to 68 K in the range of 0.5 pF to 1 pF. This phenomenon is observed since the dielectric constant of subcooled liquid nitrogen at atmospheric pressure is greater than that of liquid nitrogen.

The measured capacitance and calculated relative permittivity are tabulated for the set operating frequency of 60 Hz and 100 Hz in Table 4 and Table 6 respectively. The dissipation factor for operating frequency 60 Hz is shown in Table 5. The dielectric loss for operating voltage of 1 V is tabulated for operating frequency 60 Hz and 100 Hz in Table 7 and Table 8 respectively. After performing the experiment, it was observed that the compression of samples in between the electrodes increases the capacitance and conductivity, resulting in higher dielectric losses.

5. Conclusions

An experimental set up was developed to measure the relative permittivity of PPLP at temperatures from 65 to 300 K at various frequencies for different layer thicknesses of PPLP. With the varying thickness of PPLP stacks, a variation of capacitance was achieved and the relative permittivity was calculated from the thickness of the used sample and cross-sectional area of the electrode.

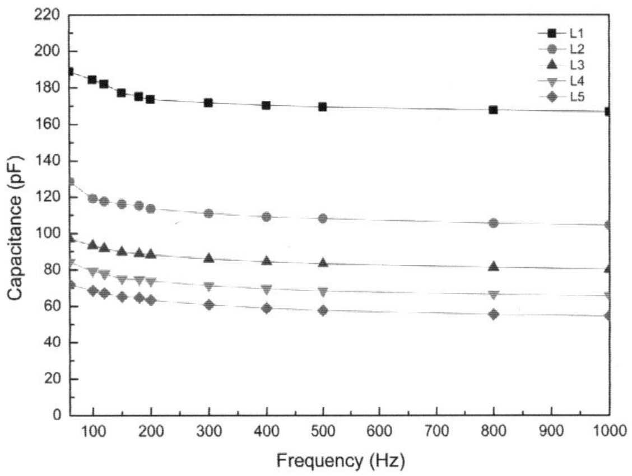

Figure 4. Capacitance vs. frequency plot for PPLP samples at 300 K.

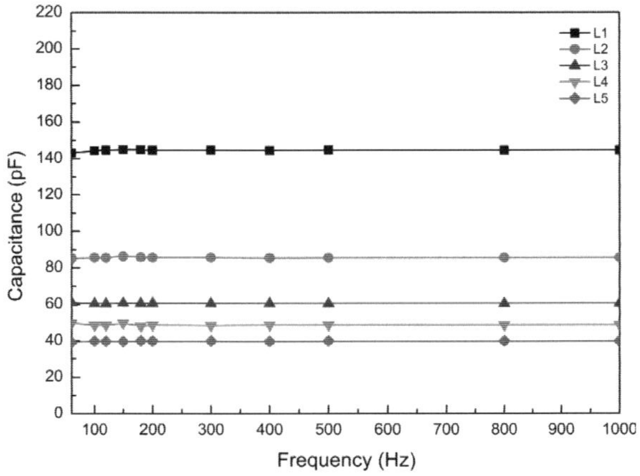

Figure 5. Capacitance vs. frequency plot for PPLP samples at 77 K.

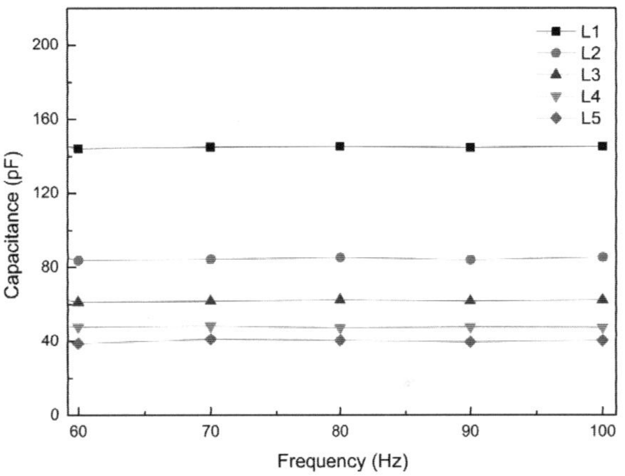

Figure 6. Capacitance vs. frequency plot for PPLP samples at 68 K.

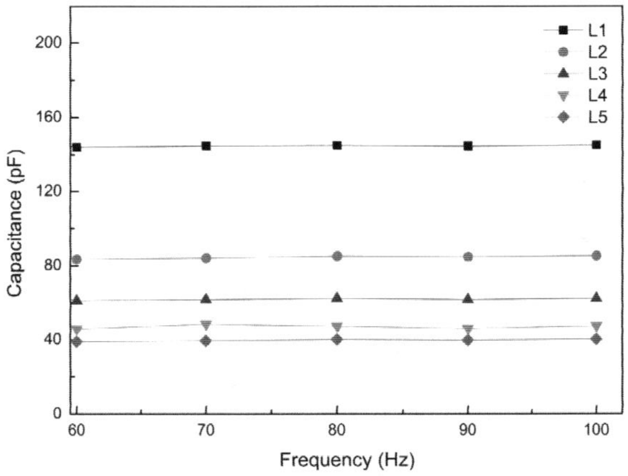

Figure 7. Capacitance vs. frequency plot for PPLP samples at 72 K.

Table 4. Measured capacitance at different temperatures.

S.No	Temp (K)	Capacitance (pF) at 60 Hz					Capacitance (pF) at 100 Hz				
		L1	L2	L3	L4	L5	L1	L2	L3	L4	L5
1.	300	188.63	128.65	97.12	84.04	71.44	178.55	119.12	93.46	79.36	68.71
2.	77	143.52	83.28	60.98	47.32	38.44	137.27	84.55	62.24	47.31	39.77
3.	72	143.93	83.43	61.21	47.45	38.79	137.98	85.23	62.40	47.45	40.21
4.	68	144.6	83.77	61.25	47.84	38.93	138.47	85.45	62.64	47.69	40.40

From the dielectric loss calculation, the loss reduces with an increase in number of layers due to increase in thickness of the samples which reduces the electric field intensity. At subcooled liquid nitrogen temperatures, the dielectric loss decreases due to a decrease in polarization of the samples. It was observed that the compression of samples in between the electrodes increases the capacitance and conductivity, which affects the dielectric loss and restricts the proper penetration

Table 5. Dissipation factor ($tan\ \delta$) at different temperatures.

S.No	Temp (K)	Dissipation factor (60 Hz) L1	L2	L3	L4	L5
1.	300	2.43E-01	2.53E-01	2.28E-01	1.29E-01	1.73E-01
2.	77	2.61E-02	4.03E-02	3.63E-02	3.03E-02	7.66E-02
3.	72	2.47E-02	4.02E-02	3.61E-02	3.02E-02	7.59E-02
4.	68	2.14E-02	4.00E-02	3.61E-02	3.00E-02	7.56E-02

Table 6. Relative permittivity at different temperatures.

S.No	Temp (K)	Relative permittivity at 60 Hz					Relative permittivity at 100 Hz				
		L1	L2	L3	L4	L5	L1	L2	L3	L4	L5
1.	300	2.38	2.97	3.57	4.07	4.31	2.33	2.75	3.44	3.84	4.12
2.	77	1.80	1.92	2.24	2.29	2.30	1.82	1.96	2.29	2.29	2.38
3.	72	1.82	1.93	2.25	2.29	2.32	1.83	1.97	2.30	2.29	2.41
4.	68	1.82	1.94	2.25	2.31	2.33	1.83	1.98	2.30	2.31	2.42

Table 7. Dielectric loss per unit volume at 60 Hz (W/m^3.)

S.No	Temp (K)	Dielectric loss per unit volume				
		L1	L2	L3	L4	L5
1.	300	1.34E-01	5.18E-02	2.22E-02	8.26E-03	7.67E-03
2.	77	1.09E-02	5.35E-03	2.22E-03	1.09E-03	1.81E-03
3.	72	1.04E-02	5.35E-03	2.22E-03	1.09E-03	1.81E-03
4.	68	9.00E-03	5.35E-03	2.22E-03	1.09E-03	1.81E-03

Table 8. Dielectric loss per unit volume at 100 Hz (W/m^3.)

S.No	Temp (K)	Dielectric loss per unit volume				
		L1	L2	L3	L4	L5
1.	300	5.04E-01	4.81E-02	2.67E-02	1.87E-02	1.31E-02
2.	77	1.71E-03	1.77E-03	2.56E-04	1.06E-03	6.43E-04
3.	72	1.71E-03	6.62E-04	2.56E-04	1.06E-03	6.43E-04
4.	68	7.37E-04	1.77E-03	2.56E-04	1.06E-03	6.43E-04

of liquid nitrogen into the samples. However, during the operation of high voltage HTS cable, any uneven voltage distribution on the dielectric material can lead to formation of hot spot. This rise in temperature reduces the critical current of the HTS cable and acts as a load on the cryogen.

From the experimental results, we conclude that these data are helpful for calculation of AC losses for the HTS cable at subcooled condition. For PPLP the dielectric loss per unit volume at liquid nitrogen temperature is 11 mW/m^3 and at room temperature is 134 mW/m^3 at operating frequency of 60 Hz. At the subcooled temperature, the effective relative permittivity increases marginally however there is reduction in tan delta loss which in turn reduces the dielectric loss and is useful for HTS based AC cables. This will also help to select the number of dielectric layers optimally which will ensure that with minimum number of PPLP layers will provide sufficient electric field gradient and electrical insulation at subcooled temperatures. The minimum number of dielectric layers will also reduce the diameter of the HTS cable, which results in reducing the diameter of cryostat and/or may reduce the pressure drop for the cryogen flow which in turn will reduce the pumping cost of cryogen to cool HTS cable.

6. References

[1] Dondapati R S and Rao V 2012 *IEEE transactions on applied superconductivity* **23** 5400604–5400604

[2] Hayakawa N, Ishida K, Matsuoka T and Kojima H 2016 *IEEE Transactions on Applied Superconductivity* **27** 1–4

[3] Nagao M, Kurimoto M, Takahashi R, Kawashima T, Murakami Y, Nishimura T, Ashibe Y and Masuda T 2011 Dielectric breakdown mechanism of polypropylene laminated paper in liquid nitrogen *2011 Annual Report Conference on Electrical Insulation and Dielectric Phenomena* (IEEE) pp 419–422

[4] Chang H M and Ryu S H 2011 *Cryogenics* **51** 132–138

[5] Suzuki H, Takahashi T, Okamoto T, Akita S and Yasuda K 2003 Electrical insulation property of liquid nitrogen/synthetic paper composite electrical insulation layer for high temperature superconducting cable *Proceedings of the 7th International Conference on Properties and Applications of Dielectric Materials (Cat. No. 03CH37417)* vol 3 (IEEE) pp 1182–1185

[6] Hata R, Yoshino Y and Shimizu T 2000 Application of pplp to ehv and uhv class underground and submarine cables *2000 IEEE Power Engineering Society Winter Meeting. Conference Proceedings (Cat. No. 00CH37077)* vol 1 (IEEE) pp 676–680

[7] Kwon I S, Hwang J S, Koo J H, Na J Y and Lee B W 2016 *IEEE Transactions on Applied Superconductivity* **26** 1–5

[8] Gerhold J 1998 *Cryogenics* **38** 1063–1081

[9] Jiao C and Xia Z 2014 *Journal of Superconductivity and Novel Magnetism* **27** 1813–1818

[10] Ahmad Z 2012 Polymer dielectric materials *Dielectric material* (IntechOpen)

[11] Meyers R A 2002 *Encyclopedia of physical science and technology* (Academic)

[12] Bulinski A and Densley J 1999 *IEEE Electrical Insulation Magazine* **15** 14–22

[13] Suda Y, Itoh M, Sakai Y, Matsuura K, Honma N and Kimura T 1996 *Cryogenics* **36** 567–571

[14] Park C, Wei J, Singh S, Narra S, Graber L and Imperatore M 2018 The characteristics of film capacitors at room temperature and in liquid nitrogen *2018 AIAA/IEEE Electric Aircraft Technologies Symposium (EATS)* (IEEE) pp 1–7

[15] https://www.aplab.com/test-measurement/component-testers.html

Acknowledgement

The Authors would like to thank Central Power Research Institute (CPRI), Bengaluru, Karnataka, India for their funding of the HTS cable project at Indian Institute of Technology, Kharagpur.

Thermal conductivity of niobium and thermally sprayed copper at cryogenic temperature

M. Fouaidy

IJCLab UMR9012 – CNRS / Université Paris-Saclay / Université de Paris 91405 Orsay Cedex, France

mohammed.fouaidy@ijclab.in2p3.fr

Abstract. We have developed a test facility dedicated to thermal conductivity k(T) measurements at low temperature. These experimental devices allow the test of four samples simultaneously during each run. The measurements are performed using steady-state axial heat flow method with a careful control of heat leaks to the surrounding. The apparatus was successfully used for measuring k(T) (Temperature T in the range: 1.5 K - 10 K) of niobium and other materials used for the fabrication of Superconducting Radio Frequency (SRF) cavities. We have tested the following samples: 1) niobium samples from different supplier (as received or/and after various high temperature Heat Treatments (HT)), 2) thermally sprayed copper coatings. The resulting experimental data are presented and compared to the experimental results previously reported by other groups and theoretical values. As expected heat treatment have a strong impact on k(T) and in particular on the phonons peak at T~2K. More precisely, HT @ 1200°C with Ti gettering improves the Nb Residual Resistance Ratio (RRR) by a factor of 3 and consequently k(T). Finally, the correlation between the niobium RRR and the thermal conductivity at T=4.2 K is confirmed in good agreement with the Wiedemann-Franz law

1. Introduction

Superconducting Radio Frequency (SRF) cavities (resonant frequency f_0 range: 88MHz-1500 MHz) are used in many particle accelerators: 1) facilities in operation (e.g. Continuous Electron Beam Accelerator Facility (CEBAF) [1] at Thomas Jefferson National Accelerator Facility (TJNAF), European X-Ray Free-Electron Laser Facility (European XFEL) [2] at Desy, Système de Production d'Ions RAdioactifs en Ligne de 2ème generation (SPIRAL2) [3] at Grand Accélérateur National d'Ions Lourds (GANIL), 2) facilities under construction (e.g European Spallation Source (ESS) [4], Multi-purpose hYbrid Research Reactor for High-tech Applications (MYRRHA) [5]). These SRF cavities are made of high purity (i.e Residual Resistance Ratio or RRR≥300) bulk niobium (Nb) sheets. However, a part from few tests with optimized 1-cell β=1 ILC type cavities, at large scale the maximum achieved accelerating field E_{acc}^{max} in SRF resonators are still below intrinsic limit. This theoretical limit, is given by the superheating magnetic field B_{sh} of niobium: B_{sh}= 190 mT@ T = 2.0 K resulting in E_{acc}^{max} ~ 50 MV/m β =1 ILC cavities. Furthermore, at high gradient (i.e $E_{acc} \geq$ 15 mV/m for β=1 SRF resonators) both E_{acc} and the unloaded quality factor Q_0 are limited by two anomalous RF losses: 1) Electron Field Emission (EFE), 2) Thermal Breakdown (TB) or quench. Thanks to the R&D effort, cavity preparation and processing recipes (i.e careful cleaning (e.g. Buffered Chemical Polishing (BCP), Electro-Polishing (EP), High Pressure Water

Rinsing (HPWR)), handling in clean room) developed since 3 decades, the onset threshold of EFE was pushed towards E_{acc}= 20 MV/m-30 MV/m. Finally, according to experimental observations, quench is triggered by anomalous RF losses due to strong Joule heating of localized normal conducting (nc) defects (i.e metallic inclusions in the Nb): the typical values of the defect diameter and surface resistance are respectively D ~1µm-100µm and R_D ~5 mΩ. As example, consider a 352 MHz spoke cavity [6] operating at E_{acc}= 9 MV/m and cooled by superfluid helium (He II) at T=2 K, the heat flux density q_D due to Joule heating of a nc defect (D=10µm, R_D =5 mΩ), located in the high surface magnetic field region is 6.1MW/m^2: this value is ~6 orders of magnitude higher than the heat flux q_{SRF} in the superconducting region. The heat transfer from the defect to He II is controlled by two parameters: 1) niobium thermal conductivity, 2) the Kapitza resistance at Nb-He II interface [7-9]. Consequently, in order to achieve accelerating fields close to the E_{acc}^{max}, we have to overcome and find remedies to quench by two approaches: 1) Reduce defects size and density on the RF surface, 2) Thermally stabilize the remaining defects. In this paper, we will present and discuss our thermal conductivity experimental data: they are analyzed and compared with previously reported data.

2. Motivations for measuring the thermal conductivity of niobium

The accelerating gradient of SRF cavities is very demanding: E_{acc}=33 mV/m is needed for ILC. Furthermore, Nb thermal conductivity $k_{Nb}(T)$ at low temperature (i.e T < 4 K) is the main parameter which controls the cavity thermal quench limit: for a quench due to a defect, $E_{acc}^{max} \sim (k_{Nb})^{1/n}$ with n≅ 2. For the high purity industrial niobium sheets, the values of the RRR and k_{Nb} at T=4.2 K are respectively 300 and $k_{4.2K}$= 60 W/m.K [8] . Moreover, bulk niobium cavity fabrication and preparation process consist of several steps: 1) deep drawing hydroforming or spinning of Nb sheets resulting in a plastic deformation, 2) Electron Beam Welding where some of the area is being heated and melted, 3) cavity processing and tuning (i.e Buffered Chemical Polishing (BCP), Electro-Polishing (EP), plastic deformation for tuning, heat treatment for hydrogen outgassing). All these processes have strong impact on niobium properties including $k_{Nb}(T)$ (i.e modification of the thermal phonons peak at T~2 K, see next sections) and could lead to a reduction of the quench field (Fig. 1).

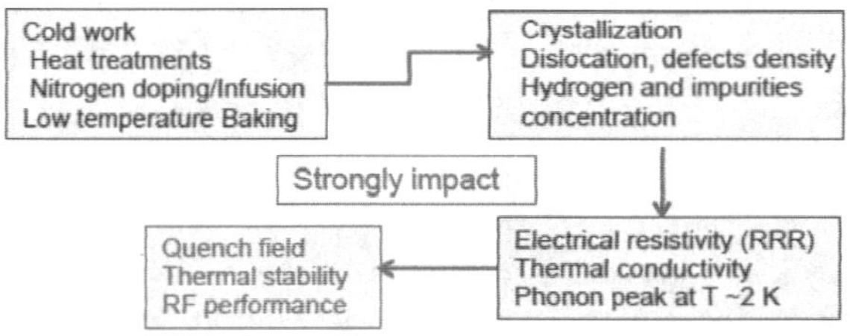

Figure 1. 1 Impact of cavity fabrication and preparation processes on the transport properties and quench field.

Consider an ILC SRF 3.8 mm thick niobium cavity with a nc defect (D=20µm, R_D =5 mΩ) on the RF surface and cooled by superfluid helium at T=2 K. The niobium RRR and thermal conductivity at 2K are respectively 250 and 10 W/m.K, while the Kapitza conductance is H_K= 8340 W/m^2.K. We have modeled the SRF cavity as a niobium disc with a nc at its center and obtained an analytical solution of steady-state of the heat equation [10]. We then computed the temperature at the center of the defect versus accelerating field: the resulting theoretical quench field is 30. 5 MV/m. Finally, it was recently [11-14] demonstrated that nitrogen doping (N-Doping) and infusion (N-infusion) allows reduction of RF losses in SRF cavities by a factor ~3. However, when SRF resonators are subjected to N-doping, a systematic reduction of E_{acc}^{max} of was observed in different laboratories [11-13]. It should be stressed that the following Heat Treatment (HT) processes modify [15-22] niobium thermal conductivity and some of its mechanical properties: 1) High temperature annealing under vacuum at 800°C (3h00) or 650 °C

(10h00) for hydrogen outgassing [15, 23-24], 2) Low temperature baking, 3) N-Doping and N-infusion. For all the aforementioned arguments, it is necessary to get reliable and high precision thermal conductivity experimental data of niobium and other materials commonly used in SRF cavities fabrication.

3. Experimental set-up and test procedure

3.1 Test cell description

The principle of the measurement is the steady state axial heat flow method performed at low heat flux (i.e q~50W/m^2), but with some refinements so as to insure a good reliability of the measurement and improve its sensitivity and accuracy. The test-cell shown in Fig. 2, was described elsewhere [19-22].

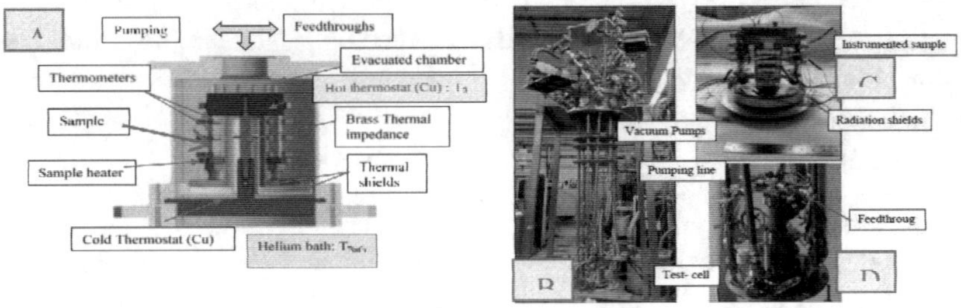

Figure 2. A) Thermal conductivity test-cell diagram, B) Test-cell on the cryogenic insert prior to the test. C) Samples mounted on the hot thermostatic Cu block. D) Close view to the Test-cell on the cryogenic insert.

This cell allows precise measurement of low temperature (i.e 1.5 K - 60 K) thermal conductivity k(T) of materials .The test-cell includes 5 major components, which are enclosed in a stainless-steel vacuum insulating jacket : 1) A temperature (T_0) controlled hot thermostatic OFE copper block, 2) A cold Cu block cooled by the (LHe bath, temperature: T_{bath}), 3) a thermal impedance (brass rod) between the heat sink and the cold source, 4) Four test-sample (length = 55 mm, width = 10 mm, thickness = 0.3 mm to 4 mm depending on the specimen) attached to the hot thermostat at their upper extremity, 5) Two gold coated OFE Cu radiation shields surrounding the heat sink and the samples. Each sample is equipped (Fig. 2) with a removable heater attached to its lower extremity and 3 calibrated (temperature range 1.5 K – 300 K) Cernox resistors. The main features of the thermal conductivity test stand are: 1) Steady-state axial heat flux method, 2) careful control of heat leaks, 3) Simultaneous tests of four samples, 4) Temperature range: 1.55 K- 30 K, 5) Precision: +/-12%, 6) Sensitivity over a wide range of k(T).

3.2 Experimental Procedure

Thermal conductivity measurements are performed in several steps: 1) Set and regulated T_o to the desired value, 2) Precise measurement of T_0 by a reference thermometer absolutely calibrated to within ± 10 mK. This thermometer is used for referencing all the thermometers at without heating (e.g q=0). 3) Measurement of the temperature profile (4-probes method with sensing current (I_{th}=10 μA) reversal so as to eliminate thermoelectric voltages) without heat flux (i.e q=0). 4) Subject each sample to a finite heat flux (q) and measure q and the temperature profile along the four samples. Note that the steps #4 is repeated for 6 values of q. Then the set point T_0 is changed and the whole above process is repeated from the step #1. In order to improve accuracy of measurement and avoid the impact of systematic errors in temperature measurement, we have eliminated the calibration and/or fit errors and/or eventual thermostat temperature drift during the run by using the following relationship for the net heating of the sensor 'k':

$$\Delta T_k(q) = (T_k(q) - T_0(q)) - (T_k(0) - T_0(0)) \quad (1)$$

4. Results and discussion

4.1 Thermal conductivity of Niobium

Several samples of niobium supplied by different companies were either tested as received or/and subjected to various Heat Treatment (HT) prior to the experiment then tested. In particular, we studied the effect of both the material initial purity (i.e RRR), which was also either measured and/or calculated from the material impurity content (O, N, H, Ta...) analysis, and the subsequent HT on the sample thermal conductivity.

4.1.1 Reactor grade niobium. In order to qualify niobium sheets produced by different suppliers, we have tested four sample: a) As received reactor grade Nb sample from Plansee, b) As received reactor grade Nb sample from Cabot, c) A heat treated (i.e 1200°C with Ti gettering) reactor grade Nb from ATI specialty materials, former Wah-Chang, d) As received Nb sample from Heraeus. The experimental data (Fig. 3) lead to the following observations: 1) The thermal conductivity versus temperature data depend strongly on the supplier resulting in a large difference between Cabot (calculated RRR= 50-63) and Plansee (calculated RRR= 45-56) samples for temperatures lower than 3.5 K and a factor ~ 4 at the phonon peak (T=2K), 2) For temperatures lower than 2.5 K, the thermal conductivity of Plansee (RRR~50) sample is higher than thermal conductivity of Heraeus (RRR~180) sample with higher RRR, 3) as expected [21-23], annealing at 1200°C with Ti gettering increases significantly the thermal conductivity of ATI specialty materials, former Wah – Chang niobium (See Fig. 4 for as received thermal conductivity data) : for this sample k is improved by a factor 2.5 and 3 respectively at 2 K (phonons peak) and 4.2 K.

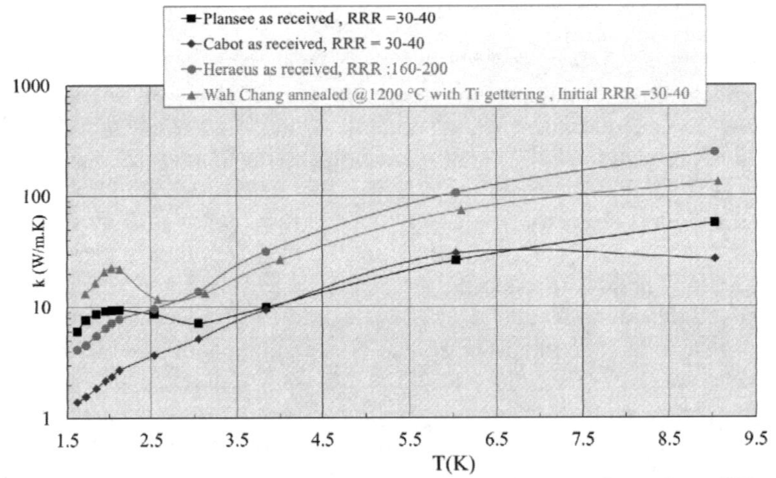

Figure. 3 Thermal conductivity versus temperature of niobium from four different suppliers.

4.1.2 Effect of high temperature vacuum annealing. We have investigated the effect of High Temperature (HT) annealing at 800 °C, 1200 °C and 1400°C of niobium samples from three different suppliers (ATI specialty materials, former Wah – Chang, Heraeus and Tokyo Denkai). The samples were tested as received or after HT annealing. Note that annealing at 1200°C and 1400°C were performed with Ti gettering. The experimental results (Fig. 4) call for several remarks: 1) as expected, high temperature annealing under vacuum at 800 °C increases significantly the thermal conductivity of niobium for temperatures lower than 4.2 K; this $k_{Nb}(T)$ increase is more important as T is decreased, 2) The phonon peak at T~2 K depend strongly [21-23] on the heat treatment to which niobium is subjected, 3) The dependence of the phonons peak of thermal conductivity according to RRR is not monotonous (i.e low RRR niobium could have higher phonon peak), 4) As the quench field due to a resistive defect (size 10-50 µm) depends strongly on the phonon peak, a niobium with low RRR could have a higher quench field as compared to a high RRR (e.g> 200); For T= 2 K SRF cavity operation one should use high RRR niobium but from thermal stability against quench due to normal defect the important parameter is the phonon peak.

The experimental data of Fig. 3 and Fig. 4 lead to the following main conclusions: 1) Shape of niobium thermal conductivity vs temperature (e.g phonon peak at 2K) depends strongly on supplier, production process (raw material, rolling, annealing, ...) and post processing, 2) High temperature annealing at 800 °C for hydrogen outgassing increases the thermal conductivity, 3) The phonon peak at T~2 K depends on impurities content including H and plastic deformation, 4) Post-purification with Ti gettering à T>1200 °C (Reduction of C, O and N content in Nb) increases the thermal conductivity, 5) Reduction of quench field subsequent to heat treatment and/or baking could be attributed to reduction of the phonon peak at T~ 2K.

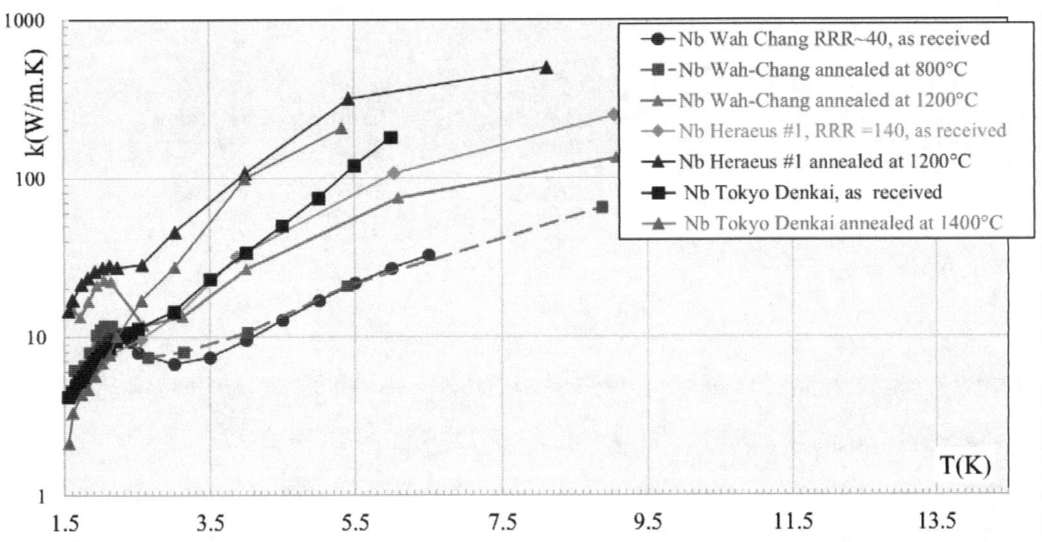

Figure 4. Effect of vacuum annealing at 800 °C, 1200 °C and 1400°C on the thermal conductivity on niobium supplied by ATI specialty materials, former Wah – Chang, Heraeus and Tokyo Denkai.

4.1.3 Relationship between the thermal conductivity of niobium at 4.2 K and RRR. The RRR is the ratio of the electrical resistivity at the ice point ρ(T=273 K) and the so-called residual resistivity ρ_R at the normal boiling point of LHe: ρ_R depend strongly on the purity of the metal and its heat treatment. The total electrical resistivity ρ(T) has four contributions: 1) the intrinsic phonons ($\rho_{Ph}(T)$) contribution, 2) The impurities part (ρ_{imp}), 3) the dislocation part (ρ_{disloc}), 4) the grain boundary contribution (ρ_{GB}). The resulting resistivity is then:

$$\rho(T) = \rho_{Ph} + \rho_{imp} + \rho_{disloc} + \rho_{GB} \tag{2}$$

For niobium at T<25 K, the phonons and impurties contribution are given by :

$$\rho_{Ph}(T) = 6.510^{-4}T^3\ \Omega.\text{m} \tag{3}$$

$$\varrho_{imp} = \Sigma_j C_j \left(\frac{\partial \varrho}{\partial c}\right)_j \tag{4}$$

Where C_j and $\left(\frac{\partial \varrho}{\partial c}\right)_j$ are respectively the concenration of the impurity 'j' in Nb and the sensitivity of ρ to this impurity [24]. Using the above relationship and the impurities concentration given by niobium suppliers, we have calculated (Table 1) the theoretical value of RRR of four samples. Moreover, Table

1 includes also the RRR values deduced from our experimental thermal conductivity data and those calculated from RRR using the Wiedemann-Franz law: for not too large RRR (i.e RRR <250), the niobium thermal conductivity at 4.2 K is proportional to RRR. Moreover, for as received samples (e.g samples #2 - #4) the theoretical RRR value estimated from Niobium impurities content are in good agreement with measured electrical resistivity data. Note that both thermal conductivity and RRR of the sample #1 were increased due to annealing at 1200°C with titanium gettering: we confirm that this process is effective in reducing C, O and N concentrations in niobium resulting in the observed increase of k(T) and RRR of niobium.

Table 1 Comparison of computed, measured electrical and thermal conductivities at T=4.2K.

Sample #	1	2	3	4
Thickness (mm)	0.36	1.94	0.92	2.05
Supplier	Wah-Chang	Plansee	Heraeus	Cabot
Initial RRR (Spec.)	30-40	30-40	160-200	30-40
RRR = a.$k_{4.2}$ with a= 4 - 5 m.K/W	36-46	56	166	41
RRR (calculation : impurities conc.)	122-153	45-56	164-205	50-63
k4.2 (W/m.K) Measurement	30.5	11.2	41	12.5
Treatment	HT@ 1200 °C (*)	As received	As received	As received

(*) Heat Treatment with Ti gettering: process aimed at reducing impurities such as C, O an N.

4.2 Thermal conductivity of copper coating

4.2.1 Motivation for testing copper coatings. Due to their high unloaded quality factor Q_0~10^{10} SRF cavities are very sensitive to small perturbations (Lorentz force, vibrations, ...). For example, consider the 1.3 GHz ILC cavity operated in a pulsed mode at E_{acc}=33 MV/m the Lorentz detuning is $\Delta f_{Lpulsed}$= 544 Hz. In order to keep this detuning below the cavity bandwidth, there are two options: 1) stiffening of the cavity by thermally sprayed copper coating, 2) dynamic compensation of the detuning by means of a Piezo tuner. As few experimental thermal and mechanical properties experimental data of sprayed copper at cryogenic are available in the literature, we have performed thermal conductivity measurement on some samples.

4.2.2 Thermal conductivity of copper coatings produced by thermal spraying. Briefly the process for producing coating with a spraying gun is the following: 1) a powder of the material to be sprayed is injected into the gun, 2) plasma is produced (gas carrier) and gives the kinetic and thermal energy for melting and spraying the particles solidify on the substrate located in front of the gun. We have measured the thermal conductivity of three samples produced by two out of the four following spraying methods: a) Atmospheric Plasma Spraying (APS), b) Vacuum Plasma Spraying (VPS), c) Inert Gas Plasma Spraying (IGPS or CAPS), d) High Velocity Oxy-Fuel Spraying (HVOF).Our experimental data presented in Fig. 5 lead to the following observations: 1) the spraying method have a strong impact on the thermal conductivity of the copper coatings, resulting in large variations of k(T) up to a factor ~10 at a given temperature, 2) this low thermal conductivity is attributed to the porosity of the coatings and the oxidation of the sprayed copper particles during their flight from the gun to the substrate, 3) the thermal conductivity the depend strongly on the temperature for all the samples tested, 4) The thermal conductivity of sprayed copper $k_{CuCoating}$ coatings are much lower than thermal conductivity of niobium k_{Nb} : k_{Nb}/ $k_{CuCoating}$ =6 -15 at 2 K and k_{Nb}/ $k_{CuCoating}$ = 3 -7 at 4 K.

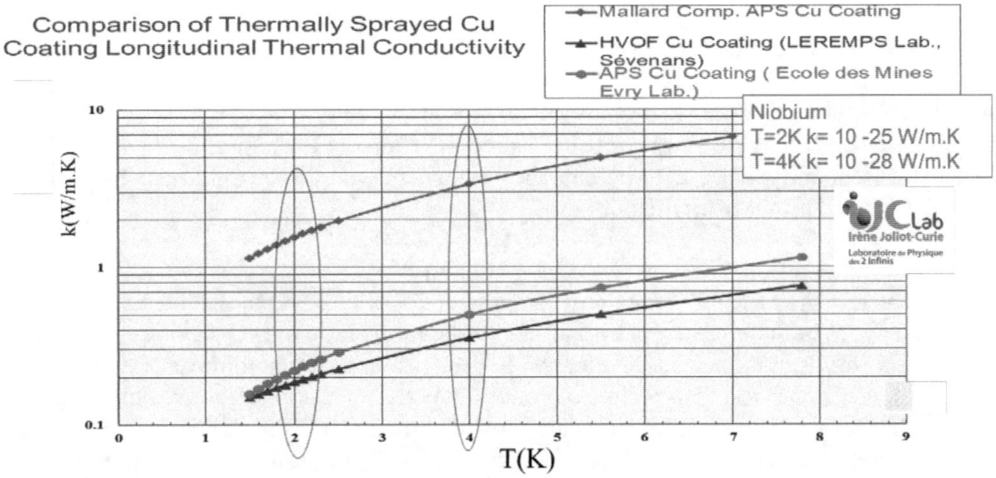

Figure 5. Thermal conductivity versus temperature of the 3 different thermally sprayed copper coatings.

In order to analyse the copper coating (APS Cu coating) experimental thermal conductivity data, we have developed a simple model called series-parallel network model (Fig. 6)

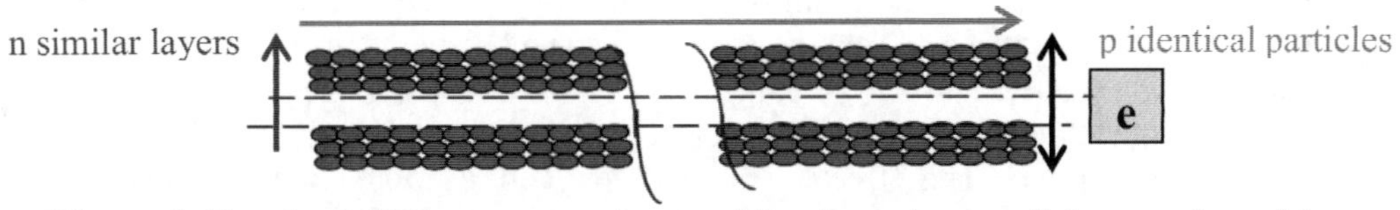

Figure 6. Sketch of APS copper coating used for the series-parallel network model.

The main assumptions of this simple thermal model are: 1) the thermally sprayed coating of thickness e, consists of an homogeneous lamella structure with n similar layers of identical particles in the transverse direction and p particles in the longitudinal direction, 2) simple series parallel network model where n and p are assumed to be large (i.e particle diameter d ≈50 µm, n ≈50, p ≈1000), 3) only solid conduction is considered and the transverse plan is isothermal. The resulting network resistance is:

$$R_T = \frac{p(R_1+R_2)}{n} \qquad (5)$$

Where $R_l \approx d/k_{Cu}$ (k_{Cu} : bulk copper thermal conductivity) is the particles specific thermal resistance and $R_2 \approx 1/h_{oxide}$ is the interfacial resistance, which dominated by the oxide thermal conductance h_{oxide}. For h_{oxide}, we used the literature experimental data [26] obtained with copper tubes oxidised under ambient atmosphere (temperature: 200 °C) during a period of 15 or 30 min. Notice that the presence of oxides has a stronger effect on the interfacial thermal conductance h_{oxide} as compared to the influence of contact surfaces roughness. The theoretical results at T = 4 K, according to the above model are summarized in table 2.

Table 2 Thermal resistance of an oxidized Cu coating according to network model.

RRR	k_{Cu} (W/m.K)	R_l (K.m²/W)	R_2 for t =30 min (K.m²/W)	R_L (K.m²/W)	R_T (K.m²/W)
3	18	27.3 10^{-7}	5 10^{-3}	5.5	0.12
60	366	1.4 10^{-7}	5 10^{-3}	5.5	0.12

As expected the overall Cu coating thermal resistance R_T is dominated by the oxide thermal conductance h_{oxide}. Moreover, the measured axial (i.e along the lamellae direction) thermal conductivity at T=4 K is $k_{APSCoating}$ = 3.4 W/m.K resulting in an overall experimental thermal resistance R_T = e/ $k_{APSCoating}$ 1.6 10^{-2} K.m²/W. There is a fairly good agreement (factor 7.5) with experimental data if one considers the simplicity of the model (i.e very crude assumptions...) and the fact that the particles are probably more oxidised. Indeed the temperature of the melted particles is much higher than 200 °C resulting in higher oxidation because the chemical reactions rate increases dramatically with temperature ($K_{oxidation} \propto \exp(-\Delta H/R.T)$), even if the time of flight of these particles is much shorter than 30 minutes. Consequently, one has to operate the plasma spraying process under inert atmosphere (IGPS) and this is also important from metallurgical considerations.

5. Conclusion

A test facility dedicated to the measurement at low temperature (1.5 K- 10 K) of the thermal conductivity as function of temperature of materials commonly used for the fabrication of SRF cavities and ancillaries (e.g. power couplers, cryostat, …) was developed. Several samples of niobium sheets supplied by different companies were either tested as received or/and subjected to various Heat Treatment (HT) prior to the experiment then tested. The experimental data show that the phonons peak around T=2 K depends strongly on the following: 1) niobium sheets production process, 2) impurities including hydrogen and heat treatment with and without titanium gettering. Heat treatment purification of Niobium with titanium gettering at temperature from 1200 °C to 1400 °C impacts strongly k(T). This effect is due to reduction of dislocations density in Niobium and purification (i.e reduction of oxygen content). Heat treatment at 800 °C (hydrogen outgassing) impacts also k(T): it confirms dedicated experiment results [27] where niobium samples were heat treated at 800 °C then saturated by hydrogen prior to the test. The reduction of quench field subsequent to baking and/or nitrogen doping could be attributed to reduction of k(T) the phonons peak. Finally, as expected according to the well-known Wiedemann-Franz law, the correlation (e.g $k_{4.2} \propto$ RRR) between RRR and $k_{4.2}$ is confirmed. The thermal conductivity of sprayed copper $k_{CuCoating}$ coatings are much lower than thermal conductivity of niobium k_{Nb} : $k_{Nb}/ k_{CuCoating}$ =6-15 at 2 K and $k_{Nb}/ k_{CuCoating}$ = 3-7 at 4 K. The effective thermal conductivity of thermally sprayed copper are limited by the material porosity and the oxides: the plasma spraying process should be performed under inert atmosphere (IGPS) and with process parameters allowing to reduce the material porosity.

6. References

[1] L. Harwood and C.Reece 'CEBAF at 12 and 25 GeV', Proc. 10th Workshop on RF Superconductivity, 2001, Tsukuba 332-336
[2] W. Decking et al 'A MHz-repetition-rate hard X-ray free-electron laser driven by a superconducting linear accelerator' Nature Photonics. 14. 1-7. June 2020 DOI:10.1038/s41566-020-0607-z
[3] A. Mosnier, 'SPIRAL2: A High Intensity Deuteron and Ion Linear Accelerator for Exotic Beam Production', Proc. PAC 2003, Portland, USA, pp 595-597
[4] R. Garoby et al., 'Progress on the ESS Project Construction', presented at IPAC'17, Copenhagen, Denmark, May 2017, paper MOXBA1
[5] MYRRHA, http://myrrha.sckcen.be
[6] P. Duchesne, S. Bousson, S. Brault, P. Duthil, G. Olry and D. Reynet, 'Design of the 352 MHz, beta 0.50, double-spoke cavity', Proc. of SRF Workshop, Paris, France, 2013
[7] K. Mittag 'Kapitza conductance and thermal conductivity of copper, niobium, and aluminium in the range from 1.3 K to 2.1 K', Cryogenics 13,94 (1973)
[8] M. Fouaidy et al., 'Kapitza conductance and thermal conductivity of materials used for SRF cavities fabrication', Proc. of the 9th workshop on RF superconductivity, Santa Fe, Nov. 99, pp. 263 – 266
[9] M.R. Smith, T. Zhang,Y. Xiang, S.W. Van Sciver, National High Magnetic Field Lab. (Florida-USA), J.G. Weisend II, P. Schmueser (Desy-Hamburg), M. Fouaidy (IPN Orsay), 'Experimental investigation of the thermal resistance in niobium samples for superconducting RF cavities', Advances in Cryogenic Engineering, Vol. 45A, pp 899-904, 1999
[10] M. Fouaidy, IPN Orsay, 'Heat transfer in a niobium plate heated locally in superfluid', Internal Report IPNO 90-03 (In French, "Transfert de chaleur au sein d'une plaque de niobium

IOP Conf. Series: Materials Science and Engineering 1241 (2022) 012012 doi:10.1088/1757-899X/1241/1/012012

localement chauffée dans l'hélium superfluide", Rapport Interne IPNO 90-03
[11] P. Dhakal, 'Status of Infusion Studies at Jlab' TTC2020 CERN, 4 Febru-ary 2020
[12] K. Umemori 'New results of KEK N-infusion and mid-T bake', TTC2020 CERN, 4 February 2020
[13] M. Checchin 'New Insight on Nitrogen Infusion' TTC Workshop, INFN Milano, 8 Feb. 2018
[14] F. Zhu et al, 'The effect of nitrogen infusion on a 162.5 MHz low beta superconducting HWR cavity', NIMA 2019, Vol 937, Sept. 2019
[15] M. Fouaidy et al. 'Commissioning of Vacuum Furnace and First Successful Heat Treatment of SRF Bulk Nb Cavities at IPN Orsay'," IEEE Transactions on Applied Superconductivity, Vol., N°4, June 2018
[16] G. R. Myneni and P. Kneisel, Proc. of the 6th workshop on RF superconductivity (1993), Newport News, USA, pp 210-244
[17] P. Dhakal et al. 'Effect of high temperature heat treatments on the quality factor of a large-grain superconducting radio-freqency niobium cavity', Phys. Rev. Special Topics – Accelerators and Beams (2013), Vol. 16, p42001-1
[18] Myneni, Ganapati Rao et al. 'Elasto-Plastic Behavior of High RRR Niobium: Effects of Crystallographic Texture, Microstructure and Hydrogen Concentration', AIP Conf.Proc. 671 (2003) 227-242 JLAB-ACT-03-02
[19] S. Bousson, M. Fouaidy T. Junquera, J. Lesrel, N. Hammoudi, J.C Le Scornet, J. Lesrel (IPN Orsay), 'Kapitza conductance and thermal conductivity of materials used for SRF cavities fabrication', Proc. of the 9th workshop on RF superconductivity, Santa Fe, Nov. 99, pp. 263 – 266
[20] F. Koechlin and B. Bonin, 'Parametrization of the niobium thermal conductivity in the superconducting state', Supercond. Sci. Technol. 9 (1996) 453–460
[21] F. Koechlin and P. Dolégiéviez 'Measurements of thermal conductivity of industrial niobium', Proc. SRF1991, Desy, Hamburg, Germany (1991)
[22] M. Fouaidy et al., 'Low temperature thermal conductivity of niobium and materials for SRF cavities', MOP040, Proc. SRF2019, Dresden, Ger-many, 2019
[23] M. Fouaidy et al. 'Status of high temperature vacuum heat treatment program at IPN Orsay TUP019, Proc. SRF2019, Dresden, Germany, 2019
[24] K.K. Schulze, Journal of Metals, May 1981, p33
[25] M. Fouaidy et al. 'Recent results of High Temperature Vacuum Heat Treatment program of SRF Resonators at IJCLab', IEEE Transactions on Applied Superconductivity Vol. 31, Issue 5 http://dx.doi.org/10.1109/TASC.2021.3062788
[26] S. Van SCIVER et al. in Advances in Cryogenic Eng., Vol. 34, pp 443-450
[27] S.K. Chandrasekaran et al. 'Hydrogen saturation and the thermal conductivity of superconducting niobium', Proc. of SRF2013, Paris, France, pp. 589-592

Acknowledgments
This research was supported by IN2P3/CNRS grant. We thank J.C. Lescornet and N. Hammoudi for their help and technical assistance in the preparation of test stand and for their valuable participation to the tests presented in this paper.

Thermal Property Measurements of Al-Alloy for Space Cryogenic Missions

Keisuke Shinozaki[1], Tadahito Mizutani[2], Takenori Fujii[3], Takashi Onaka[4], Takao Nakagawa[5] and Hiroyuki Sugita[2]

[1] R & D Directorate, JAXA, 3-1-1, Yoshinodai, Chuo-ku, Sagamihara city, Kanagawa, 252-5210, Japan
[2] R & D Directorate, JAXA, 2-1-1, Sengen, Tsukuba, Ibaraki, 305-8505, Japan
[3] Cryogenic Research Center, Univ. of Tokyo, Japan
[4] Department of Physics, Meisei Univ. Japan
[5] Institute of Space and Astronautical Science, JAXA, 3-1-1, Yoshinodai, Chuo-ku, Sagamihara city, Kanagawa, 252-5210, Japan

E-mail: `shinozaki.keisuke@jaxa.jp`

Abstract. In space cryogenic missions, the thermophysical properties of materials must be studied and measured to establish a highly reliable cooling chain. An Al-alloy is a primary material in cryogenic thermal structural designs and is particularly used for providing high thermal conductance of thermal shields and thermal straps. We measured low-temperature properties (the thermal conductivity and the electrical resistivity) of several Al-alloys and found that A6063 had a lower thermal conductivity than that reported previously. We also found that the Al-alloy ST-60, which has equivalent thermal conductivity as pure aluminum, has higher thermal conductivity than A6061 and A6063, and the material is a good candidate for the space cryogenic design at the temperature range between 4K and 300K. These results are critical for the thermal study of upcoming next-generation space astronomy missions, such as LiteBIRD, SPICA and ATHENA.

1. Introduction

In space cryogenic missions, it is necessary to optimize thermal and structural designs for reducing the heat load at low temperatures and for reducing the mass of the payload with respect to the mechanical environment during launch. In some cases, however, data of the thermal and mechanical properties at low temperatures are insufficient. Additionally, several papers suggest that significant variations in thermal conductivity at low temperature were observed even with the same alloy and tempering process [1][2][3]. Hence, the material sample must be manufactured and measured in an early phase of the mission study [4].

The Space Infrared Telescope for Cosmology and Astrophysics (SPICA) is to be launched into the second Lagrangian point (L2) in the Sun-Earth system, and a 2.5 m-class large IR telescope is cooled below 8 K in combination with effective radiant cooling and a mechanical cooler system in the payload module[5][6][7]. In the conceptual study, all critical materials proposed to use in the payload module design were listed and thermophysical properties of these materials were searched in the literature. It was found that several materials had unreliable values in thermophysical properties at low temperatures which had to be measured. Particularly, it was found that the Al-alloy A6063 was measured in only one paper prior to our study. The A6063 sample needed to be measured and compared with the reference.

This paper describes the results of measuring the thermal conductivity and electrical resistivity of Al-alloy samples proposed to use in the spacecraft cryogenic payload, and comparison of these results with the references. This study allows a highly reliable thermal design for the low-temperature payload modulew. Our results are valuable to upcoming space science missions, such as the LiteBIRD (Lite (Light) satellite for the studies of B-mode polarization and Inflation from cosmic microwave background Radiation Detection) [8][9] and the ESA X-ray observatory ATHENA (Advanced Telescope for High Energy Astrophysics)[10].

2. Proposed materials to be studied

An Al-alloy is mainly used as a thermal structure with an appropriate thermal conductance at cryogenic temperatures. Additionally, high purity aluminum is usually used for thermal straps because of its high thermal conductivity.

Table 1. Proposed Al-alloys in the thermal design of SPICA payload module.

Material	Application	Temperature range	Number of samples
Al-alloy A1050H24	Thermal shield, structure	10 ~ 150 K	1
Al-alloy A6061T651	Thermal shield, structure	10 ~ 150 K	1
Al-alloy A6063H14	Thermal shield, structure	10 ~ 150 K	1
Al-alloy ST-60[1)]	Thermal shield, structure	10 ~ 150 K	2

1) The ST-60 is the original brand and was developed by SDK (Showa Denko).

Table 1 shows Al-alloy materials proposed to be used in the low-temperature payload module of SPICA. These Al-alloys are mainly used for the shield panels, scientific instrument's structures and the face sheet of the honeycomb structure of the thermal shields.

When the thermal shield design is considered, it is better to have higher thermal conductance for more effective radiative cooling and lower mass. A1050 has a high thermal conductivity at low temperatures and the Al-alloy is a candidate for the use of the face sheet of the open honeycomb structure, but it has relatively lower mechanical strength. The high mechanical strength is also important for the shield panels, because these panels are not honeycomb structures but thin metal sheets. A6063 is relatively high thermal conductivity and good strength in literature. ST-60 is an original brand that was developed by SDK (Showa Denko). ST-60 is categorized as A6063 in chemical compositions, but high thermal conductivity with high strength is expected at low temperatures. Therefore, A6063 and ST-60 are good candidates as the shield panels of lower mass, while the relatively low thermal conductivity Al-alloy A6061 is used as the baseline of the shield panel for the reliable design.

We note that FRP (Fiber Reinforced Plastic) materials and non-metal materials proposed for the thermal design were also listed and thermal property measurements of FRPs have already been published [4].

3. Thermal property measurement

Low-temperature thermal properties of Al-alloy samples were measured with the Physical Property Measurement System (PPMS, Quantum Design, Inc.) in collaboration with the cryogenic research center at the University of Tokyo. The samples were manufactured to be a thin wire of $0.25 \times 1.3 \times 17.5$ mm as the main shape with thin arms for the experimental setup in the PPMS so that both thermal conductivity and electrical resistivity were measured in parallel. Two samples were measured for ST-60, while one sample was measured for each of the other Al-alloys. The measurement error was typically 5∼10 %, which was dominated by a systematic error of the data reduction in the temperature measurement, temperature calibration accuracy, and the heat capacity of the sample base plate [4].

4. Results of thermal conductivity measurements

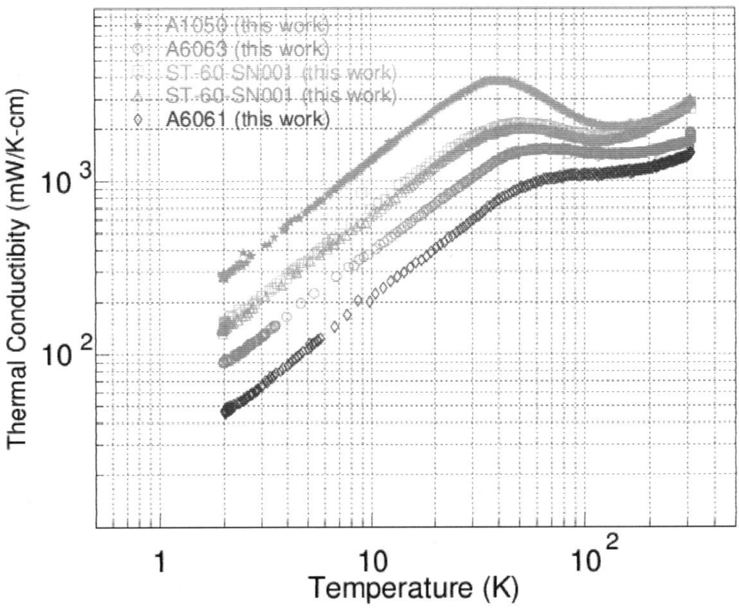

Figure 1. Measured thermal conductivity of five samples, A1050 (green stars), ST-60 (yellow open squares and orange open triangles), A6063 (gray open circles) and A6061 (purple open diamonds).

All five samples were successfully measured between 2 K and 300 K as shown in Figure 1.

One of the most important issues in this study is to compare the measured thermal conductivity with that in literature. The measured thermal conductivity of the A6063 sample in this work was apparently lower than that in references as shown in Figure 2. In particular, the thermal conductivity of the A6063 in references is about 2.5 times higher than the measurement result of the A6063 sample between 10 K and 50 K. The measured thermal conductivity of two ST-60 samples are almost the same, and about 1.5 times higher than the A6063 sample between 10 K and 50 K, but lower than that in references.

The A1050 and the A6061 samples were also measured and compared with those in literatures. The measured thermal conductivity of the A1050 sample was compared with that of A1100 in references and they are roughly compatible with each other as shown in Figure 3 left. On the other hand, the measured thermal conductivity of the A6061 sample between 4 K and 60 K was close to that in Barucci et al.[1] which were about 1.3 times higher than that in National Institute of Standards and Technology (NIST)[12] as shown in Figure 3 right.

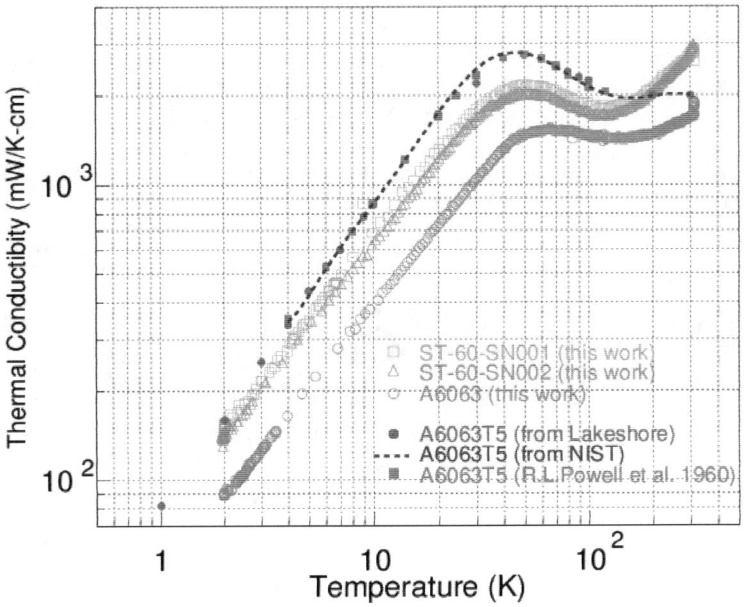

Figure 2. Comparison between the measured thermal conductivity of the A6063 samples and references. ST-60 samples in this work (yellow open squares and orange open triangles), A6063 sample in this work (gray open circles), reference data in Lakeshore (red circles)[11], reference curve in NIST (blue dashed line)[12] and reference data in Powell et al. (pale purple squares)[13].

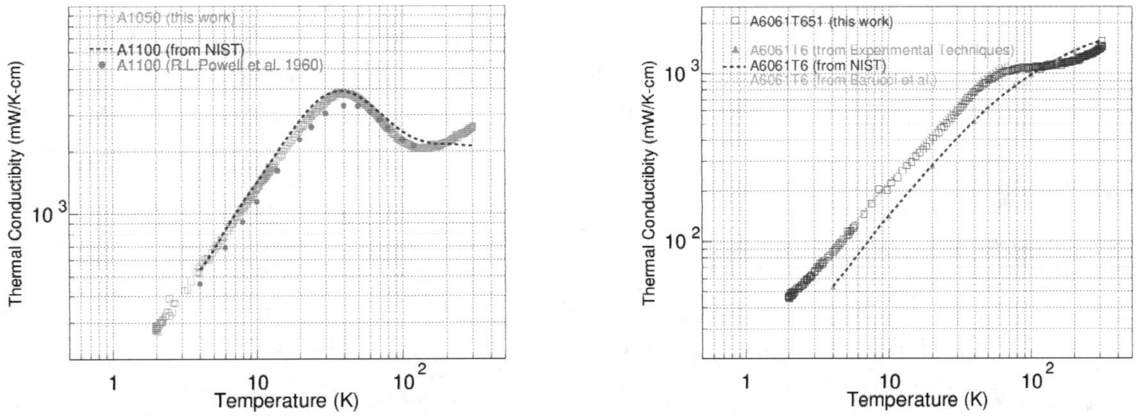

Figure 3. Comparison between the measured thermal conductivity and references for A1050 (left) and A6061 (right). The A1050 sample in this work (green open squares), reference curve in NIST (blue dashed line)[12] and reference data in Powell et al. (pale purple circles)[13] in left figure, and the A6061 sample in this work (purple open squares), reference data in the Handbook (green triangles) [14], reference curve in NIST (blue dashed line)[12] and reference curve in Barucci et al. (yellow dotted line)[1] in right figure.

IOP Conf. Series: Materials Science and Engineering 1241 (2022) 012013 doi:10.1088/1757-899X/1241/1/012013

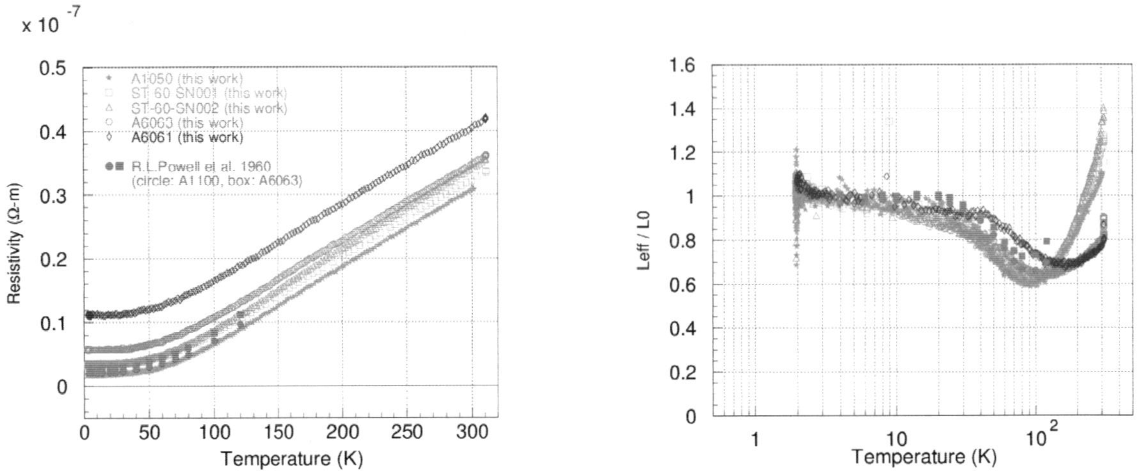

Figure 4. Measured electrical resistivities (left) and L_{eff}/L_0 (right) of five samples, A1050 (green open stars), ST-60 (yellow and orange open squares), A6063 (gray open circles), A6061 (purple open diamonds) and reference data in Powell et al. (pale purple circles for A1100 and squares for A6063)[13].

5. Results of electrical resistivity measurements

In this measurement, electrical resistivities of all five samples were also measured in parallel with thermal conductivity. Figure 4 left shows the measured electrical resistivities of all the samples. The ST-60 samples in this work have values close to the A6063 results from other literatures, while the A6061 sample has slightly higher resistivity.

In the Wiedemann-Franz law, there is a relation between electrical resistivity ρ, thermal conductivity k, and temperature [15] with the following equation.

$$\rho k = L_0 T \tag{1}$$

L_0 is the Lorenz number (approximately equal to 2.44×10^{-8} W/Ω/K^2). Then, the effective Lorenz number L_{eff} was computed by inserting the measured electrical resistivity ρ_{mea} in place of ρ and the measured thermal conductivity k_{eff} in place of k in equation 1. Lorenz ratios L_{eff}/L_0 for five samples are shown in Figure 4 right. The L_{eff}/L_0 measured in other paper is also shown, and all five samples in this work have similar L_{eff}/L_0 with those in other paper between 4 K and 100 K[13].

6. Discussion

6.1. The chemical compositions of the A6063

We searched other papers and handbooks in which the thermal conductivity of A6063 is given. We found that A6063 was measured in only one paper by Powell et al.[13] and other handbooks refer to this paper [12][16][17][18]. In handbook Brookhaven[18], the reference for A6063 is "Powell et al. 1958 to be published". However, we could not find it and it is assumed to be the same paper of Powell et al. published in 1960 [13].

The chemical compositions of the Al-alloy samples and references are shown in Table 2. The standard chemical composition specified in JIS (Japanese Industrial Standard) is also shown in Table 2, while these numbers in JIS are the same as those in ASTM International[19]. It can be confirmed that there is no apparent difference between the A6063 sample in this work and that in Powell et al.[13] in chemical composition, and both are categorized to be A6063.

Table 2. Chemical compositions of A6063 samples and standards in weight percent.

Material	Si	Fe	Cu	Mn	Mg	Cr	Zn	Ti	Others
A6063 H14 (this work)	0.41	0.15	0.0	0.04	0.47	0	0	0.01	0.00
A6063 T5 (Powell et al.)[13]	0.38	0.1	0.01	0.1	0.65	0.01	0.01	0.01	Ga 0.1, V 0.01, Ca 0.001, Pb 0.001
A6063 (JIS)1)	0.2~ 0.6	max 0.35	max 0.1	max 0.1	0.45 ~0.9	max 0.1	max 0.1	max 0.1	max 0.05 per each, max 0.15 in total
A6061 (JIS)1)	0.4~ 0.8	max 0.7	0.15 ~0.4	max 0.15	0.8~ 1.2	0.04~ 0.35	max 0.25	max 0.15	max 0.05 per each, max 0.15 in total
ST-602)	0.2~ 0.6	max 0.35	max 0.1	max 0.1	0.45 ~0.9	max 0.1	max 0.1	max 0.1	max 0.05 per each, max 0.15 in total

1) The chemical composition specified in JIS (Japanese Industrial Standard).
2) The specification by SDK.

Table 3. Mechanical properties of A6063 samples and standards.

Material	Tensile strength (N/mm^2)	Proof strength (N/mm^2)
A6063 H14 (this work)	237	226
A6063 T5 (Powell et al.)[13]	N/A	N/A
A6063 T4 (JIS)1)	170	90
A6063 T5 (JIS)1)	185	145
A6063 T6 (JIS)1)	240	215
A6063 T83 (JIS)1)	255	240
ST-60 (SDK)2)	200~300	180~280

1) Typical number in JIS (Japanese Industrial Standard).
2) The specification by SDK.

The mechanical properties as well as the thermal properties of the Al-alloy are influenced by the temper state in the heat treatment process during manufacture. The A6063 sample in this work has the cold work hardening as H14, and the measured mechanical properties of the A6063 sample is compatible with those of T6 in JIS standard as shown in Table 3. Because the A6063 sample in Powell et al. has the temperature treatment of T5, there is a possibility that the discrepancy of the thermal conductivity between the A6063 sample in this work and the sample in Powell et al. is caused by the temperature treatment. The mechanical properties of the sample in Powell et al. were not available and we cannot confirm this possibility. Another possibility is that the A6063 sample in Powell et al. may have had higher purity, because of a peak at around 45K and a similar behavior with A1050.

Two ST-60 samples have the same range of chemical compositions to A6063 as shown in Table 2, while the thermal conductivity of these ST-60 samples has 1.5 times higher than the A6063 sample at 40K. It suggests that there are large variations with respect to the thermal conductivity even in the same category of A6063.

6.2. Lorenz ratio

As shown in Figure 4, all five samples in this work have similar Lorenz ratios L_{eff}/L_0 with that in other paper[13] between 4 K and 100 K. The total thermal conductivity of metal is assumed to be an approximate sum of the electron thermal conductivity and the lattice thermal conductivity, and the Lorenz ratio is unity if the lattice thermal conductivity can be ignored in the total thermal conductivity. Therefore, it can be explained that the Lorenz ratio of these Al-alloy samples is close to unity below 40 K, because the lattice thermal conductivity changes as $\propto T^3$ and is negligible at low temperatures. Powell[13] indicates that the conduction electrons are not scattered elastically at intermediate temperatures and that the effective Lorenz number decreases considerably (if the lattice thermal conductivity is negligible). The Lorenz ratios computed for Al-alloy samples in this work are also consistent with this interpretation.

7. Conclusion

In this study, low-temperature properties (the thermal conductivity and the electrical resistivity) of several Al-alloys were measured and the following results were obtained.

- It was found that the A6063 in this work has a lower thermal conductivity than other paper even when there is no apparent difference between the A6063 sample in this work and that in other paper in chemical compositions. There is a possibility that the discrepancy is caused by the temperature treatment. Another possibility is that the A6063 sample in other paper had higher purity.
- The Al-alloy ST-60 has a higher thermal conductivity than the A6061 and A6063 samples, while the ST-60 samples have the same range of chemical compositions to A6063. It suggests that there are large variations with respect to thermal conductivity even in the same category of A6063.
- Each effective Lorenz numbers L_{eff} of each sample were also estimated between 2 K and 300 K, and they have behaviors similar to the results reported in other papers.

These results are critical and provides the high reliability and feasibility for the thermal study of upcoming space science missions, such as LiteBIRD and ATHENA.

8. References

[1] Barucci, M and Risegari, L and Olivieri, E and Pasca, E and Ventura, G 2004 *World Scientific* Astroparticle, Particle and Space Physics, Detectors and Medical Physics Applications pp 541–545
[2] Woodcraft A L 2005 *Cryogenics* **45** 421–431
[3] Comber B and Glazer S 2012 42nd International Conference on Environmental Systems aIAA 2012-3560
[4] Shinozaki K, Mizutani T, Fujii T, Onaka T, Nakagawa T and Sugita H 2015 *Physics Procedia* **67** 270–275
[5] Roelfsema P R, Shibai H, Kaneda H, Sauvage M, Giard M, Najarro F, Bradford C M, Yamada T, Tauber J, Audard M, Dieleman P, Doi Y, Elbaz D, Griffin M, Helmich F, Heske A, Honda M, Jellema W, Jin S, Juvela M, Kamp I, Kerschbaum F, Kiss C, Kohno K, Krause O, de Lange G, Larsson B, Martin-Pintado J, Matsuhara H, Nakagawa T, Nagao T, Naylor D, Nomura H, Ogawa H, Onaka T, Rouquet Jacques Spinoglio L, Szczerba R, van der Tak F, Vandenbussche B, Wang S Y and Withington S 2020 *Proc. SPIE, 11443, Space Telescopes and Instrumentation 2020: Optical, Infrared, and Millimeter Wave* vol 11443 p 1144325
[6] Nakagawa T, Goto K, Kaneda H, Matsuhara H, Matsumoto J, Mizutani T, Ogawa H, Saijo M, Sato Y, Sawada K, Shibai H, Shinozaki K, Sugita H, Takeuchi S, Tokoku C, Tirolien T and Uchida H 2020 *Proc. SPIE, 11443, Space Telescopes and Instrumentation 2020: Optical, Infrared, and Millimeter Wave* vol 11443 p 1144328
[7] Shinozaki K, Sato Y, Tokoku C, Saijo M, Ogawa H, Nakagawa T, Sawada K, Sugita H, Mizutani T, Matsuhara H, Tsunematsu S, Yoshida S, Kanao K, Okabayashi A, Narasaki K and Shibai H 2020 *Proc. SPIE, 11443, Space Telescopes and Instrumentation 2020: Optical, Infrared, and Millimeter Wave* vol 11443 p 1144329

[8] Sekimoto Y *et al.* 2018 *Proc. SPIE, 10698, Space Telescopes and Instrumentation 2018: Optical, Infrared and Millimeter Wave* vol 106981Y
[9] Hasebe T, Sekimoto Y, Dotani T, Mitsuda K, Shinozaki K and Yoshida S 2019 *Journal of Astronomical Telescopes, Instruments, and Systems* **5** 044002
[10] Barret D *et al.* 2018 *Proc. SPIE, 10699, Space Telescopes and Instrumentation 2018: Ultraviolet to Gamma Ray* vol 106991G
[11] 2016 *Temperature Measurement and Control Catalog* (Lake Shore Cryogenics, Inc.: Lakeshore)
[12] NIST Cryogenics home page https://trc.nist.gov/cryogenics/materials/materialproperties.htm
[13] Powell R L, Hall W J and Roder H M 1960 *Journal of Applied Physics* **31** 496–503
[14] Ekin J W 2006 *Experimental Techniques for Low-Temperature Measurements* (OXFORD University Press)
[15] Pobell F 2007 *Matter and Method at Low Temperatures* (Springer: Springer)
[16] for Basic Standards (US) Cryogenics Division I, Mann D and Association A G 1977 *LNG Materials and Fluids* (U.S. National Bureau of Standards, Boulder, Colorado.)
[17] Touloukian Y, Powell R, Ho C, Klemens P, Thermophysical and in E P I A C L 1970 *Thermophysical Properties of Matter - The TPRC Data Series, Volume 1, Thermal conducitivity - Metallic Elements and Alloys* (IFI / Plenum Data Publishing Corporation, New York, USA)
[18] Jensen J, Tuttle W, Stewart R, Brechna H and Prodell A 1980 *Brookhaven National Laboratory Selected Cryogenic Data Notebook Volume 1* (Brookhaven National Laboratory, Associated Universities, Inc.)
[19] Designation: B221-05a standard specification for aluminium and aluminium-alloy extruded bars, rods, wire, profiles, and tubes Tech. rep. ASTM International

Acknowledgments

The authors deeply appreciate all members of the SPICA team and the cryogenic group in Sumitomo Heavy Industry. Ltd., Niihama works for their support of this work.

IOP Conf. Series: Materials Science and Engineering 1241 (2022) 012014 doi:10.1088/1757-899X/1241/1/012014

Fatigue life characterization of hand folded and vacuum formed Kresling origami bellows at 77K

Francis Dunne, Kjell Westra, Caleb Mattox, and Jacob Leachman

Hydrogen Properties for Energy Research (HYPER) Laboratory, School of Mechanical and Material Engineering, Washington State University, Pullman, WA, 99164-2920, USA

Email: jacob.leachman@wsu.edu

Abstract. Propellant migration in microgravity is a critical challenge for in-space fuel storage and transfer. Origami fuel bladders are a new design solution to prevent cracking in polymeric materials within the cryogenic regime. This paper explores manufacturing techniques for producing origami bladders as well as the limits of possible thicknesses for origami bladders. Origami bladders are manufactured from polyvinylidene fluoride (PVDF) and polyethylene terephthalate glycol (PETG). Hand folded bladders follow a Kresling bellows origami pattern while the thermoformed samples are prepared using a mold of the Kresling bellows and an EZFORM LV 1827 thermoformer is used to shape the polymeric film. Compressive fatigue testing is carried out within liquid nitrogen to confirm the sample can survive deformation at cryogenic temperatures. Hand folded origami bellows specimens survived 3300+ cycles with minimal signs of degradation, while similar PVDF specimens cracked in fewer than 10 cycles.

1. Introduction

Bladders offer a potential solution to cryogenic fuel management issues that have persisted in aerospace for decades [1]. Most specifically propellant migration in microgravity. Microgravity causes fuel to become de-stratified and migrate within the ullage as fuel is expelled [2]. In contrast, bladders work by collapsing or crumpling to expel fuel. As a bladder compresses, the internal volume is also reduced. This reduces the chances for propellant to migrate or form bubbles. Recently the HYPER lab at Washington State University demonstrated a polymeric origami fuel bladder to solve this issue [3].

Polymeric materials have historically seen limited cryogenic application due to brittle mechanical properties below glass transition temperatures. Embrittlement is associated with loss of polymer chain mobility as temperature decreases. Generally speaking, though, plastics offer potentially important advantages for fuel bladders when compared to metal or ceramic materials including low density, high specific strength, low thermal conductivity and diffusivity, electronic resistivity, low Young's moduli, and high ductility [4]. However, the biggest advantage polymers may have been the novel manufacturing techniques recently developed for these materials.

To expand upon our preliminary results for the origami bladders [3] this work seeks to identify suitable bladder manufacturing methods and materials. The suitability of the manufactured bladders is tested via repeated mechanical cycling in liquid nitrogen (LN2). The results offer insight into both the fatigue characteristics of origami bladders to thousands of cycles, as well as investigates possible manufacturing techniques to produce scalable and leak-tight structures.

2. Materials and Experimental Methods

The bladders formed in the initial proof of concept work were manufactured via hand folding following a pattern [3]. Increasing the repeatability and scalability of bladder manufacture is important for the long-term viability of this concept. The best manufacturing methods depend on material and thickness, which were both un-investigated. To address this issue, a simplified single fold geometry was tested while varying film thickness. This most simplified case was then compared to a hand-folded full bladder origami structures that served as a repeatability test to the initial proof of concept. 3D printing and vacuum forming were then tested to compare both single the single fold and the full bladder origami structures.

Increasing the thickness of fuel bladders has the added advantage of maintaining higher pressure differentials before loss of structural integrity. To reduce the manufacturing time and inconsistencies of the sample's, simplified creases were made to study the effects of thickness. The pattern applied for sample preparation utilizes triangular indentions along the edges to stiffen the panels. This helps to limit the amount of panel stretching due to shear forces (see Fig. 1). The triangular folds help isolate the limits of the crease. The dashed lines indicate valley folds while the solid lines indicate mountain folds.

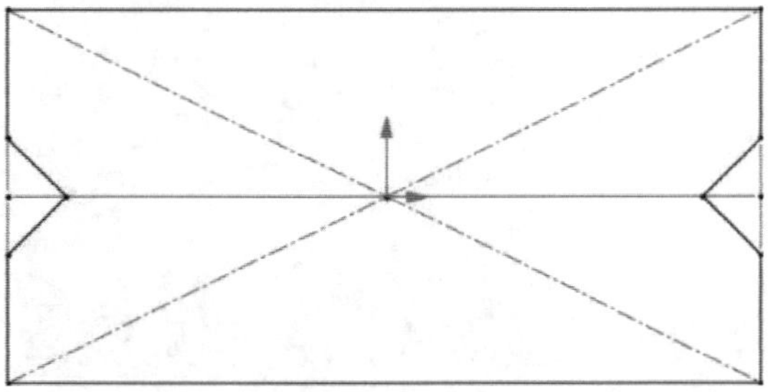

Figure 1. *Crease pattern for thickness study*

Fully origami bladder specimens were manufactured for this work for comparison to the proof of concept [3] and as a reference for repeatability. For fuel bladder applications, the Kresling pattern is promising due to large internal volumes compared to other bladder geometries. A total of 5 bladders were constructed using hand folding techniques and sealed by using Mylar tape along the open seam. The pattern used was a slightly modified Kresling pattern which was originally taken from Schenk M. et al [5]. The blue lines denote valley folds in which the vertex of the folds are recessed when viewed straight on. The red lines denote mountain folds in which the vertices are protruding from the surface. The modification made to the original pattern was alternating the fold direction halfway up the bladder (measured from the vertical axis, see Fig. 2). For extended fatigue cycling, bladders were constructed out of three different material thicknesses: 76.2 μm, 127 μm , and 177.8 μm Mylar. To hand fold the bladder, first the template was printed out and fixed to the table. Then, a sheet of bi-axially oriented Mylar fixed onto the surface. The transparency of Mylar allows simple tracing of pattern lines onto the Mylar using a straight edge. Next, the valley folds (solid lines) are folded individually until none remain. The material is then flipped, and all the mountain folds (dashed lines) are then folded. This process may need to be repeated multiple times to get stiff creases in the material. This process results in the material forming the shape of the desired bellows. Finally, metallized Mylar tape is placed along the seam to hold the structure in place.

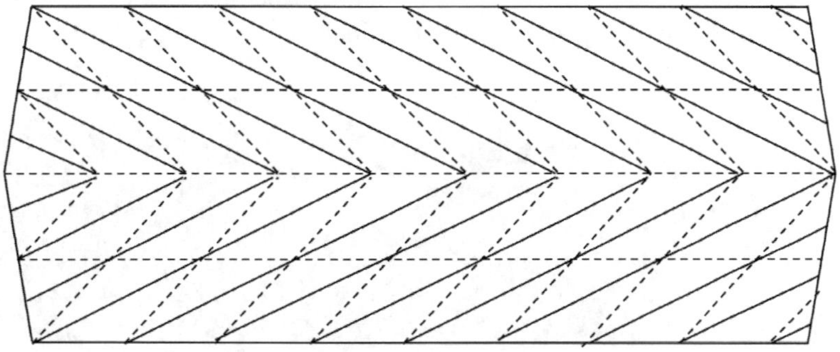

Figure 2. *Modified Kresling Pattern, solid lines indicate valley folds and dashed lines indicate mountain folds*

3D printing and vacuum forming are evaluated as alternative manufacturing methods to improve the repeatability, scalability, and cost of bladder manufacture. Various 3D printing processes were explored such as selective laser sintering (SLS) and MultiJet fusion (MJF). Polyurethane (TPU), polyamide (PA), and polylactic acid (PLA) were among the candidate materials for the 3D printing experiments. Three-point bending specimens with a living hinge geometry (see Fig. 3) were produced along with a Kresling pattern bladder.

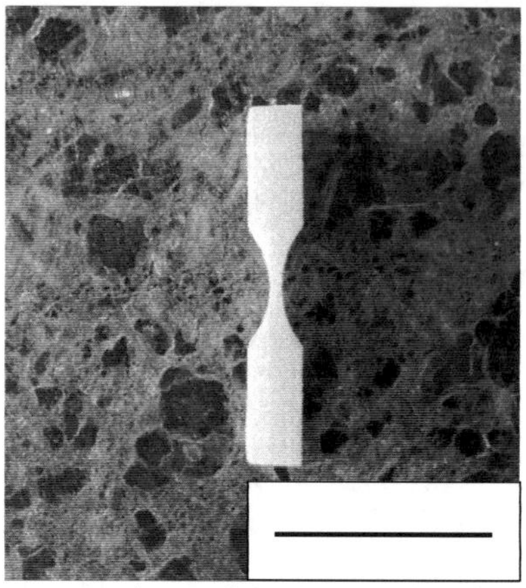

Figure 3. *3D printed living hinge sample used for 3 point bending experiment (1/4" thick on the ends, 0.03" thick in center, 1 inch scale bar)*

Vacuum forming involves heating a polymeric film close to its melting temperature and then forming the film around a mold. Vacuum is used to help pull the heated film around the mold to improve the definition of more complex parts. An EZFORM LV 1827 was used as the vacuum former for the experiment. Previously Wang et al demonstrated a vacuum forming approach to forming origami bellows [7]. A different vacuum forming method was used in which the entire structure was formed at one time rather than formed in halves and then spliced together. To ensure that the mold and part could be separated after forming, a 5-part, modular mold was 3D printed out of PLA (see Fig 4). The mold is comprised of a center cylinder which has 4 slots that the remaining pieces slide into place with. After the forming process is complete, the center piece can be pressed out and then the outer diameter parts collapse in towards the center to free the part from the mold.

Vacuum forming shows promising results for several reasons. First, vacuum forming is a natural fit for working with thin films. Thin films take very little time to heat to their respective melt temp and are easy to form once heated. In addition, the vacuum forming process leaves a part that is fully sealed around the circumference as well as fully sealed on the one end of the bladder (see Fig. 4). This leaves one face still to be sealed on the bladder. Figure 4b shows the final part from using the modular mold with 25.4 μm thick PVDF.

Figure 4. *a) Modular mold and b) Vacuum formed bellows*

3. Experiment

To explore the limits of the bellows and the various manufacturing techniques within the cryogenic temperature regime, an automated fatigue testing setup was adapted from an existing gelbo-flex testing apparatus. To automate the fatigue testing apparatus, a stepper motor was controlled using an Arduino Uno. This Arduino enabled control over the distance and rate at which the actuator traveled. Samples are fixed to a mandrel which is driven by a stepper motor to compress the sample at a precise rate (0.5 Hz) and distance while fully submerged in LN_2. It is expected that thickness is the most critical metric for determining polymer fatigue life in cryogenic conditions, due to the increased tensile and compressive loading on the creases [9]. To confirm this theory, the simplified creases (see Fig. 5) were affixed to the testing apparatus and repeatedly flexed until failure was observed.

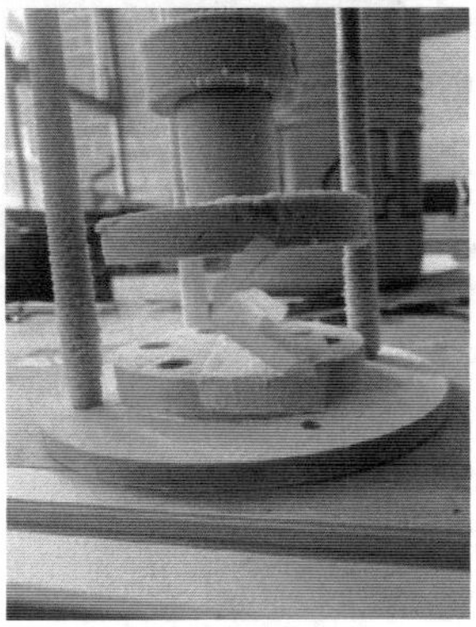

Figure 5. *76.2 thick Mylar Crease Experiment after testing in* LN_2

Testing the Kresling bellows structures followed a similar procedure as used in the simplified crease test. The bellows are fixed to the bottom of the LN_2 flask (see Fig. 6 b), and the mandrel is lowered onto the specimen. One cycle is defined to be a compression and expansion of the bellows; 90% compression for each cycle was achieved via markings on the testing shaft. A cycle rate of 0.5 Hz was kept consistent throughout the testing procedure. It has been documented that strain rate is not a critical factor in the mechanical properties of polymers at cryogenic temperatures and this variable was not controlled beyond the 0.5 Hz cycle frequency [10]. Due to the lack of visibility within the testing chamber, there is some uncertainty with the precise amount of compression being achieved. In addition, the lack of visibility also requires that the tests are stopped periodically to inspect the status of the sample. Inspection of the bellows was typically done every 400 cycles, but there was some slight deviation from this standard periodically. Any visible cracking or hole formation is considered a failure given that these structures are potential fuel tanks.

Figure 6. *a) CAD model of the automated fatigue testing design and b) Automated Testing setup and LN2 Flask with a 3" diameter Kresling bellows fatigue sample*

Finally, to test the 3D printing methods and material, a testing apparatus, which followed ASTM standard D7774, was machined to perform a 3-point bending tests in LN_2. One notable change from the ASTM standard involved the addition of the upper portion of the fixture to prevent the specimens from shifting while immersed in the LN2. The sample is tested by lowering the point load (see Fig 7. b) onto the sample. Then a computer controls the number of cycles and displacement that the living hinge samples undertake.

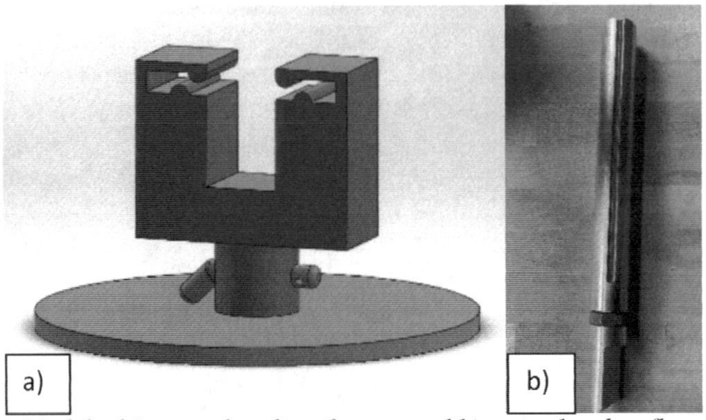

***Figure** 7. a) Modified 3-point bending fixture and b) point load to flex the samples*

4. Results and Discussion

4.1. Thickness Study

Table 1 shows the compression fatigue results for the Mylar simplified creases. The Mylar results confirm how critical thickness is to determining bellows survivability at cryogenic temperatures. Doubling the thickness from 177.8 μm to 355.6 μm thick Mylar reduced the lifespan of the samples by two orders of magnitude. Given that the cycling goal of the fuel bladder is 100 cycles, the next step should be to construct a Kresling bellows out of 254 μm thick Mylar and confirm survival beyond 100 cycles.

The PVDF simplified creases performed much worse when compared to the Mylar samples. The 76.2 μm samples were not able to survive more than 5 cycles before failure. This data helped guide fatigue testing of the Kresling bellows shown in table 2, in which only very thin PVDF of 25.4 μm was tested because thicker samples would fail. Fatigue failure is a unique phenomenon compared to maximum stress. However, in previous work, D.H. Weizel et al reported that the maximum bending radius for plastics is defined by $r_m = \frac{E_y t}{2\sigma_u}$, where E_y is the Young's modulus, t is thickness and σ_u is the ultimate tensile strength. This suggests that ultimate tensile strength and young's modulus are the two most important material properties for long lasting, and robust bellows structures [9].

***Table 1.** Simple Mylar Crease Fatigue Results*

Mylar Thickness Study							
Thickness		177.8 μm		254 μm		355.6 μm	
Sample #	Cycles Before Failure	#1	1000+	#1	90-110	#1	0-10
		#2	1000+	#2	140-160	#2	0-10
		#3	1000+	#3	140-160	#3	20-30

4.2. Bellows Fatigue Testing

The Mylar bellows fatigue results are tabulated in Table 1. It is critical to note that each of the Mylar samples tested did not break due to cycling. All the samples are still intact and could be put through additional cycling to find their fatigue limits. The bellows start to show some signs of crazing marks, indicated by increased opacity, along the crease lines. However, it is unknown how many more cycles the bellows could undergo. These preliminary results were the foundation for studying how thickness effects the survivability. Given that the initial cycle goal was for the bellows to survive 100 cycles at cryogenic temperatures, the Mylar bellows exceed expectations. The PVDF had drastically different results compared to the Mylar samples. Further work is needed to understand what material properties are critical for determining survivability at cryogenic temperatures.

Table 2. *Kresling Bellows Fatigue testing results (ND= Not Done)*

	Thickness			
Hand folded Material	25.4 μm	76.2 μm	127 μm	177.8 μm
Hand folded Mylar	ND	1500	3300	1500
Hand folded PVDF	ND	ND	0	ND
Vacuum Formed PVDF	100	ND	ND	ND

4.3. 3D Printed Living Hinges (3-Point Bending)
The 3D printed samples had several manufacturing constraints including thickness and surface finish. Regardless of the material, the 3D printed samples failed with deformation as small as 2.5 mm from resting position. The minimum thickness of 3D printed specimens is primarily to blame. The SLS printed samples had a lower limit of 508 μm thick while the MJF samples had a lower limit of 762 μm thick. For reference, that is an order of magnitude greater than the films that have shown success flexing in liquid nitrogen.

In addition to the issues with print thickness, each method of 3D printing had an additional flaw. MJF printed parts are not watertight which makes them incompatible for fuel bladder applications. SLS printing showed very poor thermal cycling performance, most likely caused from voids left in the sample. After 2 thermal cycles from 77K to room temperature, the samples would crack without the presence of any load. These various issues in addition to challenges associated with scaling the size of bellows, quickly deterred further research in this manufacturing process.

5. Conclusion
Various bellows fabrication techniques such as 3D printing and vacuum forming were explored and compared. Three point bending tests were performed on the 3D printing samples to test their flexural performance. Through this experiment, it was found that 3D printing cannot produce thin enough parts to survive in LN_2. Vacuum forming seems to be a promising manufacturing technique for future fuel bladder applications due to the compatibility with thin films as well as the ability to produce a sealed bellows. A simple study on how thickness affects fatigue performance was performed and verified the hypothesis that small changes in material thickness drastically impact the fatigue life. Additionally, the simplified crease test confirmed that material properties such as ultimate tensile strength and elastic modulus are important factors in fatigue life. Finally, extended fatigue testing was carried out into the thousands of cycles in which the Mylar samples showed promising results. In future work, the HYPER lab plans to connect this relationship back to physical theory to develop a predictive design solution for polymeric thin films at cryogenic temperatures.

IOP Conf. Series: Materials Science and Engineering 1241 (2022) 012014 doi:10.1088/1757-899X/1241/1/012014

6. References

[1] Pope D, Penner J. Development of cryogenic positive expulsion bladders, NASA-CR-72115; 1968.

[2] J.W. Hartwig Propellant management devices for low-gravity fluid management: past, present, and future applications J Spacecraft Roxck, 54 (4) (2017)pp. 808-824, 10.2514/1.A33750

[3] Kjell Westra, F. Dunne., Stasia Kulsa, Mathew Hunt, Jacob Leachman (2020). "Compliant polymer origami bellows in cryogenics." Cryogenics Volume 114.

[4] Brandrup, J., et al., *Polymer handbook.* Vol. 89. 1999: Wiley New York.

[5] Schenk M, Viquerat AD, Seffen KA, Guest SD. Review of inflatable booms for deployable space structures: packing and rigidization. J Spacecraft Rock 2014;51 (3):762–78. https://doi.org/10.2514/1.A32598.

[6] ASTM International. D7774-17 Standard Test Method for Flexural Fatigue Properties of Plastics. West Conshohocken, PA; ASTM International, 2017. doi: https://doi.org/10.1520/D7774-17

[7] Wang, H., et al. *Structural Design and Forming Method of Single-Curved Surface Structure Based On miura-Origami.* in *ASME 2016 International Design Engineering Technical Conferences and Computers and Information in Engineering Conference.* 2016.

[8] Howell, Larry L. "2.2 Stiffness and Strength ." *Compliant Mechanisms*, John Wiley & Sons, 2002, pp. 22–26.

[9] Weitzel, D. H., R. F. Robbins, and P. R. Ludtke. *ELASTOMERIC SEALS AND MATERIALS AT CRYOGENIC TEMPERATURES. PART 2.* NATIONAL BUREAU OF STANDARDS BOULDER CO BOULDER United States, 1965.

[10] Hartwig G. Polymer properties at room and cryogenic temperatures. New York: Plenum Press 1994. p. 1–17, 117–172.

An investigation of internal heat flux in SiC/AlN/Epoxy composites by mesoscopic simulation

Z R Zhou[1,2], R J Huang[*1,2], T Wang[1], S Y Xie[1,2], Y L Zhao[1,2], S S Wu[1,2], W Y Zhao[1,2], Z X Wu[1], H M Liu[1] and L F Li[*1,2]

[1] State Key Laboratory of Technologies in Space Cryogenic Propellants, Technical Institute of Physics and Chemistry, Chinese Academy of Sciences, Beijing 100190, China

[2] Center of Materials Science and Optoelectronics Engineering, University of Chinese Academy of Sciences, Beijing 100049, China

huangrongjin@mail.ipc.ac.cn; laifengli@mail.ipc.ac.cn

Abstract. Polymer-matrix composites play an important role as packaging material in superconducting equipment. The research on thermal conductivity of composites by combining computer numerical simulation and reference data can not only save the expenses but also shorten the period of material development. Herein, the internal heat flow of silicon carbide/aluminum nitride/Epoxy composites was simulated by COMSOL finite element software. And the thermal conductivity calculated by the finite element software is compared with the existing thermal conductivity mechanism. Meanwhile, the influence of the fraction, gap distance, interfacial thermal resistance, etc. on the thermal conductivity of composites was analyzed. In short, it can provide a reference for the thermal conductivity design of polymer composites.

1. Introduction

As the carrier material of electronic components, polymer plays an important role in many fields, such as microelectronics, aerospace, and superconducting electrical power equipment [1-2]. However, polymer materials generally have poor thermal conductivity (TC), and their TC is generally about 0.2 W/(m·K) [3]. To ensure the long-term reliable normal operation of devices, the heat dissipation of electronic components has become an urgent problem to be solved [4]. Therefore, the development of thermally conductive polymer composites has become a research hotspot.

Polymer composites with thermally conductive fillers have been widely studied and applied due to their cheap, diverse, and flexible designability. In general, many factors are affecting the TC of composites, such as the shape, TC, the volume fraction of the filler, the dispersion of the filler in the matrix, and the interface thermal resistance (TR) of the filler and the matrix [5]. Therefore, there are many design schemes for composites. All the composites are designed by experimental methods, which will consume a lot of manpower and material resources. The research method combining computer numerical simulation and reference data can not only save the funds but also greatly shorten the cycle of material development. Therefore, finite element analysis (FEA) is used to predict the TC of the thermally conductive polymer composites, and analyze the relationship between the TC of the composites, and the shape, size, and distribution of the thermally conductive filler [6]. It is of guiding significance to the design of polymer thermal conductive composites.

Herein, the thermal conductivity and heat capacity at constant pressure of silicon carbide (SiC), aluminum nitride (AlN), and Epoxy resin (Epoxy) were measured from 100 K to 300 K. Based on these parameters, the FEA is used to simulate the internal heat flow of SiC/AlN/Epoxy composites, and their TC was analyzed by combining the existing thermal conductivity mechanism. Furthermore, the relationship between the TC and the gap distance (between the particles), the interfacial thermal resistance (between the particles and the matrix), the relative size of the particles, the inclination Angle (between particles) are explained.

2. Results and discussion

2.1. Model description

Inspired by the ball-and-stick models in physics, as shown in Figure 1A. The good matching effect between the spherical particles and the columnar particles can make the composites have better TC. Because the columnar filler can be used as the main structure of the heat conduction network, and the spherical particles effectively connect the columnar particles and fill the gap in the main structure of the heat conduction network. For the composites with cylindrical and spherical fillers uniformly distributed in the polymer matrix, the internal heat flow was simulated and analyzed by the FEA. The influence of particle size, interfacial thermal resistance, and spatial distribution on the internal heat flow and effective TC of the composite was explored to provide a corresponding reference for designing the polymer composites. Herein, COMSOL finite element software is used for modeling and analysis.

According to the above analysis, a microscopic Representative Volume Element (RVE) model was established, as shown in Figure 1B. The RVE model consists of three parts: fiber, spherical particle, and polymer matrix. To facilitate the simulation, two different particles are controlled to be filled at an equal ratio of 1:3. Among them, Figure 1B1 is the most ideal case, in which particles are closely connected in the most ideal way and the gap value between particles is ignored. In practice, there is a certain gap between particles, and the polymer matrix is filled in the gap, as shown in Figure 1B2. In addition, there is an inclination Angle between particles, as shown in Figure 1B3.

In nanoscale RVE, the interfacial thermal resistance (TR) between particles and polymer matrix needs to be taken into account [7-8]. Figure 1C shows the TR of spherical particles filled in the polymer matrix. TR can be defined as:

$$R = \tau \frac{\nabla T}{J} \tag{1}$$

Where, τ is the thickness of the interface layer; J represents the heat flow in a certain direction; ∇T is the temperature gradient between the matrix and the filler. To facilitate the calculation of interface thermal resistance, the above formula is simplified to obtain:

$$R = \frac{\tau}{K_s} \tag{2}$$

As mentioned above, there is interfacial TC (K_s) between the filler and the matrix. According to previous work, when the equivalent TR between the filler and the polymer matrix is about 10^{-8} $m^2 \cdot K \cdot W^{-1}$, the TR is small, and the heat transfer efficiency is high. However, when 10^{-4} $m^2 \cdot K \cdot W^{-1}$ is assumed to be the equivalent TR between the filler and the polymer matrix, the TR is larger, and the heat transfer efficiency is low [9-10].

Due to the various components of composites, the effective prediction of the TC (Mark it as K or k) of composites can greatly reduce the workload and provide a theoretical basis for the preparation of higher TC of composites. The reconstructed Maxwell model (RM model, Equation (3)) can fit the TC of the composite with a variety of fill rates and different packing sizes [11].

$$\frac{K - K_m}{2K + K_m} f_m + \frac{K - K_{ef}}{2K + K_{ef}} f = 0 \quad (3)$$

$$\frac{1}{K_{ef}} = \frac{1}{K_p} + \frac{1}{K_s} \quad (4)$$

In Equation (3) and (4), K_m and K_p is the effective TC of the matrix and the fillers, respectively; K_{ef} is the effective TC of composites; f_m and f is the volume fraction of the matrix and the filler, respectively.

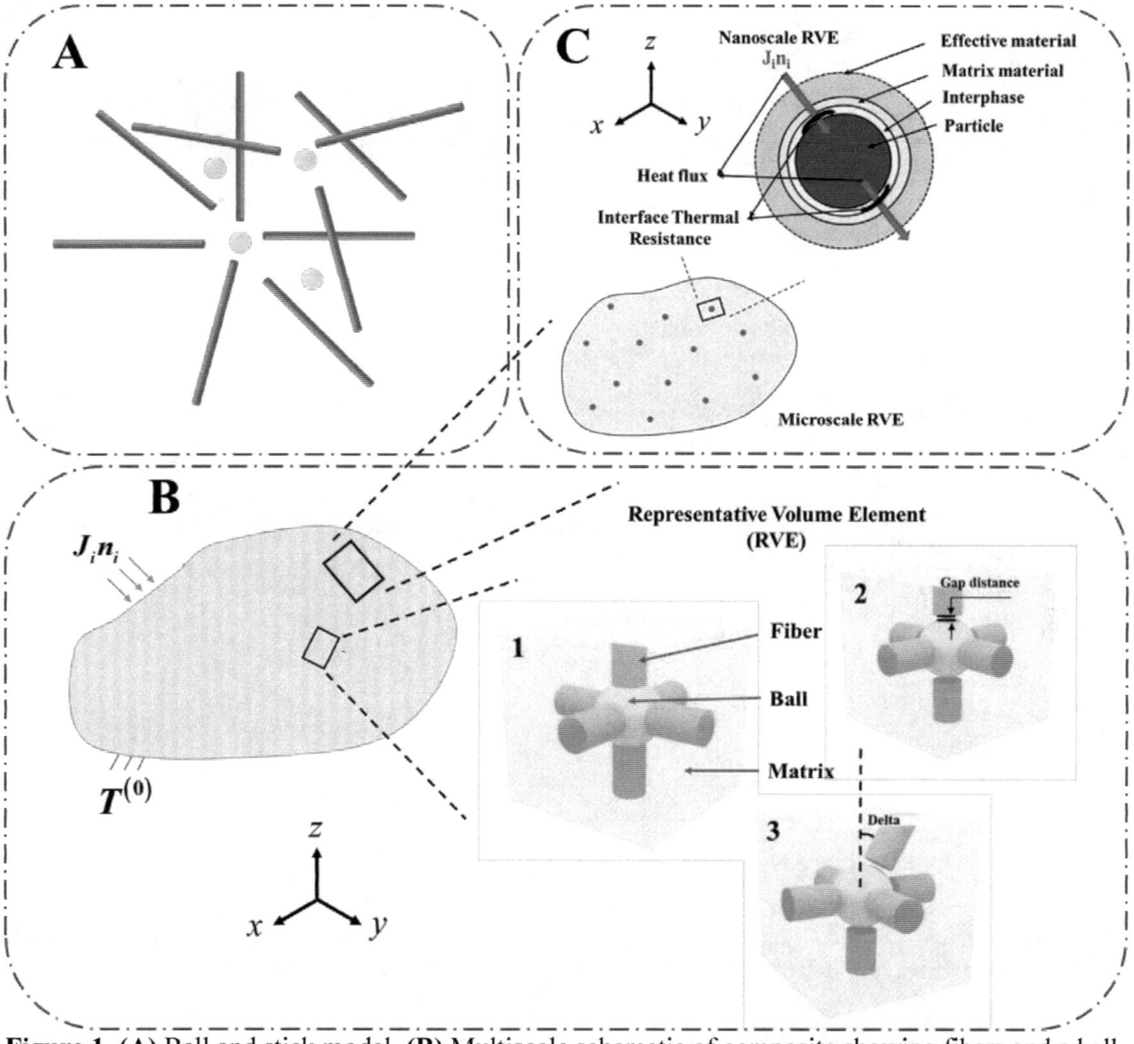

Figure 1. (A) Ball and stick model. **(B)** Multiscale schematic of composite showing fibers and a ball embedded in a matrix as an RVE. **(C)** The microscopic diagram of the interface between the particle and the matrix.

2.2. Parameters of the fillers and matrix

SiC and AlN are commonly used as thermally conductive fillers. AlN has the advantages of high intrinsic TC, non-toxic, stable crystal structure, and low price [12]. SiC also has the advantages of stable chemical properties and high TC [13]. In the model, the parameters of cylindrical particles refer to those of SiC fiber, and those of spherical particles refer to those of AlN particles in the model. Epoxy resin (Epoxy), as a thermosetting material, is often used as the matrix of polymer composites because of its excellent insulating properties, chemical stability, and mechanical properties [14]. Therefore, Epoxy is selected as the polymer matrix. The specific parameters of filler and polymer matrix from 100 K to 300 K are shown in Figure 2.

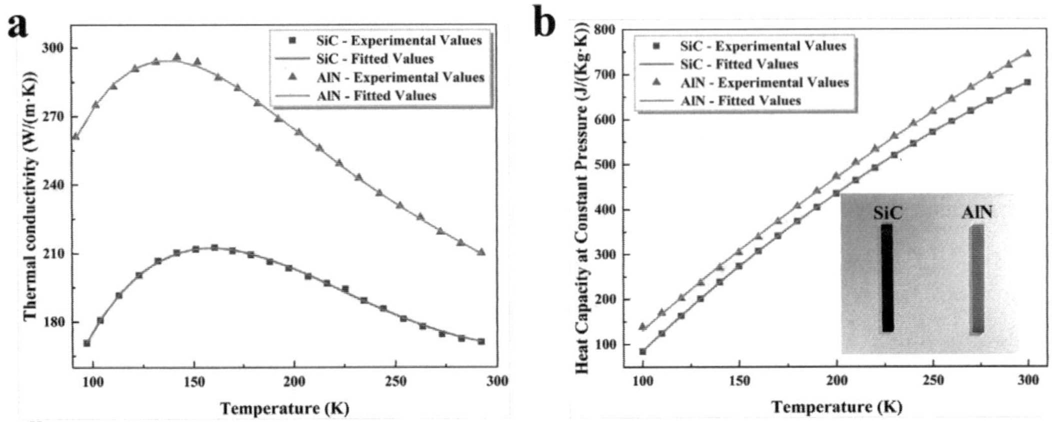

$$K_{AlN} = -243.58192 + 10.62655 \times T - 0.07311 \times T^2 + 2.05695 \times 10^{-4} \times T^3 - 2.11422 \times 10^{-7} \times T^4$$

$$K_{SiC} = -242.97224 + 7.82829 \times T - 0.04691 \times T^2 + 1.14461 \times 10^{-4} \times T^3 - 9.92559 \times 10^{-8} \times T^4$$

$$HCCP_{AlN} = -271.4992 + 4.34091 \times T - 0.00316 \times T^2$$

$$HCCP_{SiC} = -366.86205 + 5.0366 \times T - 0.00514 \times T^2$$

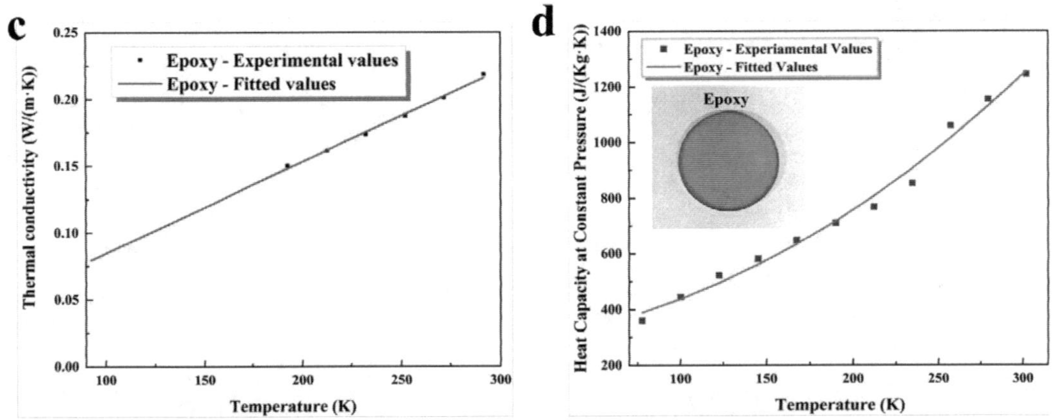

$$K_{Epoxy} = 0.01605 + 6.84965 \times 10^{-4} \times T \qquad HCCP_{Epoxy} = 286.79563 + 0.66137 \times T + 0.00843 \times T^2$$

Figure 2. (a) The TC and **(b)** the HCCP of SiC and AlN from 100 K to 300 K. **(c)** The TC and **(d)** The HCCP of Epoxy from 100 K to 300 K. The insets of **(c)** and **(d)** are the digital photos of SiC, AlN and Epoxy, respectively.

In practice, the TC and heat capacity at constant pressure (HCCP, C_p) of the material will vary with the temperature. Herein, the K and HCCP of the SiC, AlN, and Epoxy were measured by Physical Property Measurement System (PPMS), as shown in Figure 2. The fitting curve are shown in the corresponding figure, and the polynomial formula is displayed directly below the corresponding fitting curve. Density (ρ) changes little with temperature, so density is regarded as a constant, and its value is shown in table 1. In the model, it is assumed that the average diameter of the fibers obtained is 8 um, the length is long enough. And the average diameter (Mark it as D) of the spherical particles is 4 or 8 or 12 um. The above data will be used to set the parameters of the model.

Table 1. The density of silicon carbide, aluminum nitride, and epoxy resin at room temperature [15].

Material	6H-SiC	AlN	Epoxy
Density (Kg•m^{-3})	2980	3240	1170

2.3. Distribution of Heat flux

Considering that the function of x and y fibrous filler is the same, and for the convenience of analysis, the fibrous filler in the direction of x is removed and the fibrous filler in the direction of y is retained in the analysis. The initial temperature of RVE was set at 293.15 K, that is, $T^{(0)} = 293.15\ K$. The upper surface was the heat inflow interface at a constant temperature of 300 K, that is, T_u=300 K, while the lower surface was the heat outflow interface at a constant temperature of 100 K, that is, T_l=100 K, as shown in Figure 3a. Except for the upper and lower boundaries, all the other boundaries are thermal insulation, that is, $\sum \vec{n}_i \cdot \vec{J}_i$=0. The principle of the model is based on the differential equation of unsteady heat conduction, as shown below:

$$\rho C_p \frac{\partial T}{\partial t} + \nabla(-k\nabla T) = Q \tag{5}$$

Where, $T = T(t, x, y, z)$ is the distribution function of temperature; Q is heat energy per unit volume. Therefore, the distribution of heat flux of RVE is mapped, and the respective roles of spherical and fibrous fillers in the heat transfer process are analyzed, as shown in Figure 3b-f. The arrows in Figure 3b-f represent the flow direction of the heat flow, whose magnitude is normalized.

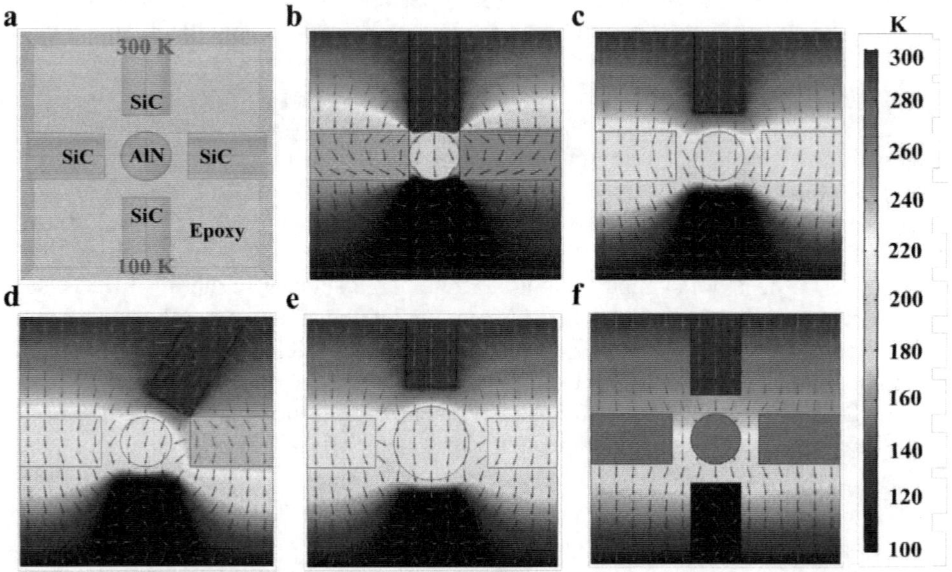

Figure 3. The temperature distribution of the cross-section in the steady-state RVE. **(a)** A brief schematic of the model. **(b)** The radius of spherical particles is 4 um without gap distance (GD). **(c)** The radius of spherical particles is 4 um with a 3 um GD. **(d)** The radius of spherical particles is 4 um with a 3 um GD and the included angle between the upper fiber and the right transverse fiber is 60°. **(e)** The radius of spherical particles is 8 um with a 3 um GD. **(b-e)** The radius of columnar particles is 4 um and the equivalent TR between the spherical particles and the polymer matrix is 10^{-8} m^2·K·W^{-1}. **(f)** The radius of spherical particles is 4 um, and the radius of columnar particles is 4 um, and the equivalent TR between the spherical particles and the polymer matrix is 10^{-2} m^2·K·W^{-1} with a 3 um GD.

2.4. Calculation of thermal conductivity of composites

Herein, the temperature field within the RVE satisfies the differential equation in the steady-state heat conduction problem:

$$\frac{\partial}{\partial}\left(k_x\frac{\partial T}{\partial x}\right)+\frac{\partial}{\partial y}\left(k_y\frac{\partial T}{\partial y}\right)+\frac{\partial}{\partial z}\left(k_z\frac{\partial T}{\partial z}\right)+Q=0 \tag{6}$$

Where, k_x, k_y and k_z are the thermal conductivity in three principal directions in Cartesian coordinate. When heat flows in and out of a sample in a specific direction (set as the z-axis), the Fourier law is simplified as follows:

$$q_z=-k_z\frac{\partial T}{\partial z}=-k_z\frac{T_u-T_l}{L} \tag{7}$$

Where, q_z is the total heat flux along the z-axis, which is the convergent value calculated by finite element software. L is the distance between the upper and lower surfaces. In this model, the thermal conductivity (TC, $k_x=k_y=k_z=k$) is calculated by **Equation (7).**

Figure 4 shows the relationship between the TC of composites and the parameters of the model. In Figure 4a, the TC decreases significantly with GD than without GD. As the GD increases, the TC decreases. Moreover, the smaller the spherical particles, the greater the TC. Because the fibers are longer with the smaller the spherical particles in RVE, the axial heat transfer distance of the fibers is increased, as shown in Figures 3c and e. The tilt of the columnar particles has little effect on the TC. But when the particles are reduced to a certain extent, the upper fiber will contact the transverse fiber, increasing the TC.

Assuming $1/K_s$=0.13 m·K·W^{-1}, the purple curve between the TC of the composite and the filling loading of particles is obtained by using the RM model, as shown in Figure 4b. Through the comparison with the data of reference, it can be seen that the theoretical simulation value is in good agreement with the actual value [12, 16]. As shown in Figure 4b, the TC increases with the increase of filler loading, which is much larger than the TC values simulated by the RM Model. As the filling loading increases, the GD gradually decreases, which increases the heat transfer distance of the axial fibers and further improves the TC of the composites.

In Figure 4c, only when the TR1 varies between 10^{-4} m^2·K·W^{-1} and 10^{-6} m^2·K·W^{-1}, the TC increases significantly with the increase of TR1. When the TR1 is TR1 is greater than 10^{-4} m^2·K·W^{-1} or less than 10^{-6} m^2·K·W^{-1}, the TC does not change with the change of TR1, because there is little or sufficient heat exchange between the filler and the matrix. Moreover, if the TR1 is small enough, the TC of the composite is lower than that of pure Epoxy. And only when the TR1 is large enough, the differences in TC of the composites filled with different fillers' sizes will become obvious. As for TR2, the change of TR2 has little effect on the TC of the composites, indicating that the contribution of spherical particles to the TC of the composites is limited, as presented in Figure 4d.

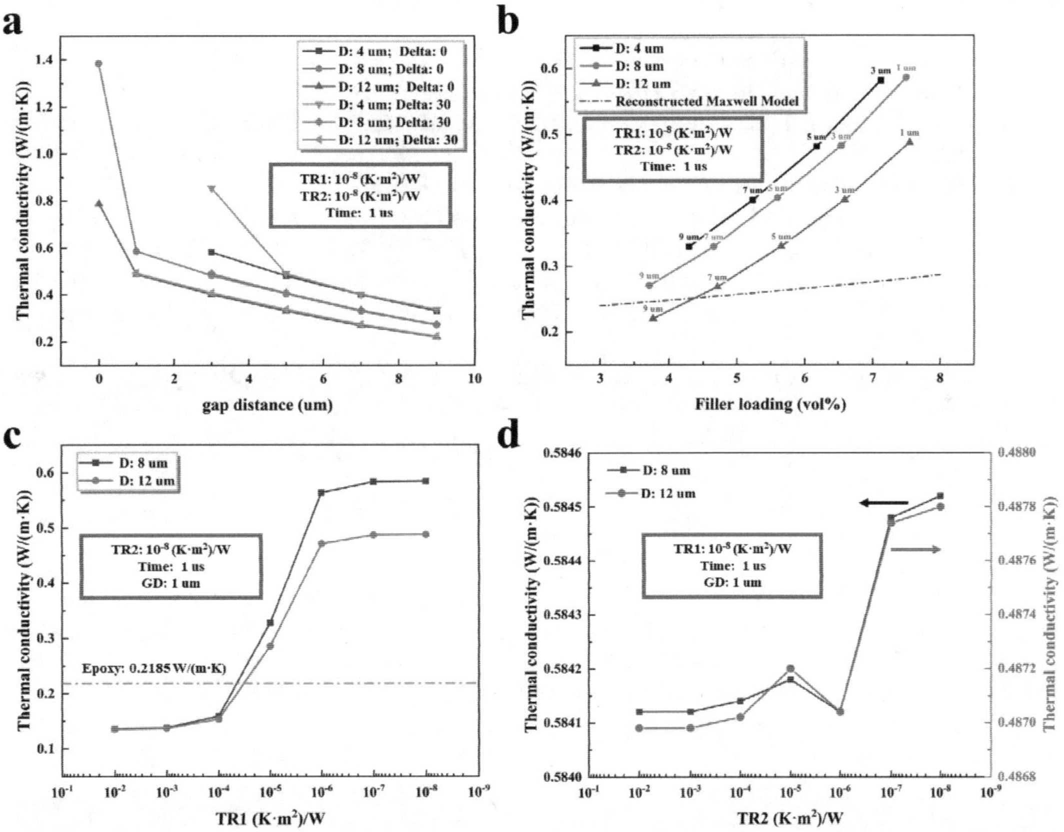

Figure 4. (a) TC of composites with different GD, diameter of the spherical particle, and tilt Angle. **(b)** Diagram of the TC of composites with the filler loading. (The values on the data point are the GD corresponding to filler loading) **(c)** Values of TC of composites with different TR between cylindrical particles and matrix (Mark it as TR1). **(d)** TC of composites with different TR between spherical particles and matrix (Mark it as TR2). The parameters in the blue box are fixed values. Time (t) is the time elapsed from the initial state, and the average temperature of RVE is about 290K at 1 us.

3. Conclusion

The contribution of spherical particles and fibrous particles filled with Epoxy to the TC of the composite was analyzed by mesoscopic simulation. As the loading of AlN or SiC increases, the TC of AlN/Epoxy and SiC/Epoxy composites increases. And when the filler is connected, the heat flow can flow rapidly from the hot end to the cold end through the filler. However, if the tremendous TR between the filler and the matrix exists, the heat flow through the interface layer and matrix will decrease obviously. Above all, increasing the fraction of fillers appropriately and reducing the gap as well as the TR between fillers can increase the TC of composites effectively. In addition, to improve the contribution of fiber to the TC of composites, the fiber should be highly oriented to the direction of heat flow. In all, it is of guiding significance to the design of polymer thermal conductive composites as packaging material.

4. References

[1] Vadivelu M A, Kumar C R and Joshi G M 2016 *Compos. Interfaces* **23** 847-72
[2] Kausar A 2020 *POLYM-PLAST TECH MAT* **59** 895-909
[3] Huang X, Jiang P and Tanaka T 2011 *IEEE Electr. Insul. Mag.* **27** 8-16
[4] Moore A L and Shi L 2014 *Mater. Today* **17** 163-74

[5] Burger N, Laachachi A, Ferriol M, Lutz M, Toniazzo V and Ruch D 2016 *Prog. Polym. Sci.* **61** 1-28
[6] Gardea F, Naraghi M and Lagoudas D 2013 *ACS. Appl. Mater. Inter.* **6** 1061-72
[7] Warzoha R J and Fleischer A S 2014 *Nano Energy* **6** 137-58
[8] Roy A K, Farmer B L, Varshney V, Sihn S, Lee J and Ganguli S 2012 *ACS. Appl. Mater. Inter.* **4** 545-63
[9] Hill R F and Supancic P H 2004 *J. Am. Ceram. Soc.* **87** 1831-5
[10] Hansson J, Nilsson T M J, Ye L and Liu J 2017 *Int. Mater. Rev.* **63** 22-45
[11] Xu J Z, Gao B Z and Kang F Y 2016 *Appl. Therm. Eng.* **102** 972-9
[12] Huang X, Iizuka T, Jiang P, Ohki Y and Tanaka T 2012 *J. Phys. Chem. C* **116** 13629-39
[13] Fu J F, Lu Q, Zong P S, Chen L Y, Ding G J, Dong X, Yu W Q and Shi L Y 2017 *Key Eng. Mater.* **727** 553-7
[14] Chen C, Wang H, Xue Y, Xue Z, Liu H, Xie X and Mai Y-W 2016 *Compos. Sci. Technol.* **128** 207-14
[15] Singh A K, Panda B P, Mohanty S, Nayak S K and Gupta M K 2017 *Polym Plast Technol Eng* **57** 903-34
[16] Bae J-W, Kim W, Cho S-H and Lee S-H 2000 *J. Mater. Sci.* **35** 5907-13

Acknowledgments

This work was supported by Strategic Priority Research Program of Chinese Academy of Sciences (No. XDB25040300), National Key Research and Development Program of China (No. 2019YFA0704904), Key Research Program of Frontier Sciences of Chinese Academy of Sciences (No. QYZDB-SSW-JSC042), President's International Fellowship Initiative (No. 2019VEA0017).

IOP Conf. Series: Materials Science and Engineering 1241 (2022) 012016 doi:10.1088/1757-899X/1241/1/012016

Numerical simulation and experimental investigation of the thermal insulation performance of multi-layer insulation material at 20-298 K

Y C Zhao[1, 2], X Li[1, 2], F Z Shen[1, *], H C Zhang[1], and L F Li[1, 2]

[1]Key Laboratory of Cryogenics, Technical Institute of Physics and Chemistry, Beijing 100190, China;

[2]University of Chinese Academy of Sciences, Beijing 100049, China;

E-mail: shenfuzhi@mail.ipc.ac.cn

Abstract. Thermal insulation plays a major role in the use of cryogenic systems. Due to its superior thermal insulation property, Multi-layer insulation materials (MLI) are widely used in various cryogenic systems, such as superconducting magnets, space propellant storage, satellites and so on. Thus, it is crucial to estimate and measure the insulation performance of it. This paper has carried out a numerical simulation of the performance of one kind of MLI through a modified Layer-by-Layer model. The thermal insulation propriety of the test specimen is also measured by a newly designed measurement system, which is based on the steady-state axial heat flux method. Finally, the simulation and test results are compared and discussed.

1. Introduction

Multi-layer insulation (MLI) system is a kind of thermal insulation system, which normally consists of several metal foils or metal coating polymeric films with high reflectance and thin spacers with low thermal conductivity [1, 2]. Due to the excellent property in thermal insulation, MLI systems are widely used in cryogenic systems such as cryogenic liquid transportation, spacecraft and superconducting system to reduce heat leakage.

The insulation performance of MLI is affected by the material of spacer and metal coating, layer density, the temperature of cold and warm boundary, etc. Thus, it is essential to measure the thermal property of MLI at different conditions and improve the thermal performance of the MLI system.

As a traditional thermal property testing method of MLI at cryogenic temperature, measurement by evaporation rate is usually carried out with cryogenic liquid [3, 4]. Because of the difficulty of temperature regulation, evaporation has a limited boundary temperature. Moreover, it is expensive to use liquid helium and dangerous to use liquid hydrogen. Therefore, cryocoolers have been applied as cold sources of the MLI measurement systems. This method has many advantages such as simple temperature regulation and low cost of measurement, etc.

In this paper, the thermal property of one kind of MLI has been studied through a kind of MLI measurement system cooled by a G-M cryocooler. By using a modified layer-by-layer model, the property is also simulated and compared with the result of the experiment.

2. Experiment

2.1. Apparatus and test specimen

The measurement system is shown in Figure 1. It is cooled by an RDK-415D two-stage G-M cryocooler and has two thermal shields made of OFHC (oxygen-free high-conductivity copper). The outer shield function as the warm boundary of the MLI measurement system, is thermally attached to the 1st stage cold head of cryocooler through some copper braids. While the inner shield functions as the cold boundary, and is thermally attached to the 2nd stage cold head through a support plate and a calibrated rod. The temperature of the cold and warm boundary is controlled by a temperature controller.

The heat flow through MLI is absorbed by the cold boundary and transmitted to the 2nd stage cold head through the calibrated rod. By placing the rhodium-iron resistance thermometer on the calibration rod, the temperature gradient on the calibration rod is measured. Comparing the test data to the calibration data of the calibrated rod, the heat flow can be calculated. It was calculated that the total measuring error is about 9.47%, which is better than the error of the traditional measurement systems of MLI.

The detailed description and measurement principle of the device is described in the article [5] .

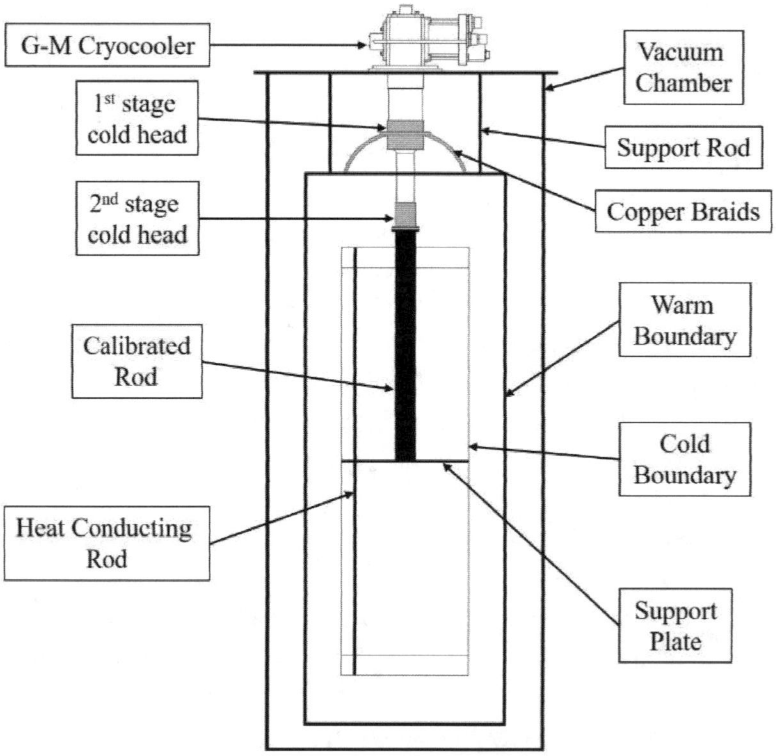

Figure 1. Schematic diagram of the MLI measurement system

The test specimen is a commercial MLI material (provided by Wuhan Yang-Ge-En Technology Co., Ltd.) composed of double aluminized-Mylar (reflector) and polyester non-woven fabric (spacer). The size of it is 1220mm×700mm, and the layer density is 29.4 layers/cm. The number of layers is 30 and the area of the cold boundary is 0.76m^2.

2.2. Experiment result

Set the temperature of the cold boundary as 20K, and the warm boundary as the ambient temperature. Record the temperature of the upper and lower location of the calibrated rod when the system reached thermal equilibrium. The temperatures of these two locations during the thermal equilibrium are shown in Figure 2.

It was calculated that the temperature difference between the upper and lower calibrated rod is 0.921K when the system reaches thermal equilibrium, and the vacuum degree is 3.4×10^{-4} Pa. According to the calibration data of the calibrated rod, the heat flow is 1.07 W. Treating the area of cold boundary as the heat transfer area of the MLI, the heat flux is 1.41W/m^2 by calculation.

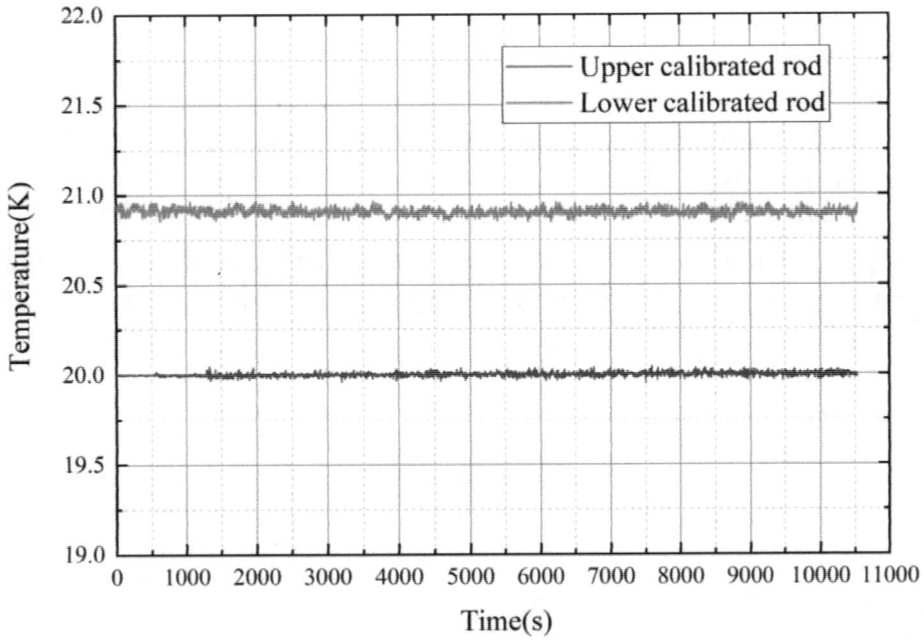

Figure 2. Temperature histories of the upper and lower calibrated rod

3. Calculation

3.1. Modified calculation model

The Layer-by-Layer model is a classical calculation model of thermal properties of the MLI materials [6]. It divided the heat transfer through MLI into three parallel parts, including solid conduction (k_s), gas conduction (k_g) and thermal radiation (k_r) [7]. The total thermal conductivity can be expressed as:

$$k = k_s + k_g + k_r \tag{1}$$

3.1.1. *Solid conduction.* The solid conduction of MLI mainly from the thermal contact between spacers and reflectors, and it can be calculated through the following equation:

$$k_s = \frac{C_1 fk}{d} \tag{2}$$

Where,

k_s – solid conduction, $[W/m^2 \cdot K]$;

C_1 – an empirical constant, 0.008 for the specimen;

f – relative density of the spacer compared to the solid material, 11.6 for the specimen;

k – thermal conductivity of the spacer, $0.17 + 7 \times 10^{-6}(800 - T) + 0.0228 \cdot \ln(T)$ $[W/(m \cdot K)]$ for the specimen [7], T represents the temperature, K;

d – thickness of a piece of spacer, 1.8 [μm] for the specimen.

3.1.2. *Gas conduction.* The thermal conductivity of the residual gas conduction can be written as follows:

$$k_g = C_2 p\alpha \tag{3}$$

Where,

k_g – gas conduction, $[W/m^2 \cdot K]$;

C_2 – $[(\gamma + 1)/(\gamma - 1)] \cdot [R/8\pi MT]^{1/2}$, $\gamma = c_p/c_v$, $R = 8.314$ J/mol · K, M represents the molecular weight of the gas, [g/mol]. For this experiment, $C_2 = 1.1666$ [7];

p – gas pressure, [Pa];

α – accommodation coefficient, 0.9 for air [6];

In this experiment, the pressure of the vacuum chamber under cryogenic temperature reaches 10^{-5} Pa, considering that the gas pressure of the interlayers is higher than the pressure of the chamber, the gas pressure of the interlayers can be considered to be 10^{-3} Pa.

3.1.3. *Thermal radiation.* The coefficient of thermal radiation is as the following equation:

$$k_r = \frac{\sigma(T_{N+1} + T_N)\bullet(T_{N+1}^2 + T_N^2)}{\frac{1}{\varepsilon_{N+1}} + \frac{1}{\varepsilon_N} - 1} \tag{4}$$

Where,

k_r – coefficient of thermal radiation, $[W/m^2 \cdot K]$;

σ – Stefan-Boltzmann coefficient, $5.67 \times 10^{-8} [W/m^2 \cdot K^4]$;

T_N, T_{N+1} – temperature of layer N and layer N + 1, [K];

$\varepsilon_N, \varepsilon_{N+1}$ – emissivity of layer N and layer N + 1;

Because the temperature difference between adjacent layers is small, the emissivity of adjacent layers can be considered the same, thus $\varepsilon_{N+1} = \varepsilon_N$.

According to the previous studies, emissivity of the double aluminized-Mylar has an approximate positive correlation with temperature at cryogenic temperature, and it can be described by the following equation roughly [8]:

$$\varepsilon = 0.0131 + 4.643 \times 10^{-5} T \tag{5}$$

3.2. Calculation procedure

The relation between k and T_N, T_{N+1} (temperature of layers) is established through the above mathematical processing. Based on the above equations, the temperature of each layer of MLI can be calculated through an iterative process. The calculation process is shown in Figure 3.

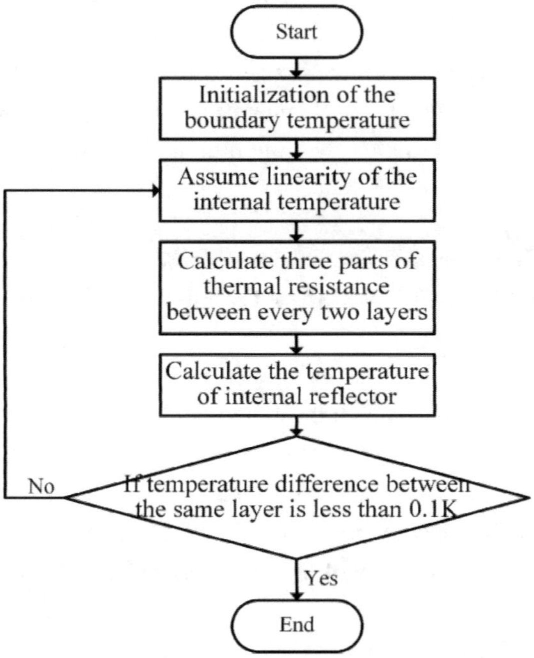

Figure 3. Calculation flow chart

3.3. Results and discussion

The temperature of layers of MLI has been calculated, and compared to the result of measurement. The comparison is shown in Figure 4.

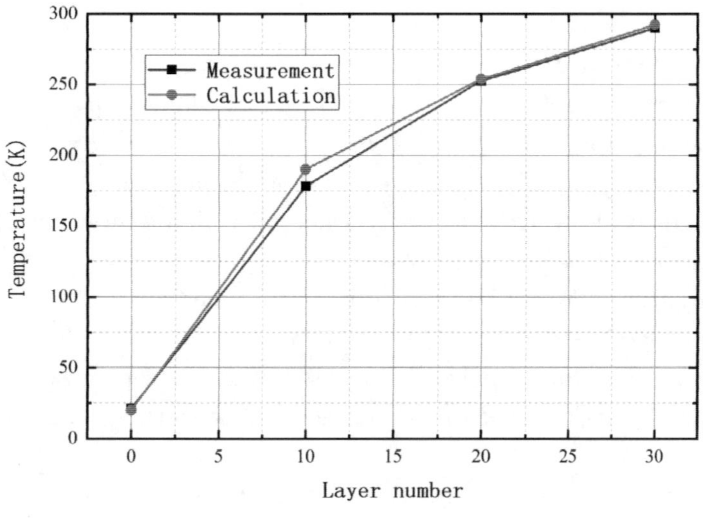

Figure 4. Comparison of the temperature of layers between measurement and calculation

As can be seen in Figure 4, the calculation result of the difference in temperature between the first ten layers is larger than that in the measurement. Due to the larger preload at the internality of the MLI in the measurement, the internal layer density is higher, thus the solid thermal conduction is higher than that in the calculation. Past studies have shown that leakage of solid conduction and thermal radiation account for different share of total heat leakage [9]. The change of temperature has more influence on thermal radiation. While at the lower temperature zone, the solid conduction accounts for the major part of the total heat transfer, and the thermal radiation accounts for the major part of the total heat transfer at the higher temperature zone. Thus, the temperature difference between the measurement and calculation results is larger at the internal layers, and diminishes at the outer layers.

4. Conclusion

By using the MLI measurement system cooled by a G-M cryocooler, the heat flux of one kind of commercial MLI material has been measured when the temperature of cold boundary is 20 K and warm boundary is 298 K in this article.

The temperature of layers of MLI is also calculated by using a modified layer-by-layer model, and the calculation result has been compared with the experiment result. The difference between experiment and calculation result has also been analyzed in this article.

On one side, the obtained result is instructive to the design and construction of cryogenic cryostat. On the other side, the analyze for the contribution of different heat transfer mode in different temperature range is also important for the optimization of the high-performance MLI.

5. Future

In order to provide guidance for the MLI application, the heat flux of MLI when the temperature of the cold boundary and the warm boundary changed will be measured in the future, and the changing law will also be summarized. The thermal property of more different kinds of MLI materials will be tested to find better MLI materials. In addition, on the basis of the measurement and calculation analysis, new MLI materials will be developed.

6. Reference

[1] Zhu H L, Yao C, Li Y, and Pan H L 2016 Heat Transfer Analysis of Compound Multi-Layer Insulation for Cryogenic Tank under Different Service Conditions *Conf.* IOP Conference Series: Materials Science and Engineering 129 p 012053

[2] Mesforoush H, Pakmanesh M R, Esfandiary H, Asghari S, and Baniasadi E 2019 *J.* Experimental and Numerical Analyses of Thermal Performance of a Thin-Film Multi-Layer Insulation for Satellite Application *Cryogenics.* **102** 77-84

[3] Fesmire J E and Johnson W L 2018 *J.* Cylindrical Cryogenic Calorimeter Testing of Six Types of Multilayer Insulation Systems *Cryogenics.* **89** 58-75

[4] Wang B, Huang Y H, Li P, Sun P J, Chen Z C, and Wu J Y 2016 *J.* Optimization of Variable Density Multilayer Insulation for Cryogenic Application and Experimental Validation *Cryogenics.* **80** 154-63

[5] Li X, Xu D, Shen F Z, Liu H M, and Li L F 2019 Design of a Multi-Layer Insulation (Mli) Measurement System with a G-M Cryocooler *Conf.* IOP Conference Series: Materials Science and

Engineering 502 p 012075
[6] Huang Y, Wang B, Zhou S, Wu J, Lei G, Li P, and Sun P 2017 *J.* Modeling and Experimental Study on Combination of Foam and Variable Density Multilayer Insulation for Cryogen Storage *Energy.* **123** 487-98
[7] Mcintosh G E 1994 *J.* Layer by Layer Mli Calculation Using a Separated Mode Equation *Advances in Cryogenic Engineering.* **39B** 1683-90
[8] Frolec J, Králík T, Musilová V, Hanzelka P, Srnka A, and Jelínek J 2018 *J.* A Database of Metallic Materials Emissivities and Absorptivities for Cryogenics *Cryogenics.* **97** 85-99
[9] Zheng J P, Chen L B, Cui C, Guo J, Zhu W X, Zhou Y, and Wang J J 2018 *J.* Experimental study on composite insulation system of spray on foam insulation and variable density multilayer insulation *Applied Thermal Engineering.* **130** 161-168

Acknowledgments
This work was supported by the National Key R&D Program of China (Grant No.: 2017YFE0301405), and CAS Key Laboratory of Cryogenics, Technical Institute of Physics and Chemistry (Fund No.: CRYOQN201904 and Fund No.: CRYOQN202104).

Niobium rod quality and its impact on the production of Nb_3Sn strand for the Divertor Tokamak Test Facility toroidal coils

D. Smathers[1], J. Flanigan[1], J. Kim[2] and K. Sim[2]

[1] H. C. Starck Inc./Materion, 45 Industrial Place, Newton, MA, 02164 USA
[2] Kiswire Advanced Technology Co. Ltd., 223 Techno 2-ro, yuseong-gu, 34026 Daejeon City, South Korea

E-mail: david.smathers@hcstarcksolutions.com

Abstract. Niobium Tin cable is being used in the high field magnets in the Divertor Tokamak Test Facility under construction by ENEA in Italy. Kiswire Advanced Technology was selected to provide 55 tonnes of high-performance Niobium Tin wire for the 18 Toroidal Field Coils. The conductor design is based on a 0.82mm Nb_3Sn strand similar to that provided for the ITER TF coils. H. C. Starck Solutions provided the pure Nb rod and Ta sheet for their conductor design. The enhanced strand requirement has both a minimum current (320 A at 12 Tesla) and maximum hysteresis loss (± 3T cycle 1000mJ/cc) and a Cu/Non-Cu of 1:1. To achieve the current and hysteresis loss reliably, the niobium rod must maintain its shape throughout the drawing process. The n value requirement (>20) is a good indication of niobium filament uniformity while the RRR requirement (>100) is a good indication of tantalum barrier performance. The HCSS rod supplied to KAT provided consistently high Ic (>320A at 12 Tesla), low hysteresis loss and the Ta sheet provided very good RRR values. We report here the consistency of more than 11 tonnes of HCSS Nb rod properties for chemistry, hardness, grain size and wire performance.

1. Introduction

In the European Research Roadmap to the Realization of Fusion Energy[1] a number of milestones must be achieved in order to develop a production fusion reactor. Beyond ITER, the goal is to build a net positive power machine called DEMO by 2050. One of the missions is to develop capable heat exhaust systems to allow the highest possible performance of the core plasma. There is a concern that the heat exhaust system for ITER may not extrapolate to the higher heat loads in DEMO so an alternative exhaust strategy needs to have a maturity sufficient for the DEMO Engineering Design to move forward. If such a promising alternative exhaust concept emerges it will be implemented in the Divertor Test Tokamak (DTT) facility that is also a joint European collaboration.

The DTT magnet system will be built with superconducting cable in conduit conductors cooled by forced flow Supercritical Helium at 4.5K[2]. The Central Solenoid and the Toroidal Field coils will use Nb_3Sn strand in the cables and consume more than 55 metric tons of wire. The high performance Nb_3Sn wire is being produced by Kiswire Advanced Technology Co, Ltd. (KAT). The wire design is based on the 0.82mm strand supplied to ITER[3] but with higher expectations for critical current and AC loss. The DTT wire design must exceed 320 amps at 12 Tesla, have a ±3 Tesla AC Loss lower than 1000 mJ/cc and maintain a RRR of greater than 100 with a Cu:Non-Cu ratio of 1:1. The niobium rod used in the KAT process must have consistent properties in order to achieve the high critical current and low loss required as well as long piece lengths for cable production. H. C. Starck Solutions (HCSS) was selected to provide the niobium rod for the filament and tantalum sheet for the diffusion barrier. HCSS provided more than 11 metric tons of reactor grade niobium (ASTM B392) with a maximum hardness of 85 HV0.5 and a grain size finer than ASTM 6. As the total engineering strain will exceed 15 (3.4 million to 1 area reduction), the mechanical properties and material homogeneity are critical.

2. Kiswire Advanced Technology Manufacturing Process

The KAT process for making Nb_3Sn wire and various design choices has been described earlier[3] and some improvements have been made to the design to achieve higher critical current density while maintaining low losses.

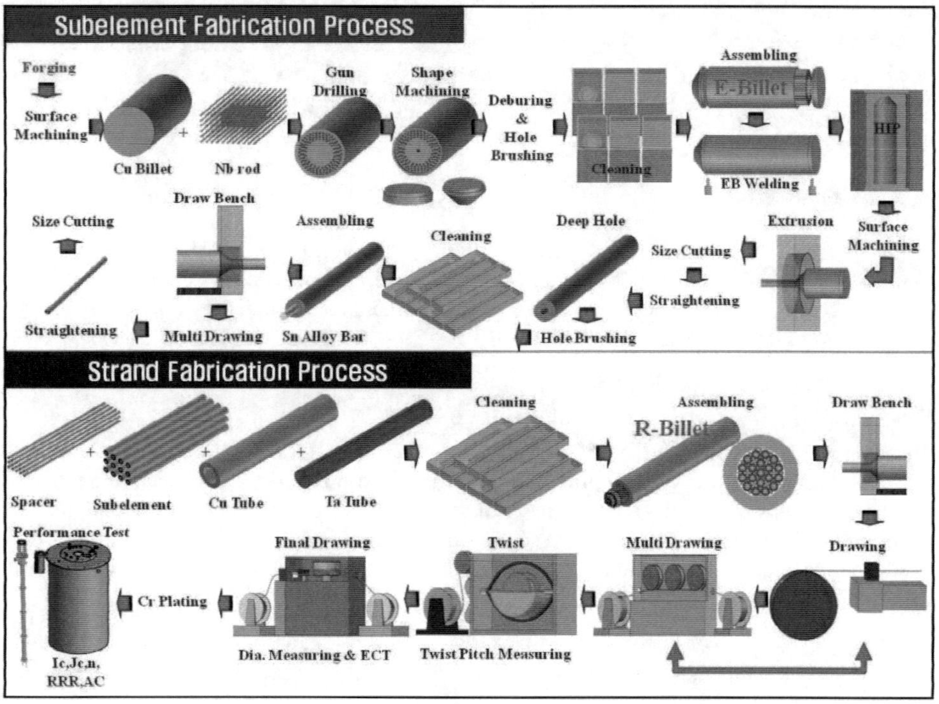

Figure 1. Internal tin manufacturing process layout.

The sub-element fabrication process involves inserting niobium rods into drilled copper holes. Both the copper and the niobium are fully soft so the straightness and diameter tolerance are critical to avoid damage to the rods and holes during the assembly process. After assembly, the billet is electron beam welded and HIPped to remove void space prior to extrusion. Once extruded, the bar is deep hole bored and the tin-titanium alloy core added and drawn to the strand assembly diameter. The DTT strand is improved relative to the ITER strands by decreasing the copper fraction of the matrix area. Additionally, KAT distributed the tin containing spacers through the cross section of the wire. The distribution of the tin elements aids in the movement of the tin during the solid state reaction to form the Nb_3Sn. The DTT wire has 18 extra tin spacers distributed in the cross section compared to the ITER wire. As a result, the heat treatment recipe was dramatically shortened relative to the ITER strand from a total of 478 hours to 323 hrs. The final recipe was 210C/50hr+450C/50hr+650C/100hr plus ramp times.

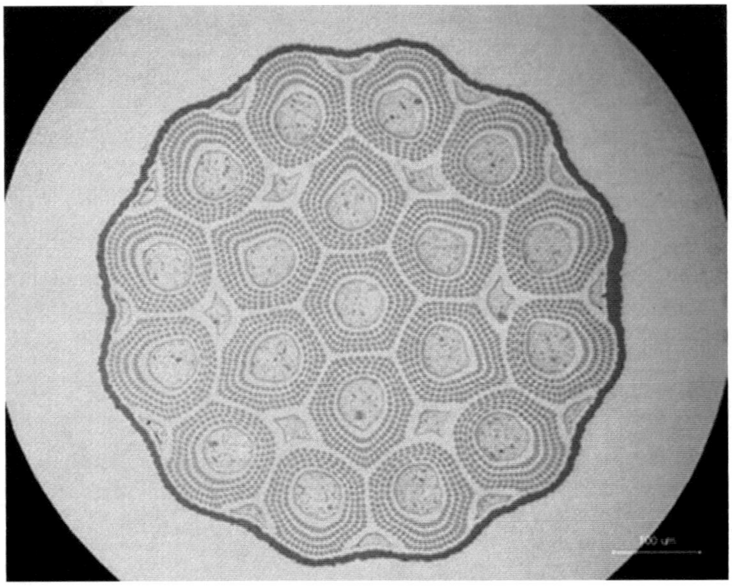

Figure 2. DTT wire cross section at 0.82mm. Cu:Non-Cu ratio 1:1. Wire design consists of 19 Sub-elements of 180 Nb filaments, each with a Sn-Ti core, 18 Sn-Ti spacers between the sub-elements surrounded by a pure Ta diffusion barrier.

3. H. C. Starck Solutions Manufacturing Process

The Niobium rod was produced from ingots electron beam melted at the Hermsdorf, Germany facility of HCSS. The ingots are melted a minimum of two times at controlled rates in order to control the ingot microstructure consistency. The input material was blended to provide the desired reactor grade chemistry with this process. After melting, the niobium was extruded and worked to the rod diameter specified by the customer. The final anneal was chosen to provide the best combination of hardness and grain size.

3.1. General Chemistry

The material was specified as ASTM B392 reactor grade niobium rod[4]. The average chemistry over the history of the production of 11 metric tons is given in the table below. Most impurities are quite low due to the electron beam melting process. The ASTM specification does not require the use of Glow Discharge Mass Spectroscopy (GDMS) but this is standard practice at HCSS.

Table 1. Average chemistry for production ingots compared to the ASTM B392 limits. All values in ppm by weight using GDMS except for elements marked with * values in ppm by weight using LECO.

Element	**Fe**	**Si**	**Ni**	**Zr**	**Ti**	**Hf**	**W**
Max	50	50	50	200	200	200	300
Mean	0.04	0.9	0.01	0.04	0.01	3.0	20
Element	**Mo**	**Ta**	**O***	**N***	**C***	**H***	**Nb**
Max	100	1000	150	100	100	15	Bal
Mean	10	492	50	7.0	5.5	2.0	Bal

The range and distribution of tantalum will be discussed later regarding wire performance. The only element that has any correlation to rod mechanical properties is oxygen but generally, 50ppm or less oxygen does not impact the performance very much.

3.2. Hardness vs Grain Size

The rod specification included a tight diameter tolerance, a limit of 85 Vickers Hardness (HV0.5), an ASTM E112 grain size of 6 or finer and a strict straightness requirement. For very pure niobium, the main factor controlling the grain size is cold work and annealing temperature[5]. It can be a fine line between keeping a fine grain size and not letting the hardness get too high. The data in figure 3 shows that the relationship is not a strict function and it was possible with experience to move the grains size finer without driving the hardness higher. All the batches supplied had a mean intercept grain size between 17 microns (ASTM 8.5) and 34 microns (ASTM 6.5). It is difficult to get consistent hardness readings in soft, pure metal. Part of the vertical spread in data is due to the gauge capability for Vickers hardness. Within a batch, the data for each sample hardness consisted of an average of 9 points on a grid and grain size consisted of three locations from center to outside. In Figure 3, data from four samples of each batch are averaged for hardness and grain size.

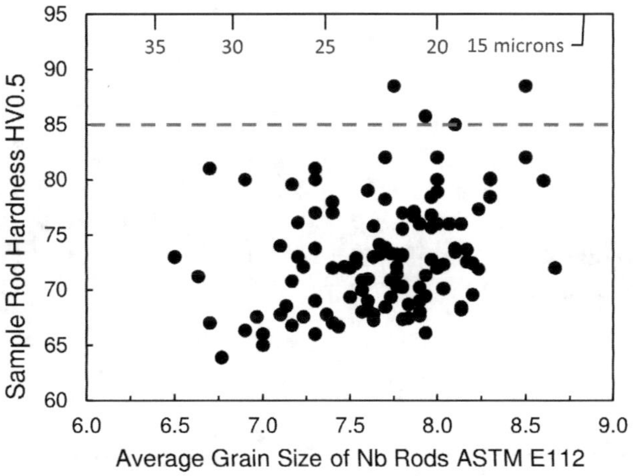

Figure 3. A sample rod was measured by both HCSS and KAT for every batch supplied to KAT. A general trend can be observed that finer grain size comes with increased hardness. Lots with hardness greater than 85 would be rejected. The equivalent intercept grain size is shown at the top of the chart.

A single HCSS batch of niobium rods may make several KAT primary extrusions (Niobium/Copper stacks of 180 rods each) and some extrusions will have mixed batches of niobium rods. Each primary extrusion will then produce multiple final wire assemblies.

3.3. Tantalum impurity

The key elemental impurity that separates the various grades of niobium is tantalum[4]. Reactor grade has the lowest maximum limit for tantalum, 1000 ppm by weight. Tantalum is a naturally occurring impurity in niobium ore and cannot be removed during electron beam melting. The tantalum content of the starting oxide determines the final content of virgin ingot through the thermite reductio process. Depending on the ore source, the tantalum content of oxide can be less than 50ppm to over 2000 ppm. Niobium sourced from tantalum production is usually very low tantalum while niobium sourced from pyrochlore usually contains much higher tantalum. Oxide pricing can be related to tantalum content while the higher tantalum containing oxides are more available and not tied to tantalum production. To manage availability and cost of reactor grade niobium ingot, an average of 50% clean solids (recycle) was mixed with virgin material from thermite in the manufacture resulting in a range of tantalum contents, all kept below 1000ppm. The mean tantalum content of the production was 492 ± 135 ppm, with a maximum of 787 ppm and a minimum of 196 ppm. HCSS found no correlation to either grain size or hardness over this range.

4. Correlation of Rod Attributes to Wire Results

A subset of the data in figure 3 was tracked through wire manufacture, heat treatment and testing. The critical current at 12 Tesla for all batches ranged from 305 to 365 Amps with a mean of 335 Amps and the ±3 T Loss ranged from 575 to 1050 mJ/cc with a mean of 820 mJ/cc. In the subset for correlation, the I_c range was 312 to 360 Amps and the loss ranged from 608 to 950 mJ/cc. The niobium rod hardness ranged from 65 to 90 while the grain size ranged from 6.7 to 8.5. The only rod property that correlates to the final wire current performance is the starting niobium grain size. The loss had a weaker correlation to grain size but showed a reverse trend, getting lower as the grain size became finer. As the I_c is increasing as the grain size becomes finer, a lower loss is counter intuitive. If the effective filament size is determined by the wire architecture, a higher I_c should result in a higher loss but the opposite is observed. The effective filament size is larger than the actual filament size when local filaments bridge together and respond to an alternating field as a larger filament.

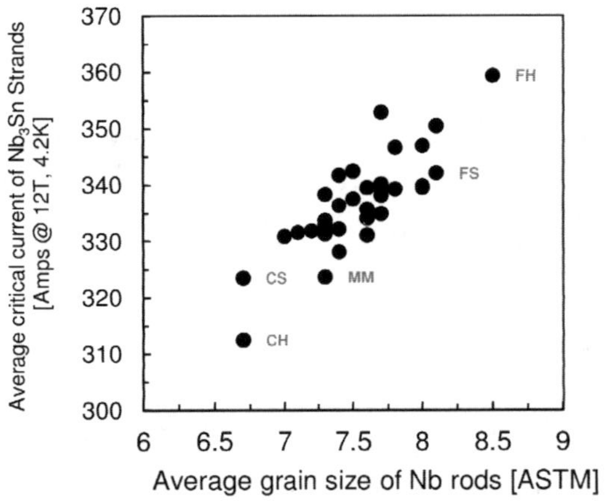

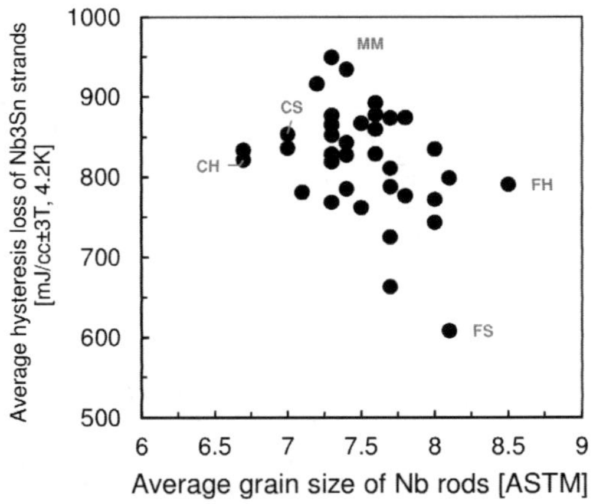

Figure 4. Left is I_c at 12 Tesla vs initial grain size of the niobium rod. Right is the ±3T Loss vs initial grain size of the niobium rod. Indicated points from Table 2. The grain size is the average of four rods measured by KAT. Designations of the five points are described in Table 2.

Table 2. Selected samples from the edges of the grain size – hardness distribution. FH is fine grain/harder, FS is fine grain/softer, CS is bigger grain/softer, CH is bigger grain/harder and MM is from the middle of the distribution. Hardness and grain size are the average of four samples from a batch.

Sample	Hardness HV0.5	ASTM G. S.	Ic at 12T Amp	±3T Loss mJ/cc
FH	88	8.5	359	791
FS	72	8.1	342	609
MM	77	7.3	324	950
CS	81	6.7	313	822
CH	67	6.7	324	834

It is clear from Figure 4 that the critical current is affected more by the grain size than the hardness. Even thought the hardness ranges overlap, the seperation is complete by grain size for both critical current and AC loss. It has been observed in the past that the roughness of the niobium-copper interface in co-drawn materials is dependent on the initial grain size of the niobium[6]. As the niobium is body centered cubic and the copper is face centered cubic, the difference in slip systems causes some niobium grains to rotate prior to deforming to get into the easy direction. So, then it is not surprising that the reacted wire properties are in fact related to the initial grain size of the niobium. The wire design has four rows of closely spaced niobium filaments surrounding the main tin source. The tin must diffuse between the filaments in the first row to reach the second and third row. The addition of the extra tin sources between elements aids in the tin distribution to the outer niobium rows but the main source of tin is the core. If the niobium filaments distort too much, the spacing between the filaments may be reduced and slow the diffusion of the tin during the final heat treatment. The effective filament size is much larger than the individual filaments and bridging may be enhanced when the niobium ridges are larger due to bigger starting grain size. Qualitatively the interface roughness can be seen for four of the samples at the final assembly stage. An attempt is being made to quantify the interfacial roughness to make a more definitive relationship with the starting niobium rod grain size.

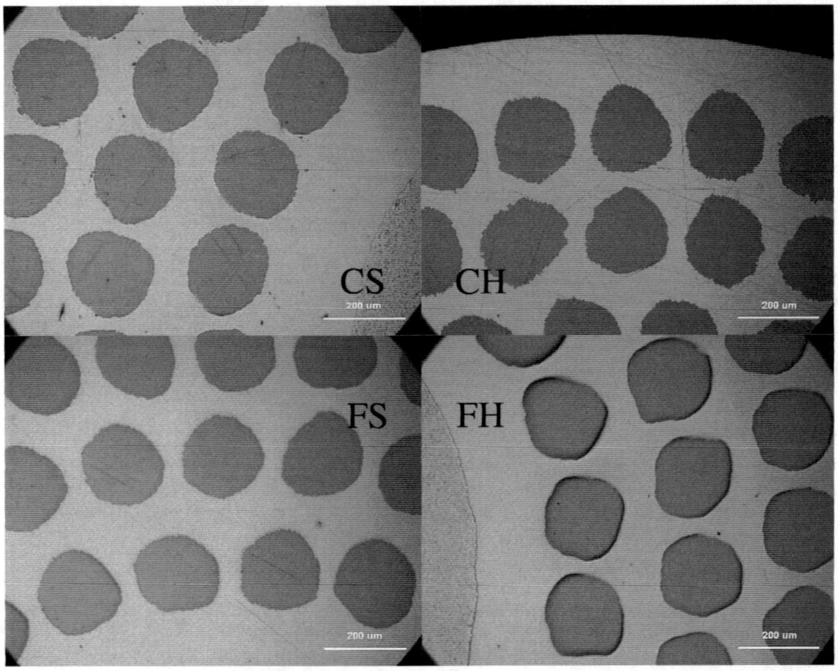

Figure 5. Micrographs of the Cu-Nb-Tin element prior to assembly into the final product. The interfacial roughness between the Cu and Nb gets smoother as the grain size is reduced. FH is fine grain/harder, FS is fine grain/softer, CS is bigger grain/softer and CH is bigger grain/harder (see Table 2). 200μm scale bar

Another indicaton of the quality of the entire process including the niobium homogeneity is the n value of the critical current transition. The DTT specification is an n value greater than 20. The KAT process produced wires with a mean (average) n value of 50 with a standard deviation of 3.5 at 4.2K, 12 Tesla.

The critical current and AC loss can be evaluated as a function of tantalum content in the rods. The same final wire properties shown in figure 4 are plotted for the rod tantalum content in figure 6.

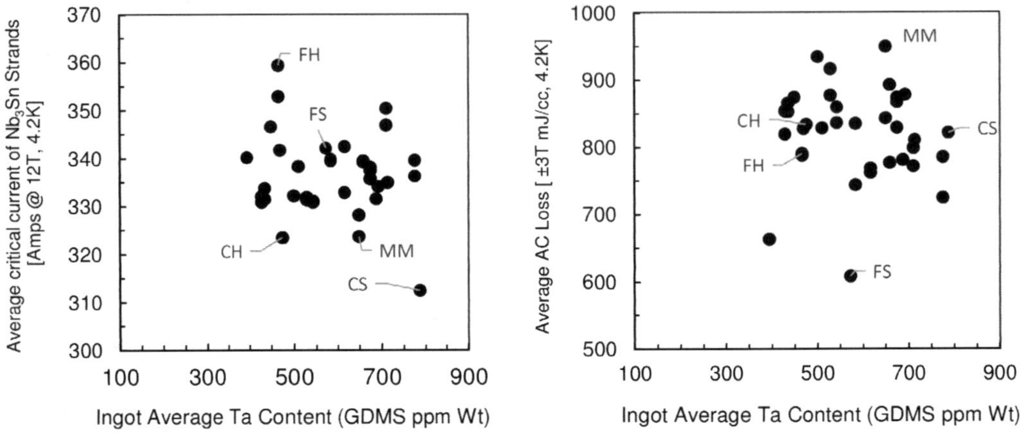

Figure 6. Left is I_c at 12 Tesla vs tantalum content of the Nb rod. Right is the ±3T Loss vs tantalum content of the Nb Rod. The mean value of I_c (335 Amps) and ±3T Loss (820mJ/cc) is present for all values of tantalum. Letter designations correspond to data in Table 2.

HCSS also supplied the tantalum sheet for the diffusion barrier (seen in figure 2). The RRR values of all the reacted wires met the specification of 100:1 (R@273K/R@20K). This is another testiment to the quality of the starting materials, the wire design and the processing.

5. Summary

Kiswire Advanced Technology Co, Ltd. has continued to demonstrate a high level of competence in the manufacture of Internal Tin Wire for large programs such as ITER and DTT. HCSS has demonstrated high volume capability for niobium and tantalum materials to support niobium tin conductor manufacture. In the collaboration described here, we have shown how the niobium properties affect the wire performance. The desirable product is a fine grain, low hardness niobium consistent from rod to rod and batch to batch. The acceptable range of these attributes will be a function of the actual wire design and processing route. Once established through interaction of the rod producer and the wire producer, the attribute ranges can be used to control the quality of the wire output. Capability of the measurement process should be taken into account. Tantalum content of the reactor grade niobium has no influence on the final properties of the rod or the wire processing, heat treatment or final attributes. For future large projects, high volume production will require sourcing niobium from various ores and consume recycle, making the insensitivity to tantalum good news. In over 11 metric tons of niobium supplied for the production of 55 metric tons of wire, the tantalum ranged from 196 to 787 ppm without impacting performance at any stage of the processing. While all the range of hardness and grain size produced very good wire performance, KAT preferred an ASTM grain size above 7 and a hardness below 85.

The KAT wire design has 28% higher performance than the wire designed and delivered for the ITER program with consistently high critical current, high n value and high RRR. This has been achieved through the adjustment of matrix copper fraction and the improvement of the tin-titanium cores in the distributed tin-containing spacers. A lower limit to the Nb_3Sn layer J_c, calculated based on the niobium volume in the wire, is 3,400 A/mm^2 at 12T, 4.2K. Compared to the US-DPC conductor (18 element, Nb0.8Ti, Sn3.5Mg) the layer J_c is 60% higher while the effective filament sizes are quite similar[7,8]. An even higher layer J_c is possible if the niobium content and tin content are increased but potentially at the expense of the effective filament size.

6. References

[1] Donne T and Morris W 2018 , European research roadmap to the realisation of fusion energy-long version, *EUROfusion*, www.euro-fusion.org/eurofusion/roadmap, ISBN 978-3-00-061152-0

[2] Albanese R, Crisanti F, Martin P, Martone R and Pizzuto A 2019 Divertor Tokamak Test Facility Interim Design Report –*ENEA* DTT_IDR_2019_WEB.pdf (dtt-project.it) ISBN: 978-88-8286-378-4

[3] Park P and Kim K 2008 Status of Nb3Sn strand development in Korea *Cryogenics* **48** pp 347-353

[4] ASTM B392 -18 2018 Standard specification for niobium and niobium alloy bar, rod, and wire *ASTM international, West Conshoshoken, PA*, www.astm.org

[5] Smathers D 1988 Production scale niobium ingots for radio frequency cavity application *Journal of Less Common Metals* **139** pp 61-66

[6] Balachandran S, Barber R, Huang Y, Miao H, Parrell J, Hong S, Griffin R and Hartwig K 2010 Fine grained Nb for internal tin Nb_3Sn conductors *Adv. In Cryogenic Eng*. **1219** pp 216-223

[7] Smathers D and Siddall M 1990 Manufacture and Evaluation of Tin Core Modfied Jelly Roll Cables for the US-DPC Coil *Adv. In Cryogenic Eng*. **36** pp 131-137.

[8] McKinnell J, Siddall M, O'Larey P and Smathers D 1994 Increased superconducting critical current density in internal tin niobium-tin (Nb3Sn) composite wires by magnesium doping, 1994 *Adv. In Cryogenic Eng* **40** pp 945-952.

IOP Conf. Series: Materials Science and Engineering 1241 (2022) 012018 doi:10.1088/1757-899X/1241/1/012018

Neutron irradiation effect on critical current and critical temperature of Nb_3Sn wire

Arata NISHIMURA[1,2] and Yoshimitsu HISHINUMA[1]

[1] National Institute for Fusion Science. 322-6 Toki, Gifu 5095292 Japan.
[2] Technical Institute of Physics and Chemistry, CAS. 29 Zhongguancun East Rd, Haidian, Beijing 100190 China

E-mail: nishimura.arata@toki-fs.jp

Abstract. The ITER project is ongoing in France, and it is expected that a lot of high energy neutrons will be generated, and some neutrons will reach to superconducting magnets. To keep the safe and stable operation of the superconducting magnets, the neutron irradiation data must be piled up and the mechanisms of the irradiation effect must be investigated. A 15.5 T superconducting magnet and a variable temperature insert were installed in a radiation control area at Oarai centre in Tohoku University, and the superconducting properties of the irradiated Nb_3Sn wires have been evaluated. The critical current was improved once and drastically dropped after irradiation of 10^{23} n/m^2, and the critical temperature degraded monotonically, and the critical magnetic field decreased around 17 T after the irradiation of 10^{23} n/m^2. The mechanism of the irradiation effect was discussed based on a flux pinning concept.

1. Introduction

The fusion energy is anticipated as a new energy to reduce the greenhouse effect in the world. The ITER was designed, and the construction has been continued from 2007 [1]. Since the fusion energy will be earned from the fast neutrons which have 14 MeV kinetic energy after the deuterium and tritium (D-T) reaction, the fusion device is a kind of a huge neutron source. All elemental components which make up a fusion reactor will be irradiated by the neutrons. Therefore, the neutronics on the component materials is very important and the basic data and the fundamental understanding of the neutron irradiation must be piled up as far as possible.

The superconducting magnets are applied to the fusion devices, as the magnets can produce strong magnetic field in the plasma vacuum vessel. Roughly saying, when the magnetic field becomes twice, the confinement time becomes twice as long. Therefore, the strong magnetic field is anticipated, and the large-scale superconducting magnets are designed and fabricated in the ITER project.

In the plasma vacuum vessel, the blanket modules will be installed, and the fast neutrons will be captured by the blankets and the kinetic energy would be changed to the thermal energy to generate the electricity. Since the blanket cannot capture all neutrons perfectly, some neutrons will stream out of the vessel and reach to the superconducting magnets. The superconducting magnet consists of the superconducting wires, the electric insulation materials, and structural materials. It is well known that the superconducting wires are weak against the neutron irradiation and degrade the superconductivity with relatively small number of the neutron fluence.

To keep the safe and stable operation of the superconducting magnets in the fusion device, the precise database of the irradiation effect of the superconducting wires must be established, and the

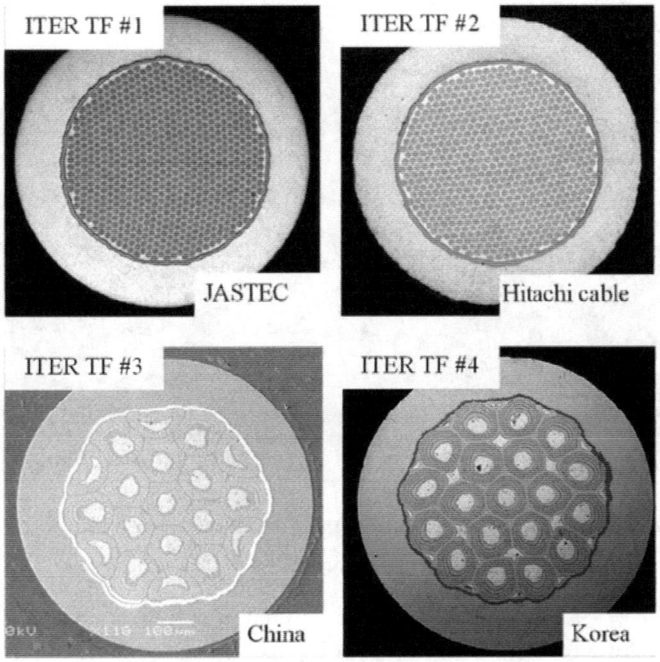

Fig. 1 Cross sectional views of tested Nb_3Sn wire. ITER TF #1 and #2 are bronze route wires. ITER TF #3 and TF #4 are internal tin processed wires.

irradiation mechanisms on the superconducting wires must be clarified. In this study, a bronze route Nb_3Sn wire and an internal tin processed Nb_3Sn wire were tested, and the neutron irradiation effect of these wires on the critical current and the critical temperature were investigated. Some superconducting data of the post irradiated tests will be presented, and the mechanisms of the irradiation effect will be discussed.

2. Test samples and test procedures

A bronze route Nb_3Sn wire and an internal tin processed Nb_3Sn wire were prepared for the study. All wires were developed for the ITER TF coil conductors. The heat treatment was carried out with the standard condition for the ITER TF coil wire (240 hours at 650 °C). The cross-sectional views of the tested wires are shown in Fig. 1. The bronze route wires have very fine filaments, and the internal tin processed wires have big tin rods surrounded by the fine Nb filaments. These wires were produced in around 2000 and the improved wires were used for the fabrication of the ITER TF coil conductors. The diameter is 0.8 mm, and the electric barrier was plated on the surface of the wire which was removed before the heat treatment was performed.

After the heat treatment, the Nb_3Sn wires were cut to about 35 mm long and sent to a fission reactor, BR2, in Belgium to carry out the neutron irradiation. The samples were bundled loosely and covered with aluminum foil and then put into an aluminum alloy capsule. The samples were free in the capsule and there was no restraint. The inside of the capsule was filled with helium gas and the cap was TIG welded. The capsule was kept in the BR2 for a certain time to be irradiated to the targeted fluence without any special shielding. The cooling water temperature was controlled at about 50 °C during the irradiation. The neutron fluence in this study ranges from $4.90\text{x}10^{22}$ n/m^2 to $1.71\text{x}10^{23}$ n/m^2 (> 0.1 MeV), and these values exceed the ITER design value ($2.0\text{x}10^{22}$ n/m^2) and were expected in the range of the fusion DEMO design. The corresponding neutron fluence for energy >1 MeV ranges between $1.8\text{x}10^{22}$ n/m^2 and $8.73\text{x}10^{22}$ n/m^2. For an easier comparison with previous data, the values of neutron fluence of over 0.1 MeV were used. After the irradiation, the capsule was sent back to Oarai center in Tohoku University and kept in the hot cell in the radiation control area. The residual

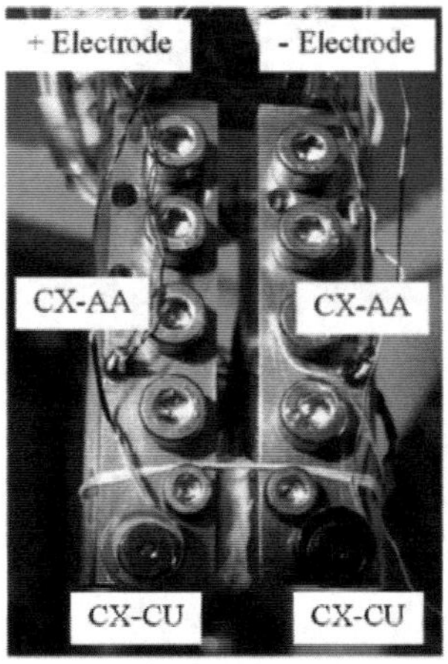

Fig. 2 Sample holder of test wire.

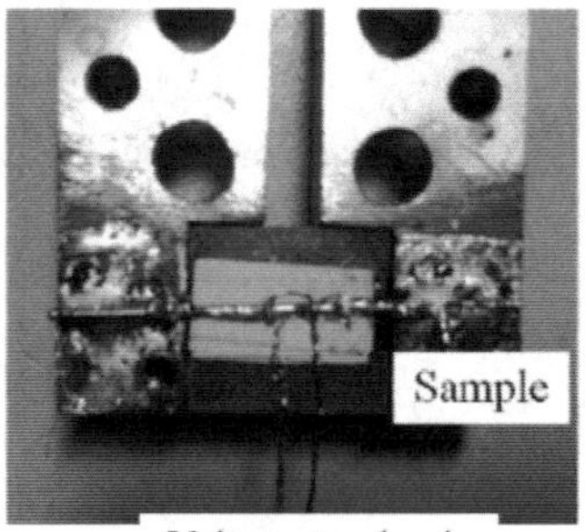

Fig. 3 Soldered sample and voltage taps.

radioactivity was measured, and the post irradiation experiments were conducted after confirming that the measured radioactivity was below the reference value.

To carry out the post irradiation test of the superconducting wires, a 15.5 T superconducting magnet and a variable temperature insert were installed in the radiation control area at Oarai center [2]. The 15.5 T superconducting magnet is cooled by a GM refrigeration and has 52 mm room temperature bore in the vertical direction of the magnet. The variable temperature insert with conduction cooling was installed to measure the critical current and the critical temperature of the superconducting wires. The sample holder is shown in Fig. 2. The bus bars are contacted with a superhigh purity aluminum rod which is connected to the second stage of the GM refrigerators. The radiation heat load and the joule heating of the copper bas bars are transferred to the GM refrigerators through the aluminum rod.

The sample was soldered on a sample holder as shown in Fig. 3. A low melting point solder was used to reduce the damage on the sample. Two sets of voltage taps were soldered on the sample. The central voltage was used for the determination of the critical current and both-ends voltage was used for the interlock of the sample current supply system.

The relation between the sample current and the voltage was measured, and the regression curve was found using a power function of $V = C \times I^n$, where C and n are constants. Then the current at 20 μV/cm in the regression curve was defined as the critical current of the sample under the given condition. Because of the high n-value in Nb_3Sn, no significant difference is expected for the different criteria. The critical current of non-irradiated sample was measured in liquid helium using 27 T hybrid magnet in Tohoku University, and the 10 μV/cm was adopted as the criteria.

The sampling rate of the V-I measurement was 10 Hz, and the ramp rate of the sample current was 150 A/s to reduce the joule heating of the copper holder. But in case of the lower critical current, the 50 A/s was adopted to get more data points [3,4].

3. Test results and discussions

The critical currents of the non-irradiated and the irradiated samples are shown in Fig. 4 to Fig. 6. These data were obtained in the bronze route Nb_3Sn wires, ITER TF #1. The general tendency is the same for other samples. In these figures, the critical current on the non-irradiated wire, the critical current of the irradiated wire and the ratio of the critical current after irradiation to that of non-irradiated sample (I_C/I_{C0}) are presented. Figures 4, 5 and 6 show the results when 4.90×10^{22} n/m^2, 7.92×10^{22} n/m^2 and 1.71×10^{23} n/m^2 of over 0.1 MeV neutrons were irradiated.

It is very clear that the critical current was improved by the irradiation in a case of 4.90×10^{22} n/m^2. However, when the neuron fluence increases, the critical current decreases and it dropped drastically in a case of the 1.71×10^{23} n/m^2 irradiation. When the critical current is linearly extrapolated, the critical magnetic field would be expected to around 19 or 20 T for this result. Generally, the critical

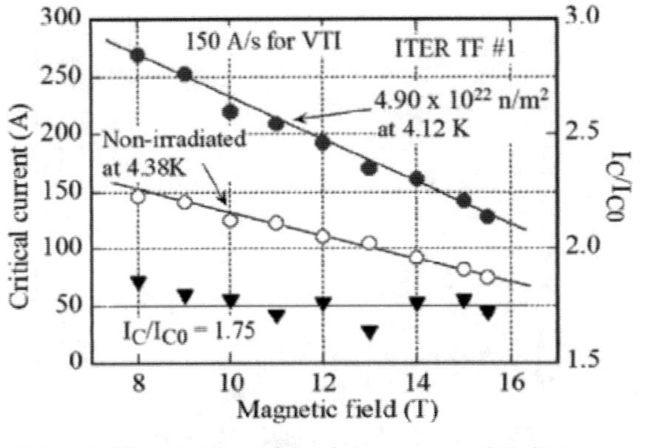

Fig. 4. Change in critical current and I_C/I_{C0} against magnetic field. Bronze route Nb_3Sn wire. Neutron irradiated to 4.90 x 10^{22} n/m^2.

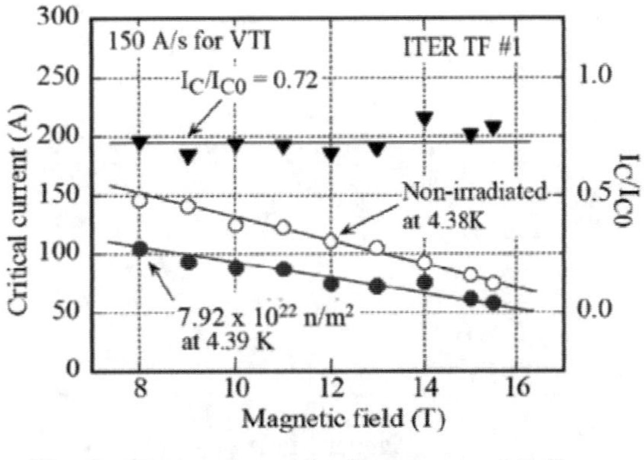

Fig. 5. Change in critical current and I_C/I_{C0} against magnetic field. Bronze route Nb_3Sn wire. Neutron irradiated to 7.92 x 10^{22} n/m^2.

current of non-irradiated Nb_3Sn wires is over 24 T, it is obvious that the critical magnetic field degrades because of the irradiation. In cases of 4.90 x 10^{22} n/m^2 and 7.92 x 10^{22} n/m^2 irradiations, the I_C/I_{C0} is almost constant in the range of 8 T to 15.5 T, but it gradually goes down in case of the 1.71 x 10^{23} n/m^2 irradiation.

The dependence of I_C/I_{C0} on the neutron fluence of over 0.1 MeV is shown in Fig. 7. The obtained data in the study were plotted near the data written by Kuroda et. al. [5,6], which were shown with round open symbols. The I_C/I_{C0} shows a peak at around 5 x 10^{22} n/m^2 and goes down as the fluence increases. The plots near the fluence of 1.8 x 10^{23} n/m^2 have bars, because the I_C/I_{C0} comes near to zero in the higher magnetic field. The general tendency is the same for both samples of the bronze route and the internal tin processed. Since the data showed a wide scatter, the reason will be explored in the future work.

The critical temperatures of the irradiated samples against the magnetic field are shown in Fig. 8 to Fig. 10. Figures 8 and 9 show the results of the bronze route Nb_3Sn wires and Fig. 10 presents the results of the internal tin processed Nb_3Sn wire. When the wires were irradiated to 4.90 x 10^{22} or 7.92 x 10^{22} or 1.71 x 10^{23} n/m^2, the critical temperatures come down comparing with the critical temperatures of 8.23 x 10^{21} n/m^2 irradiated samples. It should be noticed that the larger neutron fluence makes the critical temperature lower in all cases. When the amount of the temperature decrease is focused, the ITER TF #1 showed the small degradation among three samples at the lower magnetic field, and the relative degradation becomes smaller in the higher magnetic field. The critical

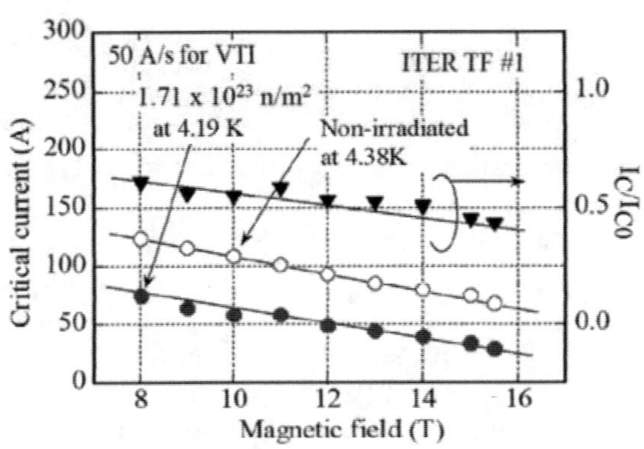

Fig. 6. Change in critical current and I_C/I_{C0} against magnetic field. Bronze route Nb_3Sn wire. Neutron irradiated to 1.71 x 10^{23} n/m^2.

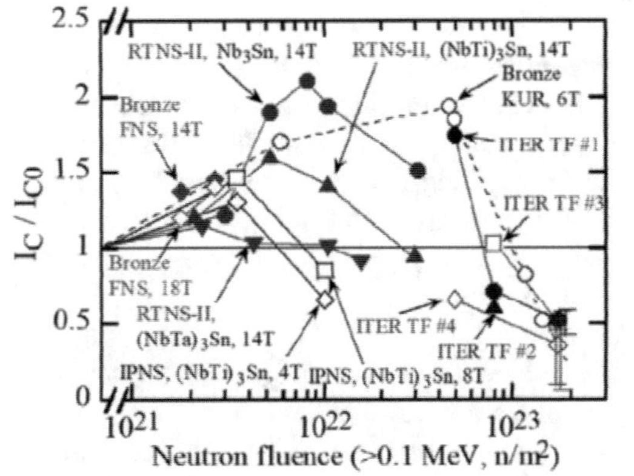

Fig. 7. Change in I_C/I_{C0} against neutron fluence of over 0.1 MeV. Published data were plotted [5,6].

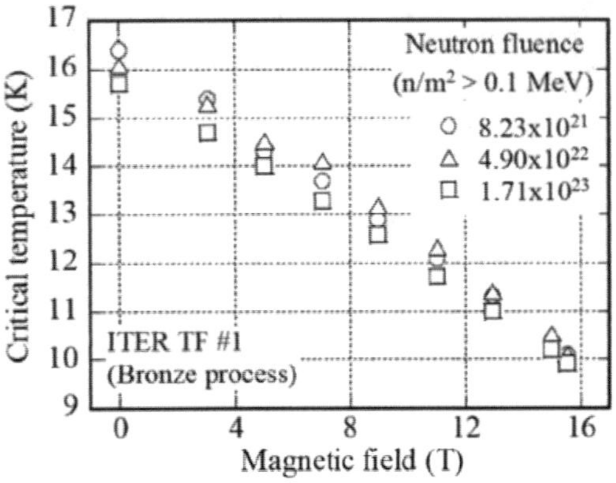

Fig. 8. Change in critical temperature of irradiated wires against magnetic field. Bronze route Nb_3Sn wire, ITER TF #1.

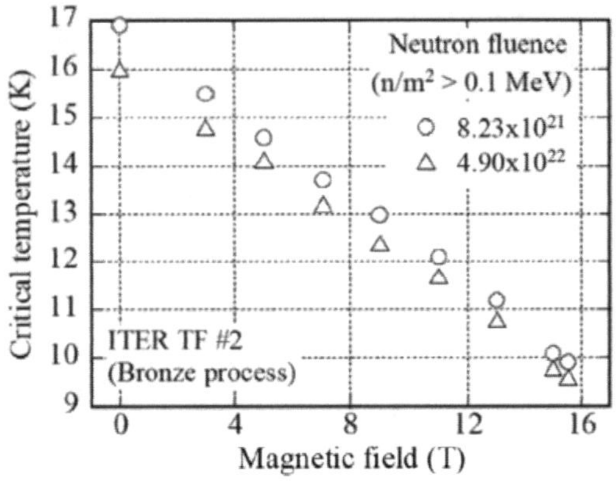

Fig. 9. Change in critical temperature of irradiated wires against magnetic field. Bronze route Nb_3Sn wire, ITER TF #2.

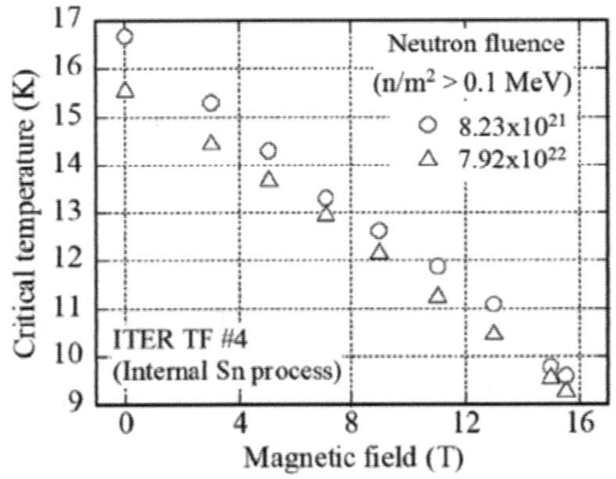

Fig. 10. Change in critical temperature of irradiated wires against magnetic field. Internal tin processed Nb_3Sn wire, ITER TF #4.

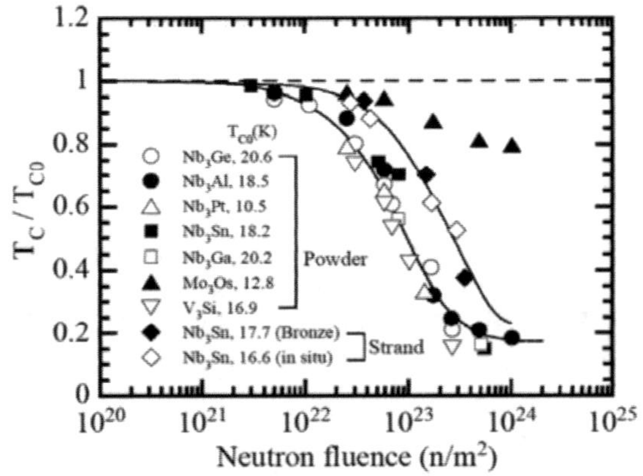

Fig. 11. Change in T_C/T_{C0} against neutron fluence of over 0.1 MeV. Published data are plotted [6,7].

temperature measurements are continuing, and the quantitative discussion would be expected after the further investigations.

In 1990s, Kuroda et. al. reported the change in the critical temperature of the irradiated Nb_3Sn strands at self-field [6]. In 1970s, A. R. Sweedler et. al. published the irradiation effect of seven kinds of A15 type compound powders on the critical temperatures [7]. These results are shown in Fig. 11. The horizontal axis is the neutron fluence of over 0.1 MeV. The vertical axis shows the ratio of the critical temperature after irradiation (T_C) to the critical temperature of non-irradiated sample (T_{C0}). The data of the Nb_3Sn strands are plotted with open and solid diamond symbols.

There is a trend that the A15 powders show harder degradation than the Nb_3Sn strands. The data of the strands show that the T_C/T_{C0} goes down within several percentages after 10^{22} n/m² irradiation. The data shown in Fig. 8 to Fig. 10 suggest that the T_C/T_{C0} obtained in this study are less degraded than the published results, especially in the higher magnetic field. But it must be pointed out that the critical temperature decreases by about 10 % at about 5 x 10^{22} n/m² neutron fluence where the critical current shows the peak of I_C/I_{C0} value. And at about 2 x 10^{23} n/m² neutron fluence where the I_C/I_{C0} becomes small and B_{C2} would be expected to be degraded significantly, the T_C/T_{C0} value would become down by about 40 %.

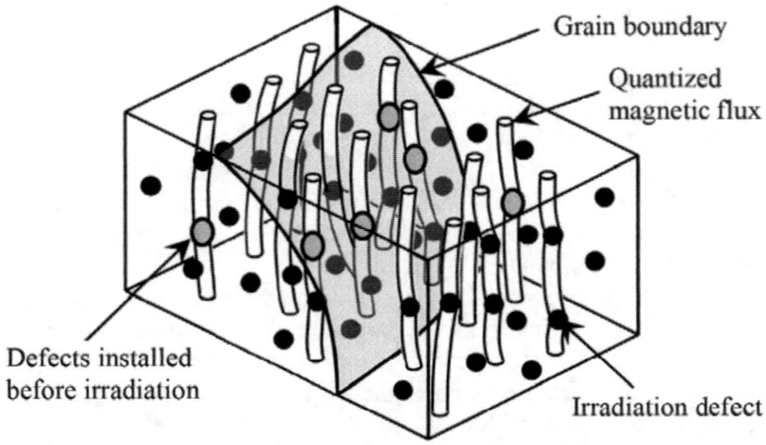

Fig.12 Image of irradiation effect on magnetic flux pinning. Irradiation defects strengthen pinning force but break crystalline of Nb_3Sn structure at the same time.

From the above experimental results, the mechanisms of the neutron irradiation effect on the superconductivity of the Nb_3Sn wire would be considered as follows. Figure 12 shows an image of the magnetic flux pinning in the Nb_3Sn matrix. After the heat treatment to produce Nb_3Sn crystals, some inclusions or precipitates would remain on the grain boundary and in the matrix. When the Nb_3Sn filaments are cooled down to below the critical temperature, the magnetic field becomes quantized flux and pinned by the defects installed before the irradiation such as the inclusions. The pinning force is stronger when the filaments are cooled down to the lower. During the neutron irradiation, the fast neutrons hit the atoms and make cascades and voids, so-called irradiation defects. Although the size of these defects is very small comparing with the inclusions, these defects strengthen the pinning force, and the I_C/I_{C0} would increase. The B_{C2} would be expected to be increased at the same time. But the increment of the I_C/I_{C0} would stop and begin the degradation as the neutron fluence increases. When the fluence becomes above ca 2×10^{22} n/m^2, the I_C/I_{C0} begins to decrease and comes near to zero at the higher magnetic field because the crystalline of A15 structure would be destroyed, which means that B_{C2} drops drastically.

Regarding the critical temperature, it does not increase like the critical current. It decreases monotonically even when the I_C/I_{C0} increases. The degradation degree of the critical temperature is considered not so large at around 2×10^{22} n/m^2. It should be noticed that the irradiation damage would make the T_C lower. The details will be investigated in further experiments.

4. Summary

The ITER project has been carried out from 2007 and the fusion DEMO to generate the D-T neutrons continuously is under consideration. Although the first plasma of the ITER would be delayed and it is not clarified when it will be completed at present, the basic and the practical investigations and R&Ds on the fusion DEMO design activities have been performed steadily and continuously. Since the fusion DEMO cannot avoid the neutron irradiation, it is necessary to investigate the irradiation effect on the superconductivity of the candidate superconducting wires. The post-irradiated test facilities were installed in the radiation control area in Tohoku University and the hard experiments have been conducted. In this report, the critical current and the critical temperature of the bronze route and the internal tin processed Nb_3Sn wires were focused, and the mechanisms of the irradiation effect was discussed based on the magnetic flux pinning concept.

The main results are summarized as follows:

On the diagram between the I_C/I_{C0} and the neutron fluence, the I_C/I_{C0} shows the peak at around 2×10^{22} n/m^2, and the I_C/I_{C0} decreases as the fluence increases. At about 2×10^{23} n/m^2 of the irradiation fluence, the I_C/I_{C0} comes near to zero at 15.5 T, and it means that the B_{C2} drops significantly by the neutron irradiation.

As to the irradiation effect on the T_C/T_{C0}, it decreases monotonically as the fluence increases. The degree of the T_C/T_{C0} degradation is within several percentages at around 2 x 10^{22} n/m^2 of the fluence, and it is expected that the decrease of the T_C/T_{C0} would be about 40 % at about 2 x 10^{23} n/m^2 under the self-field. At 15.5 T, the degradation of the T_C/T_{C0} would become smaller than that at zero outer field. Although the current sharing temperature (T_{CS}) of the irradiated wires has not been evaluated at present, the T_{CS} margin must be considered taking account of the irradiation effect when the superconducting magnets for the fusion DEMO are designed.

The pinning force would be strengthened by the irradiation defects, though the size of those defects would be very small comparing with the inclusions. At the same time, the crystalline of Nb_3Sn structure would be broken by the irradiation resulting in the drastic drop of B_{C2}.

There must be still a lot of unclarified parameters in the neutron irradiation research for the superconducting magnets of the fusion DEMO. Many effects such as the mechanical strain during the irradiation and the dose rate on the superconductivity have not been clarified yet. The further investigations are continued, and more detailed discussion will be made in near future.

5. References

[1] ITER web site. https://www.iter.org/
[2] A. Nishimura, et. al., J. Nucle. Mat. 417 842 (2011).
[3] A. Nishimura, et. al., Supercon. Sci. Technol. 32 (2019) 024004. https://doi.org/10.1088/1361-6668 /aaf903.
[4] A, Nishimura, Mat. Sci. Eng. 756 (2020) 012013. https://doi.org/10.1088/1757-899X/756/1/ 012013.
[5] A. Nishimura, et. al., AIP Conf. Proc. 1435, 217 (2012). https://doi.org/10.1063/1.4712099
[6] T. Kuroda, et. al., J. of Atomic Energy Society Japan. 37 652 (1995) (in Japanese).
[7] A. R. Sweedler,et. al., J. of Nucle. Mat. 72 50 (1978).

Acknowledgments
The authors would like to express their sincere thanks to Prof. Tatsuo Shikama at Hachinohe Institute of Technology, Mr. Masanori Yamazaki at Tohoku University and Prof. Laifeng Li at Technical Institute of Physics and Chemistry for their encouragement and contributions.

This work was supported by the NIFS Collaborative Research Program (NIFS19KECF029) and Grant-in-Aid for Scientific Research (19K03804). Also, it was supported by Chinese Academy of Sciences President's International Fellowship Initiative, Grant No. 2019VEA0017.

This work was performed at the International Research Centre for Nuclear Materials Science in the Institute for Materials Research, Tohoku University.

IOP Conf. Series: Materials Science and Engineering 1241 (2022) 012019 doi:10.1088/1757-899X/1241/1/012019

Current sharing and stability in an extremely low AC-loss MgB_2 conductor

C J Kovacs[1], T J Haugan[2], M D Sumption[3], M Tomsic[1], M Rindfleisch[1]

[1] Hyper Tech Research Inc., 539 Industrial Mile Road, Columbus, OH, USA
[2] AFRL-RQQM, Wright-Patterson AFB, Dayton, OH, USA
[3] Department of Materials Science and Engineering, The Ohio State University, Columbus, OH, USA

ckovacs@hypertechresearch.com, timothy.haugan@us.af.mil, sumption.3@osu.edu, mtomsic@hypertechresearch.com, mrindfleisch@hypertechresearch.com

Abstract. In the push to develop high power electric aircraft, superconducting technology promises to significantly reduce mass and volume of motors and generators. However, challenges related to AC-loss and thermal management are a significant factor in preventing the proliferation of aerospace superconducting technologies. Increasing the resistance of the metal matrix stabilization has only gone so far in reducing coupling currents for higher frequency applications. In this research, Multiphysics simulations of a single composite filament were used to investigate stability decreases when using very high thermal conductivity electrical insulator (CsI) or metal-to-insulator transitioning material (V_2O_3) to replace the slightly resistive metal matrix typically used for a low AC loss MgB_2 composite wire. The insulators separate the MgB_2 filaments entirely, only allowing transient current sharing to occur with the high purity Nb diffusion barrier or with the metallic state V_2O_3. These simulations show that for these very low AC-loss composites at 20 K, instability will become a major issue due to reductions in current sharing. With higher electrical conductivity metal-to-insulator materials, higher thermal conductivity impregnation materials, and thicker metallic diffusion barriers it may be possible to find a reasonable balance between AC-loss and stability.

1. Introduction

Single aisle and twin aisle large commercial aircraft presently contribute approximately 1.9 to 2.3% of global CO_2 emissions [1]. While this appears to be a small portion, decreasing global greenhouse gas emissions will require a many pronged approach. Of the many initiatives and programs to develop remarkably high efficiency commercial aircraft by 2035 [2–5], a later generation NASA N+3 vehicle will require roughly 45 MW of electrical power [6]. At powers this high, nearly every electrical component will require significant breakthroughs in light weighting and performance [7]. There is substantial focus on increasing the efficiency and performance of motors and generators, and some design teams have proposed the use of cryogenic Hydrogen cooling [8] and superconducting composites in armature windings [9]. While it is true superconducting composites generate zero DC losses, under AC conditions there are still eddy current, magnetization, and coupling loss components [10]. Analytical descriptions exist, shown in equations 1 and 2, which describe loss components of a filamentary metal-superconductor composite wire under a DC transport current and alternating magnetic field.

$$\widetilde{P_{mag}}=f\left(\frac{8}{3\pi}\right)B_0\widetilde{J_c}d_{eff} \qquad [W/m^3] \qquad (1)$$

Where "$\widetilde{P_{mag}}$" is magnetization loss averaged over a field cycle, "J_c" is the critical current density of the superconductor, "f" is frequency, "B_0" is the alternating field amplitude, and "d_{eff}" is an effective filament diameter.

$$P_{coup\text{-}eddy}=\left(\frac{r_r}{r_o}\right)^2\left(\frac{dB}{dt}\right)^2\left(\frac{L_p}{2\pi}\right)^2\left(\frac{1}{\rho_{ms}}\frac{r_o^2-r_r^2}{r_o^2+r_r^2}+\frac{1}{\rho_{tf}}\frac{r_r^2-r_c^2}{r_r^2}+\frac{1}{\rho_{mc}}\frac{r_c^2}{r_r^2}\right)+\frac{1}{4\rho_{ms}}\left(\frac{dB}{dt}\right)^2\left(\frac{r_o^4-r_r^4}{r_o^2}\right) \qquad [W/m^3] \qquad (2)$$

Where "$P_{coup\text{-}eddy}$" is the sum of eddy and coupling loss, "dB/dt" is equal to $4B_0f$, "r_{xxx}" refers to different radii within the composite (r = outer filament region, c = filament free core, o = strand), "ρ_{xxx}" refers to different resistivities within composite (ms = outer sheath, mc = core, tf = interfilamentary region, and "L_p" is the twist pitch of superconducting filaments. From equations 1 and 2 it is clear that minimizing filament size and increasing the metal matrix resistivities are ways to reduce AC-loss.

One of the promising superconducting composites for armature windings is metal-MgB_2. MgB_2 superconducting composites can be fabricated as an isotropic wire conductor with very fine filaments (~20 μm) embedded in a somewhat resistive metal matrix [11]. Doped high-J_c MgB_2 composites generally have a transition temperature, T_c, near 36 K which allows operation in LH_2 with a 16 K temperature window [12]. MgB_2 composites are cheaper than similar current carrying REBCO high temperature superconducting coated conductors and BSCCO composite wire at low fields. While attempts have been made to reduce AC-loss of MgB_2 conductor, the AC-loss remains too large for higher frequency (200 Hz and up) proposed next generation aerospace motors [13]. Because higher frequency operation is generally desired to meet the aggressive power density targets, new steps have to be taken to reduce AC-Loss, and particularly coupling losses which become dominant at higher frequencies.

Interfilamentary metal could in theory be removed, greatly reducing higher frequency losses, but interfilamentary metal provides an important role in "stabilizing" the superconducting composite from thermal perturbations, such as magnetothermal instabilities [14]. Without metal stabilizer and the ability to current share with neighbouring superconducting filaments, when an isolated superconductor filament transitions to the normal state, there will likely be irreversible damage due to excess Joule heating. The metal stabilizer also conducts heat away from thermal perturbations so the superconducting filaments can rapidly thermally recover back to the normal state before critical instability limits are reached. Metal stabilizer directly increases minimum quench energy "MQE", i.e. the thermal energy required to initiate normal zone propagation ("NZP") which continue to expand and do not shrink and recover. Normal zone propagation velocity "NZPV" is the growth rate of irrecoverable normal zones along a conductor once instability has been reached.

The purpose of the simulations in this paper is to investigate the possibility of total electrical separation of superconducting filaments, greatly reducing high frequency AC-loss, but maintaining enough stability to be operable. One investigated method was using a high thermal conductivity impregnatable ceramic, CsI, to isolate MgB_2 filaments which individually have minimal metal stabilization via the Nb barrier. If the electrically isolating matrix has high enough thermal conductivity, it may be possible to have only a moderate reduction in MQE even without filament-to-filament current sharing. From a realistic perspective, depending on the composite construction, it is possible to chemically etch the stabilizer of a metal-MgB_2 composite and leave only Nb sheathed separated filaments. CsI melt impregnation can occur near 630 °C, i.e. safe temperatures for Nb_3Sn, MgB_2, and BSCCO conductors. The use of CsI in stabilizing superconducting wires and magnets has previously been proposed, due to its high thermal conductivity and dielectric strength [15–17].

The other investigated method to isolate filaments is using a thin sheath of metal-to-insulator transitioning material, V_2O_3 which behaves as electrical isolation at low temperatures but allows current sharing at higher temperatures experienced during normal zone growth [18]. For both of these approaches to remove coupling losses, MQE will definitely be reduced. The question is whether the reductions in stability are acceptable for the enabling of higher frequency operation.

2. Details of simulations

A single MgB_2 filament, Nb diffusion barrier, and surrounding CsI, V_2O_3, or Cu was first created using Onshape® cloud-based computer-aided design software. The components were then imported and created into a three-dimensional assembly in COMSOL Multiphysics® version 5.3.0.223 with 64-bit Windows 10. The "Heat Transfer" and "AC/DC" modules were utilized along with the "Electromagnetic Heating" and "Temperature Coupling" interfaces [19–20]. All COMSOL simulations were transient solutions with start time (t_0) of 975 μs, end time (t_f) of 2.7 ms, with a continuously increasing step size described as the equation below.

$$t_{step}=2^{(-10 \text{ stepping by } 0.05 \text{ to } -8.5)} \quad [s] \qquad (3)$$

An actual low-loss MgB_2 conductor, an example mesh for V_2O_3, and the geometry of the three situations examined are shown in figure 1 below from left to right respectively. The single filament composite was 1 mm long, the MgB_2 filament diameter was 20 μm, the Nb diffusion barrier was 7.5 μm thick, and the outer diameter was 55 μm in all cases.

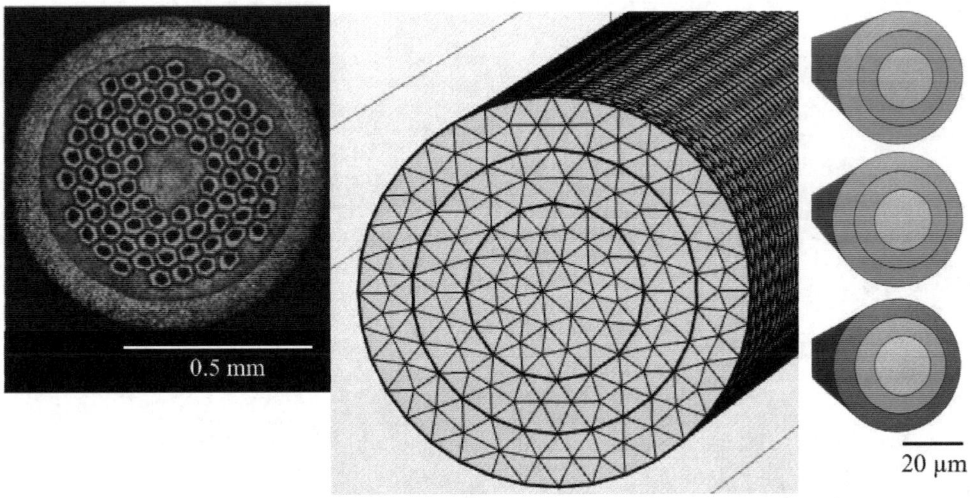

Figure 1. (Left) a low AC-Loss MgB_2 composite wire from Hyper Tech Research [11], (middle) the mesh refinement chosen for the (Cu or CsI or V_2O_3)-Nb-MgB_2 simulation, and (right) the different geometries simulated. Green, gray, orange, pink, and blue refer to MgB_2, Nb, Cu, CsI, and V_2O_3 respectively.

The material properties required for the Multiphysics simulations were heat capacity "C_p(T)", thermal conductivity "κ(T)", electrical conductivity "σ(T)", and mass density. The sources for the temperature dependent properties of the five materials used (polycrystalline MgB_2, Nb, Cu, V_2O_3, CsI) are shown in table 1. The material properties were assumed to be magnetic field independent. Data points were taken from literature plots using Graph Grabber v2.0, and were input into COMSOL Multiphysics. For continuous material property data from 20 K to elevated temperatures, a linear interpolation was used between a fine data point collection and a constant extrapolation equal to the highest temperature data point was used. The data utilized from literature went up to at least 300 K. The highly non-linear electrical conductivity data for V_2O_3 was simulated directly from literature data. The highly non-linear electrical conductivity data for MgB_2 was simulated by overlapping a continuous function onto a transition region curvature from literature. Below 36 K, MgB_2 was assumed to have a conductivity described by the inverse of the continuous equation 4 which matched well with literature data. Above 36 K, MgB_2 was assumed to have a constant conductivity equal to the inverse of equation 4 with T equal to 36 K.

$$\rho_{MgB_2} = \frac{E_0+E_0\left(\frac{T-T_b}{T_{cs}-T_b}\right)^n}{J_0} \quad [\Omega m] \tag{4}$$

In the equation above, $E_0 = 1 \times 10^{-5}$ V/m, $T_b = 20$ K, $T_{cs} = 25$ K, n = 20, and J_0 was the starting current density in the MgB_2 filament for that simulation in A/m^2.

Table 1. Literature sources for heat capacity, thermal conductivity, and electrical conductivity used for the Multiphysics simulations of a polycrystalline MgB_2 composite filament. Material properties which have multiple sources were needed to span as much of the temperature range of interest as possible, 20 K to greater than 300 K. This data is presented in graphs in the appendix.

	C_p(T)	κ(T)	σ(T)
Poly-MgB₂	Wang 2001 [21]	Bauer 2018 [22]	Kulich 2013 [23]
Nb	Chou 1958 [24] Arblaster 2017 [25]	Powell 1954 [26]	Hall 1968 [27]
Cu	Simon 1992 [28]	Simon 1992 [28]	Simon 1992 [28]
CsI	Taylor 1963 [29]	Gerlich 1982 [30] Lawless 1984 [17] Wei 2019 [31]	"Perfect Insulator"
V_2O_3	Keer 1976 [32]	Andreev 1978 [33]	Morin 1959 [34]

Starting at 1 ms into a transient simulation, a 0.5 W, 20 μm diameter, 100 μs spherical thermal perturbation was initiated at the geometric center of the composite filament, within the MgB_2. This was performed to simulate a magnetothermal instability. A fixed temperature of 20 K was on the outer surface of the filamentary composite and current terminals, located at the ends of the geometry, injected current into the MgB_2 cores. Transient simulations were performed with fixed thermal perturbation pulse parameters and different current densities until generated normal-zones no longer recovered and continued to grow towards the ends of the geometry, i.e. at MQE for the given current density. Temperature probes were located near the thermal perturbation, and because of the lengthy time of simulations, any simulation which had internal temperatures (after the thermal perturbation) greater than 1500 K were deemed unrealistic and terminated. Early terminations had no impact in determining MQE or NZPV.

3. Results from simulations

For the first series of simulations, an outer sheath of Cu with a resistivity ratio ($RR_{273/20K}$) of 10 was used to simulate an isolated Cu-Nb-MgB_2 filament with the interfilamentary region of a low AC-loss MgB_2

composite. With the thermal perturbation parameters and thermal boundaries chosen, it was not possible to generate enough Ohmic losses to nucleate a critical sized normal zone even with current densities in the MgB_2 filament reaching ~1.8 x 10^5 A/mm^2 or 90 A in the single filamentary composite. Experiments with more aggressive thermal perturbation parameters still resulted in recovery, even when peak normal zone temperatures reached 1500 K. Because the baseline situation resulted in excess cooling and an accompanying unrealistic, almost unquenchable stability condition, it was deemed that these simulations should be used qualitatively instead of quantitatively in describing changes in stability between different composite filaments.

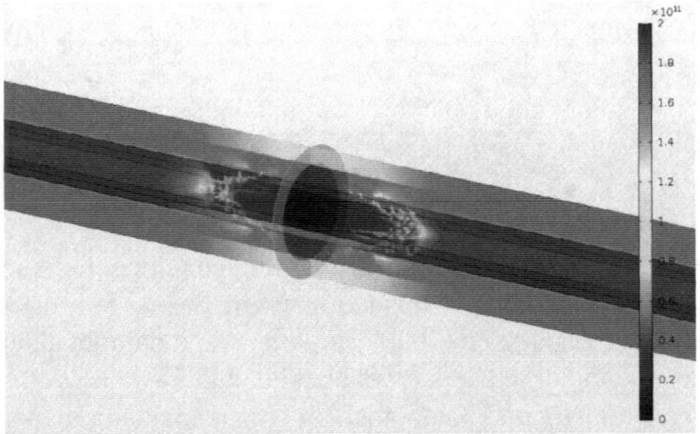

Figure 2. Current density (scale in A/m^2) around recoverable normal zone for composite with a 10RR Cu outer sheath around the Nb barrier, and with a 0.5 W, 20 μm diameter, 100 μs spherical thermal perturbation.

The CsI-Nb-MgB_2 composite filament was simulated next. After generating multiple recoverable normal zones, it was possible to find the approximate current density in which (for the fixed thermal perturbation parameters listed previously) unrecoverable NZP started to occur. Unrecoverable NZP started to occur at ~3 x 10^4 A/mm^2 in the MgB_2 or 15 A in the filamentary CsI-Nb-MgB_2 composite. Figure 3 shows the temperature gradients and current sharing during the thermal perturbation for the 12.5 A recoverable normal zone situation. Figure 4 shows the temperature gradients and current sharing that occurs after the thermal perturbation for the 15 A propagating normal zone situation. From the 15 A results, the NZPV was ~70 cm/s, which is on par with strand level NZPV experimental data [12]. The peak temperatures of the generated normal zone, close to 1500 K, are above the melting point of CsI and Cu, and again support the opinion the results of this study examining stability should be kept qualitative and not quantitative.

Lastly, the V_2O_3-Nb-MgB_2 composite filament was simulated. Again, the switch from recoverable to unrecoverable NZP was monitored and images of the results are shown in figure 5. The V_2O_3-Nb-MgB_2 composite became unstable to the thermal perturbation at 12.5 A and had a NZPV of ~50 cm/s. The peak temperatures of the generated normal zone were less than 400 K.

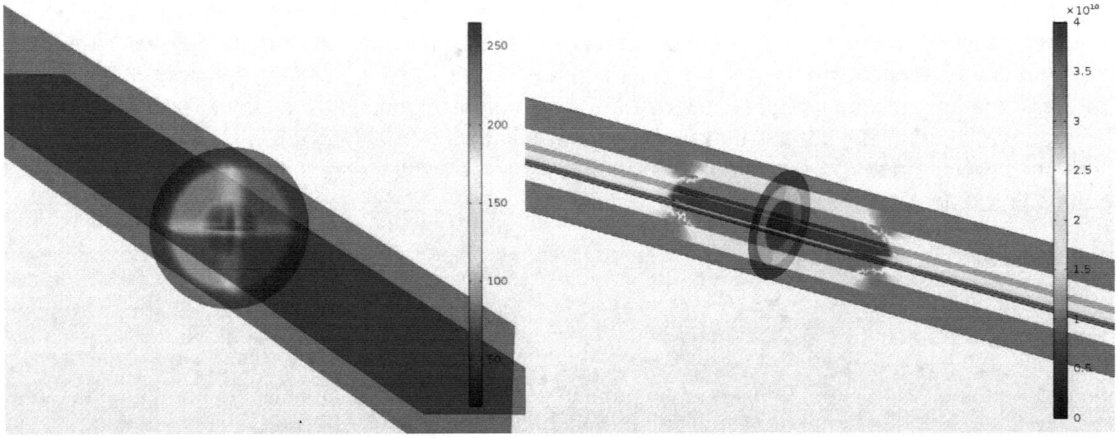

Figure 3. (Left) Temperature (scale in Kelvin) around recoverable normal zone for composite with a CsI high thermal conductivity, electrically insulating outer sheath. (Right) Current sharing (scale in A/m^2) around a recoverable normal zone (~0.15 mm long) for a composite with a CsI outer sheath, it is clear the Nb barrier performs some role in stabilization and CsI does not carry any current.

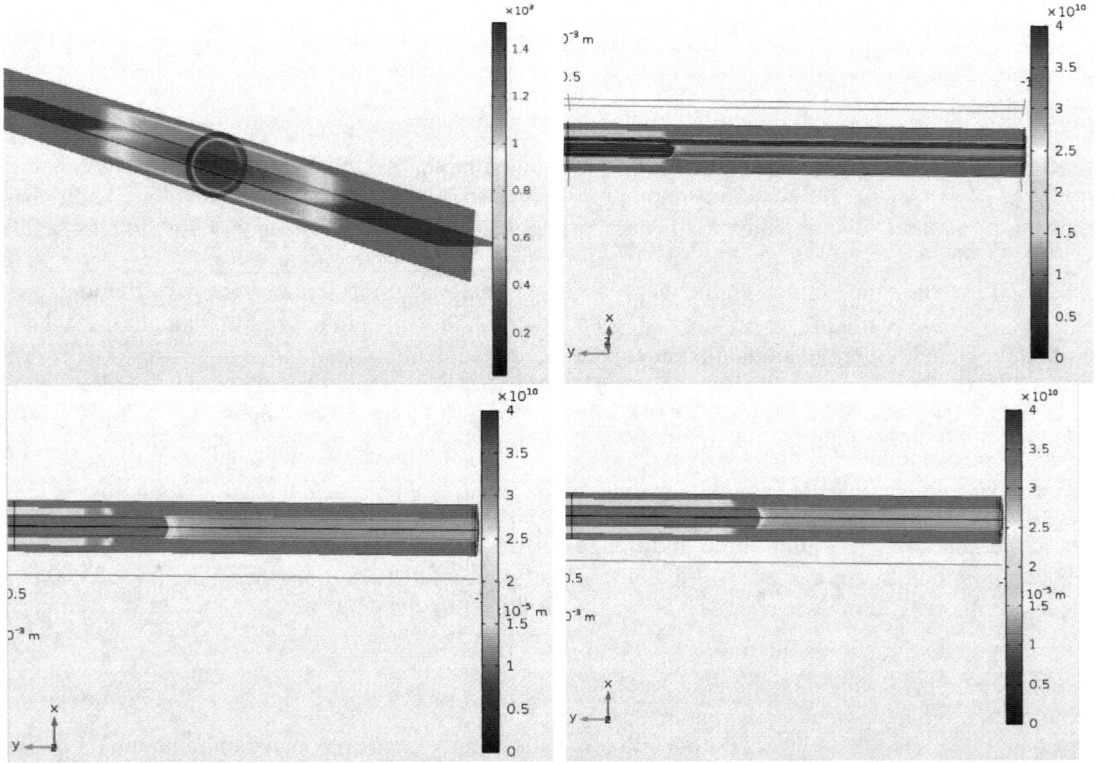

Figure 4. (Top left) Temperature (scale in Kelvin) around the propagating normal zone for the composite with a CsI high thermal conductivity, electrically insulating outer sheath. (Top right, bottom left, and bottom right) current sharing, scale in A/m^2, around a propagating normal zone at 3 different times 122 μs, 202 μs, and 334 μs after the beginning of the heat pulse. The NZPV was determined to be ~70 cm/s.

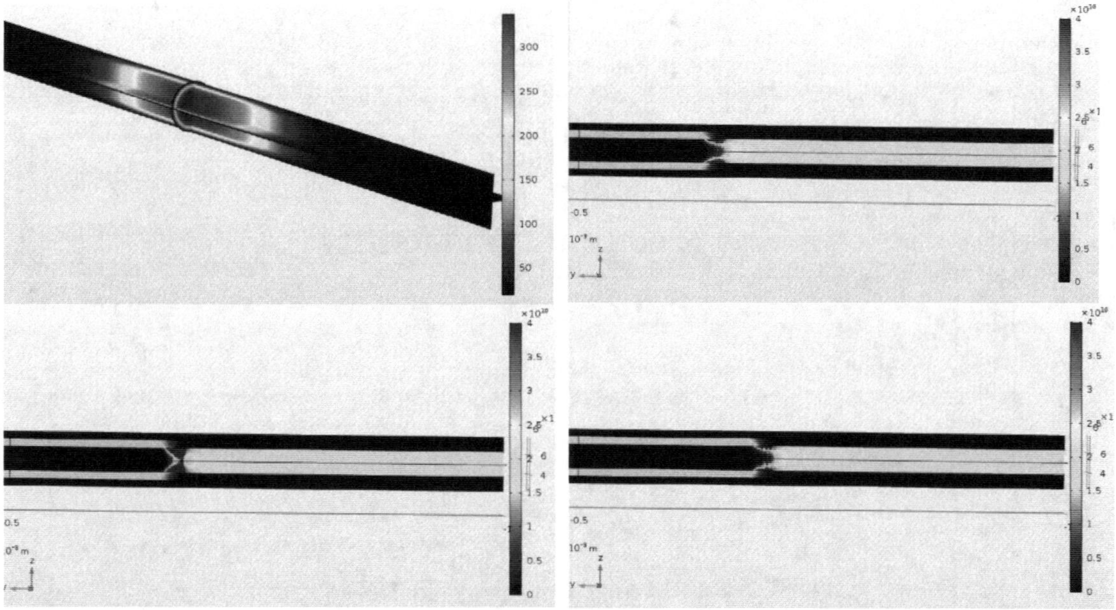

Figure 5. (Top left) Temperature (scale in Kelvin) around the propagating normal zone for the composite with a V_2O_3 metal-to-insulating transitioning outer sheath. (Top right, bottom left, and bottom right) current sharing, scale in A/m^2, around a propagating normal zone at 3 different times 122 μs, 202 μs, and 334 μs after the beginning of the heat pulse. The NZPV was determined to be ~50 cm/s.

4. Discussion of results

Replacing the metal matrix of simulated filaments into a material with poorer thermal and electrical conductivity greatly reduced stability as measured by relative ease of a fixed thermal perturbation creating an irrecoverable normal zone at different current densities. The importance of current sharing for normal state recovery was clearly highlighted. However, the stability of CsI and V_2O_3 composites may not be as poor at lower temperatures where CsI becomes much more thermally conductive and with a metal-to-insulator transition material with a higher conductivity metallic state respectively. In the case of the V_2O_3 composite simulation, it would be ideal to only have a thin (not thick) layer of V_2O_3 sandwiched between the Nb barrier and a Cu outer sheath. These possibilities are the motivation for future investigations.

By separation of filaments with CsI or insulating V_2O_3, equation 2 would change to the eddy current losses of something more familiar, that of a Litz wire. It may be application dependent whether the reduction in stability is acceptable when continuous thermal loads from AC-loss are rampant.

5. Conclusion

The demand to have higher operational frequencies for superconducting composites has called for novel methods because composites with resistive metal matrices are still too lossy, driven strongly by coupling losses at high frequencies. This computational investigation shows while it may be possible to remove coupling currents, the stability from transient thermal disturbances will be greatly reduced. At a simulated temperature of 20 K, a CsI high thermal conductivity, electrically insulating matrix may outperform a V_2O_3 metal-to-insulator transitioning matrix. Future simulations will investigate the performance of these novel matrices under more idealized conditions and more optimized geometry.

6. Appendix

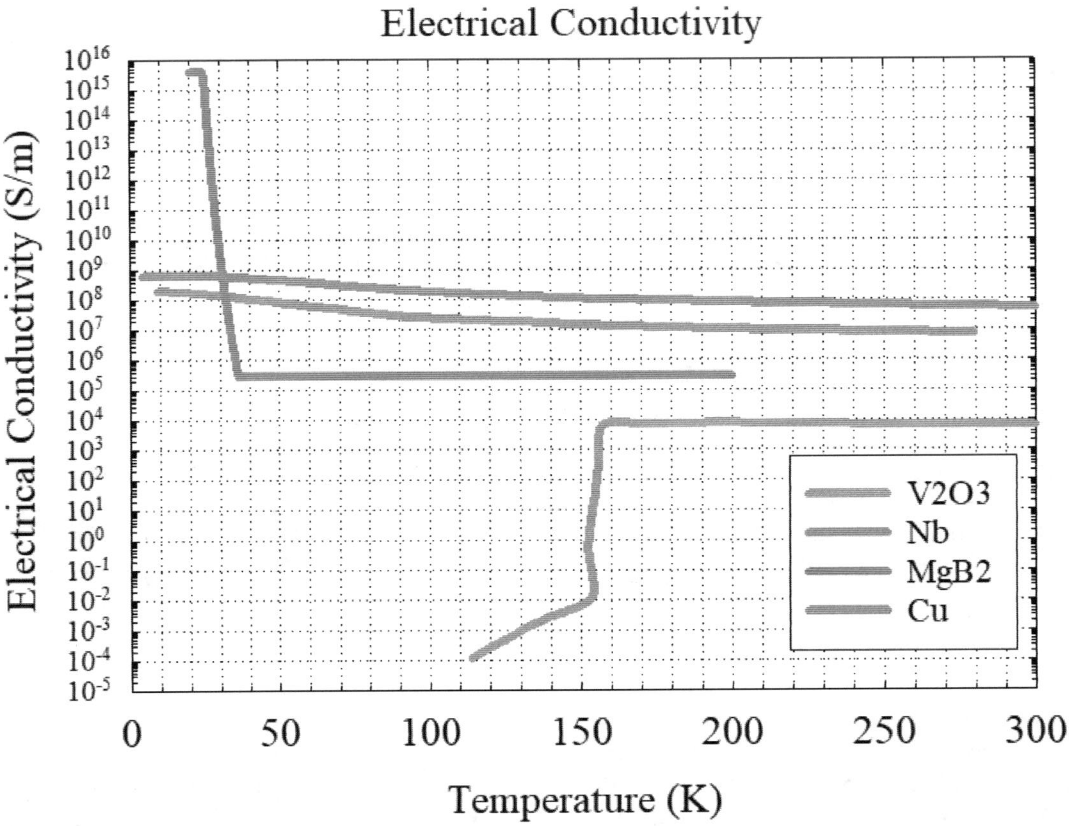

Figure 6. Electrical conductivity data for the materials used. CsI was assumed to be a perfect insulator [23,27,28,34].

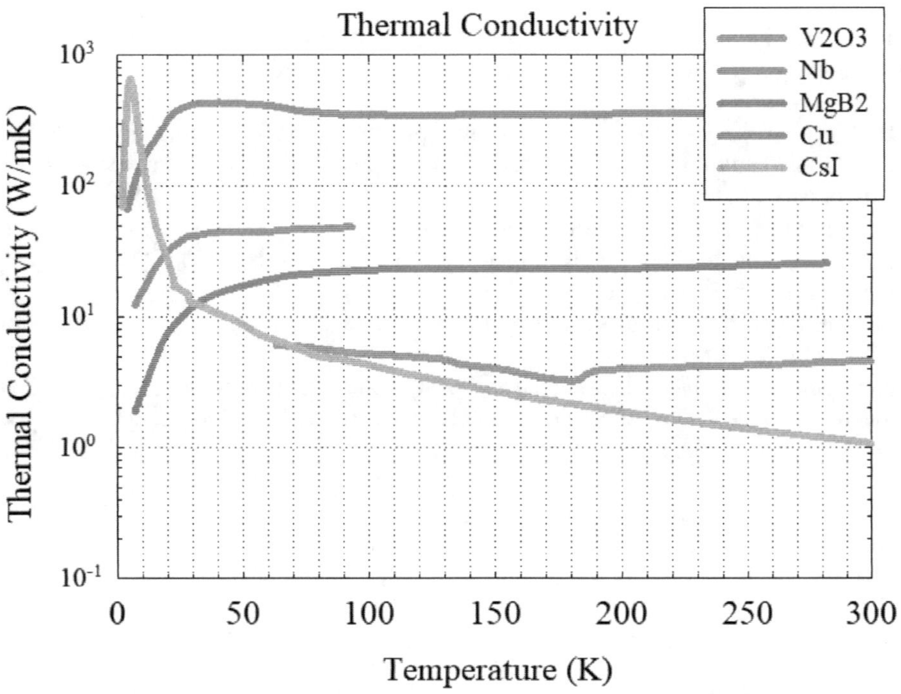

Figure 7. Thermal conductivity data for the materials used [17,22,26,28,30,31,33].

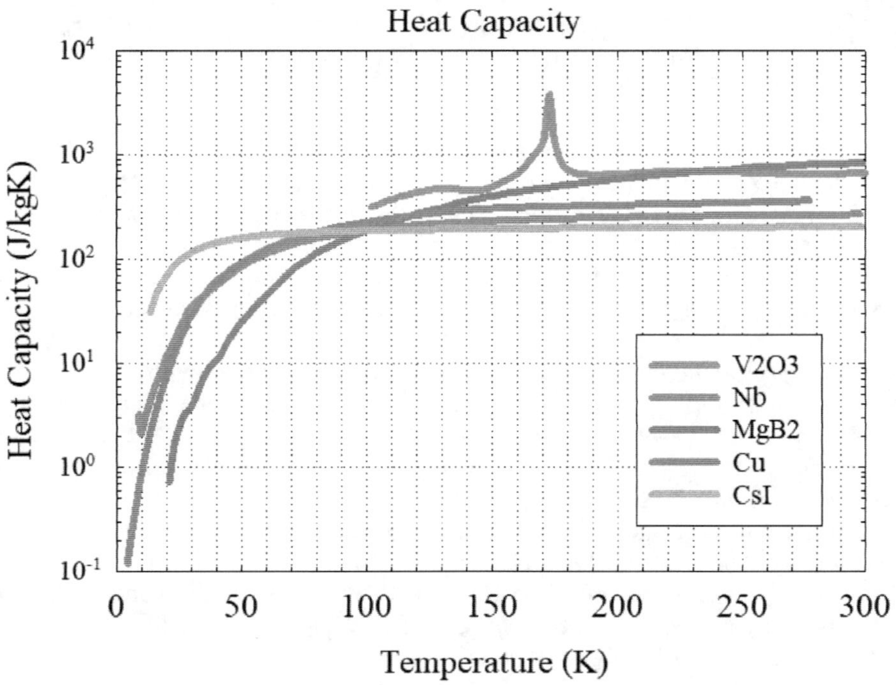

Figure 8. Heat capacity data for the materials used [21,24,25,28,29,32].

7. References

[1] Yutko B M 2011 *Approaches to Representing Aircraft Fuel Efficiency Performance for the Purpose of a Commercial Aircraft Certification Standard* (Cambridge: Massachusetts Institute of Technology) p 44

[2] Sept 21 2020 Airbus reveals new zero-emission concept aircraft *Airbus: Press Release* <https://www.airbus.com/newsroom/press-releases/en/2020/09/airbus-reveals-new-zeroemission-concept-aircraft.html.

[3] Scheelhaase J, Maertens S, Grimme W and Jung M 2018 EU ETS versus CORSIA – a critical assessment of two approaches to limit air transport's CO_2 emissions by market-based measures *J. Air Transp. Manag.* **67** 55–62

[4] 2020 FutPrInt50 future propulsion and integration: towards a hybrid-electric 50-seat regional aircraft *EASN Press Release* <https://www.easn.net/files/news/Press_Release_v02_FUTPRINT50.pdf>

[5] 2015 NASA Aeronautics advanced air transport technology (AATT) *Applied Modelling and Simulation (AMS) Seminar Series* <https://www.nas.nasa.gov/assets/pdf/ams/2015/AMS_20150521_DelRosario.pdf>

[6] Haran K S et al. 2017 High power density superconducting rotating machines – development status and technology roadmap *Supercond. Sci. Technol.* **30** 123002

[7] National Academies of Sciences, Engineering, and Medicine 2016 *Commercial Aircraft Propulsion and Energy Systems Research: Reducing Global Carbon Emissions* (Washington D. C.: The National Academies Press)

[8] Podlaski M et al. 2018 Multi-domain modelling for high temperature superconducting components for the CHEETA hybrid propulsion system *2021 AIAA Propulsion and Energy Forum (Virtual Event)* <https://doi-org.wrs.idm.oclc.org/10.2514/6.2021-3302>

[9] Kalsi S, Hamilton K A and Badcock R A 2018 Superconducting rotating machines for aerospace applications *2018 AIAA Propulsion and Energy Forum (Cincinnati)* <https://doi-org.wrs.idm.oclc.org/10.2514/6.2018-4796>

[10] Sumption M D, Wan F, Rindfleisch M and Tomsic M 2019 AC loss of superconducting materials – refined loss estimates of MgB_2 wires for superconducting motors and generators *2018 AIAA Propulsion and Energy Forum (Indianapolis)* <https://doi-org.wrs.idm.oclc.org/10.2514/6.2019-4495>

[11] Haran K S et al. 2017 High power density superconducting rotating machines – development status and technology roadmap *Supercond. Sci. Technol.* **30** 123002

[12] van Weeren H 2007 *Magnesium Diboride Superconductors for Magnet Applications* (Enschede: PrintPartners Ipskamp)

[13] Sumption M D, Murphy J, Susner M and Haugan T 2020 Performance metrics of electrical conductors for aerospace cryogenic motors, generators, and transmission cables *Cryogenics* **111** 103171

[14] Wilson M N 2002 *Superconducting Magnets* (New York: Oxford University Press)

[15] Wilson M N 1991 Stabilization, protection and current density: some general observations and speculations* *Cryogenics* **31** 499–503

[16] Lawless W N, Clark C F and Arenz R W 1983 Enthalpy improved dielectric insulation for NbTi *IEEE Trans. on Magnetics (MAG-19)* **3** 432–434

[17] Lawless W N 1984 Specific heat and thermal conductivity measurements on cesium and thallous halide crystals at low temperatures *Phys. Rev. B* **30** 10 6057–6066

[18] Hyeon C J et al. 2018 Quench behavior of 2G HTS coils with polyimide film and MIT material under over pulse current *IEEE Trans. on Appl. Supercond.* **28** 4 4603906

[19] COMSOL 5.3 2017 *AC-DC Module User's Guide*

[20] COMSOL 5.3 2017 *Heat Transfer Module User's Guide*

IOP Conf. Series: Materials Science and Engineering 1241 (2022) 012019 doi:10.1088/1757-899X/1241/1/012019

[21] Wang Y, Plackowski T and Junod A. 2001 Specific heat in the superconducting and normal state (2-300 K, 0-16 T), and magnetic susceptibility of the 38 K superconductor MgB_2: evidence for a multicomponent gap *Physica C: Superconductivity* **355** 3–4 179–193
[22] Bauer E et al. 2001 Thermal conductivity of superconducting MgB_2 *J. Phys. Condens. Matter* **13** 22 L487–L493
[23] Kulich M, Flukiger R L, Senatore C, Tropeano M and Piccardo R 2013 Effect of cold high pressure deformation on the properties of ex situ MgB_2 wires *Supercond. Sci. Technol.* **26** 105019
[24] Chou C, White D and Johnston H L 1958 Heat capacity in the normal and superconducting states and critical field of niobium *Physical Review* **109** 3 788–796
[25] Arblaster J W 2017 The thermodynamic properties of niobium *J. Phase. Equilib. Diffus.* **38** 707–722
[26] Powell R L and Blanpied W A 1954 *Thermal Conductivity of Metals and Alloys at Low Temperatures: A Review of the Literature* (Washington D. C.: U. S. Government Printing Office)
[27] Hall L A 1968 *Survey of electrical resistivity measurements on 16 pure metals in the temperature range 0 to 273 K* (Washington D. C.: U. S. Government Printing Office)
[28] Simon N J, Drexler E S and Reed R P 1992 *NIST Monograph 177: Properties of Copper and Copper Alloys at Cryogenic Temperatures* (Washington D. C.: U. S. Government Printing Office)
[29] Taylor Jr. A R, Estelle Gardner T and Smith D F 1963 Thermodynamic properties of cesium chloride and cesium iodide from 0 to 300 K *United States Department of the Interior, Bureau of Mines*
[30] Gerlich D and Andersson P 1982 Temperature and pressure effects on the thermal conductivity and heat capacity of CsCl, CsBr, and CsI *J. Phys. C: Solid State Phys.* **15** 5211–5222
[31] Wei B et al. 2019 Low-temperature anharmonicity and the thermal conductivity of cesium iodide *Phys. Rev. B* **99** 184301
[32] Keer H V et al. 1976 Heat capacity of pure and doped V_2O_3 single crystals *J. Solid State Chem.* **19** 95–102
[33] Andreev V N, Chudnovskii F A, Petrov A V and Terukov E I 1978 Thermal conductivity of VO_2, V_3O_5, and V_2O_3 *Phys. Stat. Sol. (a)* **48** 2 K153–K156 <https://doi.org/10.1002/pssa.2210480257>
[34] Morin F J 1959 Oxides which show a metal-to-insulator transition at the Neel temperature *Phys. Rev. Lett.* **3** 1 34–36

Acknowledgements
This work was supported by AFRL/AFOSR LRIR #18RQCOR100, NASA University Learning Initiative (ULI) #80NSSC19M0125, and the AFRL/RQ Aerospace Systems Directorate.

IOP Conf. Series: Materials Science and Engineering 1241 (2022) 012020 doi:10.1088/1757-899X/1241/1/012020

Development of multi-filament textured-powder Bi-2212/Ag wire with enhanced local area ratio

G D May[1,2], C Coats[1], T Elliott[1], P M McIntyre[1,2], C T Ratcliff[1], J S Rogers[1,2], and M Besser[3]

[1]Dept. of Physics & Astronomy, Texas A&M University, USA
[2]Accelerator Technology Corp., College Station, TX USA
[3]Ames Laboratory, Ames, IA USA

gdmay@tamu.edu

Abstract. The performance and cost of Bi-2212/Ag wire is limited by the large area fraction of Ag matrix (~3:1) that is required in the conventional oxide-powder-in-tube fabrication process. An alternative process is being developed in which fine-powder Bi-2212 is uniaxially compressed to form a bar tetrahedral bar with a Ag-alloy foil cladding. The fine powder naturally textures under compression (aligns the a-b planes perpendicular to the direction of compaction). Earlier work demonstrated that a symmetric billet of trapezoidal-cross-section textured-powder (TP) bars draws conformally so that the local area ratio is preserved.

1. Introduction

BI-2212 is the only Type II high-temperature superconducting ceramic that has been successfully fabricated as a multi-filament round wire [1]. Dramatic progress has been made in the performance of Bi-2212/Ag round wire over the past decade: invention of overpressure processing (OP) [2] to densify the cores; and two new methods [3] have been proven for synthesis of the fine powder that yield even higher performance than was obtained using Nexans' eutectic route [4].

Bi-2212/Ag round wire is conventionally manufactured using the oxide-powder-in-tube (OPIT) process [1]. The manufactured cost of the wire is however driven by the dominance of Ag alloy in the billet composition for in the OPIT process. The local area ratio (LAR) = Ag:Bi-2212 in the wire is limited to ~3:1 by the microstructure of the fine powder, which inhibits the drawing of tubes in the OPIT process. Bi-2212 is micaceous: the grains have a platelet geometry which complicates re-arrangement of grains in a Bi-2212 fine powder during the filling of tubes and drawing of a multi-sub-element billet. As a result the cost of the Ag matrix is a significant fraction of the final wire cost.

Previously [5], [6], it was found that Bi-2212 fine powder can be textured by uniaxial compression. The compression naturally orients the *a-b* planes of the micaceous grains so that they are normal to the compression axis – it produces a compacted textured-powder (TP) bar. Also like mica paper, textured fine-powder Bi-2212 has a much lower modulus to rearrange grains under plastic deformation of a TP bar during extrusion and drawing. In a previous study [7] the assembled geometric structure and local area ratio is approximately preserved in past tests using uniaxial compressed TP bars.

In the present work, TP bars have been fabricated and a favorable billet geometry has been devised for development of multi-sub-element wire, with a goal of optimizing the composition so that LAR~1:1 can be preserved during extrusion and drawing to fine wire.

2. Background/Previous work

2.1. Billet geometry

The billet geometry must be axially symmetric so that the forces during extrusion and drawing do not deform the registration of the constituent bars. In particular the texturing process defines the micaceous plane in which area reduction produces sliding with minimum sliding friction. The texturing plane of all bars should therefore be oriented in the azimuthal direction in the billet geometry. This consideration led a choice of tetrahedral cross-section for all bars except the central core. In each layer of a multi-layer billet, the number and size of the tetrahedral bars can be chosen so that all bars have comparable area.

2.2. Tetrahedral TP bars

Figure 1 shows the procedure used to fabricate each tetrahedral TP bar. A two-component die set is used to compress the powder in the required geometry. The trough in the female die confines the powder on three sides of the tetrahedral bar; the male die slides within a slot in the female die and compresses the top face of the bar. The female die is lined with a Ag foil strip (50 μm thick), then a

Figure 1. Fabrication of TP bars: a) loading a uniform filling of Bi-2212 powder into Ag foil-lined channel; b) compression of the foil-wrapped powder into TP bar. [7]

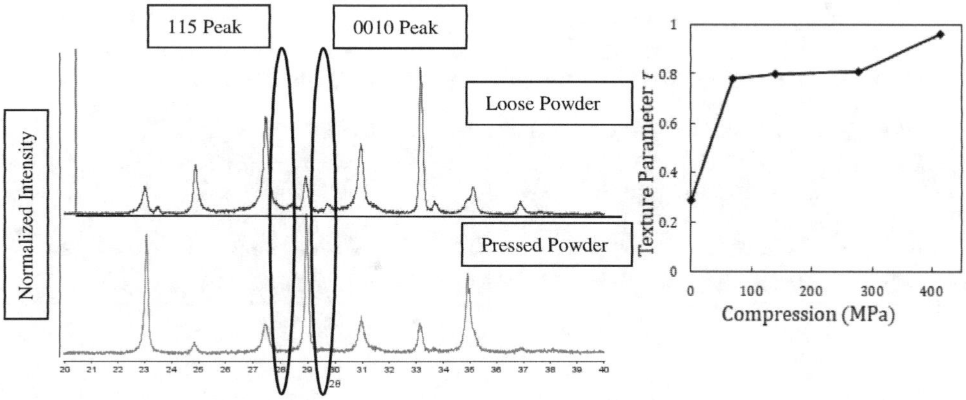

Figure 2. XRD measurement of texture parameter in cleaved faces within a TP bar: (a) diffraction spectra from a cleaved face, showing 115 peak(c-axis) and 0010 peak (a-b plane); b) texture parameter as function of compression.

supply manifold is used to uniformly fill the trough with fine-powder Bi-2212. Then the flaps of the foil are folded over the powder filling and the male die is inserted in the top slot. The die set is located in a hydraulic press and compressed to the desired compression to form a Ag-clad TP bar.

The degree of texture among the powder particles is measured by cleaving a texture parameter within a TP bar was measured by shearing the bar at several locations along its tapered sides. X-ray diffraction (XRD) is performed using the Bragg-Brentano method [8],[9] to measure the texture parameter τ which is extracted from the signal strengths of two peaks, weighted by their oscillator strengths. Figure 2a shows the XRD spectrum of one sample, in which specific diffraction directions are identified that are produced by diffraction from crystallites with an ordered *a-b* plane (0010) and with completely disordered *a-b* planes (115): $\tau = \frac{I(0010)-0.25I(115)}{I(0010)+0.75I(115)}$. Figure 2b shows the dependence of τ on compression. A compression of 200 MPa is used for all TP bars.

2.3. *Single-layer billet studies*

A key challenge in the development of multi-filament TP wire is to conformally draw a billet containing an array of TP bars to fine wire while maintaining the LAR as well as the registration of both the TP bars and the silver alloy sheaths. This requires that the textured-powder cores should have plastic flow modulus comparable to that of the silver-alloy components.

The drawing properties were developed using single-layer billets made with eight TP bars assembled to form a coaxial array around a central core [7]. The array is inserted as a loose fit into a Ag sheath tube, and the sheath tube is drawn down to wire. It is intended to utilize an alloy Ag0.2wt%Mg for the sheath, the core, and the foil wraps applied to each TP bar. For the single-billet studies delivery of Ag alloy materials was delayed, so pure-silver was used for cores and sheath tubes. This work was reported in Ref. 7 and constituted the starting point for the development of the revised billet configuration that is reported in this paper. Figure 3 shows the cross sections of two example single-layer billets at three stages of drawing. In each case the LAR of Ag/Bi-2212 and the aspect ratio H/W of the cores is given for each of the three stages of drawing.

In Billet 2, the TP bar was clad with 100 μm Ag foil, no wrap was applied, and a hollow tube was used for the central core. As the billet is drawn down, the geometry of each bar and the LAR were preserved, but the cladding foil folded into the powder filling in a problematic way. Note also that, as the billet is drawn down the center hole in the core tube closes. Once it is closed the solid center reduces more slowly that the TP bars so the LAR increases and the bar shape deforms.

In Billet 3, each TP bar was clad with 50 μm Ag foil. The foil-clad bar was inserted into a 3-sided channel of 200 μm Ag alloy foil, the side flaps were folded over on the top face of the bar, and the subassembly was compressed a second time. The bars were assembled around a solid Ag core rod. The folding of the inner cladding was cured, but the solid Ag core had stiffer modulus than the TP bars and so the bar geometry was deformed during drawing.

Two conclusions were reached from the series of single-layer billet studies:

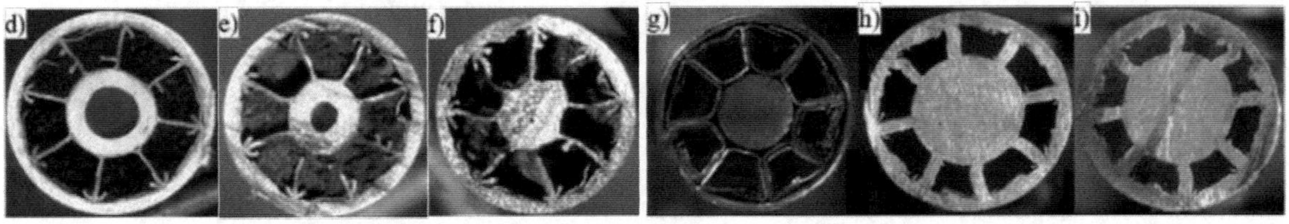

Figure 3. Single-layer billets used to study drawing properties [7]:

Billet 2	100 μm clad, no spacers, tube core	d) 8.40 mm OD: LAR = 1.17; H/W = 0.95	e) 6.20 mm OD, LAR = 1.06; H/W = 1.15	f) 4.06 mm OD, LAR = 1.07; H/W = 1.19
Billet 3	50 μm clad, 200 μm wrap, solid core	g) 12.70 mm OD, LAR = 2.32; H/W = 0.97	h) 9.60 mm OD: LAR = 2.40; H/W = 0.70	i) 6.76 mm OD: 2.49 LAR; H/W = 0.69

- Each TP bar should be Ag-clad is it is compressed to form texture, then wrapped with overlapped 200 μm Ag foil and re-compressed (as in Billet 2);
- The Ag-wrapped TP bar has an effective modulus under the plastic deformation of drawing that is *less than* that of annealed pure Ag. That is good news for the prospects of developing a billet design that preserves LAR ~1:1.
- The central core cannot be made of pure Ag, nor can it be left empty. It must have a composition with an effective modulus that closely matches that of the wrapped TP bars.

3. Three-layer billet studies

The lessons learned from the single-layer billet studies led to a revised billet design that is being used to develop three-layer billets. The billet core consists of an array of eight *triangular* TP bars, shown in Figure 4a. (layer 1). Succeeding layers of the billet contain 16 tetrahedral bars (Figure 4b) and 24 tetrahedral bars ((Figure 4c) in which the vertex angles in each case provide for compact assembly. All bars have comparable area and comparable LAR. Figure 5 shows an eight-layer billet containing 288 sub-elements, embodying the above strategy.

Figure 6 shows a three-layer billet that was assembled as the first 3 layers. The bars are arranged in layers of 8-16-24, with wedge angles that produce a conformal stacking. The alloy Ag sheath tubes surrounding layers 1 and 2 are 0.7 mm thick; the outermost sheath is 1.5 mm thick. The LAR of the overall billet is approximately 1.3:1. At the time of this paper, this 3-layer billet is awaiting extrusion in the hydrostatic extrusion press at Ames Laboratory.

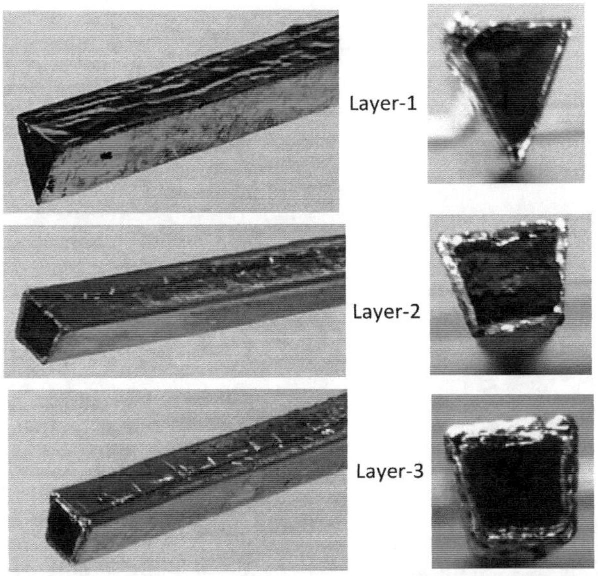

Figure 4. TP bars used in three-layer billet studies.

Figure 5. 8-layer billet (72 mm diameter; 288 sub-elements; LAR = 0.95).

Figure 6. Three-layer billet with triangular-cross-section TP bars in its core.

4. Conclusion

Single layer billets have been manufactured and shown to approximately preserve LAR while being drawn. Multi-sub-element billets are being fabricated, with the objective of developing a next-generation fine-filament Bi-2212 wire with *LAR* ~1:1. Wire architecture has been determined around an approximate Bi-2212 core of triangular bars - given area ratios and bar dimensions during past experiments - and Bi-2212/Ag fine wire will be further developed. Once the extrusion and drawing properties are optimized, process development of fine wire with OP heat treatment, and cabling experiments will be undertaken in Rutherford cable and SuperCIC geometries.

5. References

[1] Miao M *et al.*,2014 J. Phys: Conf. Series vol. 507, p 022020
[2] Matras M R *et al.* 2016 Supercond. Sci. Technol. vol 29, p 105005
[3] Jiang J *et al.*, ; 2019 IEEE Trans. Appl. Superconduct. vol 29, p 6400405
[4] Boch J, Ehrenberg J, and Rikel M, 2004 Patent US8946126B2
[5] McIntyre Peter M, Damborsky Kyle C, and Pogue Nathaniel J, 2012 Patent US20120028809A1
[6] Kellams Joshua N *et al* 2015 IOP Conf. Ser.: Mater. Sci. Eng. vol. 102 012028
[7] Rogers J S, Elliott T, May G D, McIntyre P M, Ratcliff CT, and Sattarov A, 2021 IEEE Trans. Appl. Superconduct. vol 31, p 6400605
[8] Speakman S A, 'Basics of X-ray powder diffraction', training lecture at the Center for Materials Science and Engineering at MIT,
[9] Guinebretière R *et al.*, 'Instrumental aspects in X-ray diffraction on polycrystalline materials', 2005 Powder Diffraction vol 20, p 294-305

Temperature dependent pinning efficiency in multilayer and single layer BZO/YBCO nanocomposite films

M Panth[*1], V Ogunjimi[1], M A Sebastian[2,3], D Zhang[4], B Gautam[1], J Jian[4], J Huang[4], Y Zhang[4], T Haugan[2], H Wang[4], and J Wu[1]

[1]*Department of Physics and Astronomy, the University of Kansas, Lawrence, Kansas 66045, USA*
[2]*U.S Air Force Research Laboratory, Aerospace Systems Directorate, WPAFB, OH 45433, USA*
[3]*University of Dayton Research Institute, Dayton, OH 45469, USA*
[4]*School of Materials Engineering, Purdue University, West Lafayette, IN 47907, USA*
*Corresponding author E-mails: panthm@ku.edu, jwu@ku.edu

Abstract: The $BaZrO_3/YBa_2Cu_3O_7$ (BZO/YBCO) interface has been found to affect the vortex pinning efficiency of one-dimensional artificial pinning centers (1D-APC) of BZO. A defective BZO/YBCO interface due to a lattice mismatch of ~7.7% has been blamed for the reduced pinning efficiency. Recently, we have shown incorporating $Ca_{0.3}Y_{0.7}Ba_2Cu_3O_{7-x}$ spacer layers in BZO/YBCO nanocomposite film in multilayer (ML) format can lead to a reduced lattice mismatch ~1.4% through the enlargement of lattice constant of YBCO via Ca diffusion and partial Ca/Cu replacement on Cu-O planes. In this work, the effect of this interface engineering on the BZO 1D-APC pinning efficiency is investigated at temperatures of 65-81 K through a comparison between 2 and 6 vol.% BZO/YBCO ML samples with their single-layer (SL) counterparts. An overall higher pinning force (F_p) density has been observed on the ML samples as compared to their SL counterparts. Specifically, the peak value of F_p ($F_{p,max}$) for the 6% BZO/YBCO ML film is about ~ 4 times of that of its SL counterpart at 65 K. In addition, the location of the $F_{p,max}$ (B_{max}) in the ML samples shifts to higher values as a consequence of enhanced pinning. For the 6% BZO/YBCO ML sample, a much smaller "plateau-like" decrease of the B_{max} with increasing temperature was observed, which is in contrast to approximately linear decrease of B_{max} with increasing temperature in the 6% SL film. This result indicates the importance of restoring the BZO/YBCO interface quality for better pinning efficiency of BZO 1D-APCs especially at higher BZO doping concentration.
Keywords: BZO/YBCO interface, pinning efficiency, strain field, lattice mismatch, vortex pinning

1. Introduction:

High temperature superconducting (HTS) wires based on $YBa_2Cu_3O_7$ (YBCO) coated conductors are a critical element to realize a broad variety of applications including power transmission cables, transformers, fault current limiters, high field magnets, etc. [1-4]. The temperature and field regimes in which these devices are operated are different, ranging in a few to tens of Teslas, depending on specific applications. Improved magnetic vortex pinning via the addition of nanoscale impurity artificial pinning centers (APCs) into the YBCO matrix has provided a viable approach towards achieving high critical current density (J_c) at applied magnetic fields and temperatures [5, 6]. Various APCs of different morphologies, including c-axis aligned one-dimensional nanorods (1D-APCs), ab-plane aligned two-dimensional planar defects (2D-APCs), and three-dimensional nanoparticles (3D-APCs) have been reported through the addition of secondary dopants like $BaZrO_3$ (BZO), $BaHfO_3$ (BHO), $BaSnO_3$ (BSO), $YBa_2(Nb/Ta)O_6$, Y_2O_3, etc. in YBCO (or other rare-earth variation of YBCO) to form APC/YBCO nanocomposites films [7-15].

The self-assembly of the c-axis aligned 1D-APCs in the YBCO matrix via vertical epitaxy is driven by a strain field that originates from the 1D-APC/YBCO interface due to their lattice mismatch at the interface (in the direction of the c-axis of YBCO perpendicular to the film) [8, 16, 17]. In the case of the BZO 1D-APC/YBCO nanocomposite films, the larger lattice constant of the BZO results in a positive lattice mismatch of ~7.7% at the BZO 1D-APC/YBCO interface and hence a tensile strain on the c-axis of YBCO [18]. This leads to a highly defective BZO/YBCO interface for reduction of the interface strain [19]. In contrast, an approximately coherent BHO 1D-APC/YBCO interface with a much reduced interface defect concentration was observed and attributed to a smaller lattice mismatch of ~6.7% and more adaptive elastic properties of BHO [13]. The higher pinning efficiency of BHO 1D-APCs, as reflected in the more than twice higher maximum pinning force density $F_{p,max}$ ~182 GNm^{-3} at 65 K than that of BZO 1D-APCs, suggests a close correlation between the pinning efficiency of 1D-APCs and the 1D-APC/YBCO interface quality [13]. Among others, BZO 1D-APCs have been intensively studied after the pioneering work by Driscoll et al [7]. Despite a large BZO/YBCO lattice mismatch, BZO 1D-APCs can form at fairly high concentrations proportional linearly to the BZO doping approximately up to 10 vol.%, which is important to enhance pinning at high applied magnetic fields [20-26]. In contrast, the concentration of the BHO 1D-APCs exhibits a nonlinear dependence on the BHO doping and could not seem to be further increased above BHO doping of 4-5 vol.% [12, 13, 27]. In fact, when BHO 1D-APCs form coherent interfaces with YBCO, the accommodation of the BHO and YBCO lattices leads to not only a large strain that hampers formation of high concentration BHO 1D-APCs, but also degrades T_c of the BHO/YBCO nanocomposite films. At 6 vol.% BHO doping, T_c decreases to ~78 K, which results in much reduced J_c in the temperature range of 65-77 K [13].

Considering the advantages of BZO in generating high-concentration 1D-APCs in BZO/YBCO nanocomposite films, repairing the BZO/YBCO interface may recover the pinning efficiency of the BZO 1D-APCs. In a recent exploration, we have developed a multilayer (ML) approach by inserting two $Ca_{0.3}Y_{0.7}Ba_2Cu_3O_{7-x}$ (CaY-123) spacer layers in BZO/YBCO nanocomposite film for improving the BZO/YBCO interface after BZO 1D-APCs form in BZO/YBCO nanocomposite films. In the ML films, Ca from the spacer layers can diffuse dynamically during the film growth to YBCO layers. While replacements of Y, Ba and Cu with Ca are all possible, the Ca/Cu replacement is most favorable, as compared to the other two options on tensile strained YBCO lattice (in c-axis due to formation of BZO 1D-APCs). This is because Ca/Cu replacement can lead to elongated c-axis lattice constant of YBCO up to 1.24 nm when the Cu cations of smaller size (by 30%) on the Cu-O planes are partially replaced with larger Ca cations. This prevents formation of interface defects, such as dislocations, and leads to a coherent BZO/YBCO interface owing to the reduced lattice mismatch up to ~1.4%. Considerably enhanced pinning efficiency of BZO 1D-APCs has been observed [28]. It should be noted that the Ca diffusion and Ca/Cu replacement occur dynamically during the growth of the BZO 1D-APCs with a minimum impact on the nucleation (or the concentration) of the BZO 1D-APCs. In this paper, a systematic study of the pinning efficiency of the BZO 1D-APCs in 2% and 6% BZO/YBCO ML samples as a function of temperature in the range of 65-81 K is presented using single-layer BZO/YBCO (SL) samples as a reference. Significant J_c enhancement is obtained in the temperature range of 65-81 K as illustrated in the overall higher pinning force density (F_p) and reduced temperature dependence of pinning in ML samples as compared to their SL counterparts', which can be attributed to the formation of coherent BZO 1D-APC/YBCO interface and thereby restoration of the pristine pinning efficiency of the BZO 1D-APCs.

2. Experimental

In this work, 2 and 6 vol.% BZO doped YBCO nanocomposite films were deposited on (100) $SrTiO_3$ (STO) single crystal substrates. One set of samples were made as reference samples, which included single-layered YBCO films doped with 2 and 6 vol.% BZO (BZO/YBCO-SL). These samples are referred as 2% SL and 6% SL, respectively, in the rest of the paper. The other set of samples is the same except incorporation of two calcium containing $Ca_{0.3}Y_{0.7}Ba_2Cu_3O_{7-x}$ (CaY-123) spacer layers inserted in each BZO doped YBCO nanocomposite film. These samples have a multilayer structure with three ~50 nm thick BZO doped YBCO layers separated by two alternating ~10 nm thick spacer layers and will be regarded as BZO/YBCO-ML and which are denoted by 2% ML and 6% ML in rest of the paper. All samples were made using pulsed laser deposition (PLD) technique. A KrF excimer laser was employed for PLD at the laser wavelength ~248 nm, with a pulse energy ~450 mJ. The optimal PLD repetition rates were 8 Hz and 2 Hz for BZO/YBCO and the CaY-123 spacer layers respectively. During the sample deposition, the oxygen pressure was maintained at ~300 mTorr whereas the sample temperature was set at 825°C [29]. After the PLD deposition, the films were annealed at 500 °C in 1 atmospheric oxygen pressure for 30 minutes. The details of the BZO/YBCO SL film fabrication have been reported previously [27, 30, 31]. The film thicknesses were measured using a Tencor P-16 profilometer. The thicknesses of SL and ML samples are around 150 nm and 170 nm respectively. Crystallinity and lattice parameters were determined by x-ray diffraction (XRD) utilizing a Bruker D8 Discover diffractometer. In both SL and ML samples, c-axis aligned BZO 1D-APCs have been confirmed using transmission electron microscopy [28] with a diameter ~5 nm and comparable concentrations at the given BZO doping.

In order to study the electrical transport properties, two parallel microbridges with length of 500 μm and width of 20 or 40 μm were patterned on each film using standard photolithography. The details of the patterning and sample wiring for the transport measurement can be found in our previous works [27, 30, 31]. Briefly, Ag contact pads of ~ 120 nm in thickness were deposited through a shadow mask using DC magnetron sputtering at the deposition rate of approximately 0.07 nm/second under the argon pressure of 30 mTorr. The electrical connection to the microbridges were made by attaching platinum wires of 50 μm in diameter on Ag pads using Indium. The $J_c(B)$ measurement were taken at $\theta = 0°$ ($B//c$-axis) and at temperatures of 65-81 K. Quantum Design Ever-Cool II Physical Property Measurement System equipped with a 9 T magnet was used for the electrical transport measurement. Standard four-probe technique was used in the measurement of the transport properties including resistance-temperature (R-T) curves from which T_c can be determined, and $J_c(B)$ curves. The J_c values were determined by applying a standard criterion of 1 μV/cm. The pinning force density (F_p) was calculated from the equation $\boldsymbol{F_p} = \boldsymbol{J_c} \times \boldsymbol{B}$. The maximum pinning force density ($F_{p,max}$) and its location (B_{max}) were determined from the $F_p(B)$ curves.

3. Results

Figure 1a and 1b compare the XRD θ-2θ spectra of the 2% and 6% SL and ML films. The SL films are denoted by black and ML in red respectively. The appearance of the YBCO (00l) peaks (#) indicates the c-axis orientation of the BZO/YBCO nanocomposite films on the (100) STO substrates (+). Additionally, both SL and ML spectra show the major BZO (001) peaks at $2\theta \sim 46.5°$. The c-lattice constants estimated from the YBCO (00l) peaks for the 2% and 6% SL films are 11.74 Å and 11.81 Å, respectively. The larger lattice constant in the 6% SL film confirms the increased tensile strain with increased BZO doping most probably attributed to the increased strain field overlap at higher BZO doping [16, 32, 33]. Meanwhile the same c-lattice constants of 11.76 Å is observed on both the 2% and 6% ML films. This may be attributed to the combined effects from interfacial strain and partial Ca/Cu replacement as shown in our recent works. Basically, the replacement is mostly happening around the BZO/YBCO interface where maximum tensile strained occurs. This has been confirmed via High resolution transmission electron microscopy images with the elongated c-axis lattice constant of YBCO up to 1.24 nm as described in the previous work [28]. A coherent BZO/YBCO interface was formed owing to the reduced lattice mismatch up to ~1.4%.

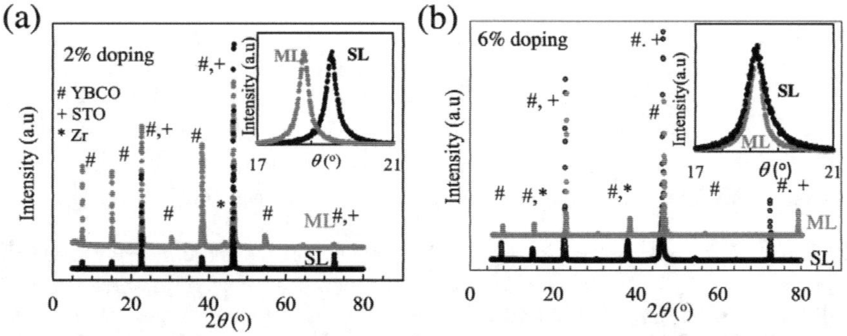

Figure 1: X-ray diffraction. (a) θ-2θ scans taken on 2% SL (black) and 2% ML (red) with the inset showing rocking curves of the YBCO (005) peak for the 2% SL (black) and ML (black) films. (b) θ-2θ scans taken on 6% SL (black) and 6% ML (red) with the inset showing rocking curves of the YBCO (005) peak for the 6% SL (black) and ML (red) films. The spectra were generated using Cu-kα radiation of wavelength 1.54 Å. The symbols #, + and * represent the YBCO (00l), STO substrate (100) and major BZO (001) peaks respectively.

This Ca/Cu replacement on YBCO's Cu-O planes resulting in strain reduction is evidenced in the microstrain evaluation in XRD data shown in figure 1. In the 6% ML film, the c-lattice constant decreases as compared to its SL counterparts. This can be explained via higher reduction of microstrain as revealed in its FWHM values evaluated from (005) YBCO rocking curves. These are shown in the inset of figure 1a and 1b for 2% and 6% SL and ML (SL as black and ML as red curves) sample respectively. This has been confirmed from the smaller full width at a half maximum (FWHM) of YBCO (005) rocking curves on the 2% and 6% ML samples. The FWHM values for the 2% SL and ML samples are 0.35° and 0.32° respectively, which correspond to ~ 10% reduction in the microstrain in the latter as opposed to the former. In the 6% SL and ML samples, the FWHM values are 0.59° and 0.38° respectively. This means that a reduction of 35% of microstrain is obtained in the 6% ML film as compared to its counterpart. In addition, the c-lattice constant of 6% ML sample is the same as that of the 2% ML sample. This result therefore suggests the benefit of the ML approach is more significant in BZO/YBCO nanocomposites with higher BZO doping concentrations, in which strain field overlap is more significant and the impact on the nanocomposite performance is more severe.

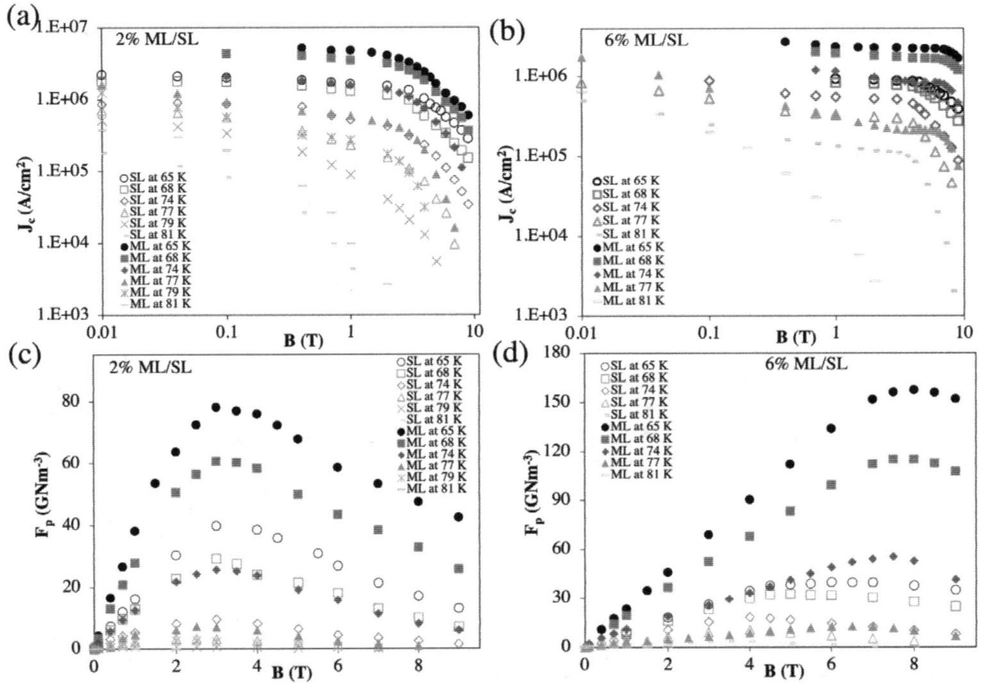

Figure 2: (a) Comparisons of $J_c(B)$ curves of the 2% SL with 2% ML nanocomposite films at *B//c*. (b) Comparisons of $J_c(B)$ curves of the 6% SL with 6% ML nanocomposite films at *B//c*. $F_p(B)$ plots at different temperature for (c) 2% SL and ML films (d) 6% SL and ML films.

The T_c values are determined to be 89.2 K, 87.5 K from the R-T curve measurement on the 2% SL and ML samples. Similarly, the T_c values of 86.9 K and 84.0 K, respectively, were observed on the 6% SL and ML samples. The slightly reduced T_c values in the ML samples indicates a minor Ca diffusion is happening in the YBCO matrix due to the Ca diffusion from the CaY-123 spacers [34, 35]. Figure 2a compares the $J_c(B)$ curves of the 2% SL and ML films at *B//c* in the field range of 0-9 T and temperature range of 65-81 K. The ML samples exhibit overall higher J_c values in high field range despite a lower T_c (by ~1.0 K) in the ML samples. For example, at 65 K, a J_c enhancement of 2.9, 2.4, 2.1, and 2.1 times is observed on the 2% ML samples at 1 T, 3 T, 5T and 9 T respectively over the 2% SL samples. At 77 K, the J_c enhancement factor of 2.5, 3.1 and 2.2 is observed at 1 T, 3 T and 5 T respectively. Similarly, at 81 K, the 2% ML sample has 1.7, 2.2 times higher J_c at self-field and at 1 T. Furthermore, the $J_c(B)$ curve for the 2% ML sample exhibits much reduced *B*-field susceptibility as compared with the 2% SL sample at all temperature study conducted. These results suggest the benefits of ML approach in enhancing the J_c via restoration of pristine pinning efficiency of BZO 1D-APCs. Figure 2b compares the $J_c(B)$ curves of the 6% SL and ML films at $B//c$ in the field range of 0-9 T and temperature range of 65-81 K. A similar trend of enhanced pinning has been observed in the 6% ML sample as compared to its SL counterpart. At 65 K, the ML samples show J_c enhancement factors of 2.5, 2.6, 2.9, and 4.3 times over its SL counterparts at 1 T, 3 T, 5 T and 9 T respectively. At 77 K, ML sample has 0.95, 0.72, and 1.2 times higher J_c than SL counterparts at 1 T, 3 T and 5 T respectively. Similarly, at self-field and 1 T, J_c enhancement of 1.6 and 0.11 is observed for ML samples at 81 K. It should be noted that the higher J_c values for the 6% ML sample were observed in the entire field range at almost all temperatures. It is seen from $J_c(B)$ plots, magnetic field susceptibility for 2% ML sample is higher than 6% ML samples. This suggest that a larger number of BZO 1D-APCs are efficient for pinning in the 6% ML sample, which is expected from the higher concentration of BZO doping in BZO/YBCO nanocomposite. In addition, the $J_c(B)$ enhancement increases with the applied B fields. For example, at a temperature of 65 K and 68 K, the J_c value for the 6% ML sample is about 4.3 times of that of the 6% SL counterparts at 9 T, which is in contrast to ~2.1 in the 2% case at the same condition.

The enhanced $J_c(B)$ leads to an enhanced $F_p(B)$ in the 2% ML sample in the entire temperature range of measurements as shown in figure 2c. The $F_p(B)$ curves are of a characteristic bell shape for both SL and ML samples and the peak value of F_p (F_{pmax}) locates at B_{max}. For the entire temperature and field range higher F_p values can be observed for ML sample. At 65 K, the F_{pmax}'s for the ML and SL samples locate almost at the same locations of B_{max} ~3.0 T despite an enhanced F_{pmax} value by approximately 96% from ~39.8 GN/m^3 for the 2% SL sample to ~78.1 GN/m^3 for the 2% ML sample. At 1 T, 5 T and 9 T, an enhancement factor of 2.3, 2.1, and 3.2 times has been observed on the ML sample over its SL counterpart's. At 77 K, approximately ~2 times higher F_p at 1 T, 3 T and 5 T observed for 2% ML sample as compared with SL counterparts. In comparison with 2% case, the 6% ML sample shows much more enhanced F_p which is shown in figure 2d. At 65 K and fields of 1 T, 5 T, and 9 T ML sample has F_p values 2.5, 2.9 and 4.1 times of SL counterparts respectively. In addition, an overall higher B_{max} values can be seen in 6% ML sample as compared to the 6% SL sample. At 65 K, the B_{max} ~8 T is approximately 33% higher than that of the 6% SL sample. At 77 K, F_p values are comparable and little better for ML samples compared to SL counterparts. For example, F_p enhancement of 0.95, 0.72 and 1.2 at 1 T, 3 T and 5 T respectively observed at 77 K. The F_p's enhancement are consistent with the J_c's enhancement observed in figure 2a and 2b.

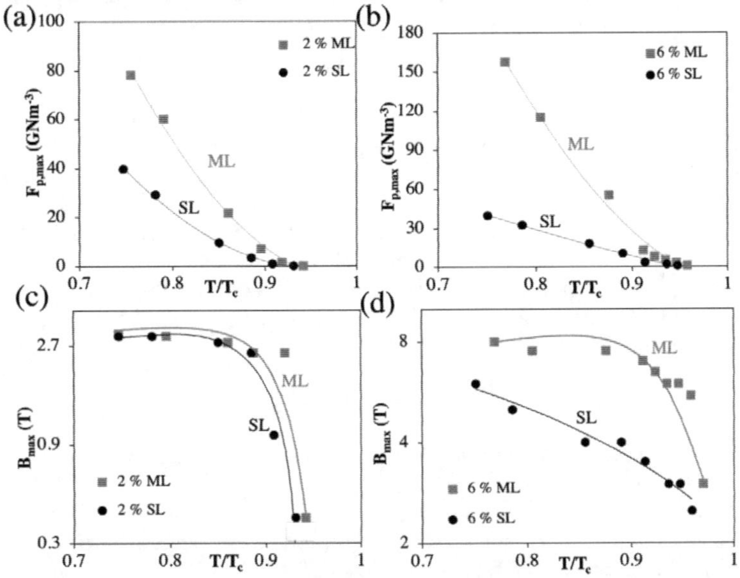

Figure 3: Pinning force density ($F_{p,max}$) as a function of reduced temperature (T/T_c) plots for (a) 2% SL and ML sample (b) 6% ML and SL sample. B_{max} vs T/T_c plots for (c) 2% SL and ML sample (d) 6% ML and SL sample.

Figure 3a and 3b compare the $F_{P,max}$ values as a function of reduced temperature (T/T_c) for the 2% and 6% SL and ML samples, respectively. Overall, high $F_{p,max}$ values are observed on ML samples as compared to their SL counterparts'. At T/T_c = 0.755, the $F_{p,max}$ values ais 78.1 GNm^{-3}, and at T/T_c = 0.769, $F_{p,max}$ values is 157.7 GNm^{-3} for 2% and 6% ML samples, which are 1.9 and 3.9 times of their SL counterparts'. At T/T_c = 0.91, the ratios between $F_{p,max}$ values of ML and SL samples are 1.6 and 3.6 for 2% and 6% samples. Although the SL and ML samples exhibit a similar temperature dependence of $F_{P,max}$, the higher $F_{P,max}$ values in the ML samples imply an extension of the temperature range for applications of the BZO/YBCO nanocomposites.

Figure 3c and 3d compare the B_{max} as a function of T/T_c for the 2% and 6% samples, respectively. At the lower BZO doping of 2 vol.%, the B_{max} values for the SL and ML samples are comparable in the lower T/T_c range up to ~0.85, while higher B_{max} values are observed in the ML sample. This illustrates the benefit of the BZO/YBCO interface engineering using the multilayer approach. Specifically, the removal of the defective YBCO interfacial layer would lead to reduced degradation of the T_c value at the interface. A sharper BZO/YBCO interface is critical to the pinning efficiency of BZO 1D-APCs [36]. Specifically, the higher B_{max} values in the 2% ML sample implies more BZO 1D-APCs remain efficient at elevated temperatures. A similar but more dramatic trend can be observed in 6 vol.% samples. In fact, the B_{max} vs T/T_c curves for SL and ML samples differ qualitatively, besides much higher B_{max} values for the ML sample over the entire range of T/T_c. At T/T_c ~0.76, the B_{max} ~8 T for the ML sample is 1.3 times of that for the SL sample. At T/T_c ~0.95, this ratio is increased to 2.2. In contrast to the approximately linear decreases of the B_{max} vs T/T_c in the SL sample, a more or less plateau can be seen in that for the ML sample at lower T/T_c, followed by a sharp decrease of the B_{max} when approaching T/T_c =1. For example, in 2% ML, B_{max} drop sharply to 0.4 T at T/T_c ~1 from B_{max} of 2.5 T when T/T_c equals to 0.91. Similarly in 6% ML film, B_{max} drops to 3 T at T/T_c ~1 from 5.5 T at T/T_c ~0.95. This observation suggests that the BZO 1D-APCs lose their pinning efficiency at elevated temperatures in SL samples, while they remain efficient pins in a wider temperature range in the ML sample. In contrast, the more dramatic enhancement of the pinning in the 6% ML sample is not surprising considering the expected benefits of improved BZO/YBCO interface and reduced strain field overlap at high concentration of the BZO 1D-APCs.

4. Conclusions

A newly developed ML scheme by inserting two ~10 nm thick CaY-123 spacer layers in 2% and 6% BZO 1D-APCs/YBCO nanocomposite films has shown to enhance the pinning efficiency of BZO 1D-APCs by reducing the BZO/YBCO interface lattice mismatch and hence interfacial defect concentration. Significant J_c enhancement has been observed on ML samples as compared to their SL counterparts in the temperature range of 65-81 K. For example, a J_c enhancement of 4.3 times in 6% ML film has been observed at 65 K and at magnetic field of 9 T in comparison with 6% SL counterparts. The overall J_c enhancement has led to enhanced $F_{p,max}$ value by approximately 96% for the 2% ML and 296% for the 6% ML samples, respectively. The pinning enhancement in the ML samples is also illustrated in their higher B_{max} values in the temperature range of 65-81 K. Furthermore, a qualitatively different temperature dependence of B_{max} has been observed on 6% ML sample as compared to the 6% SL one. Specifically, a much smaller decrease of the B_{max} values with increasing temperature has been observed in the ML sample, suggesting that the majority of BZO 1D-APCs remain efficient in pinning at temperatures close to T_c. This result indicates the importance of the BZO/YBCO interface on pinning efficiency of BZO 1D-APCs.

5. References

[1] Alecu G *Crystal structures of some high-temperature superconductors* Romanian Reports in Physics 2004 56(3) p 404-412.
[2] Mourachkine A *Room-temperature superconductivity* 2004 Cambridge Int Science Publishing.
[3] Obradors X and T Puig *Coated conductors for power applications materials challenges* Superconductor Science and Technology 2014 27(4) p 044003.
[4] Kalsi S S *Applications of high temperature superconductors to electric power equipment* 2011 John Wiley & Sons.

[5] Matsumoto K and P Mele *Artificial pinning center technology to enhance vortex pinning in YBCO coated conductors*. Superconductor Science and Technology 2009 23(1) p 014001.
[6] Matsumoto K et al *Enhancement of critical current density of YBCO films by introduction of artificial pinning centers due to the distributed nano-scaled Y_2O_3 islands on substrates* Physica C Superconductivity 2004 412 p 1267-1271.
[7] MacManus-Driscoll J et al *Strongly enhanced current densities in superconducting coated conductors of $YBa_2Cu_3O_{7-x}$+ $BaZrO_3$* Nature materials 2004 3(7) p 439-443.
[8] Wu J and J Shi *Interactive modeling-synthesis-characterization approach towards controllable in situ self-assembly of artificial pinning centers in RE-123 films* Superconductor Science and Technology 2017 30(10) p 103002.
[9] Miura S et al *Strongly enhanced irreversibility field and flux pinning force density in $SmBa_2Cu_3O_y$-coated conductors with well-aligned $BaHfO_3$ nanorods* Applied Physics Express 2017 10(10) p 103101.
[10] Chen, S., et al., *Generating mixed morphology $BaZrO_3$ artificial pinning centers for strong and isotropic pinning in $BaZrO_3$–Y_2O_3 double-doped YBCO thin films*. Superconductor Science and Technology, 2017. 30(12): p. 125011.
[11] Mele P et al *Ultra-high flux pinning properties of $BaMO_3$-doped $YBa_2Cu_3O_{7-x}$ thin films (M= Zr, Sn)* Superconductor Science and Technology 2008 21(3) p 032002.
[12] Gautam B et al *Transformational dynamics of BZO and BHO nanorods imposed by Y_2O_3 nanoparticles for improved isotropic pinning in $YBa_2Cu_3O_{7-\delta}$ thin films* AIP Advances 2017 7(7) p 075308.
[13] Gautam B et al *Probing the effect of interface on vortex pinning efficiency of one-dimensional $BaZrO_3$ and $BaHfO_3$ artificial pinning centers in $YBa_2Cu_3O_{7-x}$ thin films* Applied Physics Letters 2018 113(21) p 212602.
[14] Wu J, B Gautam, and V Ogunjimi *Pinning Efficiency of Artificial Pinning Centers in Superconductor Nanocomposite Films* in *Superconductivity* 2020 Springer p 29-52.
[15] Ogunjimi V et al *The angular range of effective pinning by one-dimensional artificial pinning centers in $BaZrO_3/YBa_2Cu_3O_{7-x}$ nanocomposite films* AIP Advances 2019 9(8) p 085110.
[16] Horide T et al *Structural evolution induced by interfacial lattice mismatch in self-organized $YBa_2Cu_3O_{7-\delta}$ nanocomposite film* ACS nano 2017 11(2) p 1780-1788.
[17] Wu J Z et al *The effect of lattice strain on the diameter of $BaZrO_3$ nanorods in epitaxial $YBa_2Cu_3O_{7-\delta}$ films* Superconductor Science and Technology 2014 27(4) p 044010.
[18] File P D *Joint Committee on Powder Diffraction Standards (JCPDS)* ASTM, Philadelphia PA 1997.
[19] Cantoni C et al *Strain-driven oxygen deficiency in self-assembled, nanostructured, composite oxide films* Acs Nano 2011 5(6) p 4783-4789.
[20] Wee S H et al *Engineering nanocolumnar defect configurations for optimized vortex pinning in high temperature superconducting nanocomposite wires* Scientific reports 2013 3 p 2310.
[21] Horide T et al *Influence of matching field on critical current density and irreversibility temperature in $YBa_2Cu_3O_7$ films with BaMO3 (M= Zr, Sn, Hf) nanorods*. Applied Physics Letters 2016 108(8) p 082601.
[22] Horide T et al *Elastic strain evolution in nanocomposite structure of $YBa_2Cu_3O_7$+ $BaZrO_3$ superconducting films* Japanese Journal of Applied Physics 2014 53(8) p 083101.
[23] Xu A et al *Strongly enhanced vortex pinning from 4 to 77 K in magnetic fields up to 31 T in 15 mol.% Zr-added (Gd, Y)-Ba-Cu-O superconducting tapes*. Apl Materials 2014 2(4) p 046111.
[24] Goyal A et al *Irradiation-free, columnar defects comprised of self-assembled nanodots and nanorods resulting in strongly enhanced flux-pinning in $YBa_2Cu_3O_{7-\delta}$ films* Superconductor Science and Technology 2005 18(11) p 1533.
[25] Peurla M et al *Effects of nanocrystalline target and columnar defects on flux pinning in pure and $BaZrO_3$-doped $YBa_2Cu_3O_{6+x}$ films in fields up to 30 T* Physical Review B 2007 75(18) p 184524.
[26] Wu J et al *Pinning efficiency of one-dimensional artificial pinning centers in $YBa_2Cu_3O_{7-x}$ thin films* IEEE Transactions on Applied Superconductivity 2019 29(5) p 1-5.

[27] Sebastian M A P et al *Study of the flux pinning landscape of YBCO thin films with single and mixed phase additions* $BaMO_3$+ *Z: M= Hf, Sn, Zr and Z=* Y_2O_3, *Y211* IEEE Transactions on Applied Superconductivity 2017 27(4) p 1-5.

[28] Ogunjimi V et al *Enhancing Magnetic Pinning by* $BaZrO_3$ *Nanorods Forming Coherent Interface by Strain-Directed Ca-doping in* $YBa_2Cu_3O_{7-x}$ *Nanocomposite Films* Superconductor Science and Technology 2021.

[29] Baca F J et al *Interactive growth effects of rare-earth nanoparticles on nanorod formation in* $YBa_2Cu_3O_x$ *thin films* Advanced Functional Materials 2013 23(38) p 4826-4831.

[30] Chen S et al *Enhancement of isotropic pinning force in YBCO films with* $BaZrO_3$ *nanorods and* Y_2O_3 *nanoparticles* IEEE Trans. Appl. Supercond 2017 27(4) p 4-8.

[31] Wang X et al *Eliminating thickness dependence of critical current density in* $YBa_2Cu_3O_{7-x}$ *films with aligned* $BaZrO_3$ *nanorods* Journal of Applied Physics 2010 108(11) p 113911.

[32] Khan M Z et al *Enhanced flux pinning isotropy by tuned nanosized defect network in superconducting* $YBa_2Cu_3O_{6+x}$ *films* Scientific reports 2019 9(1) p 1-12.

[33] Samoilenkov S et al *Anisotropic strain of* $BaZrO_3$, $BaCeO_3$ *and* Y_2O_3 *nanoinclusions in a* $YBa_2Cu_3O_{7-x}$ *epitaxial film matrix and its relation to the oxygen content of the superconductor* Superconductor Science and Technology 2011 24(5) p 055003.

[34] Hammerl G et al *Enhanced supercurrent density in polycrystalline* $YBa_2Cu_3O_{7-\delta}$ *at 77 K from calcium doping of grain boundaries* Nature 2000 407(6801) p 162.

[35] Schmehl A et al *Doping-induced enhancement of the critical currents of grain boundaries in* $YBa_2Cu_3O_{7-\delta}$ EPL (Europhysics Letters) 1999 47(1) p 110.

[36] Blatter G et al *Vortices in high-temperature superconductors*. Reviews of modern physics 1994 66(4) p 1125.

Acknowledgement
This research was supported in part by NSF contracts Nos: NSF-DMR-1508494 and 1909292 and NSF-ECCS-1809293, the AFRL Aerospace Systems Directorate, the Air Force Office of Scientific Research (AFOSR) LRIR #14RQO8COR and LRIR #18RQCOR100. D.Z. and H.W. acknowledge the support from the U.S. National Science Foundation for the high-resolution STEM effort at Purdue University (DMR-1565822 and DMR-2016453).

Interface Engineering for Enhanced Magnetic Vortex Pinning by 1D-BZO APCs in a Wide Angular Range

Victor Ogunjimi[1*], Mary Ann Sebastian[2,3], Di Zhang[3], Bibek Gautam[1], Jie Jian[4], Jijie Huang[4], Yifan Zhang[4], Timothy Haugan[2], Haiyan Wang[4], and Judy Wu[1]

[1]*Department of Physics and Astronomy, the University of Kansas, Lawrence, Kansas 66045, USA*
[2]*U.S Air Force Research Laboratory, Aerospace Systems Directorate, WPAFB, OH 45433, USA*
[3]*University of Dayton Research Institute, Dayton, OH 45469, USA*
[4]*School of Materials Engineering, Purdue University, West Lafayette, IN 47907, USA*

*Corresponding author E-mails: victorogunjimi@gmail.com, jwu@ku.edu

Abstract: Microstructural analysis of the $BaZrO_3$ (BZO)/$YBa_2Cu_3O_7$ (YBCO) interface has revealed a highly defective and oxygen deficient 2-3 nm thick YBCO column surrounding the BZO one-dimensional artificial pinning centers (1D-APCs). The resulting semi-coherent interface is the consequence of the ~7.7% BZO/YBCO lattice mismatch and is responsible for the low pinning efficiency of BZO 1D-APCs. Herein, we report an interface engineering approach of dynamic Ca/Cu replacement on YBCO lattice to reduce/eliminate the BZO/YBCO lattice mismatch for improved pinning at a wide angular range of the magnetic field orientation. The Ca/Cu replacement induces a local elongation of the YBCO c-lattice near the BZO/YBCO interface, thereby ensuring a reduction in the BZO/YBCO lattice mismatch to ~1.4% and a coherent BZO/YBCO interface. This has resulted in enhanced pinning at B//c-axis and a broad angular range of B-field orientation. For example, the 6 vol.% BZO/YBCO film with interface engineering exhibits F_p ~158 GN/m^3 at 65 K and B//c-axis, which is 440% higher than the ~36.1 GN/m^3 for the reference 6% BZO/YBCO sample, and enhanced J_c and F_p in a wide angular range up to ~ 80°. This result illustrates a facile scheme for engineering 1D-APC/YBCO interface to resume the pristine pinning efficiency of the 1D-APCs.
Keywords: BZO/YBCO interface, pinning efficiency, strain field, lattice mismatch, vortex pinning

1. Introduction

Artificial pinning centers (APCs) self-assembled in $YBa_2Cu_3O_{7-x}$ (YBCO) demobilize magnetic vortices to enhance critical current density needed for high temperature superconductor (HTS) applications. One-dimensional (1D) c-axis aligned APCs like $BaZrO_3$ (BZO)[1-5], $BaHfO_3$ (BHO)[6-10], $BaSnO_3$[11, 12] $YBa_2(Nb/Ta)O_6$[13, 14] etc. have been reported to effectively pin vortices for magnetic field oriented along the c-axis. The APC material lattice mismatch with YBCO affects the pinning efficiency of the APCs. BZO 1D-APC/YBCO interface has been reported to be semi-coherent with high defect density a consequence of the ~ 7.7% BZO/YBCO lattice mismatch [15-17]. Meanwhile, a high-resolution electron microscopy study of the BZO/YBCO interface, has reported a cylindrical YBCO shell with a thickness of a few nm surrounding the BZO 1D-APCs. This region has been shown to contain a large concentration of defects such as dislocations with much reduced superconductivity due to oxygen deficiency [18]. Since the pinning force per unit length of the BZO 1D-APCs is proportional to the radial derivative of the pinning energy at the BZO/YBCO interface [19], the reduced superconductivity in this YBCO shell at the BZO/YBCO interface is expected to degrade the pinning efficiency of the BZO 1D-APCs [18]. Experimentally, considerably higher pinning efficiency, illustrated in the higher critical current density (J_c) and pinning force density (F_p), have been observed on BHO 1D-APCs that form a coherent and almost defect-free interface with YBCO [16]. Specifically, this is illustrated in a peak of $J_c(\theta)$ centered at θ=0, or B//c-axis (θ represents the tilt angle of the B field away from the c-axis in the plane perpendicular to J_c) in the 1D-APCs/REBCO nanocomposites. A more recent study in probing the effect of the enhanced pinning efficiency of 1D-APCs on the angular range of the enhanced pinning with the pinning efficiency, through a comparison of the angular dependence of the peak pinning force density (F_{pmax}) and its location (B_{max}) on the BZO 1D APC/YBCO and BHO 1D-APC/YBCO nanocomposite films [20, 21], has revealed a broader angular range of enhanced pinning of θ>67° in the latter, suggesting the further benefit of a coherent 1D-APC/YBCO interface in reducing the J_c anisotropy with respect the B field orientations.

In order to achieve a coherent BZO/YBCO interface, this work explores reduction of the BZO/YBCO lattice mismatch via enlarging the c-axis constant of YBCO locally near the BZO 1D-APCs. Specifically, we explore a partial replacement of smaller Cu+2 ions on the Cu-O planes of YBCO lattice with larger Ca+2 ions (by 30%) for enlarged c-axis lattice constant locally through formation of planer defects or stacking faults. In order to control the amount of Ca in YBCO in the BZO 1D APC/YBCO nanocomposite films, two $Ca_{0.3}Y_{0.7}Ba_2Cu_3O_{7-x}$ (CaY-123) spacers of only 10 nm in thickness were inserted into the 2 vol.% (and 6 vol.%) BZO 1D APC/YBCO nanocomposite thin films of ~ 150 nm in thickness. In these multilayer (ML) samples consisting of three 50 nm thick BZO 1D-APC/YBCO and two CaY-123 layer each of 10 nm in thickness stacking alternatively, diffusion of Ca from the spacers into YBCO increases the c-axis lattice of the YBCO dramatically to allow coherent BZO 1D APC/YBCO interface[22] to form. This coherent interface has a profound effect on the pinning efficiency of the BZO 1D-APCs as illustrated in the almost double to quadruple of the F_p values in the ML samples as compared to the reference single-layered BZO 1D-APC/YBCO samples of the same BZO doping. In addition, significantly reduced pinning anisotropy was observed especially at high magnetic fields.

2. Experimental

Pulsed laser deposition (PLD) was used to deposit 2-6 vol.% BZO doped YBCO nanocomposite films on (100) $SrTiO_3$ (STO) single crystal substrates[1, 23, 24]. To isolate the Ca doping effect, one set consisting of single-layer (SL) YBCO films doped with 2% - 6% BZO in the range of 2-6 vol.% (2-6% BZO/YBCO-SL) were fabricated as reference samples. The other set consists of two additional Ca-containing spacer layers inserted in each BZO doped YBCO nanocomposite film to form a multi-layer (ML) structure. The ML structure has three 50 nm thick BZO doped YBCO layers separated by two ~ 10 nm thick $Ca_{0.3}Y_{0.7}Ba_2Cu_3O_{7-x}$ (CaY-123) spacers (Figure 1). The CaY-123 spacers were deposited with a PLD repetition rate of 2 Hz. All samples were deposited in 300 mTorr oxygen (O_2) at a substrate temperature of ~825 °C[1]. The BZO/YBCO layers were deposited at a repetition rate of 8 Hz in both sets of the samples. For the ease of discussion, the SL and ML samples are aliased as xSL and xML respectively where x is either 2% or 6%. Post deposition, the samples were annealed for about 30 minutes dwell time at 500 °C in one atmosphere O_2 pressure. The film thicknesses were measured using a Tencor P-16 profilometer. For electrical transport measurement, Ag contact pads were sputtered on the samples through a shadow mask. The samples were then patterned using standard photolithography (Leica) to create two microbridges of length ~500 μm and width of 20 and 40 μm respectively. The details of the patterning and sample wiring for the transport measurement can be found in our previous works.[25, 26] The samples were mounted on a oxygen-free Cu stage using Ag paste and resistance–temperature (R-T) and current-voltage (I-V) characteristic curves were measured as function of temperature T (65K-77 K) and the magnetic field B (up to 9.0 T) applied in the c-axis of the BZO/YBCO films in a Quantum Design Ever-Cool II Physical Property Measurement System (PPMS). To minimize the heating at high currents, a pulsed current source (Keithley 2430 1KW Pulse Source Meter) was adopted with the pulse width of ~ 500 ms for the I-V curves. J_c was determined by applying 1 μV/cm standard criterion.[26-28] The microstructural images of the films were acquired via Cs-corrected transmission electron microscopy (TEM) and scanning transmission electron microscopy (STEM). The TEM and STEM results were achieved using a 1.6 Å resolution Thermo Fisher Scientific TALOS F200X TEM. All STEM images were captured using a high angle annular dark field mode (HAADF). Furthermore, a new generation Thermo Fisher Scientific microscope was used to acquire high-resolution STEM (HRSTEM) images with a resolution ~ 63 pm.

3. Results and Discussions

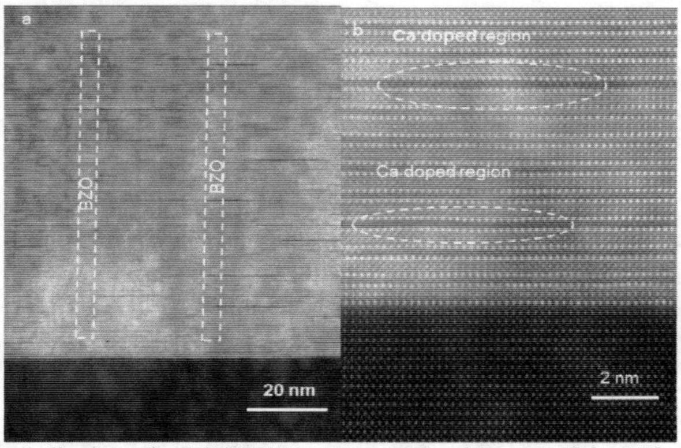

Figure 1. HRSTEM cross-section images of the ML film. Figure 1a indicate the growth of c-axis aligned BZO 1D APCs through the film's thickness. The BZO/YBCO interface lattice distortions around the APCs suggesting a coherent BZO/YBCO interface. This coherent BZO/YBCO interface occurs because of the increase in YBCO c-lattice parameter as a result of the formation of stacking faults in the Ca rich regions (enclosed in ellipse Figure 1b) due to the dynamic Ca/Cu replacement.

Figure 1 displays the HRSTEM cross-section images of the ML films. Figures 1a shows c-axis aligned BZO 1D-APCs embedded in the films. This is consistent with the expectation that Ca/Cu replacement takes place after the APCs are formed and hence would not interrupt the epitaxial growth of the APCs. Nevertheless, Figure 1a indicates that the APCs in the ML are segmented by the CaY123 spacers but the relative thinness (~ 10 nm) of the spacers may ameliorate some of the pinning detriments associated with truncated APC growth in multilayer YBCO structures with much thicker YBCO spacers [29-31]. Furthermore, significant YBCO lattice distortions are largely absent in the ML. This distortion in the YBCO lattice of SL film, which have been reported previously [32-36], arises from the strain around the BZO/YBCO interface as a result of the ~ 7.7% BZO/YBCO lattice mismatch. The absence of these defects in the ML may be attributed to the stacking faults formed in the Ca rich region (enclosed in ellipse in Figure 1b) of the ML film. The formation of the stacking faults, via Ca/Cu replacement, locally extends the YBCO c-lattice parameter thereby reducing the BZO/YBCO lattice mismatch around the BZO/YBCO interface and hence the interfacial strain.

Figure 2 compares the XRD θ-2θ spectra of 2% doped (a and b) and 6% doped(c and d) films. The black and red lines represent the SL and ML films respectively. The appearance of the YBCO (00l)

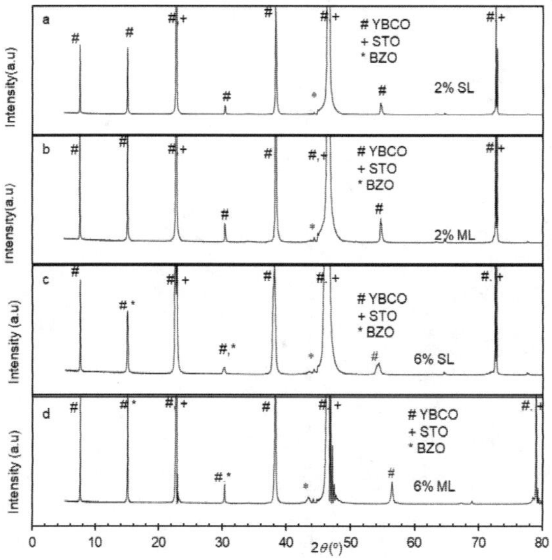

Figure 2. θ-2θ scans of SL (a and c) and ML (b and d) films at BZO doping of 2% (a and b) and 6% (c and d). The ML films were fabricated with structure or/and PLD growth conditions of: t_{spacer} =10nm, and R_{spacer} =2Hz . Cu-kα radiation of wavelength 1.54 Å was used. The symbols #, + and * represent the YBCO (001), STO substrate (100) and major BZO (001) peaks respectively.

peaks (#) confirms the c-axis orientation of the BZO/YBCO nanocomposite films on the (100) STO substrates (+). Also, a major BZO (001) peak at ~ 46.5° appears in all three samples. The c-lattice constants were estimated from the YBCO (00l) peaks. The c-lattice constants were 11.746 Å and 11.819 Å for the 2% and 6% SL films respectively, confirming the increase in tensile strain as BZO doping concentration is increased. This increase in tensile strain is most probably due to the increased strain field overlap [37, 38]. Conversely, the same c-lattice constants of 11.766 Å observed on both the 2% and 6% ML samples may indicate a tempering down of the BZO doping induced strain field overlap effect by the ML scheme. Clearly, the enlargement of the YBCO c-axis lattice constant occurs only locally near BZO 1D-APC/YBCO and YBCO/STO interfaces due to formation of stacking faults induced by Ca/Cu replacement on Cu-O planes as revealed in the HRSTEM study in Figures 1b.

Figure 3 is the J_c *(B)* and F_p (*B*) curves of the 2% doped films (3a-3c) and 6% doped films (3d-3f) along various orientations of *B* at 65 K. The SL films (open symbols) are compared to the ML films (solid symbols) at *B*//c (Figures 3a-3b and 3d-3e) and at other *B* orientations (Figures3c and 3f). Comparisons of J_c *(B)* and F_P *(B)* curves along *B//c* confirm enhancement of pinning for magnetic field applied along the c-axis. The F_P *(B)* data taken along other *B* orientations confirm the extension of the pinning enhancement beyond *B*//c. For the 2% doped films, J_c enhancement can be observed over the entire *B* field range up to 9.0 T. For example, at 1.0 T, 5.0 T and 9.0 T, the corresponding J_c values of 4.39 MA/cm^2, 1.58 MA/cm^2 and 0.59 MA/cm^2 for the ML sample are respectively about 1.7, 1.8 and 2.7 times of that of the SL film. Qualitatively, both F_P *(B)* curves exhibit an inverted bell shape with a peak value of F_{pmax} located at B_{max} (Figure 3b). Interestingly, the 2% ML and 2% SL samples have comparable B_{max} values of 3.0 T, which is smaller than the accommodation field (~5.0 T) calculated from $B^* = n^*\Phi_o$, based on the flux quantum Φ_o ~2.07*10^{-15} Wb and areal density n^* of the BZO 1D-APCs acquired from TEM characterization[1, 16]. Therefore, the insertion of the CaY-123 spacer layer does not seem to affect the self-assembly of the BZO 1D-APCs in the 2% ML samples. The F_{pmax} along *B*//c of ~ 97.7 GNm^{-3} for the 2% ML sample at 65 K is 1.7 times of the F_{pmax} of ~ 57.1 GNm^{-3} in the 2% SL sample and surpasses the ~ 80 GNm^{-3} of the 2% BHO-YBCO SL sample [16]. It should be noted that the F_{pmax} values for the 2% SL sample are comparable to that reported in literature [5, 39-41]. Indeed, Ca2+/Y3+ substitution could be energetically preferred in strain-free or low-strain conditions at low BZO doping. This means that Ca/Cu and Ca/Y substitutions may occur simultaneously in the 2 vol.% BZO doped BZO/YBCO nanocomposite films with the former near the BZO 1D-APC/YBCO interface as confirmed in Fig. 1 while the latter, away from the interface. The Ca/Y substitution is well known to overdope YBCO as illustrated typically in the T_c decrease with the

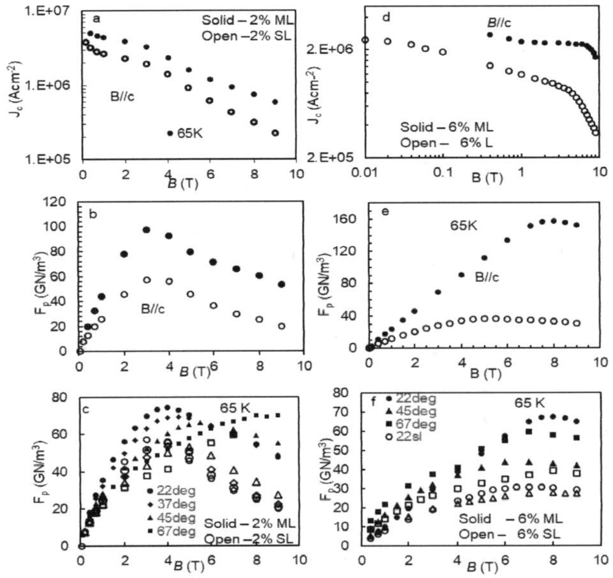

Figure 3. Transport data on the 2% doped (3a-3c) and 6% doped (3d-3f) films at 65K. The SL films (open symbols) are compared to the ML films (solid symbols) at *B*//c (3a-3b and 3d-3e) and at other *B* orientations (3c and 3f). Comparisons of J_c *(B)* and F_p *(B)* curves along *B//c* confirm enhancement of pinning for magnetic field applied along the c-axis. The F_p *(B)* data taken along other *B* orientations confirm the extension of the pinning enhancement beyond *B*//c.

substitution. Based on the moderately lower Tc of 87.5 K which is ~ 1 K lower than that of its SL counterpart's, we suspect the Ca/Y substitution induced overdoping effect (negative) is minor. This argument is consistent with the in-field pinning enhancement in 2 % ML over the 2% SL counterpart.

Therefore, the improved pinning by the BZO 1D-APCs in the 2% BZO-YBCO ML sample could be attributed to the coherent BZO/YBCO interface of the ML film. The data indicate that the enhancement of pinning in the 2% ML film clearly extends beyond 0°. The pinning improvement in the 2% ML sample is illustrated in the overall higher F_P values almost over the entire B field range (Figures 3b-3c). The F_{pmax} along 22° - 67° in the 2% ML samples are larger than the corresponding values in the 2% BZO-YBCO SL film. For example, along 22° F_{pmax} is ~ 74.3 GNm^{-3} for the 2% BZO-YBCO ML and only 55.3 GNm^{-3} for the 2% BZO-YBCO SL sample. Also, along 67°, the F_{pmax} of ~ 70.0 GNm^{-3} measured on the 2% BZO-YBCO ML is ~70% more than the value measured on the 2% SL film. This pinning enhancement close to the ab-plane may be largely attributed to the ab-plane aligned stacking faults generated by the Ca/Cu replacement in YBCO. The extension of pinning enhancement beyond $B//c$ is not limited to the 2% ML as can be seen in the J_c (B) and F_P (B) data on the 6% doped films in Figures 3d-3f. Similar to the 2% doping case along $B//c$, 6% ML sample exhibits enhanced $J_c(B)$ over the entire B field range of 0-9 T as compared to their 6% SL film and the enhancement increases with the applied B fields (Figure 3d). For example, an enhancement factor of 2.0 and 5.0 can be observed at B= 1.0 T and 9.0 T respectively (Figure 3d). Clearly, $J_c(B)$ curve for the 6% ML sample exhibits much reduced B-field susceptibility as compared with the 2% ML film and hence more enhanced pinning in a larger B field range. This is expected as the 6% ML film has three times the concentration of the BZO 1D-APCs as the 2% ML sample. The improved $J_c(B)$ leads to an improved F_p (B) in the ML samples as shown in Figure 3e. While the B_{max} is similar in the 2% doped films, the enhanced B_{max} ~ 8.0 T along $B//c$ in the 6% ML sample is 60% more than that of the 6% SL sample. Furthermore, the enhanced F_{pmax} ~ 158 GN/m^3 in the 6% ML sample (which is more than 4 times that of the 6% SL sample) is the best reported so far in BZO 1D-APC/YBCO nanocomposites films[5, 17, 42-45]. Figure 3f indicates that, similar to the 2% ML, pinning benefits in the 6% ML extends well beyond $B//c$ and up to ~ 67°. As mentioned above, this extension of benefits close to the ab plane may be attributed to the formation of 2D planar defects induced by the Ca doping.

As a further confirmation of the extension of improved pinning beyond $B//c$ in ML films, Figure 4a-

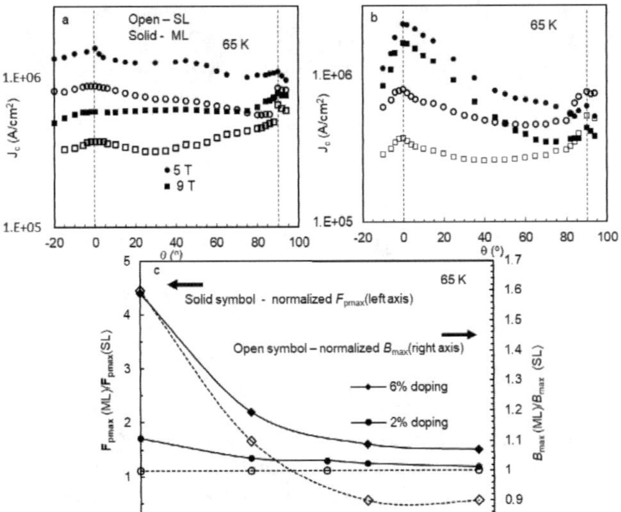

Figure 4. The pinning properties of the films at 65K. J_c (θ) data on 2% doping (a) and 6% doping (b) in 5 T (circle) and 9 T (square). Open and solid symbols represent single-layer (SL) and multilayer (ML) films respectively. (c) The normalized (the ML film data normalized to that of the SL film) $F_{P,max}$ (solid symbols and left axis) and normalized B_{max} (open symbols and right axis) as a function of (θ). Circle and diamond represent 2% doped and 6% doped film respectively.

4b compares the J_c $(\theta$, 65K) data for the samples in Figure 3. The 2% (Figures 4a) and 6% (Figures 4b) are compared at 5.0 T (circle) and 9.0 T (square). The solid and open symbols correspond to the ML and SL films respectively. Essentially, each ML film has higher J_c than its corresponding SL sample over the entire θ range. Case in point, the maximum $J_c(\theta \leq 85°)$ of 2% ML sample is almost double and about 1.5 times that of the 2% SL film at 5.0 T and 9.0 T respectively. J_c (θ) enhancement is even more pronounced at a higher BZO doping of 6 vol.%. In the range $\theta \leq 85^{o}$, enhancement factors as high as 2.8 and 4.5 are seen at 5.0 T and 9.0 T respectively. Furthermore, the $F_{p,max}$ values of the ML were normalized to that of the SL counterpart at different θ angles. In Figure 4c the normalized $F_{p,max}$ values (solid symbol and left axis) at 65K were plotted as function of θ for 2% (circle) and 6% (diamond). The data shows a monotonic decrease in the normalized $F_{p,max}$ with θ and the highest enhancement factors along B//c of 1.7 and 4.4 for the 2% ML and 6% ML respectively. When B is oriented to ~ 67° these factors are reduced to ~ 1.0 and 1.5 respectively. Similar to the $F_{p,max}$, the normalized B_{max} (open symbol and right axis) was calculated and plotted on the same figure (Figure 4c). In contrast to $F_{p,max}$, the B_{max} enhancement is largely insignificant (enhancement factor mostly ~ 1.0) for both BZO doping levels. The only significant B_{max} enhancement is the 1.6 value observed along B//c for the 6% ML film. The dramatic enhancement $F_{p,max}$ and B_{max} along B//c in the 6% ML sample is not surprising considering the expected benefits of improved BZO/YBCO interface and reduced strain field overlap at high concentration of the BZO 1D APCs. This angular enhancement in ML films is comparable to what has been reported on the BHO/YBCO nanocomposite films in which the 1D-APC/YBCO interface is coherent[20, 21].

4. Conclusions

In summary, a Ca doped BZO/YBCO nanocomposite ML scheme has been employed to extend angular width of pinning to ~ 67° - 85° for 2% and 6% BZO doped YBCO nanocomposites by promoting the formation of stacking faults and coherent BZO/YBCO interfaces. The Ca source is provided from the two ~10 nm thick CaY-123 spacers inserted into the 2% and 6% BZO/YBCO nanocomposite films of 150 nm in total thickness to form the ML structures. The coherent interface was induced by a reduction in the lattice mismatch at the BZO/YBCO interface as a result of the formation of the stacking faults. Microstructural analysis using HRSTEM confirms the formation of the stacking faults and the coherent BZO/YBCO interface. Remarkably, the BZO 1D-APCs with a coherent interface with YBCO matrix exhibits wide angular pinning enhancement as demonstrated in the higher transported J_c (θ) and F_p (θ) at 65 K. For example, at B = 5 T and 9 T, the J_c data in the ML films are significantly higher than those of the SL films up to ~ 85°. Furthermore, $F_{p,max}$ data show pinning enhancements up to at least ~ 67°. Similarly, along B//c the $F_{p,\ max}$ enhancements are 1.7 and 4.4 for the 2% ML and the 6% ML respectively. The 6% ML also has 60% bigger B_{max} than the B_{max} of ~ 5.0 T of the 6% SL film. These results, together with the remarkable enhancement in pinning along B//c, demonstrates not only the critical importance of the BZO/YBCO interface in specific pinning efficiency of 1D-APCs, but also points to a viable approach to achieve it.

5. References

[1] Baca, F.J., et al., *Interactive Growth Effects of Rare-Earth Nanoparticles on Nanorod Formation in $YBa_2Cu_3O_x$ Thin Films.* Advanced Functional Materials, 2013. **23**(38): p. 4826-4831.
[2] Chen, S., et al., *Enhancement of isotropic pinning force in YBCO films with $BaZrO_3$ nanorods and Y_2O_3 nanoparticles.* IEEE Trans. Appl. Supercond, 2017. **27**(4): p. 4-8.
[3] MacManus-Driscoll, J., et al., *Strongly enhanced current densities in superconducting coated conductors of $YBa_2Cu_3O_{7-x}$+ $BaZrO_3$.* Nature materials, 2004. **3**(7): p. 439.
[4] Matsumoto, K. and P. Mele, *Artificial pinning center technology to enhance vortex pinning in YBCO coated conductors.* Superconductor Science and Technology, 2009. **23**(1): p. 014001.
[5] Wee, S.H., et al., *Engineering nanocolumnar defect configurations for optimized vortex pinning in high temperature superconducting nanocomposite wires.* Scientific Reports, 2013. **3**: p. 2310.
[6] Gautam, B., et al., *Microscopic adaptation of $BaHfO_3$ and Y_2O_3 artificial pinning centers for strong and isotropic pinning landscape in $YBa_2Cu_3O_{7-x}$ thin films.* Superconductor Science and Technology, 2018. **31**(2): p. 025008.
[7] Matsushita, T., *Flux pinning in superconducting 123 materials.* Superconductor Science and Technology, 2000. **13**(6): p. 730.

[8] Miura, S., et al., *Characteristics of high-performance $BaHfO_3$-doped $SmBa_2Cu_3O_y$ superconducting films fabricated with a seed layer and low-temperature growth.* Superconductor Science and Technology, 2015. **28**(6): p. 065013.
[9] Sebastian, M.A.P., et al., *Study of the flux pinning landscape of YBCO thin films with single and mixed phase additions $BaMO_3$+ Z: M= Hf, Sn, Zr and Z= Y_2O_3, Y211.* IEEE Transactions on Applied Superconductivity, 2017. **27**(4): p. 1-5.
[10] Tobita, H., et al., *Fabrication of $BaHfO_3$ doped $Gd_1Ba_2Cu_3O_{7-\delta}$ coated conductors with the high Ic of 85 A/cm-w under 3 T at liquid nitrogen temperature (77 K).* Superconductor Science and Technology, 2012. **25**(6): p. 062002.
[11] Mele, P., et al., *Ultra-high flux pinning properties of $BaMO_3$-doped $YBa_2Cu_3O_{7-x}$ thin films (M= Zr, Sn).* Superconductor Science and Technology, 2008. **21**(3): p. 032002.
[12] Varanasi, C., et al., *Thick $YBa_2Cu_3O_{7-x}$+ $BaSnO_3$ films with enhanced critical current density at high magnetic fields.* Applied Physics Letters, 2008. **93**(9): p. 092501.
[13] Feldmann, D., et al., *Improved flux pinning in $YBa_2Cu_3O_7$ with nanorods of the double perovskite Ba_2YNbO_6.* Superconductor Science and Technology, 2010. **23**(9): p. 095004.
[14] Opherden, L., et al., *Large pinning forces and matching effects in $YBa_2Cu_3O_{7-\delta}$ thin films with $Ba_2Y(Nb/Ta)O_6$ nano-precipitates.* Scientific reports, 2016. **6**: p. 21188.
[15] Foltyn, S., et al., *Materials science challenges for high-temperature superconducting wire.* Nature materials, 2007. **6**(9): p. 631.
[16] Gautam, B., et al., *Probing the effect of interface on vortex pinning efficiency of one-dimensional $BaZrO_3$ and $BaHfO_3$ artificial pinning centers in $YBa_2Cu_3O_{7-x}$ thin films.* Applied Physics Letters, 2018. **113**(21): p. 212602.
[17] Wu, J., et al., *Pinning Efficiency of One-Dimensional Artificial Pinning Centers in $YBa_2Cu_3O_{7-x}$ Thin Films*. Vol. 29. 2019. 1-5.
[18] Cantoni, C., et al., *Strain-driven oxygen deficiency in self-assembled, nanostructured, composite oxide films.* Acs Nano, 2011. **5**(6): p. 4783-4789.
[19] Blatter, G., et al., *Vortices in high-temperature superconductors.* Reviews of modern physics, 1994. **66**(4): p. 1125.
[20] Ogunjimi, V., et al. *The effect of APC/YBCO interface on the angular range of effective pinning by one-dimensional artificial pinning centers in $YBa_2Cu_3O_{7-x}$ nanocomposite films.* in *IOP Conference Series: Materials Science and Engineering*. 2020. IOP Publishing.
[21] Ogunjimi, V., et al., *The angular range of effective pinning by one-dimensional artificial pinning centers in $BaZrO_3/YBa_2Cu_3O_{7-x}$ nanocomposite films.* AIP Advances, 2019. **9**(8): p. 085110.
[22] Ogunjimi, V., et al., *Enhancing Magnetic Pinning by $BaZrO_3$ Nanorods Forming Coherent Interface by Strain-Directed Ca-doping in $YBa_2Cu_3O_{7-x}$ Nanocomposite Films.* Superconductor Science and Technology, 2021. 34 (10): p. 104002
[23] Baca, F.J., et al., *Control of $BaZrO_{(3)}$ nanorod alignment in $YBa_{(2)}Cu_{(3)}O_{(7-x)}$ thin films by microstructural modulation.* Applied Physics Letters, 2009. **94**(10): p. 3.
[24] Sebastian, M.A., et al., *Comparison Study of the Flux Pinning Enhancement of $YBa_2Cu_3O_{7-\delta}$ Thin Films With $BaHfO_3$ + Y_2O_3 Single- and Mixed-Phase Additions.* IEEE Transactions on Applied Superconductivity, 2019. **29**(5): p. 1-5.
[25] Wang, X., et al., *Eliminating thickness dependence of critical current density in $YBa_2Cu_3O_{7-x}$ films with aligned $BaZrO_3$ nanorods.* Journal of Applied Physics, 2010. **108**(11): p. 3911.
[26] Chen, S., et al., *Enhancement of Isotropic Pinning Force in YBCO Films With $BaZrO_3$ Nanorods and Y_2O_3 Nanoparticles.* IEEE Transactions on Applied Superconductivity, 2017. **27**(4): p. 1-5.
[27] Gautam, B., et al., *Microscopic Adaptation of $BaHfO_3$ and Y_2O_3 Artificial Pinning Centers for Strong and Isotropic Pinning Landscape in $YBa_2Cu_3O_{7-x}$ Thin Films.* Superconductor Science and Technology, 2017.
[28] Emergo, R.L.S., et al., *The effect of thickness and substrate tilt on the BZO splay and superconducting properties of $YBa_2Cu_3O_{7-\delta}$ films.* Superconductor Science and Technology, 2010. **23**(11): p. 115010.
[29] Horide, T., M. Ishimaru, and K. Matsumoto, *Observation of inhomogeneous depinning in $YBa_2Cu_3O_7$ composite multilayers.* Superconductor Science and Technology, 2019. **32**(8): p. 085001.

[30] Horide, T., et al., *Hybrid artificial pinning centers of elongated-nanorods and segmented-nanorods in $Yba_2Cu_3O_7$ films.* Superconductor Science and Technology, 2016. **29**(10): p. 105010.
[31] Malmivirta, M., et al., *Enhanced flux pinning in YBCO multilayer films with BCO nanodots and segmented BZO nanorods.* Scientific reports, 2017. **7**(1): p. 1-8.
[32] Horide, T., et al., *Structural evolution induced by interfacial lattice mismatch in self-organized $YBa_2Cu_3O_{7-\delta}$ nanocomposite film.* ACS nano, 2017. **11**(2): p. 1780-1788.
[33] Shi, J.J. and J.Z. Wu, *Structural transition of secondary phase oxide nanorods in epitaxial $YBa_2Cu_3O_{7-\delta}$ films on vicinal substrates.* Philosophical Magazine, 2012. **92**(34): p. 4205-4214.
[34] Wu, J. and J. Shi, *Interactive modeling-synthesis-characterization approach towards controllable in situ self-assembly of artificial pinning centers in RE-123 films.* Superconductor Science and Technology, 2017. **30**(10): p. 103002.
[35] Wu, J.Z., et al., *The effect of lattice strain on the diameter of $BaZrO_3$ nanorods in epitaxial $YBa_2Cu_3O_{7-\delta}$ films.* Superconductor Science and Technology, 2014. **27**(4): p. 044010.
[36] Yoshida, Y., et al., *Approaches in controllable generation of artificial pinning center in $REBa_2Cu_3O_y$-coated conductor for high-flux pinning.* Superconductor Science and Technology, 2017. **30**(10): p. 104002.
[37] Khan, M.Z., et al., *Enhanced flux pinning isotropy by tuned nanosized defect network in superconducting $YBa_2Cu_3O_{6+x}$ films.* Scientific reports, 2019. **9**(1): p. 1-12.
[38] Samoilenkov, S., et al., *Anisotropic strain of $BaZrO_3$, $BaCeO_3$ and Y_2O_3 nanoinclusions in a $YBa_2Cu_3O_{7-x}$ epitaxial film matrix and its relation to the oxygen content of the superconductor.* Superconductor Science and Technology, 2011. **24**(5): p. 055003.
[39] Huhtinen, H., et al., *The effect of BZO doping concentration and thickness dependent properties of YBCO films grown by PLD on buffered NiW substrates.* Physica C: Superconductivity, 2012. **472**(1): p. 66-74.
[40] Huhtinen, H., et al., *Influence of $BaZrO_3$ dopant concentration on properties of $YBa_2Cu_3O_{6+x}$ films in magnetic fields up to 30 T.* Journal of Applied Physics, 2010. **107**(5): p. 053906.
[41] Jha, A.K., et al., *Tailoring the vortex pinning strength of YBCO thin films by systematic incorporation of hybrid artificial pinning centers.* Superconductor Science and Technology, 2015. **28**(11): p. 114004.
[42] Goyal, A., et al., *Irradiation-free, columnar defects comprised of self-assembled nanodots and nanorods resulting in strongly enhanced flux-pinning in $YBa_2Cu_3O_{7-\delta}$ films.* Superconductor Science and Technology, 2005. **18**(11): p. 1533.
[43] Gutierrez, J., et al., *Strong isotropic flux pinning in solution-derived $YBa_2Cu_3O_{7-x}$ nanocomposite superconductor films.* Nature materials, 2007. **6**(5): p. 367-373.
[44] Horide, T., et al., *Geometric and compositional factors on critical current density in $YBa_2Cu_3O_{7-\delta}$ films containing nanorods.* Superconductor Science and Technology, 2018. **31**(6): p. 065012.
[45] Wang, X., et al., *Eliminating thickness dependence of critical current density in $YBa_2Cu_3O_{7-x}$ films with aligned $BaZrO_3$ nanorods.* Journal of Applied Physics, 2010. **108**(11): p. 113911.

Acknowledgement
This research was supported in part by NSF contracts Nos: NSF-DMR-1508494 and 1909292 and NSF-ECCS-1809293, the AFRL Aerospace Systems Directorate, the Air Force Office of Scientific Research (AFOSR) LRIR #14RQO8COR and LRIR #18RQCOR100. D.Z. and H.W. acknowledge the support from the U.S. National Science Foundation for the high-resolution STEM effort at Purdue University (DMR-1565822 and DMR-2016453).

IOP Conf. Series: Materials Science and Engineering 1241 (2022) 012023 doi:10.1088/1757-899X/1241/1/012023

Low cost, simpler HTS cable conductors for fusion energy systems

Y Zhai[1], A Otto[2] and M Zarnstorff[1]

[1]Princeton Plasma Physics Laboratory, Princeton, NJ 08540
[2]Solid Material Solutions, LLC, Chelmsford, MA 01863

Email: yzhai@pppl.gov

Abstract. The cost and complexity of large, high-field superconducting magnet modules and related subsystems comprise 30% to 60% of the fusion reactor core capital cost. The strategic plan for the U.S. burning plasma research, the Fusion Energy Sciences Committee Report (FESAC) "Power the Future: Fusion and Plasmas", and 2021 NASEM report "Key Goals and Innovations needed for a U.S. Fusion Pilot Plant" recommends that the U.S. pursue innovative science and technology to enable construction of a Fusion Pilot Plant (FPP) that produces net electricity from fusion at reduced capital cost. To achieve this, a novel combination of lower-cost high temperature superconductors (HTS) in cable configurations with co-wound reinforcement for higher current density are being investigated using a simplified construction strategy to produce compact stable coils. They would be capable of generating 20 T at up to 10-20 K. Small-scale, inexpensive test coils and prototypes will help develop each feature and validate cabled conductor design models. The near term goal is to validate engineering approaches, scientific models and fabrication capabilities applicable to fusion reactor development such as U.S. fusion nuclear science facility (FNSF), sustained high-power density tokamak facility (SHPD) and FPP designs. The design options include lower-cost, high-strength, quench resistant REBCO or Bi-2212 cables in an all metal coil design that simplifies HTS coil construction and quench protection system, with co-wound reinforcements that integrate stress management in HTS cable design and provides thermal mass to help prevent quench damage.

1. Introduction

The fusion energy sciences advisory committee (FESAC) recently reported that 2G high-temperature superconductors (HTS) are transformative for the U.S. to pursue that could promote efficient advances towards fusion energy, building on burning plasma science and technology [1]. The National Academies also released a new report in 2021 on bringing fusion to the U.S. Grid with the committee recommendations on the key goals and innovation needed for the U.S. Fusion Pilot Plant (FPP) [2]. At least two different magnetic configurations are under consideration [3-5]. Tokamaks, with significant time-changing fields and thus coil voltages in central solenoid (CS), poloidal field (PF) and control coils, may require high current cables or novel insulations, while stellarators, where magnetic fields are in steady state and there are several strategies for making the 3D field shaping, can be more suitable for no insulation (NI) coils, a new enabling coil technology recently developed [6] and validated for high field magnet applications [7-9]. Significant progress has been made in the past years on the maturity of HTS for high field magnet applications [10-16], including prototyping and testing of NI coils, testing of high current HTS cables and fusion relevant model coils at privately funded startup companies such as ST40 experiment at Tokamak Energy [13] and SPARC at Commonwealth Fusion (CFS) [12,14].

The state-of-the-art low temperature superconducting (LTS) magnet technology such as Nb_3Sn cable-in-conduit conductors (CICCs) for coil winding adopted by ITER is costly (over \$5000 per meter for CICCs only) with achievable engineering current density of ~50 A/mm^2 in CICCs and <20 A/mm^2 in the toroidal field (TF) coil winding pack. Higher current densities are required for compact fusion [3-5]. The rapid advances of HTS cables to achieve high field and high current density (>100 A/mm^2) for fusion motivated the consideration of low aspect ratio compact fusion energy systems [3]. HTS is promising in reducing size and cost by increasing the allowable magnetic field B, current density, and operating temperature. These HTS characteristics are high-leverage since the maximum fusion power density scales as B^4 and the fusion core mass and cost scale with device size R roughly as R^3. Of particular importance to the tokamak design is the ability of the CS magnet to quickly discharge from a high current level, which provides sufficient magnetic flux swing that induces a toroidal plasma current for starting plasma. The result of the Fusion Energy Systems Studies (FESS) showed that a high current density of over 70-80 A/mm^2 in the TF and CS coil winding pack is required to achieve a satisfactory compact tokamak system design [4,10].

The TF, CS and PF coils have the largest cost of all subsystems in a magnetic-confinement fusion energy system. A cost model analysis for REBCO coated conductors showed a linear scaling for building parallel plants that is unlikely to achieve the desired goal of 20\$/kA-m at >20 T and 10-20 K for fusion energy systems, despite the high critical current density achieved. BSCCO (2223 and 2212) is the HTS that can be fabricated into both tape and either round or rectangular multi-filamentary wire, and has been commercially proven by Sumitomo Electric Industries, Ltd. with low cost and reliable 2223 tape fabrication [17]. BSCCO HTS conductors are mature with well over 1000 km in shipped total length, for very large orders. Bi-2223 wire demonstrated low cost characteristics and it was shown that low loss Bi-2212 can be more cost-effective than REBCO coated conductors for fast ramp fields in central solenoids [10,17-18]. Bi-2212 with loss reduction features [18] can provide the best combination of attributes and cost effectiveness. Unlike REBCO coated conductors with a slow normal zone quench propagation of mm/s, BSCCO has a faster normal zone propagation velocity (cm/s), which is critical for high field magnet quench detection and protection.

Other system integration challenges in fusion magnet engineering include nuclear effects on the coil winding pack such as neutron damage of superconductors, degradation of insulators and the residual resistance ratio (RRR) in copper and silver stabilizers, as well as volumetric nuclear heating of cryogenic TF magnets. Operating modes such as pulsed versus steady state will also impact reactor system radial build and the magnet design concept. No insulation (NI) coil technology has been developed and a number of small coils were tested for small bore high field HTS magnet applications. Despite these key advantages, NI coil designs have not been extended to large coil applications. In fusion nuclear environment, advantages of NI coils for TF application include acceptable long charging time for long pulse or steady state plasma operations, and advantages for TF coil organic insulation removal includes: 1) improved radiation resistance as insulation has lower neutron fluence limits, a design driver; 2) mitigate mechanical strain effects due to insulation soft sponginess and delamination due to insulation normal tensile strain; and 3) high fabrication cost of vacuum pressure impregnation (VPI) and error-prone VPI process. Many fusion coil failures (conventional fabrication approach) can be traced back to insulation mechanical failure due to mechanical strain effect.

2. High temperature superconductors

Commercially available REBCO coated flat-tape conductors and Bi-2212 multifilamentary wires in either round or rectangular form are the most promising HTS types for a potentially near term fusion reactor design. The high critical current density and critical temperature makes HTS the only option for the TF and CS coil design of low cost fusion reactors in >20 T and 10-20 K liquid helium-free operating environments.

2.1 REBCO coated conductors

YBCO coated conductors are promising in present fusion application due to their proven high critical current density and mechanical tensile strength (>500 MPa longitudinal stress limit and ~0.5% irreversible strain compared to 0.2~0.3% in Nb_3Sn LTS for ITER) [19-21]. In addition, YBCO coil fabrication can be simpler than LTS coil, since it does not require heat treatment and vacuum pressure impregnation.

As a result of fast-growing private investment, SuperOx in 2019 manufactured over 300 km of 4 mm wide YBCO coated conductor, mostly delivered to CFS (over $10M ordered). This was the largest completed order publicly known for the REBCO wire industry. The wire was fabricated on a strong Hastelloy C276 substrate with a buffer layer architecture based on textured MgO using ion beam assisted deposition and with the HTS layer grown by pulsed laser deposition (PLD) [16].

Despite this success in fusion magnet application developments, challenging issues remain on the maturity of YBCO for fusion magnet systems. These include slow normal zone propagation (NZP) for quench detection and coil protection, large deleterious screening current and related local strain effects, excessive AC losses during transient field ramp and potential delamination as a result of intrinsically low through-tape tensile and shear stress strengths in the coated conductor design [19].

2.2 Bi-2223 and Bi-2212

Bi-2212 wires provide an attractive combination of attributes and cost effectiveness for some coil types, primarily because Bi-2212 can be processed into high J_e wires with loss reducing features and flexible shapes. These wires can be used as sub-elements to fabricate low-loss, large cross-section, reinforced wire forms that provide the high operating currents required for low inductance in large, ramped field coils.

The most challenging requirement for an HTS wire or cable in a central solenoid for inductive plasma startup is achieving acceptably low AC losses during high field ramp rates. This requires innovative features in conductors. In multifilament wires, time-varying magnetic fields induce voltages that drive induced current loops within filaments, filament-to-filament and within the matrix, with corresponding dissipative energy losses. Losses have been widely studied with LTS and MgB_2 wire, and to a lesser extent with Bi-2223 and the REBCO tape. An early study at Oxford Superconducting Technology (OST) and National High Magnetic Field Laboratory (NHMFL) Applied Superconductivity Center, Florida State University (ASC/FSU) has shown that, even with considerable filament merging, AC loss in standard 2212 wire is reduced about two-fold by axial twisting [22]. More recently, a near-full set of loss reducing features for 2212 has also been developed by Solid Material Solutions, LLC (SMS), building on earlier work on the development of lower AC loss wires for transformer applications [23]. These features can be readily integrated into the development of low-loss 2212 conductors for fusion applications, such as the CS coils for fusion nuclear science facility (FNSF) and sustained high power density (SHPD) plasma startup that PPPL is developing. The outer diameter of the CS in a compact ST such as SHPD can be approximately 0.4 meter, due to the higher fraction of bootstrap currents in a compact Tokamak reactor design [24-26].

Hysteresis loss within the superconducting material is reduced by a multifilament structure with negligible filament merging along the wire length and small cross-sectioned filaments. Eddy current loss is typically a negligible contributor within the matrix material only. Filament coupling loss spans the matrix from one filament to the next, and is reduced by smaller wire diameters and proportionally reducing filament count. Coupling loss is also reduced by axially twisting the wire to the shortest practical pitch length at the selected wire diameter and increasing filament to filament resistances.

In the case of the CS coil designed for 5 T per second ramp rate (equivalent to the FNSF CS coil operation [4, 11]), a larger, higher filament count wire can be used, still with a relatively tight axial twist, and assembled pitch length with a fully transposed architecture. Several 2212-based wide wire designs are shown in Figure 1, including a CS coil type design. The size of the sub-element wire will be about 0.5 mm wide and with a thickness in the 0.2 to 0.3 mm range.

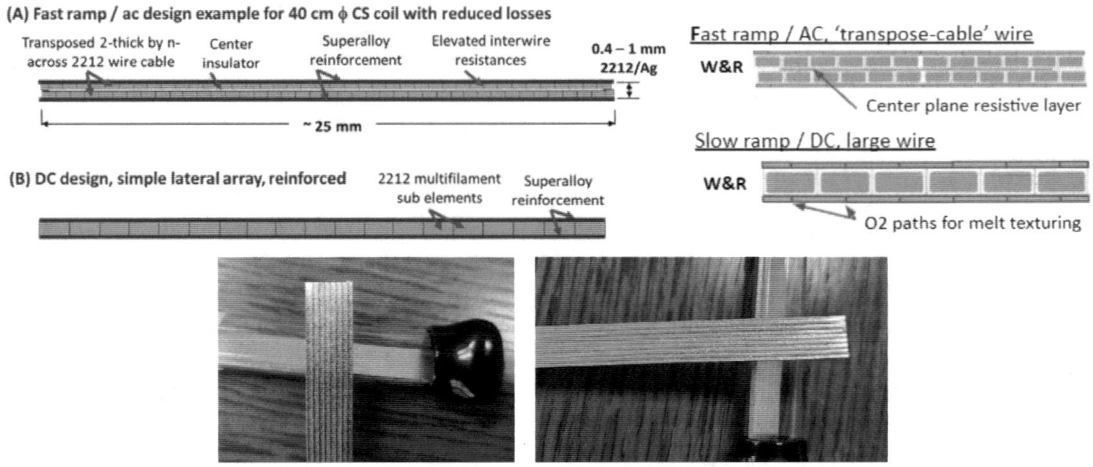

Figure 1. Cross sections of a range of specific wire design variation. (A) Fully transposed 2-layer thick (0.4-1 mm), N across (~25 mm wide) rectangular wire or cable conductor for fast ramp CS coils, (B) single layer, multifilament simple lateral array wire design with superalloy reinforcement for TF coils

3. HTS cables

A number of HTS cables based on stacked or layered tapes of REBCO coated conductors such as CroCo [27], the Conductor on round core (CORC) [28], and twisted stacked-tape cable (TSTC) [29] were developed for high field magnets including tokamak fusion applications. Due to the high critical current density achieved in coated conductors, CORC cable is one of the most promising cables being developed and tested in terms of HTS maturity for some large-scale high field magnet applications. The first CORC cable based model coil solenoid was recently tested at the National High Magnetic Field Laboratory (NHMFL), in the 160 mm large bore high field user magnet facility [15]. This is the first model coil wound with REBCO HTS cable tested. Another 15.2 cm diameter model coil solenoid using a 6 mm diameter CORC cable, consisting of 16 4 mm wide YBCO tapes was also developed between ACT and PPPL. The 2 layer, 12 turn HTS model coil is being tested at NHMFL on relevant parameters for the next step ST OH operations. The dimension and performance of the 2-layer model coil solenoid is directly scalable to the plasma startup operations needed for FNSF and SHPD.

A design using wide 2212 wire (see Figure 1(A)) for CS coils will have a 2 x N Rutherford build, with a central high resistivity superalloy strip to reduce layer to layer filament coupling, and higher resistance material between the wire sub-elements. The wide wire will be roll-consolidated and superalloy strip bonded to one or both surfaces, followed by reaction to form high J_c 2212, followed by testing and utilization of suitable pieces to wind the coil in the react-and-wind (R&W) mode. Wire coupling loss is analogous to filament coupling loss and is reduced by similar factors. Generally speaking, similar to the ITER CICC, cabling pitch is analogous to wire axial twist, with shorter pitch reducing loop size, and AC loss. Optimally increased wire-to-wire contact resistance reduces loss while allowing required current sharing [30]. Higher resistivity superalloy reinforcement strips are also applied to the center plane of 2 high by N wide low loss cables to provide strengthening plus controllable and optimal resistance increases between the top and bottom layer wires in the cable. The analysis technique established for Bi-2223 and 2212 identifies optimal conductor features and trade-offs that enable low loss even in high current cable design.

4. Coil designs for FNSF and SHPD

Fusion magnets are very different from high field magnets in high energy physics, magnetic resonance imaging (MRI) and nuclear magnetic resonance (NMR). They are uniquely challenging in requiring high fields in large coil-apertures or bore openings for sufficient plasma volume to produce a high neutron flux in steady state operation. The neutron irradiation limit for coil insulation and copper stabilizer can be lower than that of the superconductors. A fast current discharge is also required in the ohmic heating coil to provide sufficient flux swings needed for plasma startup. Table 1 presents the TF coil design parameter for FNSF.

Table 1. FNSF design parameters

	ITER	FNSF
TF conductors	Nb_3Sn CICC	6-around-1 CORC
Major/Minor radius (m)	6.2/2	4.8/1.2
Plasma current (MA)	15	8.0
Plasma center field (T)	5.3	9.3
TF operating current (kA)	68	64
TF coil max B field (T)	11.8	20.3
Winding pack current density (A/mm^2)	17	39
TF A-Turns (MA)	9.1	14.0
# of Turns	134	218
# of TF Coils	18	16
TF winding pack IB/OB radius (m)	2.78/10.75	2.2/10.5
Centering force (MN/m)	54	142-153
Vertical self-force/in-board leg (MN)	103	244-265
Length of coil centerline (m)	34.1	33.1
Winding pack (WP) volume (m^3)	12.4	11
Inductance per coil (H)	1.0	2.5
Total stored energy (GJ)	41	80
Fusion power (MW)	500	450

4.1 Toroidal Field Coil System

High current superconducting cables are typically needed to reduce coil inductance and allow quench protection in large scale insulated fusion magnets. This reduces the discharge voltage in coil insulation (i.e., limited to be within 20-30 kV during normal and fault conditions, for example as defined in ITER electrical design criteria [31-32]). For the cable-in-conduit (CICC) approach, it is unrealistic to model detailed performance and critical current behavior of each tape used in a HTS cable (stacked tape or others). Following similar approaches for qualifying LTS cables for ITER magnets, R&D programs are needed to validate HTS cable performance under the static load and cyclic fatigue load conditions. A

6-around-1 CORC cable-in-conduit conductor was shown to be promising for the FNSF TF coil winding pack in recent studies [10], see Figure 2. It is not a constant tension Princeton D shape, where the bending stress can be eliminated. The vertical location of the PF coils limits the transition of coil shape from the in-board to outboard leg, so a constant D shape is impractical. Figure 3 presents the stress distribution in the winding pack during the full field normal operation.

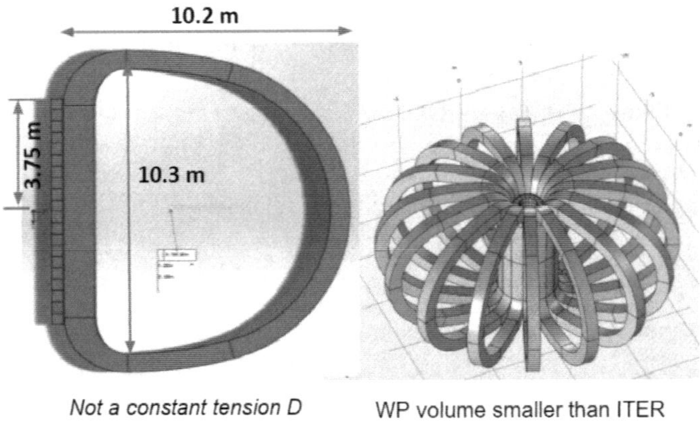

Figure 2. TF coil system for FNSF (left); shape of TF coil from inboard to outboard transition (right)

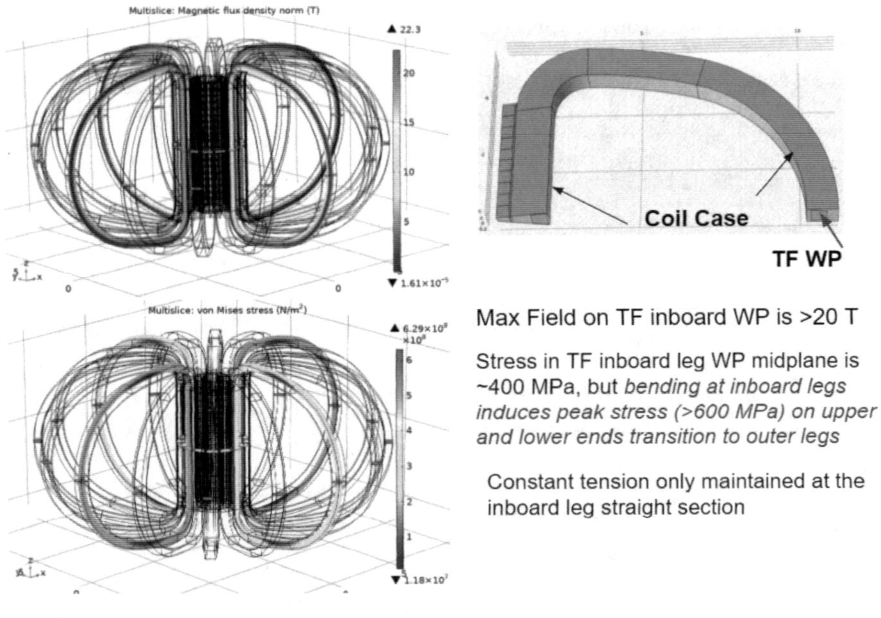

Figure 3. The magnetic field and von-Mises stress distribution of FNSF TF winding pack

4.2 Central Solenoid Coils

4.2.1 CS Coils for FNSF

A Bi-2212 conductor was considered and analyzed for the CS coil winding pack. Although there is no time limit in ramping up the CS before starting a plasma, ramping becomes complicated due to AC losses and power supplies once the plasma is formed, with ~100 seconds of ramp time used here for FNSF based on ITER. The startup is divided into two main phases, a breakdown and quick early ramp up of plasma current I_p, and then a slower ramp up to flattop. The CS and PF coils are connected across resistors initially to get a hard ramp very early for breakdown and early I_p burst, and then they are connected to power supplies for the slower ~100 second ramp up. The very early phase is about 1-5 seconds, and gets the I_p up to 100-500 kA. In the ~100 second ramp up phase, too slow of a ramp will waste volt-second in resistive losses in the plasma. Detailed calculations will be performed to quantify the impact of ramp up time to minimize the plasma resistive losses. Figure 4 presents the magnetic field and von Mises stress distribution for the FNSF central solenoid during the DC back bias loading condition prior to plasma startup with a uniform conductor current of 15.75 kA in each of the CS coil modules. The peak stress on the coil mid-plane is 630 MPa, less than the 660 MPa stainless steel 316 design stress limit. Unlike ITER long pulse plasma operations, the FNSF is designed to be operated in a steady state of days or weeks for each plasma run. Therefore, fatigue effects are not design driving factors.

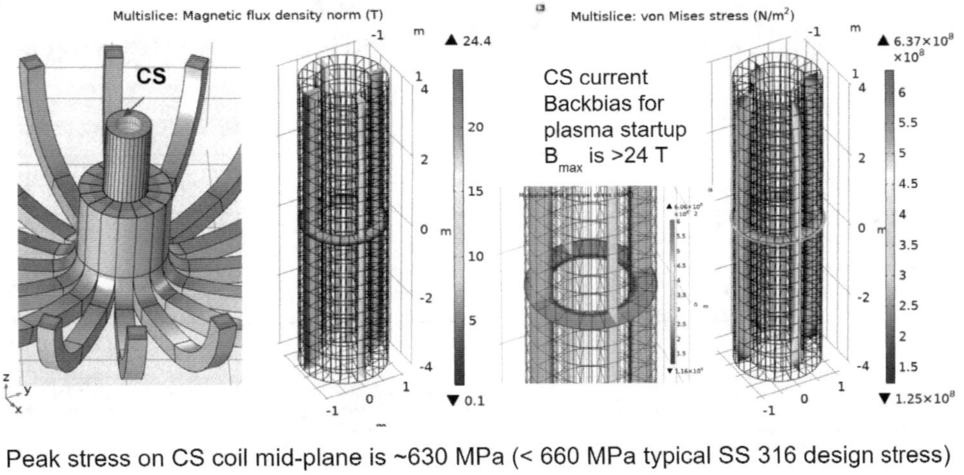

Figure 4. The magnetic field and winding pack stress analysis for FNSF central solenoid coil

4.2.2 CS Coils for SHPD

Our analysis results in Table 2 as well as in Figure 5 for the SHPD 40 cm bore diameter CS coil illustrate the effects of using the new, simple, lower cost Bi-2212 wide array design (Figure 1), with varied thicknesses, on this coil's characteristics, namely its radial build, outer diameter, required superalloy reinforcement and wire length, as well as its operating current and current density. The trade-off analysis in Table 2 is based on a reinforced Bi-2212 array of 25 mm width and thickness in the 0.4 mm to 0.8 mm range that is co-wound with high strength superalloy to keep total maximum strain in the 2212 conductor below 0.35%, and with a maximum operating current at 80% of Ic in the 4.2 K -10 K temperature range. It is seen that the lower cost, technically preferable react and wind (R&W) approach becomes feasible even as 2212/Ag thickness is increased (along with operating current) to a maximum level that still allows R&W coil winding to 40 cm diameter, without giving up too much strain margin so as to still have enough axial strain tolerance to support the Lorentz force induced hoop stress at the inner turns. This analysis of the wide array 2212 design has been found to enable R&W coil fabrication with this wide array 2212 wire for operating currents above 5 kA with present standard wire Je and in excess of 10 kA with best demonstrated Je levels. With this design, hoop stresses are also fully supported by the winding, negating the need for additional reinforcement.

Reacted 2212 can only be strained in tension during winding to below its critical strain of about 0.5%. In winding onto the smallest diameter layer, tensile strain on the outermost 2212 filaments must not exceed this critical strain in order to avoid Ic degradation. However, even for larger bore / wind diameters, where the outermost filaments remain below their critical strain levels, part of their strain tolerance budget is used up in winding, leaving only the remaining amount to their critical strain available to accommodate the Lorentz force-induced hoop strain arising when the coil is energised. As a result, as 2212/Ag thickness is increased, a greater amount of its tensile strain is used up in winding, and a relatively greater amount of reinforcement is required to support the Lorentz force, which in turn increases coil radial build and outer diameter while reducing its operating current density. In order to attain the required high operating current levels, for example > 5 kA and preferably > 10 kA, the conductor width can be increased, as well as its thickness. In recent work at Solid Material Solutions, 2212/Ag wire as wide as 25 mm has been fabricated by a low cost technique, and this width has been selected for the analysis presented.

Table 2. Low loss, high current 2212 wire options for PPPL SHPD 40 cm bore CS coil winding

Generic Features	Operating Temperature		4.2 K-10 K	Max strain for Bi-2212		0.35%
Wire with transposed low loss components	B at I_{op} in long solenoid		20 T	2212 modulus at 4.2 K		42 GPa
Strips thin enough to conform to coil diam	B at I_c in long solenoid		25 T	Superalloy modulus		270 GPa
Turn resistance material qty negligible	Wire width		25 mm	Superalloy CTE		13.5e-6 / K
Coil made of stack of pancakes	Coil inner diameter		40 cm	Assumed cooling		Conduction
React & wind /wind & react also an option	Field max ramp rate		5 T/s			
Attributes	Present 2212/Ag J_{op} ~420 A/mm^2			J_{op} ~700 A/mm^2, lower base cost*		
2212/Ag thickness in reinforced wire Max I_{op} of wire in coil (at 0.8 x I_c)	0.4 mm, 4.2 kA	0.6 mm, 6.3 kA	0.8 mm, 8.4 kA	0.4 mm, 7 kA	0.6 mm, 105 kA	0.8 mm, 14.0 kA
Not graded, same reinforcement everywhere						
Radial build (m)	0.140	0.155	0.182	0.125	0.143	0.170
Coil estimated outer diameter (m)	0.680	0.7088	0.764	0.650	0.686	0.740
Coil J_{op} (at 0.8 x I_c)	116	102.8	88.7	127.9	112.1	95.6
Avg. reinforcement thickness (mm)	1.049	1.851	2.987	1.790	3.148	5.062
Wire length per pancake (m)	163	110	88	94	64.8	51.9

Additional design features are added to the trade-off and coil / conductor analysis in Figure 5, focused on the fast ramped design, and specifically investigating wind and react (W&R) as compared to R&W coil fabrication modes, the key tradeoff being the benefits of higher operating current with thicker 2212/Ag versus the need for increased thickness of reinforcement in the less costly and lower risk R&W mode, as well as single thickness reinforcement as compared to graded reinforcement designs using three thicknesses for three radial build zones and different hoop loads from the local Lorentz forces. Present current densities and earlier development stage J_{op} levels were both analyzed.

IOP Conf. Series: Materials Science and Engineering 1241 (2022) 012023 doi:10.1088/1757-899X/1241/1/012023

The estimated outer diameters of the coil for different builds versus 2212/Ag thickness in Figure 5 (left) illustrate that the 65 cm outer diameter (OD) target can be achieved by both present (420 A/mm^2) and higher (700 A/mm^2) J_{op} wires under development by R&W at 2212/Ag thickness of 0.6 mm or less, with graded coils, while for W&R the target can be met with graded W&R even at 420 A/mm^2 and all 2212/Ag thicknesses. Figure 5 (right) presents coil ODs estimated for two different coil fabrication modes, two operating current densities and two reinforcement modes showing the regimes where a 65 cm OD or less is attained. The case of W&R for the 700 A/mm^2 graded (G) and non-graded (NG) cases are not shown, but they both come in well under the 65 cm diameter target in the full 0.4 - 1.2 mm 2212/Ag thickness range analyzed. The results show that flexibility of 2212 wire design can meet CS coil operation requirements, without the need for a CICC cable construct.

Figure 6 presents the magnetic field and stress analysis of the TF-CS coil system for the SHPD using the Bi-2212 wire co-wind structural reinforcement. The peak stress in the co-wind reinforcement is less than the 660 MPa SS 316 design allowable. The strain in the 2212 wire is within the 0.3% limit. Therefore the 2212 wire is well protected by the structural reinforcement in the coil winding pack.

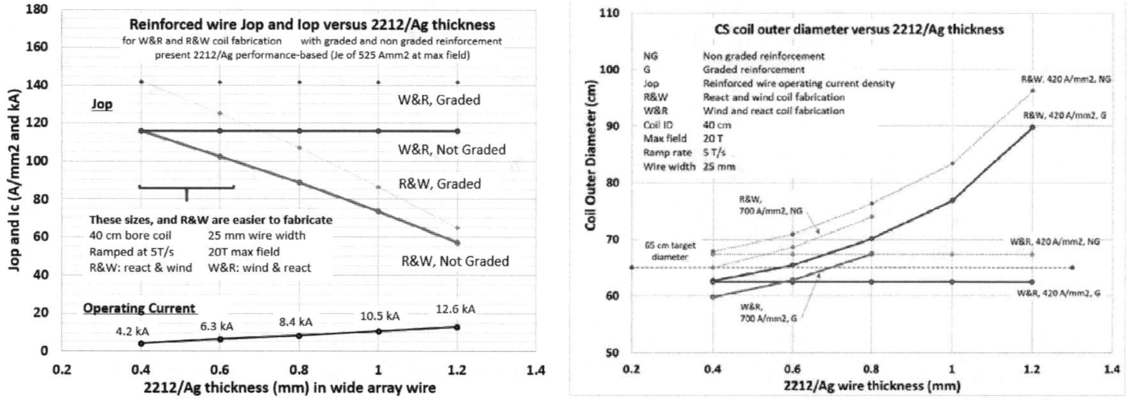

Figure 5. Impact of 2212 wire design to CS magnet radial build. Operating current and current density (J_{op}) of reinforced wide 2212 wire versus 2212/Ag thickness (left), illustrating that J_{op} with the W&R approach stays constant with respect to wire thickness (and I_{op}) increasing, while due to the need for increasingly more reinforcement with R&W, J_{op} decreases as the 2212/Ag gets thicker. This example is for present J_e levels in standard 2212 wires at 20 T (525 A/mm^2 J_e, 420 A/mm^2 J_{op}). Coil ODs for two different fabrication modes, two operating current densities and two reinforcement modes show regimes where a 65 cm OD or less is attained (right).

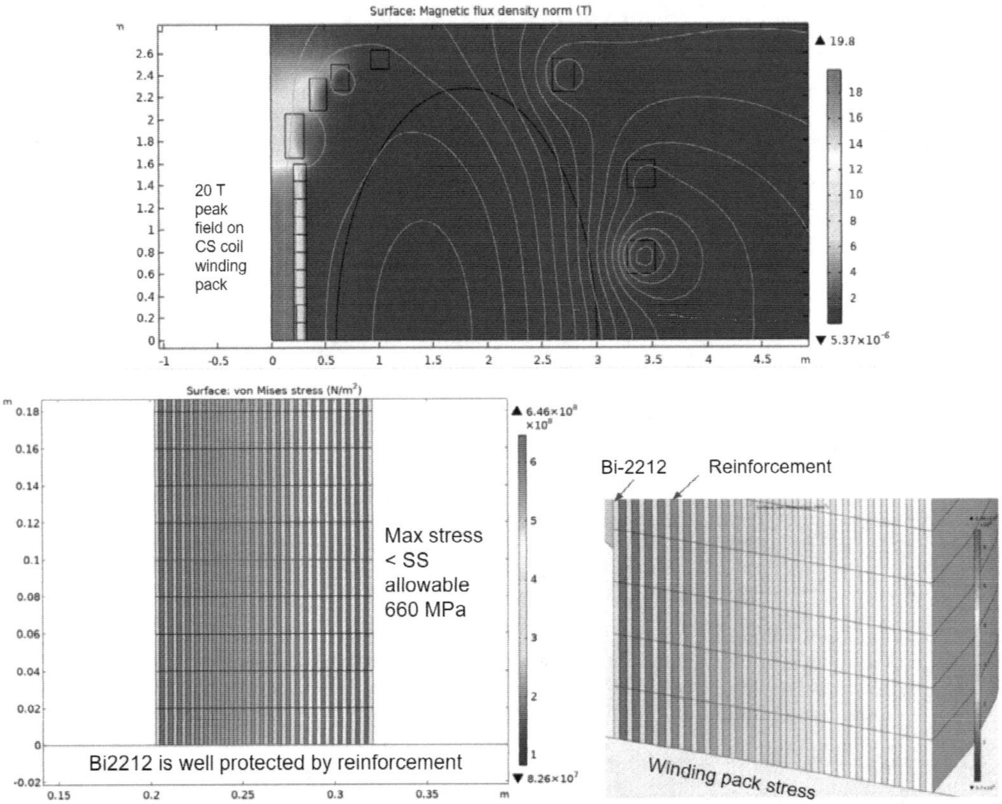

Figure 6. The 20 T CS coil winding pack analysis for SHPD PF-CS coil system

5. Cost considerations

Lowering HTS conductor cost for building a commercial fusion magnet system is critical for next step studies for FPP. At present, the cost for commercial 4-6 mm wide YBCO coated conductors is over \$60 per meter (~\$50 per kA-m) and this will lead to over \$5000 per meter for 50 tape cables. The estimated conductor cost alone for winding the CORC based CS coil for FNSF will be over \$200M based on the total cable length needed. HTS conductor manufacturing scale up, cost-reduction technical developments and yield improvements at large throughputs are needed to drive cost down. For YBCO based fusion coils, building parallel plants will not drastically reduce the overall cost as equipment cost for coated conductor fabrication will linearly scale up with the number of production lines.

Table 3. Cost considerations

	Nb_3Sn CICC* ITER	Bi-2212** FNSF	YBCO CORC
Max B field (T)	13	20-25	20-25
Unit cost (\$/m)	4480	~500	~5000
Cost/Performance (\$/kA-m)	43-88	40-70	469

*ITER CS/TF coils - Nb_3Sn CICC at (4.2 K, 12 T); YBCO CORC at (4.2 K, 20 T) performance for FNSF; ITER cost includes R&D and conductor testing; YBCO CORC cost does not include jacketing, QA and conductor testing

**Bi-2212 at (4.2K, 20T) of ~10 kA in large volume for R&W large cross-sectioned wire (~25 mm wide by 0.6 mm thick) for FNSF/SHPD

The cost numbers for 2212 conductors are based on the pricing of Sumitomo Electric's 2223 tape. Although those data are not published they can be reduced to cost components and explained. The cost data basis is also consistent with the current pricing of 2223 tapes sold by Sumitomo Electric - namely for larger orders it is in the 20 - 30 $ / m range. This is based on the very reasonable assumption that Sumitomo Electric is unlikely to sell HTS tape or any of its other commercial products at below cost. The cost of BSCCO is a key driver for its likely commercial utility and success in some applications, and in turn its lower cost in part derives from the simplicity of its manufacture and resulting much higher manufacturing yields. The additional cost estimates for building the magnet are based on actual, but as yet limited cost data we have for building HTS coils.

6. Conclusions

It is important for next step fusion reactor design to reduce superconducting coil costs and simplify construction processes. Required HTS cost reductions, coil fabrication and build simplifications as well as high operating current and current densities can be achieved using multifilamentary Bi-2212 conductors with practically accessible design variations around combinations of optimum space-filling cross-sections, low loss configurations based on proven LTS designs, and high modulus reinforcement. The more readily customizable Bi-2212 wire design is also integrated directly with the coil winding pack layout and the TF-CS magnet system design concept for performance, coil build and cost benefits.

7. References

[1] "A long range plan to deliver fusion energy and to advance plasma science," February, 2021. A report to Fusion Energy Sciences Advisory Committee (FESAC), Powering Future Fusion & Plasmas

[2] "Bringing Fusion to the U.S. Grid," The National Academies of Sciences, Engineering, Medicine, 2021.

[3] J.E. Menard et al., "Fusion nuclear science facilities and pilot plants based on the spherical tokamak," *Nucl. Fusion*, vol. 56, 2016.

[4] C.E. Kessel et al., "The Fusion nuclear science facility, the critical step in the pathway to fusion energy," *Fusion Sci. Technol.*, 68:2, March 2017, 225-236

[5] J.E. Menard, "Compact steady-state tokamak performance dependence on magnet and core physics limits, " *Phil Trans. R. Soc.* A377:20170440, 2019.

[6] S. Hahn et al., "HTS pancake coils without turn-to-turn insulation," *IEEE Trans. Appl. Supercond.*, vol.21, no.3, June 2011, 1592-1595.

[7] S. Hahn et al., "No-insulation (NI) winding techniques for premature-quench-free NbTi MRI magnets," *IEEE Trans. Appl. Supercond.*, vol.22, no.3, June 2012, 4501004.

[8] Y. Zhang et al., "Numerical simulation of a no-insulation BSCCO toroidal magnet applied in magnetic confinement fusion," *Science and Technology of Nuclear Installations*, vol. 2018, 2914036.

[9] Y. Zhai et al., "Design, construction and testing of no-insulation small subscale solenoids for compact tokamaks," *Supercond. Sci. Technol.*, vol.34, 105003, August 2021.

[10] Y. Zhai et al., "Conceptual design of HTS magnets for fusion nuclear science facility," *Fusion Eng. Des.*, vol. 168, 112611, July 2021.

[11] Y. Zhai et al., "Conceptual design of central solenoid for fusion nuclear science facility," *IEEE Trans. Appl. Supercond.*, 10.1109/TASC.2021.3063056, March 2021.

[12] E.E. Salazar et al., "Fiber optic quench detection for large-scale HTS magnets demonstrated on VIPER cable during high-fidelity testing at the SULTAN facility," *Supercond. Sci. Technol.*, vol.34, 035027, February 2021.

IOP Conf. Series: Materials Science and Engineering 1241 (2022) 012023 doi:10.1088/1757-899X/1241/1/012023

[13] M. Windridge, "Smaller and quicker with spherical tokamaks and high-temperature superconductors," *Phil. Trans. R. Soc.*, A337:20170438, 2019.

[14] Z. Hartwig et al., "VIPER: an industrially scalable high-current high-temperature superconductor cable," *Supercond. Sci. Technol.*, vol.33, no. 11, October 2020.

[15] D C Van Der Laan et al, "A CORC cable insert solenoid: the first high temperature superconducting insert magnet tested at currents exceeding 4 kA in 14 T background magnetic field," *Supercond. Sci. Technol.*, vol.33, 05LT03, 2020.

[16] A. Molodyk et al., "Development and large volume production of extremely high current density $YBa_2Cu_3O_7$ superconducting wires for fusion," *Sci. Rep.*, 2021; 11:2084.

[17] K. Hayashi, "Commercialization of Bi-2223 superconducting wires and their applications," SEI Technical Review, n91, pp.68-74, October 2020.

[18] A. Otto, "Transposed high current cables made with strong, rectangular Bi-2212 superconductor wire," *The 13th European Conference on Applied Superconductivity (EUCAS 2017)*, Geneva, Switzerland, Sept. 2017.

[19] H. Maeda and Y. Yanagisawa, "Recent developments in high-temperature superconducting magnet technology (Review)," *IEEE Trans. Appl. Supercond.*, vol.24, no.3, 4602412, June 2014.

[20] D.W. Hazelton and V. Selvamanickam, "SuperPower's YBCO coated high-temperature superconducting (HTS) wire and magnet applications," Proceedings of the IEEE, vol. 97, no.11, pp. 1831-1836, November 2009.

[21] N. Cheggour and L.F. Goodrich, "Characterization of Nb_3Sn strand for ITER," Final report, DE-AI05-07OR23274, NIST, May 2012.

[22] H. Miao et al., "Bi-2212 round wire development for high field applications," *J. Phys. Conf. Ser.* 507, 022020, 2014.

[23] A. Otto, "Multifilament superconducting wire with high resistance sleeves," US patent 9859039, February, 2016.

[24] G.L. Jackson et al., "Non Inductive plasma initiation and startup in the DIII-D tokamak," *Nucl. Fusion*, 51, 083015, 2011.

[25] T. Brown et al., "The development of a Sustained High Power Density (SHPD) facility," *Fusion Eng. Des.*, vol. 168, 112214, July 2021. https://doi.org/10.1016/j.fusengdes.2020.112214

[26] J.E. Menard et al., "Fusion pilot plant performance and the role of a sustained high power density tokamak," *Nucl. Fusion*, 62, 036026, February, 2022.

[27] M. Wolf, et al., "High temperature superconductors for fusion applications and new developments for the HTS CroCo conductor design," *Fusion Eng. Des.*, vol. 172, 112739, June 2021.

[28] D.C van der Laan et al, "Status of CORC cables and wires for use in high-field magnets and power systems a decade after their introduction," *Supercond. Sci. Technol.*, vol.32, 033001, 2019.

[29] M. Takayasu et al., "Investigation of HTS twisted stacked tape cable conductor for high field, high current fusion magnets," *IEEE Trans. Appl. Supercond.*, 10.1109/TASC.2017.2652328, 2017.

[30] K.Yagotintsev et al., "AC loss and contact resistance in REBCO CORC, Roebel, and stacked tape cables," *Supercond. Sci. Technol.*, vol.33, 085009, 2020.

[31] C. Neumeyer, "Magnet Superconducting and Electrical Design Criteria," ITER_D_22GRQH v1.2, Sep. 2009.

[32] R.P Reed and N. Martovetsky, "Electrical insulation systems for the ITER CS modules," *AIP Conference Proceedings* 1574, 354 (2014), https://doi.org/10.1063/1.4860648

Acknowledgments

This work was supported by the U.S. DOE contract No. DE-AC-02-09CH11466.

Superconducting liquid cryogen insulated power cables for medium voltage applications

H Lopez[1,2], P Mensah[1,2], P Cheetham[1,2], C H Kim[1], S Pamidi[1,2]

[1]Center for Advanced Power Systems, Florida State University, Tallahassee, FL, 32310, USA
[2]FAMU-FSU College of Engineering, Department of Electrical & Computer Engineering, Tallahassee, FL, 32310, USA
cheetham@caps.fsu.edu

Abstract. Liquid Nitrogen (LN2) is used as the cryogen and is part of the dielectric insulation system in most high temperature superconducting (HTS) power cables. However, in electric propulsion systems for transportation, using LN2 is not feasible due to asphyxiation concerns in the event of a leak in a confined space. Gaseous helium (GHe) has been proposed as an alternate cryogen to relieve this asphyxiation concern. GHe has benefits such as a wider operating temperature range and versatility for a centralized cryogenic cooling system for multiple devices which could potentially reduce the overall weight of the superconducting system. The drawback is that GHe has a lower dielectric strength compared to LN2, which limits its use to low-medium voltage applications. A hybrid cryogen HTS cable was proposed using LN2 as the dielectric media and GHe as the cryogen of an HTS cable to solve the limitations and issues associated with each cryogen and explained in this paper. A 1-m prototype cable that used LN2 as the cryogen and dielectric was fabricated and dielectric breakdown measurements were performed on it to establish a baseline on what to be expected from the hybrid cryogen HTS cable.

1. Introduction

Electric propulsion systems for transportation belong to the most recent large-scale power systems. Electric ship and aircraft require novel ideas and designs to meet the increasing power demands. Besides rising power demands, there are also targets for energy efficiency, power density, and resiliency. These demands add constraints to size and weight while at the same time calling for higher power densities to meet the power demands. High temperature superconducting (HTS) power cables are being explored for electric transport (electric ships and aircraft) applications. These applications call for a current carrying capacity of 10 times that of conventional technology [1],[2]. This ampacity target could be achieved by using a conventional conductor with a larger cross-section; however, cable designs for the transportation sector have strict weight and space restrictions. HTS power cables have been proposed as a means of achieving power-dense, resilient, and efficient electrical transmission targets. HTS power cables have a higher current density (typically 50-100 A/mm^2) when compared to conventional conductors such as copper (1-2 A/mm^2). HTS power cables can achieve higher power transmission ratings without having to operate at higher voltages, thus can reduce the size and weight of electrical components within a power system.

Choosing the right cryogen and dielectric is typically one of the major challenges of HTS power applications. The coupling of electrical and thermal properties of HTS power cables dictates that the cryogen used to achieve the desired operating temperature has an influence over the voltage and current ratings [2]. Liquid Nitrogen (LN2) is one of the most used cryogens for HTS applications due to its low cost, abundance, high dielectric strength, and inertness in chemical reaction. Using LN2 as a coolant is not always practical or feasible and there are times when an alternate cryogen must be used. One reason for this is the presence of nitrogen gas bubbles. LN2's breakdown strength has been reported to decrease by a factor of two when nitrogen gas bubbles are present [7]. These bubbles are usually minimized by sub-cooling and pressurizing LN2; however, the formation of these nitrogen bubbles cannot be totally prevented [8].

LN2 also provides a risk of asphyxiation in applications where there are confined spaces, such as electric ships and aircrafts, in the event of a cryogenic leak. This is mainly due to its large volume-expansion ratio when boiled off into a gas. LN2 expands to over 700 times its liquid volume when converted to its gas phase replacing oxygen in air [2]. Because of these reasons, it is necessary to use alternate cryogens to achieve the desired operating temperature while reducing the safety concerns [9]. Gaseous helium (GHe) is the preferred cryogen for shipboard applications and has been proposed as a solution to the risk of asphyxiation. GHe has a wider operating temperature range (4 - 80 K) compared to LN2 (60 – 80 K). Being able to operate at lower temperatures allows for higher current densities, thus increasing the power density [10]. The use of GHe also allows for multiple HTS devices to be connected to a single, centralized, cryogenic closed loop. This is advantageous for the electrification of transportation because it could reduce weight compared to using multiple cryogenic systems for different devices [2].

One major drawback of utilizing GHe as a cryogen is its low dielectric strength, which restricts it to low-medium voltage applications [10]. Alternative insulation designs, such as mixing small mole percentages of hydrogen (H2) balanced with GHe, have been proposed to improve the voltage rating of GHe cooled HTS cables [11]. It is possible to reduce the concerns associated with using LN2 and GHe as cryogens, and that is by using both at the same time: LN2 as the dielectric and GHe as the cryogen. By doing so, there is the potential to take advantage of each cryogen's benefits while mitigating the drawbacks such as asphyxiation when using LN2 and the low dielectric strength of GHe.

A project called Best Paths proposed the idea of using LN2 as dielectric and GHe as cryogen [13]. The study conducted in [2] investigates the practicality of a hybrid HTS cable design by utilizing commercially available HTS conductors and cryostats. Both studies focused on using LN2 and GHe as dielectric medium and cryogen, respectively. However, these studies, and the designs that will be presented in this paper, are also applicable for liquid hydrogen (LH2), which is considered as a potential cryogen for electric aircraft applications. LH2 has been considered as the cryogen and fuel source for NASA's N3-X aircraft since it can be stored in the fuel tanks, thus reducing the need to have cryocoolers. Once used as the cryogen for motors and cables, it will be gaseous hydrogen which can then be used as a fuel for the aircraft.

This paper reviews the hybrid HTS cable design considerations and proposes using a superconducting liquid cryogen insulated power cable to establish a baseline of what to expect from a hybrid cryogen HTS cable.

2. Superconducting power cable with hybrid cryogens

As it was stated before, the use of LN2 as a cryogen has been discouraged given the asphyxiation concerns present when implementing HTS cables for electric transport applications. However, it is possible to minimize the volume of LN2 by using LN2 solely as the dielectric media and GHe as the cryogenic media for HTS power cables. The concept of a hybrid cryogen HTS cable has been discussed before in [13]. However, the idea is to develop a hybrid cryogen HTS cable that is smaller and lighter to maximize the number of cables that can fit within commercially available cryostats, thus reducing space used and weight.

2.1. *Hybrid HTS cable design*

The main goal when designing a hybrid HTS cable is to minimize the total volume of LN2, and it must be done in a way so that the desired voltage rating can be achieved. One way to reduce the total volume of LN2 within a cryostat is to use some sort of solid insulation as part of the design, such as the lapped tape insulation used in LN2-cooled HTS cables.

Figure 1 below shows LN2 filled butt gaps which are created on purpose as part of the insulation design. These butt gaps ensure mechanical flexibility of the HTS cable at cryogenic temperatures and reduce the mechanical stress at the HTS layer and electrical insulation interface [14]. These butt gaps also cause partial discharge (PD) when the maximum electric field exceeds the partial discharge inception voltage (PDIV). PD occurs in the butt gaps as it is filled with LN2, and it has a lower relative permittivity than the tape and creates a local enhancement of the electric field [14]. The PDIV depends on if the LN2 within the butt gap remains as a liquid or if a phase change occurs and it becomes a gas [2]. In the event of nitrogen gas being trapped in the butt gaps, a lower PDIV is expected since it has a lower dielectric strength [14]. Polypropylene laminated paper (PPLP) is typically used as insulation tape for the lapped tape design; however, it is porous, and this allows LN2 to impregnate the tape [2]. Other insulation tape materials such as Kapton and Cryoflex®, which are non-porous, could be utilized to further reduce the total amount of LN2 within the cryostat.

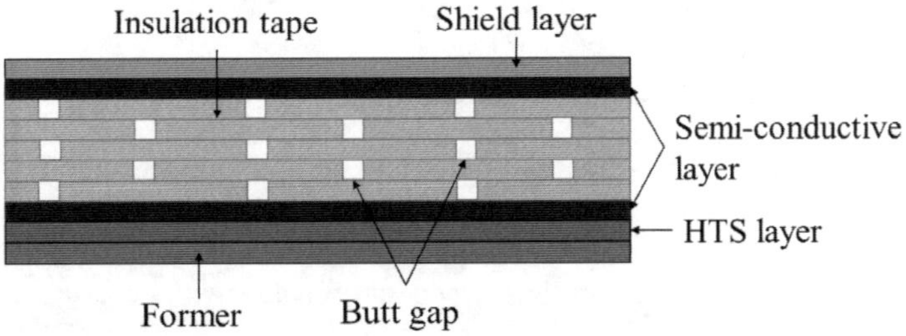

Figure 1. Butt gaps in a lapped tape cable

Utilizing LN2 as part of the dielectric medium would limit the HTS cable to operate with a temperature range of 64-80 K to ensure a phase change does not occur to the nitrogen. This results in a lower current density of the HTS cable when compared to GHe-cooled HTS cables operating at lower temperatures. However, the 65-70 K temperature range of the hybrid cable could still provide a high enough current density. For DC HTS cables, with lengths less than 100m, the heat loads typically come from the terminations and heat leak from the cryostat. For the terminations, the heat loads are typically 100 – 300 W and the heat load along the HTS cable due to cryogenic leaks is approximately 1 W/m [15],[16].

For the hybrid cryogen HTS cable design, it was decided to use Conductor On Round Core (CORC®) HTS conductors because of the higher current density and smaller geometry offered by this technology when compared to conventional copper conductors. These conductors are capable of operating at 5-10 kA while only having an outer diameter of 3-7 mm [17]. Based on the study conducted in [2], it is possible to achieve voltages between 15-20 kV using a dipole configuration with CORC conductors within a 39 mm commercially available cryostat. This allows to achieve higher power densities and flexibility by using two or more CORC® conductors within a single cryostat.

Utilizing commercially available cryostats and HTS CORC® conductors, a Superconducting Power Cable with Hybrid Cryogens (SPCHC) is developed.

2.1.1. Novel design of a SPCHC

For this design, the W7 CORC® conductor developed by Advanced Conductor technologies was selected. It has a diameter of 4.5 mm and a critical current (I_C) of 4.4 kA at 76 K. Typically, operating currents of HTS cables are between 40-60% of I_C and for this design, 50% of I_C was selected [2]. With this in mind, and using the W7 CORC® conductor, an operating current of 2.2 kA is envisioned, and higher currents can be achieved by operating CORC® conductors in parallel. Figure 2 shows a proposed HTS cable design of the SPCHC. As seen in this figure, six conductors are placed into individual chambers that are to be filled with LN2. The conductors are being kept concentric with respect to the chamber using standoff spacers, and all of this is placed into a GHe-filled 39 mm cryostat.

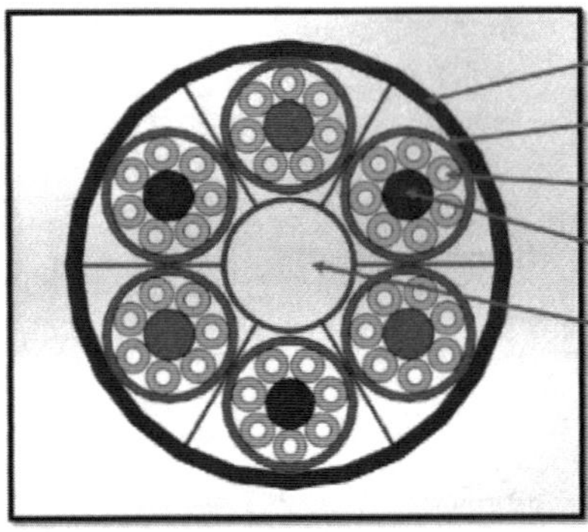

Figure 2. Novel SPCHC 6-conductor design model

This design ensures that the good dielectric properties of LN2 will help to achieve higher voltage ratings compared to GHe. LN2 also has higher specific heat capacity which gives it superior cooling capabilities. This is beneficial in the event of an electrical breakdown occurring within the insulation to help with heat dissipation of the associated plasma arc [2].

3. Superconducting liquid cryogen insulated power cable

A 1-m prototype of the SPCHC cable was developed out of commercial off the shelf (COTS) parts. In this prototype, a 4.78 mm outer diameter (OD) copper tube was used as the conductor to represent the HTS conductor. The conductor chamber was a 19.05 mm OD stainless steel tube, which is part of a 33.78 mm conflat flange (CF) full nipple from LDS Vacuum and was grounded. For this prototype, LN2 was used as both the dielectric medium and cryogen, therefore this prototype cable was called a superconducting liquid cryogen insulated power cable prototype. Six 4.78 mm OD GPO-3 insulator spacer rods were used to keep the conductor concentric with respect to the grounded chamber to avoid any field enhancements in areas where the conductor is closest to the ground. Kapton tape was wrapped around the insulator rods and the conductor to hold the insulator rods in place.

This cable was installed into the setup seen in figure 3. This figure also shows a cross sectional cut of the prototype cable. The setup consists of a 33.78 mm CF full nipple, two 2.75" CF tees (also from LDS Vacuums), and two 69.85 mm 5.75 kV bushings from CeramTec. One end of the conductor was connected to the high voltage bushing and the other end of the conductor was connected to the other high voltage bushing. A stress sphere was screwed to one of the bushings to reduce the electric field enhancement at the end of the cable; an adaptor for high voltage connection was screwed to the other bushing. The complete chamber, which consisted of the CF nipples and tees, was grounded. It is worth mentioning that this setup also serves as a pressure vessel. The setup was then placed in a chamber that is filled with LN2 to keep it at 77 K.

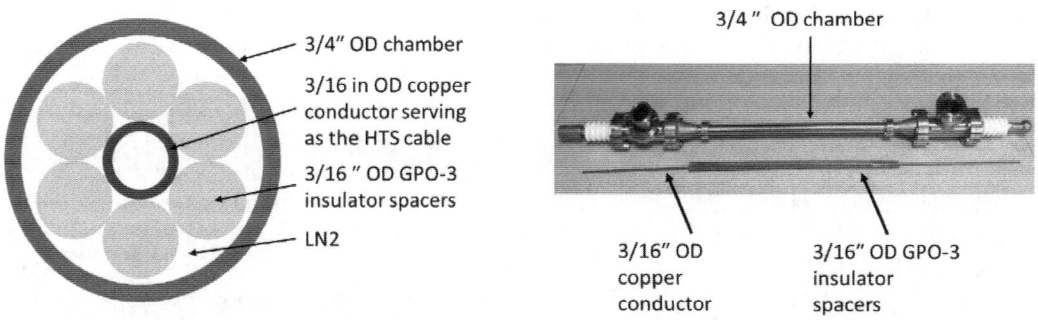

Figure 3. Superconducting liquid cryogen insulated power cable prototype

4. Experimental setup and results

For this prototype cable, two tests were performed: an AC dielectric breakdown test in an LN2 open bath environment and an AC dielectric breakdown test with pressurized and subcooled LN2. One of the high voltage bushings was connected to a 100 kV, 7.5 kVA AC transformer, and the other end (the other high voltage bushing) had a stress sphere to avoid any field enhancements.

Figure 4 shows a diagram of the experimental setup. For the open bath test, the top parts of the tees were left open to allow LN2 to be boiling inside the cable at 1 atm. For the pressurized case, connections pipes were connected to the tees for: LN2 inlet, GN2 inlet, LN2/GN2 outlet, and pressure gauge. For the pressurized case, a vacuum and purging cycle was performed to ensure there were no impurities in the setup. An LN2 dewar was connected to the inlet line to fill the setup with LN2 and constantly pressurized at 0.24MPa to prevent LN2 from forming bubbles. This dewar was always kept inside the Faraday Cage during measurements under being submerged in LN2.

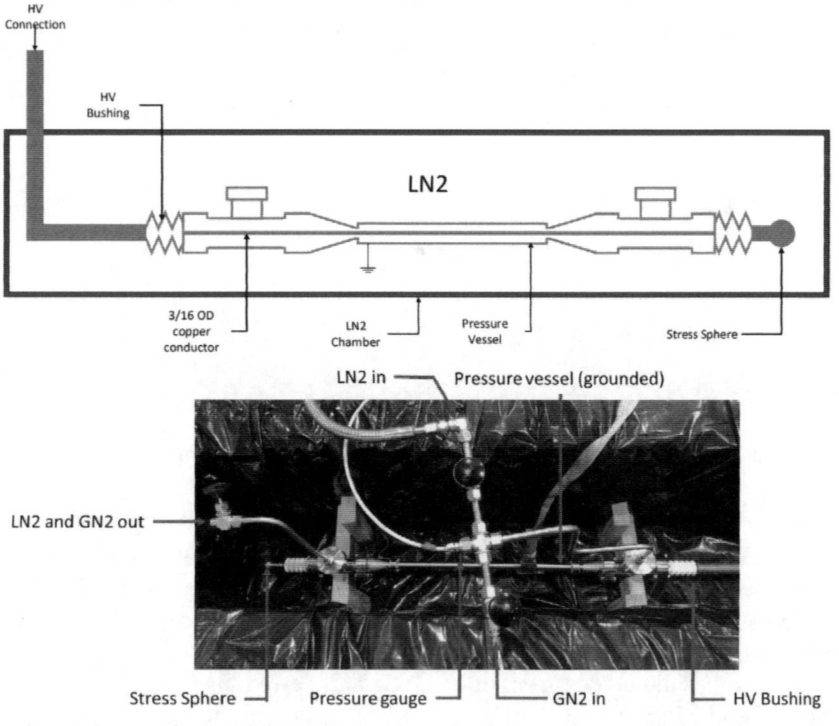

Figure 4. Experimental setup for breakdown measurements (for open bath case, omit the pipe connections)

The AC dielectric breakdown tests were performed for both cases and 10 measurements were performed for each case. The results are summarized and presented in Table 1. Figure 5 shows the dielectric breakdown voltages measured in both tests. There is a variation in the results obtained from the open bath case. This may be attributed to the formation of nitrogen bubbles trapped in the setup during the experiment. There was a 40.8% increase in the average breakdown voltage when the setup was pressurized and subcooled. Also, the data seems to be more consistent when compared to the open bath case. This is mainly because the LN2 was pressurized to 0.24MPa and this brought its boiling point to approximately 86 K. The setup was then subcooled to 77 K which helped prevent bubble formation. The data presented in Figure 6 and table 1 is presented as AC root mean square (RMS) voltage. The average peak AC breakdown voltage can be used to estimate the DC breakdown voltage while the cable is in its transient time period given that, conductivity is a function of temperature. The DC cable could ideally behave like an AC cable for the first couple of hours when energized and it is calculated as $\sqrt{2} * V_{AC\ rms}$. Additional measurements are needed to understand the steady state conditions which behaves as a resistive field. The breakdown voltage of this design could be improved by using higher pressures to pressurize the LN2. Also, the formation of bubbles could be further reduced by using GHe as the coolant to sub cool the setup even further.

Table 1. Summary of dielectric breakdown tests performed on cable prototype (AC rms)

Case	Breakdown voltage (kV)		
	Maximum	Minimum	Mean
Open bath	23.6	14.6	19.1
Pressurized and subcooled	29.8	24.3	26.9

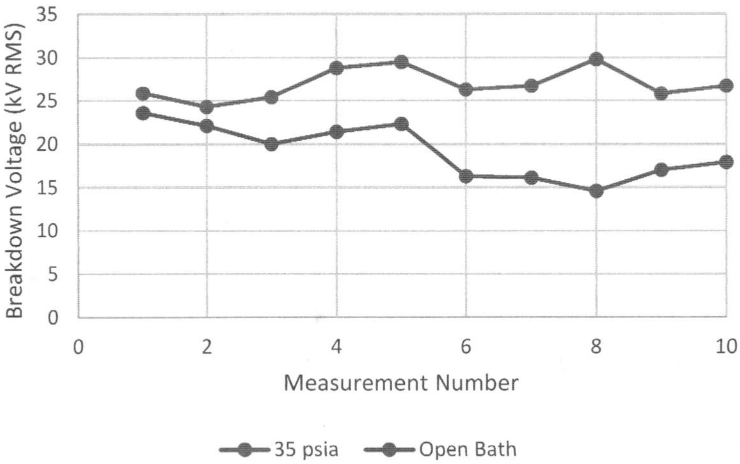

Figure 5. Results of AC dielectric breakdown tests on SPCHC prototype cable.

5. Discussion

The goal of this study was to develop an HTS power cable design that addressed the issues involved with using LN2 and GHe as a cryogen and dielectric medium respectively for HTS cables. It also worked to establish a baseline on what to expect not only for a hybrid cable that uses LN2 and GHe, but for any hybrid cryogen cable, such as ones that could use liquid hydrogen, hydrogen gas, etc. The SPCHC cable design presented allows for higher voltage ratings when compared to conventional HTS power cable designs due to the higher dielectric strength of LN2. The design also reduces the risk of asphyxiation that is present in confined spaces. It does so by reducing the volume of LN2 within the cryostat. One major challenge involved with this prototype was preventing the formation of nitrogen bubbles, which reduce the dielectric strength of LN2, thus having a negative effect on the breakdown voltage. This can be solved by pressurizing the LN2 at higher pressures to increase its boiling point and using GHe as the cryogen to keep it at a low enough temperature such that it will not boil.

Further experimental investigations are required to determine the cable's performance with a different solid insulation, such as a lapped tape insulation environment. Future work includes studying the effect of having both solid and liquid nitrogen as the insulation, which occurs at temperatures lower than 63 K. Also, it is important to design the cryogenic network of this cable. For this, it would be important to understand how the cable behaves in an environment with circulating LN2. Another future work involves conducting a study to measure the LN2 volume that is required for the design. The SPCHC design has the potential to be used to replace conventional copper cables in all electric aircraft and ships, such as when HTS degaussing system was installed on a United State Navy San Antonio class (LPD-17) which resulted in a 75% reduction in weight including the cryogenic system and 80% reduction in cable length [9].

6. Conclusion

This paper discussed the challenges of using either LN2 or GHe as the sole cryogen and dielectric medium in HTS cables. The proposed hybrid cryogen HTS cable design addresses these challenges by taking advantage of each cryogen's dielectric, thermal, and system level benefits. The proposed hybrid cable design is also applicable when using LH2 as the dielectric for electric aircraft applications. A prototype cable, called a superconducting liquid cryogen insulated power cable, was built to establish a baseline on what to expect from an SPCHC. The experiments performed on this prototype shows promising results and further improvements can be made on the design to meet dielectric, thermal, space, and weight requirements for electric transport applications. Further studies are required to explore the feasibility of using the SPCHC power cable design in electric transportation applications, where it must interface with a cryogen circulation system.

7. References

[1] Doerry N and Amy J 2015 DC voltage interface standards for naval applications *2015 IEEE Electric Ship Technologies Symposium, ESTS 2015*, 318–325

[2] Stamm T, Cheetham P, Kim C H, and Pamidi S V 2020 Superconducting Power Cable Design with Hybrid Cryogenic Media - Gaseous Helium for Cooling and Liquid Nitrogen for Dielectric Insulation *2020 IEEE Electrical Insulation Conference, EIC 2020*. 22–26.

[3] Pamidi S V 2015 Research and Development Efforts at Center for Advanced Power Systems in the area of Superconducting Power Devices

[4] Felder J L, Kim H D, and Brown G V 2009 Turboelectric Distributed Propulsion Engine Cycle Analysis for Hybrid-Wing-Body Aircraft *47th AIAA Aerospace Sciences Meeting Including the New Horizons Forum and Aerospace Exposition, January*. AIAA 2009-1132.

[5] Chalfant J 2017 ESRDC Notional Ship Data. May

[6] Stemmle M, Merschel F, Noe M, and Hobl A 2013 AmpaCity - Installation of advanced superconducting 10 kV system in city center replaces conventional 110 kV cables *2013 IEEE International Conference on Applied Superconductivity and Electromagnetic Devices, ASEMD 2013*. 323–326

[7] Sauers I, James R, Ellis A, Tuncer E, Polizos G, and Pace, M 2011 Effect of bubbles on liquid nitrogen breakdown in plane-plane electrode geometry from 100-250 kPa *IEEE Transactions on Applied Superconductivity*. **21**1892–5

[8] Pamidi S, Kim C H, and Graber L 2015 High-temperature superconducting (HTS) power cables cooled by helium gas *Superconductors in the Power Grid: Materials and Applications* pp. 225–260

[9] Fitzpatrick B K, Kephart J T, and Michael Golda E 2007 Characterization of gaseous helium flow cryogen in a flexible cryostat for naval applications of high temperature superconductors *IEEE Transactions on Applied Superconductivity*. **17** 1752–1755.

[10] Cheetham P, Kim C H, Graber L, and Pamidi S 2017 Practical Considerations for the Design of a Superconducting Gas-Insulated Transmission Line for Shipboard Applications *IEEE Electrical Ship Technical Symposium.*
[11] Cheetham P, Kim W, Kim C H, Graber L, Rodrigo H, and Pamidi S 2017 Enhancement of Dielectric Strength of Cryogenic Gaseous Helium by Addition of Small Mol% Hydrogen *IEEE Transactions on Applied Superconductivity.* **27** 1–5.
[12] Felder J L, Kim H D, and Brown G V 2009 Turboelectric Distributed Propulsion Engine Cycle Analysis for Hybrid-Wing-Body Aircraft *47th AIAA Aerospace Sciences Meeting Including the New Horizons Forum and Aerospace Exposition, January*, AIAA 2009-1132.
[13] Marian A and Bruzek C E, Advancing superconducting links for very high power transmission 2018.
[14] Cheetham P, Kim W, Kim C H, Pamidi S V, Graber L, and Rodrigo H 2017 Use of partial discharge inception voltage measurements to design a gaseous helium cooled high temperature superconducting power cable *IEEE Transactions on Dielectrics and Electrical Insulation.* **24** 191–199
[15] Al-Taie A, Telikapalli S, Cheetham P, Kim C H, and Pamidi S V 2020 Liquid Nitrogen Cooled Superconducting Power Cable with No Solid Insulation *IOP Conference Series: Materials Science and Engineering.* **756** 012033
[16] Demko J A A 2015 High-temperature superconducting cable cooling systems for power grid applications In *Superconductors in the Power Grid: Materials and Applications.* pp 261–280 Elsevier
[17] Weiss J D, Mulder T, ten Kate H J, and van der Laan D C 2017 Introduction of CORC ® wires: highly flexible, round high-temperature superconducting wires for magnet and power transmission applications *Superconductor Science and Technology.* **30** 014002

Acknowledgments
This work is supported by the Office of Naval Research (ONR) through the following grants: N00014-18-12693 and N00014-16-1-2956

IOP Conf. Series: Materials Science and Engineering 1241 (2022) 012025 doi:10.1088/1757-899X/1241/1/012025

Design of warm dielectric terminations and electrical breaks for high-temperature superconducting power cables

P Mensah[1, 2], H Lopez[1, 2], P Cheetham[1, 2], C H Kim[1] and S Pamidi[1, 2]

[1]Center for Advanced Power Systems, Florida State University, Tallahassee, FL, 32310, USA

[2]FAMU-FSU College of Engineering, Department of Electrical & Computer Engineering, Tallahassee, FL, 32310, USA

Email: pm20dq@my.fsu.edu

Abstract. Large electric transportation systems such as electric aircraft and electric ships have been designed with multiple high temperature superconducting (HTS) power cables as part of superconducting power distribution network, and a good electric insulation for the power cables is a critical design requirement needed to be considered. This is because terminations in HTS cable represent a location of significant electrical and thermal stresses and needs to be addressed to reduce the substantial heat load that might be introduced into the cryogenic system. The use of a warm dielectric termination is seen as a potential solution to reduce the electrical standoffs required as part of the design and allow for a high-power dense termination design. A warm dielectric termination design using commercially off-the-shelf ceramic breaks as electrical breaks is explored and withstand and partial discharge measurements are performed to determine the viability of the design concept.

1. Introduction

The utilization of high temperature superconducting (HTS) power cables in the electrical power grid to help solve rapidly growing power demands is steadily being accepted by electric power utilities as demonstrated in China [1] and Germany [2]. Terrestrial HTS power cables predominantly use liquid nitrogen to keep the cables at cryogenic temperatures [3]. Liquid nitrogen being inexpensive and having good dielectric properties is used as the cryogen for HTS power cables in electric power grid applications which typically consist of a single HTS cable [4]. A growing area of interest is the use of HTS cables for large-scale electric transportation applications which includes ships and aircraft [1] . For transport applications, short lengths of HTS cables need to terminate at several discrete locations where they either transition to conventional devices at room temperature or to be coupled to other HTS devices.

Using liquid nitrogen as the cryogen in such enclosed spaces in transport poses safety challenges as nitrogen can displace oxygen and lead to asphyxiation in the event of a leak [6]. The electrical power system for electric transport system would also require HTS machines (motors and generators) which have an operating temperature of 40 – 50 K which cannot be served by liquid nitrogen [7]. An additional cryogen together with the liquid nitrogen needs to be utilized to operate the HTS machines at their required operating temperatures, which demands additional weight to the power system and reduces its overall power density. Utilizing one cryogen with much lower operating temperatures to cool the entire power system components makes it ideal and eliminates the extra weight introduced by an additional cryogenic storage as part of the system. Gaseous helium (GHe) has a practically wide operating range of 20-80 K, which makes it feasible to be used as cryogen for all of HTS devices within the power system. While GHe is seen as a replacement for liquid nitrogen in transport applications, it has some challenges such as low dielectric properties and low heat capacity.

HTS devices are being developed for electric ship and aircraft applications, however, this paper focuses on the development of HTS cable termination for electric ship applications. The anticipated power rating of electric ships is 100 MW and an operating voltage in the medium voltage range is required. Numerous HTS cables are required within the power system to achieve the power demands from various components and overall safety constraints of the system. An efficient power system is one that is resilient and offers redundancy. In a transport system, failure of a component or a section of the system should not result in the loss of the whole power system. The occurrence of a fault in a partially superconducting power system could be the result of a number of factors from both the electrical and cryogenic components of HTS devices. Designing parallel cable routes which terminate at common locations within the power system allows for reconfiguration and resiliency in the event of a cable failure, as the unaffected HTS cables could accommodate the rated current [8].

For this design to achieve high power density, efficient solutions are required for both the HTS cable and termination. High power-dense HTS cable terminations represent notable challenges and if the electrical, thermal, and mechanical constraints of HTS cables are unaccounted for, it could result in the failure of the entire power system [9]. During the cooling of the HTS cable, mechanical stresses are introduced at the terminations which need to be accounted for due to the thermal contractions of the cable [12]. For HTS cables, the majority of the heat loads come from the terminal lead in the termination which transitions from room to cryogenic temperature and is made from a conventional conductor. Typical thermal load in the termination was estimated to be 350 W [13]. This leads to an added difficulty in removing the heat generated and maintaining the required cryogenic temperature for the HTS cable. There have been previous termination designs to help mitigate these problems, but these designs are not for large continuous thermal loads, and the complexity and size of terminations make them undesirable in an efficient and compact cryogenic system [13].

There are two dielectric insulation designs for HTS power cables - warm and cold dielectric insulations. The cold dielectric insulation is where the electrical insulation is directly applied on the HTS cable and is in the cryogenic environment inside the cable cryostat. Ongoing research is being conducted on the use and application of cold dielectric termination with the mismatch in coefficient of thermal expansion between the HTS conductor and polymeric insulation being a major challenge [14]-[16]. With the warm dielectric insulation, the electrical insulation employed has no direct contact with the HTS cable and it is on the exterior of the cable cryostat, hence, the insulation is at room temperature [17]. This allows for traditional techniques to apply electrical insulation, such as extrusion, to be utilized. A termination design where warm dielectric insulation concept is utilized is beneficial as it seeks to offer reduced insulation complexity, negligible heat load generation, and significant reduction of the size of the terminations [5].

The warm dielectric insulation design should be limited only for the cable terminations, but not for the whole cable system interconnections. This is because a number of cables in cryostats have various phases and voltage potentials and warm dielectric would require all separate cryostats for each cable, which results in additional weight added to the system and components. This paper presents a modern warm dielectric termination design concept with electric breaks for high-temperature superconducting power cable, which shows the feasibility to provide a high power dense termination design. The design concept is explored using finite element modeling and the dielectric properties of prototype warm dielectric termination utilizing helium gas is characterized.

2. Termination of Cables in a Cryogenic system

In electric transport applications, multiple cables are expected to terminate at a common point. Therefore, it is imperative to understand the basic structure of the termination as shown in Figure 1. For simplicity, Figure 1 depicts a single HTS cable but this design concept can be extended to develop a cryogenic node as previously discussed in [18]. Designing the warm dielectric insulation termination for helium gas applications should be compatible with a closed-loop cryogenic circulation system where a cryocooler and fan are used to continuously circulate the cryogen. Figure 1 also depicts how a warm dielectric is used for the termination design while a cold dielectric insulation is applied for the cable interconnections. To allow the components within the cryogenic system to be at different voltage potentials, electrical breaks are required on either side of the termination. Maintaining the cable insulation with cold dielectric design helps increase the overall power density of the system compared to utilizing a warm dielectric for the entire system.

For a cold dielectric termination, a single cable termination will be made of cables having multiple voltage potentials which need to have adequate clearance between one another as well as the cryostat which is at ground potential. The clearance required by the terminal leads and the cryostat needs to account for the mechanical stress applied by the HTS cable during thermal contractions as moving the terminal lead off-axis can lead to significant electric field enhancement. For the warm dielectric termination, the terminal lead and the inner wall of the cryostat are at the same voltage potential which makes the installation and design of the terminal lead simpler. The premise behind this design is that for GHe cooled MVDC HTS cables two warm dielectric terminations provide a higher power-dense solution compared to a single cold dielectric termination. An electric field analysis was completed to compare the feasibility of the cold and warm dielectric designs.

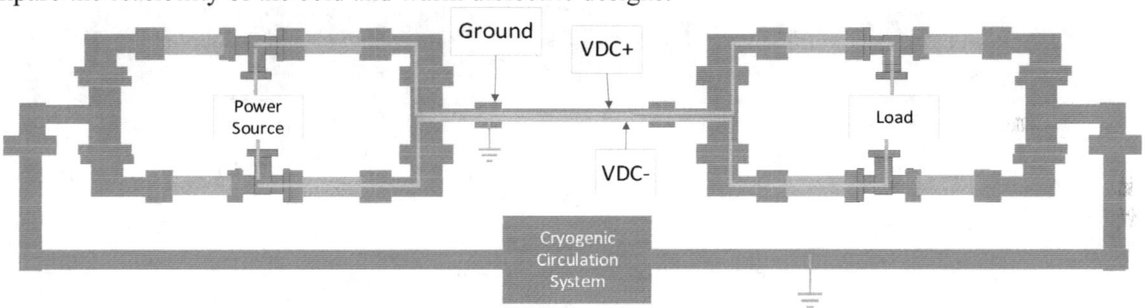

Figure 1. A closed-loop cryogenic cable termination

2.1 Electric Field Analysis of Warm and Cold dielectric Insulation designs

An electric field study was conducted to understand and compare the differences in electric field enhancement that will be introduced in both warm and cold dielectric terminations using the finite-element method. In setting the parameters for the use of the finite-element method, the same geometry was employed for both the warm and cold dielectric terminations (Fig 2.a) and the determined boundary conditions were changed by changing the voltage potentials of the cryostat to achieve a cold dielectric (Fig 2.b) and warm dielectric (Fig 2.d).

The dimensions for a commercial off-the-shelf cryostat were used, a current rating of 2-4 kA was assumed for the high-temperature superconducting cable and approximately 2000 mm^2 was chosen as the cross-sectional area of the terminal lead.

For the warm dielectric termination, in conducting the electric field analysis, the thickness of the polyethylene used was assigned a value of 3 mm which was the actual thickness specified in the manufacturer's specification sheet. The polyethylene jacket is typically used as mechanical protection for the cryostat, however, polyethylene is typically used as extrusion insulation for power cables. The highest electric field between the outer cryostat wall and the ground layer on a 60 mm cryostat is approximately 0.2 kV/mm for an applied nominal voltage of 1 kV. Conventional high voltage cable utilizing cross-linked polyethylene insulation with a thickness of 3 mm is used for medium voltages in the range of 20-30 kV. Juxtaposing the thickness of the XPLE to that of the polyethylene validates electric field calculations of allowable operating voltages of 20 kV.

For the cold dielectric termination, it was observed that the highest electric field enhancement occurred between the terminal lead and the inner wall of a 60 mm cryostat and was about 0.2 kV/mm for an applied nominal voltage of 1 kV. While this is the same maximum electric field shown for the warm dielectric design, the insulation medium is GHe which has significantly lower dielectric strength. The required gap distance in an actual application to achieve the required power densities makes the 5mm gap distance between the terminal lead and cryostat wall too small and not practicable. This is because the terminal lead and cryostat wall need to be aligned precisely to prevent an increase in electric field intensity which will result in lower withstand voltages. To experimentally validate this theory an experiment setup was developed which consisted of warm and cold dielectric terminations, and a cold dielectric cable.

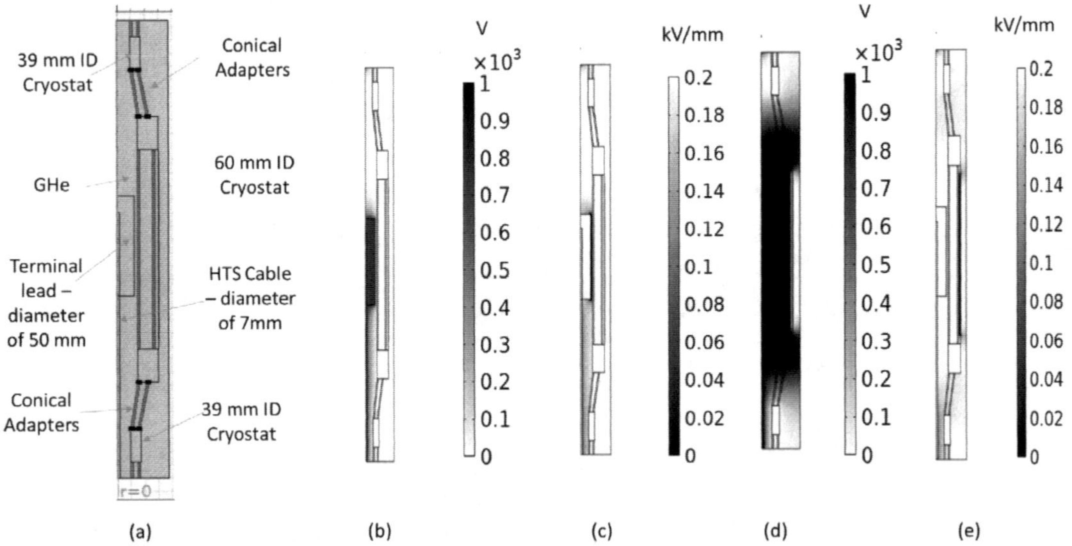

Figure 2. (a) geometry of a cable termination (drawn to scale), (b) electric potential when utilizing a cold dielectric design, (c) electric field when utilizing a cold dielectric design, (d) electric potential when utilizing a warm dielectric design, (e) electric field when utilizing a warm dielectric design.

3. Experimental Setup

The experimental setup was fabricated using commercial off-the-shelf 2.75-inch Conflat (CF) hardware. An aluminum conductor was used to represent a cold dielectric HTS cable in performing the experiment. A total of seven (7) glass polymer laminate (GPO-3) insulation rods with an outer diameter of 9.5 mm were sheathed at the ends of the aluminum conductor. Kapton tape was used to keep the GPO3 firmly wrapped around the conductor keeping it concentric and also from parts of the walls of the CF hardware which will be at ground potential throughout the high voltage measurements. A copper mesh was wrapped around the mid-section of the conductor that was going to be at the warm dielectric termination compartment of the setup. The conductor with the wrapped mesh was inserted into a brass tube with an outer diameter of 12.7 mm. This arrangement of having the mesh on the aluminum conductor and in the brass tube was to allow for the high voltage terminal lead to make contact with the model cable acting as the HTS cable.

Two ceramic breaks rated for 65 kV as shown in fig 3 (a) were used as electrical breaks to allow for the CF hardware to be at different voltage potentials. The ceramic breaks are hollow to allow for the flow of the cryogen. The ends of the aluminum conductor were connected to two high voltage bushings and stress spheres capped on the bushings as demonstrated in the schematic. The high voltage was applied to the sample and a check for continuity was performed to ensure the intended voltage potential of the experimental setup was achieved. After setting up the experiment, a series of purging cycles involving vacuuming and filling the setup with nitrogen gas were performed to eliminate all contaminants that were present. The setup was filled with helium gas and pressurized at 2 MPa and placed in a liquid nitrogen bath at 77 K.

AC partial discharge and withstand voltage measurements were performed on the setup using a 100 kV, 5kVA, 60 Hz AC (RMS) power supply. The partial discharge measurements were performed first with the voltage being increased with increments of 300 V to allow for the examination of activity, and care was taken to prevent the cable from breaking down during the experiments. This was done to ensure the integrity of the cable was not compromised before withstand voltage measurements were performed. A ramping voltage of approximately 300 V/s was set and used during the intrinsic withstand voltage measurement. Due to the design of the experiment, ground wires were attached to different ends of the ceramic breaks to ensure both ends of the cable were at ground potential. The experimental setup used in performing both measurements is shown in Fig 3(b).

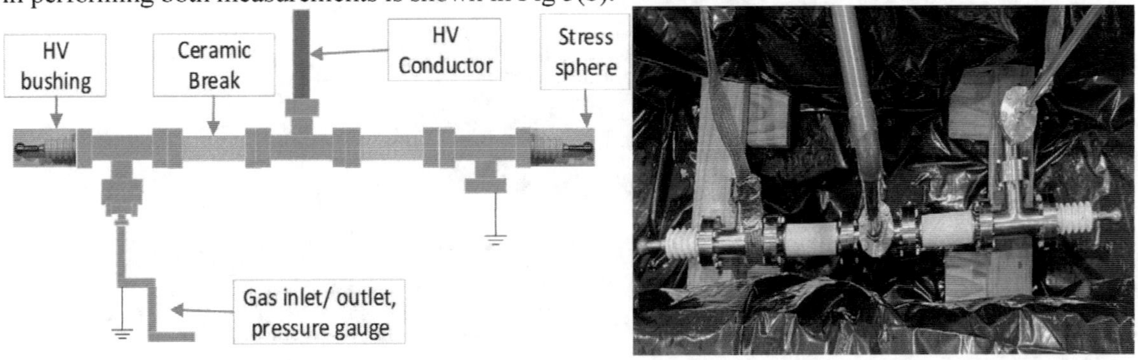

Figure 3. (a) schematic for the warm dielectric termination. (b) Actual experimental setup for the warm dielectric termination.

4. Results

The data presented in Fig 4 shows the withstand voltage and the partial discharge inception voltage (PDIV) values. An average value of 18 kV was obtained as the withstand voltage for the warm dielectric termination after seven (7) measurements. A surface flashover occurred between the high voltage bushing pin and the ceramic which left a surface track on the ceramic as shown in Fig 5.

A summary of the PDIV data collected for the warm dielectric termination with pressurized GHe at 2 MPa was plotted on the graph as shown in Fig 4. At a voltage of 12 kV, an apparent charge of 20 pC averagely was recorded which demonstrated the occurrence of a partial discharge activity.

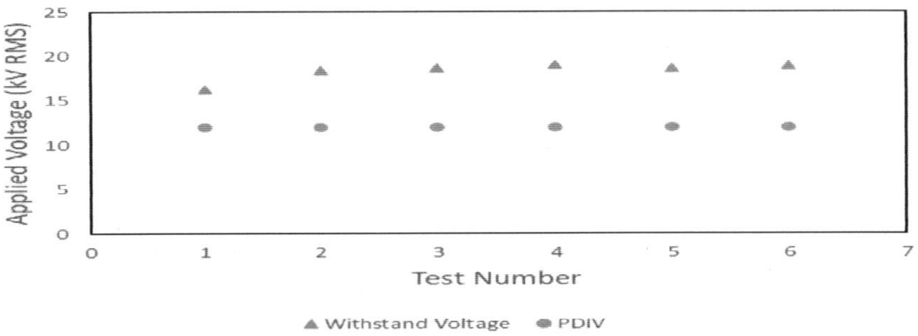

Figure 4. Withstand voltage and PDIV values for multiple tests performed

Figure 5. Surface flashover track found on the ceramic around the HV bushing pin.

5. Discussion

An electric field analysis was conducted using the finite element method to characterize and understand the electric field enhancement that ensues in an event of misalignment where the bushing pin is not held concentric and touches the ceramic around the pin as shown in Fig 6(c) The relative permittivity of the ceramic was set at 21 and 1 for helium gas. A notional 1 kV is applied to the bushing pin and the inner wall of the CF hardware is set to ground potential. A triple point is created as soon the bushing pin is not concentric and makes contact with the ceramic. The mechanism of breakdown likely changes when the triple point is formed from intrinsic breakdown through the helium gas to surface flashover along the ceramic bushing. Given the clearances between the bushing pin and the ceramic, only a small amount of mechanical stress from the cable is enough for the pin to sit off its axis. There is an electric field enhancement as seen in Fig 6(c) which creates the potential for a surface flashover to occur and, subsequently does around 18 kV.

In Fig 6(c), the color scale was set to a maximum of 0.4 kV/mm to allow the reader to see the change in the electric field. The maximum electric field within this model was greater than 1 kV/mm at the triple point between bushing pin, helium gas, and ceramic. The field enhancement at the triple point could explain the reason why PDIV started around 12 kV and the cable eventually breaks down at 18 kV. Similar measurements have been performed in the past by our team and for the geometry utilized for this experiment, a withstand voltage between 20-30 kV was expected [19]. When the bushing is held concentric in Fig 6(b), breakdown occurs through the helium. The electric field for this arrangement is uniform and can be characterized by the coaxial electric field equation. The maximum electric field is on the surface of the conductor and is 0.4 kV/mm when a notional 1 kV is applied.

The bushing pin making contact with the ceramic and creating high electric field enhancement is one of the challenges that the experimental setup highlights when designing terminations. Another challenge is that the high voltage bushing is rated for 4 A and it would not be applicable for the rated current carrying capabilities of HTS cables. The cold dielectric bushing used in the experiment was a limiting factor in our experimental setup. As it could be seen in the results in Fig 5 where the breakdown was occurring around the high voltage bushing. In an actual cold dielectric termination design, where the current demands are in the range of 2-4 kA, the geometry of the terminal lead would need to be larger to accommodate the substantial increase in cross-sectional area of the terminal lead and bigger bushings will be needed.

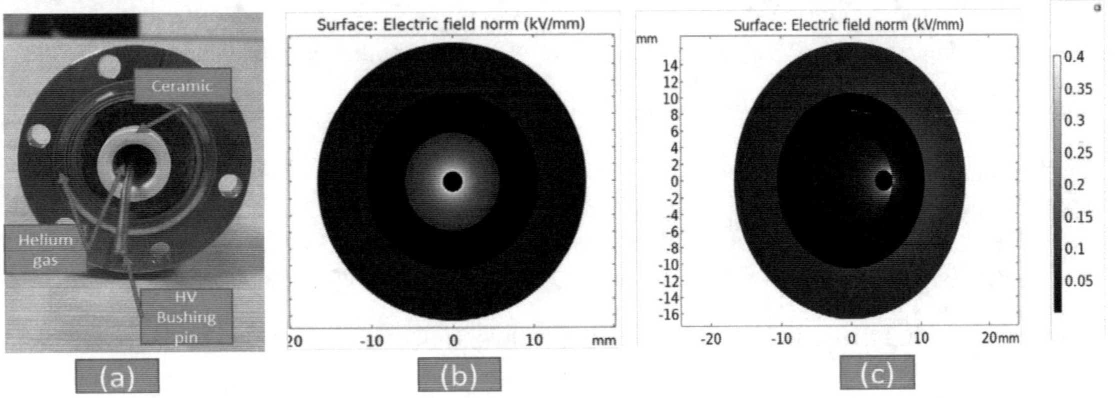

Figure 6. (a) HV bushing rated for 4 A (b) Electric field analysis of the HV bushing pin held concentric (c) Electric field analysis of the HV bushing pin making contact with the ceramic.

6. Conclusion

This paper explored a warm dielectric termination design with electrical breaks for high-temperature superconducting power cables in electric transportation systems. The basic structure of a termination where a warm dielectric is used for the termination design while a cold dielectric insulation is applied for the cable in a closed-loop cryogenic circulation system was discussed. The electric field enhancement for both cold and warm dielectric terminations was studied to understand the electric field enhancement that might be generated with each termination design. The experimental data on the withstand voltage and the partial discharge measurements on the warm dielectric termination design concept showed potential and an improved design concept might need to be studied and developed to reduce the possibility of surface discharge occurring, hence, increasing the withstand voltage.

7. References

[1] J. X. Jin *et al.*, "Enabling high-temperature superconducting technologies toward practical applications," *IEEE Trans. Appl. Supercond.*, vol. 24, no. 5, 2014, doi: 10.1109/TASC.2014.2346496.

[2] M. Stemmle, F. Merschel, M. Noe, and A. Hobl, "AmpaCity - Advanced superconducting medium voltage system for urban area power supply," *Proc. IEEE Power Eng. Soc. Transm. Distrib. Conf.*, pp. 1–5, 2014, doi: 10.1109/tdc.2014.6863566.

[3] B. K. Fitzpatrick, J. T. Kephart, and E. Michael Golda, "Characterization of gaseous helium flow cryogen in a flexible cryostat for naval applications of high temperature superconductors," *IEEE Trans. Appl. Supercond.*, vol. 17, no. 2, pp. 1752–1755, 2007, doi: 10.1109/TASC.2007.897763.

[4] W. J. Shin, J. S. Hwang, J. K. Seong, J. G. Lee, and B. W. Lee, "Breakdown characteristics of liquid nitrogen for transmission-class superconducting electric equipment," *Proc. 2012 IEEE Int. Conf. Cond. Monit. Diagnosis, C. 2012*, no. September, pp. 766–769, 2012, doi: 10.1109/CMD.2012.6416259.

[5] S. Pamidi, C. H. Kim, and L. Graber, "High-temperature superconducting (HTS) power cables cooled by helium gas," in *Superconductors in the Power Grid: Materials and Applications*, 2015, pp. 225–260.

[6] C. M. Rey, "Superconductors in the Power Grid: Materials and Applications," *Superconductors in the Power Grid: Materials and Applications*. pp. 1–438, 2015, doi: 10.1016/C2013-0-16436-7.
[7] J.-G. Kim, M. A. Salmani, L. Graber, C. H. Kim, and S. V. Pamidi, "Electrical characteristics and transient analysis of HTS DC power cables for shipboard application," in *2015 IEEE Electric Ship Technologies Symposium (ESTS)*, Jun. 2015, vol. 1, pp. 376–381, doi: 10.1109/ESTS.2015.7157922.
[8] S. P. Transmission, P. J. Ferrara, M. A. Uva, J. Nowlin, and A. P. Side, "Naval Ship-to-Shore High Temperature," vol. 21, no. 3, pp. 984–987, 2011.
[9] L. Graber, J. Kim, C. H. Kim, and S. V. Pamidi, "Thermal Network Model for HTS Cable Systems and Components Cooled by Helium Gas," *IEEE Trans. Appl. Supercond.*, vol. 26, no. 4, pp. 1–5, Jun. 2016, doi: 10.1109/TASC.2016.2545102.
[10] N. Suttell, C. H. Kim, J. Ordonez, J. Kephart, L. Graber, and S. Pamidi, "Three-Dimensional Finite-Element Analysis of Terminations for Gaseous-Helium-Cooled High-Temperature Superconducting Power Cables," *IEEE Trans. Appl. Supercond.*, vol. 26, no. 4, pp. 10–14, 2016, doi: 10.1109/TASC.2016.2526633.
[11] B. Nemeth, I. Kiss, and R. Cselko, "Examination of a Cable Terninal Failure," pp. 47–49, 2008.
[12] G. M. Hathaway, A. E. Davies, and S. G. Swinglert, "ECl," pp. 84–87.
[13] N. Suttell *et al.*, "Investigation of solid nitrogen for cryogenic thermal storage in superconducting cable terminations for enhanced resiliency," *IOP Conf. Ser. Mater. Sci. Eng.*, vol. 278, no. 1, p. 012019, Dec. 2017, doi: 10.1088/1757-899X/278/1/012019.
[14] C. H. Kim, J. Kim, and S V Pamidi, "Cryogenic Thermal Studies on Cryocooler-Based Helium Circulation System for Gas Cooled Superconducting Power Devices," *Int. Cryocooler Conf.*, pp. 504–513, 2014.
[15] C. H. Kim, S. K. Kim, L. Graber, and S. V. Pamidi, "Cryogenic thermal studies on terminations for helium gas cooled superconducting cables," *Phys. Procedia*, vol. 67, pp. 201–207, 2015, doi: 10.1016/j.phpro.2015.06.035.
[16] L. Graber, C. H. Kim, S. V Pamidi, and D. Knoll, "Dielectric Design Validation of a Helium Gas Cooled Superconducting DC Power Cable," no. June, pp. 157–161, 2014.
[17] P. Cheetham, W. Kim, C. H. Kim, S. V. Pamidi, L. Graber, and H. Rodrigo, "Use of partial discharge inception voltage measurements to design a gaseous helium cooled high temperature superconducting power cable," *IEEE Trans. Dielectr. Electr. Insul.*, vol. 24, no. 1, pp. 191–199, 2017, doi: 10.1109/TDEI.2016.005909.
[18] P. Cheetham, P. Mensah, C. H. Kim, and S. V. Pamidi, "Design of Warm Dielectric Terminations and Nodes for High Temperature Superconducting Power Cables." in IEEE Electrical Insulation Conference 2021 (Accepted)
[19] P. Cheetham et al., "Development of a High-Temperature Superconducting Gas-Insulated Power Cable," in IEEE Transactions on Applied Superconductivity, vol. 30, no. 6, pp. 1-7, Sept. 2020, Art no. 7700207, doi: 10.1109/TASC.2020.2976606.

Acknowledgments
This work is supported by the Office of Naval Research (ONR) through the following grants: N00014-18-12693 and N00014-16-1-2956.

The calculation and measurement of electromagnetic energy flows for investigation of critical current estimation for high-temperature superconducting coil by means of pick-up coils

Junya Omura, Go Sato, and Akifumi Kawagoe

(Kagoshima University, 1-21-40 Kohrimoto, Kagoshima City, Kagoshima, Japan)

E-mail: k9090225@kadai.jp

Abstract. For practical operations of high-temperature superconducting (HTS) devices, we have proposed a monitoring and diagnosing system to detect the degradations of superconducting properties before normal transitions occur locally in the HTS winding. The proposed system estimates the distributions of critical current indirectly from measurement results of the active power components of the Poynting vector around the monitoring HTS devices conducting ac currents by means of pick-up coils pairs mounted surrounding the HTS devices. For this purpose, it is necessary to solve the inverse problem. In order to estimate the critical current distribution, therefore, it is first necessary to be able to calculate the measurement signal distribution from the critical current distribution of the monitoring coil. This paper first proposes the calculation method of signal distributions measured by our system, next reports the results of the experiments and analysis, which were carried out to verify the validity of the calculation method. A pancake coil wound with Bi-2223 multifilamentary tape exposed to a uniform AC magnetic field was assumed as a monitoring coil in the experiments and analyses. The good agreement between the measured signals obtained from the experiments and the numerical analysis demonstrates the effectiveness of the proposed calculation method of measured signals.

1. Introduction

Research and development of devices using high-temperature superconducting (HTS)wires is underway [1]. On these HTS devices, extremely stable device operations are expected because the specific heat of the materials increases significantly at higher operating temperatures.

In the case of HTS wires, when a normal transition area occurs in the winding of the HTS coils due to the heat propagation properties of the HTS wires, the area remains within a narrow range and may become a hot spot. It has been pointed out that the degradation of superconducting properties caused by mechanical damage applied to the HTS wire during the winding process of the coil, its cooling, and operations trigger these hot spots [2, 3].

If these abnormalities exist in the HTS coil, it is hazardous to continue the operation of an HTS coil without noticing the abnormality in the winding. The abnormality mentioned above may trigger the thermal runaway of the coil, even if the coil current is below the critical current. If they remain undetected, the coil may burn out in the worst case.

However, the abnormality and the thermal runaway triggered by it are difficult to detect during operation. Even if the abnormality is detected, it is often already too late to continue the operation of the devices. Therefore, the coil protection process must be provided immediately after the detection.

Providing protection means the HTS devices' circuit needs to be dumped. Unexpected interrupting the circuits, such as the power distribution system, including superconducting transformers, is a serious problem. In addition to protecting superconducting devices, such a serious accident must be prevented as much as possible. For this purpose, the abnormality monitoring method of superconducting coils used in the past, i.e., the quench detection system [4-8], is insufficient. Therefore, for practical HTS devices, the monitoring and diagnosing system is required to detect the degradation of superconducting properties before the normal transition locally occurs and diagnose the degree of degradation.

We have proposed and been studying the new system to detect and diagnosing the abnormality of the HTS winding [9-14]. In this system, active power components of Poynting vectors around the HTS devices are measured by means of the pickup coil pairs. The active power components are the flow of electromagnetic energies corresponding to the AC losses in the HTS windings. The components are hereinafter referred to as P. When an anomality locally occurs in the HTS winding and the critical current degrades, in this area, the magnetization process and the hysteresis losses changes. Therefore, measured P_{EF} also changes. And the magnitude of the P_{EF} might be almost proportional to the AC losses, corresponding critical currents in the area. If it is clearly the relation between critical currents and measured P_{EF}, critical current estimations of the HTS windings from measured P_{EF} might be possible by solving the inverse problem, which calculate critical currents from measured P_{EF}.

We have already shown in the previous papers that P_{EF} can be measured and that changes in the critical current can be detected [11, 12]. In order to solve the inverse problem to find the critical current from the measured P_{EF}, we first need to establish a calculation method of the measured P_{EF} from both magnetic environment and critical currents of the HTS windings.

In this paper, firstly, the principle of our proposed diagnosing system is explained. Next, the method to calculate the P_{EF} measured by the system are proposed. Finally, the experiments and numerical analyses are described in detail. In this experiment and analysis, we assumed a pancake coil wound with Bi-2223 multifilament tape exposed to a uniform AC magnetic field as the monitoring coil.

2. Principle of the diagnostic system

This diagnostic system measures the Poynting vector, which corresponds to the AC loss of the coil under test, in order to diagnose abnormalities in the superconducting coil. Figure 1 shows a schematic diagram of the diagnostic system. Pickup coils for electric field measurement (pickup coil-E: PC_E) and magnetic field measurement (pickup coil-H :PC_H) are placed around the coil to be measured. By measuring these pickup coils as signals, the Poynting vector can be measured. However, the measured Poynting vector includes not only the loss of the winding but also the electromagnetic energy exchanged between the power supply and the coil. Therefore, by integrating the Poynting vector over one AC cycle, it is possible to measure only the loss component of Poynting vector (P_{EF}), which is corresponding to the AC loss of the winding. In other words, The value of P_{EF} is expressed as follows. The electric field measured by PC_E is E_θ, the magnetic field measured by PC_H is H_z, and one cycle time is T.

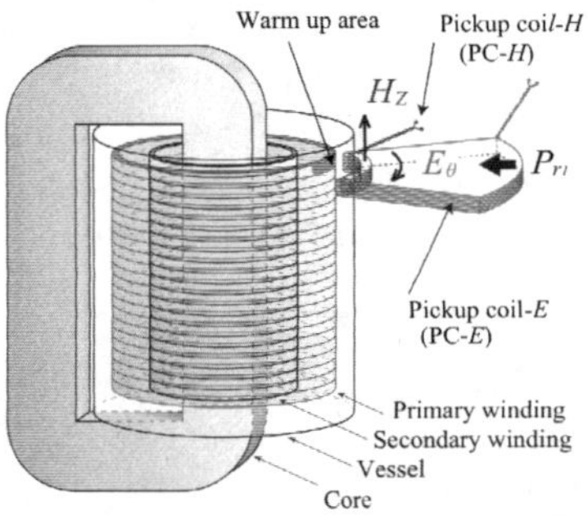

Figure 1. A schematic illustration of one application example of the proposed diagnosing system to the superconducting transformer [9]. In practical case, multiple pick-up coil pairs are placed around the monitoring HTS coils for measuring the distributions of P_{EF} around the HTS coils.

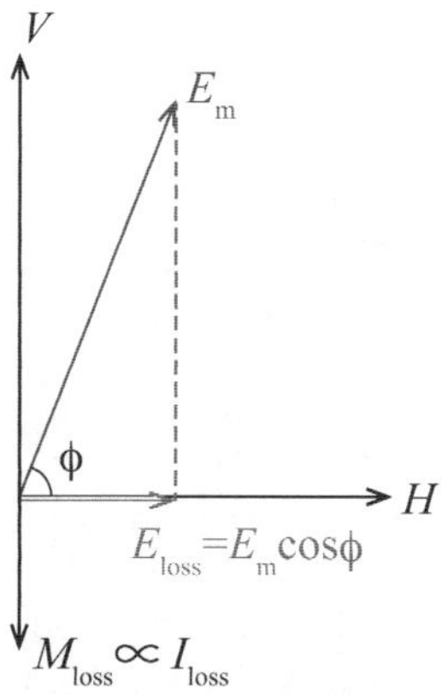

Figure 2. Phasor diagram of measured signal by the proposed diagnosing system.

$$P_{\mathrm{EF}} = \frac{1}{T}\int_T E_\theta \times H_z \, dt \tag{1}$$

P_{EF} obtained equation (1) corresponds to local ac losses in the winding of the HTS coil. In general, AC losses are mainly hysteresis losses in the HTS coil. Therefore, P_{EF} includes the information of local critical current of the winding. By measuring this P_{EF} with high accuracy and back-calculating the critical current of the winding, the winding can be constantly monitored and diagnosed. In practical case, multiple pick-up coil pairs are placed around the monitoring HTS coils for measuring the distributions of P_{EF} around the HTS coils. The distributions of local critical currents are obtained from the distributions of P_{EF}.

The advantages of this diagnostic system are as follows.

- No direct contact with the coil windings is required, resulting in superior safety.
- This system has higher sensitivity and reliability because it is an electrical measurement.
- A sensor installed in a space at room temperature can diagnose abnormalities in the coil.
- Local critical current degradations can be estimated by measured P_{EF} s at multiple points.

3. Proposal of the calculating method

The method for calculating P_{EF} is described below: P_{EF} can be obtained by integrating the cross-product of the electric field E_m measured by PC_E and the magnetic field H_m measured by PC_H in one cycle T. So P_{EF} can be calculated as

$$P_{\mathrm{EF}}=\frac{1}{T}\oint_T \boldsymbol{E}\times\boldsymbol{H}\,\mathrm{d}t=\oint_T EH\,\mathrm{d}t = \frac{1}{2}H_{\mathrm{m}}E_{\mathrm{m}}\cos\phi = \frac{1}{2}H_{\mathrm{m}}E_{\mathrm{loss}} \tag{2}$$

E_{loss} is assumed to be generated by the loss component of magnetization M_{loss} caused by the AC loss in the coil. In case of HTS coil, ac losses are mainly hysteresis losses. This M_{loss} has only a fundamental frequency and is a component whose phase is 90 degrees behind the magnetic field. Figure 2 shows the phasor diagram corresponding to each component of electric and magnetic field. The true magnetization M in the superconducting tape is out of phase with the applied magnetic field H_e and has harmonics components corresponding to non-linear properties of the superconductivity. However, when external magnetic fields with only the fundamental frequency are applied to the HTS tapes, AC losses are determined by the fundamental frequency components of the magnetizations, M_0. In addition, if the phase difference between this magnetization and the applied magnetic field is ϕ, then the magnetization and the applied magnetic field are respectively expressed as follows.

$$M_0 = M_{\mathrm{m}}\sin(\omega t + \phi) \tag{3}$$

$$H_{\mathrm{e}} = H_{\mathrm{m}}\sin\omega t \tag{4}$$

Since the main component of the AC losses is the hysteresis loss, the loss W_h is expressed as follows:

$$W_h = \mu_0 \oint M \, dH_e = \mu_0 \oint M \frac{dH_e}{dt} dt = \pi\mu_0 H_m M_m \sin\phi = \pi\mu_0 H_m M_{loss} \quad (5)$$

Thus, the AC loss can be determined by considering only the loss component of magnetization, M_{loss}. A variant of equation (5) is expressed as follows.

$$M_{loss} = \frac{W_h}{\pi\mu_0 H_m} \quad (6)$$

Therefore, E_{loss} can be obtained by calculating M_{loss} from the AC loss and calculating the electromagnetic field distribution generated around the coil by the M_{loss}. P_{EF} can be obtained multiplying E_{loss} and H_m.

This calculation method is an approximate method. In reality, when the coil under test is exposed to an AC magnetic field, the internal magnetic flux penetration process is very complex. And, in order to analyze them accurately, it is necessary to use the finite element method, which is time-consuming. Therefore, the proposed method assumes that the only loss component of the magnetization of the coil under test that is necessary for P_{EF} calculation is M_{loss}, which can be calculated from the AC loss value without using the finite element method.

4. Verification of the calculating method

In chapter 4, we show the effectiveness of the P_{EF} calculation method proposed in Chapter 3. For this purpose, P_{EF} are obtained from experiments and numerical analysis and compared in case when only external magnetic fields are applied to the HTS tapes. So AC losses are only magnetization losses. In section 4.1, the sample coil for the demonstration test is described. In section 4.2, the details of the analytical model used in the numerical analysis are described. In section 4.3, the results of the demonstration test are reported. Finally, in section 4.4, a comparison between the analytical results and the demonstration test results is presented. In this paper, comparison between experiment and calculation

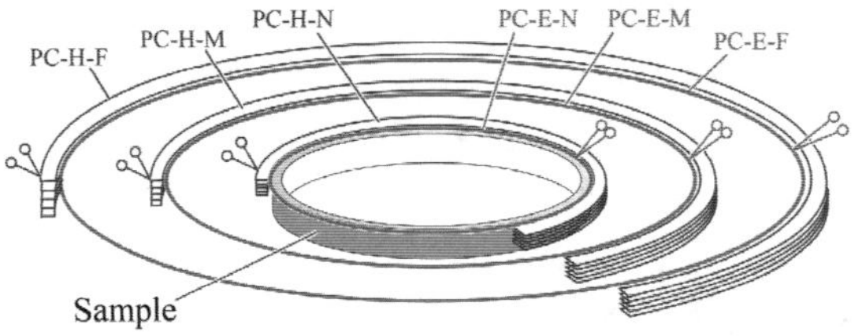

Figure 3. The schematic diagram of the experimental sample coil and pick-up coils for measuring electromagnetic energy flows. The sample coil was a single pancake coil with a 9.5 turn. PC_E and PC_H represent pick-up coils for measuring electric fields and magnetic fields, respectively. The additional words to PC_E or PC_H -N, -M, and -F represent the measuring position.

Table 1.
Parameter of the sample coil and the winding tape.

Tape	
Filament material	Bi-2223
Dimension	4.2mm × 0.22 mm
Critical Current	115A @77.3K
Coil	
Form	Pancake coil
Turn number	9.5
Inner diameter	70mm

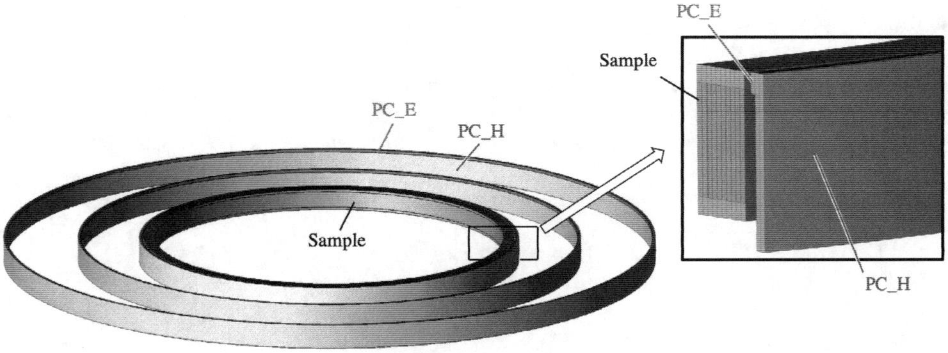

Figure 4. Analysis model and its enlarged figure. The sample coil consists of ten rings, each with two rings encapsulated inside. Round-trip current is applied to these two rings to account for the magnetization, M_{loss} of the sample. PC_E was the simple circularly formed pick-up coil. PC_H was the non-inductive pick-up coil consisting of two circular pick-up coils with different diameters connected.

is carried out in case of only external magnetic fields applied to the winding. Because, in practical coil, ac losses are determined by external magnetic fields, i.e., magnetization losses are dominant. It is important to consider the applicability of this calculation method to more general conditions as well. In principle, it can be considered that there is a loss component of magnetic moment in the HTS tapes when transport current is applied to the tapes as well as case when magnetization losses are generated in HTS tapes under external magnetic fields unless HTS tapes is not in flux flow conditions. The validity of above approximated calculation method will be discussed as future work.

4.1. Sample coil for the verification
The sample coils for the P_{EF} measurement are described below: Bi-2223 multifilamentary tape was used to make a pancake coil sample with an inner diameter of 70 mm, and a uniform AC magnetic field was applied to the sample. The direction of the applied field is parallel to the flat face of the Bi-2223 tape. For accurate measurements, the sample coil was multi-turn. Figure 3 shows the schematic diagram of the sample coil, and Table 1 shows its specifications. PC_E and PC_H are pick-up coils for measuring electric and magnetic fields, respectively. PC_E was the simple circularly formed pick-up coil that measures the electrical fields in the circumferential direction along the wire of the PC_E. PC_H was the non-inductive pick-up coil consisting of two circular pick-up coils with different diameters connected and measures local magnetic fields linked between two pick-up coils constituting PC_H. The direction of the measured magnetic fields is parallel to the sample coil axis. Three pairs of pick-up coils were set up around the sample coil. These pick-up coil pairs measured P_{EF} at distances of 1 mm, 14 mm, and 29 mm from the sample coil. These distances are named N (near), M (middle), and F (far), respectively. The additional words to PC_E or PC_H -N, -M, and -F represent the measuring position. These pick-up coils were made of copper wire with ϕ0.08 mm in diameter (ϕ0.1 mm including insulator).

4.2. Analysis model
The analytical model used to calculate the P_{EF} was created under the same conditions as the experimental sample coil. Figure 4 shows the analytical model and its enlarged figure. The sample coil consists of ten rings, each with two rings encapsulated inside. Round-trip current is applied to these two rings to account for the magnetization, M_{loss} of the sample, corresponding to magnetization losses. To perform the

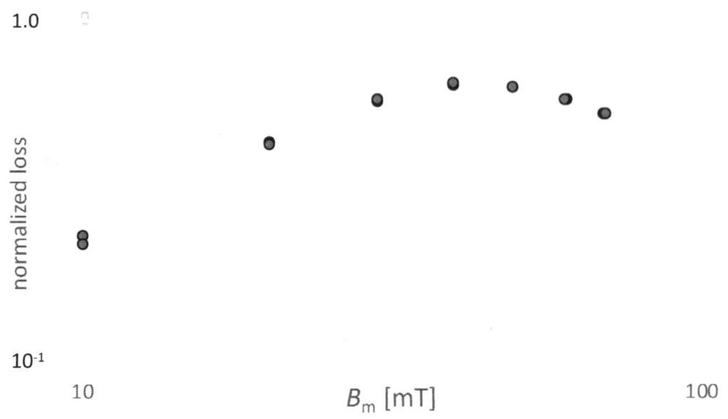

Figure 5. Measured magnetization losses in a short sample of Bi-2223 tapes applied magnetic fields in direction of parallel to the flat face of the tape. The normalized losses are that magnetization losses per one cycle of the magnetic fields, per unit volume of the sample are divided by B_m^2/μ_0.

proposed calculation method, we first obtain M_{loss} from equation 6. For this purpose, the measured magnetization loss in the short sample of the same Bi 2223 tape as the winding tape for the sample coil was used as W_h in equation 6. Figure 5 shows the measured losses at 10 Hz. In figure 5, the horizontal axis is the applied magnetic field, and the vertical axis is the normalized losses that magnetization losses per one cycle of the magnetic fields, per unit volume of the sample divided by B_m^2/μ_0, B_m is amplitude of applied magnetic fields.

Based on the M_{loss} obtained from the magnetization loss properties in the short sample, the magnetic moment m_{loss} is given to the analytical model. Assuming that the elliptical cross section of the sample coil is S and the total length of the winding tape of the sample coil is l, the magnetic moment m_{loss} is expressed as follows.

$$m_{\text{loss}} = M_{\text{loss}} S l \tag{7}$$

Secondly, voltages of pick-up coils are calculated when sample coils have a magnetization of M_{loss}. The calculations are carried out by finit element method, FEM, software JMAG®. By using the obtained voltages of the pick-up coils, energy flows can be calculated.

4.3. Experimental result

The P_{EF} is measured using the sample coil and pickup coil pairs shown in figure 3. The sample coil was wound by Bi-2223 multifilamentary tape, whose critical currents is 115 A at 77K, in self-field. The direction of the applied magnetic fields is the parallel to the surface of the tape. Amplitude of the fields were 6 – 70 mT. The frequency of the fields were 50 Hz for Near, 10 Hz for Middle and Far.

Figure 6 shows the experimental results of the magnetic field dependence of P_{EF} when the measurement distance is varied. The vertical and horizontal axes represent the normalized energy flow per cycle of the applied magnetic field, that P_{EF} are divided by by B_m^2/μ_0, and the applied magnetic field, respectively. From the top, the results are shown for the measurement distances Near (N), Middle (M), and Far (F). From the results, we can see that the measured energy flow is similarly to the magnetization loss of the superconducting tape. The absolute values of the energy flow at different measurement distances are different, but the dependence on the magnetic field is the same at different distances.

4.4. Comparison of experiment with numerical analyses

To confirm the effectiveness of the proposed calculation method, a comparison between the numerical results and the experimental results mentioned above was conducted.

A comparison is then made when the measurement distance is varied: the loss component of voltage V_{loss} measured by PC_E is expressed as follows:

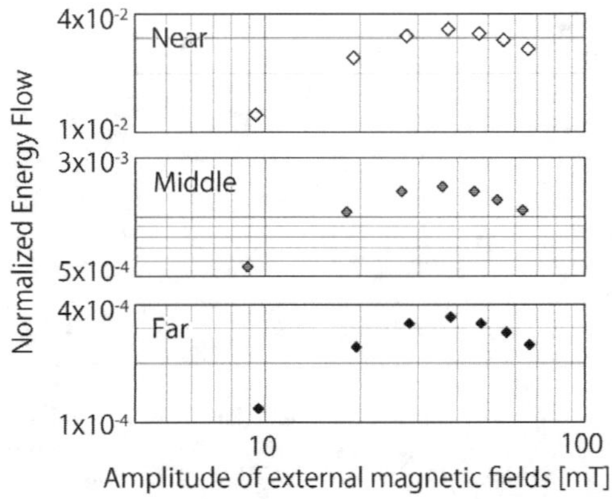

Figure 6. Magnetic field dependence of P_{EF} as the measurement distance was varied in N, M, and F. The vertical axis is the normalized energy flows that P_{EF} are divided by by B_m^2/μ_0.

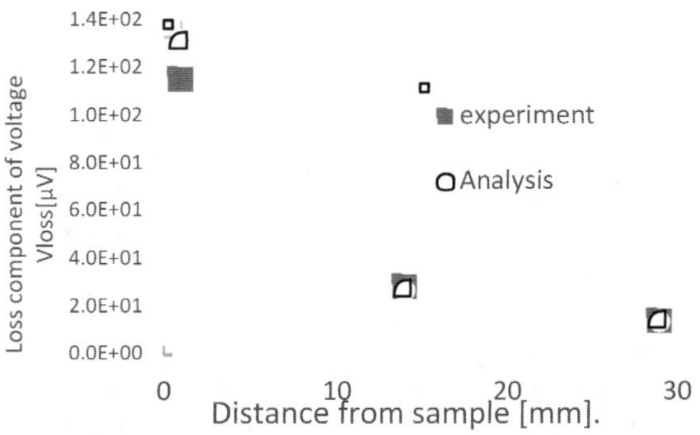

Figure 7. Comparison of the loss components of voltages between experiments and calculations at different measurement distances.

$$V_{\mathrm{loss}} = \frac{2P_{\mathrm{EF}}}{H_{\mathrm{e}}} \cdot l_{\mathrm{PC\text{-}E}} \tag{8}$$

Here, l_{PC_E} is the circumference of PC_E. Since P_{EF} is the value per turn of PC_E, V_{loss} is also the value per turn. Figure 7 shows a comparison of the analytical and experimental loss component of voltages when the measurement distance is varied. The vertical and horizontal axes represent the loss component of voltage V_{loss} [μV] and the measurement distance from the sample, respectively. The errors between the experimental and analytical values are in good agreement: 14% at a distance of 1 mm from the sample, 6% at a distance of 14 mm, and 4% at a distance of 29 mm. These results clearly show the effectiveness of the proposed P_{EF} calculation method.

Figure 8 shows the magnetic field dependence of the energy flow obtained from the analysis and experiment when the measurement distance is Far. The vertical and horizontal axes represent the normalized energy flow and the applied magnetic field, respectively, as in figure 6. This result shows that the experimental and analytical results are in good agreement. It indicates that our porposed method can calculate the electromagnetic nergy flows around superconducting coil.

In practical coil, the method to estimate local critical currents using our system and the calculation method is as follows: Firstly, based on the properties of ac losses of HTS tapes for the coil and design of the HTS coils, ac losses in the winding locally are calculated, and then the distribution of loss components of magnetizations in the tape are also calculated. The distributions of electromagnetic

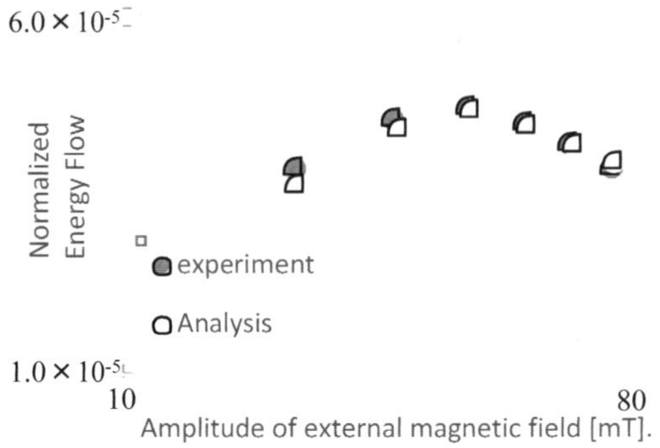

Figure 8. Comparison of the magnetic field dependence of the energy flow obtained by analysis and experiment at Far.

energy flows can be calculated around the HTS coils under operation from the distribution of loss components of magnetizations. The obtained distributions of PEF are in case that the winding soundnesses are in good conditions. Secondly, during operations, the distributions of PEF around the coil measured by our system are compared with the predicted values mentioned above. Then actual loss components of magnetization under actual operations are calculated. Furthermore, degradations of the critical current are estimated from the calculated loss component of magnetization. In actual PEF, all magnetizations in the winding tapes of the coil are superpositioned. Therefore, to calculate the loss component of magnetization, inverse-problems are need to be solved. The study of the solving method is future work.

5. Conclusion

In order to be able to estimate critical currents of the HTS coils before normal transitions occur in the coils, the calculation method of electromagnetic energy flow have been proposed, and the experiments and numerical analysis have been carried out to verify the validity of the method.

This calculation method is an approximate method in which only the loss component of the magnetization in the winding is considered. The experimental and analytical results agreed well when the measurement positions were some distance from the winding. The validity of the of the calculation method have been confirmed from both of experimental and analyses.

6. References

[1] Yuh Shiohara, Masateru Yoshizumi, Yuji Takagi, Teruo Izumi, *Physica C: Superconductivity* Vol. 484, pp. 1-5, 2013
[2] H. Maeda, Y. Yanagisawa, *IEEE Trans. Appl. Supercond.,* vol. 24, no. 3, Art. Num. 4602412, 2014.
[3] T. Takematsu, R. Hu, T. Takao, Y. Yanagisawa, H. Nakagome, D. Uglietti, T. Kiyoshi, M. Takahash, H. and Maeda, *Physica C*, Vol. 470, pp. 674-677, 2010
[4] M.N. Wilson : Clarendon Press Oxford, 1983, pp. 200-232.
[5] K. Takeuchi, S. Murase, S. Kim, G. Nishijima, K. Watanabe, T. Takeuchi, N. Banno, A. Kikuchi, and N. Nanato, *Cryogenics*, vol. 48, no. 3-4, pp.148-153, 2008
[6] T. Ishigohka, T. Tsuchiya, Y. Adachi, A. Ninomiya, N.Yanagi, K. Seo, H. Sekiguchi, S. Yamada, S. Imagawa and T. Mito, *IEEE Trans. Appl. Supercond.,* vol. 15, no. 2, pp. 1423-1426, 2005
[7] T. Ogitsu, A. Devred, K. Kim, J. Krzywinski, P. Radusewicz, R. I. Schermer, T. Kobayashi, K. Tuchiya, J. Muratoe, and P. Wanderer, *IEEE Trans. Magnet,* vol.30, no. 4, pp. 2273-2276, 1994.
[8] A. Siemko, J. Bilian, G. Gerin, D. Leroy, L. Walckiers, and R. Wolf, *IEEE Trans. Appl. Supercond.,* vol. 5, no. 2, pp. 1028-1031, 1995

[9] H. Hiwatashi, R. Mizoue, A. Kawagoe and T. Eguchi, *IEEE Trans. Appl. Supercond.,* vol. 24, no. 3, Art. Num. 9001104, 2014
[10] K. Habu, S. Kaminohara, T. Kimoto, A. Kawagoe, F. Sumiyoshi, H. Okamoto, *Physica C*, vol. 470, no. 20, pp. 1860-1864, 2010
[11] A. Kawagoe, H. Moribe, K. Kakiyama and F. Sumiyoshi, *IEEE Trans. Appl. Supercond.,* vol. 23, no. 3, Art. Num. 9000804, 2013
[12] A. Kawagoe, R. Kosaka, K. Hamada, and H. Okamoto, *TEION KOUGAKU(J. Cryo. Super. Jpn.)*, vol. 47, no. 5, pp. 268-275, 2012
[13] K. Sakemoto, T. Kubo and A. Kawagoe, *Physics Procedia*, vol. 81, 2016, pp. 199-202
[14] A. Kawagoe, K. Hosoda and J. Tsuruda, *IEEE Trans. on Appl. Supercond.,* vol. 29, no. 5, Art. Num. 5501504, 2019

IOP Conf. Series: Materials Science and Engineering 1241 (2022) 012027 doi:10.1088/1757-899X/1241/1/012027

Approaches for High Performance and Thermally Optimized Flexible Cryogenic Microwave Ribbon Cables

Sherman E. Peek[1], **Archit Shah**[1], **Bhargav Yelamanchili**[1], **Vaibhav Gupta**[1], **John A. Sellers**[1], **David B. Tuckerman**[2], **Michael C. Hamilton**[1]

[1]Auburn University, Auburn AL

[2]Tuckerman & Associates, Inc., Lake Stevens, WA

E-mail: sep0022@auburn.edu, mch0021@auburn.edu

Abstract. Design, fabrication, and testing of cryogenic cables with thermal performance considerations is presented in this work. Cables were designed with commercial off-the-shelf materials for low-GHz operation in a small form factor. Thermal and microwave simulations were performed during the design of cables. Fabrication processes were developed to create features in Cu-cladding and to successfully deposit thin-film metal in large step-height regions. Post-fabrication processes such as laser machining and connector assembly were used to cutout and interface with fabricated cables. Microwave performance was measured at room temperature and at 4.2 K with comparable results to simulations performed. Cables designed and fabricated in this work provide a solution to thermal and spatial limitations while achieving targeted microwave performance at cryogenic temperatures.

1. Introduction

Many electronic cables that are used in today's cryogenic closed-cycle systems impart large volumes within the system. Minimizing interconnect volumes is paramount to realize more dense and complex systems. A solution to minimizing interconnect volumes that is heavily pursued is the fabrication of flexible cables [1–5]. Though these cables are a solution to physical dimension constraints, electrical and thermal aspects must be considered in the designs of cables and cable assembly procedures that are intended for use in dense cryogenic systems. In terms of thermal considerations, passive, active, and radiative loads must be managed [6]. Each solution is typically tailored towards specific applications leading to various approaches of thermal management being explored [7–9]. Approaches developed and demonstrated in this work seek to alleviate challenges for designing solutions to electrical, thermal, and spatial constraints by creating plug-and-play cryogenic microwave ribbon cables for cryogenic performance by using commercial off-the-shelf substrates along with conventional microelectronic fabrication and assembly techniques.

2. Cabling Design Approach

Cabling design approach followed four primary criteria: 1) microwave performance intended for low-GHz clock signals, 2) minimize thermal contributions from hot-stage to cold-stage of a hypothetical cryogenic system, 3) use commercial off-the-shelf products when possible, and 4) implement in a small form factor. The result of the outlined criteria is a simple coplanar waveguide (CPW) fabricated from commercially available DuPont™ Pyralux® Cu-cladded Kapton with deposited Ti/Cu thin films interfaced to surface mount sub-miniature push-on SMPS connectors.

2.1. Thermal Simulations of Cabling Design

The start of the design process consisted of thermal simulations. Thermal simulations were performed in Ansys Icepak by modeling two SMPS connector footprints with a transition connecting the CPW's ground-signal-ground configuration from one footprint to the remaining footprint. Two design cases were studied in the thermal simulations performed: 1) uniform Cu thickness (18 µm, thickness of Cu-cladding) across the entire transmission line and 2) decrease in Cu thickness (0.28 µm, thickness of thin film deposition) in the transition region between SMPS footprints. The thickness of 0.28 µm was determined to achieve a targeted DC resistance. Thermal conductivity of Cu at 4.2 K in the model was assumed to be 100 W/m-K for both 18 µm and 0.28 µm thick metals based on thermal conductivity residual-resistance ratio values from Hust [10]. Ambient temperature was defined as 4.2 K. Radiation contributions were neglected. Thermal sources were used on each edge of the SMPS footprints where the "hot-side" footprint thermal source was defined as 77 K and the "cold-side" footprint thermal source was defined as 4.2 K. A representative model used for thermal simulations and simulation results are shown in Figure 1. Steady-state heat flow was calculated along with direct temperature monitoring of the "cold-side" SMPS footprint. Table 1 summarizes the results of the thermal simulations. Simulations completed demonstrated improved thermal performance from hot-side to cold-side when decreasing the thickness of metal in the transition region between SMPS footprints.

Table 1: Results from thermal simulations.

Case	Cu-Cladding (18 µm)	Thin Film (0.28 µm)
Temperature At Thermal Monitor	8.91 K	6.00 K
Steady-state Heat Flow	45 µW	9 µW

2.2. Structural Design and Microwave Simulations of Cables

Once thermal simulations were completed, cables were designed and simulated to achieve desired microwave performance. Figure 2 presents critical design dimensions, drawn using Keysight Technologies' PathWave Advanced Design System. Loopback cabling structures were designed as a test structure to evaluate designed cables. CPW conductor gaps are designed to be 30 µm more narrow than desired due to fabrication considerations for chemical wet etching, which will be discussed in Section 3. Simulations to evaluate microwave performance were completed in Ansys HFSS using a 3D model with one view shown in Figure 3. Results from the simulations performed are shown in Figure 4. Microwave results of the simulations performed meet target requirements for this particular cable design.

IOP Conf. Series: Materials Science and Engineering 1241 (2022) 012027 doi:10.1088/1757-899X/1241/1/012027

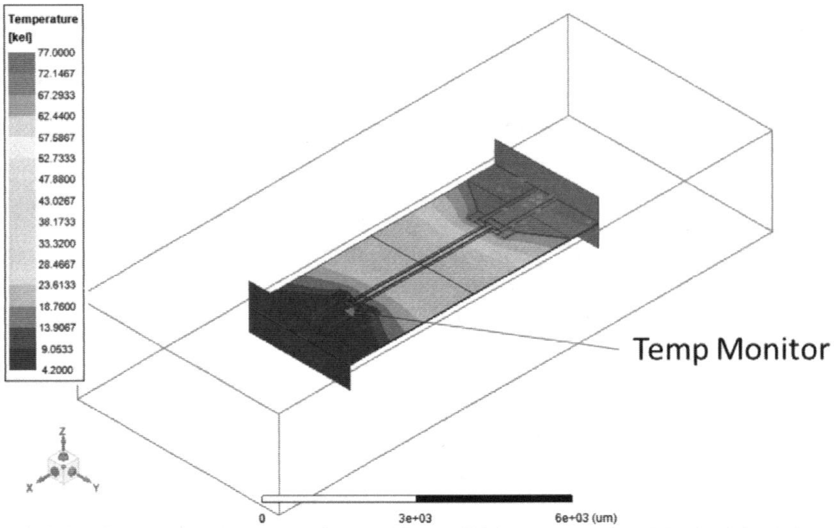

Figure 1: Representative model and simulation results of a symmetric thru line used to compare thermal impact of metal thicknesses between SMPS footprints.

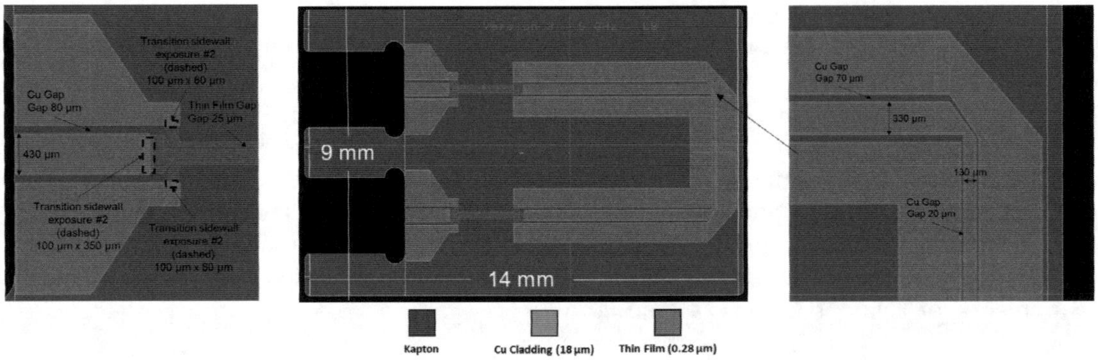

Figure 2: Critical dimensions of the designed loopback cable presented in this work. Cu cladding gaps were designed to be 30 µm more narrow than desired to comply with fabrication methods. The smallest gap in Cu-cladding is 20 µm. The gap in thin film is 25 µm. Dashed lines denote locations that experience a second exposure before developing, as discussed in Section 3.

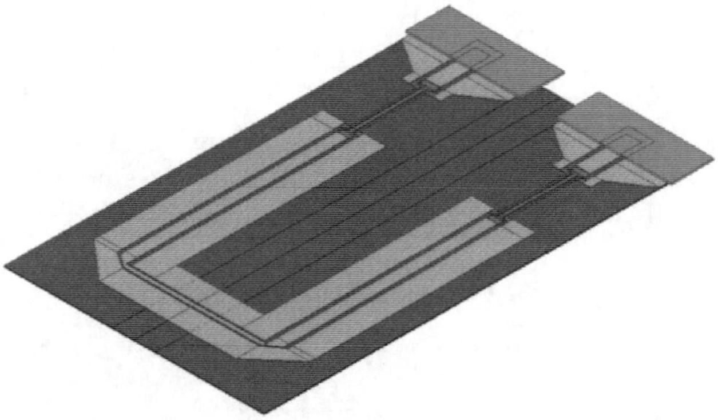

Figure 3: A 3D model of the designed cable used to simulate microwave performance in Ansys HFSS.

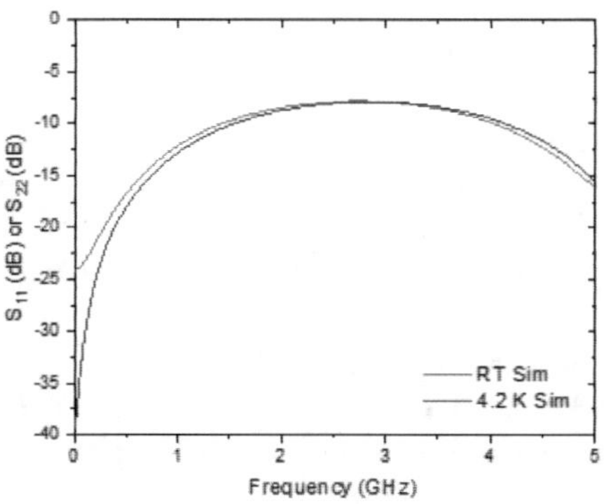

(a)

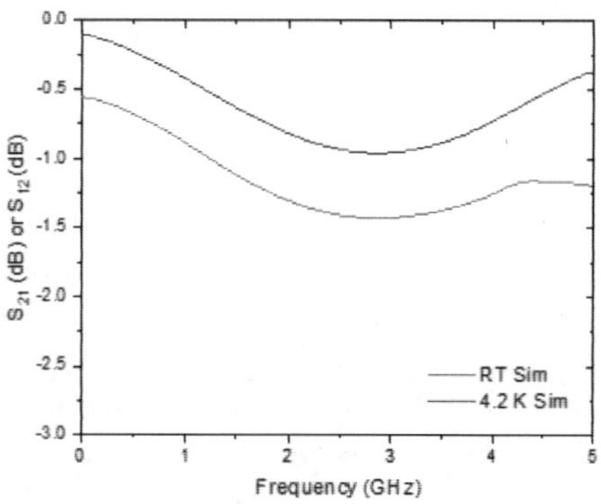

(b)

Figure 4: Simulation results extracted from Ansys HFSS for room temperature (RT) and 4.2 K operation. (a) S_{11} or S_{22} and (b) S_{21} or S_{12}.

3. Fabrication of Cables

This section details fabrication processes used to realize the designed cables presented in this work. A schematic cross-section of the fabrication flow to be used as reference for this section is found in Figure 5. Many processes used in this work have been developed from our previous work [11–13].

3.1. Thin-Film Fabrication

Fabrication began with mounting Cu-cladded Kapton sheets to fabrication rings to maintain relative flatness for the duration of the fabrication process. Neglecting to use the fabrication rings resulted in inconsistent fabrication yields. Figure 6a shows a substrate when mounted to fabrication rings. After the substrate was mounted to the fabrication rings, a conventional positive-tone photolithography process was completed to define the gaps in the Cu-cladding for the CPW transmission lines. After the photoresist was developed, Cu-cladding was etched using a 2:1 mixture of DI water and Transene 49-1 Cu etchant. We found during this process that an additional photolithography step may be necessary (i.e., photolithography, etch, remove

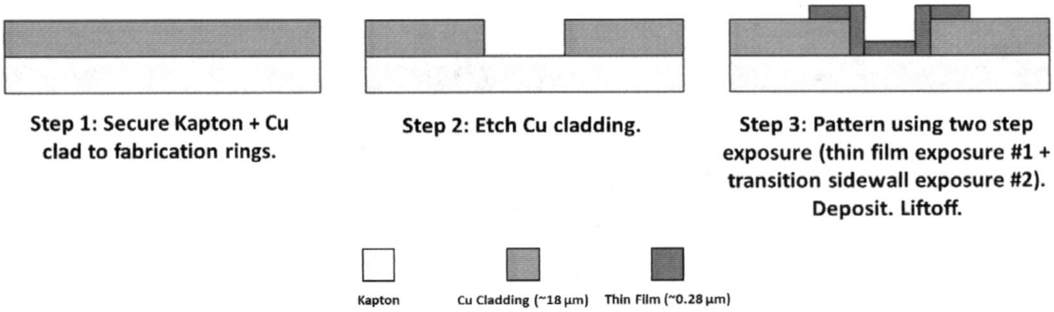

Figure 5: Schematic cross-section of fabricated cables using commercially available off-the-shelf Cu-cladded Kapton sheets as the base substrate with etched features and added resistive thin films for thermal optimization.

photoresist, photolithography, etch) to better protect the desired Cu-clad due to degradation of the resist layer during etching. Structures etched using a single photolithography step were observed to contain mosaic-like defects on the top surface of the Cu-cladding where the resist layer began to crack as shown in Figure 6b. It is important to note that due to the isotropic etching process, designs must be biased, as previously mentioned in Section 2.2, such that the lateral etch amount of the Cu reflects the desired geometry upon completing the vertical etching. After etching the Cu-cladding, thin film deposition was performed. A two-step exposure was used when performing positive-tone photolithography in the Cu-cladding to thin film transition regions (shown by dashed lines in Figure 2) due to a large thickness step between thicker Cu-cladding and regions for thin film Cu. The first exposure step exposes the entirety of the thin film pattern in the desired locations but has shown insufficient photoresist removal in the transition regions after resist development, as shown in Figure 6c. The second exposure in the two-step process is completed directly after the first exposure using a separate mask reducing the exposed areas to only the transition regions. This step enhanced the resist removal in areas that experience large step transitions, as shown in Figure 6d. A 2 minute ion milling process was performed to clean the surface of the deposition sites and to roughen the surface for better metal adhesion, followed by a deposition of Ti (50 nm) and then Cu (230 nm) as the signal layer. The deposited metals then underwent a lift-off process by way of acetone immersion to complete the thin film deposition step.

3.2. Post-Fabrication Processing

After the thin-film fabrication processes were completed, the next step in realizing the designed cables was to cut individual cables from the substrate. A Newport Laser μfab™ system was used in conjunction with a Spectra-Physics Spirit® femtosecond laser system to perform each cable cutout. Due to lack of planarity of the substrate as the laser processing begins/continues, a low magnification lens was used to ensure the entire depth of the cable cutout remained inside the depth of focus of the laser. Parameters used to perform the cable cutouts with minimal Kapton carbonization and heat-affected zone (HAZ) damage are shown in Table 2 and a resulting cutout is shown in Figure 7a.

After each cable was cut out, SMPS connectors were attached using Sn63/Pb37 solder paste applied onto the connection regions of the cables. Connectors were reflowed with 350 °C hot air on a Cu heat sink. To protect the connectors and cables from handling damage, Masterbond EP29LPSP epoxy was deposited onto connector locations using a "glob top" technique and cured at 80 °C for 14 hours. A resulting cable is shown in Figure 7b.

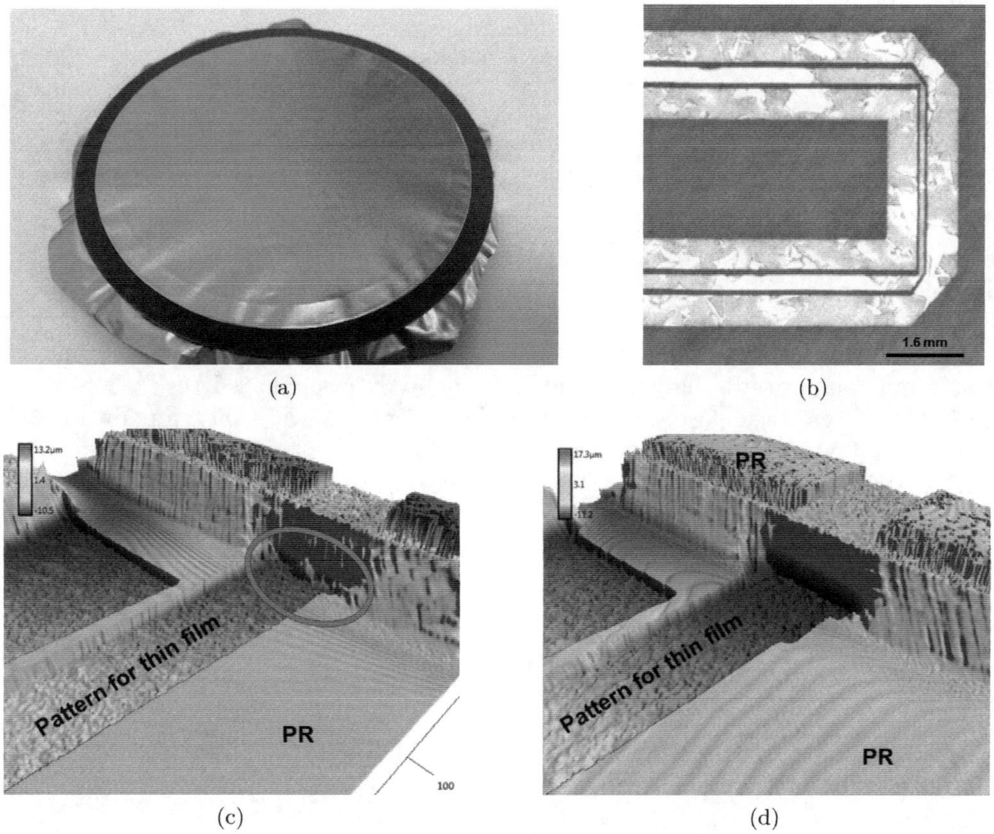

Figure 6: Collection of images during the thin-film fabrication process. (a) Image of Cu-cladded Kapton sheet mounted to fabrication rings to ensure consistent fabrication alignment during processing. (b) Image of top surface of Cu-cladding after Cu etching procedure. Photoresist during the etching procedure began to crack, exposing the Cu-cladding to the etchant resulting in top surface etching in a "mosaic-like" pattern. (c) 3-D profilometry of a thin film pattern without using a two-step exposure technique with residual photoresist (PR) outlined in red in a step transition region. (d) 3-D profilometry of the thin film pattern using a two-step exposure technique showing fully developed photoresist (PR) in a step transition region.

IOP Conf. Series: Materials Science and Engineering 1241 (2022) 012027 doi:10.1088/1757-899X/1241/1/012027

4. Testing and Results

After fabrication of the cables, testing of the fabricated cables was performed in a cryogenic dewar probe by directly submersing into liquid helium to achieve temperatures as low as 4.2 K. Cables of the testing infrastructure were connected to the fabricated cable by using a custom assembly jig to assist cable connectivity. Measurements were performed by using an Agilent N5227A PNA. The testing infrastructure included two 36" coaxial SMA cables leading from the PNA to the top-side of the probe. The probe consisted of two 120" coaxial cables connected to two 6" coaxial SMPS cables. The coaxial SMPS cables interface with the fabricated cable. Calibration was performed at both room temperature and at 4.2 K to the end of the coaxial SMPS cables. Automatic Fixture Removal (AFR) was used to remove coaxial SMPS cable contributions. A representative schematic is provided in Figure 8. The source power from the PNA for each measurement was +20 dBm. Figure 9 presents measured performance of the cables at room temperature and at 4.2 K as compared to simulated results. Simulated and measured results compared closely when comparing S_{11} and S_{22} at each temperature. Simulated and measured differences that are observed in insertion loss performance are likely caused by fabrication defects, as mentioned in Section 3. Even so, fabricated cables are proven functional at temperatures as low 4.2 K.

Table 2: Laser machining parameters for cable cutouts using Newport Laser µfab™ system.

Parameter	Value
Lens Objective	4x/NA: 0.1
Max Power (Prior To Lens)	970 mW
Wavelength	520 nm
Pulse Duration	420 fs
Repetition Rate	200 kHz
Pulse Picker Divider	100
Repetitions Per Cable	3
Scan Speed	1 mm/sec

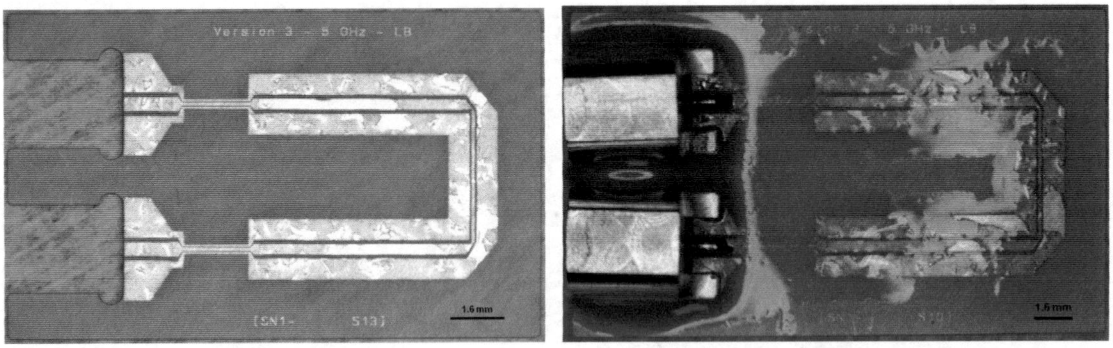

(a) (b)

Figure 7: (a) Image of a single cable after laser cutout with minimal HAZ damage and carbonization of the Kapton around the perimeter of the cable. (b) Image of a single cable after attaching SMPS connectors and curing protective epoxy coating.

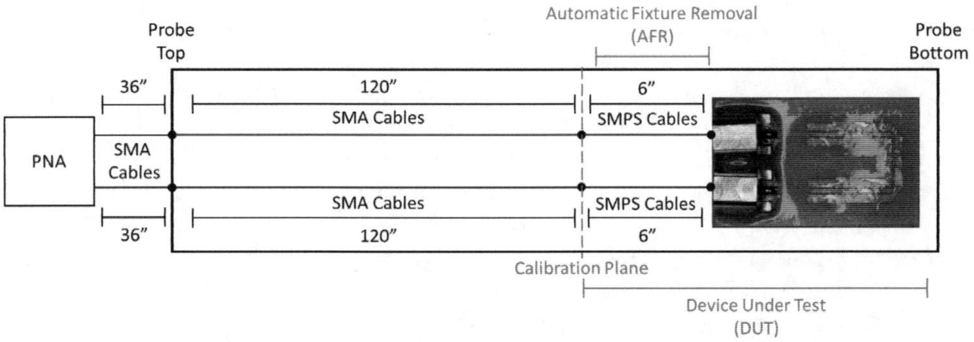

Figure 8: Representative schematic of the testing infrastructure detailing the lengths of cabling, calibration plane, and location of the Automatic Fixture Removal (AFR).

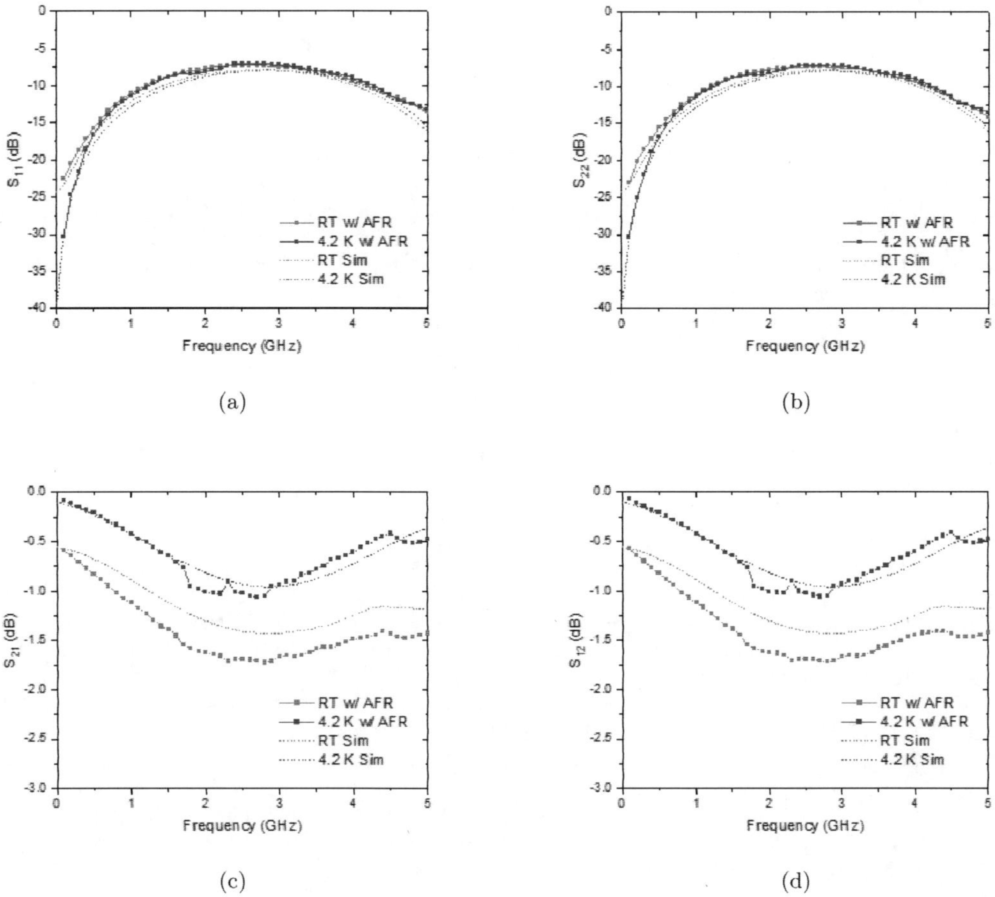

Figure 9: Measurements results for fabricated cables at room temperature and 4.2 K by direct submersion in liquid helium compared to simulated performance. (a) S_{11}, (b) S_{22}, (c) S_{21}, and (d) S_{12}.

5. Conclusion

In this work, we describe approaches to realize a cable fabricated from commercial off-the-shelf substrates with thermal optimization considerations. Fabrication processes used when creating the presented cables use largely conventional microelectronic fabrication techniques with developed techniques such as a two-step exposure process for large step transition features. Post-fabrication processing steps were developed to cutout cables with a femtosecond laser as well as to attach small form factor connectors to fabricated cables. Cables fabricated were proven functional at room temperature and at 4.2 K with comparable performance to simulated results. Differences observed when comparing simulated to measured performance are likely attributed to fabrication defects. We presume that by optimizing the etching process, simulated and measured performance will have a closer comparison. Additionally, the impedance matching of the designed cable for the measured frequency range can be improved beyond this work's targeted performance. Additional work to optimize the microwave performance at the transition region from Cu-cladding to thin-film is necessary to achieve better impedance matching. Aspects regarding the approaches used in this work to realize a thermally optimized cable solution can be refined, however results presented are encouraging for future iterations of cables.

Acknowledgments

We gratefully acknowledge funding and technical guidance from Microsoft for this work. We thank the Alabama Micro/Nano Science and Technology Center (AMNSTC) for providing access to fabrication and characterization facilities used in this work.

References

[1] Shah A, Gupta V, Yelamanchili B, Peek S E, Chowdhury M F F, Rashid A S, Goteti U S, Sellers J A, Tuckerman D B and Hamilton M C 2021 *IEEE Transactions on Applied Superconductivity* **31** 1–6

[2] Yelamanchili B, Gupta V, Zou S, Christiansen R, Cichon M C, Sellers J A, Park M, Mahjouri-Samani M, Tuckerman D B and Hamilton M C 2021 *IEEE Transactions on Applied Superconductivity* **31** 1–5

[3] Gupta V, Adams M L, Sellers J A, Niedzwiecki N, Rush N, Tuckerman D B and Hamilton M C 2021 *IEEE Transactions on Applied Superconductivity* **31** 1–4

[4] Schmidgall E R, Griggio F, Thiel G H, Peek S E, Yelamanchili B, Shah A, Gupta V, Sellers J A, Hamilton M C, Tuckerman D B and d'Hollosy S 2021 Reliability characterization of a flexible interconnect for cryogenic and quantum applications *2021 IEEE International Reliability Physics Symposium (IRPS)* pp 1–7

[5] Zou S, Yelamanchili B, Gupta V, Sellers J A, Isaacs-smith T, Tuckerman D B and Hamilton M C 2019 *Superconductor Science and Technology* **32** 075006 URL `https://doi.org/10.1088/1361-6668/ab1825`

[6] Krinner S, Storz S, Kurpiers P, Magnard P, Heinsoo J, Keller R, Lütolf J, Eichler C and Wallraff A 2019 *EPJ Quantum Technology* **6** 2 ISSN 2662-4400, 2196-0763 URL `https://epjquantumtechnology.springeropen.com/articles/10.1140/epjqt/s40507-019-0072-0`

[7] Melhuish S J, McCulloch M A, Piccirillo L and Stott C 2016 *Review of Scientific Instruments* **87** 104706 ISSN 0034-6748, 1089-7623 URL `http://aip.scitation.org/doi/10.1063/1.4964475`

[8] Demko J, Lue J, Gouge M, Lindsay D, Roden M, Willen D, Daumling M, Fesmire J and Augustynowicz S 2003 *IEEE Transactions on Appiled Superconductivity* **13** 1930–1933 ISSN 1051-8223 URL `http://ieeexplore.ieee.org/document/1211989/`

[9] Long H M and Notaro J 1971 *Journal of Applied Physics* **42** 155–162 ISSN 0021-8979, 1089-7550 URL `http://aip.scitation.org/doi/10.1063/1.1659553`

[10] Hust J G and Lankford A B 1984 Thermal conductivity of aluminum, copper, iron, and tungsten for temperatures from 1 K to the melting point Tech. Rep. NBS IR 84-3007 National Bureau of Standards Gaithersburg, MD edition: 0 URL `https://nvlpubs.nist.gov/nistpubs/Legacy/IR/nbsir84-3007.pdf`

[11] Gupta V, Zou S, Hamilton M, Sellers J and Ellis C 2016 Barrier Layers to Protect Nb Superconductivity in Thin Film Superconducting Flexible Cables *Applied Superconductivity Conference 2016 (ASC 2016)*

[12] Gupta V, Sellers J, Ellis C, Zou S, A Hernandez G, Bai R, Cao Y, Tuckerman D and Hamilton M 2016 *Additional Conferences (Device Packaging, HiTEC, HiTEN, & CICMT)* **2016** 2075–2094

[13] Peek S, Gupta V, Sellers J, Tuckerman D, Yelamanchili B, Shah A and Hamilton M 2021 Fabrication and flip-chip assembly processes for cryogenic applications using indium-indium and sac-indium bump connections on flexible interconnects *GOMACTech 2021, Virtual*

Mechanical stability of SIS100 Bypass Line busbars clamping system under AC current load

A Iluk[1], M Chorowski[1], J Poliński[1], T Eisel[2], B Streicher[2], R Trockel[2] and P Grzegory[3]

[1] Wroclaw University of Science and Technology, Wyb. Wyspanskiego 27, 50 – 370 Wroclaw, Poland
[2] GSI Helmholtzzentrum für Schwerionenforschung GmbH, Planckstraße 1, 64291 Darmstadt, Germany
[3] KrioSystem, Wrocław, Poland

E-mail: artur.iluk@pwr.edu.pl

Abstract. This paper presents the lessons learned from tests of first module of superconducting cryogenic bypass line (BPL), a part of the international Facility for Antiproton and Ion Research (FAIR) SIS100 cryogenic system, currently under construction in Darmstadt, Germany. Design, manufacturing, and installation of the superconducting cryogenic bypass line is a part of a Polish in-kind contribution to the FAIR project, realized by the Wroclaw University of Science and Technology. The main goal of the tests was to check the superconducting, Nuclotron type busbar system containing four pairs of busbars, transferring 13.2 kA pulsing current with the ramp rate of 28 kA/s. The mechanical stability of the busbars, especially at the connection region, was investigated with the use of vibration sensors and cameras located inside the vacuum space. The tests revealed insufficient mechanical stability of the busbars in the connection area due to pulsing Lorentz forces, and necessity of additional supports and clamps. Results of the tests were presented and discussed. The conclusions can be significant not only for the bypass line design, but also for design of the busbar connections in the superconducting magnets.

1. Introduction

Facility for Antiproton and Ion Research (FAIR) is the particle accelerator facility currently being under construction by at GSI Helmholtzzentrum für Schwerionenforschung, Darmstadt, Germany [1]. It will be unique particle accelerator facility which will use ions, antiprotons and the special isotopes (new atomic nuclei) for research in the many fields, among of others for particle physics, plasmas physics, atomic and anti-matter physics as well as for bio, medical and material science.

FAIR layout is schematically presented in the figure 1. It will compose of the accelerator infrastructure already existing at GSI site, as UNILAC linear and SIS18 circular accelerators, but also of the infrastructure provided within the FAIR project, which main elements are the SIS100 superconducting accelerator ring, Superconducting Fragment Separator (Super-FRS), storage rings and experimental stations.

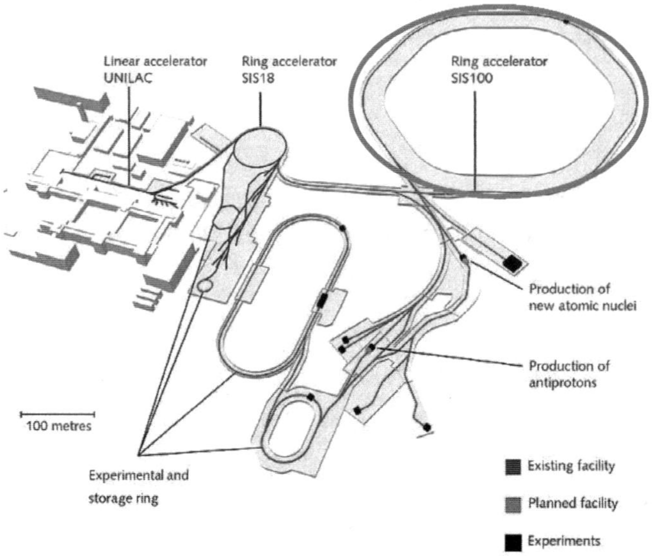

Figure 1. Layout of the FAIR accelerator facility and location of the SIS100 synchrotron [2]

The primary, ion beam will be produced and pre-accelerated up to 20% of the speed of light in the UNILAC and then up to 90% of the speed of light in the SIS18 accelerators. Next the beam will be directed to the SIS100 ring for further acceleration up to 99% of the speed of light or transferred directly to the ions research experiment stations or the string rings. The ions beam, by the collision of with dedicated targets, will be also used for production of antiprotons or special isotopes secondary beams. Before a dedicated experiment, the isotope beams will have to be filtered out of particles that are not of interest to the experiment. Sorting of particles according to their mass and charge will be done in the Super-FRS.

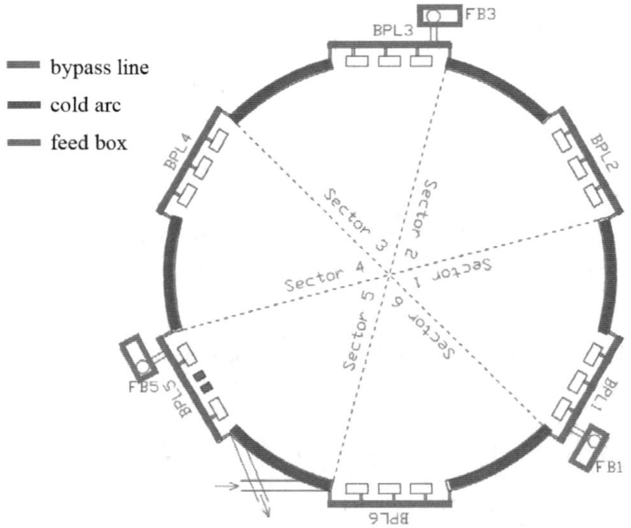

Figure 2. Schematic representation of the SIS100 accelerator showing the six Bypass Line (BPL) sections. The cold arc sections and BPL sections are for clarity not shown proportional to each other. FB1, FB3, FB5 – feed boxes

Wroclaw University of Science and Technology (WUST), in close collaboration with GSI Helmholtzzentrum für Schwerionenforschung, is responsible within Polish In-Kind contribution to

FAIR project for design, production, installation and commissioning of the local cryogenic system combined with the power transfer system for the SIS100 synchrotron including 6 Bypass Lines, 3 Feedboxes (FB) and 6 Current Lead Boxes (fig. 2).

2. Cryogenic Bypass Line

The BPLs and FBs are dedicated to transferring liquid helium and AC electric current between SIS100 arc sections and superconducting quadrupole magnets located in warm straight sections of the synchrotron [3, 4]. A main innovative feature of the cryogenic bypass line is transferring the electric current and liquid helium in one vacuum vessel, while in other similar projects, namely, the Large Hadron Collider at CERN (CH) [5] or the Tevatron at FermiLab (USA) [6], those functions were separated.

The view of the BPL module is presented in figure 3. The endcap areas 1 and 2 the interconnection to the subsequent BPL modules closed for test bay the test Endcaps. The feedbox connection was connected during the test to the test Feedbox providing LHe/GHe and electrical current. The general cross section of the Bypass Line is shown in figure 4. The design and operational parameters of the process pipes are presented in the table 1.

The coexistence of superconducting busbars and liquid helium process pipes in one limited space was a source of the serious design and production challenges.

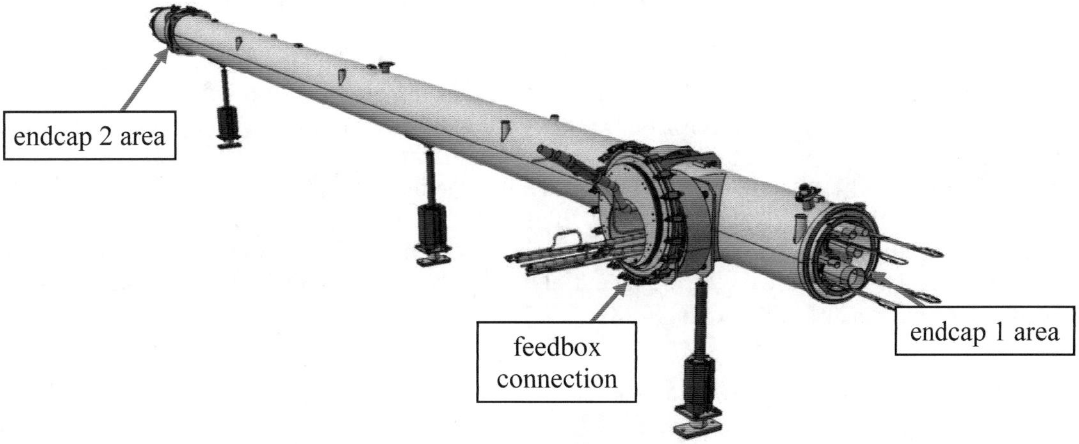

Figure 3. General view of the SIS100 Bypass Line (BPL) module.

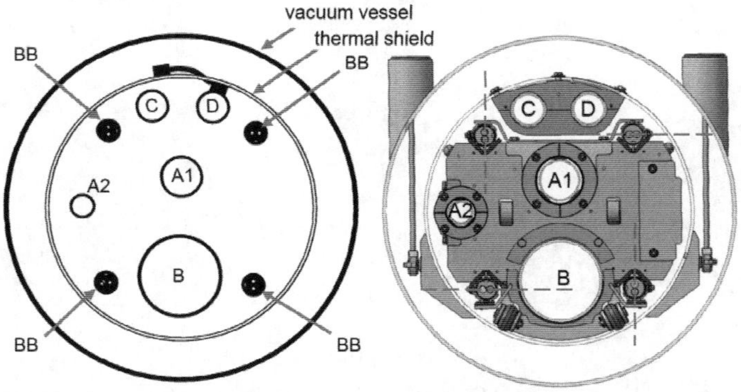

Figure 4. General cross-sectional layout of SIS100 bypass line: without supports (left) with supports and axes of busbar pairs (right). A1 – magnet supply line, A2 – vacuum chamber supply line, B – magnet and vacuum chamber return line , C –shield supply, D – shield return, BB – superconducting busbars (pairs)

Table 1. Operational and design parameters of the BPL process pipes

	Headers & Shells	O. Diameter & Thickness [mm]	Operating temperature [K]	Operating pressure [bar(a)]	Design pressure [bar(a)]	Test pressure [bar(a)]
A1	He supply magnets	54x2	4.5	3	20	28.6
A2	He supply vacuum chamber (VC)	32x2	4.5	3	20	28.6
B	He return magnets +VC	108x3	4.3	1.1	20	28.6
C	He supply shield	42.4x2	50	18	20	28.6
D	He return shield	42.4x2	80	17	20	28.6

The direct consequence of combining the helium process pipes and superconducting busbars into a common straight vacuum vessel was the necessity of providing thermal compensation on the busbars, with similar function as the axial compensators on the process pipes, within very limited space. It brings several challenges in the design:

- the busbars routing in the compensation loop area should minimize the electromagnetic cross-talk between busbar pairs,
- support system able to resist the significant Lorentz forces between busbars in the pair,
- prevention of cyclic movement of the busbars caused by AC current cycles.

Especially the electromagnetic cross-talk - the current induced in one cable by the electromagnetic field generated by the close, high-current cable – is a significant problem influencing the precise control of the fast-ramping SIS100 magnets [4]. For this reason the busbar pairs were located as far as possible from each other and two pairs were rotated by 90 deg (fig. 4 right, visible axes of BB pairs) [8].

The cross section of the NbTi superconductor is shown in fig. 5. The outer superconducting wires are cooled by the helium flowing through central CuNi tube. The details and mechanical properties of the cable components and are given in [7, 8].

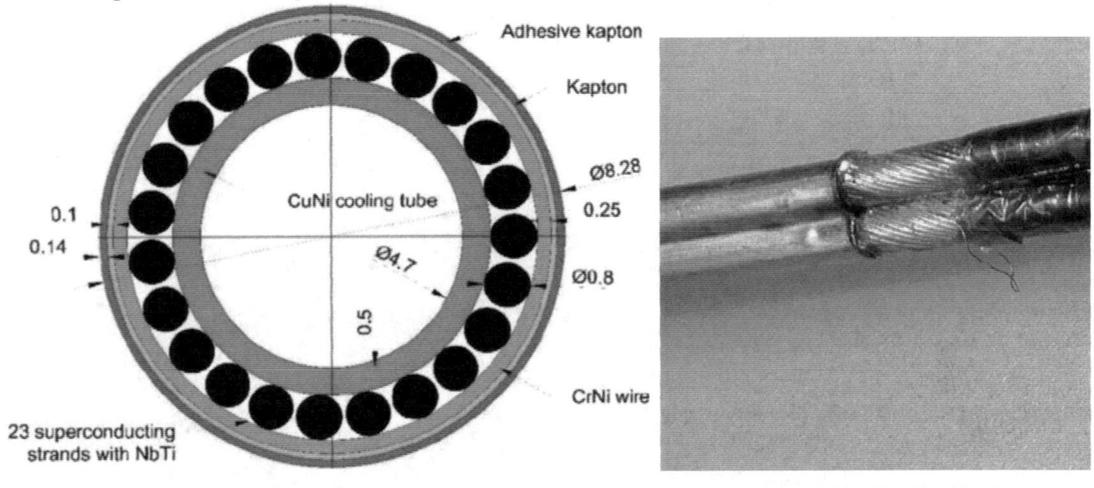

Figure 5. Cross-section of the Nuclotron type NbTi superconducting cable for the busbars of SIS100 (left) and the real superconductor uninsulated and soldered (right)

3. Cold electrical tests

The objective of the work was investigation of the mechanical stability of the busbar system installed inside BPL module and loaded by AC current. The high current flowing through the busbars results in a significant Lorentz forces applied to the system. The BBs with its support system was tested so far successfully for 3 BPL pieces manufactured by company KrioSystem. The tests were conducted in the conditions similar to the real one, with cooling by cold helium The modules was tested at 13.2 kA pulsing current with the ramp rate of 28 kA/s, i.e. with 1 Hz.

Figure 6. The location of the acceleration sensor on the support structure of the busbars

The mechanical stability of the busbars, especially at the region of modules interconnection, was investigated with the use of vibration sensors (fig. 6) and cameras located inside the vacuum space (fig. 7). For vibration measurements, the PCB Piezotronics ceramic shear ICP accelerometers PCB-352B has been used, with sensitivity 1000mV/g and frequency range 2 to 10000 Hz. The sensors were fixed to the busbar support structures in axial direction on both end the module. The tests revealed insufficient mechanical stability of the busbars in the connection area due to pulsing Lorentz forces, and necessity of additional supports and clamps shown in figures 8 and 9.

Figure 7. Camera installed inside the cold space of interconnection to observe busbar movement and additional clamps (arrows) used for stabilization of the busbars

Figure 8. Additional clamps in BPL necessary to prevent excessive BB movement under AC current load

Figure 9. Additional clamps in voltage tabs area (connections to detect the quenches of BB) necessary to prevent excessive BB movement under AC current load

The final electrical scheme of connection used for cold tests of the BPL module is shown in figure 10. The current flow directions are shown by arrows. The helium circuits for the module itself (Helium In and Helium OUT) as well as the helium supply for additional small busbar piece making the shortcut on the right side of the scheme (Helium IN/OUT) were presented. It should be mentioned, that for initial tests the different connection scheme was where not all busbars were powered. When only single busbars in the pair was powered, the unbalanced Lorentz forces caused strange side effects, like very noisy work of the module or even observed movement of external metal equipment next to working BPL module. Powering all of the busbars solved these issues.

IOP Conf. Series: Materials Science and Engineering 1241 (2022) 012028 doi:10.1088/1757-899X/1241/1/012028

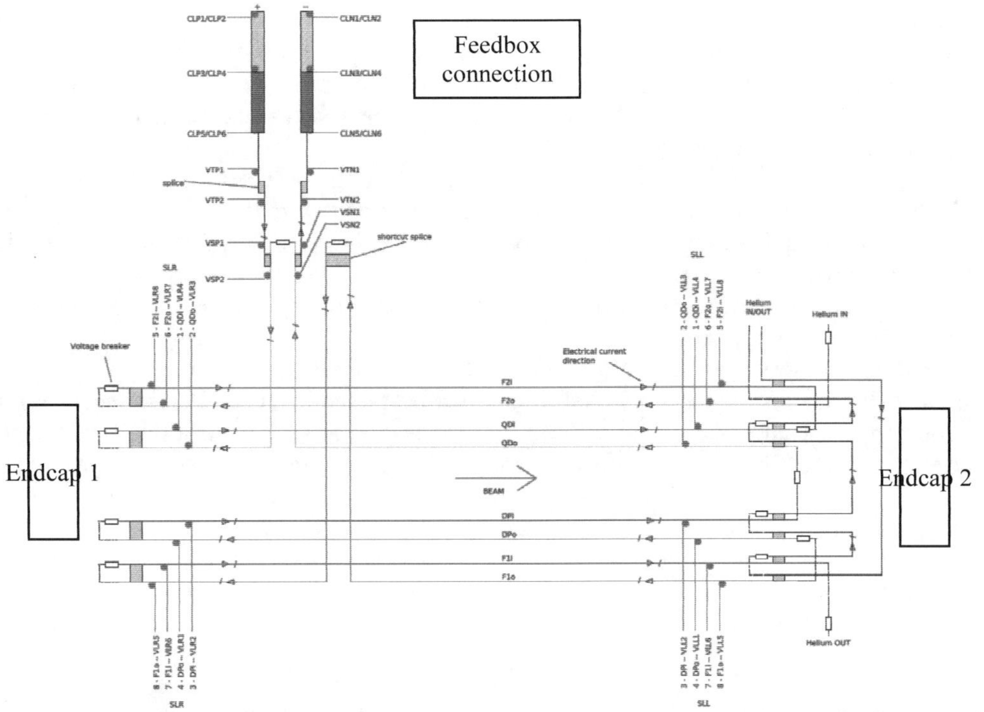

Figure 10. Electrical scheme of the BPL module cold test: VLxx – voltage tabs for quench detection, CLxx – current leads, F1x, F2x, QDx, DPx – busbars

The busbar cyclic movement due to Lorentz forces under AC current is an issue not only in the Bypass Lines. Similar situation is in the SIS100 synchrotron magnets. In figure 11 (left) the busbar pair in the superconducting magnet is shown. On the right hand side there is visible the BB clamp. The movement of the free parts of the busbars was captured by camera and analyzed by the photogrammetry software TEMA 3.4 [9].

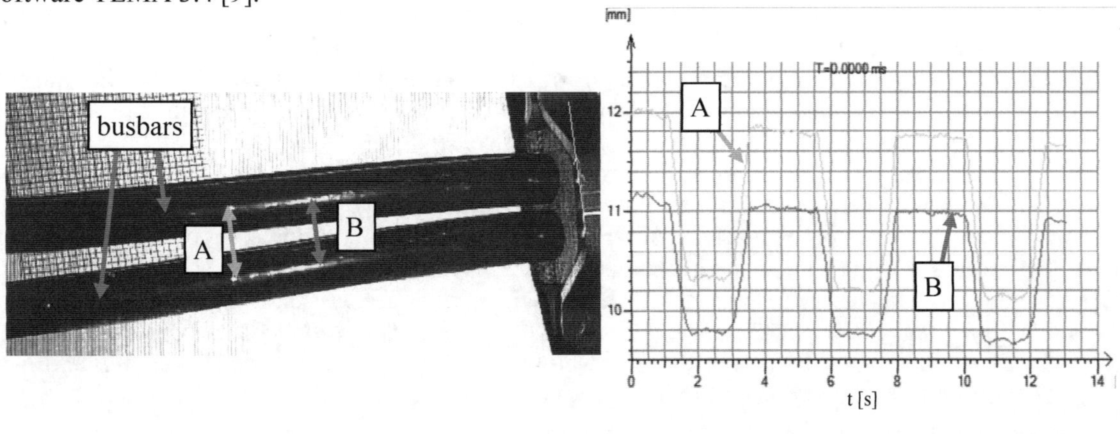

Figure 11. Busbar pair in the superconducting magnet of SIS100 synchrotron, the single busbar diameter equal to 8.3 mm, A, B – location of measured distances (left), the distance [mm] between busbars in the pair under AC current pulses (right)

In figure 11 (right) the example results of the photogrammetric measurements for magnet busbar pair shown in figure above are presented. Maximum detected amplitude of the busbar movement due to repelling forces during ramping was equal to 1.6 mm for initial busbar distance 10 mm.

4. Summary

The electrical cold test of the superconducting Bypass Line of the SIS100 synchrotron were conducted for 3 BPL modules manufactured in Wrocław (Poland) by company KrioSystem. Initial test revealed a few issues connected mainly with electrical test connection scheme and insufficient clamping of the busbars in the test connection area.

During the electrical test of the high current superconducting lines, especially for AC current, always both BB in pair should be powered, because unbalanced Lorentz forces can damage the support system designed for balanced forces. The BBs of a pair shall be powered with the same current in opposite direction to achieve a stress compensation by the clamping system. The correction of the powering system to symmetrical one and application of additional clamps to stabilize busbars in the test endcap connection areas solved the problems.

The vibration sensors have been used to evaluate/compare the movement of the entire cold mass of the module. The sensors are not very useful in detection of stabilization issues, because provide little information about location of the problem. Much more useful tools were the optical cameras used during the cold at critical locations. The cameras and analysis by photogrammetric software provide very useful and precise information about stability of the busbar and actual thermal movements, especially in the crowded environment of the combined electrical/hydraulic transfer line.

Special attention should be paid to the stabilization system of the at the interconnection areas, not only during the test, but also in the interconnections between modules at final assembly location. In case of AC current, insufficiently stabilized busbars can undergo significant cyclic strains and fatigue damage. An implication of the results presented in this paper will be the extended investigation of the Lorentz force induced strain of superconducting cable components for a wide range of the cable routing and various designs of clamps in the interconnection areas of modules and magnets.

5. References

[1] Spiller et al., The FAIR Heavy Ion Synchrotron SIS100, Journal of Instrumentation 15 (2020)
[2] R. Lalik "Krakowski wkład w badania naukowe prowadzone w ośrodku FAIR w Darmstadt." Foton 142 (2018).
[3] B. Streicher *"Detailed specification od the Local Cryogenic By-Pass System for SIS100"*, F-DS-K-20e SIS100 Bypass System v1.6.pdf
[4] Eisel, T., et al. "Local Cryogenics for the SIS100 at FAIR." IOP Conference Series: Materials Science and Engineering. Vol. 101. No. 1. IOP Publishing, 2015.
[5] Th. Goiffon et al. "Conceptual design of the Cryogenic Electrical Feedboxes and the Superconducting Links of LHC." Proceedings of the Twentieth International Cryogenic Engineering Conference (ICEC20). Elsevier Science, 2005.
[6] M. Geynisman et al. Cryogenic system for the Tevatron. No. FNAL/C-96/296; CONF-9605237-1. Fermi National Accelerator Lab., Batavia, IL (United States), 1996.
[7] A. Iluk "Investigation of mechanical strains in thermal compensation loop of superconducting NbTi cable during bending and cyclic operation" Materials, 2021, 14 (5)
[8] A. Iluk, K. Malcher, W. Słomski, M. Chorowski, J. Poliński, T. Eisel, B. Streicher, P. Spiller "*Design of the Cryogenic Bypass Line for SIS100 synchrotron*", Applied Science, 2020, 10 (22)
[9] V. Gampala et al. "Enhancement of resolution and image reconstruction in digital image correlation." Materials Today: Proceedings (2021).

Acknowledgments

The work has been realised within the GSI – WUST agreement no. 4005/0001/21 and supported from statutory funds the Polish Ministry of Science and Higher Education grant no 8211104160. The device described in the publication (apparatus, cryogenic equipment, etc.) was made for the FAIR GmbH Center as part of the in-kind contribution made by the Polish Shareholder in FAIR GmbH (Jagiellonian University), declared by the Polish government in the international FAIR Convention and financed from the state budget.

Strategies for conformal REBCO windings

J S Rogers[1], P M McIntyre[1,2], T Elliott[2], G D May[1] and C T Ratcliff[1]

[1]Texas A&M University

[2]Accelerator Technology Corp.

E-mail: jsr12e@tamu.edu

Abstract. A high-field winding can be fabricated from a cable of non-insulated REBCO tapes stacked face-to-face without twisting. If the cable is oriented within each turn of a winding so that the tape face is closely parallel to the magnetic field at its location, the supercurrent capacity of that cable is enhanced ~3x greater than in a transverse or twisting orientation. This concept for a conformal winding was presented in a previous paper pertinent to the body winding of a REBCO based high-field dipole. Strategies are presented and simulated for how the same orientation can be sustained in the flared ends of a high-field hybrid dipole.

1. Introduction

REBCO tape has remarkable properties for use in superconducting magnets. It can operate with useful current density up to liquid nitrogen temperature and it can produce very high magnetic field at temperatures of 20-40 K. The manufactured tape is ready to use as supplied and does not require a final heat treatment after winding into a magnet winding – an important benefit compared to Nb_3Sn and Bi-2212.

On the other hand, REBCO has undesirable properties that have limited its usefulness in high field magnets. REBCO is ruinously expensive, making it infeasible for use in many applications. Additionally, it is a highly anisotropic superconductor [1] with a critical current that is strongly dependent on the magnitude and direction of the magnetic field in which it is operating (Figure 1). The critical current is approximately 3x greater when the magnetic field at the tape is parallel to the tape face ($B_{\|}, \theta = 90°$) than when it is perpendicular ($B_{\perp}, \theta = 0°$).

In a previous paper, a conformal winding method was presented in which a REBCO winding was configured in a dipole such that the tapes within the winding are everywhere oriented parallel to the field at the tape, enabling the winding to use the full capacity of the REBCO films. In this work we use a REBCO insert winding as the core of a hybrid dipole which as obeys this conformity with the field lines. Figure 2 shows an example field design for such a dipole. The insert winding consists of a tape-stack cable in which 25 REBCO tapes are stacked face-to-face, and each turn of the cable is oriented closely parallel to the field at that location.

Clustering REBCO tapes has been used by several authors including Roebel cable [2], CORC [3], cable-in-conduit [4], and stacked tapes [5]. In most cases REBCO tapes are stacked in a face-on cluster and then multiple clusters are cabled around a solid copper core with a twist pitch so that each cluster

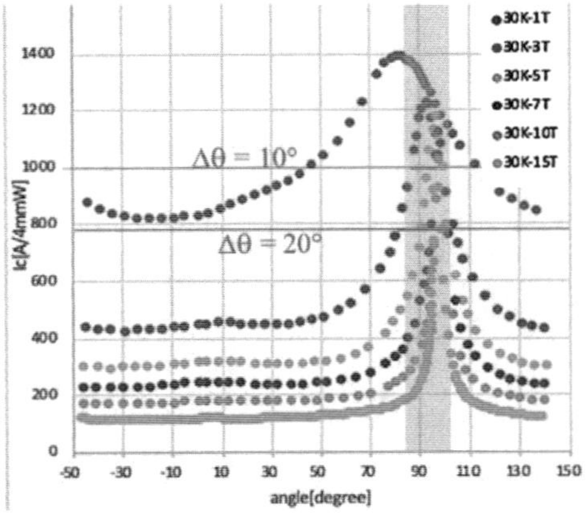

Figure 1. Dependence of I_c on the angle between the tape face normal and the magnetic field, for various field strengths (30 K temperature, from [1]).

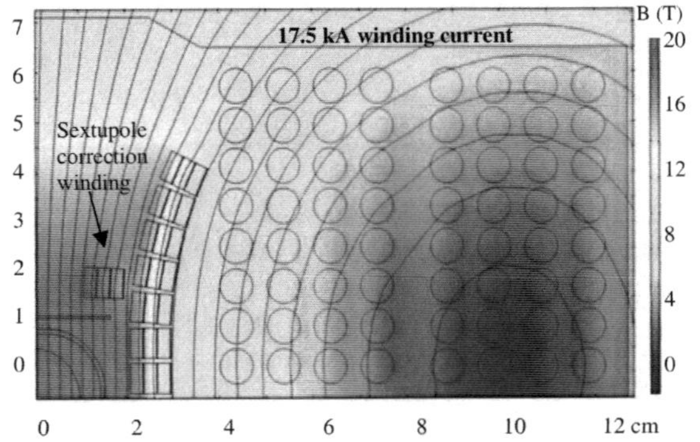

Figure 2. Field distribution in one quadrant of an 18 T dipole conformal REBCO winding. The field magnitude is shown in color while black lines show contours of the A_z component of the vector potential, ie. "lines of force"

spends equal length on the inside and outside of the cable in a winding. Yagotintsev *et al.* [6] review the measurements of AC loss and contact resistance in these several forms of clustered-tape conductors. All the present approaches to tape clustering involve twisting the cluster and thus forego any possibility to sustain the $B_{\parallel}$ condition that would allow the REBCO tape to be used at its full potential; this is the goal of the conformal winding method.

There are two challenges in making the conformal winding realizable:

- Can uniformly high critical current be sustained in the end windings of the dipole insert (where the magnetic field and the winding must both be flared) and in the leads, as it is in the body?
- Can the tape-cluster cable support dynamic current-sharing within each cable turn so that inductive forces within the non-insulated cable do not concentrate its current in outer tapes and drive premature quench?

2. Optimization of conformal winding

The hybrid dipole shown in Figure 2 contains an insert winding made from rectangular REBCO tape-stack cable and an outsert winding made from Nb_3Sn Cable-in-Conduit 'SuperCIC' [7]. Table 1 summarizes the main parameters of both windings. The tape-stack cable contains 25 6-mm-wide REBCO tapes with 100 μm Cu clad to one side. The SuperCIC cable contains 17 0.85 mm-diameter wires of Hi-Lumi-class 108/127 Nb_3Sn/Cu wires, spiral-wrapped as a single layer around a thin-wall perforated center tube, then pulled to a loose fit through a bronze sheath tube and drawn to compress the

Table 1. Main parameters of the 18 T hybrid dipole.

Bore field @ 4.2K short-sample	18.3	T
Cable current (windings in series)	17.5	kA
Aperture: horizontal, vertical	40, 30	mm
REBCO: #tapes/cable x #cables/shell x #shells	25x 16 x 3	
Nb_3Sn CIC: #wires/CIC x #cables/layer x #layers	17x 16 x 8	
B_{max} in REBCO, CIC	20.0, 11.8	T
Sextupole @ full field, injection field	-5, -21	units

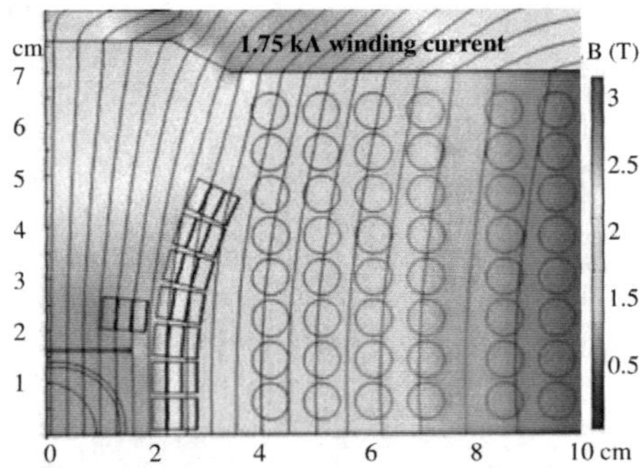

Figure 3. Field distribution at injection field.

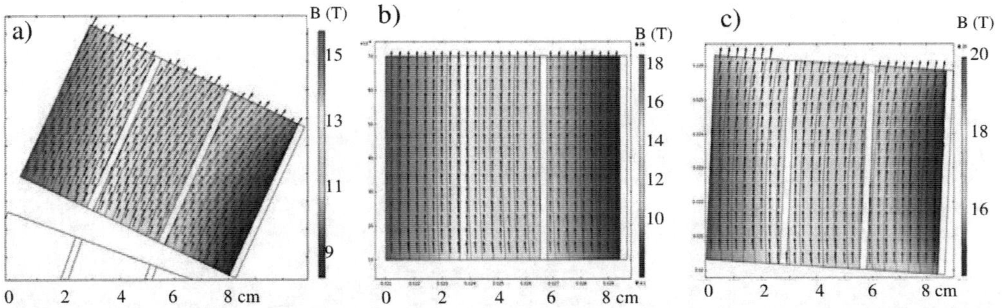

Figure 4. Detailed simulation of the fields on the tape faces for a) top right; b) mid-plane; and c) sextupole 3-turn blocks of tape-stack cables in the REBCO insert winding. Shown in color are the field magnitudes while black arrows represent the direction of the field vectors

wires against the center tube to immobilize them. The winding is housed in a steel flux-return shown as the outer rectangular boundary in Figure 2.

The windings are operated in series, and the number of wires in each winding are chosen so that a cable current of 17.5 kA produces ~18 T bore field and corresponds approximately to the critical current in each winding. The REBCO insert can operate at up to 20K, while the Nb_3Sn winding operates at ~5K.

2.1. Conformal geometry at injection field

The field distribution in the winding region is significantly different at injection field than at collision field, due to progressive saturation in the flux return. Figure 3 shows the field distribution at injection field (1.75 kA winding current). There is a significant angle θ between the tape face and the field direction in several blocks, but the value of $|B|$ is low enough that the critical current $I_c(|B|, \theta)$ is sufficient that there is no risk of quenching provided that current sharing is dynamically stable.

2.2. Body field simulations

The rectangular REBCO tape stack cables are oriented along a curving contour to conform with the flaring behavior of the magnetic field in the region of the insert winding. The cylindrical CIC cables are oriented in a rectangular block-coil array in the outsert winding. There is also a 'sextupole correction' sub-winding of REBCO cables located inside and above the main REBCO insert, which is operated in series to correct sextupole field in the bore. Immediately above the bore tube is a steel flux plate that suppresses multipoles at injection field to suppress the effects of persistent currents and snap-back at injection field. The REBCO insert contributes ~8T and the CIC winding ~10T to the bore field.

Figure 4 shows a detailed cross-section of 3 particular 3-turn blocks within the REBCO insert winding: the top-right block, the mid-plane block, and the sextupole block. These blocks exhibit variously the highest field or the maximum deviation from the $B_{\parallel}$ orientation for some of the constituent tapes. The current capacity of the *m*th tape in each cluster can be estimated by extracting the local sheet critical current density $K_m(u)$ as a function of the location u across the width of that tape, using the local values for $|B(x, y)|$ and θ, and adding the contributions for segments spanning the entire width w of that tape: $I_{cm} = \int_0^w K_m(u)du$. The total cable current capacity is then obtained by adding the maximum current capacity of the 25 tapes in that cable segment.

2.3. End field Simulations

An additional challenge is to design a winding strategy for the magnet ends that does not compromise the operational current capacity. In the 3D end regions of the dipole, fields flare and bend in a disorderly fashion raising concerns about the critical current of the cables. The flared ends of the tape-stack winding are formed using a method first pioneered by Willy Sampson (BNL) in the 1970s. Figure 5 shows a flared-end quadrupole containing a stacked-tape winding of dip-process Nb_3Sn tape. Each tape-stack cable is twisted about its axis as it is flared vertically to form a catenary. All tapes remain as a stack but

Figure 5. Example of flared ends in a tape-stack cable: Sampson's flared-end quadrupole using dip-coated Nb_3Sn superconducting tape.

no tape is bent in the hard direction. With this approach each tape face naturally follows closely the flaring of the magnetic field in the end.

This twisted-flare end has been modeled in a 3-D CAD design, and the fields have been modeled using 3-D Magnetic Field simulations in COMSOL. Figure 6 shows the results in three example cross sections through the end region: a) a y/z cross section through the vertical midplane; b) an x/y cross section at the transition from body to flared end; and c) an x/y cross section at a location 3 cm into the body from that transition. The $I_c(|B|, \theta)$ for each tape is estimated using the data of Figure 1 and then is summed across all tapes to obtain the I_c in each tape-stack cable at that location.

There is a compensating interplay that sustains I_c with little or no degradation through the flared ends: the inner tapes are well-aligned to with the field direction but have the strongest value. Although the outer tapes flare at a significant angle with respect to the field, the field strength is low enough that the angular dependence of $I_c$is also broadened. *The flared-end regions do not present a 'weak link' limit to the I_c of the cable in any turn of the winding.*

3. Dynamic current-sharing in the conformal winding

The current sharing strategy in the cable is based on the interplay of Lorentz forces acting on currents and contact resistance between stacked-tape faces during operation. Figure 7 shows a 3-turn block of cables detailing the 25 layers of stacked tape in each cable, insulation between cable turns, and a laminar spring that provides uniform compression of tape stacks. Here we employ the current sharing strategy proposed in our previous work [8] in which a conformal winding for a 3.5T dipole with only REBCO (no outsert winding) was designed. In summary, as the dipole is ramped to increase the bore field from its value at injection energy to that at collision energy, the total current entering each tape-cluster cable is increased. As shown in Figure 8, the magnetic field at each cable produces a Lorentz force $\vec{F}$ that redistributes the current density flowing within each tape-cluster cable away from the dipole bore. This implies that as the cable current is increased the outermost tape would quench prematurely while the current in neighboring tapes of that cable was still only a fraction of their critical current.

However, there is a countervening effect as discussed by Trillaud *et al.* [9]. As the current in each tape approaches its limit, the internal dynamics within its REBCO layer shifts from flux creep to flux flow before reaching the normal state (as demonstrated in Figure 1 from [9]). The local electric field in the tape increases exponentially with current:

$$E_z = E_0(I/I_c)^n \quad (1)$$

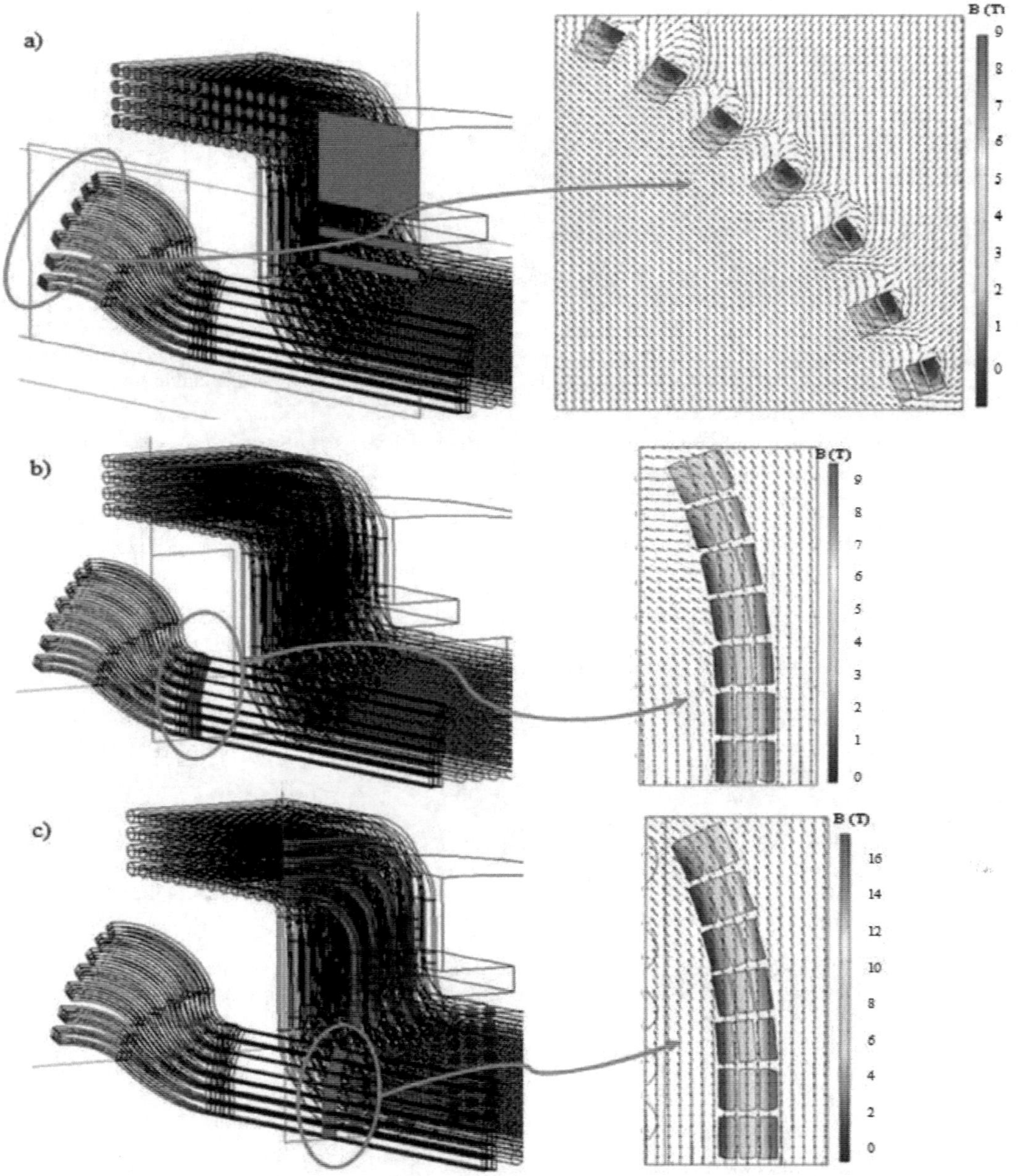

Figure 6. Three sections through the REBCO end winding of the hybrid dipole: a) section through the vertical midplane; b) x/y cross section at the transition from body to end winding; c) x/y cross-section 3 cm into the body winding. Black arrows indicate field directions.

where $E_0 = 10^{-6}\frac{V}{m}$ is the quench criterion, $I_c(|B|, \theta, T)$ is the critical current for the conditions (B, θ, T) for that tape, and $n{\sim}30$ is the index that characterizes the power-law dependence of the superconductor-normal transition for REBCO. As the current I in a given tape approaches I_c, the generated E_z causes

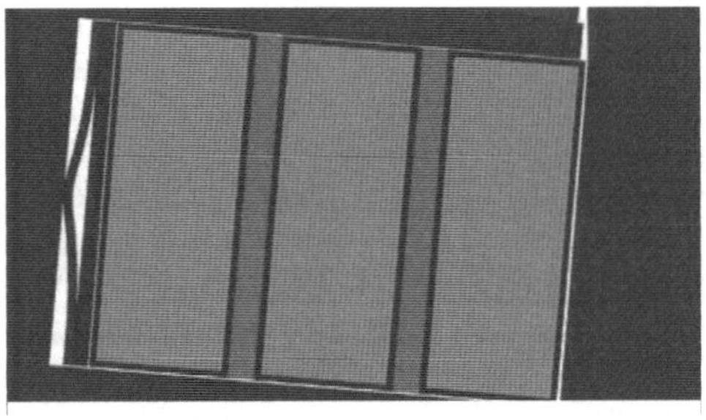

Figure 7. Detail showing a block of 3 tape-stack tapes, each containing 25 tapes, with a laminar spring to provide uniform ~1 MPa compression to all cables and insulation space between cable turns.

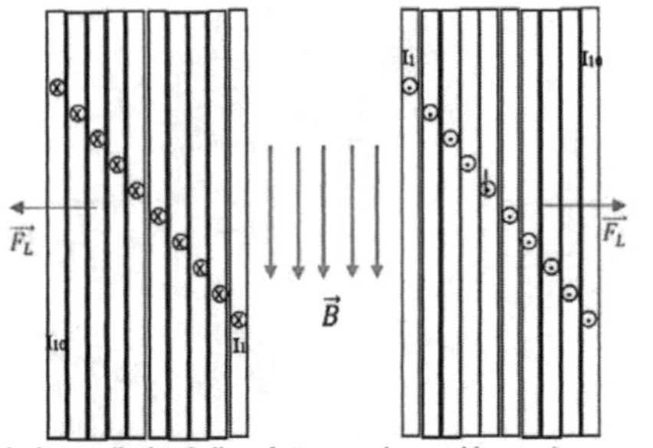

Figure 8. Schematic illustration of current-sharing dynamics in a conformal winding of tape-stack REBCO cable.

the current in that tape to redistribute into the neighboring tapes where $I \ll I_c$. Since REBCO can operate at temperatures >20 K where the heat capacity of the tape (~T^3) is much greater than that in low-temperature superconductors, it is feasible to remove the generated heat so that a local 'soft-quench' in a tape segment drives re-distribution of current within the cable rather than propagating. This pattern repeats for neighboring tapes as the current is ramped.

The 'soft quench' approach has been used to good effect in 'non-insulated' (NI) pancake windings for high-field solenoids [10] but never in a dipole. The dynamics of quasi-equilibrium current sharing among the tapes within a non-transposed cable follows nearly identically to those detailed in [8]. The model uses the approach for NI REBCO pancake coils by Noguchi [11] and relies on tape face-to-face contact resistance that decreases with compression as measured by Lu *et al.* [12] shown in Figure 9.

There is interesting physics in the current-sharing among NI REBCO tapes that are face-aligned with $\vec{B}$. The above qualitative model lays the groundwork to develop a multi-scale model connecting the normal transition at nanoscale to the redistribution of currents and forces among the tapes and will be the focus of future work.

4. Field homogeneity for collider requirements

Field homogeneity is of particular concern for the dipole magnets of an accelerator or collider. The sextupole harmonic can be selectively canceled by placement of one correction turn in the winding, at the location shown in Figure 2. This magnetic design shown has been optimized to produce nearly pure dipole field over a dynamic range of field 0.2-4 T in which the amplitudes b_n are all less than 10^{-4}.

Current sharing poses a further challenge for field homogeneity. However, at injection field, the current in each tape cluster is located mainly in the outermost tape; at collision field, the current is approximately equally shared, so the 'current location' for that cable turn is shifted inwards by half the cluster width.

Conventionally, the nth multipole of a dipole field distribution is defined at a location $\vec{r} = r(\hat{x}cos\theta + \hat{y}sin\theta)$ by the expansion:

$$B_x + iB_y = \sum b_n e^{-ni\theta} \left(\frac{r}{R}\right)^{n-1} \quad (2)$$

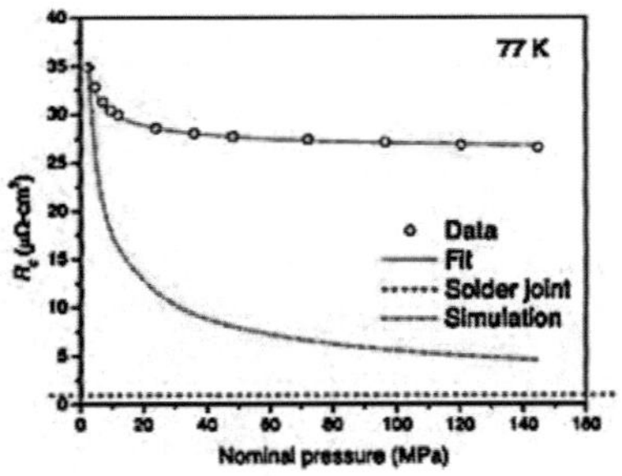

Figure 9. Contact resistance between two copper-clad REBCO tapes as a function of compression.

Here the multipole coefficients b_n are defined as,

$$b_n = \frac{1}{\pi B_0}\left(\frac{R_{ref}}{r}\right)^n \int_0^{2\pi} B_y(\theta)\cos(n\theta)\,d\theta \quad (3)$$

where R_{ref} is a reference radius that is generally taken to be ~66% of the aperture. It is convenient to take $r = R_{ref}$ and to follow the convention that $b_0 = 1$, thereby implying:

$$B_0 = \frac{1}{\pi}\int_0^{2\pi} B_y\,d\theta \quad (4)$$

The multipoles have been evaluated for the magnetic design for the maximum and injection field limiting cases. The difference in the calculated multipoles is $\Delta b_n < 0.5 \times 10^{-4}$ for all multipoles! This remarkable result is a consequence of the conformal design strategy: since each tape cluster is oriented with its face closely parallel to $\vec{B}$, the field distribution is insensitive to the horizontal position of the 'current location' of that cluster.

Prestemon *et al.* [13] modeled current transfer in a stacked-tape cable and found that a region of cable with coupling resistance $R{\sim}10\ n\Omega$ is effective in equalizing current among its tapes. The laminar spring sustains $1\ MPa$ compression giving a contact resistance $R_c{\sim}35\mu\Omega \cdot cm^2$. For a tape of width $w = 6mm$ the characteristic length of cable for stability is

$$L = \frac{R_c}{R{\cdot}w} \sim \frac{35\mu\Omega{\cdot}cm^2}{10n\Omega{\cdot}6mm} = 60\text{m} \quad (5)$$

This is conveniently the approximate length for one turn of a collider dipole so ramping current should stabilize turn-by-turn.

5. Conclusion

A strategy is presented by which a cable containing multiple REBCO tapes may be configured in an insert magnet winding in such a way that the favourable $B_{\|}$ orientation is sustained everywhere in the body of the winding such that the REBCO tapes could perform to their maximum potential. An Nb_3Sn SuperCIC outsert winding shapes the fields on the tapes and adds to the total bore field. A method for dynamic current sharing is discussed, in which current would naturally re-distribute among the tapes of each cable turn as windng current is increased without inducing premature quench. Dynamics and field multipoles at different stages of operating current were studied, as well as stabililty during the ramping process.

6. References

[1] http://www.superpower-inc.com/content/2g-hts-wire
[2] Goldacker W, Grilli F, Pardo E, Kario A, Schlachter SI and Vojenciak M 2014 *Supercond. Sci. Technol.* **27** 093001
[3] Van der Laan DC: 2015 Patent WO2015006350A3
[4] Shun Y *et al.*, 2017 Patent CN107564623A
[5] Iwasa Y, Bascuñán J and Hahn S 2018 Patent US10079092B2
[6] Yagotintsev K, Anvar VA, Gao P, Dhalle MJ, Haugan TJ, van der Laan DC, Weiss JD, Hossain M S A and Nijhuis A 2020 *Supercond. Sci. Technol.* **33** 085009
[7] Breitschopf J, Chavez D, Elliott T, Kellams J, McIntyre P and Sattarov A 2020 *IOP Conf. Ser.: Mater. Sci. Eng.* **756** 012031

[8] McIntyre P, Rogers J and Sattarov A 2021 *IEEE Trans. Appl. Superconduct.*, **31**, 5, 1-5, 10.1109
[9] Trillaud F, dos Santos G, Sotelo G 2021 *Materials,* **14**, 1892, 10.3390/ma14081892
[10] Choi S, Jo HC, Hwang YJ, Hahn S and Ko TK 2012 *IEEE Trans. Appl. Superconduct.* **2**, 3, 4904004
[11] Noguchi S 2019 *IEEE Trans. Appl. Supercond.* **29**, 5, 4602607
[12] Lu J, Goddard R, Han K and Hahn S 2017 *Physics: Instruments and Detectors* 1701.00447
[13] Martinez ACA, Ji Q, Prestemon SO, Wang X and Mary-Cuna GHI 2020 *IEEE Trans. Appl. Superconduct.* **30**, 4, 1-5, 6600605

IOP Conf. Series: Materials Science and Engineering 1241 (2022) 012030 doi:10.1088/1757-899X/1241/1/012030

3D Modelling and Validation of the Optimal Pitch in Commercial CORC Cables

M. U. Fareed, M. Kapolka, B. C. Robert, M. Clegg, & H. S. Ruiz

College of Science and Engineering & Space Park Leicester, University of Leicester, Leicester LE1 7RH, United Kingdom

E-mail: muf2@le.ac.uk ; mk652@le.ac.uk ; hsrr1@le.ac.uk

Abstract. Conductor on a rounded core (CORC®) cables with current densities beyond 300 A/mm^{-2} at 4.2 K, and a capacity to retain around 90 % of critical current after bending to a diameter of 3.5 cm, make them a strong candidate for high field power applications and magnets. In this paper, we present a full 3D-FEM model based upon the so-called H-formulation for commercial CORC® cables manufactured by Advanced Conductor Technologies LLC. The model presented consists of tapes ranging from 1 up to 3 SuperPower 4mm-width tapes in 1 single layer and at multiple pitch angles. By varying the twist pitch, local electromagnetic characteristics such as the current density distribution along the length and width are visualized. Measurements of macroscopical quantities such as AC-losses are disclosed in comparison with available experimental measurements. We particularly focused on the influence of the twist pitch by comparing the efficiency and performance of multiple cables, critically assessing the optimal twist pitch angle.

1. Introduction

The future updating of large-scale superconducting magnets for high-energy applications, such as those involved in the High-Luminosity Large Hadron Collider (HL-LHC) project [1], and fusion energy projects such as ITER [2], both require the optimal designing of high current capacity cables for minimizing the number of Ampere-turns in the manufacturing of superconducting coils. Thus, in the last few years a very promising design of high temperature superconducting cables has started to dominate the landscape for high field applications [3] by following the simple concept of winding Coated Conductors On a Round Core, nowadays known as CORC® cable [4]. These coated conductors are also known as the second generation of high temperature superconducting (2G-HTS) tapes, which when helically wounded on a cylindrical core offer superior advantages over other cabling technologies such as, a low manufacturing cost, a relatively easy production, and enhanced engineering current density [5]. They also offer the possibility to have a full transposition of the tapes without affecting their physical properties along the cable length, it because of the relatively high elasticity module of the 2G-HTS tapes, and the use of flexible formers with a small diameter. It allows the cable structure to bend in any direction without irreversible degradation [6]

The potential of CORC®cables for high field magnets have been gradually investigated since their inception, performing viability studies on one to multiple layer cables at up to 20 T, and at different temperatures [7, 8]. These studies have demonstrated engineering current densities beyond the 300 Amm^{-2} at 4.2 K and up to 20 T [9, 10], proving the concept of CORC®cables as a suitable candidate for future high field magnets. Moreover, it has been recently reported that CORC®cables with low tape windings on strong elastic cores are indeed suitable for the manufacturing of 40-60 T field magnets, allowing an effective protection against the high operating stresses (¿600 MPa) and strains associated with high-field magnet operation [11]. Nevertheless, by reducing the tape winding angle, the amount of tape necessary to manufacture a determined length of a conductor can increase substantially and furthermore, it is already known that variations of the angle with which each one of the tapes is wound onto the former could severely change the magnetization measurements of the CORC®cable [12,13]. Thus, this can lead to the possibility to encounter robust increments on the AC-losses of the cable once in operation. Nevertheless, it is known that the magnetization losses of the overall cable can be reduced by optimal twisting (pitch angle) of the multiple tapes at a single layer, which should be first computationally assessed in order to reduce the prototyping and manufacturing costs.

Due to the rising complexity and demand of CORC®cables with applications pressing for a higher number of layers and 2G-HTS tapes, a deep understanding and sound computational model for reproducing the electromagnetic characteristics of commercially available CORC® cables, and in particular its AC-losses, results are of utter importance. Nevertheless, the high aspect ratio of the helically twisted tapes at the CORC®cable, makes the devising of finite element models computationally demanding. Although with the continuous improvement of software and hardware, every year this becomes less of a concern. Certain simplifications can still be useful as long as the electromagnetic quantity of interest is not the AC-losses, as the level of accuracy demanded in this case is commonly the highest. This is added to the fact that the change on the power density that leads to the superconducting energy losses cannot be fully captured by 2D models, as it is the motion of the profiles of current density inside the superconducting tapes what causes the losses. In this sense, 2D models of CORC®cables [14] do not consider the twist pitch, and hence at short twist pitch lengths they can suffer of significant errors. By using symmetry conditions at the full 3D geometry whose approaches can render to computations over a 2D domain, these drawbacks can be addressed. Assuming that the current only flows along the 2D domains following the helicoidal trajectories, a scalar power-law for the resistivity of the superconductor makes the eddy current approach feasible for the solution of the 3D CORC®cable but within a 2D formulation for the electromagnetic variables [15]. Similarly, by neglecting any significant hysteretic process across the thickness of the 2G-HTS tapes, i.e., by assuming infinitely thin superconducting tapes where all relevant pinning dynamics manifest only at their surface, 3D geommetry models such as the ones based on the so-called T-A formulation result in the computational modelling of 2D domains [16]. These models although represent a significant improvement in terms of the computing time when compared with fully 3D numerical formulations [17,18], still might suffer of the sufficient level of accuracy for determining subtle changes in the AC-losses of the CORC®cables by the change of the twist pitch. Which can in-turn actually modify the vortex orientation inside the SC material, i.e., along the thickness of the SC tape. In other words, 2D-domain simplifications assume that all vortex inside the SC tape are like fluxons with a solid symmetry axis, whilst in the reality these fluxons can be intertwine within the SC material [19]. Therefore, despite fully 3D-formulations based on the H-formulations being computationally more expensive, these are certainly needed to reconstruct the full electrodynamics of the CORC®cables, in particular when the task ahead is to validate the AC-loss measurements of the early commercial designs of CORC®cables [20]. All other models up to date have focused mostly on the validation of their computational principles through comparison of in-house lab-made CORC®cables, which has left the door open for investigating further improvements on the full 3D modelling of commercially made CORC®cables, to make them affordable with relatively inexpensive workstations.

Therefore, in this paper we present the AC-losses study of the single layer commercial CORC®cable [20], assessing whether their choice of the twist pitch length provides the minimum

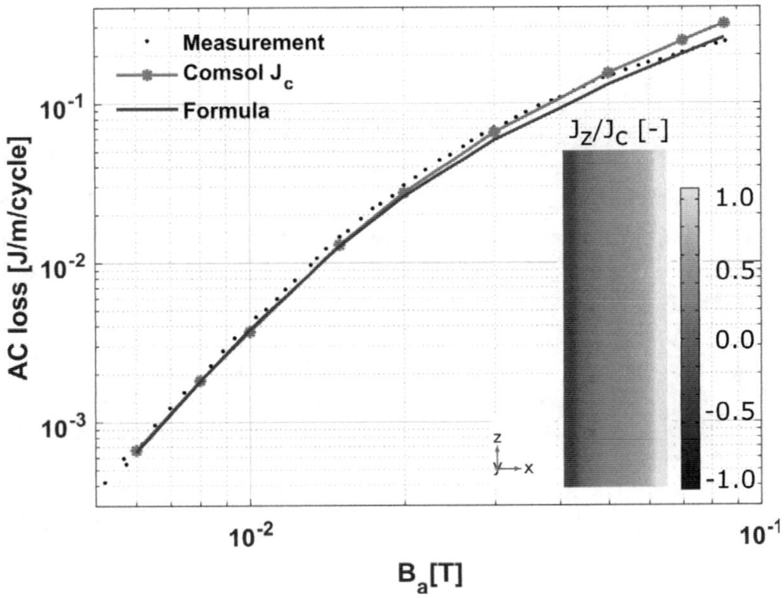

Figure 1. The AC losses comparison of the 12 mm straight tape measurements, the COMSOL model and analytical solution shown perfect agreement and confirmed correctness of the model.The inset shown current distribution of the screening current at the peak of the 20 mT magnetic field.

amount of hysteresis losses. The computational method is based on the H-formulation, enabling a full 3D modelling of the electromagnetic variables and helicoidal geometry of the cable, for which the model itself is validated first with the experimental measurements and analytical solution for a straight 2G-HTS tape in Sec. 2 of this manuscript. Then, the CORC®cable design is explained with the different twist pitch lengths, where we reveal a direct comparison between our numerical model and the AC-losses measurements reported in the literature. Our calculations show how the optimum twist pitch length can be estimated, by considering different factors such as the intensity of the AC-losses and the amount of 2G-HTS tape involved. In this sense, we disclose the window of change for the twist pitch angle which could offer similar AC-losses at different intensities of an applied magnetic field, confirming the field-ranges for which the pitch angle chosen for the manufacturing of the CORC®cables reported in the literature are certainly the most optimal.

2. Computational model and validation

In this section, we explain and validate our computational model for analysing the AC-losses produced by superconducting wires and cables with a 3-dimensional topology. The model is based on the general PDE formulation of Comsol Multiphysics, which allows a robust implementation of Maxwell equations through 3D Finite Element Methods (FEM), with the state variables defined in what is commonly known as the H-formulation. This formulation allows us to adopt the magneto quasi-steady (MQS) approach of the critical state theory [19] and neglect the time dependence of the electric displacement vector $\mathbf{D}$ at the general definition of Ampère's law, $\nabla \times \mathbf{H} = \partial_t \mathbf{D} + \mathbf{J}$. Assuming the magnetic permeability of any superconductor to be isotropic and equal to the magnetic permeability of free space, $\mu_0 \approx 4\pi \times 10^{-7}$ H/m. Due to no intrinsic magnetization vector $\mathbf{M}$, the magnetic flux density can be simply written as $\mathbf{B} = \mu_0 \mathbf{H}$, with $\mathbf{H}$ the magnetic field strength vector, and Faraday's law defined by $\nabla \times \mathbf{E} = -\mu_0 \partial_t \mathbf{H}$.

Therefore, if no other material is introduced where the condition $\mathbf{B} = \mu_0 \mathbf{H}$ could not be ensured, Faraday's law can be written for the entire space as the vectorial PDE:

$$\left| \begin{array}{c} \partial_y E_z - \partial_z E_y \\ \partial_z E_x - \partial_x E_z \\ \partial_x E_y - \partial_y E_x \end{array} \right| = -\mu_0 \left| \begin{array}{c} \partial_t H_x \\ \partial_t H_y \\ \partial_t H_z \end{array} \right| , \tag{1}$$

where E_x, E_y, and E_z are the Cartesian components of the electric field $\mathbf{E}$, and H_x, H_y, and H_z are the corresponding components of the magnetic field $\mathbf{H}$. Analogously, we can express Ampère's law in the same vectorial notation as follows:

$$\left| \begin{array}{c} J_x \\ J_y \\ J_z \end{array} \right| = \left| \begin{array}{c} \partial_y H_z - \partial_z H_y \\ \partial_z H_x - \partial_x H_z \\ \partial_x H_y - \partial_y H_x \end{array} \right| , \tag{2}$$

where J_x, J_y, and J_z define the cartesian components of the current density vector $\mathbf{J}$.

Whilst the conductive properties of any non-superconducting material can be introduced by the linear function of Ohm's law, $\mathbf{E} = \sigma \mathbf{J}$, with σ the electric conductivity of the media, in the case of superconductors the material law governing the electrical conductivity of these materials follows a highly non-linear relation. For the sake of simplicity and numerical convergence, this is introduced as an $E - J$ power law. As PDE formulations do not allow instantaneous transitions of state, i.e., current states where any variation of the magnetic field leads to an instantaneous transition of $J = 0$ to $J = \pm J_c$ when $E \neq 0$. Hence, to tackle the non-linearity of Bean's concept [21], and to incorporate the physical principles of the critical state theory [19], a standard $E - J$ power law can be introduced,

$$E = E_c \frac{\mathbf{J}}{|\mathbf{J}|} \left(\frac{|\mathbf{J}|}{J_c} \right)^n , \tag{3}$$

with E_c being the standard critical electric field criterion of 1μV/cm, from which the parameters J_c and n can be experimentally measured [22]. For commercial REBCO tapes at 77 K, n is commonly assumed to be greater than 25 with critical current densities (J_c) around 25 GA/m^2. J_c and n can also be as isotropic or anisotropic parameters depending on the requirements of the experimental observations, as it is known that these parameters can depend on external quantities such as magnetic field and temperature. However, it is already known that unless the operational temperature is demanded to change, like in the case of superconducting fault current limiters [23], isotropic approaches commonly come up with sufficiently accurate results for high transport currents [24]. For this reason, within this paper we will demonstrate that the isotropic approach is sufficient for the optimization of CORC®cables at different twist pitch angles, and due to it's importance within the cryogenic industry, an accurate estimation of the AC-losses.

As an initial validation of our model, we have confirmed our numerical results through direct comparison with the experimental measurements of the AC losses in a straight YBCO tape under applied magnetic field [20]. The dimensions of the measured tape were 12 mm width, 25 cm length, and a 1 μm thick superconducting layer with a critical current density of 2.5 GA/m^2 at self-filed conditions and 77 K, i.e., with an $I_c = 300$ A. Then, for tackling the computational challenges associated with the large aspect ratio of the superconducting cross-section, the thickness of this layer and consequently its J_c has been renormalized by a factor of 50. This allows to reduce the meshing density around the superconducting tape, increasing the numerical convergence of the PDE solver whilst reducing its computing time.

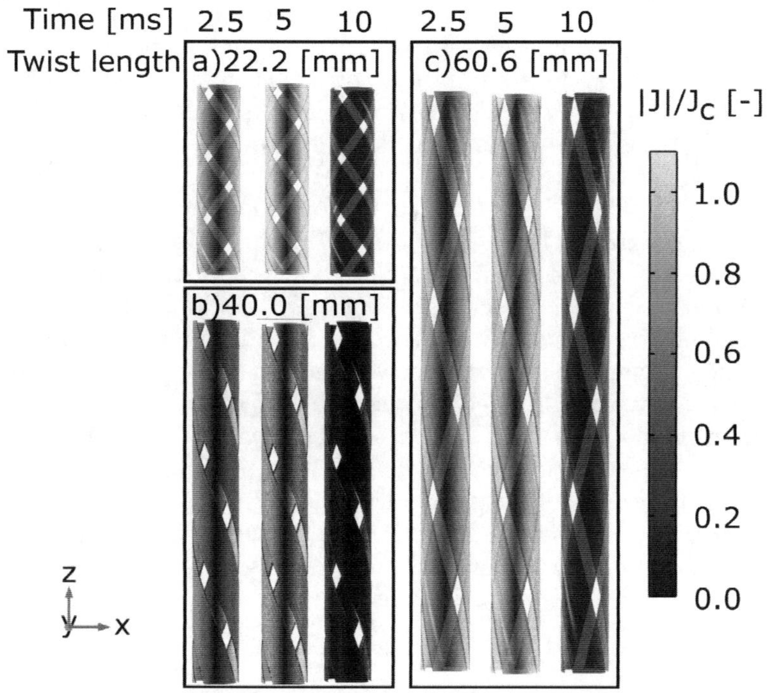

Figure 2. Magnetization currents in CORC cables with twist lengths of (a) 22.2 mm, (b) 40 mm, and (c) 60.6 mm, at half-peak (2.5 ms), first peak (5 ms), and zero-field (10 ms) of an external magnetic field applied in the y-direction $B_{ext} = B_a \sin(\omega t)$.

Likewise, our numerical results have been compared not only with the experimental measurements (see Fig. 1), but also confirmed via the analytical solution of Halse [25],

$$W = \frac{8\mu_0 J_c^2 w^2}{\pi} \left[\ln \left(\cosh \left(\frac{\pi H_p}{J_c} \right) \right) - \frac{\pi H_p}{2J_c} \tanh \left(\frac{\pi H_p}{J_c} \right) \right], \tag{4}$$

where w it the width of the tape and H_p is the peak value of the applied (AC) magnetic field. The calculated AC losses by Halse's analytical formula confirms the correctness of our numerical model, allowing us to use it confidently in the modelling of more sophisticated 3D structures such as the CORC®cables, whilst gaining an insight of the local distribution of current density along the superconducting tape, as it is shown in the inset of Fig. 1.

3. Results & discussion

In this section, we focused at first on the modelling of the AC-losses of experimentally measured 1-layer CORC®cable [20], validating the effectiveness of our model. Then, a comprehensive study on how the twist pitch length influences the calculated AC-losses has been pursued, concluding whether the manufactured cable design can be considered as optimal or not.

The full geometry of a 3D 1-layer CORC®cable with different twist pitch lengths has been modelled with 3 tapes twisted along the z axes on an stainless-steel former of 4.96 mm diameter, shown in Fig. 2. Since the experimental measurements have shown negligible eddy current in the stainless-steel former [20], this can be assumed as an electrically insulated material in the computational model, with the same properties as the surroundings of the superconducting tapes. The tapes in the cable are 4 mm wide with a critical current of $I_c = 25$ A per mm width, measured at 77 K and self-field conditions. In the model we applied the oscillating magnetic field in the y direction, with amplitudes ranging from 20 mT to 50 mT based on the original experimental conditions. The tapes are initially arranged with a twist pitch angle of 68°, i.e., for building a CORC®cable of 40 mm along the $z-$length. Our simulations show an excellent agreement with the experimental measurements (Fig. 3), with slightly higher losses at fields greater than 40 mT. It has been reported in [20] that their experimental measurements beyond this field magnitude might have been affected by some heating effects, which may indicate a decrease in the J_c and consequently the AC-Losses. For this reason, at the field range when full penetration of the magnetic field occurs, our numerical simulations can be considered reliable for not only the the AC-losses, but also because it confirms the negligible influence of eddy currents at the stainless steel former. It is also worth checking the distribution of magnetization currents due to their influence on the occurrence of power losses. The current distribution of this case is shown in the CORC®cable at the left bottom pane of Fig. 2, where the screening currents spread into the tapes in a similar way to what can be seen in a straight tape (see Fig. 1) or a bulk wire [26, 27], i.e, from the outer edges of the cable (perpendicular to the direction of the applied magnetic field) towards its center. Notice that in Fig. 2, it is the normalized magnitude of the total current density $|\mathbf{J}|$ what is shown. Although the dominant component of the current density is J_z, the helix structure of the CORC®cable does not allow a greater comparison with the above mentioned '2D' geometries as the one shown here.

Having verified our numerical model for the 1-layer 3 tapes CORC®cable, its twist pitch angle has been systematically changed from $\sim 35°$ to $\sim 75°$. This has helped us to determine whether the twist pitch angle of the experimentally measured cable with twist length $\Gamma_l = 40$ mm, and diameter $D = 4.96$ mm, i.e, $\beta = \arctan(\Gamma_l/(\pi D)) = 68.72°$, can be considered as an optimal CORC®cable design or not. Therefore, the results presented in Fig. 3 are for showing the AC-losses of different CORC®cables considering a maximum rate of change of 10% the pitch angle reported in the experiments. This is equivalent to a study of at least 7 twist lengths, with pitch angles of approximately 34.36°, 41.23°, 48.10°, 54.97°, 61.85°, 68.72°, and 75.59° respectively. Likewise, the spiral length, i.e, the approximate amount of coated conductor used for a given twist pitch length can be calculated by $CC_l = \sqrt{\Gamma_l^2 + (\pi D)^2}$, which results in significant differences in the amount of of tape required for each of the twist pitch angles analyzed, or being more specific, three times 18.88 mm, 20.72 mm, 23.34 mm, 27.15 mm, 33.01 mm, 42.93 mm, and 62.6 mm, respectively.

Based on this study we can conclude that for the broad range of twist pitch lengths studied, the distribution and dynamics of magnetization currents in the 1-layer 3-tapes CORC®cable manufactured by Advanced Conductor Technologies (see Fig. 2), and consequently its AC-Losses (see Fig. 3), do not significantly change as a function of the twist pitch angle. However, by shortening the twist pitch angle, i.e., by reducing the cable-length, also called the twist pitch length, the AC-losses at high field (50 mT) increase in about a 15%. Whilst at low field (20 mT) they tend to decrease in the same ratio, at least for $\Gamma_L < 40$ mm. Despite a reduction of the AC-losses at low field being beneficial, the reduction in the distance covered by the CORC®cable is of at least 27% to 73% the length covered by the manufactured cable with $\Gamma_l = 40$ mm. Therefore, in order to cover the same 40 mm distance, the amount of tape required by the CORC®cable with $\beta \simeq 34.36°$ and $\Gamma_l \simeq 10.65$ mm, would be of at least 2.27 times the amount of tape required for the standarized 40 mm cable with $\beta = 68.72°$.

The slight reduction in the the AC-losses at low magnetic fields and low pitch angles is due to the reduced path in the loops of current density that form along the $z-$axis, i.e, screening the magnetic field applied along the $y-$direction. However, as the applied magnetic field increases, for shorter loops of the magnetization currents, i.e., for lower twist pitch angles, larger becomes the magnetic field penetration rendering to higher AC-losses. Therefore, a certain balance must be found between the twist pitch length and the AC-losses of the CORC®conductor, such that

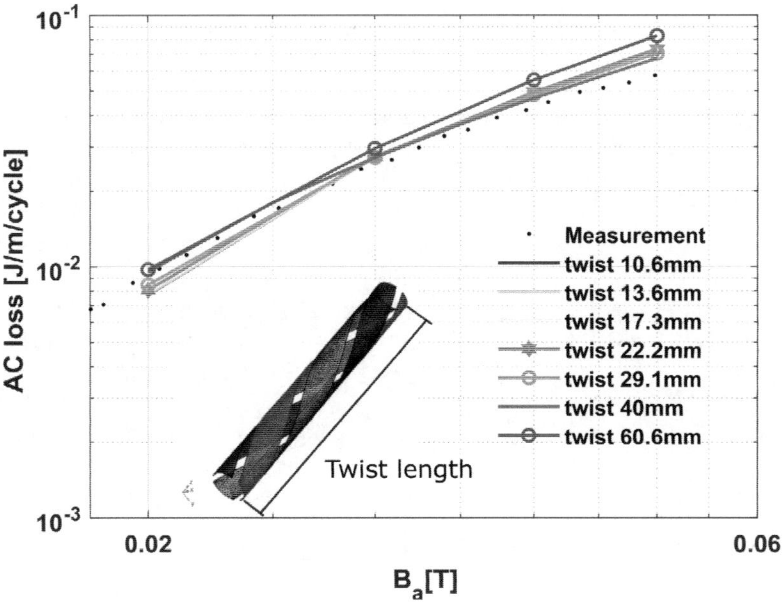

Figure 3. The AC losses of the CORC®cable with 40 mm twist pitch length shown nice agreement with measurement. The study of different twist pitch confirmed the optimum twist length of 40 mm or twist angle 68.72° with slight increase and reduction of the AC losses by 15% outside of this length. The inset shows the twist pitch length in the cable.

for long cable lengths and high magnetic fields, no excessive amount of 2G-HTS is demanded. As the AC-losses of the CORC®cable for $\Gamma_L > 40$ mm and applied magnetic fields greater than 40 mT start to show a noticeable increment (see Fig. 3), we have found that the optimal twist pitch length for the 1-layer 3-tapes CORC®cable manufactured by Advanced Conductor Technologies is indeed 40 ± 5 mm. In fact, as the amount of conductor required for $\Gamma_l = 40$ mm is of just 42.93 mm, no more than a 3% increment in the cabling cost will be expected (excluding the cost of the SC-former) in comparison with an equivalent cable composed by straight tapes, making this dimensions ideal for the inception marketing of CORC®cables in high power systems applications.

4. Conclusion

Based on the possibility to validate the H-formulation in a fully 3D model of straight SuperPower 4-mm width tapes (SCS4050) either from experimental observations or by fully analytic methods, we expanded our model to present a comprehensive study on the twist pitch angle of CORC®cables. Here we shown a CORC®cable model with 3 tapes in 1-layer, following the exact manufacturing dimensions and physical properties of the cables assembled by Advanced Conductor Technologies LLC. We have proven how our method reproduces with good accuracy the experimental measurements of AC-losses when the cable is exposed to a transverse magnetic field, it without the need for reducing the dimensionality of the superconducting tapes from a finite thickness layer to a strip approach. Seven CORC®cables with different twist pitch lengths have been studied in order to confirm their optimum design. Twist pitch lengths in the range of $\Gamma_L = 10.6 - 60.6$ mm show a slight increase of the AC-losses by a maximum of 15% at high magnetic fields, and a similar reduction by 15% at low magnetic fields. This reduction of the AC-losses at low fields and $\Gamma_L < 40$ mm is caused by shorter screening currents loops. With an increasing magnetic field, the AC-losses escalated above the measured AC-losses value in the 40 mm cable case. This was due to higher amount of the HTS tapes per the cable length. In the case of the twist length 60.6 mm, the AC-losses increases due to the current loops being too long. This study has confirmed the optimum twist pitch length is indeed the one in the experimentally measured cable, i.e., of 40 mm. In fact, as by shortening the twist pitch length the tape length required increases, then we have found that an optimal twist pitch length of 40 ± 5 mm is encountered by balancing the AC-losses at high and low field encountered, with the need to have a sufficiently long twist pitch length for the manufacturing and commercial production of the CORC®cable. Further studies on multiple layer CORC®cables are required to confirm whether the above conclusions stand as a principle of generality, as although from the electromagnetic point of view the physical model will remain unaltered, the aggregated computational burden might present significant challenges for the uncovering of new physics on these designs.

5. References

[1] Béjar Alonso I, Brüning O, Fessia P, Rossi L, Tavian L and Zerlauth M 2020 High-luminosity large hadron collider (hl-lhc): Tech. rep. (CERN Yellow Reports: Monographs cern-2020-010

[2] Oliva A B, Aprili P, Bellesia B, Rebollo E B, Boutboul T, de Sousa P C, Lino M P C, Cornelis M, Gavouyere-Lasserre P, Harrison R, Hernandez A, Jimenez M, Loizaga A, Martinez M, Readman P, Romano G, Rossi D, Sborchia C, Valente P, Viladiu E, Batista R, Casarin V, Kostopoulos C, Lo Bue A, Pompa E, Pozuelo-Segura E, LIM B S, Koczorowski S, Luongo C, Mitchell N and Poncet L 2020 *IEEE Transactions on Applied Superconductivity* **30** 1–13 ISSN 1051-8223

[3] Wang X, Caspi S, Dietderich D R, Ghiorso W B, Gourlay S A, Higley H C, Lin A, Prestemon S O, van der Laan D and Weiss J D 2018 *Superconductor Science and Technology* **31** 45007 ISSN 0953-2048

[4] van der Laan D C 2009 *Superconductor Science and Technology* **22** 065013 ISSN 0953-2048

[5] Šouc J, Gömöry F, Vojenciak M, Solovyov M, Seiler E, Kovác J and Frolek L 2017 *Superconductor science abd technology* **30** 105014 ISSN 0953-2048

[6] van der Laan D C and Ekin J W 2007 *Applied physics letters* **90** 052506–052506–3 ISSN 0003-6951

[7] van der Laan D C, Weiss J D, Noyes P, Trociewitz U P, Godeke A, Abraimov D and Larbalestier D C 2016 *Superconductor Science and Technology* **29** 055009

[8] Mulder T, Weiss J, van der Laan D, Dudarev A and Kate H t 2020 *IEEE Transactions on Applied Superconductivity* **30** 1–5

[9] Mulder T, Weiss J, van der Laan D, Dhallé M and Kate H T 2018 *IEEE Transactions on Applied Superconductivity* **28** 1–4

[10] van der Laan D C, Weiss J D and McRae D M 2019 *Superconductor Science and Technology* **32** 033001

[11] van der Laan D C, Radcliff K, Anvar V A, Wang K, Nijhuis A and Weiss J D 2021 *Superconductor Science and Technology* **34** 10 ISSN 0953-2048

[12] Takeuchi K, Amemiya N, Nakamura T, Maruyama O and Ohkuma T 2011 *Superconductor Science and Technology* **24** 085014 ISSN 0953-2048

[13] Goo J, Han J W, Lee S, Kim W S, Choi K and Lee J K 2021 *IEEE Transactions on Applied Superconductivity* **31** 1–5

[14] Li Q, Tan H and Yu X 2014 *IEEE Transactions on Applied Superconductivity* **24** 1–4 ISSN 1051-8223

[15] Stenvall A, Siahrang M, Grilli F and Sirois F 2013 *Superconductor Science and Technology* **26** 45011 ISSN 0953-2048

[16] Wang Y, Zhang M, Grilli F, Zhu Z and Yuan W 2019 *Superconductor Science and Technology* **32** 25003 ISSN 0953-2048

[17] Sheng J, Vojenciak M, Terzioglu R, Frolek L and Gomory F 2017 *IEEE Transactions on Applied Superconductivity* **27** 1–5 ISSN 1051-8223

[18] Tian M, Yang J, Shen B, Ozturk Y, Ma J, Shah A, Patel I, Li J and Coombs T A 2021 *IEEE Transactions on Applied Superconductivity* 1–1 ISSN 1051-8223

[19] Badía-Majós A, López C and Ruiz H S 2009 *Phys. Rev. B* **80**(14) 144509

[20] Majoros M, Sumption M D, Collings E W and van der Laan D C 2014 *Superconductor Science and Technology* **27** 125008

[21] Bean C P 1962 *Phys. Rev. Lett.* **8**(6) 250–253

[22] Zhang X, Zhong Z, Ruiz H S, Geng J and Coombs T A 2017 *Superconductor Science and Technology* **30** 025010
[23] Ruiz H S, Zhang X and Coombs T A 2015 *IEEE Transactions on Applied Superconductivity* **25** 5601405 ISSN 1051-8223
[24] Robert B C, Fareed M U and Ruiz H S 2019 *Materials* **12** 2679
[25] Halse M R 1970 *Journal of Physics D: Applied Physics* **3** 717
[26] Ruiz H S, Badía-Majós A, Genenko Y A, Rauh H and Yampolskii S V 2012 *Applied Physics Letters* **100** 112602
[27] Ruiz H S and Badía-Majós A 2013 *Journal of Applied Physics* **113** 193906

Acknowledgements
This work was supported by the UK Research and Innovation, Engineering and Physical Sciences Research Council (EPSRC), Grant Ref. EP/S025707/1 led by H.S.R. All authors acknowledge the use of the High Performance Computing facility ALICE at the University of Leicester.

IOP Conf. Series: Materials Science and Engineering 1241 (2022) 012031 doi:10.1088/1757-899X/1241/1/012031

Computational Modelling of Russia's First 2G-HTS Triaxial Cable

M. Clegg, M. U. Fareed, M. Kapolka & H. S. Ruiz

College of Science and Engineering & Space Park Leicester, University of Leicester, Leicester LE1 7RH, United Kingdom

E-mail: mlc42@leicester.ac.uk ; hsrr1@le.ac.uk

Abstract. A better understanding of the interaction between three phases is required when developing superconducting cables for high voltage AC systems. With a particular focus on the energy losses of real power transmission cables, in this paper we utilize the so-called H-formulation of Maxwell equations to devise a 2D model for superconducting triaxial cables. The major aim of this model is to comprehend and reproduce the experimental observations reported on the first triaxial prototype cable developed by SuperOx and VNIIKP. The computationally modelled and prototyped cable is made of up to 87 tapes of 4 mm width SuperOx tape arranged across the three phases. Our computational results are compared to the experimental measurements performed by VNIIKP with the electrical contact method, showing a high degree of accuracy over the outer phase of the cable, whilst revealing technical issues with the experimental measurements at the inner phases. Thus, in consultation with VNIIKP it has been concluded that for the actual experimental measurement of the AC losses at the inner phases, and consequently of the overall cable, a sophisticated calorimetric setup must be built. Still our model is capable to provide an independent assessment of the VNIIKP-SuperOx cable design, by investigating the magnetic profiles per phase in the time domain. In this sense, we confirm that the unbalanced arrange of currents and distancing between the phases affirmatively lead to no magnetic leakages, and therefore to an adequate balance of the cabling inductance.

1. Introduction

Improving the current density capacity of power transmission is important as the electricity demand across the world continues to increase. Accommodating for this pressure is possible with high temperature superconducting (HTS) cables that allow for high levels of current density while also maintaining a low level of energy losses produced by the system. HTS cables have been investigated over many years as numerous projects have already been installed and tested within several power grids, showing the potential of these cables when applied in this sector [1]. However, a downfall in the HTS cable production is the capital and operational cost of the HTS wire and refrigeration system [2]. Therefore when constructing a HTS cable, further thought needs to be put into the configuration of the cable with the three concentric phases resulting in the most compact design and therefore lowest material consumption [3]. This is especially important when considering a three phased system due to the added energy losses imposed by the alternating currents, which although could be significantly reduced by considering DC currents, depending on the amount of current being transmitted [4–7], all major transmission applications and nearly all distribution systems operate under AC conditions. Thus, for the development of three-phase networks aided by superconducting cables, one configuration of great interest is the

Table 1. Measured dimensions and critical currents for the triaxial prototyped cable in [9].

Phase	Inner Radius (mm)	Number of Tapes	Critical Current (A)
1	18	27	127
2	19.6	29	121
3	21.7	31	116

designing of HTS triaxial cables [8–14], capable to offer a high critical current density with a low amount of HTS wire, whilst maintaining the power system balance.

The possibility of triaxial HTS cables in three-phase power networks has been actively researched since the beginning of this century [8, 15], and since then multiple projects have been conducted such as the Bixby project at Columbus, Ohio [16] and the AmpaCity project at Essen, Germany [17] both using the first generation (1G) of Bi-2223 tapes for the manufacturing of their cables. The AmpaCity cable was in fact the first HTS cable to be integrated into a real power system of a major city, and has been the precursor of many major developments of HTS AC cables for power applications [18]. Nevertheless, more recently the development of triaxial cables has been focused on the use of the so-called second generation (2G) of HTS tapes, with 2G HTS triaxial cables being produced in South Korea [13] and Russia [9]. These projects have displayed the capabilities of producing a large scale three phase transmission system using the latest developments on the manufacturing of superconducting tapes, and the possibility to balance their total currents for a null magnetic field outside of the conductor, as the different radii of the three phases produces non-uniform current distributions due to the magnetic induction suffered by the type-II superconductors [19].

A study of the first Russian triaxial HTS cable prototype was published in 2017 by the Russian Scientific Research and Development Cable Institute also known as VNIIKP, in collaboration with the 2G-HTS manufacturing company SuperOx [9]. This paper produced two prototypes; one being a 10 m 1G Bi-based HTS cable and the other being a 4 m ReBCO based 2G-HTS cable. The VNIIKP team inspected the AC-losses at 50 Hz of the 4 m cable with respect to each phase/layer, leading to some results which needed to be investigated from the computational point of view. In fact, it was made evident how the recording of the AC-losses in the first and second phase, i.e., the inner phases in the triaxial design, could not be trusted due to an unforeseen technical failure in the electrical method used at these phases. Nevertheless, it was agreed that the reported measurements for the AC-losses at the third phase (outer-phase in the triaxial cable) should not have been affected, and therefore it is the purpose of this paper to validate or disregard such observation, starting with a two-dimensional (2D) numerical approach. In this sense, it is to be mentioned that although the 2D approach does not allow to have a faithful representation of the twist pitch for each one of the cable phases, by knowing the experimentally measured critical current at each one of the phases for the prototyped cable, these can be assumed to intrinsically account for the effect of the magnetic mutual inductance of the other phases. Therefore, global quantities such as the magnetization or AC losses of the cable should be reproduced by considering the real critical current density measured at each one of the phases of the triaxial cable (see Table 1), these instead of the self-field critical current of a single planar tape that is commonly defined in the classical 2D-modelling. Still, a completely accurate measuring of the magnetic field in the space surrounding the SC tapes cannot be ensured, but as this also depends on the mesh quality invoked, even the 3D models could render to slight deviations from the experimental results. Therefore, a 2D modelling supported by the experimental measurements for the critical current density in the real geometry of the cable for

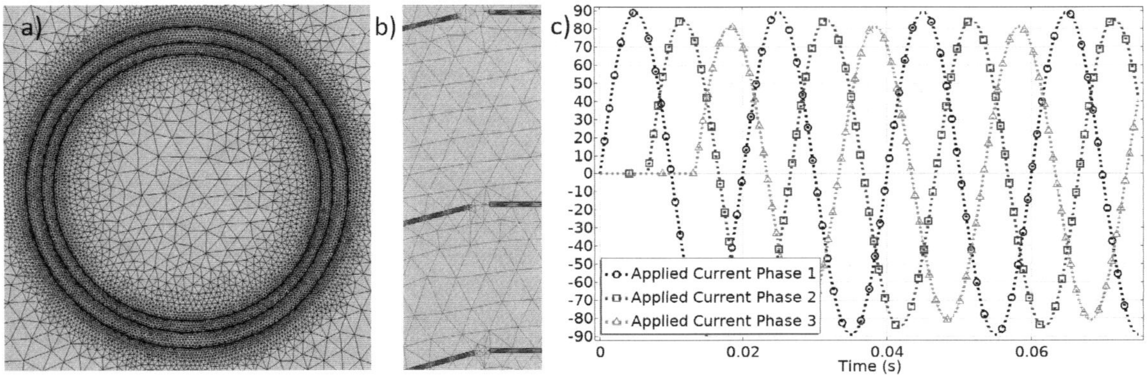

Figure 1. a) 2D representation of the meshed cross-section for the ReBCO triaxial cable prototyped in [9]. b) A zoomed-in of the meshing and relative distance between neighbouring tapes in the triaxial arrangement. c) The unbalanced AC distribution of transport current applied to the three phases of the cable as informed by the experiments and computational needs.

which the AC losses are to be determined, is considered by us as the most sensible approach allowing to avoid the high computational costs associated with the 3D modelling.

Thus, in this paper we present the numerical estimation of the AC-losses produced by the first Russian triaxial 4 m ReBCO cable prototype, by the means of a detailed 2D model based on the so-called H-formulation previously used by our group and other researchers in other geometries [20–26]. The triaxial cable contains 87 SuperOx 4 mm width tapes with all their physical properties and dimensions being reported in [27,28]. This is done with particular focus on the magnetic profiles of the three phase system and a further interest in the AC-losses within the different phases of the model. From the research conducted, there is an interest in the magnetic interaction between the three superconducting layers (phases), with particular focus on the events occurred at the current peaks such that the correct hysteretic cycle is determined. Our numerical simulations have yield to a good agreement between our AC-loss calculations and those measured in the triaxial cable, proving the reproducibility and reliability of our method, as well as of the low hysteretic losses produced by the SuperOx triaxial cable by means of its characteristic feature of no magnetic leakage.

2. Geometry of the triaxial cable and H-Formulation method

This model features what in the computational jargon for electromagnetic modelling is called as the "air" domain, i.e., a no magnetic highly resistive media, it surrounding the cross-section of the triaxial cable which is composed by three concentric rings of finite superconducting tapes, each defining a phase of the conductor. With the first or inner phase radius being 18 mm forming a "ring" made of 27 tapes, the number of tapes at each phase is then increased by 2 tapes within a no-symmetric arrangement between their radii. The values of the radii and number of tapes are stated in table 1, together with the average intensity of the critical current experimentally measured at each one of the tapes per phase [9]. Each tape is composed by a superconducting layer of 4 mm width and 1 μm thickness with the critical current density dependence on the magnetic field and angle, $J_c(B, \theta)$, as reported in Ref. [27].

In Figure 1, it is to be noted that the thickness of the superconducting layer within the SuperOx tape has been increased by a factor of 50. This allows to overcome the numerical difficulties that can be caused by the large aspect ratio of the 2G-HTS tapes, whilst maintaining the integrability of the physical quantities by renormalizing the critical current density J_c as it is customarily done in the numerical modelling of 2G-HTS tapes [22, 23, 29]. Likewise, as none of the other layers composing the SuperOx tapes are known to be magnetic nor carry any of the transport current which is to flow only on the superconducting layer, these can be assumed to have the same electrically insulating and non-magnetic properties of the so-called "air" domain, what allows to neglect also the occurrence of eddy currents. Same assumptions apply to the electrically insulated stainless steel cable former, as no magnetization signal nor eddy currents have been reported in this case [9]. The varied distance between each phase can be seen in Fig. 1 a) and Fig. 1 b), which is used to provide a sufficient accuracy for the simulations. Fig. 1 c) shows the three-phase unbalanced applied current with phase 2 and 3 being applied from a third of a cycle and two thirds of a cycle, respectively. This is to have a zero-field solution at the initial time instant ($t = 0s$), enabling the time-dependent solution and convergence of the PDE system. This implies to perform the numerical computation for at least 3.75 cycles of the applied current to the first phase (see Fig. 1 c)), with the AC losses for the entire cable being calculated within 0.05166 s and 0.07166 seconds, i.e., being this the first encountered hysteretic cycle after a magnetic relaxation period of 0.05166 s.

Using the H-formulation in a two-dimensional setting implies to have only two dependent variables, which are the magnetic field in both the x and the y direction. This two-dimensional arrangement implies that the current density flows only perpendicularly to the magnetic field, i.e., along the z-axis. These factors are implemented into COMSOL using the general Partial Differential Equations (PDEs) module, with the magnetic field and the current density being calculated through Faraday's law:

$$\begin{vmatrix} \partial_y E_z \\ -\partial_x E_z \end{vmatrix} = -\mu \begin{vmatrix} \partial_t H_x \\ \partial_t H_y \end{vmatrix} , \tag{1}$$

with the Ampere's law implemented as a constitutive equation for

$$J_z = \partial_x H_y - \partial_y H_x \tag{2}$$

The electrical behaviour of the superconducting material is defined using the conventional $E - J$ power law model [23],

$$\boldsymbol{E} = E_0 \frac{\boldsymbol{J_z}}{|\boldsymbol{J_z}|} \left(\frac{|\boldsymbol{J_z}|}{J_c} \right)^n , \tag{3}$$

where the standard electric field criterion of 1 μV/cm has been is adopted. An n-value of 34.4 is used according to the experimental measurements stated earlier for the critical current density dependence $J_c(B, \theta)$ for the 4 mm SuperOx tapes [27].

Special interest has been paid onto the hysteretic AC-losses of the superconducting tapes, which are calculated by the integration of the local density of power dissipation ($\mathbf{E} \cdot \mathbf{J}$) across the superconducting tapes (S), and then over the time span for a full hysteretic cycle ($f.c$), i.e,

$$Q = \omega \int_{f.c} dt \int_S \boldsymbol{E} \cdot \boldsymbol{J} dS . \tag{4}$$

However, in the case of a triaxial cable with an apparently unbalanced distribution of transport current (see Fig. 1 c)), the correct time span for defining a fully hysteretic behaviour has to be computationally determined. The transport current applied to each one of the phases is included within the COMSOL platform as pointwise constraints for each one of the tapes, satisfying that the integral function for the variable J_z (Eq. 1) over the cross section of the superconducting domains (per phase) minus the applied current per phase equals zero.

Finally, it is to remind that the rescaling of the superconducting layer thickness from 1μm to 50μm, and consequently the critical current density, is not only for tackling the large aspect ratio problem of the superconducting tape cross-section, but to include a sufficiently fine mesh across the thickness where the profiles of current density in a could flow within the Bean's approach. As finer is the mesh as accurate is the motion of the flux front profile inside the superconductor, and therefore as accurate becomes the calculation of the AC-losses. However, as finer is the mesh as higher is the random access memory (RAM) and computational time demanded. In this regard, the numerical calculations shown at Figs. 2 & 3 are for a mesh of seven rectangular elements (segments) across the thickness of each one of the 87 superconducting tapes, whose width has been split into a free-triangular mesh with about 28 elements across each one these segments, i.e., each superconducting layer being defined by approximately 196 finite elements within the most general Lagrange shape function with a quadratic element order, this to avoid mathematical reductions that could lead to infeasible physical solutions [30]. The total number of finite elements (1012516), vertex elements (1772), and boundary elements (18821) at each one of our models has been reduced by removing mesh control entities, which are for instance, unnecessary edges within the material domains which are automatically created by the CAD assembly [30]. Therefore, consideration of only 352 vertex elements and 3703 boundary elements, with a minimum element quality of 0.016, led to affordable computing times ranging from 2.7 hours at $I/I_c = 0.1$ and 3.3 hours at $I/I_c = 1$ per cycle. A standard PC with a clock-speed per core-processor of 3.6 GHz and 64 GB RAM has been used, with only one core active per computation.

3. Results

To analyse the magnetic interaction between the superconducting tapes and the three none equally distanced phases, particular attention has been played towards the magnetic field produced between each phase at both the peak current value and when the applied current is equal to 0 A. This is due to these values showing both extremes of the electromagnetic response produced by the triaxial cable, that being when either the maximum or zero transport current condition is being applied.. Therefore, In Fig. 2, the hysteretic magnetic $B-$field distribution for 6 cases are displayed. At the top, each phase reaches its peak applied current, starting from phase 1 (Fig. 2 a)) to phase 3 (Fig. 2 c)), respectively. The bottom pane shows the corresponding $B-$field profiles when each phase current equals 0 A. All subplots show no field leakage outside the triaxial cable proving the correct inductive balancing of the same.

In Fig. 2 a), i.e., for the 1st phase peak current of 127 A, the intensity of the magnetic field B between phase 1 and 2 is (in average), about twice the $B-$field between phase 2 and 3, with the maximum field produced being about 45 mT. Then, as it can be seen in Fig. 2 b), i.e., for the following peak value of the 2nd phase peak current of 121 A, a relatively even balance for the $B-$fields between phases is seen. Finally, in Fig. 2 c), i.e., for the 3rd phase peak current of 116 A, it can be seen an opposite behaviour to what was observed in Fig. 2 a), but showing a slightly lower maximum-field of 40 mT caused by the imbalance in the applied current and the different spacing between the phases.

Similar response is shown in the bottom pane of Fig. 2 d) to f) when each phase reaches its zero transport current condition ($I_{tr} = 0$ A), with the other phases at $I_{tr} \neq 0$ A for $\phi = \pm 120^\circ$ the phase reference. Thus, when no current is in phase 1 (Fig. 2 d), nearly no magnetic field is seen in between phases 1 and 2 but a very small field leakage at phase 2 is observed around the tapes. On the other hand, when there is no transport current in phase 2 (Fig. 2 e)), nearly no screening of the magnetic field is provided by this phase, but the compensation of the field-phase outside the cable demonstrates its current balancing by the continuous absence of magnetic leakage, which agrees with the original optimization method used at [9] for the triaxial cable designing. This is also confirmed by Fig. 2 f), where no B-field is observed around the phase 3 at the instant

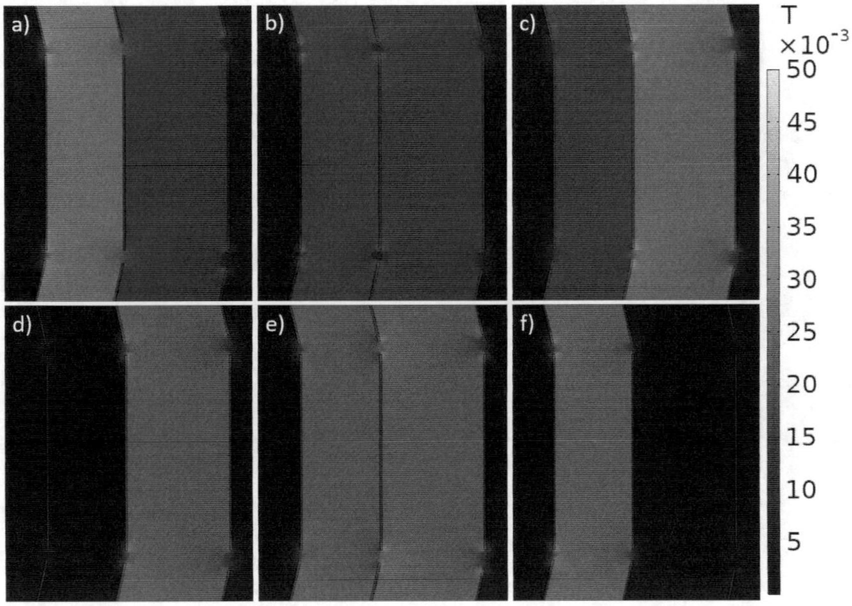

Figure 2. The magnetic profiles of the cable across a hysteretic time period with (a), (b) and (c) being when phase 1, 2 and 3 are individually at the peak of the respective phases applied current. The bottom pane of (d), (e) and (f) are when the applied currents of phase 1, 2 and 3 equal 0 A, respectively.

of time when only this phase has null transport current.

On the other hand, initially the AC-losses caused a concern due to the difficulties in replicating the losses produced within the phase 1 and 2 of the cable. When looking at Fig. 3 a), the calculated AC losses for third phase provides an adequate explanation of the experimental results, but it shows an acute difference at low transport currents for the phases 2 and 3, which both show a very similar but unexpected high losses behavior in the experimental measurements for the applied current range of $0.1I_c$ to $0.6I_c$, whilst the numerical estimation for currents higher than $0.8I_c$ show a good resemblance with the experimental observations. Therefore, despite a proper current balancing of the system and no magnetic leakage was previously demonstrated, it resulted necessary to investigate what could be the potential issues that may have caused these differences with the experimental observations, before concluding the occurrence of technical issues in the experimental setup which were duly discussed with the VNIIKP team. Thus, from the computational perspective, the first attempt was to reduce the n-value within phase 1 from 34.4 down to 8 to see if this would have a drastic change within the AC-losses (see Fig. 3 b)). With phase 2 being the highest AC-losses for most of the applied current range at the default values, by reducing the n-value, this caused the relationship between phase 1 and 2 to become closer at lower applied currents. Unfortunately, it also reduced the AC-losses within phase 2 very slightly, instead of increasing them to justify the large difference. Consequently, degrading the n-value of the superconducting tapes cannot be the reason for the difference between the computationally assessed and measured AC losses for the inner phases of the triaxial cable.

A second theory was tested to see if there was any significant degradation of the 2G-HTS tapes used in the inner phases, which implied to keep the same applied current for each one of the tapes, but reducing the critical current density of phases 1 and 2 (see Fig. 3 c)). The aim here was to maintain the AC-losses of phase 3 so that they continue to reproduce the experimental observations, whilst finding an alternative critical current density which could increase the AC-

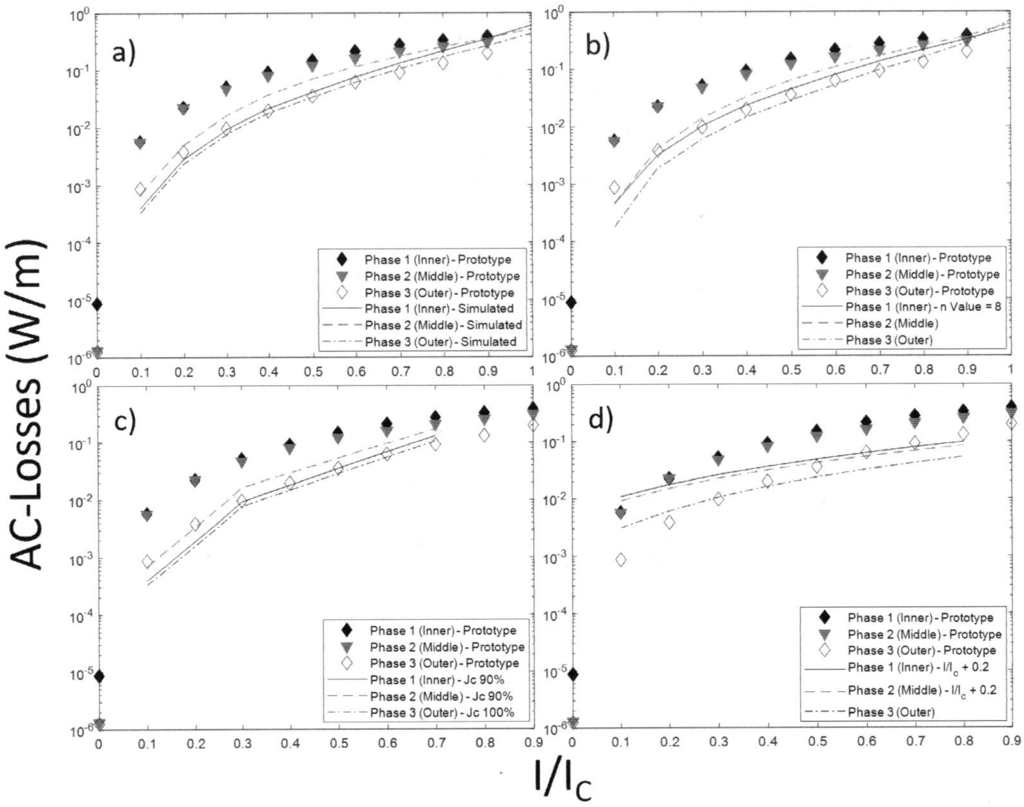

Figure 3. Calculated AC-Losses with (a) default values, (b) reduced n-values, (c) reduced Jc in phase 1 and 2, and (d) increased transport current by $0.2I_c$ for phase 1 and 2.

losses of phases 1 and 2. The values of $0.1I_c$, $0.3I_c$ and $0.7I_c$ were used as reference points so that we could compare where the largest discrepancy was for the default values. Again, these results did not provide a sound justification for the difference in the AC-losses.

The final effort to attempt explaining the differences in the AC-losses predicted for the inner phases, was to increase the applied current by an extra $0.2I_c$ within phase 1 and 2, assuming a possible but unexpected raise in the amount of current driven by the power sources (see Fig. 3 d)). In this sense, although increasing the applied current of phase 1 and 2 for low values of the stated applied currents ($0.1I_c$ to $0.3I_c$ with phase 3 as the reference) raised the estimated values for the AC losses within these phases, being now closer with the experimental results, it has the opposite behavior for the phase 3. Moreover, the discrepancies with the experimental measurements become greater as the transport current at each one of the phases increases, what allow us to conclude that the issue encountered is not related with the correct definition of the applied currents shown in Fig. 1 c). Thus, as any degradation of the J_c properties of the superconducting tapes nor a technical failure on the power sources can explain the acute raise in the AC losses observed for the inner phases, it has been concluded that the most likely reason for having the significant increase in the AC losses for the phases 1 and 2, with the phase 3 being appropriately reproduced, lies in a possible failure of the electric contacts for the voltage taps fixed at the phases 1 and 2, which could not be revised after the full assembly of the cable, as these were already under the electrical setup used for the outer layer, i.e., the phase 3 of the cable.

For future publications, we have been allowed to disclose that the VNIIKP-SuperOx team is currently preparing a fully integrated calorimetric setup, aiming to overcome the technical difficulties encountered with the measurement of the AC losses at the inner (first) and middle (second) phase of the triaxial cables. Meanwhile, our team is preparing a comprehensive study on further prototypes currently being developed at the VNIIKP facilities [14].

4. Conclusion

In this paper, we have focused on the computational modelling of the first Russian triaxial cable produced by VNIIKP and SuperOx [9]. After investigating different scenarios for the forecasting of the AC-losses from the finite elements modelling of the triaxial cable, this in comparison with the attained experimental results from the electrical probe method at each one of its phases, it has been determined that the actual AC losses at the inner phases, i.e., the phase 1 and 2 of the cable are actually lower than the originally reported. In this sense, to support this conclusion we have performed a comprehensive study on the different factors that can alter the critical current density in the different phases of the cable, assuming some degradation of its electrical properties caused by the cable manufacturing, it despite the fact that the critical current at each one of the phases/tapes was correctly measured after the cable assembly. Thus, we have concluded that no modification of the n- and J_c values in the $E - J$ power law of the ReBCO conductors could explain the unusual rise in the AC losses observed for the phase 1 and 2 of the cable, nor a failure in the calibration of the power sources. On the contrary, as a correct estimation of the AC losses has been obtained for the outer phase of the cable, i.e., its third phase, and also, it has been determined that the intensity and time dependence of the applied currents with the experimentally measured critical currents after the cable assembly, correctly leads to the proper power balancing of the cable by proving the absence of magnetic leakages, it has resulted clear for our teams that the source of the additional losses in the inner phases is not caused by the cable itself but by the voltage taps setups used for the measuring of the AC-losses in the inner two phases, which cannot be accessed after the cable assembly. In this sense, VNIIKP is currently preparing a fully integrated calorimetric setup which by default, will overcome the technical difficulties encountered with the electrical method for the measurement of the AC losses.

Alongside predicting the actual AC-losses produced by the VNIIKP-SuperOx triaxial cable, within the experimental specifications for the cable testing, we have demonstrated that an adequate magnetic field balancing of the system can be achieved, it produced by the non-isotropic arrangement of the phases that has been considered by VNIIKP. At the phase 1 peak current, these results show that the intensity of the magnetic field **B** between phase 1 and 2 is, about twice the B-field between phase 2 and 3. This behaviour was also observed when phase 3 reached its peak current, but with a slightly lower maximum-field between phase 2 and 3. When the 2nd phase reaches its peak, it produces a relatively even magnetic balance between the inner and outer phases. Likewise, at the instants of time when no transport current flows across anyone of the three phases of the triaxial cable, it has been demonstrated that no magnetic leakages are to be observed even when no magnetic screening of the secondary phase is achieved. This demonstrates the adequate current balancing on the system of transport currents considered by VNIIKP in the Russia's first commercial prototype for triaxial cables.

References

[1] Doukas D I 2019 *IEEE Transactions on Applied Superconductivity* **29** 1–5 URL `https://ieeexplore.ieee.org/document/8626522`

[2] Kalsi S S 2011 *Applications of High Temperature Superconductors to Electric Power Equipment* (Hoboken, N.J: Wiley-IEEE Press) ISBN 978-0-470-16768-7

[3] Willén D, Matheus C, Lindsay D and Gouge M 2005 The application of triaxial high-temperature superconducting power cables in distribution networks *CIRED 2005 - 18th International Conference and Exhibition on Electricity Distribution* pp 1–4
[4] Robert B C and Ruiz H S 2018 *IEEE Transactions on Applied Superconductivity* **28** 8200905
[5] Robert B C and Ruiz H S 2018 *Superconductor Science and Technology* **31** 035006
[6] Ruiz H S and Badía-Majós A 2013 *Journal of Applied Physics* **113** 193906
[7] Ruiz H S, Badía-Majós A, Genenko Y A, Rauh H and Yampolskii S V 2012 *Applied Physics Letters* **100** 112602
[8] Fisher P W, Cole M J, Demko J A, Foster C A, Gouge M J, Grabovickic R W, Lue J W, Stovall J P, Lindsay D T, Roden M L and Tolbert J C 2003 *IEEE Transactions on Applied Superconductivity* **13** 1938–1941
[9] Fetisov S S, Zubko V V, Zanegin S Y, Nosov A A, Ryabov S M and Vysotsky V S 2017 *IEEE Transactions on Applied Superconductivity* **27** 1–5
[10] Sun Q, Lv H, Gao S, Wang S, Wei K and Mauersberger M 2019 *IEEE Transactions on Applied Superconductivity* **29** 1–5
[11] Lee S J and Yang H S 2019 *IEEE Transactions on Applied Superconductivity* **29** 1–5
[12] Cao, Jingying, Chen, Jie, Yang, Liu, He, Qidi, Liu, Gang, Hu, Libin, Tan, Xiao and Li, Chenying 2019 *E3S Web Conf.* **118** 02056 URL `https://doi.org/10.1051/e3sconf/201911802056`
[13] Lee S J, Kang S Y, Park M, Won D, Yoo J and Yang H S 2020 *Energies (Basel)* **13** 2053 ISSN 1996-1073
[14] Fetisov S S, Zubko V S, Zanegin S Y, Nosov A A and Vysotsky V S 2021 *IEEE Transactions on Applied Superconductivity* **31** 1–4
[15] Politano D, Sjostrom M, Schnyder G and Rhyner J 2001 *IEEE Transactions on Applied Superconductivity* **11** 2477–2480
[16] Demko J, Nielsen C, Sauers I, James D, Gouge M, Lindsay D, Roden M, Tolbert J, Willen D and Traeholt C 2007 *IEEE Transactions on Applied Superconductivity* **17** 2047–2050 ISSN 1051-8223
[17] Stemmle M, Merschel F, Noe M and Hobl A 2014 Ampacity - advanced superconducting medium voltage system for urban area power supply *2014 IEEE PES T D Conference and Exposition* pp 1–5
[18] Malozemoff A, Yuan J and Rey C 2015 5 - high-temperature superconducting (hts) ac cables for power grid applications *Superconductors in the Power Grid* Woodhead Publishing Series in Energy ed Rey C (Woodhead Publishing) pp 133–188 ISBN 978-1-78242-029-3
[19] Nassi M 2000 *Superconductor Science and Technology* **13** 460–463 ISSN 0953-2048
[20] Brambilla R, Grilli F and Martini L 2007 *Superconductor Science and Technology* **20** 16
[21] Rodriguez-Zermeno V M, Mijatovic N, Traeholt C, Zirngibl T, Seiler E, Abrahamsen A B, Pedersen N F and Sorensen M P 2011 *IEEE Transactions on Applied Superconductivity* **21** 3273–3276 ISSN 1051-8223
[22] Baghdadi M, Ruiz H S and Coombs T A 2018 *Scientific Reports* **8** 1342
[23] Robert B C, Fareed M U and Ruiz H S 2019 *Materials* **12** 2679
[24] Robert B C, Fareed M U and Ruiz H S 2019 *Journal of applied physics* **126** 123902 ISSN 0021-8979
[25] Robert B C, Fareed M U and Ruiz H S 2019 *Engineering Research Express* **1** 015037
[26] Fareed M U, Robert B C and Ruiz H S 2019 *IEEE Transactions on Applied Superconductivity* **29** 5900705 ISSN 1051-8223
[27] Zhang X, Zhong Z, Geng J, Shen B, Ma J, Li C, Zhang H, Dong Q and Coombs T A 2018 *Journal of Superconductivity and Novel Magnetism* **31** 3847–3854 ISSN 1557-1947
[28] Lee S, Petrykin V, Molodyk A, Samoilenkov S, Kaul A, Vavilov A, Vysotsky V and Fetisov S 2014 *Superconductor science and technology* **27** ISSN 0953-2048
[29] Stenvall A, Siahrang M, Grilli F and Sirois F 2013 *Superconductor Science and Technology* **26** 45011 ISSN 0953-2048
[30] Bathe K J 2006 *Finite Element Procedures* (Ukraine: Prentice Hall) ISBN 9780979004902

Acknowledgements

This work was supported by the UK Research and Innovation, Engineering and Physical Sciences Research Council (EPSRC), through the grant Ref. EP/S025707/1 led by H.S.R. All authors acknowledge the use of the High Performance Computing facility ALICE at the University of Leicester. Special thanks are given to Vitaly Vysotsky, Sergey Fetisov, and Vasily Zubko from VNIIKP, for the fruitful discussions and data sharing on their superconducting triaxial cables.

Structural Finite Element Analysis of REBCO Tape Delamination with Solid-Shell Element under Various Loads

Zijia Zhao, Peter Moore, and Luisa Chiesa

Tufts University, Mechanical Engineering Department,

Medford, MA 02155, USA

E-mail: Zijia.Zhao@tufts.edu

Abstract. Structural Finite Element Analysis was performed to study the delamination behaviour of layer laminated REBCO tapes using a Solid-Shell element under various loading conditions, including (1) transverse tensile delamination, (2) shear delamination and (3) peeling delamination. A Solid-Shell element is well suited to simulate REBCO tape because the material and thickness of all the layers can be specified within a single element. The results obtained from the models utilizing Solid-Shell elements are compared to a more computationally intensive method known as Cohesive Zone Method (CZM), which has successfully modelled the initiation and propagation of REBCO's delamination [1]. The results show that the stress state of REBCO tape in transverse tensile and shear directions can be accurately captured by the proposed method employing Solid-Shell elements if the transverse tensile and shear delamination stress is measured experimentally for a specific REBCO and used as failure criteria. In this work, the failure criteria adopted from previous experimental studies are 50 MPa in the transverse tensile direction and 10 MPa in the shear direction. This work provides a time-efficient modelling tool to predict and mitigate delamination in REBCO tapes used in complex systems.

1. Introduction

Layer laminated Rare Earth-Barium-Copper-Oxide (REBCO) high temperature superconducting (HTS) tapes are promising materials for high magnetic field applications due to their excellent mechanical and electrical properties. Experiments have shown that REBCO tapes are very strong in the axial tensile direction but very weak in the transverse tensile direction [2]. When the tape is subjected to excessive load, the tape will become permanently damaged and its performance degraded. The strength of REBCO tapes against delamination stresses can often be the limiting factor of the performance of the conductor. Delamination has been studied through experiments and simulations, and can be roughly classified into three modes: transverse tensile delamination, cleavage or peeling delamination, and shear delamination.

Experiments measuring the transverse tensile strength have determined that the delamination usually occurs either in the REBCO layer or at the interface between the REBCO and buffer layers, but can also happen at the buffer/substrate or REBCO/silver interfaces [3]. Critical current degradation tends to occur before mechanical delamination, and does so either abruptly or gradually depending on where the delamination occurred [3,4]. The danger of delamination caused by transverse tensile stresses during thermal cooldown (particularly in epoxy-impregnated coils), thermal quenches, and Lorentz forces is well documented in literatures [5,6]. Experiments to measure the transverse tensile strength of REBCO tapes have shown that delamination can occur at stresses within the range of 10 to 100 MPa [7–9]. This wide range of strengths can be attributed to different manufacturing processes and voids or defects in the tape [10]. For example, the stress at delamination at the edges of a 4 mm wide REBCO tape that has been slitted during production, is significantly lower. Slitting can cause micro-cracks in the REBCO crystal layer, weakening its mechanical strength [7]. Delamination can also occur when the layers are peeled apart by cleavage stresses. This mode of delamination constitutes a significant risk as it causes high stress concentration. The nominal cleavage strength (defined as the cleavage force over the entire loaded area) at which delamination occurs can be as low as 0.5 MPa at the slit edge of a tape [8].

Tests measuring the behavior of tapes under pure shear stress reported delamination strength between 3.7 MPa and 7.2 MPa, depending on location of the loading edge of the tape, edge slitting, and temperature [11]. The critical current underwent abrupt and significant degradation as the stress increased past the delamination strength.

To better understand the stress distribution within the tape as well as delamination crack propagation, various numerical models of REBCO tapes subjected to delamination stresses have been developed. Earlier models focused on the yielding of individual layers without modelling delamination between layers [9]. In more recent studies, a finite element simulation technique known as Cohesive Zone Method (CZM), in which the delamination is modeled through a zero-thickness cohesive layer between two surfaces, was shown to accurately model the delamination of a REBCO tape at the REBCO/buffer interface [1]. To implement CZM, each layer of the tape has to be modeled. Due to the high aspect ratio between the width and thickness, a very refined mesh density to ensure the accuracy of the results is needed. Additionally, the plastic deformation and geometric nonlinearities associated with delamination require the use of small load-step increments in the FEA software to avoid computational errors. Given these requirements, the CZM method is computationally very expensive, and it would be hard to apply the same method to more complex multi-tape system, such as a Cable-in-Conduit Conductors (CICC).

In this work, a less computationally intensive FEA simulation technique, utilizing Solid-Shell elements in ANSYS®, has been developed to predict the delamination hazard in complex systems. The Solid-Shell element is well suited to simulate a REBCO tape, as the material and thickness of the individual layers can be specified. In previous studies, the Solid-Shell element was successfully used to reproduce the stress-strain behaviour of REBCO tapes under axial tensile load [12]. In this study, Solid-Shell elements are used to predict the delamination hazard of REBCO tapes. Three types of delamination modes including transverse tensile, peeling, and shear delamination are studied, and the results are compared to the one from the CZM method for validation.

2. Simulation Methods

To explore the options on modelling delamination with reduced computational efforts, each delamination mode considered (transverse tensile, peeling and shear) is modelled in three different ways (figure 1). The REBCO tape in this study is assumed to be 4 mm wide, 10 mm long and 0.096 mm thick. The layer composition and the corresponding material properties are listed in Table 1. The material properties at 77 K are adopted from our previous work, and gathered from the literatures [12]. The REBCO and buffer layers (ceramic materials) are modelled together with a combined thickness of 2 μm.

Table 1. Material properties and thickness of the layers in REBCO tape.

Layer	Thickness	Elastic Modulus (GPa)	Yield Strength (MPa)	Tangent Modulus (GPa)
Copper	2 x 20 μm	70	120	5
Silver	2 x 2 μm	76	65	4
Substrate	50 μm	170	980	6
REBCO/Buffer	2 μm	157	700	--

In Model 1, all the layers of a REBCO tape are modelled with individual elements and are bonded together. The initiation and propagation of the delamination process is captured by a zero-thickness CZM layer located in between the silver and REBCO/buffer layer. In Model 2, the CZM layer is removed. In both models, two elements are specified through each layer, and the element size is 0.1 mm in other two directions. The purpose of this model is to capture the deformation of the tape when the delamination initiates as predicted in Model 1. In Model 3, a single solid-shell element (SOLSH190) through the thickness of the tape is used for the model, greatly reducing the computational time. Within the solid-shell element, six layers representing each layer of a REBCO tape are defined, and each layer is assigned its own thickness and material properties as done with the other two models.

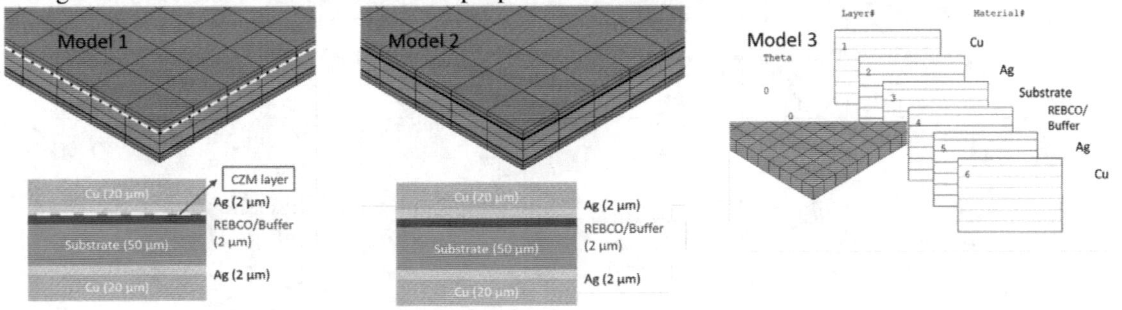

Figure 1. Three models are developed to reduce computational needs. For Model 1, all the layers of a REBCO tape are modelled, and the layers are bonded together. The initiation and propagation of the delamination process is captured by a zero-thickness CZM layer located in between the silver and REBCO/buffer layers. For Model 2, all the layers are modelled in the same manner as in Model 1, but the CZM layer is removed to capture the deformation of the tape when delamination initiates. In Model 3, the REBCO tape is modelled with only one element (Solid-Shell element) across the thickness with each layer specified.

2.1. Cohesive Zone Method

Cohesive Zone Method (CZM) is a FEA technique to model the initiation and propagation of delamination. The CZM applied in this work uses a bilinear traction-separation law to relate the relative gap of the two surfaces to the amount of stress at the interface. As shown in figure 2, in a bilinear delamination, the contact stress increases linearly with the contact gap up to the delamination strength, and then linearly decreases back to zero when complete decohesion has occurred. As loading is applied onto a layer laminated material, if the contact stress and gap are below the critical value (along line OA), the loading and deformation is fully reversible, and the material is not damaged. However, if the load exceeds the critical value the contact gap will continue to increase while the load will decrease following the AB path. If the load is removed at point B, the material is irreversibly damaged, and the contact is softened so the unloading and reloading process will follow line OB with a lower slope than OA. When the contact gap reaches point C, the material is fully delaminated. The bilinear tangential shear-sliding law could be an alternative approach to use, and it is very similar to the traction-separation law.

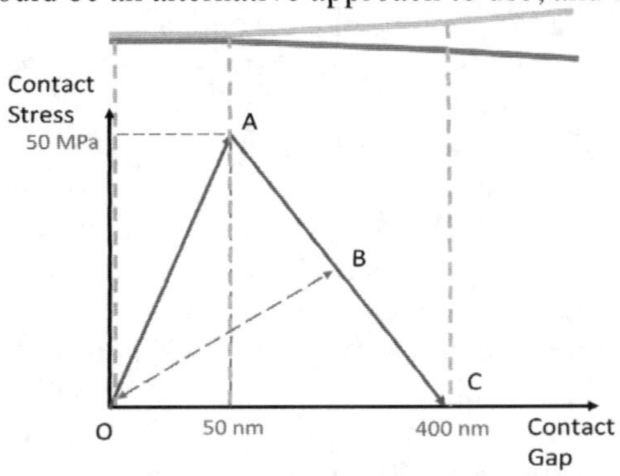

Figure 2. A demonstration of bilinear traction-separation law of Cohesive Zone Method applied in this work. For a REBCO tape, as load is applied, if the contact stress is below the critical values of 50 MPa, the loading is fully reversible. If the load exceeds the critical point A, the tape is irreversibly damaged, and the contact is softened. When unloading and reloading is applied, it will follow line OB, instead of OAB. When the contact gap reaches point C, the material is fully delaminated.

For REBCO tapes, four parameters are necessary to describe the delamination behaviour: maximum contact stress, fracture energy, and contact gaps at initiation and completion of delamination. Among those parameters, the fracture energy in both transverse tensile direction, 10 J/m^2, and shear direction, 20 J/m^2, are measured from experiments [1,10,13]. The delamination strength in transverse tensile direction measured in experiment varies from 10 to 100 MPa [7–9,14]. In this work, an average value of 50 MPa is taken as the failure criterion in transverse tensile direction, and 10 MPa is taken as the one in shear direction [11]. The contact gaps at initiation and completion of delamination in transverse tensile direction are 50 nm and 400 nm, and the interlayer sliding at initiation and completion of delamination are 50 nm and 100 nm respectively. Those parameters, gathered from literature, are listed in Table 2.

Table 2. Parameters used in the CZM layer in Model 1.

Parameter	Description	Value	Parameter	Description	Value
σ_{max}	Maximum normal tensile contact stress	50 MPa	τ_{max}	Maximum tangential contact stress	10 MPa
G_{cn}	Normal critical fracture energy	10 J/m^2	G_{ct}	Tangential critical fracture energy	20 J/m^2
$\bar{u}_n$	Contact gap at the maximum normal contact stress	50 nm	$\bar{u}_t$	Interlayer sliding at the maximum shear stress	50 nm
u_n^c	Contact gap at the completion of delamination	400 nm	u_t^c	Interlayer sliding at the completion of delamination	100 nm

2.2. Solid-Shell (SOLSH) element

In finite element analysis, a solid-shell (SOLSH) element is a 3D, 8-node structural element with all the capabilities and the topology of a solid element but it allows for layers with different material properties to be defined within it. The solid-shell element is useful for modelling laminates and high-aspect ratio structures. Previous work has shown that the stress-strain behavior in axial tensile direction obtained from solid-shell elements can reproduce very similar results to the ones obtained with experiments [12]. In addition, the use of a single solid-shell element through the thickness of the tape can reduce the computation time by more than a factor of 10 compared to using solid elements for each layer. Using solid-shell elements to represent the layered structure of the tapes opens the door to useful and accurate modeling of REBCO tape-stack in complex systems, such as CICC [12].

Since the properties of each layer (substrate, buffer, REBCO, etc.) are defined within each solid-shell element, it is not possible to model the delamination process in all its details, but this technique can provide insights on locations where delamination could occur, particularly for complex systems in which utilizing the CZM technique would be computationally prohibitive.

2.3. Loading conditions

Three loading conditions, including transverse tensile delamination, peeling delamination, and shear delamination, are studied for a 4 mm wide and 10 mm long REBCO tape and are shown in figure 3.

For all three loading conditions, the bottom surface of the REBCO tape is fixed, and displacement is applied onto a portion of the top surface. For transverse tensile delamination, transverse tensile displacement is applied onto a 2 mm by 2 mm square in the middle of the tape. Taking advantage of symmetries in the geometry, only a quarter of the tape is modelled, and symmetry boundary conditions are used. For peeling delamination, a small portion at the edge of the tape (1 mm by 4 mm) is peeled in a circular path around the peeling axis at the other edge of the tape. The peeling radius is the distance from a point in the peeling region to the peeling axis, for example, the peeling radius at the outer edge is 4 mm, and the one at inner edge is 3 mm. Since the load is applied in a circular path, the peeling force is always normal to the surface of the peeling region. The reduced mechanical strength at the edges of the tape was not considered in this study. For shear delamination, shear displacement is applied in the middle of the top surface of the tape in a 4 mm by 4 mm red square. The reduced mechanical strength at the edges was also not considered.

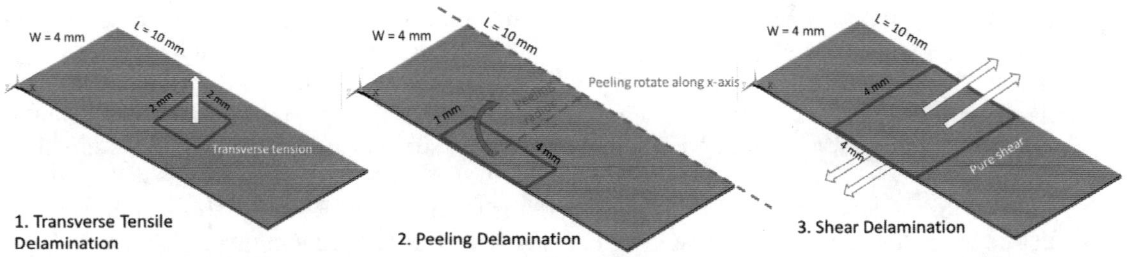

Figure 3. Three loading conditions studied in this work, including transverse tensile delamination, peeling delamination, and shear delamination. For each loading conditions, the three modelling approaches shown in figure 1 were considered.

3. Results

3.1. Transverse tensile delamination

For Model 1, the delamination process is described by the CZM layer, and the stress taken at the center of the loaded region as a function of applied transverse displacement is shown in figure 4. The vertical displacement in Model 1 is the sum of contact gap and layer deformation. Delamination initiates when the stress reaches 50 MPa and the contact gap reaches 50 nm, as expected from our definition of the CZM layer properties. As loading is applied, the layer deformation increases until the stress reaches 50 MPa. As the loading continues, the layer deformation goes back to zero when the loaded region is fully delaminated, and the vertical displacement is dominated by the contact gap behaviour. The vertical displacement and contact gap overlap when the transverse displacement reaches 400 nm due to the completion of the delamination as prescribed in the CZM layer. The non-zero stress observed at transverse displacement above 400 nm indicates the propagation of the delamination along the tape.

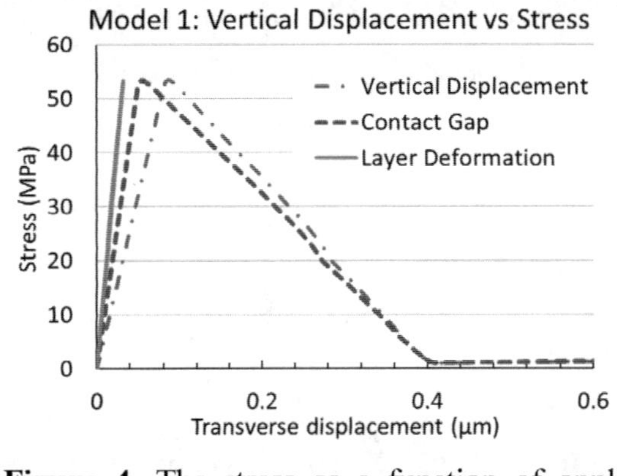

Figure 4. The stress as a function of applied transverse displacement for Model 1. The vertical displacement is the sum of contact gap and layer deformation. Delamination initiates when the stress reaches 50 MPa.

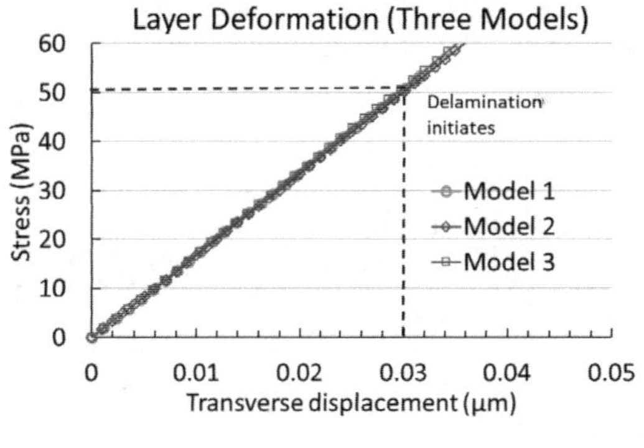

Figure 5. For Models 2 and 3, the vertical displacement is only the layer deformation. The stress as a function of layer deformation is very similar among the three models.

For Model 2 and 3, the transverse displacement is only caused by layer deformation (no physical separation is modelled). Comparing the layer deformation from all three models (figure 5) it can be observed that the stress as a function of layer deformation is very similar. In the same figure, the three models all reach a stress of 50 MPa when the delamination initiates as predicted in Model 1. The computation time for Models 1 and 2 is roughly 10 hours, while for Model 3, it is less than 20 minutes.

At the centre of the loaded region, the transverse tensile stress and strain distribution across the thickness of the tape of the three models are further compared in figure 6 when the stress reaches 50 MPa (initiation of delamination). The stress is about 50 MPa across the entire thickness of the tape for all three models, indicating the stress predicted by the Solid-Shell element agrees with Model 1 and 2 in which all individual layers are modelled. The strain distribution of Models 1 and 2 changes across the thickness of the tape as each layer has a different elastic modulus. For the Solid-Shell element (model 3), the layers are specified within one element only and therefore the strain is constant throughout the thickness (the element has nodes only at the top and bottom surfaces of the tape). The strain value is very similar to the one of the copper layer, since both the top and the bottom layers with nodes are copper. In conclusion, when the SOLSH element is adopted to model the REBCO tape, a failure criterion to predict delamination can be considered (in this example 50 MPa) reducing the complexity and the computational time needed when adopting the CZM model. Of course, if information regarding the contact gap is needed, the CZM model is the only effective method to capture the information.

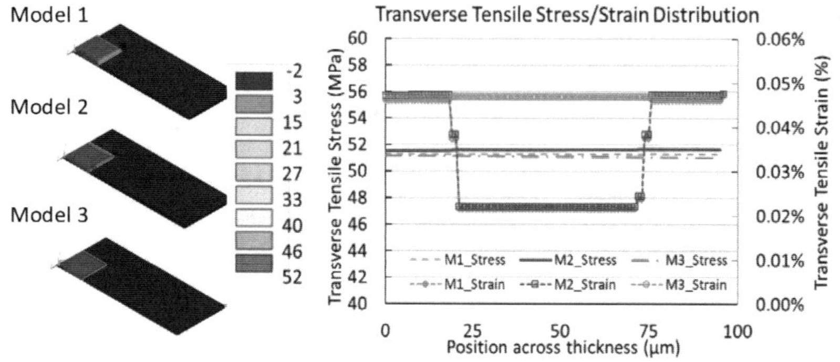

Figure 6. The transverse tensile stress distribution of the REBCO tape (contour plot), and the stress and strain distribution across the thickness at the centre of loading region are when delamination initiates as predicted in Model 1 (M1).

3.2. Peeling delamination

For peeling delamination, a small portion at the edge of the tape is peeled in a circular path according to the peeling axis at the other edge of the tape. Three locations at the peeling region are studied: the outer edge (A, peeling radius of 4 mm), the centre (B, peeling radius of 3.5 mm) and the inner edge (C, peeling radius of 3 mm) as shown in figure 7.

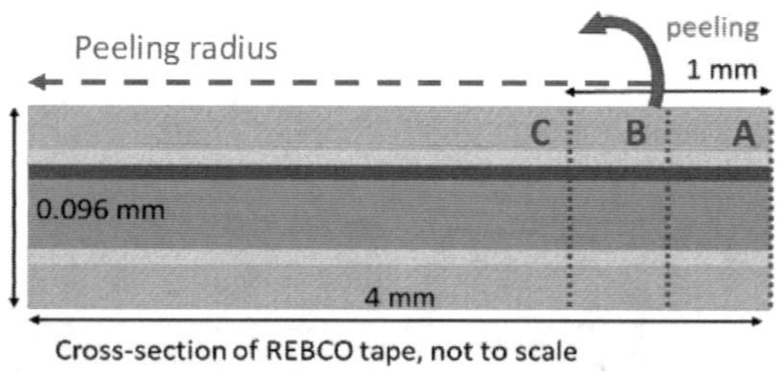

Figure 7. Three locations of the peeling region studied are shown in the cross section of the REBCO tape.

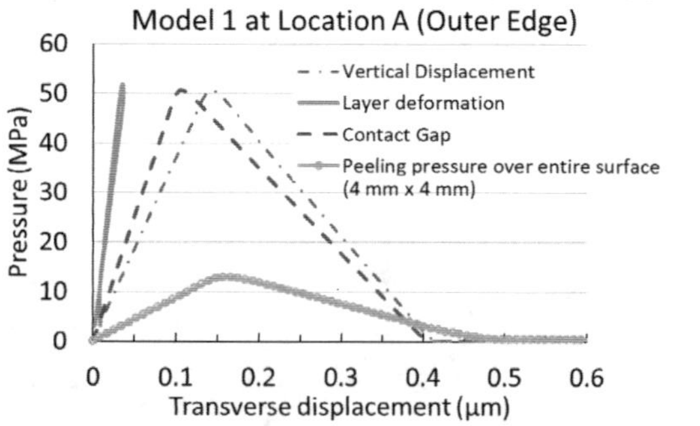

Figure 8. The stress as a function of applied transverse displacement for Model 1 at location A. Delamination initiates at a stress of ~ 50 MPa.

For Model 1, the peeling pressure and the stress of the elements at the outer edge of the tape (location A) versus transverse displacement is shown in figure 8. As in the previous loading case considered, the vertical displacement is the sum of layer deformation and contact gap. The peeling pressure is defined as the reaction force over the entire area of 4 mm by 4 mm (figure 3), and the stress is from the elements at location A. The delamination initiates when contact stress reaches 50 MPa. When the contact gap at the outer edge reaches 400 nm, the stress decreases to zero while the pressure remains a non-zero value up to 500 nm. For transverse displacement in a range from 400 to 500 nm, the outer edge is fully delaminated while the other part of the loading region is still in the delamination process.

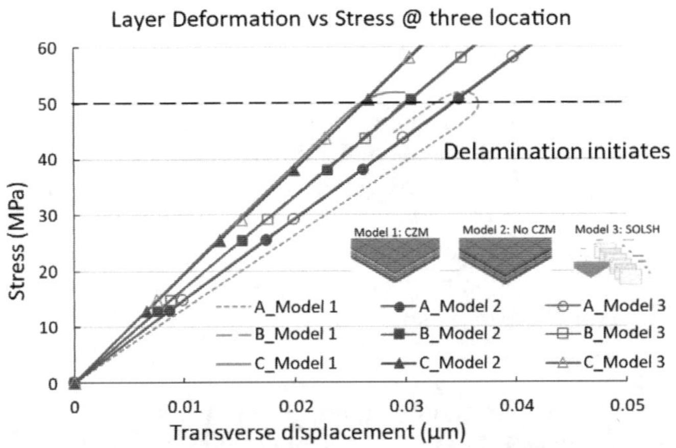

Figure 9. The comparison of the three models at three locations of the peeling region. Delamination initiates when stress reaches 50 MPa. The slopes of the stress-layer deformation curve at three locations are different. For a given stress with the same rotation angle, a higher peeling radius causes higher displacement therefore a lower slope is observed at outer edge (A). Three models show similar results at all three locations of the peeling region.

The stress vs layer deformation obtained from the three models are compared at all three locations (figure 9). As shown in the figure, delamination initiates when stress reaches 50 MPa, and the three models show similar results at all three locations of the peeling region. The slopes of the stress-layer deformation curve at three locations are different. The inner edge (C) has the highest slope, and the outer one (A) has the lowest. This is expected as, for a given stress and with the same rotation angle applied to the peeling region, a higher peeling radius causes higher displacement therefore a lower slope is observed at the outer edge of the peeling region (location A).

Since the peeling delamination in this study is dominated by the stress in the transverse tensile direction, the stress and strain distributions across the tape when delamination initiates are very similar to the ones shown in figure 7. As in the previous loading case, a transverse tensile stress of 50 MPa can be applied as a failure criterion and Model 3 can be effectively employed to predict delamination.

3.3. Shear delamination

For the case of shear load, delamination initiates when shear stress exceeds 11 MPa (figure 10). As per the previous loading conditions, the three models show similar results (figure 11).

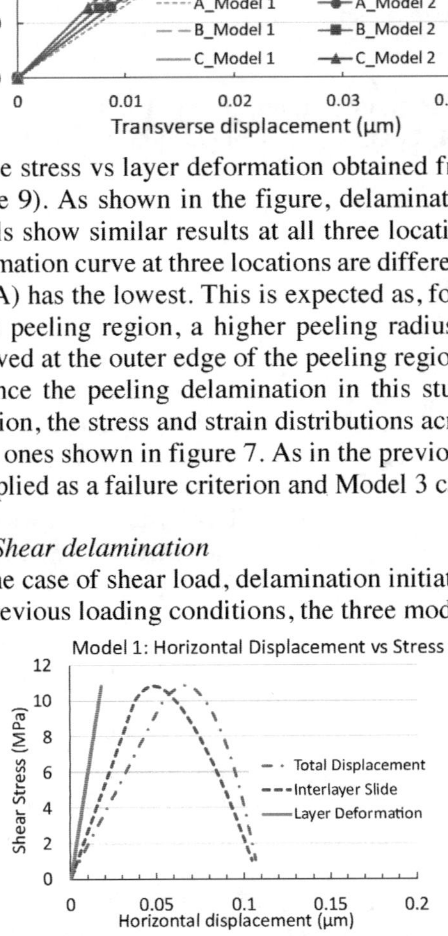

Figure 10. The shear stress as a function of applied transverse displacement for Model 1. Delamination initiates when shear stress reaches 11 MPa.

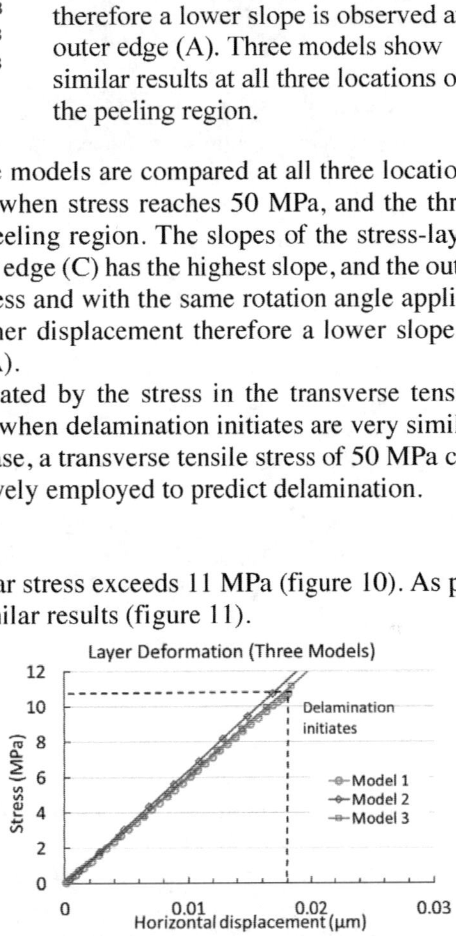

Figure 11. The three models show similar results up to the point when delamination initiates.

The shear stress and strain across the thickness at the centre of loading region was investigated, and the results show that the shear stress is about 11 MPa throughout the tape. As with the loading conditions considered earlier, a shear stress of 10 MPa could be applied as a failure criterion for Model 3 to identify possible delamination issues while reducing the complexity and computational effort of having to use a CZM layer.

4. Conclusion

To explore the options for modelling delamination with reduced computation efforts, three delamination modes of REBCO tape under transverse tensile, peeling and shear loads are modelled using three different modelling approaches. The results show that the stress state of REBCO in transverse tensile and shear directions can be accurately captured using a Solid-Shell element with specified layer structure to avoid the complex and computationally expensive CZM approach. In this work, the failure criteria used for the Solid-Shell element are taken from available experimental results (50 MPa in transverse tensile direction and 10 MPa in shear direction) but since the architecture of REBCO tapes (deposition method, material selection, layer thickness, etc.) varies significantly among manufacturers, modelling a specific tape configuration would require the knowledge of those values as part of the manufacturing specifications. The Solid-Shell element cannot capture the delamination process with exact details as with the CZM approach, but it provides critical information to predict and mitigate delamination and could provide a more efficient technique to analyse complex systems.

5. References

[1] Gao P, Chan W-K, Wang X and Schwartz J 2018 Mixed-dimensional modeling of delamination in rare earth-barium-copper-oxide coated conductors composed of laminated high-aspect-ratio thin films *Supercond. Sci. Technol.* **31** 074004

[2] Maeda H and Yanagisawa Y 2014 Recent Developments in High-Temperature Superconducting Magnet Technology (Review) *IEEE Trans. Appl. Supercond.* **24** 1–12

[3] Shin H-S and Gorospe A 2014 Characterization of transverse tensile stress response of critical current and delamination behaviour in GdBCO coated conductor tapes by anvil test *Supercond. Sci. Technol.* **27** 025001

[4] Gorospe A, Dedicatoria M J and Shin H-S 2015 Delamination Characteristics in Wide REBCO Coated Conductor Tapes Under Transverse Loading *IEEE Trans. Appl. Supercond.* **25** 1–5

[5] Yin S, Duranti M, Swenson C A, Li P, Ye L, Zhang X and Shen T 2020 Degradation of REBCO coated conductors due to a combination of epoxy impregnation, thermal cycles, and quench: Characteristics and a method of alleviation *Journal of Applied Physics* **128** 173903

[6] Kajita K, Iguchi S, Xu Y, Nawa M, Hamada M, Takao T, Nakagome H, Matsumoto S, Nishijima G, Suematsu H, Takahashi M and Yanagisawa Y 2016 Degradation of a REBCO Coil Due to Cleavage and Peeling Originating From an Electromagnetic Force *IEEE Trans. Appl. Supercond.* **26** 1–6

[7] Laan D C van der, Ekin J W, Clickner C C and Stauffer T C 2007 Delamination strength of YBCO coated conductors under transverse tensile stress *Supercond. Sci. Technol.* **20** 765–70

[8] Yanagisawa Y, Nakagome H, Takematsu T, Takao T, Sato N, Takahashi M and Maeda H 2011 Remarkable weakness against cleavage stress for YBCO-coated conductors and its effect on the YBCO coil performance *Physica C: Superconductivity and its Applications* **471** 480–5

[9] Dizon J R C, Gorospe A B and Shin H-S 2014 Numerical analysis of stress distribution in Cu-stabilized GdBCO CC tapes during anvil tests for the evaluation of transverse delamination strength *Supercond. Sci. Technol.* **27** 055023

[10] Sakai N, Lee S, Chikumoto N, Izumi T and Tanabe K 2011 Delamination behavior of Gd123 coated conductor fabricated by PLD *Physica C: Superconductivity and its Applications* **471** 1075–9

[11] Liu L, Zhu Y, Yang X, Qiu T and Zhao Y 2016 Delamination Properties of YBCO Tapes Under Shear Stress Along the Width Direction *IEEE Trans. Appl. Supercond.* **26** 1–6

[12] Allen N C, Chiesa L and Takayasu M 2016 Structural modeling of HTS tapes and cables *Cryogenics* **80** 405–18

[13] Miyazato T, Hojo M, Sugano M, Adachi T, Inoue Y, Shikimachi K, Hirano N and Nagaya S 2011 Mode I type delamination fracture toughness of YBCO coated conductor with additional Cu layer *Physica C: Superconductivity and its Applications* **471** 1071–4

[14] Majkic G, Galstyan E, Zhang Y and Selvamanickam V 2013 Investigation of Delamination Mechanisms in IBAD-MOCVD REBCO Coated Conductors *IEEE Transactions on Applied Superconductivity* **23** 6600205–6600205

Influence of cooling temperature on load-carrying performance of a radial HTS magnetic bearing

Xiang Guan[1, 2], Yincai Zou[3], Xing Bian[1], Jin Shang[1], Jihao Wu[1, 2] and Qing Li [1, 2*]

[1]State Key laboratory of Technologies in Space Cryogenic Propellants, Technical Institute of Physics and Chemistry, Chinese Academy of Sciences, Beijing, China
[2]University of Chinese Academy of Sciences, Beijing, China
[3]Southwest Institute of Technical Physics, Chengdu, China

E-mail: liqing@mail.ipc.ac.cn

Abstract. High temperature superconducting (HTS) magnetic bearing can operate in the harsh environment of low temperature and negative pressure with low loss and self-stability. Therefore, it has been gradually applied in flywheel energy storage, cryogenic liquid pump, cold compressor and some other cryogenic rotating machinery. The load-carrying performance of HTS magnetic bearing is affected by the cooling temperature of HTS directly. In this paper, the linear model is used to modify the temperature dependence to evaluate the influence of cooling temperature on load-carrying performance quantitatively, and the results are analyzed and displayed from the aspects of maximum levitation force, levitation stiffness and hysteresis effect. The research results of this paper have practical significance for the design and safe operation of HTS levitation system.

1. Introduction

High temperature superconducting (HTS) magnetic bearing can operate in the harsh environment of low temperature and negative pressure with low loss and self-stability. Therefore, it has been gradually applied in flywheel energy storage[1], cryogenic liquid pump[2], cold compressor[3] and some other cryogenic rotating machinery. At the same time, the HTS levitation system is a complex multi-physical field coupling system[4]. The cooling temperature of HTS will have a certain impact on its electromagnetic characteristics, and then it will affect the bearing capacity of the system. It is very important for the design and safe operation of HTS magnetic bearing to evaluate the influence of cooling temperature on load-carrying performance quantitatively.

Generally, the modeling and calculation of HTS levitation system is complicated, especially for the radial bearing with complex structure. Its complexity lies in solving the Maxwell equations including nonlinear electromagnetic constitutive relation of HTS and describing the changing external field of HTS under complex structure, which is equivalent to describing the relative motion of permanent magnet and HTS. Our team has proposed a simple and convenient method model for the calculation of axial force and radial force of radial HTS magnetic bearing[5, 6]. On the basis of these studies, this paper will further consider the temperature correction to evaluate the influence of cooling temperature on load-carrying performance of a radial bearing quantitatively.

2. Model establishment

2.1. Physical model

Our team has designed a radial HTS magnetic bearing[3]. The bearing adopts the form of internal permanent magnet rotor and external HTS stator. The permanent magnet rotor consists of alternating permanent magnetic rings and silicon steel sheets. The HTS stator is made of superconducting ring material produced by ATZ in Germany. Figure 1 shows the picture of permanent magnet rotor and HTS stator. The size of HTS stator is ID 50 mm × OD 80 mm × 60 mm. The size of the rotor is ID30 mm × OD48 mm × 70 mm, which consists of two permanent magnetic rings with the size of ID30 mm × OD48 mm × 2.8mm at both ends, nine permanent magnetic rings with the size of ID30 mm × OD48 mm × 5.6 mm in the middle, and several silicon steel sheets with the size of ID 30 mm × OD48 mm × 1.4 mm.

Figure 1. Picture of permanent magnet rotor and HTS stator.

2.2. Mathematical model

According to the theory of electromagnetic field, the governing equation of the solution domain is as follows when the displacement current can be ignored,

$$\nabla \times (\rho \nabla \times H) + \mu_0 \mu_r \frac{\partial H}{\partial t} = 0 \qquad (1)$$

Where, ρ and μ_r are the electrical conductivity and relative permeability of the materials in the corresponding domain. For the HTS domain, the relative permeability is 1. The conductivity is the virtual conductivity calculated according to the $E-J$ relationship. In this paper, the power exponent model is used to describe the nonlinear electromagnetic constitutive relation.

$$\rho_{sc} = \frac{E_c}{J_c} \left| \frac{J}{J_c} \right|^{n-1} \qquad (2)$$

Generally, the critical current density is only considered as a function of the magnetic flux density B. In this paper, we need to introduce temperature correction to evaluate the effect of cooling temperature. In the temperature correction model, the most commonly used is the linear model[4, 7-10], as shown in formula (3). It is worth noting that in the calculation of this paper, the heat exchange and diffusion of the HTS itself are not considered, and the temperature of the HTS is assumed to be uniformly distributed at each calculation temperature.

$$J_c(B,T) = J_{c0}\frac{B_0}{B_0+|B|}\frac{T_c-T}{T_c-T_0} \qquad (3)$$

Where, E_c is the critical electric field strength; T is the temperature; T_c is the critical temperature of HTS; T_0 is the temperature of liquid nitrogen; $J_c(B,T)$ is the critical current density; E_c, n, J_{c0}, B_0 are the characteristic parameters of HTS. Finally, we can get the electromagnetic force by integrating the HTS domain.

$$F=\iiint(J\times B)dV \qquad (4)$$

2.3. Solution method

In this paper, the Maxwell equations above (with the magnetic field intensity H as the independent variable) are solved with the help of COMSOL Multiphysics® software. It is also a common method for many scholars to solve the problem of HTS in recent years. The key of the solution process is to apply a variable magnetic field boundary to the HTS domain, which is equivalent to describing the relative motion of permanent magnet and HTS. The traditional method is to describe the external field of permanent magnet changing with time and space based on analytical formula[11, 12], or calculate the steady-state magnetic field distribution first by finite element method and then couple it with the magnetic field of HTS in two directions[13]. These methods have some limitations or disadvantages. In reference[14], the moving mesh in COMSOL Multiphysics® is proposed to solve this problem, which facilitates the modeling and solving of this kind of problem greatly. Generally speaking, this paper also adopts the method of moving mesh and solves the equations above based on H method.

The specific modeling methods we adopted are as follows. When solving the axial force of radial bearing, we establish a two-dimensional axisymmetric model. When solving the radial force of radial bearing, we establish a three-dimensional model. The displacement function is defined to describe the motion process of the permanent magnet rotor and control the deformation of the mesh. The specific setting method is shown in reference[5, 6]. The model is shown in figure 2.

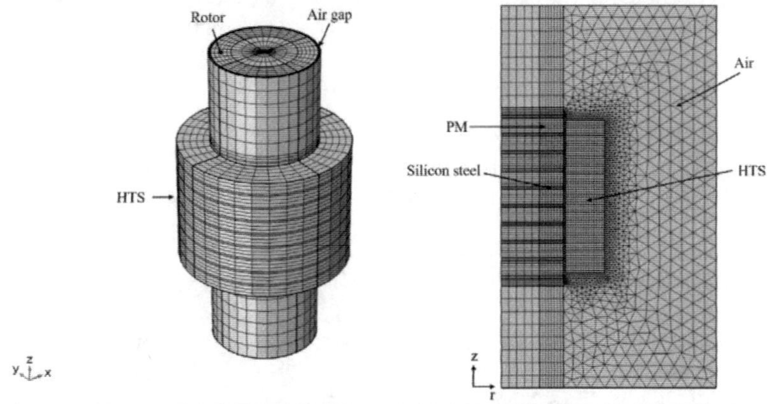

Figure 2. Modeling diagram.

3. Method validation

3.1. Verification of the levitation force calculation of radial HTS magnetic bearing

Our team has carried out some calculation and experimental research on the axial and radial force of radial HTS magnetic bearing. The specific parameters of the calculation model are introduced in reference[6]. The calculation and experimental results of radial force are also given in reference[6]. The comparison between the calculated and experimental results of axial force is shown in figure 3. In the actual working state, the displacement of the rotor of bearing is usually allowed to be in the range of tens of microns, so the calculation in this paper is limited to the range of small displacement. Within the acceptable engineering error range, we think that the calculated value of the levitation force is basically consistent with the experimental value.

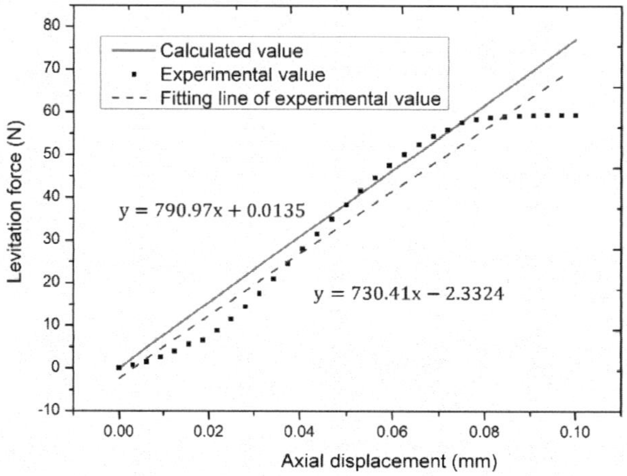

Figure 3. The comparison between the calculated and experimental results of axial force.

3.2. Verification of temperature correction model

In reference[15], the levitation force of a simple axial levitation system at different temperatures is measured experimentally. The relevant performance and experimental parameters are listed in table 1. According to the numerical method above, we simulated the experimental process. Figure 4 shows the comparison between the calculated and experimental values of the levitation force at the maximum displacement at different temperatures. In the calculation process, although some of the electromagnetic parameters are selected according to the empirical values, the calculated results are consistent with the experimental results generally. The results show that the linear temperature correction model can reflect the influence of temperature on the levitation force well within the acceptable range of engineering error.

Table 1. Performance and experimental parameters[15]

Parameter	Symbol	Value
Remanent magnetism of PM	B_r	1.05 T
Critical current criterion	E_c	$1e^{-4}$ V/m
Zero-field critical current density	J_{c0}	$2.4e^{8}$ A/m^2
Power law exponent	n	21
Critical magnetic flux density	B_0	0.37 T
Size of HTS		$\varnothing 30\times 14$ mm
Size of PM		$\varnothing 30\times 15$ mm
The displacement range of the experiment		150 mm-5 mm

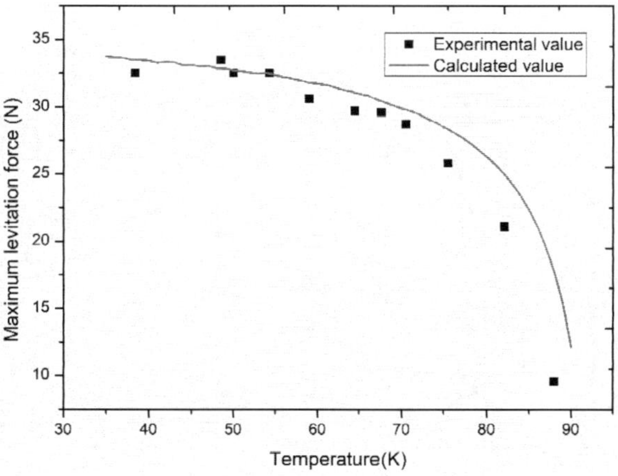

Figure 4. The comparison between the calculated and experimental values of the levitation force at the maximum displacement at different temperatures.

4. Results and discussion

On the basis of the method above, we can study the load-carrying performance of our radial HTS bearing at different temperatures quantitatively. Figure 5 and 6 show the variation curves of calculated axial force and calculated radial force under different cooling temperatures at a given stroke, which show similar laws. The value of levitation force reaches its maximum at the maximum displacement. The lower the cooling temperature is, the greater the maximum levitation force is. Taking the levitation force at 75K as the standard, we calculated the relative proportion of levitation force at different temperatures, as shown in figure 7. It can be seen that the levitation force changes sharply with temperature between 70-80K, but it changes slowly and tends to saturation below 60K. The saturation levitation force below 60K is about 1.2-1.5 times of that at 75K. In the temperature range above 80K, the levitation force decreases rapidly. This is also confirmed by the experiment in reference[16], although they are levitation systems with different sizes and structures.

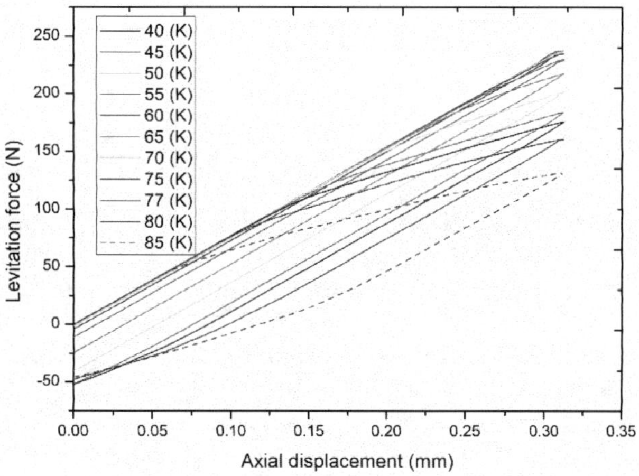

Figure 5. The variation curves of the calculated axial force of radial HTS bearing under different cooling temperatures.

IOP Conf. Series: Materials Science and Engineering 1241 (2022) 012033 doi:10.1088/1757-899X/1241/1/012033

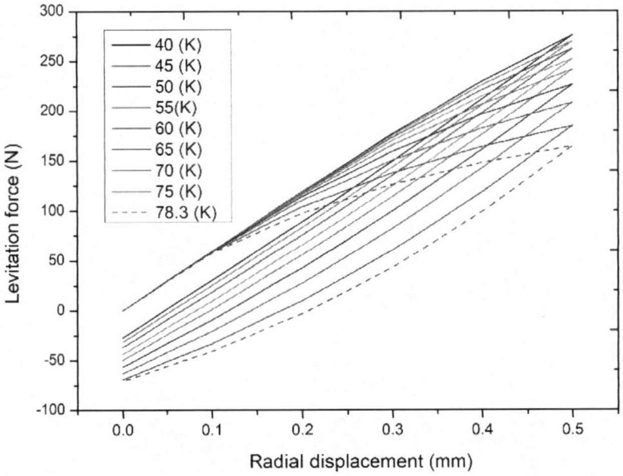

Figure 6. The variation curves of the calculated radial force of radial HTS bearing under different cooling temperatures.

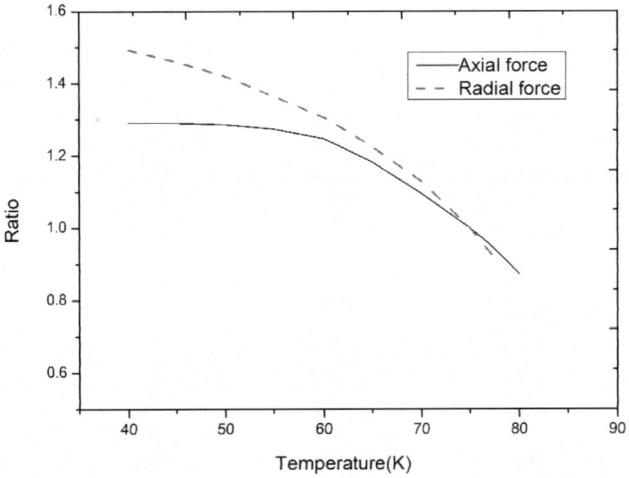

Figure 7. The calculated ratio of maximum levitation force at different temperatures.

In addition to the maximum levitation force, cooling temperature also causes the stiffness difference of levitation force. But the difference is not obvious in the small displacement range. For example, in the range of radial displacement within 0.1 mm and axial displacement within 0.05 mm, the difference of levitation force is not obvious. With the increase of displacement range, the nonlinearity of levitation force increases obviously, and the stiffness difference caused by temperature is more obvious.

Finally, it should be noted that the cooling temperature also affects the hysteresis effect of the levitation force. We calculated the hysteresis area of the reciprocating levitation force, and the results show that the higher the cooling temperature is, the larger the hysteresis area is, as shown in figure 8. It is concluded that the higher the cooling temperature is, the more obvious the hysteresis effect of the levitation force is, and the larger the corresponding loss is.

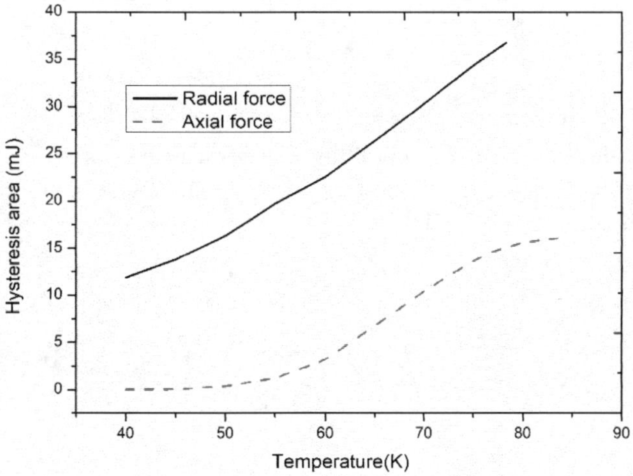

Figure 8. The hysteresis area of the reciprocating levitation force at different temperatures.

5. Conclusion

The influence of cooling temperature on the levitation force can be well described by using the linear model to modify the critical current density. It also shows that the influence of cooling temperature on the levitation force is mainly due to the influence of temperature on the critical current density.

The cooling temperature of HTS will affect the radial and axial load-carrying performance of radial bearing, which can be reflected in the maximum levitation force, levitation stiffness, hysteresis effect and so on. When the cooling temperature is lower than 60K, the bearing capacity tends to be saturated, which is about 1.2-1.5 times of that at 75K. When the cooling temperature is higher than 80K, the levitation force decreases rapidly. The lower the cooling temperature, the greater the levitation force when the same displacement occurs and the levitation stiffness will be. However, the stiffness difference caused by the temperature is not obvious in the small displacement range, but it becomes obvious beyond a certain displacement range. At the same time, the hysteresis area of reciprocating levitation force increases with the increase of cooling temperature.

6. References

[1] Miyazaki Y *et al* 2016 *Cryogenics* **80** 234-7
[2] Lin Q, Jiang D, Deng Z, Ma G, Zheng J, Wang W, Shin D, Gu X, Lin N and Shao M 2015 *J. Low Temp. Phys.* **180** 416-24
[3] Zou Y, Shang J, Guan X, Bian X, Wu J and Li Q 2020 *Cryogenics* **106** 103047
[4] Yang W, Ma G, Queval L, Li G, Ye C and Deng Y 2019 *Chinese Science Bulletin* **64** 3255-66
[5] Guan X, Zou Y, Shang J, Bian X, Wu J and Li Q 2020 *Physica C* **578** 1353752
[6] Guan X, Zou Y, Shang J, Zheng L, Bian X, Wu J and Li Q 2021 *Cryogenics* **116** 103284
[7] Alloui L, Ben Alia K, Bouillault F, Mimoune SM, Bernard L and Lévêque J 2013 *Physica C* **487** 1-10
[8] Ma GT and Rauh H 2013 *Supercond. Sci. Technol.* **26** 105001
[9] Huang Y, Zhang X and Zhou Y-H 2016 *Supercond. Sci. Technol.* **29** 075009
[10] Alloui L, Bouillault F and Mimoune S 2009 *Eur Phys J Appl Phys* **45** 20801
[11] Sass F, Sotelo GG, Junior RdA and Sirois F 2015 *Supercond. Sci. Technol.* **28** 125012
[12] Naseh M and Heydari H 2016 *IET Electric Power Applications* **11 369-77**
[13] Quéval L, Liu K, Yang W, Zermeño VMR and Ma G 2018 *Supercond. Sci. Technol.* **31** 084001
[14] Grilli F, Morandi A, Silvestri FD and Brambilla R 2018 *Supercond. Sci. Technol.* **31** 125003

[15] Jiang H, Wang J, Wang S, Ren Z, Zhu M, Wang X and Shen Xu 2002 *Physica C* **378-381** 869-72
[16] Morandi A, Fabbri M, Ribani PL, Dennis A, Durrell J, Shi Y and Cardwell D 2018 *IEEE Trans. Appl. Supercond.***28** 1-10

Acknowledgments
This work is financially supported by the fund of the State Key Laboratory of Technologies in Space Cryogenic Propellants, SKLTSCP1902, and the fund of National Research and Development Project for Key Scientific Instruments, ZDYZ2014-1.

ASCEND: The first step towards cryogenic electric propulsion

L Ybanez[1], A Colle[1], E Nilsson[1], F Berg[2], G Galla[1], M Tassisto[1], J Rivenc[3], F Kapaun[4] and G Steiner[4]
[1] Airbus UpNext/Airbus ExO Zero Emissions, Blagnac, France
[2] Airbus X-labs, Taufkirchen, Germany
[3] Airbus CRT, Blagnac, France
[4] Airbus CRT, Taufkirchen, Germany

Ludovic.ybanez@airbus.com

Abstract. If a cold source such as liquid H_2 or cryogenic fuel is available, cryogenic and high-temperature superconductivity technologies are promising to significantly increase performance of electric propulsion systems. A first study for aeronautic applications shows that the power density of electrical components could be multiplied 2 to 3 times to reach more than 30kW/kg for electric motors and power electronics, the weight of cables can be considerably reduced, and the efficiency increased by more than 50% with a voltage between below 500 V for multi-megawatt applications. With ASCEND, AIRBUS intends to demonstrate the potential and feasibility of a cryogenic and superconducting powertrain to breakthrough aircraft electric propulsion performance.

1. Introduction

The aeronautics industry is anticipating deep changes and new challenges in the near future. It is expected that air traffic will continue to grow in the next decades even taking into account the impact of COVID-19, social acceptability and long-term roadmaps such as ACARE (Advisory Council for Aeronautics Research in Europe) Flightpath 2050. The SRIA (Strategic Research and Innovation Agenda) set specific goals such as a 75% reduction of CO_2 emissions and 90% reduction of Nitrogen Oxides (NO_x) as a result of ecological and sustainable development obligations [1].

Electrified aircraft can significantly contribute to ACARE objectives [2] and ambitious projects have been launched during the past years, to demonstrate the feasibility, and to accelerate the development of fully electric or hybrid electric propulsion systems with "conventional" technologies. For high power applications, the challenge is even greater for current electric technologies to be competitive with combination of kerosene and turbofan. This is especially true for longer range needs where the propulsive power is several tens of megawatts and a large amount of energy must be carried for the long flight in the form of chemical fuel.

A promising avenue to replace conventional fuel, thanks to its high specific energy, is liquid hydrogen; whether to "burn" it in a turbofan or to use it in a fuel cell to generate electrical power. A lot of work is currently underway on fuel cell or H_2-burn propulsion but it remains essential to develop an ultra-efficient electric or hybrid propulsion system for the future aircraft. If a cryogenic source like liquid hydrogen is available on board, it can be used in two ways to increase performance of electrical components:

IOP Conf. Series: Materials Science and Engineering 1241 (2022) 012034 doi:10.1088/1757-899X/1241/1/012034

- Cryogenic technologies: by cooling "conventional" technologies and dramatically increasing electrical propulsion system performance by reducing losses (because resistivity and losses depend on system temperature) and therefore the weight of components.
- Superconducting technologies: by using high-temperature superconducting technologies for DC (direct current) distribution components, electric motors to increase performance because of no resistivity, no losses and magnetic behavior of superconducting materials.

Their potential and maturity are now demonstrated in several applications (MRI, ground distribution of electricity, scientific equipment) and promise a profound revolution in the field of high power electrical engineering in the years to come [3]. It is now essential to accelerate the adaptation of these technologies, to demonstrate their potential and feasibility with the main European institutes, SME and large groups and mature them with a flying demonstrator to be able to propose TRL6 at powertrain level before 2030 for future generations of "greener" aircraft.

Four main advantages are targeted:

- Increase the power-to-weight ratio by a factor of 2 to 3
- Enable lower voltage because of high current capabilities to reduce safety and technological issues
- Improve efficiency thanks to a decrease of Joule losses, leading to a powertrain efficiency greater than 97%, reducing weight and volume of fuel
- Propose new technologies based on superconductivity such as Fault Current Limiter to reduce the short circuit current and the oversizing of distribution network

The potential and maturity of superconducting and cryogenic technologies have been assessed but Airbus internal projects driven by CRT (Central Research and Technology), with partners and are also demonstrated on several ground applications (MRI, ground distribution of electricity, scientific equipment...). This shows a potential deep revolution in the field of high power electric engineering in the years to come.

To demonstrate the potential at powertrain and aircraft level, the aim of ASCEND (Advanced Superconducting and Cryogenic Experimental powertraiN Demonstrator) over the next three years is to develop and to test a first demonstrator detailed on Figure 1:

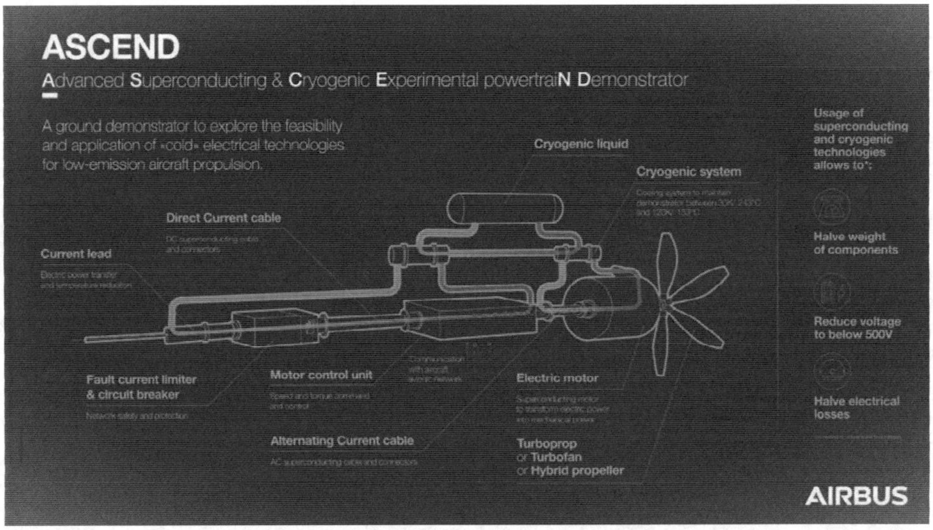

Figure 1. ASCEND demonstrator

The main objectives can be detailed as:

- Demonstrate the feasibility of cryogenic and superconducting technologies for aircraft applications with a voltage below 500 V for multi-megawatt applications

IOP Conf. Series: Materials Science and Engineering 1241 (2022) 012034 doi:10.1088/1757-899X/1241/1/012034

- Demonstrate that the propulsion system (from DC bus to shaft) can be more than two times lighter than conventional technologies and can reach 98% efficiency
- Give recommendations on the potential of these technologies on the Airbus product range and more particularly on regional and long range aircraft
- Strengthen the partnership with the best industrial and academic partners in the world and animate the ecosystem

2. Powertrain modelling and scenarios

To be able to specify detailed technical and safety requirements, to be able to assess the interest of these technologies in future aircraft applications, it is essential to improve our system modelling capabilities. Work on network modelling and simulation involving superconductors has so far been limited to individual superconducting components in a larger conventional system, e.g. [4],[5],[6]. Published modelling work for a network of connected superconducting and cryogenic components is not known to the authors. Superconducting electrical components have special behavior compared to conventional ones and thus cannot be presumed to be possible to model using conventional electrical simulation tools and standard component libraries. Superconducting components exhibit close coupling between thermal and electrical behaviors and nonlinear behavior associated with quench. Normal electrical component models often cannot accept a zero resistance value. Other impedance effects play a heightened role in the electrical behavior of the system, as well as any resistive components that make up part of the circuit. For normal electrical system modelling to be available to us in future projects, we must use the opportunity of this project to develop and validate the tools needed. This work will help to enable engineering tasks on future projects, such as:

- Short circuit analysis
- Protective equipment specification
- Motor control development
- Voltage and current specification
- Cooling system sizing
- Cooling system control development

To progress with the modelling work, all project partners contributing components to the ASCEND demonstrator will also provide validated models according to an agreed framework. The models will be integrated into a system simulation in Matlab Simulink/Simscape as illustrated in the placeholder sketch in Figure 2. The software was chosen because of its wide acceptance among the project partners and because it is increasingly a standard industrial tool for electrical, as well as multiphysics simulations and controls development.

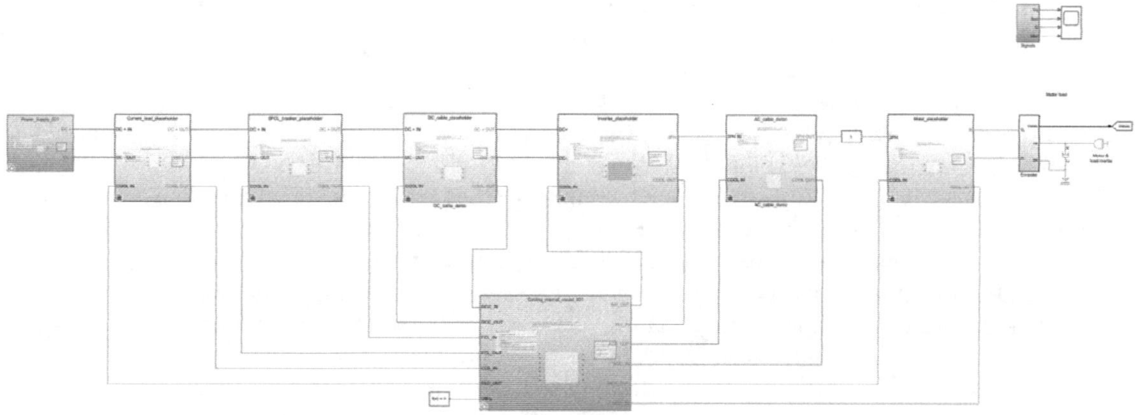

Figure 2. Placeholder model in Matlab Simulink

An important aspect of the simulation will be the use of the correct fidelity of models for the type of analysis being performed. For example, short circuit analysis requires more detailed electrical models for very short time-frame simulations (ms) and little detail in the cooling system, whereas cooling topics require simulations that run at or faster than real-time and can tolerate lower accuracy in the electrical behaviour, provided that thermal loads are well represented. In some cases, the component models must be made available in different levels of detail to accommodate these needs. The aim in the longer term is to have a library of parameterisable components that can be deployed to analyse different propulsive architectures with a known degree of confidence.

Additionally, it is our aim to build on our understanding of the application of superconducting and cryogenic technology in aircraft. To understand the impacts and interests in terms of mass, volume, efficiency and voltage level for different applications, architectural models are required to trade off the complex decisions. Simplified behavioral models, combined with sizing and mass estimation models informed by the technology bricks being developed will allow architectural trades to be carried out. This will permit us to assess the applicability of superconducting propulsion systems to different market segments and to compare the superconducting option with conventional technologies with or without LH_2 on board.

3. DC and AC distribution

A direct current distribution network is used in a propulsion chain to transfer energy from the power sources to an electrical converter, which transforms this energy into an alternating voltage/current to drive the superconducting motor. The functions fulfilled by this link are therefore energy transport functions but also protection of the network against faults such as a short circuit, arcing or other to ensure the safety and reliability of the aircraft. The ASCEND's DC link will transport power at <500 V over 10 m. The DC link is connected to an electrical converter, which transforms this energy into an alternating voltage/current to drive the superconducting motor.

However, the fault management of such a direct current distribution network is much more challenging than that of an alternating current network because of the absence of the zero-crossing of the current. This is even more challenging for a superconducting and cryogenic powertrain because the electrical system has an extremely low impedance. It is therefore essential to develop a high-power density, high-efficiency, fast-operating circuit breaker (CB) with fault current limiter and fault detection capability. The current in the DC link of the ASCEND demonstrator will be limited by a superconducting fault current limiter.

Because the DC link has a significant impact on the total mass and safety of a propulsion chain, superconducting technologies, which are mature for terrestrial applications, will allow a significant gain over the conventional technologies currently in use. The DC link in the ASCEND demonstrator will consist of a compact, flexible lightweight HTS conductor with plus and minus pole embedded in a common cryostat. The target weight for the cable (two poles) and cryostat of the ASCEND demonstrator is <2.5 kg/m. Another advantage of the superconducting distribution is the relatively small cross sectional area of the system.

The power supply will be connected to the superconductor through a current lead, optimized to carry the high current from ambient temperature to the LN_2 environment with maximal temperature of 80 K. This current leads will be dimensioned to minimize the heat leak from the warm to the cold environment [22]. Even though the current leads can be made rather compact, they will be of significant weight since they are typically made of copper. The DC distribution has also to contain connecting devices from the FCL and at the input of the power electronics. Even with these components at the extremities of the cable, the superconducting cable remains very light and the benefit of using a superconducting distribution increases with the length and the power of the distribution system.

As opposed to the DC link, the extremities of the AC link are not connected to the ambient environment. They are kept in a superconducting temperature range: at the temperature of motor control unit output in one side and at <50 K at the electric motor side.

The AC distribution of the ASCEND demonstrator consists of a 2 m long three-phase coaxial superconducting cable.

While superconducting is characterized by zero resistance in DC, the variation of AC current amplifies electromagnetic phenomena known with traditional technologies and generates losses specific to superconductivity directly related to the frequency of the current. It is essential to understand and control these effects. The high frequency of the system requires optimization of the cable layout to reduce the AC losses as much as possible.

4. Power electronics

The superconducting motor controlled by cryogenic power electronics is essential to reach the targeted performance for two main reasons:

- Increase the performance thanks to cryogenic technology
- Maintain low temperature thermal continuity between the superconducting DC distribution and the superconducting electric motors to reduce thermal losses.

Existing research has already demonstrated the feasibility and potential of cryogenic power electronics. This information is further reinforced by the results of the latest paper ([8], [9],[10],[11]) according to which the main power components (power modules, bus bars, passive components...) have better performance at lower temperatures, thus increasing the overall efficiency and power density of cryogenic power converters.

According to [8], GaN technologies present several benefits in cryogenic conditions:

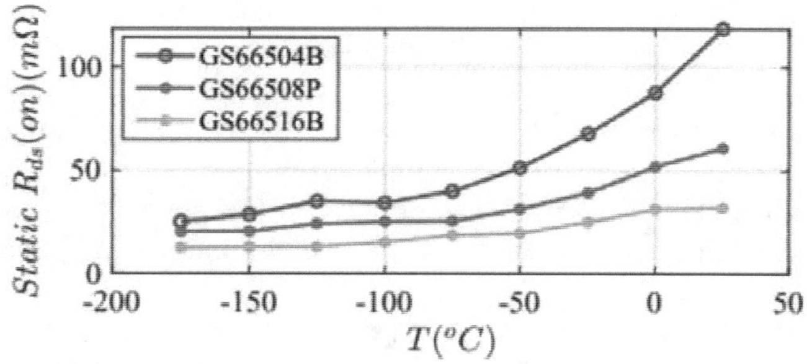

Figure 3. RDS(on) of several GaN in function of ambient temperature

A decrease by factor of 2 to 5 of RDS(on) is noticed at 77 K compared to 298K. Consequently the conduction loss also decreased by the same factor.

For the switching loss on the contrary, an increase between 25% and 50% is present. In [8], tests ran on GaN devices (GS66508P, GS66504B):

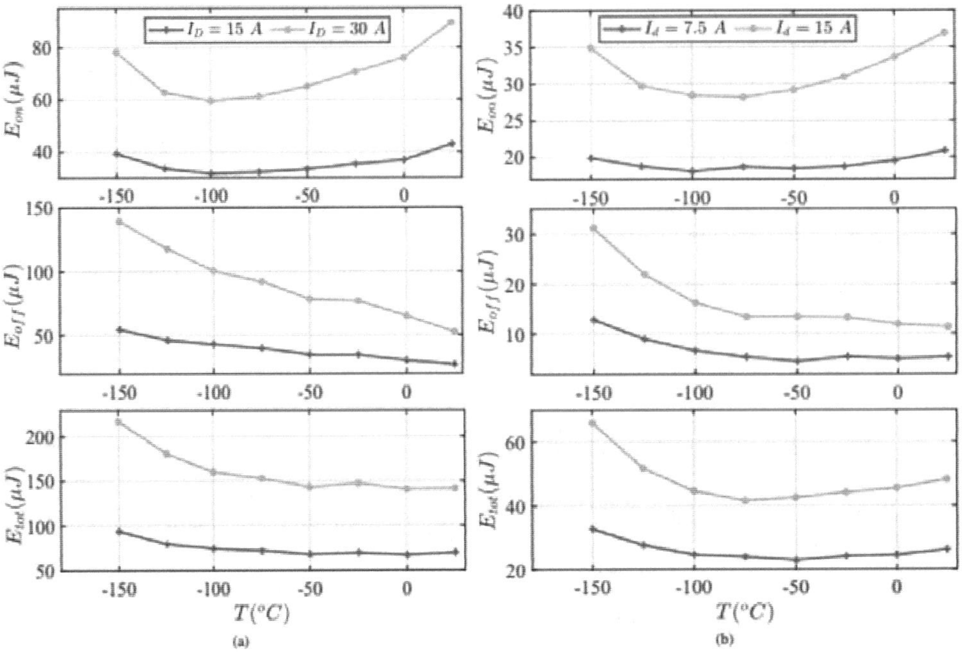

Figure 4. Turn-off, and total switching energy loses versus temperature for (a) GS66508P and (b) GS66504B

Despite the slight increase of switching loss, the conduction loss remains dominant. GaN devices show very promising performance in cryogenic conditions.

To get a complete picture, [11] focuses also on passive components such as inductors in cryogenic conditions. A nanocrystalline core combines gain of weight and low power losses compared to other core materials (amorphous, ferrite).

These studies give encouraging results to build power electronics converters at cryogenic conditions. Still several questions remain open concerning thermo-mechanical stress on the components, reliability and integration of the global system.

The targeted power range can be achieved under ambient temperature conditions with voltages up to 1200 V but in a cryogenic environment with low voltages and very high currents is more difficult because there are no standard "commercial" components characterized for this temperature range.
Currently, there is no demonstrator at the targeted power level because no application of superconductivity requires an increase in power density for converters as required for aeronautical applications.

5. Superconducting electric motor

5.1. Background

High power electrical propulsion will require electrical machines with power densities above 20 kW/kg and efficiency above 99% [12]. It is not possible to achieve such performance with conventional technologies therefore disruptive technologies will be required to achieve these values. In this context, superconducting machines have been identified as a potential solution, and therefore represent a key element of future high-power electric propulsion for aircraft applications. To reach these goals, the majority of existing superconducting machine can be classified in two groups.

5.2. *Partially superconducting motor*

Only the conductor of the rotor is made with superconductors. It corresponds to the first type of large scale machine manufactured with a high temperature superconductor [14],[15]. The biggest advantage is the use of superconducting materials with direct current leading to an absence of losses. This configuration reduces to the maximum the power/mass/volume of the cryogenic system required. The order of magnitude of the cryogenic losses is 100 W (at cold temperature) mainly due to the radiation, conduction through the torque tube and thermal losses through the current lead [15]. The choice of the operating temperature (below 77K), has not a significant impact on the losses (proportional to ΔT or the resistivity).

However, the complexity leads to extracting the losses from a rotating part. In the literature, three configurations are popular:

- Conduction with a rotating cryocooler [16]: no cryogenic seal is needed with this architecture, however high electrical power to supply the cryocooler should be provided from a static part to a rotating one. Also, the cryocooler should be designed to tolerate the vibration or other constraints when the motor is operated.
- Conduction with a thermosiphon [17]: conventionally used with liquid neon to cool the superconductors around 30K. The system was already tested on a system with rotation speed over 1krpm.
- Convection with helium loop [18]: the use of helium of gas offers a large range for the operating temperature, compared to the thermosiphon solution. The cryocooler is placed outside the motor's cryostat and static, so some cryogenic seals are needed to guide the gas to the rotating part.

Globally, a rotating cryogenic cooling is still a topic that needs to be matured, otherwise all benefits by using superconductivity on the power density improvement can be vanished.

Recently, some devices called "flux pumps" show real interest in the possibility of supplying rotating superconducting coils in a brushless way [19]. These promising devices for wound rotor superconducting machines use the particular electrical behaviour of the superconductors.

5.3. *Fully superconducting motor*

The rotor and stator are made with superconductors. The superconductors (on the stator side) are supplied by an alternating current and exposed by the rotating field of the rotor leading to transport and magnetisation losses respectively. This modification of the superconductor's magnetic state (compared to a DC electric environment) will generate some losses. The interest in this configuration is growing recently with the development of higher performance superconducting wires. For a given power, the higher the critical current of the wire (maximal allowable current before losing the superconductivity), the lower the losses will be. This leads to a reduction of the required cryogenic cooling system. The order of magnitude of the cryogenic losses of this class of motors is 1 kW (at cold temperature). Considering this configuration, some machines were designed well beyond the requirement of the electric propulsion system [20].

6. Cryogenic system

The cryogenic cooling system is necessary to maintain the powertrain at the nominal working temperature, to extract thermal losses from the electrical system and to manage faults that could occur during the operations of the powertrain.

At aircraft level two main possible cooling architectures could in principle been envisaged:

- In case a reservoir of LH_2 is embedded on board, it could be used as a "cold source" , to cool-down, directly or indirectly, the components of the powertrain
- In case LH_2 is not available on board, or if it not considered for any other use than propulsion, dedicated cooling machines shall be embedded

For the ASCEND demonstrator it has been assessed not to use liquid hydrogen as a coolant for two main reasons, i.e. to reduce the risks at demonstrator level and to explore alternative cooling methods.

In any case, in the framework of the project, liquid Hydrogen technologies are studied and assessed to explore their potential.

The design of the ASCEND cooling system has been focused not only on the efficiency, but also on the reliability and complexity balance.

ASCEND demonstrator is designed to be cooled using inert fluids, with technologies suitable for aircraft applications. The different components of the demonstrator require different working temperatures and coolants, so the cooling system is designed to cope with these requests.
In particular for:

- DC and AC distribution
- The Motor Drive
- The Electrical motor

For DC/AC distribution and for the electrical motor, a solution based on Turbo Brayton reverse cycle has been considered the best trade off taking into account the efficiency, reliability and scalability for future aircraft applications.

As stated in [13], for these kind of machine, the goal of the development is to reach a specific mass less than 3 kg/kW with efficiency of about 30 % of Carnot. Surveys show that most high efficiency cryocoolers have efficiencies of about 20 % of Carnot and specific masses of about 10 to 30 kg/kW.

Preliminary studies show that these requirements can be met by Turbo Brayton cooling machines featuring a cold capacity of the order of 1kW, as requested by ASCEND demonstrator.

7. Conclusion

Connecting all the detailed components (DC/AC distribution, cryogenic power electronic, superconducting motor, cryogenic system) in a powertrain of 500kW, observing their interactions and achieving stable operation in 3 years, is a big challenge and will be a world first; an essential step towards future full scale tests and flying demonstrators.
In addition to the hardware demonstrator itself, this demonstration will make it possible to increase the maturity of these technological bricks and the cooling options with or without liquid hydrogen. It will also make it possible, thanks to the development of thermo-electrical behavioural models and architectural trade models, to evaluate the contribution of these technologies and give strong recommendations to the whole range of future Airbus products.

8. References

[1] https://www.acare4europe.org
[2] P. J. Ansell and K. S. Haran, "Electrified Airplanes: A Path to Zero-Emission Air Travel," in IEEE Electrification Magazine, vol. 8, no. 2, pp. 18-26, June 2020
[3] Peter Cheetham, Srikar Telikapalli, Chul Han Kim and Sastry Pamidi, "Superconducting Power Devices for the Electrification of Transportation," Chapter 1 in A Comprehensive Guide to Superconductivity, Editor Rohan Morrow, (pp. 1-62), Nova Science Publishers, Inc., New York, USA, 2020.
[4] Venuturumilli S, Berg F, Zhang M, Yuan W, "Investigation of HTS cable impact on turbo-electric aircraft performance", IET Electrical Systems in Transportation, 2020.
[5] Haugan T, Sebastian M, Kovacs K, Sumption M and Tsao B-H . "Design and Scaling Laws of a 40-MW-class Electric Power Distribution System for Liquid-H2 Fuel-Cell Propulsion," AIAA 2021-3310. AIAA Propulsion and Energy 2021 Forum. August 2021.
[6] Berg F, Palmer J, Miller P and Dodds G, "HTS System and Component Targets for a Distributed Aircraft Propulsion System," in IEEE Transactions on Applied Superconductivity, vol. 27, no. 4, pp. 1-7, June 2017.
[7] Jones C E, Norman P J, Galloway S J, Burt G M, Armstrong M and Bollman A., "A pre-design sensitivity analysis tool for consideration of full-electric aircraft propulsion electrical power system architectures," 2015 International Conference on Electrical Systems for Aircraft, Railway, Ship Propulsion and Road Vehicles (ESARS), 2015, pp. 1-6.

[8] Graber L,Yin S, Mehrabankhomartash M, Cruz A.J., Saeedifard M, Evans S, Kapaun F, Revel I, Steiner G, Ybanez L, Park C "Static and Dynamic Characterization of 650 V GaN E-HEMTs in Room and Cryogenic Environments ", ECCE 2021

[9] Gui, H., et al., "Review of Power Electronics Components at Cryogenic Temperatures," IEEE Trans. Power Electron., vol. 35, no. 5, pp. 5144–5156, May 2020

[10] Mhiesan, H., M. M. Hossain, A. U. Rashid, Y. Wei, and A. Mantooth, "Survey of Cryogenic Power Electronics for Hybrid Electric Aircraft Applications," in 2020 IEEE Aerospace Conference, 2020, pp. 1–7,

[11] Graber L,Yin S, Mehrabankhomartash M, Cruz A.J., Saeedifard M, Evans S, Kapaun F, Revel I, Steiner G, Ybanez L, Park C "Characterization of Inductor Magnetic Cores for Cryogenic Applications", ECCE 2021

[12] Kiruba S Haran et al 2017 Supercond. Sci. Technol. 30 123002

[13] Ray Radebaugh "Cryocoolers for aircraft superconducting generators and motors"AIP Conference Proceedings 1434, 171 (2012)

[14] G. Snitchler, B. Gamble and S. S. Kalsi, « The performance of a 5 MW high temperature superconductor ship propulsion motor », IEEE Transactions on Applied Superconductivity, vol. 15, n° 2, p. 2206–2209, 2005.

[15] B. B. Gamble, S. Kalsi, G. Snitchler, D. Madura, et R. Howard, « The status of HTS motors », in IEEE Power Engineering Society Summer Meeting, vol. 1, p. 270–274, 2002.

[16] T. Yanamoto, M. Izumi, Y. Murase and T. Oryu, "Loss Analysis of a 3-MW High-Temperature Superconducting Ship Propulsion Motor," IEEE Transactions on Applied Superconductivity, vol. 28, pp. 1-5, 2018.

[17] R. H. Jansen, Y. De Jesus-Arce, P. Kascak, R. Dyson, A. Woodworth, J. J. Sheidler, R. Edwards, E. Stalcup, J. Whilhite, K. Duffy, P. Passe and S. McCormick, "High Efficiency Megawatt Motor Conceptual Design," in AIAA Propulsion and Energy Forum, Cincinnati, 2018.

[18] M. Frank et al., "Thermosyphon cooling system for the Siemens 400 kW HTS synchronous machine," presented at the Cryogenic Engineering Conf., Anchorage, AK, 2003.

[19] A. Colle, T. Lubin, S. Ayat, O. Gosselin, and J. Leveque, 'Test of a Flux Modulation Superconducting Machine for Aircraft', J. Phys.: Conf. Ser., vol. 1590, p. 012052, 2020.

[20] J. H. Kim et al., "Fabrication and performance testing of a 1-kW-high-temperature superconducting generator with a high-temperature superconducting contactless field exciter," Supercond. Sci. Technol., vol. 33, Jul. 2020.

[21] M. Filipenko et al, "Concept design of a high power superconducting generator for future hybrid-electric aircraft," Superconductor Science and Technology, vol. 33, no. 5, 2020

[22] R. McFee, "Optimum Input Leads for Cryogenic Apparatus", Review of Scientific Instruments 30, 98, 1959.

Electric Aircraft Fueled by Liquid Hydrogen and Liquefied Natural Gas

Srikar Telikapalli[1,2], Roberto M. Swain[1,2], Peter Cheetham[1,2], Chul H. Kim[2], Sastry V. Pamidi[1,2]

[1]FAMU-FSU College of Engineering, Tallahassee, FL, USA
[2]Center for Advanced Power Systems, Tallahassee, FL, USA

st18b@my.fsu.edu

Abstract.
The paper is a review of the opportunities and challenges of cryogenic power devices of electric aircraft, and the ongoing research and development efforts of the government agencies and the industry. Liquid Hydrogen (LH2) and Liquefied Natural Gas (LNG) are compared to support high temperature superconducting (HTS) and normal metal devices, respectively. The power devices were assumed to operate at the normal boiling point of the fuel used. The efficiencies of the electrical devices are estimated based on state-of-the-art technology. The mass flow rates and total fuel requirements for both the cryogenic fuels required to maintain the operating temperatures of the devices were simulated using thermal network models. A twin-aisle, 300 passenger aircraft with a 5.5 h flight duration was used for the models. The results show that the required masses of LH2 and LNG are 744 kg and 13,638 kg, respectively for the cooling requirement. The corresponding volumes of LH2 and LNG required are 9,760 and 30,300 L, respectively. In both cases, the estimated mass of the fuel needed for the aircraft is more than what is needed to maintain the cryogenic environment of the power devices. It was concluded that an electric aircraft with LNG cooled normal metal devices is feasible. However, an aircraft with HTS devices and cooled with LH2 is more attractive if the ongoing R&D efforts on HTS devices and LH2 infrastructure are successful. The emission reductions would be substantially higher with LH2, particularly when H2 is produced using renewable energy sources.

1. Introduction

The aviation industry is expected to grow by more than 20% in the next two decades. The expected growth would increase the demand for fuel, in the U.S alone, by 38% to 120 billion liters [1]. Hence, the emissions will increase by around 12% in the transport sector. Given these factors, ambitious targets are set to reduce the CO_2 emissions per passenger by 75%, NO_x emissions by 90%, and the perceived noise level of aircraft by 65% before 2050 [2]. The aggressive targets are not new to the aviation sector, and it has seen some of the most advanced technology and policy developments in the past two decades. Although a 20-25% decrease is expected in the fuel burn of the aircraft due to engine improvements over the next 2 decades, it is not sufficient to meet the anticipated targets [3]. Due to the limitations of height underwing, drag, and weight of the engine, further significant improvements are not possible with the conventional fuselage and wing-type designs [4]. There is a need for new progressive designs for achieving the emission reduction targets that have been set.

Blended wing body design (BWD) aided by distributed propulsion is one of the novel designs being considered [5]. In distributed propulsion, multiple propulsors are linked to one another and are situated in the most aero-dynamically ideal locations [5]. BWD with distributed propulsion offers many benefits such as a higher overall bypass ratio, advantages in the placement of propulsion systems, better payload capacity, higher fuel efficiency, and reduced noise level [6]. However, using BWD with conventional propulsion systems and aviation fuel (AV gas) is still not sufficient to achieve the targets [2].

To achieve higher efficiencies, there is a need for electric propulsion [6]. It was observed that, when most of the non-propulsion loads such as actuators, wing flaps, landing gear, and rudders were replaced from hydraulic to electric components, there was a significant improvement in the overall system efficiency of a more electric aircraft (MEA) [7]. The electric power train is lighter and is more flexible compared to an internal combustion system allowing improved overall efficiency and the prospect of adding/upgrading components in the system easily [8], [9]. Furthermore, electric motors maintain a higher efficiency over wide operating speeds and loading conditions [10]. Electric engines work efficiently even in smaller sizes unlike jet engines or propellers, thus, advancing the prospect of distributed propulsion [10], [11]. The control of the flow of electric power is more accurate and has a faster time constant when compared to energy transfer using mechanical systems [12]. This accommodates rapid changes in the loading condition of the aircraft based on the flight state. The power system will require high power ramp rates during takeoff and rapid system configurability during emergencies and contingencies [13].

Another alternative solution to achieve the target emission reduction requirements is the use of low carbon alternative high energy-dense fuels [4]. For the alternative fuel to be a viable solution, it should have several key properties. The fuel must be commercially available on a large scale, have a competitive price relative to the current AV gas, be energy-dense (both volumetric and gravimetric), be safe to ship and store, and it must be clean not only to burn but also to produce. The current AV gas meets some of the listed criteria but is not suitable in the long run due to its environmental impact in both production and burning in the engines [4].

Several large-scale developments are ongoing to advance the prospect of all-electric aircraft [14]–[18] Airbus's Zero-Emission project aims to develop hydrogen propelled aircraft by 2035, and the aircraft is envisioned to use hydrogen as a fuel for jet engines or propellers in a twin-engine configuration. The long-term goal of this project is to develop aircraft with a blended wing body design with two sets of turbo engines for generating thrust [14]. ARPA-E recently released \$10 million in funding for 6 projects as part of the Connecting Aviation by Lighter Electrical Systems project (CABLES) [15]. The project's scope is to develop high power density cables using different materials, the ancillary vacuum or cryogenic technologies, lightweight electrical insulating materials with good dielectric and thermal properties, connectors, circuit breakers, and to address the issue of partial discharge at high altitudes due to Paschen's curve. Our research group at the Center for Advanced Power Systems (www.caps.fsu.edu) is also a part of one of the ARPA-E projects focusing on the development of high power dense superconducting cables in collaboration with Advanced Conductor Technologies (www.advancedconductor.com) [16]. The research effort "Aviation-class Synergistically Cooled Electric-motors with iNtegrated Drives - ASCEND" is focused on developing high specific power (>12 kW/kg) electric machines and their thermal management systems [19]. General Electric is working on developing a MW Any-Altitude Gas Insulted Cable System (MAAGIC) for aircraft power distribution that uses carbon dioxide as the dielectric to overcome the limitations of partial discharge at high altitude [17]. NASA is funding the Center for Cryogenic High-efficiency Electrical Technologies for Aircraft (CHEETA) to examine the use of liquid hydrogen (LH2) in fuel cells to generate electric power along with cryogenic hybrid aluminum cables to improve the prospect of electric propulsion [18].

For achieving the reduction in emissions, the fuel must be clean not only to burn but also to produce. There are several efforts to increase the production of hydrogen cleanly [20]–[23]. H2@Scale is a project funded by the department of energy and aims to scale up the production of commercially available hydrogen through electrolysis, thermochemical water splitting, use of photoelectrochemical cells, and biological pathways [21]. Department of energy is also developing techniques to utilize magneto calorific effect for liquefaction of H2 gas. The process has more than 50% of the Carnot's efficiency when compared to traditional cryocoolers that work at around 40% of Carnot value [22]. Research is also being conducted to understand and reduce the boil-off rate of stored cryogenic fuel [24]. Finally, crucial studies are being conducted to understand the safety aspects of using hydrogen as a fuel in terms of onboard systems and the ancillary ground logistics with promising results [25], [26].

In this paper, we discuss the prospect of using cryogenic fuels such as LH2 and liquefied natural gas (LNG) in the framework of a BWD aircraft design using a notional electric distributed propulsion architecture. The power train consists of high temperature superconducting (HTS) devices such as HTS motors, HTS generators, and HTS cables or cryogenically cooled metallic components such as copper cables, electric motors, and generators. The results of the study are presented in the context of the total amount of cryogen used, the mass flow rates necessary to keep the system components at necessary operating temperatures over the duration of the flight, and the heat influx of different system components into the cooling loop under different operating conditions. A comparative analysis between the LH2 and LNG systems has been conducted to understand the various considerations that arise due to the operating temperature difference between the cryogenic fuels as well as the difference in their physical properties.

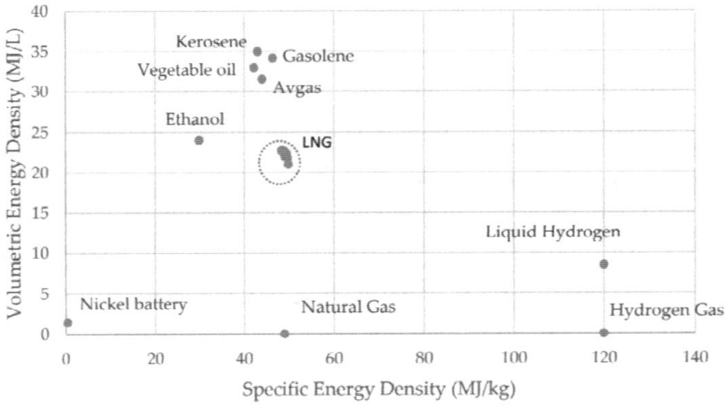

Figure 1: Volumetric vs Gravimetric energy density of different fuels [4].

2. Cryogenic Fuels

Preliminary studies show that the total CO2 emissions of future electric aircraft can be reduced by using cryogenic fuels such as LH2, LNG, and liquefied bio-methane [4]. LH2, when cleanly produced has the potential to create a zero-carbon emission transport system [14]. It was shown that LNG if burned cleanly, offers a 15% reduction in total emissions [27]. Furthermore, there is a potential to increase the power density of the electrical components in the system by operating at cryogenic temperatures using the stored cooling power available in the cryogenic fuels. The reduced total weight of the power system increases the overall efficiency and reduces the fuel consumption of the aircraft. Cryogenic fuels have a high gravimetric energy density, but a low volumetric energy density compared to conventional fuels, as shown in Fig. 1[4]. That means more space is needed to store the fuel.

Table 1 compares LH2 and LNG as fuels. An additional column for aviation fuel is also provided to give context. The takeaway from the table is that hydrogen at 15 K has 1/10th of the volumetric density of aviation fuel and about 1/5th of LNG at 110 K. However, the energy per unit mass of hydrogen is about 3 times that of AV gas. Although both LH2 and LNG are cryogenic fuels, there is a large difference in their operating temperatures. The constraints in operating a power system using LH2 and LNG are very different due to the large temperature difference. For example, the temperature of LH2 allows the use of HTS devices such as high temperature superconducting (HTS) motors, generators, and cables, while the LNG systems would work at the top end of ReBCO and BSCCO devices or with cryogenically cooled normal conductors [28]. The resistive heat load of the copper system sustained by the LNG system at high operating temperatures requires significantly higher flow rates. The latent heat of vaporization of LNG is slightly higher than LH2 requiring a smaller mass flow rate to absorb the heat from the system component. However, the specific heat capacity of LH2 is higher than LNG, therefore after the phase change, the temperature of LNG rises quicker than that of LH2. Finally, both cryogenic fuels have competitive pricing when compared to AV gas and have the potential to work as alternative fuels [29].

Table 1: Thermophysical properties of LH2 and LNG compared to AV gas [35],[36].

Property	**Aviation Fuel @ 300 K**	**LH2 @ 1 bar, 15 K**	**LNG @ 1 bar, 110 K**
Density (kg/m^3)	775-840	76.2	430-470
Volumetric energy density (MJ/L)	31	9	21-23
Specific energy density (MJ/kg)	45	120	50
Operating temperature (K)	226 – 423	15 – 20.3	91 - 115.5
Latent heat of vaporization (kJ/kg)	-	450	503
Specific heat capacity (C_p) (J/g K)	-	7.31	3.48
Cost ($/L)	0.4-0.5	0.26-0.5	0.06

There are some logistic challenges to address before cryogenic fuels can be realized in applications. Airports need to accommodate large storage or liquefaction stations of LNG or LH2 to refuel the aircraft. It is essential to maintain the purity of the cryogen storage tanks or the aircraft system components free

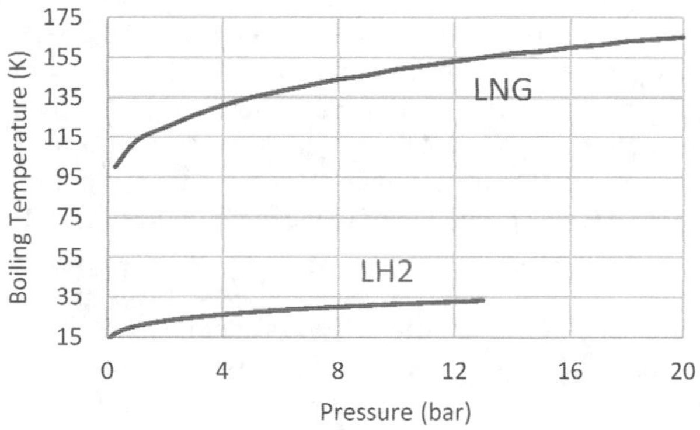

Figure 2: Pressure vs. boiling point of LH2 and LNG (Data obtained from NIST webbook).

from contamination to maintain both dielectric purposes and to avoid a reaction of LNG and water [4]. In case of leakage of LNG which is mainly composed of methane (more harmful greenhouse gas than CO2) directly into the atmosphere, the environmental benefits of low CO2 emissions are nullified. However, further benefits in emissions are realized as the sulfur content of the fuel is reduced by more than 15% [4].

Cryogenic fuels can be pressurized to attain higher energy densities and boiling points. Higher mass flow rates can be attained by pressurizing cryogen in the cryogenic circulation system to prevent phase change at operating temperature as shown in Fig. 2.

Higher operating pressure will also increase the dielectric strength of the cryogens and reduce the bubble formation if subcooled [30]. This is important as we have established that the dielectric strength of a gaseous cryogen is directly related to the density of the cryogen [31]–[33]. Furthermore, bubble formation reduces the dielectric strength of the cryogen and the partial discharge inception voltage. The dielectric constant of LH2 is about 1.24 at 1 bar. The dielectric strength and dielectric constant of LNG at 110 K and 1 bar are 27 kV/mm and 7.4, respectively. Both the cryogenic fuels can be used as a coolant and dielectric medium [31], [34]. Safety concerns related to the combustibility of the fuels in direct contact with electrical equipment need to be addressed.

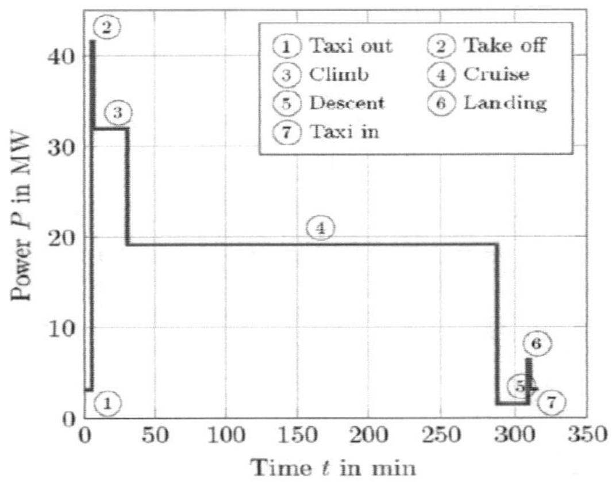

Figure 3: Typical electrical loading of a regional aircraft over the duration of flight [41].

The projected required power rating of a twin-aisle regional aircraft is 40 MW. It was found that operating in medium voltage is beneficial for electric aviation as protection equipment for high currents is very bulky [37]. Also, there is an increased potential for partial discharge inception at high altitudes due to Paschen's law. To achieve tens of MW of power rating, even at 10 kV, there is a need for cables with amperages of several kA. However, conventional copper technologies rated for kA would be heavy. There is a need for high power-dense cables and machines to obtain efficient power trains. Devices made of HTS or cryogenically cooled normal metals would offer high power densities. Cryogenic hybrid aluminum cables that support significantly higher current density compared to conventional copper technology are being investigated [38]. The resistivity of hybrid aluminum at room temperature is 11.7 times higher than at 77 K and more than 100 times higher at 20 K [38]. State of art electrical machines with a power density of around 5 kW/kg are being developed [39], [40]. The power density needs to increase to 13-40 kW/kg for long-range electric aircraft systems to be viable [39]. The machines could also be operated at higher frequencies to achieve higher power density, but as of now, it is only limited to a few hundred Hz for cryogenically cooled devices due to thermal management challenges arising from hysteresis and resistive losses [39].

HTS cables offer significantly higher current densities for a given cross-section compared to cryogenic metal conductors and provide a current limiting characteristic that could be beneficial during fault initiation [40]. HTS machines have higher overall efficiency and power density compared to their copper counterparts. When operating in DC, HTS cables have no electrical losses. However, there are several challenges in incorporating HTS technology onto onboard systems such as the higher capital cost of HTS devices, compact terminations that act as interfaces between room temperature and cryogenic systems, electrical insulation designs of HTS cables, and the behavior of HTS cable to faults in the power system.

3. Modeling Methodology

This section introduces the modeling methodology used by us to understand the cryogenic cooling opportunities and challenges for the electric power system of an electric aircraft. HTS technology or cryogenics is not suitable for all types of airplanes. Smaller aircraft would benefit more by using battery technology, direct injection of fuel, or by modifying the aerodynamic design of aircraft [39]. Cryogenic systems are viable when the total cooling power required for the system available from the fuel is considerably less than the total electrical losses incurred in the system. One more consideration is that the weight of the cryogenic system must be significantly smaller than the weight savings offered by the power system using HTS or cryogenic normal conductors.

Fig. 3 shows the power requirement of a typical aircraft during different stages of a flight. It shows the power requirement of a twin-engine, twin-aisle aircraft that can carry about 300 passengers over the duration of the flight - 330 minutes used in the models [41]. As seen from the figure, the power requirements of an aircraft are not constant and change during the flight. HTS conductors can take advantage of this fact as the size of the cable does not increase considerably by increasing ampacity. This is because the superconducting layer of an HTS conductor is only 2-3 micrometers thick and the rest of the conductor is made from supporting materials and a normal metal. Additionally, HTS cables require a former and a cryostat making the fraction of the cross-section of HTS small. Hence adding a few layers of HTS devices doesn't change the cable size mass considerably [42]–[45].

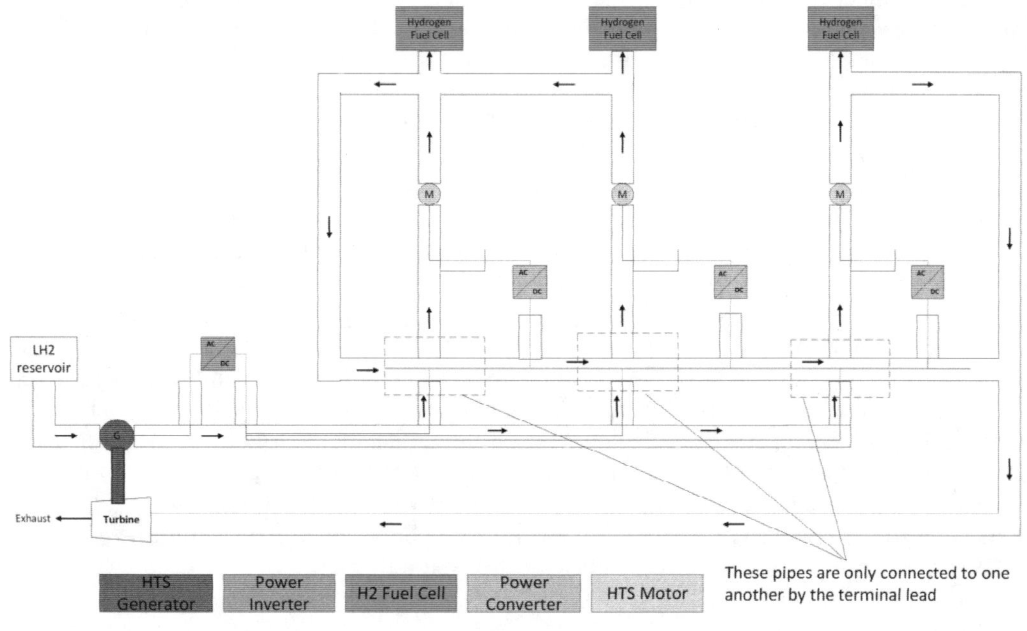

Figure 4: Notional powersystem architecture of an LH2 fuelled superconducting electric aircraft (one fourth of the system is shown)

When designing a notional power system for transportation systems such as electric aircraft, safety and redundancy are important as accidents could lead to severe damage or loss of life. Failure of one system should not lead to total failure of the power network or the essential loads such as navigation, propulsion, and pressure control. The notional power system modeled supports the power demand and has built-in redundancies for safe operation. Also, despite the difference in thermal and electrical constants, both the systems must be integrated in such a way that, during contingencies, the control systems must be able to reroute the electrical and cryogen flow with minimal downtime. The role of large heat capacities of the metallic components could be utilized to overcome the difference in time constants.

Based on these constraints discussed above, we have developed a notional architecture for the LH2 cooled aircraft system as shown in Fig. 4. Fig. 4 shows a quarter of the total system with a total cable length of 160 m. The power system consists of 2 generators, and 12 motors that are connected by power distribution cables, converters, and inverters. With this architecture, the power system can function even with the loss of one of the generators. The figure does not show the protection equipment such as relays, circuit breakers - solid-state or otherwise, disconnect switches, HTS fault current limiters needed. Additionally, other electrical components such as high-frequency transformers and other equipment are needed to connect the main bus to ancillary loads such as lighting loads, climate control, actuators, wing flaps, rudders, aero foils, and landing equipment. Multiple R&D efforts at the industry, universities, and government laboratories are developing different system components that operate at cryogenic temperature, room temperature, or in between. For example, our group is working on a two-stage cryogenic environment where the power electronics and protection equipment could be placed in dielectric fluids at 150~200 K. This would reduce the total heat introduced by the power electronics into the main cryogenic loop. Furthermore, the dielectric fluid would behave similarly to transformer oil by providing efficient heat transfer for power electronics. Finally, the energy required to maintain the devices at 150~200 K is significantly lower than the energy required to maintain the power electronics inside the main cooling loop [46].

The system engineer must consider all the components and their temperature need and understand how to integrate them into the power system. There is a need to understand the benefits and drawbacks of operating each device either in the cryogenic cooling loop or outside of it. For this set of models, we have assumed that the power electronics operate at room temperature, and we account for the heat influx through the interfaces. This is because, from our previous models described in [47], we saw that the heat loads by incorporating power electronics inside the loop were very high. The heat loads used were derived using the estimated efficiencies of power electronic components as well as HTS devices from a previous study [47]. While the heat load of the power electronic components is high, but the overall power system is still feasible. Higher heat loads can be managed by either using additional cryogenic fuel or using designs with greater efficiencies, and other system-level tradeoffs.

The nominal operating parameters of the devices such as the HTS generator, motor, and cable and their current ratings and thermal loads during various conditions were previously discussed in [48]. Table 2 shows the heat loads for an HTS system cooled by LH2 [48], based on these values, we have developed thermal loads of the cryogenically cooled copper machines and cables in an LNG cooled system (Table 3). The heat loads were calculated by assuming that they are equal to the electrical losses from each device component. The machine efficiencies used, represent the current state-of-the-art technologies. ABB reported a motor that has an efficiency of 99.05% at full load [49]. The numbers in Table 2 show that for a copper system, the heat load of a generator could be 25-50 times the heat load of an HTS generator, and it could be even higher for the motors.

Table 2: Thermal loads of HTS devices cooled by LH2 for three different levels of efficiencies [48][5].

Component	Rated power (MW)	Operating Temperature (K)	Maximum Efficiency (%)	Maximum Heat load (kW)	Medium Efficiency (%)	Medium Heat load (kW)	Minimum Efficiency (%)	Minimum Heat load (kW)
HTS Generator	20	16-40	99.98	4	99.97	6	99.96	8
HTS Motor	1.4	20-60	99.98	0.67	99.97	1	99.96	1.33
HTS Cable	-	<80	100 - DC	0.1-0.2	100 - DC	0.1-0.2	100 - DC	0.1-0.2

Table 3: Thermal loads of components of the cryogenically cooled normal metal system for three different levels of efficiencies.

Component	Rated power (MW)	Operating Temperature (K)	Maximum Efficiency (%)	Maximum Heat load (kW)	Medium Efficiency (%)	Medium Heat load (kW)	Minimum Efficiency (%)	Minimum Heat load (kW)
Generator	20	16-40	99	200	98.5	300	98	400
Motor	1.4	20-60	98	66.5	97.5	83.1	97	99.7
Cable	-	<80	I^2R losses	0.2-0.5	I^2R losses	0.2-0.5	I^2R losses	0.2-0.5

4. Thermal network models

The thermal network modeling (TNM) technique described in [50], [51] was used to simulate the power system described above. The material properties of the fuels and the power ratings and efficiencies of the devices used are listed in the tables. The physical properties of the fuels are temperature and pressure sensitive. The transition temperature and the latent heat of vaporization of the fuels are calculated in the models are based on the operating pressure of the system. The simulations conducted are for a 330-minute duration flight with different states of the aircraft as shown in Fig. 3. The heat loads and power outputs of the components in the model change based on the state of the flight.

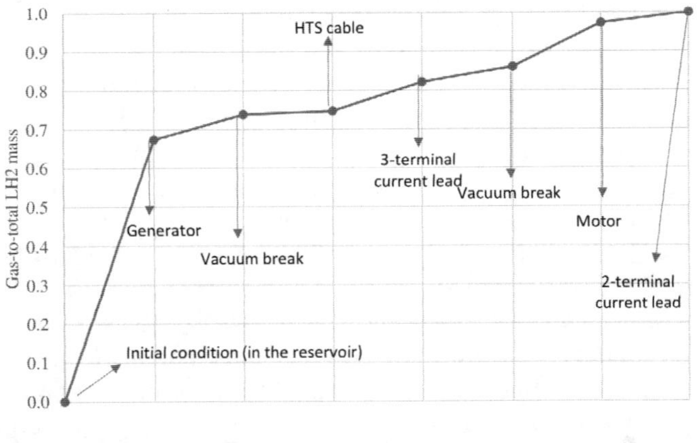

Figure 5: Ratio of mass flow rate of GH2 to total mass flow rate of the cryogen in the power system components.

The goal of the modeling effort is to use a similar system for both the cryogenic fuels, LH2 and LNG, and compare the heat loads introduced by the HTS and copper-based power system components to estimate the total mass of the cryogen needed for the duration of the flight and the required mass flow rates of the cryogen necessary to maintain the operating temperature at each device. As such, the cryogenic fuels are assumed to be stored at the lowest possible temperature as liquids with LH2 at 15 K and LNG at 91 K at 1 bar, just above their freezing point. The pressure of the system in the models is assumed to start and be maintained at 1 bar, as pressurizing the storage tanks could require heavy insulation and structural designs. Pressurization also pushes the boiling point up. The mass flow rates of the cryogen are selected to keep the temperature at T_{boil} at the exit of the last device in the cooling loop, i.e., the cryogen is in the liquid state at the start of the cooling loop and transitions to gas as it absorbs heat from the devices along the cooling loop and no liquid is left in the system when the last device is cooled (as shown in Fig. 5). Finally, the temperature of the gaseous cryogen is assumed to remain at T_{boil} until there is no liquid left in the system, although this assumption is not accurate in the real cooling process. However, the results obtained by the model still provide a close estimate in solving the steady-state solution of TNM models.

The devices in the LH2 cooled system are based on HTS technology while the devices in the LNG system are copper based. Both the HTS and copper machines are cooled by the cryogen through conduction or using secondary cooling loops with the help of heat exchangers. The HTS/copper cables would have a helium circulation system to avoid direct contact of cryogenic fuel with the cable. This is done to reduce the risk of accidents in case of fault on the cable. The circulating cryogen (helium) is run through a heat exchanger embedded in the LH2/GH2 loops. Due to the notional nature of the models, the heat exchanger efficiency is assumed to be 100%. The mass flow rates of the cryogens could be adjusted accordingly based on the actual heat exchanger efficiencies in the system.

The hydrogen system would have both liquid and gaseous hydrogen. LH2 stored in the wings would be at 15 K. It reaches the hydrogen-burning engine or fuel cells after passing through and cooling the various components. The high latent heat (443 kJ/kg) aids in maintaining the temperature of the system and keeping the required mass flow rates low. As mentioned earlier, the power electronics in the system are at ambient temperature and the heat influx into the cooling loop is considered in the models.

The LNG-based cooling system would work similarly to the LH2 system. it consists of both liquid and gaseous states flowing through the pipes. The cryogen starts in the reservoir as LNG at 95 K and 1 bar.

5. Results

In the simulations, the mass flow rate of the cryogen was selected to maintain the operating temperatures of the last device in the cooling loop at T_{boil}. T_{boil} for LH2 is 20.3 K at 1 bar and 110.7 K at 1 bar for LNG. Table 4 shows the important results of the simulations. The table shows the total heat load for different operational states of the two systems and the corresponding mass flow rates per quadrant required to keep the operating temperature of the components in all four quadrants at T_{boil}. The total heat load for each case was calculated by combining the heat loads of each component of the system (minimum efficiency case). The heat loads of the individual components are listed in Table 2 and Table 3. The efficiencies of the electrical devices were assumed to be constant throughout the flight. The assumption is reasonable as the efficiency of the electric machines stays relatively flat above 30% of the rated load and the duration of taxi and take-off where the aircraft is below 30% is small compared to the overall duration of the flight.

The main conclusions drawn from the data in Table 4 are, for a 330 min flight with the power demands listed in Fig. 3 and Table 4. the aircraft would need about 744 kg of hydrogen and 13638 kg of LNG to cool all the devices in the system. The vast difference in the amount of cryogen necessary is due to the lower efficiencies and the corresponding higher heat loads of the copper-based power devices. The masses would translate to 10,000 and 30,000 liters of LH2 and LNG, respectively. The volume of LNG is only thrice that of LH2 despite the large difference in the mass due to its higher density compared to LH2.

To give context to these values, the fuel requirement of a 300-passenger aircraft that uses AV gas for a 330-minute flight is approximately 38,000 kg [52]. Using the fuel consumption, energy density of the fuel (Table 1), and the efficiency of the current high bypass pass ratio of aircrafts, 43%, one can estimate the fuel needs of cryogenically cooled aircraft. The specific energy densities of the three fuels

– AV Gas, LH2, and LNG are listed in Table 1. We assumed the efficiency of the electric aircraft to be 75%. The calculated fuel needed for an LH2 fueled aircraft is 7,200 kg of LH2. For LNG fueled aircraft it is 17,200 kg of LNG. In terms of volumes, 94,500 liters of LH2 or 38,400 liters of LNG would be needed.

Comparing the fuel needs and the cryogenic needs of electric aircraft fueled by LH2 and LN2 that use HTS power devices or cryogenic normal metal devices, one sees that in both cases, the cryogenic fuel needs are less than the fuel needs. For the LH2 fueled aircraft using HTS devices, the cryogen needs estimated using the models described are much lower than the total fuel requirement even if additional heat loads that have not been accounted for in the models or the efficiencies of the power devices are slightly lower than the assumed as in Table 2. For the LNG system, the total cryogen is needed is only slightly lower than the total fuel needed. The significantly high gravimetric energy density of LH2 is the primary reason behind the conclusion. LNG has an energy density comparable to that of AV gas. One major challenge for the LH2 fueled aircraft design is in accommodating the large volume needed because of the low volumetric energy density.

It is worth noting that, for these models we are considering 100% heat exchanger efficiencies, the required fuel flow rates would increase depending on the heat exchanger efficiencies. Also, all the devices in the system need not operate at T_{boil} and by optimizing the mass flow rates, the total cryogen needed would be lower. We assumed all the devices to operate at T_{boil} to have comparable results between LH2 and LNG systems.

The simulations focus primarily on the mass and volume requirements of LH2 and LNG fuels for cryogenically cooled power devices. The conclusion is that an electric aircraft with LNG cooled normal metal devices is feasible. However, an aircraft with HTS devices and cooled with LH2 is attractive if the ongoing R&D efforts on HTS devices and LH2 infrastructure are successful. Additionally, the emission reductions would be substantially lower with LH2, particularly when H2 produced using renewable energies is used. The focus and investments of the government agencies and the industry on production, transportation, and delivery will accelerate the development of LH2 and HTS technologies [14]–[18], [20]–[23]. The issue of the low volumetric energy density of LH2 needs to be overcome by building highly efficient power trains and making more aerodynamic aircraft designs to reduce the overall fuel need of the aircraft.

Table 4: Comparison of required fuel consumption and total heat load at each flight state with the total amount of fuel to complete a flight for LH2 and LNG systems.

Flight state	LH2 system		LNG system	
	Mass Flowrate (g/s)	Total Heat load (kW)	Mass Flowrate (g/s)	Total Heat load (kW)
①Taxi	3	5.2	41	81.8
②Takeoff	17	29.1	419	842.9
③Climb	14	23.9	338	680.9
④Cruise	8	14.4	190	381.8
⑤Descent	2	4.6	31	61.7
⑥Landing	4	6.4	61	121.7
⑦Taxi	3	5.2	41	81.8
Total amount of cryogen needed	744 kg		13638 kg	
Volume of cryogen needed	9760 liters @ 1 bar, 15 K		30300 liters @ 1 bar, 91 K	

6. Summary

The opportunities and challenges of cryogenic power devices for electric aircraft to achieve aggressive emissions reduction targets are discussed. A summary of the ongoing research and development efforts of the government agencies and the industry is provided. The physical properties of Liquid Hydrogen

(LH2) and Liquefied Natural Gas (LNG) are compared as fuels for the aircraft and for providing a cryogenic environment to the electrical power devices. Based on the possible operating temperatures of the two cryogens, LH2 systems are assumed to support HTS devices, and LNG systems to support cryogenic normal metal devices. Thermal network models were used to simulate the mass flow rates and the total requirements for both the cryogenic fuels needed to maintain the operating temperatures of the devices. A 300-passenger twin-aisle aircraft with a flight duration of 5.5 h was used for the simulations. The results show that the required masses of LH2 and LNG are 744 kg and 13,638 kg, respectively for cooling requirements. The corresponding volumes of LH2 and LNG required are 9,760 and 30,300 L, respectively. The results show that an electric aircraft with LNG cooled normal metal devices is feasible. However, an aircraft with HTS devices and cooled with LH2 is more attractive. In both cases, the fuel masses of LH2 and LNG needed for fuel purposes are higher than the needs for cryogen purposes. It was concluded that the ongoing R&D efforts and the investments in HTS devices and LH2 infrastructure are promising. There is optimism that the emission reductions targets would be met with LH2 fueled aircraft using HTS power devices. The success of the ongoing efforts to develop large-scale production of H2 is essential for realizing the dream of zero-emission electric aircraft. The readers would benefit from several recent reviews on electric aircraft technologies, superconducting devices for electric aircraft, and the emerging liquid hydrogen and liquefied natural gas-fueled aircraft [39], [40], [53]–[59].

7. References

[1] A. L. S. Chemex *et al.*, "FAA Aerospace Forecasts Fiscal Years 2020-2040," 2019. DOI: https://www.faa.gov/data_research/aviation/aerospace_forecasts/media/FAA_Aerospace_Forecasts_FY_2020-2040.pdf.

[2] R. Alblas, "Flightpath 2050 – A long term vision for alternative fuels for aviation," Sustainable Energy Week, Brussels Belgium 2017.

[3] Air Transport Department - Cranfield University, "Fuel and Air Transport - A Report for the European Commission," p. 46, 2008, [Online]. Available: http://ec.europa.eu/transport/modes/air/doc/fuel_report_final.pdf%5Cnhttp://ec.europa.eu/transport/modes/air/internal_market/observatory_market/topic_reports_en.htm.

[4] P. Rompokos, S. Kissoon, I. Roumeliotis, D. Nalianda, T. Nikolaidis, and A. Rolt, "Liquefied natural gas for civil aviation," *Energies*, vol. 13, no. 22, 2020, doi: 10.3390/en13225925.

[5] J. L. Felder, H. D. Kim, and G. V. Brown, "Turboelectric distributed propulsion engine cycle analysis for hybrid-wing-body aircraft," in *47th AIAA Aerospace Sciences Meeting including the New Horizons Forum and Aerospace Exposition*, 2009, no. January, p. AIAA 2009-1132, DOI: 10.2514/6.2009-1132.

[6] J. Palmer and E. Shehab, "Modelling of cryogenic cooling system design concepts for superconducting aircraft propulsion," *IET Electr. Syst. Transp.*, vol. 6, no. 3, pp. 170–178, 2016, DOI: 10.1049/iet-est.2015.0020.

[7] B. Sarlioglu and C. T. Morris, "More Electric Aircraft: Review, Challenges, and Opportunities for Commercial Transport Aircraft," *IEEE Trans. Transp. Electrif.*, vol. 1, no. 1, pp. 54–64, 2015, DOI: 10.1109/TTE.2015.2426499.

[8] N. Doerry and John V. Amy Jr., "Design Considerations for a Reference MVDC Power System," *Naturwissenschaften*, vol. 30, no. 16, p. 240, 2015, doi: 10.1007/BF01486360.

[9] N. H. Doerry and H. Fireman, "Designing All Electric Ships," *Proc. Ninth Int. Mar. Des. Conf.*, p. 475, 2006.

[10] J. Martins *et al.*, "REAL-LIFE COMPARISON BETWEEN DIESEL AND ELECTRIC CAR ENERGY CONSUMPTION," pp. 209–232, 2013, [Online]. Available: https://www.novapublishers.com/catalog/product_info.php?products_id=44945.

[11] A. S. Gohardani, G. Doulgeris, and R. Singh, *challenges of future aircraft propulsion: a review of distributed propulsion technology and its potential application for the all-electric commercial aircraft*, vol. 3, no. 5. Nova Science Publisher, 2013.

[12] A. Hughes and B. Drury, *Electric Motors and Drives : Fundamentals, Types and Applications*, vol. 4th ed. Amsterdam: Newnes, 2013.
[13] N. Doerry and J. Amy, "MVDC Shipboard Power System Considerations for Electromagnetic Railguns," *6th DoD Electromagn. Railgun Work. Laurel MD, Sept 15-16, 2015.*, no. July, 2015.
[14] AIRBUS, "AIRBUS - Zero-Emission Concept Aircraft." https://www.airbus.com/newsroom/press-releases/en/2020/09/airbus-reveals-new-zeroemission-concept-aircraft.htmlhttps://www.airbus.com/newsroom/press-releases/en/2020/09/airbus-reveals-new-zeroemission-concept-aircraft.htmlhttps://www.airbus.com/newsroom/p (accessed Jul. 30, 2021).
[15] ARPA-E, "Green Car Congress - ARPA-E." https://www.greencarcongress.com/2021/02/20210227-cables.html (accessed Jul. 30, 2021).
[16] T. C. Keller, "CAPS-ACT ARPA-E project," 2021. https://www.eng.famu.fsu.edu/news/caps-act-arpa-e-grant (accessed Jul. 30, 2021).
[17] C. F. Co, "ARPA-E MAAGIC Cables." http://www.cleanfuture.co.in/2021/03/10/arpa-e-funding-electric-aircraft/ (accessed Jul. 30, 2021).
[18] L. Gipson, "NASA Aeronautics - CHEETA," 2020. https://www.nasa.gov/aero/nasa-aeronautics-selects-three-university-teams-for-research-help (accessed Jul. 30, 2021).
[19] ARPA-E, "Aviation-class Synergistically Cooled Electric-motors with iNtegrated Drives." https://arpa-e.energy.gov/technologies/programs/ascend (accessed Aug. 31, 2021).
[20] National Physcis Laboratory, "NPL -- Hydrogen Strategy Plan." https://www.npl.co.uk/news/npl-responds-to-government-hydrogen-strategy (accessed Aug. 31, 2021).
[21] Office of Energy Efficiency and Renewable Energy, "H2 @ Scale." https://www.energy.gov/eere/fuelcells/h2scale (accessed Jul. 30, 2021).
[22] J. Holladay *et al.*, "MagnetoCaloric Hydrogen Liquefaction," *DOE Annu. Merit Rev. PD131*, no. June, p. 35, 2017.
[23] banse Tom, "Hydrogen Production Plant -- Douglas County Public Utility." https://www.nwnewsnetwork.org/energy/2021-03-08/hydrogen-production-plant-groundbreaking-signals-renewed-interest-in-alt-fuel (accessed Aug. 31, 2021).
[24] T. Włodek, "Analysis of boil-off rate problem in Liquefied Natural Gas (LNG) receiving terminals.," *IOP Conf. Ser. Earth Environ. Sci.*, vol. 214, no. 1, 2019, DOI: 10.1088/1755-1315/214/1/012105.
[25] J. Leachman, "Hyper-lab Cool. Fuel." https://hydrogen.wsu.edu/ (accessed Jul. 30, 2021).
[26] ACI/ATI, "Contents Introduction 3 Hydrogen Supply Chain Aircraft / Airport Compatibility and Ground Operations Cost Considerations Gaps Identified Acknowledgments 23," 2021.
[27] J. M. Gibbs and B. Nagel, "Design , Economic Competitiveness , and Profitability of a Transportation System," *Am. Inst. Aeronaut. Astronaut.*, pp. 1–13.
[28] A. P. Malozemoff, "The power grid and the impact of high-temperature superconductor technology: An overview," *Superconductors in the Power Grid: Materials and Applications*, pp. 1–28, 2015.
[29] U. S Department of Energy, "CLEAN CITIES ALTERNATIVE FUEL PRICE REPORT April 2020 Issue" no. April, 2020, [Online]. Available: https://afdc.energy.gov/files/u/publication/alternative_fuel_price_report_april_2020.pdf.
[30] T. Stamm, "DIELECTRIC DESIGNS OF POWER CABLES FOR ENHANCED RESILIENCY IN HIGH TEMPERATURE SUPERCONDUCTING ELECTRICAL TRANSPORT SYSTEMS - Chapter 4," FAMU-FSU College of Engineering, 2020.
[31] C. Park, J. Wei, P. Cheetham, C. H. Kim, S. Pamidi, and L. Graber, "The influence of temperature on the dielectric strength of gaseous cryogens," *Proc. IEEE Int. Conf. Prop. Appl. Dielectr. Mater.*, vol. 2018-May, pp. 164–167, 2018, DOI: 10.1109/ICPADM.2018.8401185.
[32] C. Park, L. Graber, P. Cheetham, J. G. Viquez, C. H. Kim, and S. Pamidi, "A versatile

modeling technique for predicting dielectric strength improvements in gas mixtures for superconducting applications," *IEEE Transactions on Dielectrics and Electrical Insulation*, vol. 24, no. 5. pp. 2755–2764, 2017, DOI: 10.1109/TDEI.2017.006653.
[33] P. Cheetham, C. Park, C. H. Kim, L. Graber, and S. V. Pamidi, "Dielectric properties of cryogenic gas mixtures for superconducting power applications," *IOP Conf. Ser. Mater. Sci. Eng.*, vol. 278, no. 1, p. 12040, Dec. 2017, DOI: 10.1088/1757-899X/278/1/012040.
[34] P. Cheetham, C. Park, C. H. Kim, L. Graber, and S. V. Pamidi, "Dielectric properties of cryogenic gas mixtures for superconducting power applications," *IOP Conf. Ser. Mater. Sci. Eng.*, vol. 278, no. 1, 2017, DOI: 10.1088/1757-899X/278/1/012040.
[35] NIST, "NIST webBook - Hydrogen." https://webbook.nist.gov/cgi/cbook.cgi?ID=1333-74-0 (accessed Aug. 05, 2021).
[36] C. Chemicals, "LNG - Material properties and saftey." https://cameochemicals.noaa.gov/chris/LNG.pdf (accessed Aug. 05, 2021).
[37] G. F. Reed, B. M. Grainger, A. R. Sparacino, and Z. H. Mao, "Ship to grid: Medium-voltage dc concepts in theory and practice," *IEEE Power Energy Mag.*, vol. 10, no. 6, pp. 70–79, 2012, DOI: 10.1109/MPE.2012.2212613.
[38] M. D. Sumption, J. Murphy, M. Susner, and T. Haugan, "Performance metrics of electrical conductors for aerospace cryogenic motors, generators, and transmission cables," *Cryogenics (Guildf).*, vol. 111, no. July 2019, p. 103171, 2020, DOI: 10.1016/j.cryogenics.2020.103171.
[39] X. Zhang, C. L. Bowman, T. C. O'Connell, and K. S. Haran, "Large electric machines for aircraft electric propulsion," *IET Electr. Power Appl.*, vol. 12, no. 6, pp. 767–779, 2018, DOI: 10.1049/iet-epa.2017.0639.
[40] P. J. Masson, G. V. Brown, D. S. Soban, and C. A. Luongo, "HTS machines as enabling technology for all-electric airborne vehicles," *Supercond. Sci. Technol.*, 2007, DOI: 10.1088/0953-2048/20/8/005.
[41] M. Boll *et al.*, "A holistic system approach for short range passenger aircraft with cryogenic propulsion system," *Supercond. Sci. Technol.*, vol. 33, no. 4, 2020, DOI: 10.1088/1361-6668/ab7779.
[42] J. T. Kephart, B. K. Fitzpatrick, P. Ferrara, M. Pyryt, J. Pienkos, and E. Michael Golda, "High temperature superconducting degaussing from feasibility study to fleet adoption," *IEEE Trans. Appl. Supercond.*, 2011, DOI: 10.1109/TASC.2010.2092746.
[43] J. F. Maguire *et al.*, "Progress and status of a 2G HTS power cable to be installed in the Long Island Power Authority (LIPA) grid," *IEEE Trans. Appl. Supercond.*, vol. 21, no. 3 PART 2, pp. 961–966, 2011, DOI: 10.1109/TASC.2010.2093108.
[44] M. J. Gouge *et al.*, "Development and testing of HTS cables and terminations at ORNL," 2001, DOI: 10.1109/77.920333.
[45] J. D. Weiss, T. Mulder, H. J. ten Kate, and D. C. van der Laan, "Introduction of CORC ® wires: highly flexible, round high-temperature superconducting wires for magnet and power transmission applications," *Supercond. Sci. Technol.*, vol. 30, no. 1, p. 014002, 2017, DOI: 10.1088/0953-2048/30/1/014002.
[46] P. Mensah, P. Cheetham, C. H. Kim, and S. Pamidi, "Novel dielectric fluids for superconducting electric aircraft applications -- Submitted at AIAA Propulsion and Energy 2021 Forum," 2021.
[47] P. Cheetham, I. Barnola, C. Kim, and S. Pamidi, "Exploring options for integrated cryogenic circulation loop of superconducting power devices on electric aircraft," no. August, pp. 1–8, 2019, DOI: 10.2514/6.2019-4498.
[48] P. Cheetham, B. Darbha, S. Telikapalli, C. H. Kim, and S. Pamidi, "Development of power dense superconducting DC networks for electric aircraft," Aug. 2020, DOI: 10.2514/6.2020-3553.
[49] J. Rosberg, "ABB - High efficiency electric motor." https://www.abb-conversations.com/2017/07/abb-motor-sets-world-record-in-energy-efficiency/ (accessed Aug.

04, 2021).
[50] N. Suttell, C. H. Kim, J. Ordonez, D. Shah, L. Graber, and S. Pamidi, "Thermal Modeling of Gaseous Helium as a Cryogen for High Temperature Superconducting Cable Components," *IEEE Trans. Appl. Supercond.*, vol. 25, no. 3, pp. 1–5, Jun. 2015, DOI: 10.1109/TASC.2014.2374673.
[51] L. Graber, J. Kim, C. H. Kim, and S. V. Pamidi, "Thermal Network Model for HTS Cable Systems and Components Cooled by Helium Gas," *IEEE Trans. Appl. Supercond.*, vol. 26, no. 4, pp. 1–5, Jun. 2016, DOI: 10.1109/TASC.2016.2545102.
[52] Saurabh, "Fuel Consumption of Popular Aircraft." https://alliknowaviation.com/2019/12/14/fuel-consumption-aircraft/ (accessed Sep. 02, 2021).
[53] B. P. J. Ansell and K. S. Haran, "Electrified Airplanes - A path to zero-emission air travel," *IEEE Electrif. Mag.*, vol. 8, no. 2, pp. 18–26, 2020, [Online]. Available: https://doi.org/10.1109/MELE.2020.2985482.
[54] K. S. Haran *et al.*, "High power density superconducting rotating machines - Development status and technology roadmap," *Supercond. Sci. Technol.*, vol. 30, no. 12, 2017, DOI: 10.1088/1361-6668/aa833e.
[55] Cheetham Peter, S. Telikapalli, C. H. Kim, and S. Pamidi, "A Comprehensive Guide to Superconductivity - Chapter 1 - Superconducting Power Devices for the Electrification of Transportation," in *A Comprehensive Guide to Superconductivity*, 2020.
[56] J. S. Langford and D. K. Hall, "Summer 2020," *Georg. Libr. Q.*, vol. 57, no. 3, p. 1, 2020, [Online]. Available: https://ntrs.nasa.gov/api/citations/20090042355/downloads/20090042355.pdf.
[57] M. J. Armstrong *et al.*, *Architecture, Voltage, and Components for a Turboelectric Distributed Propulsion Electric Grid(AVC-TeDP)*, no. July. 2015.
[58] H. D. Kim, J. L. Felder, M. T. Tong, and M. J. Armstrong, "Revolutionary Aeropropulsion Concept for Sustainable Aviation: Turboelectric Distributed Propulsion," *21st Int. Symp. Air Breath. Engines*, pp. 1–12, 2013.
[59] G. Brown, "Space Administration Materials Aspects of Turboelectric Aircraft Propulsion."

Acknowledgments
This work is sponsored in part by the Office of Naval Research Grants and Electric Ship Research & Development Consortium: - N00014-18-1-2693 and N00014-16-1-2956

Development of Arduino Based Power Conditioning Unit for Superconducting Magnetic Energy Storage (SMES) System used as UPS for Load Leveling during Charging of Electric Vehicles

A Sarkar, A Anand, T S Datta, A S Gour and V V Rao

Cryogenic Engineering Centre, Indian Institute of Technology Kharagpur - 721302, West Bengal, India.

E-mail: abhiksarkar17011995@gmail.com

Abstract. With the advancements in Electric Vehicle (EV) technology, more and more EVs are entering into service and simultaneously the number of charging stations are also increasing. These charging stations are connected to the EVs for a small duration of time. Hence, as the number of EVs increases, the simultaneous charging of multiple EVs generates a peak active power demand for this duration. The available peak load compensation technologies such as flywheel storage, pumped hydro storage or Battery Energy Storage System (BESS) take a long time to respond and are not very efficient. With an operational efficiency exceeding 90 % and quick reaction time (< 1 second), a Superconducting Magnetic Energy Storage (SMES) system can be a viable solution for this scenario. A SMES system generally consists of a superconducting coil system being charged during the low load/no load period and discharged during the peak demand, with the help of a power conditioning unit (PCU). This paper describes the development of various sub-systems of the SMES-PCU, such as a three-phase rectifier, bidirectional chopper unit, three-phase inverter and the controllers with a description of various modes of operation and sizing of components. The paper also includes the simulation of overall SMES-PCU with suitable assumptions, to present the operational characteristics of the integrated system.

1. Introduction

Electric Vehicles (EVs) are the future of road transportation. Each year, more and more EVs are manufactured and adopted worldwide. This rise in EV population is adding to the concern of increased load on the electric power grid. The potential impact of EVs on global energy systems [1] reports a rise in the peak load by above 30% with even 25% electric vehicle penetration by 2050. This will create huge stress on the generators as they are designed to be operated at a constant maximum output power [2] [3]. This brings into consideration energy storage systems to provide power during peak loading periods.

Conventional grid level energy storage systems include pumped hydro power plants and Battery Energy Storage Systems (BESS) with a few applications of flywheel energy storage [4], super-capacitor energy storage and Compressed Air Energy Storage (CAES) [5] [6]. All these energy storage systems are having very high specific energy density (> 100kJ/kg). However, during transient power fluctuations, huge power is required to be pumped into the grid or extracted from it, which is not possible with these conventional technologies in lieu of their large actuation time [6] [8] (typically more than 1 second). Another disadvantage with these energy storage systems is their low conversion efficiency (70-80 %, except for the BESS which have 90 % efficiency [4] [6]).

Also, the penetration of EVs is not bound to be uniform as stated in McKinsey's Geospatial Analytics forecast considering Germany as an example with 10% national penetration [1]. Local pockets with significant EV populations will emerge ensuing the need to distribute the load leveling systems rather than a centralized load leveling operation. Most of the conventional energy storage systems are area specific and thus, cannot be distributed as per the EV population density.

Considering all these shortcomings, a Superconducting Magnetic Energy Storage (SMES) system is the best available option. A SMES system has a typical energy density of about 10 kJ/kg. However, it has high specific power density of (>100 kW/kg) with a power transfer efficiency of more than 95% [4] [7], making it the most suitable energy storage system for mitigating transient power fluctuations in the grid. Also, the SMES systems can be highly modular [9], adding to a distributed energy regulation capacity.

This paper describes a possible strategy for load-leveling operation of the SMES system through simulation and with suitable assumptions. Section 2 describes the overview of the proposed SMES-PCU system with its various components. Section 3 presents a theoretical approach to coil design with calculation for its various parameters, considering stacked solenoid configuration for the magnetostatic analysis using COMSOL. Section 4 describes the MATLAB Simulink model for observing the load leveling operation of the SMES system. This is followed by results and discussions on the magnetostatic analysis of stacked solenoid and the observations of the MATLAB simulation for load leveling operation of SMES system; finally ending with conclusions and bibliography.

2. Overview of SMES-PCU system

The SMES-UPS system, as shown in Figure 1 comprises of - (a) superconducting coil (b) persistent switch/action (c) type-D chopper (d) DC-link capacitor (e) bidirectional converter (f) filter circuit (g) 3-Phase transformer and circuit breakers (h) controllers for bidirectional converter and chopper circuits and (i) refrigeration system

The superconducting coil is responsible for storing the excess energy available during base load operation, from the grid in the form of magnetic flux density (B) produced due to the flow of persistent current. This trapped energy can be calculated using equation (1)

$$E = \frac{1}{2}LI^2 = \frac{1}{2\mu_o}\int_V B^2(x,y,z)\,dx\,dy\,dz \tag{1}$$

where L = inductance of superconducting coil (H), I = persistent current flowing through the coil (A) , B = volumetric flux density produced by the coil (T) and μ_o = permeability of free space = $4\pi\times\ 10^{-7}$ H m^{-1} [10]. This trapped energy is fed back to the grid during peak load when the superconducting coil discharges.

The chopper circuit is responsible to maintain a steady voltage across the DC-link capacitor by charging or discharging the coil. This is achieved by the chopper controller which accepts the DC-link voltage as feedback and accordingly switches on or off the chopper's controlled switches to maintain this voltage at a steady value [11]. The bidirectional converter converts 3-phase AC voltage from the grid to DC, charging the DC-link capacitor in rectifier mode, during the base load period. On other hand, during peak load, bidirectional converter converts the DC voltage from the DC-link capacitor and feeds back 3-phase AC voltage into the grid in inverter/regeneration mode [12]. The converter controller performs the phase synchronization

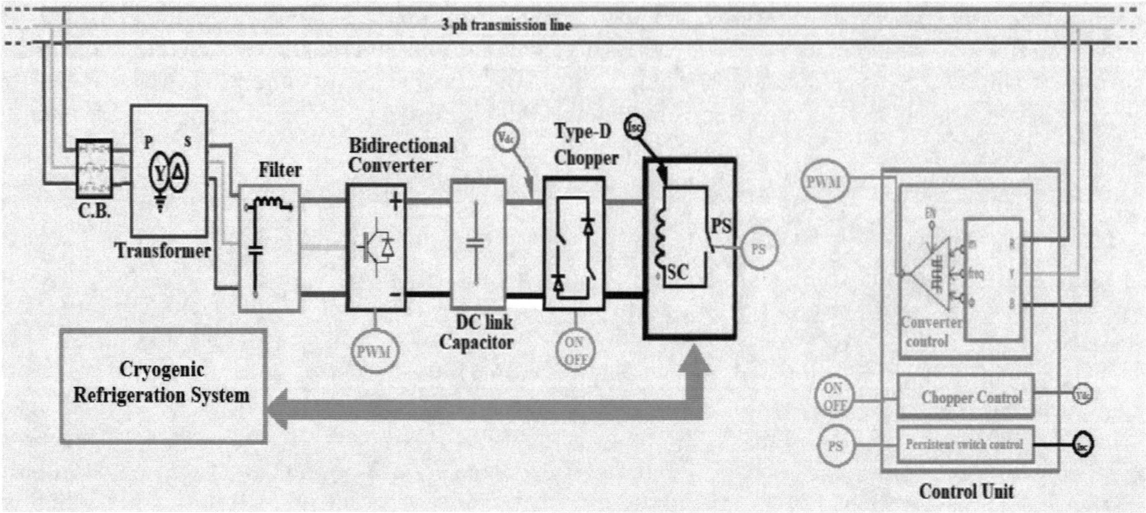

Figure 1. Schematic of SMES-UPS system

and Sinusoidal Pulse Width Modulation (SPWM) generation to trigger the converter switches at specified times to obtain 3-phase AC Pulse Width Modulated (PWM) voltages [13]. The 3-phase AC PWM voltages are then passed through a 3-phase filter to remove the harmonics from the output and a sinusoidal output is fed to the grid [14].

The 3-phase transformer is responsible for stepping down the high voltage on the grid side to the low voltage on the converter side for rectification and stepping up the low voltage on the converter side to high voltage on the grid side during regeneration. The circuit breaker isolates the SMES system from the grid in case of occurrence of any fault within the SMES system. The refrigeration system maintains the temperature of the superconducting coil at the required level to ensure the rated current flowing through the coil without quenching it.

3. Parameter estimation for a stacked solenoid coil geometry

A solenoid made up of stacks of single HTS pancake coils is considered. The coil parameters are estimated assuming each stack is made up of concentric ampere turns with incremental circumference. The input parameters are the stack's inner diameter (ϕ_i), length of HTS tape available (L), tape thickness (t), tape width (w), insulation thickness (t_i), distance between two stacks (d) and number of stacks (N_s), as shown in figure 2.

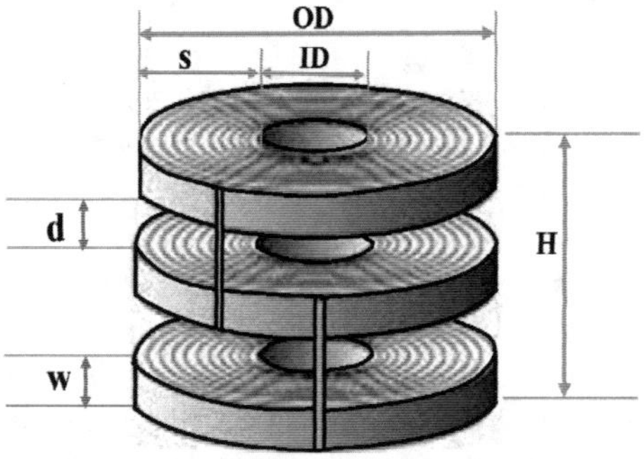

Figure 2. Parameters of stacked HTS solenoid

Arithmetic progression is applied to the incremental circumference with common difference ($t+2t_i$) and the sum of the incremental circumferences equal to (L), to obtain the number of turns (N) using equation (2)

$$N = \frac{(t+2t_i) - (\frac{\phi_i}{2}) + \sqrt{(\frac{\phi_i}{2}) - (t+2t_i)^2 - 4(t+2t_i)(\frac{L}{\pi})}}{2(t+2t_i)} \tag{2}$$

The outer diameter (ϕ_o) is calculated using equation (3)

$$\phi_o = 2[\frac{\phi_i}{2} + N(t+2t_i)] \tag{3}$$

and the height of the coil (H) is given by equation (4)

$$H = (N_s - 1)\frac{d}{2} + N_s(t+2t_i) \tag{4}$$

The difference between ϕ_o and ϕ_i gives the span of the coil (S).

4. MATLAB Simulink model for SMES-PCU

A MATLAB simulation was carried out to observe the load leveling operation of the SMES system when the grid is under overload. The Simulink model is shown in figure 3. In this, a 15 kW resistive load is connected and disconnected to the grid (3-phase 400V L-L, 50 Hz, 10 kW), between 0.02 to 0.12 seconds using timed switching of an ideal switch component, to simulate overload. The corresponding 3-phase grid voltage and current are observed under loaded as well as unloaded conditions.

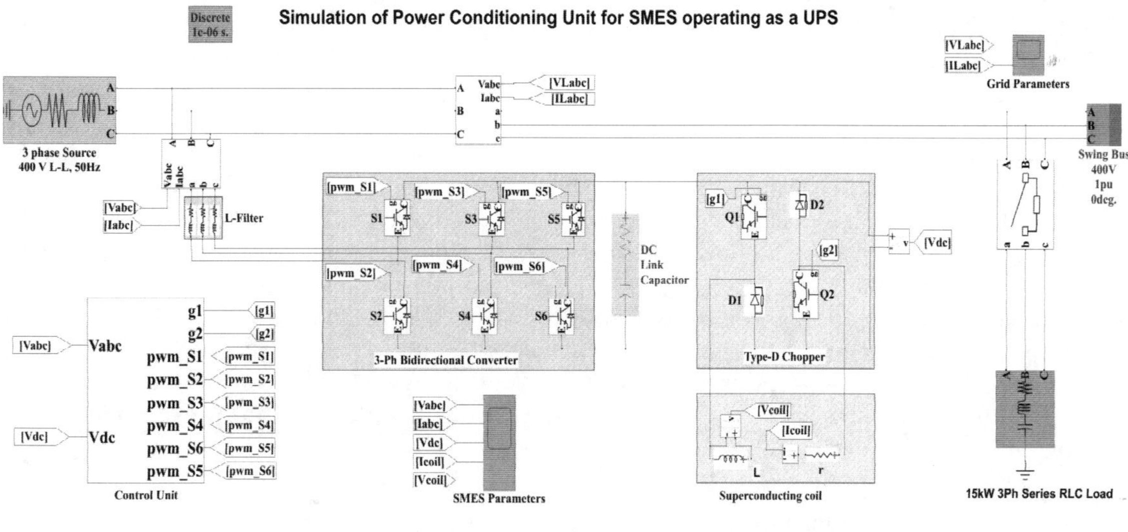

Figure 3. Image of the simulation model for SMES-UPS

The SMES system model consists of a superconducting coil modeled as an inductance (L) of 282 mH, in series with a resistance (r) of 0.01 Ω to represent the current lead and joint resistances. The coil current and voltage are measured. The type-D chopper circuit is modeled as per the schematic shown in figure 4, with S_1 and S_2 taken as IGBTs with on-state resistance (R_{on}) of 0.16 Ω and forward voltage (V_f) of 2.3 V. The diodes have on-state resistance of (R_D)of

0.03 Ω and forward voltage (V_D) of 1.19 V. The parallel RC snubber circuit for each diode has a resistance of 500 Ω and a capacitance of 250 nF. The DC-link voltage is fed back as a reference to the hysteresis band-type chopper controller where it is compared with a fixed reference voltage. If the difference between the actual voltage and the reference voltage is more than the upper threshold voltage, the IGBTs are turned on and current flows from the DC-link capacitor to the superconducting coil to charge it. Similarly, if the difference is below the lower threshold, the IGBTs are turned off and the coil discharges to charge the DC-link capacitor, provided D_1 and D_2 are forward biased. This way, the DC-link voltage is held at a steady value with a suitable tolerance for the ripple voltage.

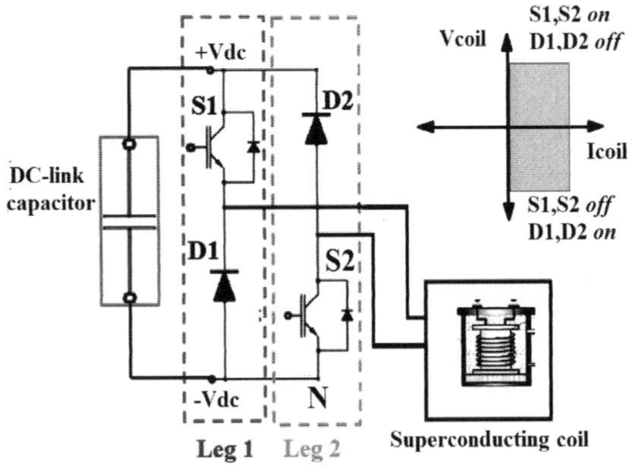

Figure 4. Schematic of type-D chopper circuit

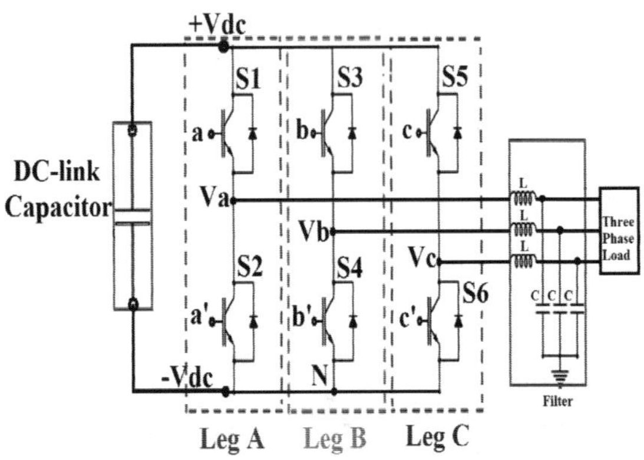

Figure 5. Schematic of Bidirectional converter circuit

A DC-link capacitance of 4700 μF with an ESR of 0.1 Ω is considered for the simulation. The bidirectional converter is modeled using IGBTs with reverse conduction diodes as shown in figure 5 and having a on-state resistance of 0.001 Ω. The controller for the bidirectional converter accepts 3-phase ac voltage from the grid. It is then passed through a Phase Locked Loop (PLL) to obtain the phase angle of the grid voltage. The phase angle from the PLL block is then sent to a MATLAB function generation block which generates 3 reference sine waves of unit amplitude each and a phase difference of 120° between any two signals. The reference waves are then passed through a modulator which essentially multiplies a gain value to the signals to produce 3-phase modulated reference signals. These modulated signals are then compared with a high frequency triangular carrier wave to produce the SPWM signals which are fed to the gate driver circuit. The output of the gate driver is fed to the upper and lower half switches of the bidirectional converter to generate 3-phase AC PWM voltage. The filter circuit is simply an L-type with an inductance of 100 mH and a series resistance of 0.1 Ω. The transformers are not considered in the simulation to reduce time.

5. Results and discussions

Stacked solenoid geometry with parameters as shown in table 1 are considered for magnetostatic analysis using COMSOL Multiphysics® v.5.3a, for two cases as shown in figure 6. It was observed that for the same geometrical parameters of solenoid, the inductance in the configuration with soft iron core and shielding was around 282 mH which is 47 times higher than 6 mH with air core. This allows for a linear increase in energy storage from 67 J to almost 3.2 kJ in the soft iron cored coil. Also, a decrease in magnetic flux penetration through the coil stacks was observed in the soft iron based configuration. This allows for an increase in the operating current of the solenoid due to enhanced J_c-B_c characteristics of the coil, which will further increase the energy storage capacity quadratically.

Table 1. Coil parameters for Magnetostatic analysis using COMSOL Multiphysics

Core Type	μ_r	ID (mm)	N	d (mm)	H (mm)	I (A)	Shielding	L (mH)	E (J)
Air	1	50	383	8	112.3	150	absent	6	67.3
Soft iron	1000	50	383	8	112.3	150	present	282	3169.5

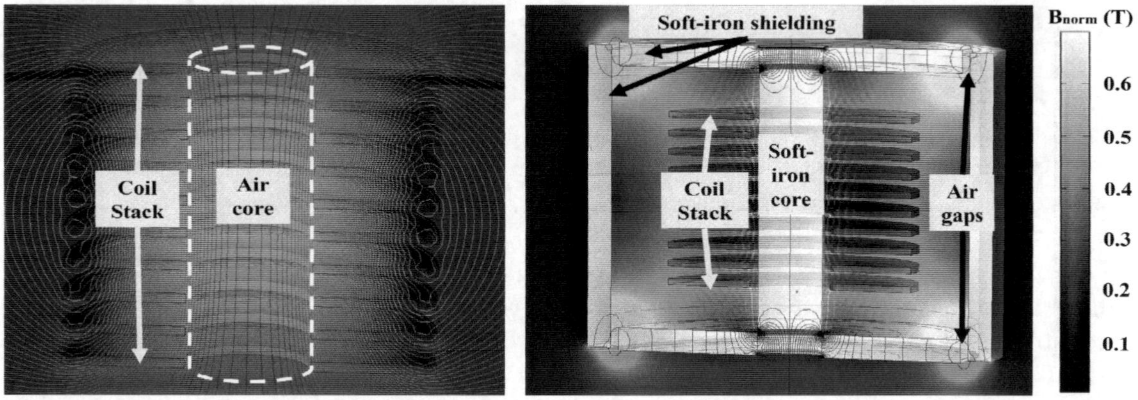

Figure 6. Magnetostatic analysis of stacked solenoid-(a) with air core (b) with soft iron core, shielding and air gaps

Further, from the MATLAB simulation of the SMES-PCU circuit, it was observed that the chopper current and voltage are following the gate pulses in the correct order for the charging and discharging cycles of the superconducting coil (figure 7). Also, a drooping characteristic of the coil current is observed, maintaining the dc link voltage at a steady level even under load.

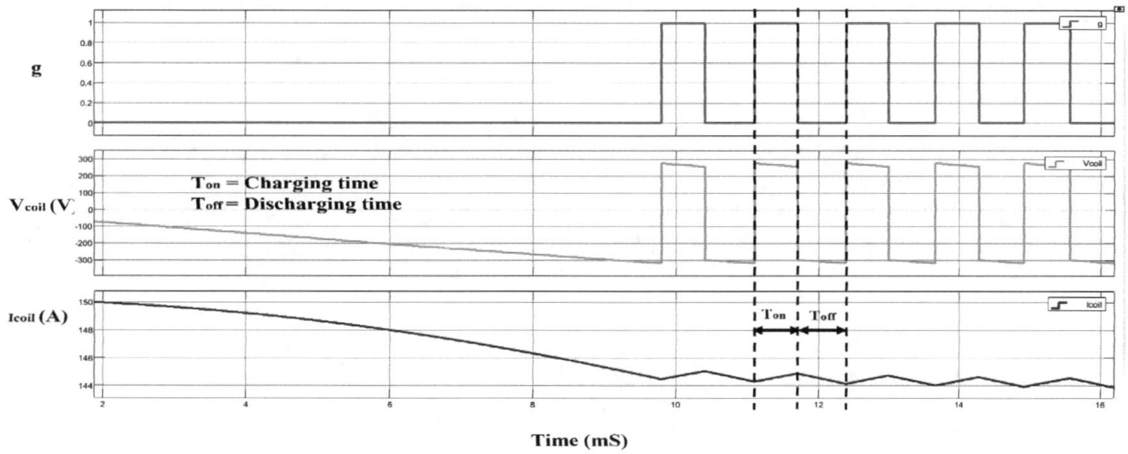

Figure 7. Charging discharging characteristics of type-D chopper and coil current

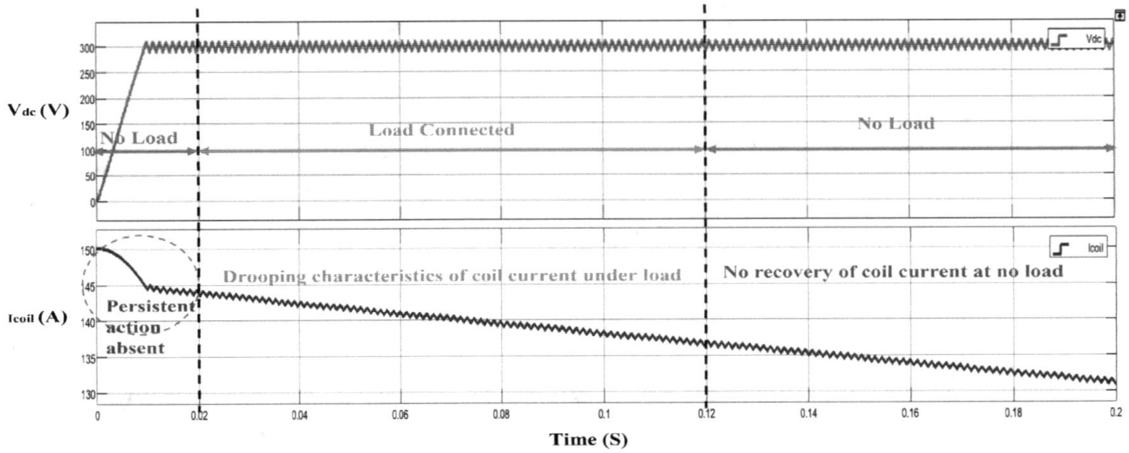

Figure 8. DC-link voltage regulation characteristics

However, there is an absence of effective dynamic regulation of DC-link voltage due to sluggish nature of hysteresis band controller for chopper control in this application. This results in limited voltage regulation as is clear from figure 9 showing load characteristics of the grid with and without SMES system. However, a decrease in the current amplitude from 45 A without SMES system to around 40 A with SMES system connected to the grid is observed. This established the reduced dependence of the load on the grid. Thus, the SMES system is participating in the load leveling operation of the loaded grid, the effectiveness of which can be improvised by inclusion of a current control algorithm in parallel with the hysteresis band controller based voltage control of the chopper circuit.

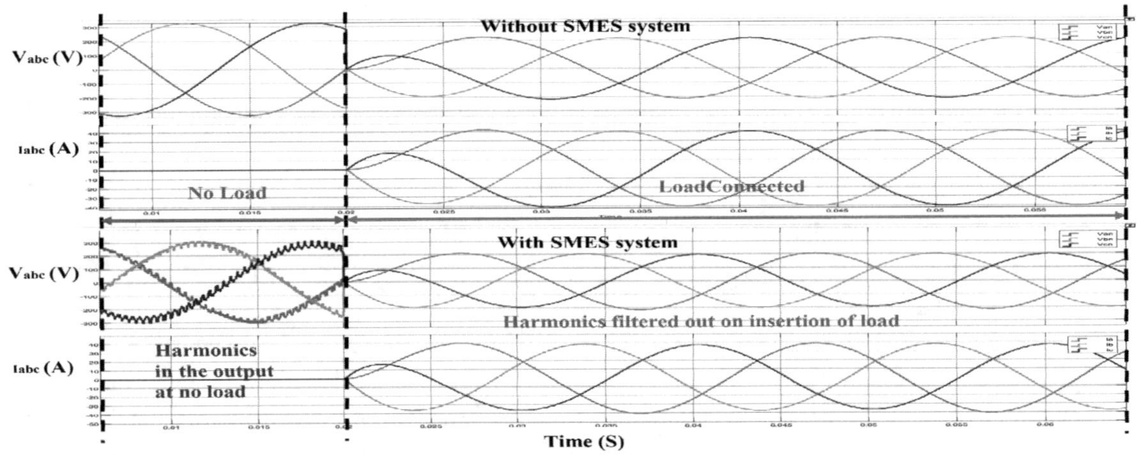

Figure 9. Load characteristics of the grid with and without SMES

Added to this, the grid voltage contains harmonics which motivates the optimization of output filter and development of a persistent switch/action to prevent the coil from discharging when the grid is not loaded.

6. Conclusions

Magneto-static analysis of stacked HTS solenoid using COMSOL shows an enhancement in the inductance and energy storage of the solenoid by 47 times with soft iron core and shielding. Also, the reduced flux penetration with this configuration further allows for more current to be passed through the superconducting coil.

MATLAB simulation of SMES system shows that coil current and voltage follow the operating modes of the type-D chopper. Drooping characteristics of the coil current was observed indicating the discharging of coil to maintain the dc link voltage. Regulation of the grid voltage under load with and without SMES system has been established with limited effectiveness. Nevertheless, a reduction in the load current amplitude by around 11.11 % is observed with SMES system integrated with the grid. This indicates that the SMES system is participating in the load leveling operation. However, it couldn't be captured effectively due to sluggish nature of the Hysteresis band controller alone. A parallel current control algorithm may enhance the performance of system, which must be taken up as a future work.

Nevertheless, grid voltage synchronization algorithm is working as per requirement. But the appearance of harmonics in grid voltage necessitates the design of a better input/output filter for the bi-directional converter and a persistent switch for the superconducting coil.

7. References

[1] www.mckinsey.com, "The potential impact of electric vehicles on global energy systems", (2018). [Online]. Available: https://www.mckinsey.com/industries/automotive-and-assembly/our-insights/the-potential-impact-of-electric-vehicles-on-global-energy-systems.
[2] T.Nitta, Y.Shirai and T.Okada (1985), "Power charging and discharging characteristics of SMES connected to artificial transmission line", IEEE Transactions On Magnetics, Vol. Mag-21, Nn. 2
[3] M. H. Ali, B. Wu and R. A. Dougal (2010), "An Overview of SMES Applications in Power and Energy Systems", in IEEE Transactions on Sustainable Energy, vol. 1, no. 1, pp. 38-47.
[4] C.K. Das, O. Bass, G. Kothapalli, T.S. Mahmoud, D. Habibi (2018), "Overview of energy storage systems in distribution networks: placement, sizing, operation, and power quality", Renew Sustain Energy Rev, 91 , pp. 1205-123. doi: 10.1016/j.rser.2018.03.068
[5] Chapter-9,"*Emergency supply equipment*",[Electrical Systems and Equipment (Third Edition)], (1992). British Electricity International, Pergamon, 748-798, ISBN 9780080405148.
[6] Moller, Kasper & Jensen, Torben & Akiba, Etsuo & Hai-Wen, Li. (2017). "Hydrogen - A sustainable energy carrier", Progress in Natural Science: Materials International. 27. doi: 10.1016/j.pnsc.2016.12.014.
[7] K. E. Nielsen and M. Molinas (2010), "Superconducting Magnetic Energy Storage (SMES) in power systems with renewable energy sources", IEEE International Symposium on Industrial Electronics, 2010, pp. 2487-2492. doi: 10.1109/ISIE.2010.5637892.
[8] McKerracher, R.D.Ponce de Leon, C. Wills, R.G.A. Shah, A.A. and Walsh, F.C., (2015) "A Review of the Iron-Air Secondary Battery for Energy Storage", ChemPlusChem, 80: 323-335. doi: 10.1002/cplu.201402238
[9] J Zheng, S S Peng, W Y Li, Y J Dai. (2019) "Magnet design of 10 MJ multiple solenoids SMES", IOP Conference Series: Earth and Environmental Science, Volume 233, Issue 3, pp. 032026.
[10] Tixador, P. (2012), "Superconducting magnetic energy storage (SMES) systems". *High temperature superconductors (HTS) for energy applications*, Woodhead Publishing Series in Energy, pp. 294-319, doi:10.1533/9780857095299.2.294.
[11] P. Mukherjee and V. V. Rao (2018), "Fuzzy Logic Controlled Superconducting Magnetic Energy Storage for Leveling Power Fluctuation of Grid Connected Wind Generator", Int. Conf. on Power Energy, Environment and Intelligent Control (PEEIC), Greater Noida, India, pp. 665-669, doi:10.1109/PEEIC.2018.8665563.
[12] M. Steurer, C. A. Luongo, P. R. Ribeiro and S. Eckroad (2003),"Interaction between a superconducting coil and the power electronics interface on a 100 MJ SMES system", in IEEE Transactions on Applied Superconductivity, vol. 13, no. 2, pp. 1806-1809, doi: 10.1109/TASC.2003.812895.
[13] Gour, Abhay. (2016). "Design and Development of Linear Moving Magnet Synchronous Motor based Twin PTC and HTS Level Sensor for LOX Recondenser", Ph.D dissertation, Indian Institute of Science, Bengaluru, Karnataka - 560012, India.
[14] M.H. Rashid (2010),"*Power Electronics Handbook*", 3rd edition, Butterworth-Heinemann, pp 183-196.

IOP Conf. Series: Materials Science and Engineering 1241 (2022) 012037 doi:10.1088/1757-899X/1241/1/012037

Test of a DC-HTS Busbar Demonstrator for Power Distribution in Hybrid-Electric Propulsion Systems for Aircraft

S I Schlachter[1], J Brand[2], S Elschner[3], S Fink[1], B Holzapfel[1], A Kudymow[1], J Willms[1]

[1] Karlsruhe Institute of Technology, Institute for Technical Physics, Hermann-von-Helmholtz-Platz 1, 76344 Eggenstein-Leopoldshafen, Germany
[2] Ingenieurbuero Brand, Karlstrasse 8, 68794 Oberhausen, Germany
[3] University of Applied Science, Paul-Wittsack-Str. 10, 68163 Mannheim, Germany

Corresponding author: sonja.schlachter@kit.edu

Abstract. In the framework of the German project TELOS (Thermo-Electrically Optimised Aircraft Propulsion Systems) a high-temperature superconducting 40 MVA DC demonstrator busbar for hybrid-electric propulsion systems for aircraft has been developed. The design current for a temperature below 25 K is 13.3 kA and the rated voltage is 3 kV. The 2-pole busbar contains 2 stacks of REBCO coated conductors which are supported by a 3D-printed structure allowing compensation of thermal length changes of the superconductor. It fits in a cryostat tube with an inner diameter of 25 mm. A special focus has been put on low-resistive joints that are necessary to connect single elements of the busbar system. The special layout of the joints allows an effective current redistribution between the different tapes in a stack. We present results for the test of the DC busbar demonstrator in liquid nitrogen at 77 K. The design current for this temperature is 3.3 kA which corresponds to a rated power of 10 MW. We applied currents up to 3.5 kA and measured the I-V characteristics and contact resistances of 90° and 180° joints in a virgin and in a strained state thus simulating thermal length changes. We also present results of Lorentz-Force tests with short AC current pulses up to 20 kA to demonstrate the viability of the design for application with currents up to 13.3 kA.

1. Introduction

The climate change caused by the increasing emission of greenhouse gases is affecting people all over the world. In 2015 the civil and military air traffic contributed 875 million tons of CO_2, i.e. about 2.5%, to the global anthropogenic CO_2 emissions [1]. The predicted yearly growth in air-travel is 4-5 %, i.e. the global flight demand doubles every 15 years [2]. In order to keep further temperature increases due to the greenhouse effect on an acceptable level, greenhouse gas emissions, e.g CO_2 or NO_x, have to be reduced drastically. In “Flightpath 2050” the European Commission has set goals for the aviation industry to be reached until 2050: 75% reduction in CO_2 emissions per passenger kilometer, 90% reduction in NO_x emissions, emission-free aircraft movements when taxiing and reduction of noise emission of flying aircraft by 65% - all numbers compared to a typical new aircraft of the year 2000 [3]. Reaching these ambitious goals just by increasing the efficiency of today’s technologies is very difficult – rather a breakthrough of new technologies for the aviation sector is needed. One promising way to achieve the goals of “Flightpath 2050” is the electrification of aircraft propulsion systems, however, batteries are too heavy to power drive trains with power in the range of several 10 MW. Hybrid-electric propulsion systems with gas turbine and generator allow generation of power in the required range on board of the aircraft. Superconductivity would substantially support this technology, because components of the drive-train like motors, generators and cables could be realized with improved power-to-weight ratios. Using cryogenic hydrogen both as coolant and fuel would result in an effective reduction of greenhouse-gas emissions. In the framework of the German project "TELOS" (Thermo-Electrically Optimised Aircraft Propulsion Systems) six partners, Airbus Group Innovations, Airbus Operations GmbH, Karlsruhe Institute of Technology, Siemens AG, Neue Materialien Bayreuth GmbH, and Technical University of Munich, joined in 2016 to develop a technological basis for a hybrid electric propulsion system for aircraft on an appropriate power level for the high power class.

A basic architecture for a cryogenic hybrid-electric propulsion system is described in detail in [4]. Two power generation units with superconducting generators which are driven by turboshaft engines are placed in the fuselage of the aircraft. The electric propulsion units (EPUs) which deliver the required thrust with superconducting motors and fans are placed in the wings of the aircraft. A DC power-distribution system with cryogenic AC/DC and DC/AC converters connect the generators and EPUs.

In this paper we focus on results obtained within the work package "Conductor Technology" in which a superconducting power distribution system for the hybrid-electric propulsion system should be developed. Originally the rated power was set to a value of 10 MW in the TELOS-project, however, to cover the requirements for the electrification of the Airbus A320 class the targeted power was later increased to a value of about 40 MW in the framework of "TELOS-Upgrade", an amendment to the original TELOS project.

The paper is arranged as follows: In section 2 the design of the two-pole DC cable is described briefly and results from a self-consistent calculation of critical currents, magnetic fields and Lorentz forces are presented. Low-resistive joints are required to connect different cable sections. Results for joints which allow current redistribution between different tapes of a single pole are described in section 3. In section 4 the test of a short demonstrator cable in liquid nitrogen is presented and test results are being discussed.

2. Power distribution system

In the framework of the TELOS project a DC power distribution system on the basis of high-temperature superconductors should be developed. The advantage of a DC architecture is the decoupling of power and thrust generation units by power electronic converters which allow running generators and motors at an appropriate speed as required by the varying phases during the flight [5]. E.g. it is possible to run the generator at high speed while the motors run at lower speed with higher torque. In case of distributed propulsion systems with many motors it is also possible to run the motors at different speed, e.g. during the cruise phase. Another advantage of DC power distribution are the reduced AC losses. Even in DC distribution systems AC losses cannot be avoided completely as the rectified signal usually contains residual ripples. However, as the ripple amplitude in the rectified signal is usually small compared to the original AC amplitude, even superconductor cable architectures like stack cables can be applied.

2.1. Cable Layout

We decided to design the DC power distribution system as a busbar system with two poles (+1.5 kV, -1.5 kV) in one cryostat. A simple stack architecture for the two poles of the cable was chosen (Figure 1 and Figure 2). This allows to change the number of tapes easily to account for changed specifications with respect to current carrying capability of the cable and also for differences in the quality of commercial superconductors. In order to minimize size and weight of the cable we decided to limit the outer dimensions of the cable assembly including superconducting tapes and 3D printed wave-guide structure to a maximum diameter of 25 mm because in TELOS we used standard steel tubes with inner diameter of 25 mm as inner cryostat tube as shown in Figure 2.

To account for simple transportation and installation, the busbar system consists of single elements with a maximum length of 6 meters. In order to connect single elements, low-resistive 180°-, 90°- and

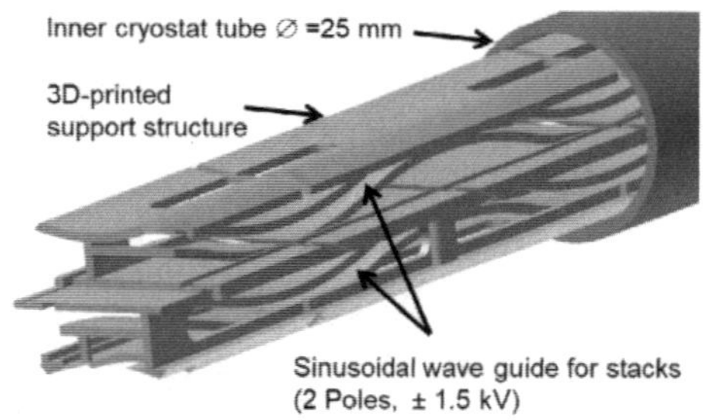

Figure 1: CAD drawing of inner cryostat tube and 3D-printed former with wave-guide for REBCO tape stacks .

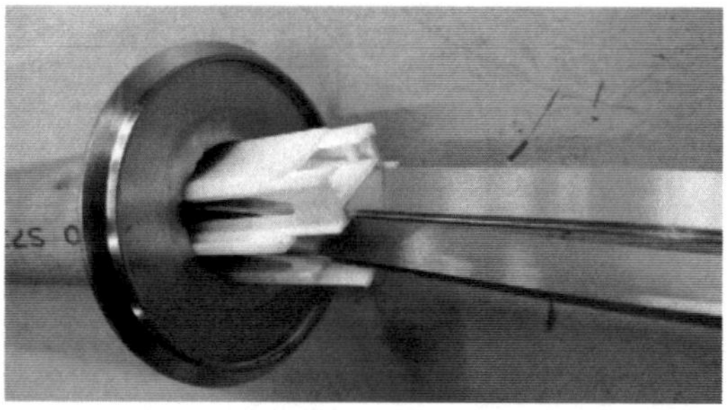

Figure 2: Photograph of inner cryostat tube with 3D-printed former and a tape stack for each pole of the cable.

T-joints had to be developed. The stack design of our two-pole cable allows face-to-face connection of the superconducting tapes so that low resistive joints could be realized.

We also decided to compensate thermal length changes of the cold cable sections within the busbar structure, avoiding compensation elements at the cable ends with a size that depends on the cable length. Compensation of thermal length changes was achieved by placing the stacks in a 3D-printed support structure made of glass-bead filled polyamide 12 with sinusoidal wave guides for the stacks. Contraction of the HTS-tapes tapes during cool-down can easily be compensated by a change of the wave amplitude. The former not only helps to compensate thermal length changes, but also ensures that the distance between the stacks is kept constant and that the REBCO stacks do not touch the inner cryostat tube.

We made several high-voltage tests to ensure that the cable-design is suitable to withstand voltages of 3 kV and more between the two poles or between one pole and the inner cryostat tube. The tests were made at room temperature in air atmosphere as a worst-case scenario. No voltage breakdowns happened at 3 kV with exception of one sample joint damaged by welding which needed repair. Voltage breakdowns appeared in air on few samples for values between 4 kV and 5 kV. However, under the foreseen operation conditions with subcooled nitrogen or liquid hydrogen the high voltage breakdowns can be expected to appear at much higher values. Fault scenarios with fluid and pressure loss would require an adapted improvement of the high voltage design.

2.2. Critical currents, magnetic fields and Lorentz forces

To calculate the current carrying capacity, the field distribution and the Lorentz forces due to the self-field for the 2-pole HTS cable, an existing MATLAB code was modified [6]. The self-consistent model uses the E-J relationship $E/E_c=(J(B)/J_c(B))^n$ for superconductors where E is the electric field, E_c the electric field criterion to determine the critical current density J_c, B is the magnetic flux density and the exponent n is a measure for the homogeneity of the material and the steepness of the transition from superconducting to normal behavior. As the current carrying capacity of REBCO superconductors is strongly dependent on temperature, magnitude and direction of the magnetic field, we used a parameter-free routine described in [7] to calculate the $J_c(B,\theta)$ dependence for a Theva ProLine conductor[1] for which experimental $I_c(B,\theta)$ data were available from the HTS database of the University of Wellington [8]. We mainly concentrated on two different application cases which represent the goals of the two project phases, TELOS and TELOS-Upgrade: distribution of a nominal power of 10 MW with an LN_2 cooled cable system (subcooled pressurized LN_2, $T \approx 70$ K) and distribution of a nominal power of 40 MW with a hydrogen cooled cable system at ($T \approx 20$-25 K). For both application cases a voltage of 3 kV (±1.5 kV) was assumed, resulting in nominal currents of 3.33 kA and 13.33 kA for electric power levels of 10 MW and 40 MW, respectively.

[1] THEVA Pro-Line 2G HTS Wire, TPL4420c, Manufacturer sample designation: ID 180152

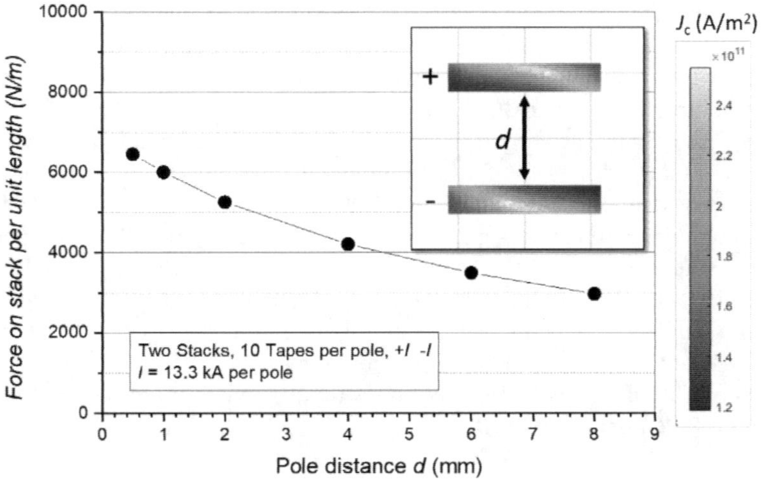

Figure 3: Calculated Lorentz forces on single stacks of a 2-pole stack cable versus stack distance for a current of 13.3 kA The inset shows the absolute values of the calculated current density distribution in the stacks of the cable with Theva-ProLine-conductor at $T = 20$ K. The calculated current density corresponds to an assumed REBCO-layer thickness of 1 µm. Real Theva conductors have REBCO layer thicknesses in the order of 3 µm. i.e. the real current densities are approx. a factor 3 lower.

In a first step we assumed having a cable with two stacks of 12 mm wide REBCO tapes, one for each pole of the cable. Each tape stack consists of 10 REBCO tapes. We calculated current densities, magnetic field distributions and Lorentz forces for different distances of the two stacks and for different temperatures. The current distribution in the two poles of the cable calculated for a temperature of 20 K is shown in the inset of Figure 3 (absolute values). The distribution is not symmetric with respect to the x-axis which is parallel to the tape surface. This is a consequence of the c-axis tilt of the superconductor layer due to the inclined substrate deposition process applied by Theva [9]. Due to the different opposite current directions in the two stacks, the magnetic field strength is high between the two stacks and decreases very fast outside the stack assembly, resulting in low stray fields outside the cryostat. The asymmetric current distribution results in Lorentz forces which do not only act in the y-direction on the tapes (perpendicular to the tape plane), but also in the x-direction, i.e. parallel to the tape plane. Therefore, the two stacks need to be stabilized not only against the repelling forces perpendicular to the tape plane, but also against forces parallel to the tape plane. In Figure 3 the Lorentz force on a single stack is shown as a function of the stack distance. If the stacks are close together the forces per unit length exceed values of 6 kN/m for a current of 13.33 kA corresponding to a power level of 40 MW at

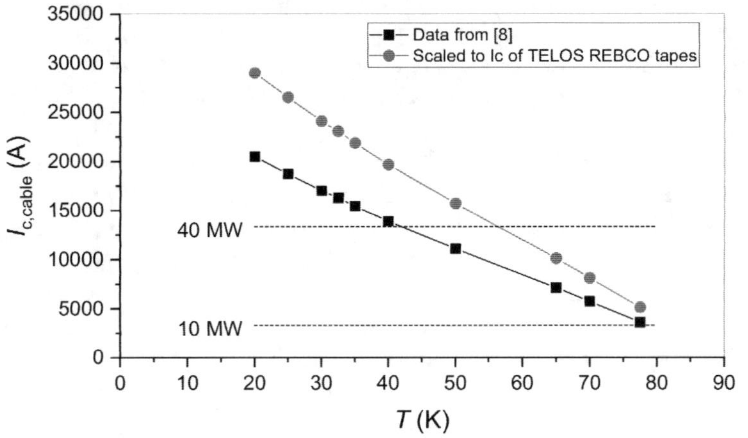

Figure 4: Calculated critical currents for a 2-pole cable with 10 Theva ProLine tapes in each stack (stack distance = 7 mm). The black squares were obtained from calculationswith $I_c(B,\theta)$ data from [8] while the red circles were obtained by scaling theses data to the higher current carrying capability of the Theva tapes used within the TELOS project.

3 kV. In order to minimize the Lorentz forces as much as possible we used a stack distance of 7 mm for our cable, resulting in Lorentz forces of 3.2 kN/m on the stacks. This allowed us to realize the wave-like guiding of the stacks for compensation of thermal length changes still keeping the diameter of the cable assembly of support structure and REBCO stacks below our limit of 25 mm.

Figure 4 shows the critical currents of the cable calculated with temperature dependent $I_c(B,\theta)$ data from [8] for different temperatures. According to the calculations, critical currents of 5.7 kA and 20.4 kA can be achieved at temperatures of 70 K and 20 K, respectively, with 2x10 12 mm wide Theva ProLine tapes. For the two application cases, power distribution of 10 MW at 70 K and 40 MW at 20 K, the cable would operate at 58% and 65% of the critical current, respectively. The Theva conductors that have been used in the TELOS-project, have an approximately 40% higher current carrying capability than the conductor from Ref. [8]. Therefore, we simply scaled the calculated data to the current carrying capability of our tapes to see up to which temperatures our cable could operate with currents of 3.3 kA and 13.33 kA. An electrical power of 10 MW could be distributed up to temperatures of ~ 80 K and a power of 40 MW up to temperatures around 55 K.

3. Joints

Since a modular structure is intended for the cable in the aircraft, it is necessary to provide couplings between such modules. 180°-connections as well as 90°- and T-joints are necessary, and in the long term any angular connections could be required. Besides cryostat and ripple losses, resistive joints are a major source of losses in DC cables. Especially at low temperatures, the cooling penalty is high due to the low Carnot efficiency and small amounts of heat can lead to strong temperature increases. In aerospace applications cooling power is usually limited because either a cryogen has to be carried on board the aircraft or space vehicle or the electric energy for mechanical coolers needs to be provided. To keep losses at a minimum, we set an upper limit for the losses of joints in our busbar system. Resistive losses of a joint in one pole should be below 1 W, i.e. a 2-pole cable coupling should have resistive losses below 2 W. For a current of 13.33 kA the resistance of a single joint thus needs to be below 5.6 nΩ. As a comparison, the heat load per unit length of commercial cable cryostats, are often in the range of 1-2.5 W/m [10].

In the past, it has been demonstrated that REBCO tapes can also be superconductively connected. Park et al. achieved a resistance $< 10^{-17}\,\Omega$ for a joint of $GdBa_2Cu_3O_{7-\delta}$ tapes [11]. However, for connecting two cable ends during installation in an aircraft, this complex process of contacting is not suitable. To achieve low-resistive joints between two REBCO tapes, face-to-face (ftf) soldering, i.e. soldering the tapes with the superconductor sides together is commonly used. Resistances of $R{\cdot}A$ (A = contact area) in the range of several 10 $n\Omega{\cdot}cm^2$ have been obtained for joints between single tapes, however, the achievable values also depend on the interlayer resistances between REBCO layer, silver layer and Cu layer and depend strongly on the manufacturing process [12].

One major problem in stacks of REBCO tapes is the redistribution of currents, due to the insulating buffer layers in the REBCO tapes. While currents flowing from one tape to the other can cause resistive losses especially in AC applications, DC applications can benefit from the possibility of current redistribution in case of single tape failure (e.g. due to local heating, mechanical damage or varying magnetic fields along the cable length). As the tapes in the stacks of the TELOS cable are only loosely

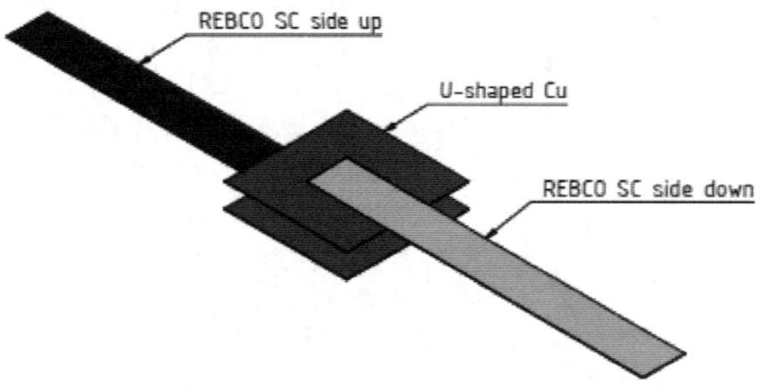

Figure 5: Face-to-Face soldering of ith tape of stack S1 to i^{th} tape of stack S2 (i=1-10). U-shaped Cu plates were added to facilitate transverse current flow between tapes in different layers of the stack.

Figure 6: Joints of stacks with ten 12 mm wide tapes, REBCO tape overlap area of 12 x 12 mm^2 and a total joint area of 36 x 36 mm^2 : a) 180° joint, b) the two 90° joints of the 2 poles of the TELOS cable demonstrator, and c) T-joint of two stacks.

connected throughout the cable section, a good connection of the tapes in the joint sections can facilitate current redistribution. We developed a special technique to connect the ith tape of a stack S1 with the ith tape of a second stack S2 (i: number of layer in stack, i=1...10) by direct ftf soldering to achieve low longitudinal resistances while ensuring also low transverse resistances between the single tapes. We used an overlap length of the two stacks which equals the tape width, i.e. the contact area was 12 mm x 12 mm in the case of 180° joints as well as in the case of 90° joints. In order to facilitate the transverse current flow, we added U-shaped Cu plates (size 36 mm x 36 mm) in each layer of the stack (Figure 5).

Before the final assembly of the cable demonstrator we fabricated several 180°- and T-joints and measured the longitudinal and transverse resistances. We used 12 mm wide tapes from SuperPower (50 μm substrate thickness) or Theva (100 μm substrate thickness) with 20 μm surround Cu coating. The thickness of the U-shaped Cu plates was 100 μm and 170 μm for joints with SuperPower and Theva tapes, respectively. Sn63Pb37 foil was used as solder between the single REBCO/Cu layers in the joint. The longitudinal and transverse resistances between single tapes of two different stack joints with SuperPower tapes are shown in Table 1 and Table 2.

Table 1. Measured resistances R (10^{-9} Ω) between single tapes from stack S1 layer Ln and stack S2 layer Lm for joint #1 .

R (10^{-9} Ω)	S2-L1	S2-L2	S2-L5	S2-L6	S2-L10
S1-L1	46	78	218	236	327
S1-L2	105	42	168	190	286
S1-L5	206	152	39	89	232
S1-L6	243	195	108	38	198
S1-L10	345	300	250	216	39

Table 2. Measured resistances R (10^{-9} Ω) between single tapes from stack S1 layer Ln and stack S2 layer Lm for joint #2 .

R (10^{-9} Ω)	S2-L1	S2-L2	S2-L5	S2-L6	S2-L10
S1-L1	45	93	220	232	330
S1-L2	117	44	166	180	286
S1-L5	210	151	39	77	229
S1-L6	238	186	98	43	197
S1-L10	360	313	260	228	60

Both joints were soldered simultaneously in a specially constructed soldering device to simulate the soldering process in a 2-pole cable. The longitudinal and transverse resistances between five of the tapes of each stack were determined in LN_2 by applying currents between two single tapes of the stacks and measuring the voltage drops. The longitudinal resistance values of the direct ftf transitions within a stacking plane (i.e. between layer i of stack S1 and layer i of stack S2) are in the range of 38-60 nΩ. The average values for joint #1 and joint #2 are 40.8 nΩ and 46.2 nΩ, respectively, corresponding to total longitudinal joint resistances of 4.08 nΩ and 4.62 nΩ for the parallel connection of the ten such ftf connections in the joints of stacks with 10 tapes. The resistive losses for such a joint would be below 1 W for a current of 13.3 kA, i.e. below the maximum value set as a goal for the TELOS cable. Taking the decreasing resistivities of metallic components in the joints into account when going from 77 K to 20 K the losses should become even smaller for applications in liquid hydrogen. The transverse resistances between tapes of different layers increase with increasing layer distance, being in the range of 327-360 nΩ for the outer tapes of the stack, i.e. between tapes of layer 1 and layer 10. For T- and 90°-joints we observed similar resistance values as for 180° joints. The low transverse resistances allow the redistribution of currents within the joints.

4. TELOS SC Cable Demonstrator

In the framework of TELOS-Upgrade a superconducting demonstrator cable was designed for a power transmission of 40 MW at $T = 20$-25 K (hydrogen cooling). This power corresponds to a current of 13.33 kA at a nominal voltage of 3 kV. Since it was not possible to realize hydrogen cooling within the time and budget framework of the TELOS project, it was already decided in the application phase of TELOS-Upgrade to test the demonstrator cable in a nitrogen bath at 77 K. A nominal current of $I_{nom,77K}$ = 3333 A was targeted for this test, which would correspond to a transmission power of 10 MW at U_{nom} = 3 kV. This power corresponds to the value that was originally targeted before the realignment through the TELOS-Upgrade. With the proof of a current-carrying capacity of $I_{nom,77K} = 3.33$ kA at $T = 77$ K and the assumption of an at least 5 times higher current-carrying capacity at $T = 20$ K as shown in Figure 4, indirect proof of a sufficient current-carrying capacity at 20 K can thus be provided. Besides the demonstration of sufficient current carrying capability one of the major goals was to demonstrate the realization of low-ohmic joints with resistances below 5.6 nΩ. Therefore, the 2-pole demonstrator cable was designed with 3 superconducting cable sections connected by 180°- and 90°-couplings and a short-circuit connection at one end (Figure 7). A special device for straining SC section 2 was built to demonstrate compensation of thermal length changes by wave-like guiding of the REBCO stacks. A major goal for the test of the demonstrator cable was the proof that it can withstand the high Lorentz forces acting on superconductors and support structure at I =13.33 kA. Lorentz force tests at 77 K were carried out with 50 Hz AC currents with amplitudes up to 20 kA and a limited duration of 40 ms. The maximum currents in these tests were far above the critical current of the cable, however, since the pulse duration was very short, the cable was not expected to be destroyed by the thermal load. The following series of tests was made: DC test #1 up to $I = 967$ A (test measurement, then minor modifications to the measurement program), DC test #2 up to $I = 3450$ A, DC test #3 (current ramp to 1 kA, stretching of

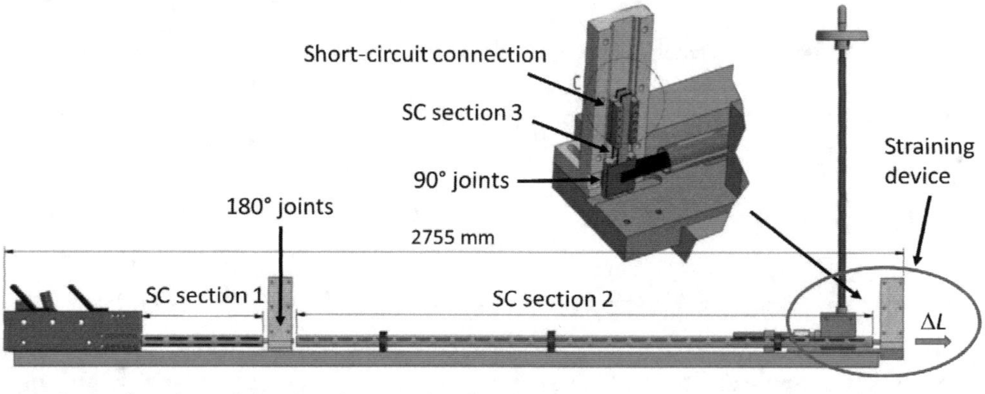

Figure 7: CAD drawing of the 2-pole TELOS demonstrator cable consisting of 3 SC cable sections connected with 180°- and 90°-joints. The two poles of the cable are connected with a short-circuit connection at one cable end. Length changes of SC section 2 (length = 1.8 m) can be performed with the straining device.

cable with ΔL=+7 mm then current ramp up to 3410 A), Lorentz force test up to 20 kA (AC, 40 ms), DC test #4 up to I = 3500 A and DC test #5 up to I=3470 A.

4.1. Initial DC test

Figure 8 shows the voltage drops measured at the 2 poles of 3 superconductor (SC) sections during DC test #2. Although the sample voltages did not show any abnormalities during pre-tests in warm conditions, it quickly became apparent that the contact to the superconductors deteriorated during the test campaign in liquid nitrogen. We suspect that in the area of the soldered contacts, flux ran between the HTS tapes by capillary action, leading to an oxidation of the tape surface. The forces occurring during the measurements may have caused a displacement of voltage taps just sticking loosely in the tape stacks. This manifested itself in an increasing, high noise. The voltage signal of pole 2 of SC section 1 was so noisy that an evaluation was not possible throughout the whole test campaign. Strongly scattering data points obviously caused by noise were masked for a clearer view. The voltage signal of pole 1 in SC section 3 shows a slight resistive increase up to 3270 A, but its values are far below the E-field criterion of 1 μV/cm. Above this current, a sharp voltage increase was initially seen, after which the values fell back to the original, slightly rising curve. Since the voltage signal was very noisy, it is not possible to say with final certainty whether the increase was merely caused by bad contacts or whether there was a current redistribution that caused the signal to increase for a short time. Above 3390 A, a stronger noise occurred, which can possibly be explained by the fact that the current was changed in smaller steps and also reduced and increased several times in between. The voltage drops at the 180°- and 90°-joints of pole 1 and 2 are shown in Figure 9 versus the applied current. The voltage drops measured at the maximum currents of 3450 A correspond to resistances of 1.86 nΩ and 1.84 nΩ for the 180° joints of pole 1 and 2, respectively and to resistances of 2.93 nΩ and 3.62 nΩ for the 90°-joints of pole 1 and pole 2, respectively. These values are well below the maximum design values of 5.6 nΩ which correspond to ohmic losses of 1 W at I=13.33 kA.

4.2. Length change

For DC test #3, the current was first increased in 50 A steps from zero to 1000 A and then kept constant. At constant current, the superconductor section 2 between the 180°- and the 90° joints of the demonstrator cable was stretched by 7 mm, i.e. by $\Delta L/L$=0.38%. A negative influence of the stretching of the cable on the voltages different sections of the cable could not be observed. Even after stretching, the cable demonstrator easily carried the nominal current of 3.3 kA.

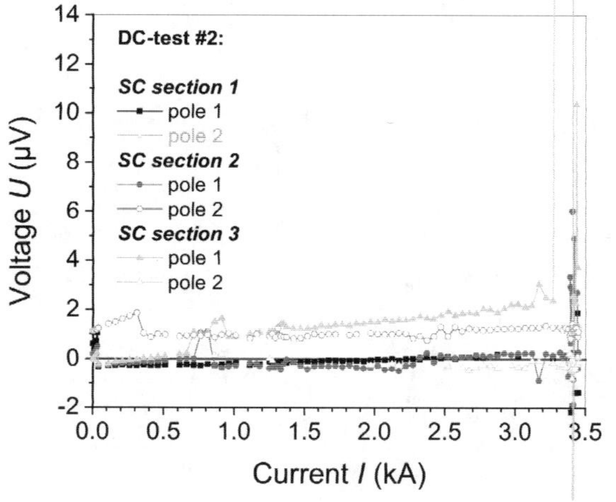

Figure 8: Voltages measured at the SC sections of the 2-pole cable versus applied current.

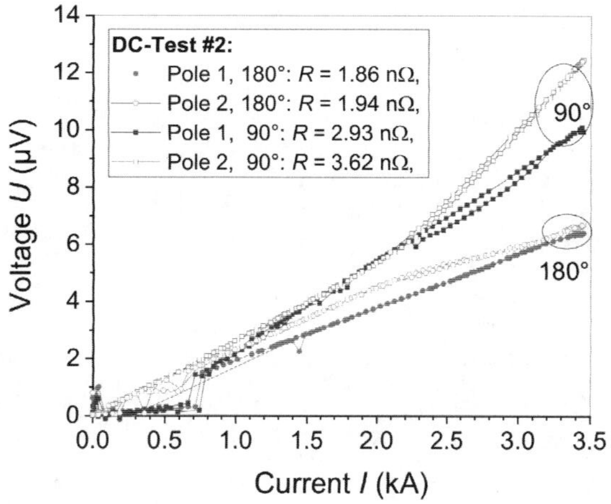

Figure 9: Joint voltages measured during DC test #2 versus current.

4.3. Lorentz force test

Following DC test #3, a Lorentz force test was carried out. The purpose of this test was to show that the cable is able to compensate the Lorentz forces occurring at a current of 13.33 kA. Since a current of 13.33 kA far exceeds the current-carrying capacity of the cable demonstrator at 77 K, AC currents of short duration (2 periods, 50 Hz) were applied in order not to heat up the conductor too much. The current amplitude was successively increased in several tests up to a maximum amplitude of 20 kA. Current and total cable voltage signals are shown in Figure 10 versus time. The maximum current amplitude of 20 kA corresponds to a maximum force per unit length of about 7.3 kN/m acting on the individual tape stacks (Figure 11). The low total voltages during the measurements with maximum currents of 0.843 KA and 1.392 kA were not measurable with the experimental setup due to the low resistivity of the cable system. Up to currents of 3.74 kA we observed a clear phase shift between current and voltage signal, indicating a non-negligible inductance of the entire circuit. In these ranges of maximum currents, the resistive part of the voltage signal is essentially caused by the resistive joints, since the resistivity of the REBCO tapes in the superconducting state is low (AC losses or incipient resistivity). In the measurements with a maximum current amplitude of 11.98 kA and higher, the phase shift becomes considerably smaller. However, the voltage curves do not show a sinusoidal behavior, but show a small shoulder. This behavior is probably due to the fact that the REBCO tapes are initially in the superconducting state at low currents, i.e. the resistivity of the system is low. As the current increases, the resistivity of the REBCO tapes increases strongly due to the critical current being exceeded, so that the voltage curve is essentially determined by the resistive component. In the measurements with the two highest maximum current amplitudes ripples appeared in the voltage signals. It is suspected that these ripples were caused by movement of the tapes near the short-circuit connection, where the tape stacks were laid over a short distance without formers.

The AC Lorentz-force test corresponds to a worst-case scenario since it is to be expected that the pulsating forces varying between zero and the maximum force represent a considerably higher load for the structure than slowly rising and falling forces occurring in the DC case.

4.4. Final DC tests

Following the Lorentz force test, further DC tests were carried out in order to rule out any damage to the superconductors. During DC test #4, a current ramp of up to 3500 A was applied to the cable. The voltages in the superconductor sections kept below the I_c-criterion although a slight increase was detected at currents above 3 kA. The voltage across the 90°-joint of pole 2 showed a steep increase at $I > 3.1$ kA, indicating a possible damage. Nevertheless, the cable was still able to carry a higher current than the design current of 3.3 kA at 77 K. In a 5th DC test, a rapid voltage increase occurred in SC section 3 at a current of 3480 A, i.e. between the 90°-joint and the short connection of pole 1 and 2, resulting in

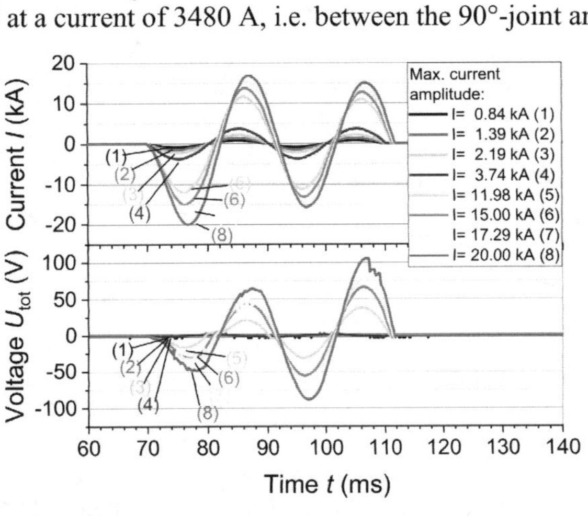

Figure 10: AC currents (50 Hz, 40 ms) and voltages versus time during Lorentz-force test.

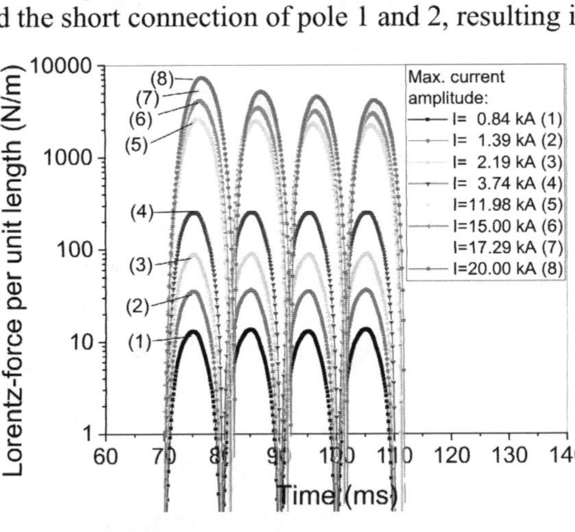

Figure 11: Calculated Lorentz forces on stacks versus time during Lorentz-force tests.

the destruction of the superconductor stack in this area. Possible causes for the occurrence of the defect include damage to individual tapes during the Lorentz force test. Damage was to be expected in SC section 3 in particular, as the SC stacks in this area were not supported by a support structure as in SC sections 1 and 2. In the area of SC section 3, only G10 supports were used to stabilize the contacts (see inset of Figure 7). The superconductor stacks themselves were free to move between the G10 supports and were strongly bent at the short connection. The screwed short-circuit connection between poles 1 and 2 proved to be another weak point. The resistance was more than 12 μΩ, resulting in a heating power of 120 W at $I = 3.5$ kA. Heating effects may have led to the destruction of SC section 3 of pole 1. The long sections 1 and 2, supported by the regular 3D printed structure, definitely were not affected.

5. Summary

We developed and tested a 2-pole DC busbar for power transmission in hybrid-electric propulsion systems for aircraft. Two application cases were considered for a voltage level of 3 kV: transmission of 10 MW electrical power with nitrogen cooling ($T \approx 70$ K) and transmission of 40 MW with hydrogen cooling at $T = 20$-25 K. We developed a 3D-printed support structure that allows compensation of thermal length changes and withstands the high Lorentz forces occurring at currents of 13.33 kA and higher. A demonstrator cable with 180° and 90° joints was tested at 77 K in LN_2. The critical current was above 3.3 kA and the resistive losses of single joints were below 1 W. A length change $\Delta L/L = 0.38\%$ at $I = 1$ kA did not deteriorate the cable properties. An AC Lorentz force test with AC currents up to 20 kA was successful in the regularly supported parts of the cable.

6. References

[1] Lee D S 2018 Available: https://assets.publishing.service.gov.uk/government/uploads/system/uploads/attachment_data/file/813342/non-CO2-effects-report.pdf
[2] Ansell P J and Haran K S 2020 *IEEE Electrification Magazine* **8** 18-26, June 2020
[3] European Commission 2011, Flightpath 2050 Europe's Vision for Aviation. Report of the High Level Group on Aviation Research, doi: 10.2777/50266
[4] Boll M, Corduan M, Biser S, Filipenko M, Pham Q H, Schlachter S, Rostek P and Noe M 2020 *Supercond. Sci. Technol.* **33** 044014
[5] Jones C E, Norman P J, Galloway S J, Armstrong M J and Bollman A M 2016 *IEEE Trans. Appl. Supercond.* **26** 3601409
[6] Zermeño V M R, Quaiyum S and Grilli F 2016 *IEEE Trans. Appl. Supercond.* **26** 4901607 HTS modeling workgroup. [Online]. Available: http://www.htsmodelling.com/?p=917
[7] Zermeño V M R, Habelok K, Stępień M and Grilli F 2017 *Supercond. Sci. Technol.* **30** 034001 http://www.htsmodelling.com/?page_id=748 - Jc_B_theta
[8] Wimbush S C and Strickland M N 2017 *IEEE Trans. Appl. Supercond.* **27** 8000105 Wimbush S and Strickland N 2016 A high-temperature superconducting (HTS) wire critical current database. figshare. Collection. https://doi.org/10.6084/m9.figshare.c.2861821.v12
[9] Stafford B H, Sieger M, Ottolinger R, Meledin A, Strickland N M, Wimbush S C, Van Tendeloo G, Hühne R and Schultz L 2017 *Supercond. Sci. Technol.* **30** 055002 (7pp)
[10] Gouge M J, Demko J A, Roden M L, Maguire J F and Weber C S 2008 *AIP Conf. Proc.* **985**, 1343-50
[11] Park Y, Lee M, Ann H, Choi Y H and Lee H 2014 *NPG Asia Mater.* **6** e98
[12] Bagrets N, Augieri A, Celentano G, Tomassetti G, Weiss K-P and della Corte A 2015 *IEEE Trans. Appl. Superc.* **25** 6602705.

Acknowledgment
The authors acknowledge funding of the TELOS project by the German Ministry of Economic Affairs and Energy under Grant-No. 20Y1516C. We dedicate this paper to our colleague Andrea Kling who passed away recently. We thank her for supporting us in mounting the cable demonstrator.

IOP Conf. Series: Materials Science and Engineering 1241 (2022) 012038 doi:10.1088/1757-899X/1241/1/012038

Investigation on Effect of Shape of HTS Pole Coil on Airgap Magnetic Field of HTS Synchronous Motor

D K Sharma[1,2], V A S Muralidhar Bathula[2], A S Gour[1], V V Rao[1]

[1] Indian Institute of Technology, Kharagpur, India
[2] BHEL Corporate R&D Division, Hyderabad, India

Email: devsharm@gmail.com

Abstract. The High Temperature Superconducting (HTS) synchronous motors offer considerably higher torque and power densities along with higher efficiencies for strategic and critical industrial applications. The most widely explored airgap based topology for HTS synchronous motors has conventional copper windings in stator and HTS pole coils in the rotor. The shape of these rotor pole coils affects various electrical parameters of the motor. In this paper, a detailed investigation has been carried out on effect of shape of HTS pole coil on airgap magnetic field and rated open circuit characteristic (OCC) of an 8 MW HTS synchronous motor along with its other parameters like pole coil excitation current, peak magnetic field inside the motor, maximum perpendicular magnetic field on HTS tape and operating temperature of HTS pole coils through FEM based electromagnetic analysis. Based on the results of electromagnetic analysis, the optimum shape of HTS pole coil has been finalised.

1. Introduction

The High Temperature Superconducting (HTS) synchronous motors have become popular choice for various strategic and critical industrial applications where high torque and power densities are essentially required along with higher efficiencies at all operating loads conditions [1,2]. Among various available topologies for HTS synchronous motors, the most widely explored one is airgap based topology which has conventional copper-based windings in stator and HTS based pole coils in rotor [3,4]. Figure 1 shows the schematic of a typical airgap topology based HTS synchronous motor. On stator side, it has conventional copper-based windings placed in a non-magnetic core supported on a cylindrical magnetic shield made of ferromagnetic material. On rotor side, it has HTS pole coils kept under cryogenic environment. The HTS rotor also has other essential sub-components like torque tubes, thermal radiation shield, vacuum sleeve, rotary coupling etc. as shown in figure 1.

One of the most essential and costliest components of a HTS rotor is the pole coil. Usually, the HTS pole coils are made in racetrack shape and are realized by a stacking few identical double pancake coils one over the other. The design and shape of these HTS pole coils play a very crucial role in determining the airgap magnetic field [5,6]. The airgap magnetic field is one of the most essential electrical parameters for determining the rated open circuit voltage [7]. The rated Open Circuit Characteristic (OCC) is a condition where the rotor spins at synchronous speed with rated DC excitation in pole coils and the voltage observed in the open circuited stator winding is equal to the rated voltage of the motor. In the present paper, the rated OCC condition for 8 MW HTS motor is studied for various design cases of HTS pole coils.

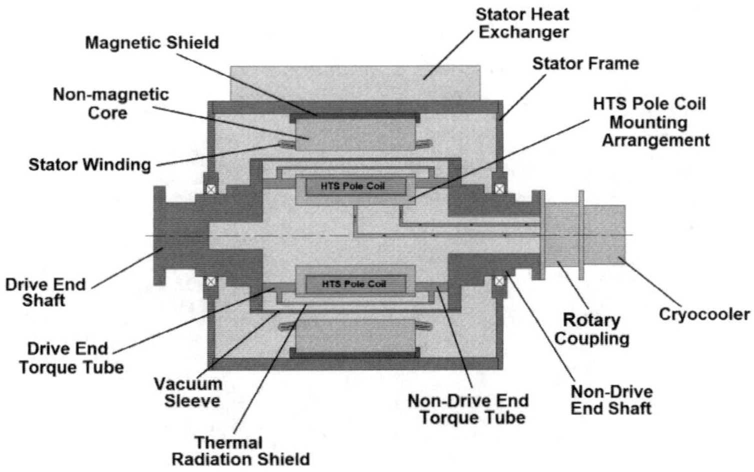

Figure 1. Schematic of a typical airgap topology based HTS motor.

2. Specifications of HTS Synchronous Motor and Pole Coil

In the present paper, a three phase, 8 MW, 500 RPM HTS synchronous motor having airgap based topology is considered. The broad technical specifications of the HTS synchronous motor are given in table 1 below. The HTS synchronous motor under consideration has a three phase, star connected conventional copper-based stator winding, whereas, the rotor has HTS pole coils. The airgap magnetic field of this HTS synchronous motor is studied for four HTS pole coil variants generated by changing the pole shoe gap i.e. the gap between the two rectilinear portions of the coil.

Table 1. Technical Specifications of HTS Synchronous Motor.

S. No.	Parameter	Value
1	Power Output	8 MW
2	Voltage (Line-to-Line)	6600 V
3	Frequency	50 Hz
4	RPM	500 RPM
5	No of Poles	12 Poles

The 8 MW HTS synchronous motor under consideration has 12 HTS pole coils in the rotor as shown in figure 2. Each single HTS pole coil in the figure is obtained by stacking five identical racetrack shaped double pancake coils having equal number of turns of HTS tape.

The shape of this HTS pole coil plays a significant role in determining the airgap magnetic field of HTS motor along with other parameters like pole coil excitation current, peak magnetic field inside the motor, maximum perpendicular magnetic field on HTS tape and operating temperature of HTS pole coils. The airgap magnetic field is one of the essential electrical design parameters which determines the rated OCC voltage of the motor for a given of HTS synchronous motor. In an HTS synchronous motor, the airgap magnetic field is created by HTS pole coils which are made of HTS tape. In the present work, the YBCO based 2G HTS tape is chosen. The technical specifications of HTS tape are given in table 2 below.

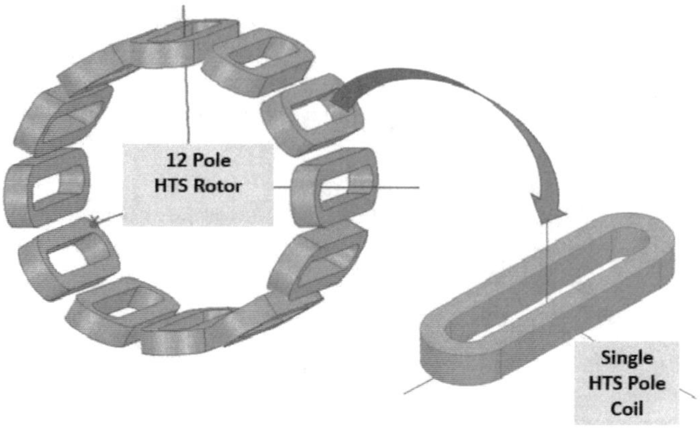

Figure 2. HTS rotor having 12 pole coils.

Table 2. Technical Specifications of HTS tape.

S. No.	Parameter	Value
1	Make	SuperPower
2	Type of HTS tape	YBCO (SCS4050-i-AP)
3	Critical Current (*Ic*)	130 A @ 77 K, sf
4	Average thickness of HTS tape	0.161 mm
5	Average width of HTS tape	4.05 mm
6	HTS tape Insulation	Polyimide
7	Stabilizer	Copper
8	Critical bending diameter	11 mm
9	Critical tensile strength	550 MPa

In general, the current carrying capacity of HTS tape selected for HTS synchronous motor is determined based on perpendicular magnetic field on it and its operating temperature. The critical current (I_c) of chosen 2G HTS tape at various perpendicular magnetic fields ($B_\perp$) for different operating temperatures are provided by manufacturer and given in figure 3 below [8].

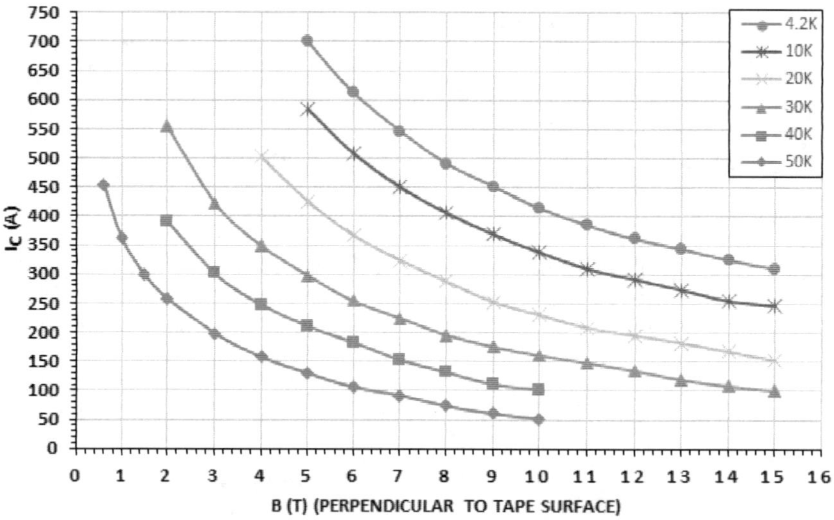

Figure 3. I_c vs $B_\perp$ characteristic of HTS tape at various operating temperatures [8].

3. Electromagnetic Analysis of various Design Variants of HTS Pole Coil

In this section, the details of electromagnetic analysis carried out for different design variants of HTS pole coil of the motor are presented. The electromagnetic analysis of 8 MW HTS synchronous motor (whose specifications are given in table 1) is carried out in ANSYS® Maxwell software for four design variants of HTS pole coil.

Figure 4 (a) and 4 (b) show the isometric view and sectional view of single HTS pole coil respectively. Figure 4 (c) shows the two-dimensional model of HTS pole coil. As shown in figure 2, the rotor of the motor has 12 HTS pole coils. Hence, during modeling of motor, 12 such pole coils have been modeled. The height of HTS pole coil and the overall width of HTS tapes within a pole coil are considered as constant during electromagnetic analysis. The gap *'g'* between two rectilinear portions of pole coil is termed as pole shoe width as shown in figure 4 (a) and 4 (c).

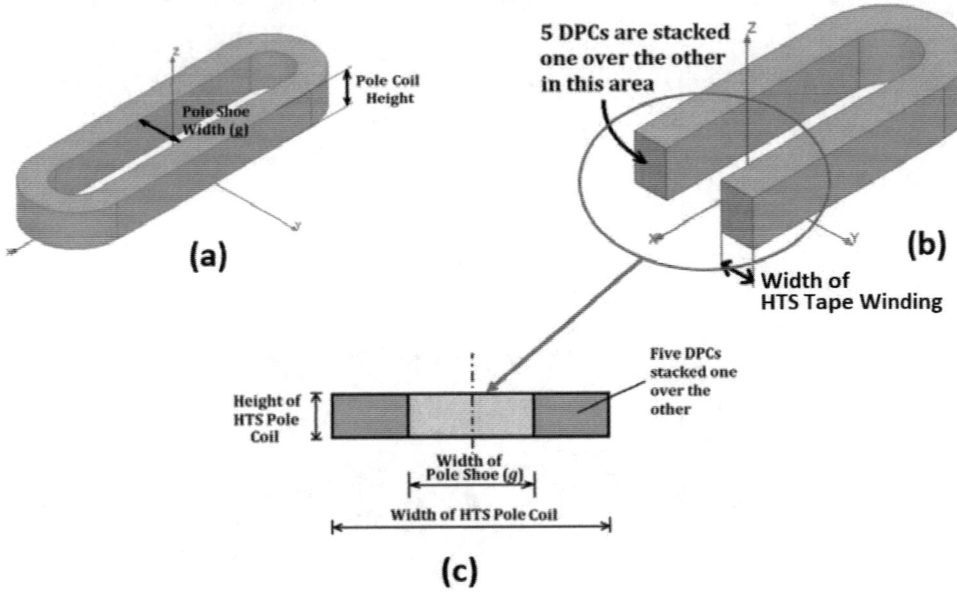

Figure 4. (a) Isometric view, (b) sectional view and (c) 2D model of HTS pole coil.

Figure 5 shows the model of 8 MW HTS synchronous motor prepared in ANSYS® Maxwell software. The conductors of all three phases (R, Y and B phases) of the stator winding along with magnetic shield are modeled. The HTS rotor is also modeled with 12 numbers of HTS pole coils.

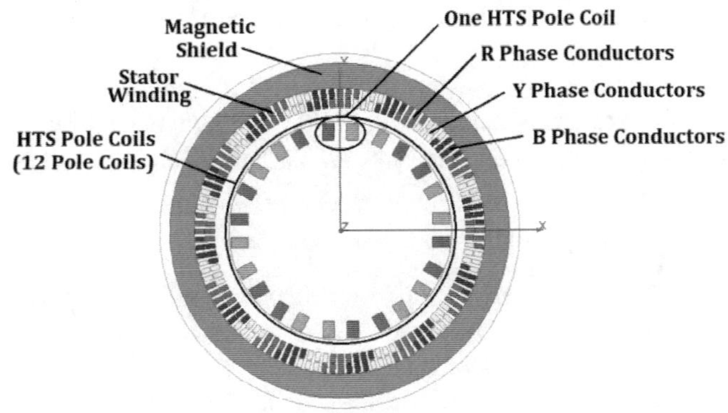

Figure 5. Model of 8 MW HTS synchronous motor developed in ANSYS® Maxwell software.

Various electromagnetic simulations are carried out to study the effect of variations of pole shoe width *'g'* between two rectilinear portions of HTS pole coil on airgap magnetic field. Table 3 shows four design variants having different shapes of HTS pole coil created by varying pole shoe width *'g'* from 40 mm to 100 mm in four regular intervals.

Table 3. Design variants of HTS pole coil.

Design variant	***Case A***	***Case B***	***Case C***	***Case D***
Pole shoe width '*g*'	*g*=40 mm	*g*=60 mm	*g*=80 mm	*g*=100 mm
	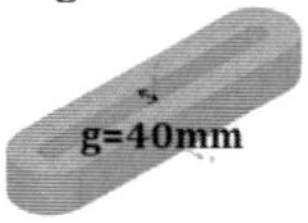	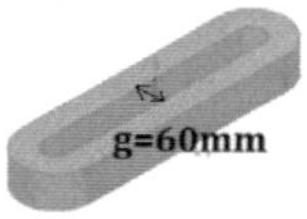	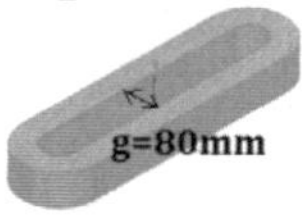	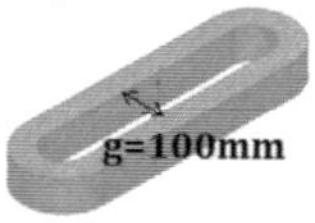

For all four cases, the HTS pole coil excitation (I_f) is varied to obtain the rated OCC condition (E_L=6600 V) and their magnetic field plots. Figure 6 shows a typical zoomed-in-view of magnetic field plot (B-plot) of HTS motor under rated OCC condition depicting the locations of the radial airgap magnetic field (B_g), the peak magnetic field (B_{peak}) and the maximum perpendicular magnetic field ($B\perp_{max}$).

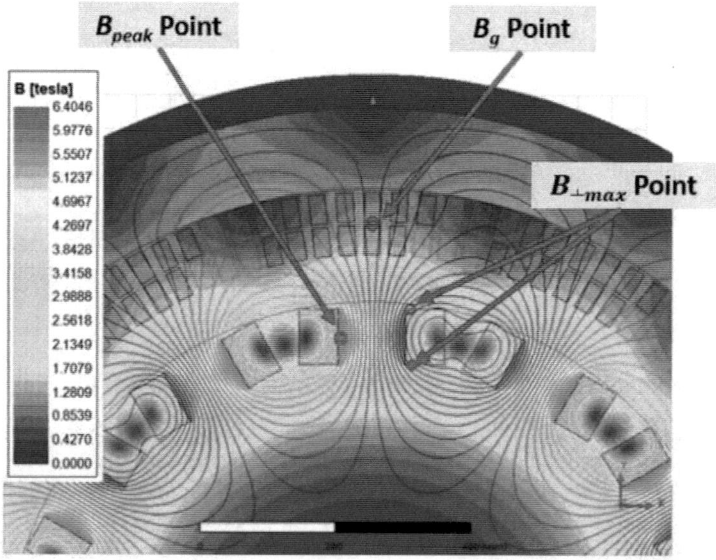

Figure 6. Location of B_{peak}, B_g and $B\perp_{max}$ in the B-Plot of *Case D*.

The effect of shape of the HTS pole coil for all these four cases at rated OCC condition on airgap magnetic field (B_g), the rotor excitation current (I_f), the peak magnetic field inside the motor (B_{peak}) and maximum perpendicular magnetic field on HTS tape ($B\perp_{max}$) are evaluated and presented in next section.

4. Results and Discussions

The rated OCC conditions are simulated for all four cases (as given in table 3). Figure 7 shows the zoomed-in-view of magnetic field plot (B-Plot) for all four cases. It can be seen from B-Plot that the radial airgap magnetic field (B_g) is nearly constant for all four cases. The peak magnetic field (B_{peak}) for *Case A* (*'g'*=40 mm) is higher. However, as pole shoe width *'g'* is increased from 40 mm to 100 mm, this peak magnetic field (B_{peak}) reduces.

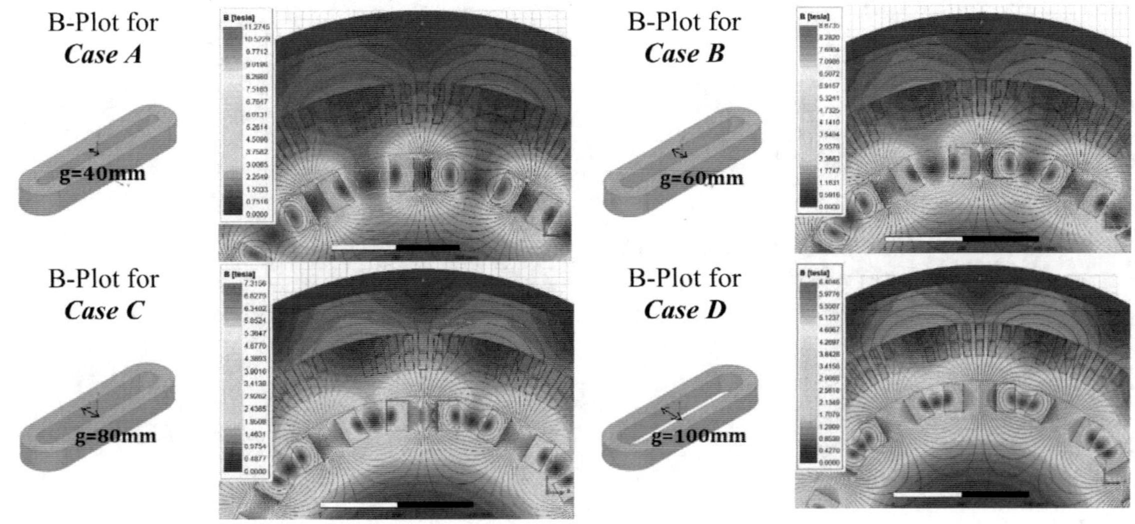

Figure 7. Magnetic field plot for all four cases.

Table 4 below shows the rotor excitation current (I_f), the peak magnetic field inside the motor (B_{peak}), the radial airgap magnetic field (B_g) and maximum perpendicular magnetic field on HTS tape ($B\perp_{max}$) for rated OCC condition for all four cases. It can be observed from the table 4 that as the pole shoe width *'g'* increases from 40 mm to 100 mm, the excitation current (I_f) as well as the maximum perpendicular magnetic field ($B\perp_{max}$) on HTS tape decreases. It can be ascertained that when the pole shoe width *'g'* is the largest (as in *Case D* with *g*=100 mm), the excitation current (I_f), the peak magnetic field (B_{peak}) and the maximum perpendicular magnetic field ($B\perp_{max}$) are lesser as compared to that for other three cases.

Table 4. I_f, B_{peak}, B_g and $B\perp_{max}$ for all four cases at rated OCC condition.

	I_f	B_{peak}	B_g	$B\perp_{max}$
Case A (*g*=40 mm)	346.1 A	11.27 T	1.28 T	6.02 T
Case B (*g*=60 mm)	282.5 A	8.87 T	1.26 T	5.01 T
Case C (*g*=80 mm)	240.7 A	7.32 T	1.24 T	4.49 T
Case D (*g*=100 mm)	211.4 A	6.40 T	1.22 T	4.18 T

Further, considering the data given in table 4, the operating temperature of pole coil with chosen HTS tape (whose specifications are given in table 2) is estimated considering a suitable margin-of-safety (in our case 67%) above operating current for all four cases. For example, the excitation current (I_f) for *Case A* (for *g*=40 mm) is 346.1 A and, with margin-of-safety, the operating excitation current is 578 A. The operating point (6.02 T, 578 A) for *Case A* is highlighted on *Ic* vs $B^{\perp}$ characteristic of HTS tape. Similarly, the operating points for other three cases are also evaluated and highlighted in figure 8. Based on the operating point of *Case A*, it can be observed from figure 8 that the next nearby temperature at which this HTS pole coil (for *g*=40 mm) can possibly be operated is 4.2 K. Similarly, the possible nearby operating temperatures for other three cases are also arrived at. The maximum perpendicular magnetic field ($B\perp_{max}$) on HTS tape, the evaluated operating current with safety margin and nearby operating temperature for all four cases are presented in table 5.

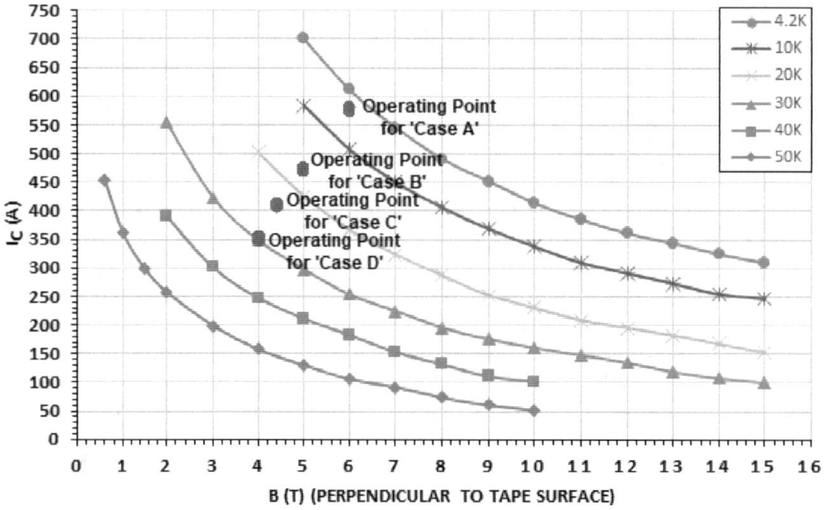

Figure 8. Operating points for all four cases on *Ic* vs *B*⊥ characteristic of HTS tape.

Table 5. Operating points and nearby temperatures for all four cases.

	$B\perp_{max}$	*Operating current with safety margin*	*Nearby operating temperature*
Case A (*g*=40 mm)	*6.02 T*	*578 A*	*4.2 K*
Case B (*g*=60 mm)	*5.01 T*	*472 A*	*10 K*
Case C (*g*=80 mm)	*4.49 T*	*402 A*	*20 K*
Case D (*g*=100 mm)	*4.18 T*	*353 A*	*30 K*

5. Conclusions

An 8 MW HTS synchronous motor is simulated at rated OCC condition for four different HTS pole coil shapes with varying pole shoe width in ANSYS® Maxwell software. Various electrical and magnetic parameters at rated OCC condition are obtained and compared for all four cases. The results of electromagnetic analyses suggest that, among all four cases, *Case D* (*g*=100 mm) gives the least operating excitation current at a reasonably good operating temperature of 30 K. Also, among all four cases, *Case D* gives the least perpendicular magnetic field on HTS pole coil and the least maximum magnetic field in the HTS synchronous motor.

6. References

[1] S S Kalsi 2004 *IEEE Power Engineering Society General Meeting* **2** 2047-2048 (Denver, CO, USA)

[2] Michael Frank et al. 2006 *IEEE Transactions on Applied Superconductivity* **16(2)** 1465-1468

[3] G Klaus, M Wilke, J Frauenhofer, W Nick and H W Neumuller 2007 *IEEE Power Engineering Society General Meeting* 1-8 (Tampa, FL, USA)

[4] Naoki Maki, Mitsuru Izumi, Masayoshi Numano, Kiyoshi Aizawa, Kagao Okumura and Katsunori Iwata 2007 *International Conference on Electrical Machines and Systems (ICEMS)* 1523-1527 (Seoul, South Korea)

[5] Y Jiang, R Pei, W Xian, Zhibin Hong and Timothy Coombs 2007 *Superconductor Science and Technology* **20(7)** 585-591

[6] Y Jiang, R Pei, Z Hong, Q Jiang and T Coombs 2008 *Journal of Physics: Conference Series* **97** 012123
[7] Victor M R Zermeno, Asger B Abrahamsen, Nenad Mijatovic, Mads Peter Sorensen, Bogi B Jensen, Niels F Pedersen 2012 *Physics Procedia* **36** 786-790
[8] https://www.superpower-inc.com/specification.aspx

Acknowledgments
Authors would like to extend their warm gratitude to the management of BHEL Corporate R&D Division, Hyderabad for giving an opportunity and extending full support to carry out this work.

Numerical Studies on Two-Phase Flow of Liquid Nitrogen to Cool HTS Power Cables

I de Souza, H K Hassan, A Anand, S K Chand, A S Gour and V V Rao

Cryogenic Engineering Centre, Indian Institute of Technology, Kharagpur, India, 721302

E-mail: isaacdesouza68@gmail.com

Abstract. Liquid nitrogen is the foremost coolant when it comes to cooling HTS power devices due to its abundance and since it is more economical than other cryogens. Sub-cooled LN_2 is preferred over saturated LN_2 as it offers substantial advantages in terms of performance. However, in the cases where it is not possible to cool HTS power cables using sub-cooled LN_2, saturated LN_2 is the only option. Additionally, even when sub-cooled LN_2 is used for long length cables, the heat-in-leaks along with other core losses result in rise in temperature and an unavoidable two-phase flow affects the cooling characteristics of the cryogen. This paper deals with the modeling and numerical analysis of a two-phase flow through an HTS power cable cryostat to understand the temperature profile, the pressure drop per unit length, and required pumping power for circulation of cryogen.

1. Introduction

High Temperature Superconductors (HTS) are employed in various power devices such as HTS power cables, Superconducting Fault Current Limiters (SFCL), Superconducting Magnetic Energy Storage (SMES), HTS motors & generators and HTS transformers as they offer zero resistance below their critical temperatures.

HTS tapes used in HTS power cables have a transition temperature of 90-110 K [1] and are often cooled using cryogens like LN_2. A majority of the HTS power cables operating and being developed all over the world are cooled using LN_2 in the operating range of 65-77 K [2].

Currently, a 1 ϕ, single pass liquid nitrogen cooled HTS power cable is being developed at the Indian Institute of Technology (IIT), Kharagpur and its schematic is shown in figure 1. The space between the inner and outer corrugated pipes is usually evacuated to sub-atmospheric pressures to minimize heat-in-leaks from the ambient via conduction and convection. At times, even Multi-Layer super Insulation (MLI) is used to minimize the heat transfer via radiation. In spite of all these provisions, the heat-in-leaks are unavoidable and for long length HTS power cables, these leaks place a huge thermal load on the cryogen. Also, in the case of an AC excitation, core losses in the form of eddy current and hysteresis loss, and dielectric losses place an additional heat load on the cryogen. These heat loads, though undesirable, may result in a two-phase flow of N_2 inside HTS power cables.

The power transmission capacity may be affected as the critical current reduces due to the rise in temperature of the cable core [3]. This temperature rise results in two-phase of cryogen which enhances probability of voltage breakdown due to the presence of bubbles [4] [5] resulting in the occurrence of faults. In addition, this also affects the thermo-hydraulic characteristics of

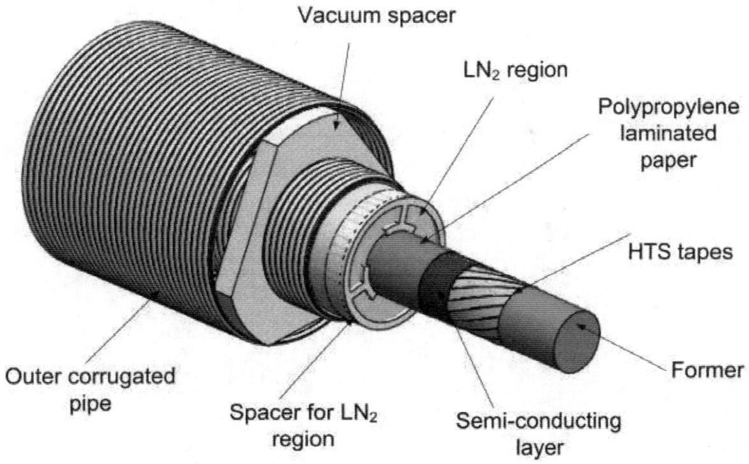

Figure 1. Schematic of a 1 ϕ, single pass, LN_2 cooled HTS power cable

cryogen which in turn leads to the quench of superconductor in HTS cable. Finally, a two-phase flow results in high pressure drops leading to a higher pumping power requirement. Though studies on cryogenic fluid flow inside HTS power cables have been reported in various numerical investigations [6] [7] [8], there is not much work reported in open literature on the analysis of a two-phase flow inside HTS power cables. This work will help in predicting the time before failure (TBF) to which the HTS cable can effectively and safely transmit rated power under the event of deterioration of vacuum level between the inner and outer cryostats.

Thus, the objective of this work is to model and numerically analyze a two-phase flow inside a 1 ϕ, single pass, LN_2 cooled HTS power cable. The volume fraction, temperature rise, velocity contours, vectors, streamlines, pressure drop and corresponding pumping power were computed. Section 2 describes the specifications of the cable being analyzed. Section 3 describes the numerical approach undertaken to analyze the problem. Section 4 lists the results and the discussions.

2. Specifications of HTS power cable

A five meter HTS power cable is being developed at IIT Kharagpur. The specifications of the HTS power cable are shown in Table 1. The dimensions of the cable are shown in Table 2. These

Table 1. Specifications of HTS power cable being developed

	Specifications
Rating	11 kV, 1 kA
HTS tape	SuNAM SLBS04150 ($I_c = 150$ A at 90 K self-field)
Dielectric	PolyPropylene Laminated Paper (PPLP)
Coolant	LN_2

dimensions have been used for the present study as shown in figure 2.

Table 2. Dimensions of HTS power cable being developed

Layer	Thickness	Outer diameter
Copper former	-	19 mm
HTS tapes	0.14 mm	19.28 mm
Semi-conducting layer	0.5 mm	20.28 mm
PPLP layer	2.5 mm	25.3 mm
Inner corrugated pipe	0.3 mm	63.4 mm
Outer corrugated pipe	0.3 mm	102.2 mm

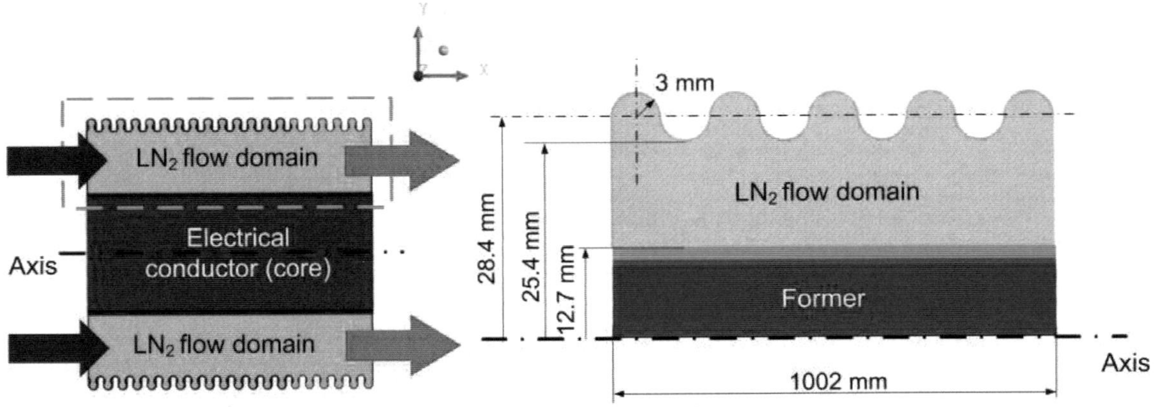

Figure 2. Cross sectional view and dimensions of HTS power cable

3. Numerical analysis

A numerical approach was undertaken to analyze the flow through the HTS power cable using ANSYS Fluent 18.2.

3.1. Computational domain

A two dimensional section at the top portion of the cable is considered for the present study. The dimensions of the fluid domain are shown in figure 2.

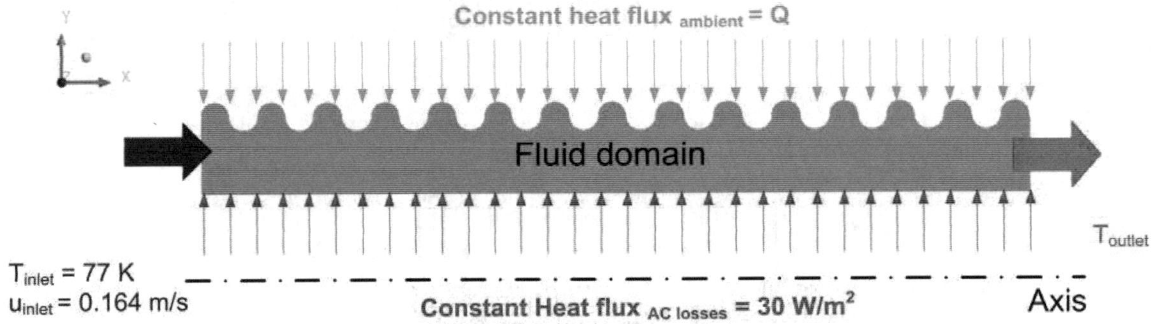

Figure 3. Fluid domain and boundary conditions

3.2. *Governing differential equations and performance parameters*

The following governing differential equations for the fluid domain are solved to capture the flow physics and compute the thermal characteristics of the two-phase flow.
Continuity equation

$$\frac{\partial \rho}{\partial t} + \bigtriangledown.(\rho\vec{\nu}) = 0 \tag{1}$$

Momentum equation

$$\frac{\partial}{\partial t}(\rho\vec{\nu}) + \bigtriangledown.(\rho\vec{\nu}\vec{\nu}) = -\bigtriangledown p + \bigtriangledown.(\bar{\bar{\tau}}) + \rho g \tag{2}$$

Since the flow domain is 2 dimensional, only the momentum equations in x and y direction are considered.
Energy equation

$$\rho C_p \frac{\partial T}{\partial t} + \rho C_p(u.\bigtriangledown T) = k \bigtriangledown^2 T \tag{3}$$

Since the flow is turbulent in nature, the standard k-ϵ model was selected.

$$\frac{\partial}{\partial t}(\rho k) + \frac{\partial}{\partial x_i}(\rho k u_i) = \frac{\partial}{\partial x_j}\left[\left(\mu + \frac{\mu_t}{\sigma_k}\right)\frac{\partial k}{\partial x_j}\right] + G_k - \rho\varepsilon \tag{4}$$

$$\frac{\partial}{\partial t}(\rho\varepsilon) + \frac{\partial}{\partial x_i}(\rho\varepsilon u_i) = \frac{\partial}{\partial x_j}\left[\left(\mu + \frac{\mu_t}{\sigma_\varepsilon}\right)\frac{\partial \varepsilon}{\partial x_j}\right] + C_{1_\varepsilon}\frac{\varepsilon}{k}(G_k) - C_{2_\varepsilon}\rho\frac{\varepsilon^2}{k} \tag{5}$$

$$\mu_t = \rho C_\mu \frac{k^2}{\varepsilon} \tag{6}$$

Volume Of Fluid (VOF) model

$$\rho = \sum_{i=1}^{n} \rho_i \alpha_i \tag{7}$$

$$\frac{\partial \alpha_i}{\partial t} + \vec{\nu}\bigtriangledown.\alpha_i = 0 \tag{8}$$

3.3. *Boundary conditions*

A uniform axial fluid velocity of 0.164 m/s, corresponding to a volumetric flow rate of 15 lit/min, was imposed at the inlet of the fluid domain. A constant temperature of liquid nitrogen at saturated condition (77 K and an absolute pressure of 1 atm) is defined at the inlet. No slip conditions (at a solid boundary, the fluid will have zero velocity relative to the boundary) are defined at the walls. At the outlet, discharge occurs to atmospheric pressure. A constant heat flux (corresponding to the heat-in-leaks from the ambient) is imposed at the inner surface of the inner corrugated pipe. Also, a constant heat flux (corresponding to AC losses) is imposed at the outer surface of the cable core. The corrugated pipe (outer wall) material is Stainless Steel (SS) while the inner former (electrical core) material is copper. The boundary conditions imposed on the fluid domain are shown in figure 3. Gravity was also considered in the present analysis in the form of acceleration defined in the negative direction of the Y axis.

3.4. Solution methodology

The governing conservative equations for the two-phase flow were solved using Finite Volume solver ANSYS Fluent 18.2. Multi phase transient simulation was performed. The governing equations for pressure, velocity and temperature were solved using the Pressure-Implicit with Splitting of Operators (PISO) algorithm. Convective terms were discretized using 2nd order upwind scheme. The flow is turbulent and standard k-ϵ model was chosen. The convergence criteria for all the conservative governing equations were set to 10^{-6} [8] [9].

4. Results and discussions

The volume fraction, streamlines, velocity vectors and contours, temperature rise, pressure drop and corresponding pumping power required to overcome the pressure drop and circulate the cryogen were computed. The liquid volume fraction at the end of the various time durations are shown in figure 4. In the figure, 1.0 corresponds to a complete liquid phase while 0.0 corresponds

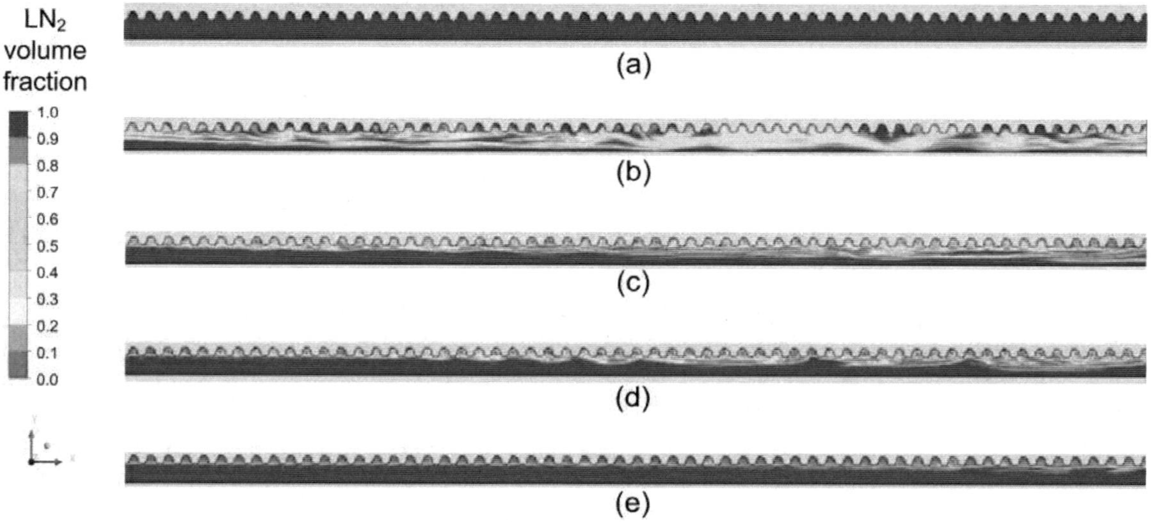

Figure 4. Liquid nitrogen volume fraction at the end of (a) 0 secs (b) 1.5 secs (c) 3 secs (d) 4.5 secs and (e) 6 secs

to a complete gaseous phase. It was observed that, as the time progressed, the flow appeared to reach steady state after approximately 6 secs. The corrugations were filled with gaseous nitrogen and formed vortices as shown in figure 6. The streamlines near the inlet of the pipe are shown in figure 5 (a). It was observed that, near the inlet, the flow converged since the width of the inlet is greater than the cross-section at the end of the first corrugation. Also, no streamlines were observed inside the corrugation. This is because vortices were observed inside each of the corrugations as shown in figure 6. Since streamlines do not intersect or overlap, no streamlines were observed inside the corrugations. Velocity contours at the inlet, mid length and the exit were plotted as shown in figure 5 (b), (c) and (d). It was observed that even after the flow seemed to be fully developed, the velocity along the length increased. This may be attributed to the pressure drop resulting from fluid acceleration (or momentum change) which is often observed in two phase flows [10]. In two phase flows, as the gaseous nitrogen (GN_2) volume fraction increases, the density of the fluid downstream decreases. For mass to be conserved, the decrease in density is compensated by an increase in velocity, thus maintaining a constant mass flow rate. A variation of the velocity along a centre line (between the inner wall of the inner corrugated pipe and the core surface) with the length is shown in figure 7 (Case C). The

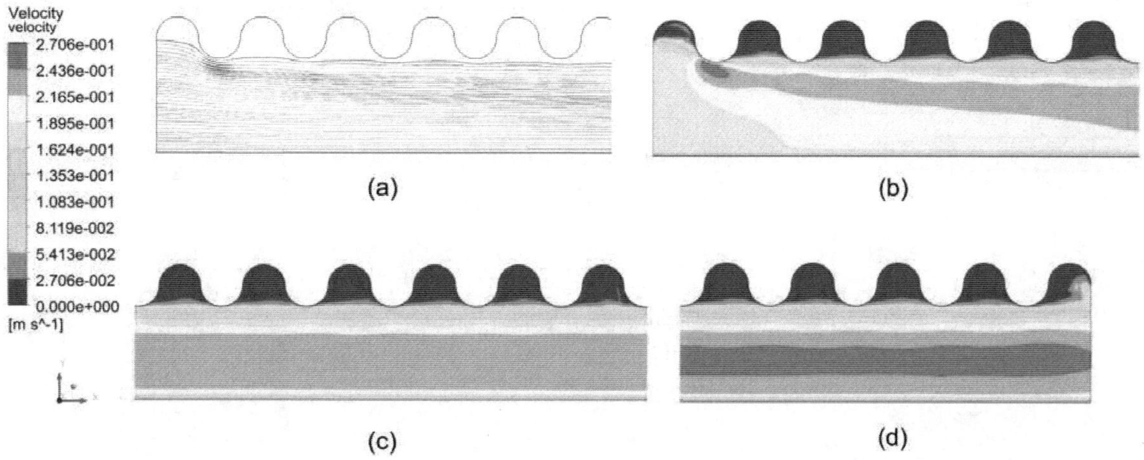

Figure 5. (a) Streamlines near inlet (b) Velocity contour near inlet (c) Velocity contour at mid length (d) Velocity contour near exit

velocity was found to fluctuate along the length of the pipe. This is attributed to the varying cross-section due to the presence of the corrugations.

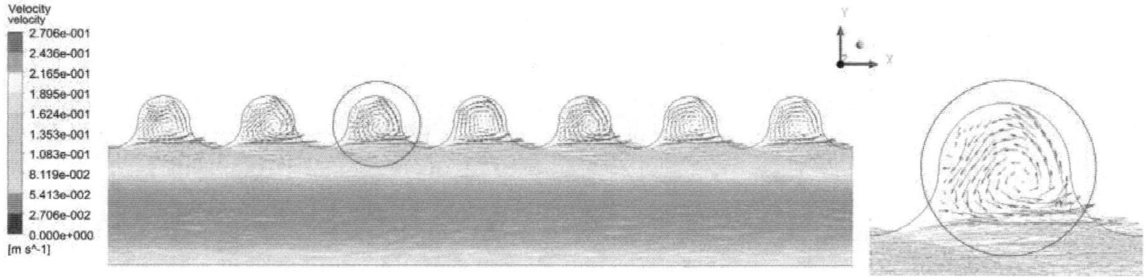

Figure 6. Velocity vectors showing vortices in corrugations for constant ambient heat flux of 260 W/m^2

Table 3. Results of parametric study

Case	Heat flux from core (W/m^2)	Heat flux from ambient (W/m^2)	Total heat load (W)	Temperature rise (K)	Velocity at outlet (m/s)	Total Pressure drop (Pa)	Required pumping power (mW)
A	30	26	9.7	0.76	0.1997	47.67	11.9
B	30	130	38.9	2.11	0.2079	54.96	14.49
C	30	260	75.4	3.21	0.2150	88.54	22.13
D	30	520	148.4	4.9	0.2398	167.71	41.92

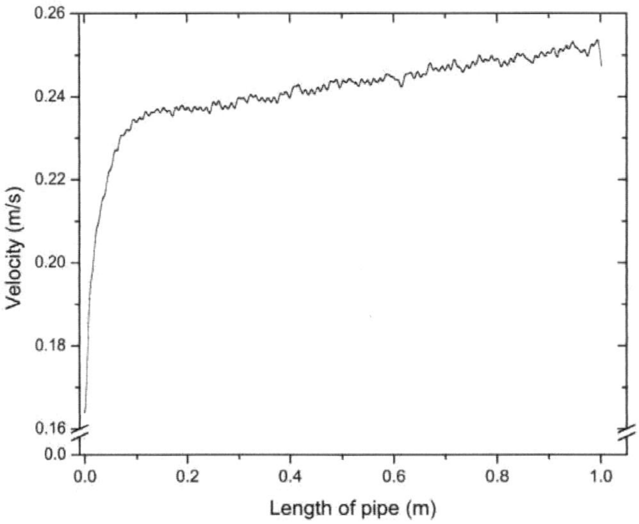

Figure 7. Variation of axial velocity along length of pipe for constant ambient heat flux of 260 W/m^2

A parametric study was performed to study the effect of varying heat-in-leaks from the ambient and the results are tabulated in Table 3. When the vacuum in the outer jacket (between the inner and outer corrugated pipe) is destroyed due to the failure of the vacuum pumping system or damage to the outer corrugated pipe, warm air from the surrounding rushes into the annular space thus placing a heat-load on the cryogen. Case A corresponds to a heat load (from core) of 1.5 W/m and a heat load (from ambient) of 2 W/m [11]. These correspond to a heat flux of 30 W/m^2 and 26 W/m^2 respectively based on the geometry under consideration. Cases B, C and D correspond to situations where the heat fluxes are 5, 10 and 20 times the reference value (2 W/m). The corresponding total heat load, outlet temperature, outlet velocity, pressure drop were computed. Since the temperature, pressure and velocity vary long the radial direction at the outlet, a mass weighted average was computed for the same. The required pumping power was computed using equation 9 [12].

$$W = q\triangle p \tag{9}$$

where, q is the volumetric flow rate, W is the required pumping power, and $\triangle p$ is the total pressure drop. The total pressure drop in a two phase flow is the summation of the frictional and the momentum pressure drops as shown in equation 10.

$$\triangle p = \triangle p_F + \triangle p_M \tag{10}$$

where, $\triangle p_F$ is the frictional pressure drop, and $\triangle p_M$ is the momentum pressure drop.

5. Conclusions

A two-phase flow through the annular flow passage of a one metre long HTS cable has been modeled and numerically analyzed. The pressure drop, velocity profiles, vectors, streamlines and the temperature rise have been computed. A parametric study has been conducted to understand the effect of heat load on the thermo-hydraulic characteristics. For a volumetric flow rate of 15 lit/min, a temperature rise of 3.12 K and a pressure drop of 88.54 Pa were observed when the total heat load was 75.4 W. Two-phase flow should be avoided to protect and effectively operate an HTS power cable.

6. References

[1] Scanlan R M, Malozemoff A P and Larbalestier D C 2004 *Proceedings of the IEEE* **92** 1639–1654

[2] Yapicioglu A and Dincer I 2019 *Sustainable Energy Technologies and Assessments* **31** 254–261

[3] Wang Y 2013 *Fundamental elements of applied superconductivity in electrical engineering* (John Wiley & Sons)

[4] Chang H, Ryu S and Kim M 2010 Performance of heat exchanger for subcooling liquid nitrogen with a gm cryocooler *AIP Conference Proceedings* **1218** (American Institute of Physics) pp 345–352

[5] Al-Taie A, Telikapalli S, Cheetham P, Kim C and Pamidi S 2020 Liquid nitrogen cooled superconducting power cable with no solid insulation *IOP Conference Series: Materials Science and Engineering* **756** (IOP Publishing) p 012033

[6] Das I and Rao V 2019 *Cryogenics* **103** 102950

[7] Kalsia M and Dondapati R S 2019 *Physica C: Superconductivity and its Applications* **567** 1353549

[8] Mubarok M H, Cater J E and Zarrouk S J 2020 *Geothermics* **86** 101801

[9] Hemidi A, Henry F, Leclaire S, Seynhaeve J M and Bartosiewicz Y 2009 *Applied Thermal Engineering* **29** 1523–1531

[10] Barron R F and Nellis G F 2017 *Cryogenic heat transfer* (CRC press)

[11] Zong X, Wei D, Han Y and Tang T 2016 *IEEE Transactions on Applied Superconductivity* **26** 1–4

[12] Thadela S, Rao V, Agarwal R and Dondapati R S 2019 *Physica C: Superconductivity and its Applications* **559** 25–31

Conceptual design of a wide band gap based cryogenically cooled MW-class inverter

Soumen Kar, Siddavatam Ravi Prakash Reddy, Kaushik Rajashekara
Department of Electrical and Computer Engineering,
University of Houston, Houston, TX 77204, USA

skar2@uh.edu

Abstract. The increasing need for higher power density and closer integration of power subsystems consisting of superconducting motors/generators, degaussing coils, energy storage modules, and cables leads one to consider the merits of refrigerating the associated power electronics down to cryogenic temperatures. High temperature superconducting (HTS) components combined with cryogenic power converters resulting in high power density power conversion systems will have a significant effect on several industrial, commercial, transportation, and renewable energy applications. Cryogenic power converters provide promising benefits over their room temperature counterparts in terms of reduced size and weight due to increased power density, improved efficiency, switching speed, and reliability. Integration could result in significant weight and space savings for the overall system. In this paper, a conceptual design study on the wide-bandgap-based (especially SiC and GaN) MW-class power inverter/converter is reported. Based on the total power loss of the designed converter, different cryogenic cooling strategies are proposed. The cooling power requirements, cooler mass, the cooler cost is evaluated based on the operating temperature of MW-class power converter. Finally, a power density comparison for different types of conventional power applications and HTS applications together with MW-class power converters is presented. The future of the cryogenically cooled power electronics system together with superconducting power devices is described for large-scale applications such as for future electric aircraft.

1. Introduction

Power electronics is an enabling technology for power conversion and control in various industrial, transportation, and commercial applications. This is an essential technology for continued development of next generation transportation vehicles' concepts like transport-class aircrafts, ships, and trains [1], which require electric power in the range of several megawatts. Traditional thermal solutions are often not capable of meeting performance requirements in harsh or challenging environments for MW to GW power level, and with high heat flux. Hence, the thermal management becomes a key design consideration [2] to address these issues. The increasing demand of high-power density in a compact and lightweight power electronic system has led to ever-growing heat fluxes that need to be effectively and quickly dissipated to maintain device temperature within safe/operable limits [1, 2-5]. Therefore, a cryogenic cooling system is needed to cool the electric motor as well as the associated power electronics in the next generation of aviation and other electrified transportation systems to achieve high efficiency. The recent trend is to investigate the use of cryogenic cooling for high power converters to achieve high efficiency, low noise, high reliability, fast switching, and integration with superconducting machines [1]. In order to further address the high temperature issues and improve the efficiency, the power converters based on wide band gap (WBG) semiconductor devices are being increasingly examined for use in various applications. These WBG semiconductors have low conduction losses, higher operating temperature, higher breakdown voltages, higher power densities (~ 50 W/in^3), and higher switching frequencies [6-8]. These WBG-based materials, particularly gallium nitride (GaN) at cryogenic temperatures have been reported to show excellent improvement in their physical properties like on-resistance, switching speed and stable breakdown voltage [7]. Mueller & Herd *et. al.* [9], Ray *et. al.* [10-13], and many other researchers have shown the enhanced efficiency of the various power converters at cryogenic temperature by immersion cooling concept [6, 14-20]. Without cryogenic fuels, immersion cryogenic cooling may not be adopted for power converters in electric aircraft with longer travel times, and cryocoolers are needed for associated MW-class power converters.

In this paper, a conceptual design study of the cryogenically cooled wide-bandgap-based MW-class power inverter/converter has been carried out and total power loss of the designed converters at 20 K and 77 K are computed and compared with their performance at room temperature. The cooling power requirements, cryocooler mass and cryocooler cost are evaluated based on 20 K and 77 K operation of MW-class power converter to obtain a selection criterion. Further, a power density comparison for different types of conventional power applications and HTS applications together with MW-class power converters is discussed. Finally, the future of the cryogenically cooled power electronics system together with superconducting power devices is described for large-scale applications such as for future electric aircrafts.

2. Design of a 1 MW converter using wide band gap semiconductor devices

2.1 Conceptual 2-level inverter

The block diagram of a 2-level inverter is shown in Figure 1. It consisits of six switching devices ($S_1 - S_6$) realised using Silicon Carbide (SiC)/ Gallium Nitride (GaN) devices. The DC link voltage is considered to be 1000 V and swicthing frequency is 150 kHz. Space vector pulse width modulation (SVPWM) technique is used for modulation of switches.. Since, the DC link voltage is 1000 V, we have considered the series connection of devices that of are of lower voltage ratings. Thus, two of the GaN devices are connected in series. For SiC devices, since 1200 V is just above 1 kV, two of these can be connected in series to match the desired voltage level. The DC link capacitor used is a polypropylene metallized film capacitor (C44HKGR6250AASJ – 1000 V DC, 250 μF). Metal film resistors are used for the design The performance of 1 MW converter is compared in terms of conduction and switching losses. The following wideband gap devices given in Table 1 are considered and analysed for their operation at temperatures of 20 K, 77 K and 300 K. We have considered SiC devices from three manufacturers and GaN device from one manufacturer. The ON-state resistance is represented by $R_{DS\text{-}ON}$, and OFF-state resistance is represented by $R_{DS\text{-}OFF}$ and diode voltage drop is represented by V_D.

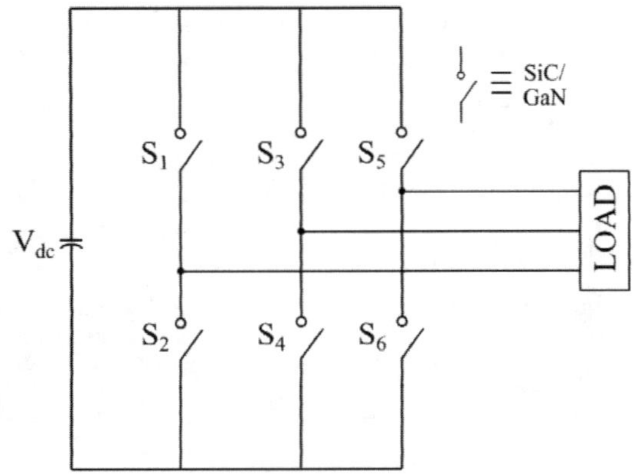

Figure 1. Configuration of 2-level inverter

Table 1. Specifications of commercially available WBG devices [21, 22]

Manufacturer		Wolfspeed (WS)	Rohm	ST	GaN systems (GS)
WBG semiconductor material		SiC	SiC	SiC	GaN
Model no.		C3M0075120K	SCT3030KL	SCT50N120	GS66540C
Max. operating junction temperature (T_{jmax})	**°C**	150	150	200	120
	K	423.15	423.15	473.15	393
Drain-source voltage (V_{DS}) (volts)		1200	1200	1200	650
Drain Current (I_D) at 25°C (A)		30	72	65	100
ON-state Characteristics of devices (R_{DS-ON}) (mΩ)	**300 K**	85.94	30.11	45.97	14.5
	77 K	279.21	48.88	80.24	3.0
	20 K	1074.9	187.88	308.92	0.768
OFF-state Characteristics of devices (V_D for SiC & R_{DS-OFF} for GaN)	**300 K**	5.2 V	4.8 V	4.7 V	202.0 mΩ
	77 K	9.5 V	10.7 V	10.6 V	2.0 mΩ
	20 K	11.1 V	11.3 V	11.2 V	0.512 mΩ

2.2 Conduction and switching loss analysis of 1 MW converter

The evaluation of semiconductor losses is presented in the following way. Let V_D and r_{DS} represent the forward voltage drop of body diode and on-state resistance of switch respectively. For GaN devices, conduction loss for a switch is expressed in terms of on-state resistance of switch itself. Thus, the average conduction loss ($P_{cond,R}$) in an inverter leg in a line cycle for say (R-phase) is given by

$$P_{cond,R} = \frac{1}{\pi}\int_{\alpha}^{\pi+\alpha}(d_R i_R^2 r_{DS} + (1 - d_R)V_d i_R)\, dwt) \tag{1}$$

Here, α depends on the power factor (cosα) of operation. The term d_R represents the duty ratio for which top switch R-phase conducts. Also, for GaN devices, as discussed above, v_D in (1) will be replaced with r_{DS}. The expressions for conduction loss in other phases can be obtained in similar terms, as given in (1).

The expression for switching loss, $P_{switching}$ (approximating to linear rise and fall of voltages and currents during switching transitions) is given by

$$P_{switching} = \frac{1}{6}\, V_{dc}\, I_L\, (t_{rise} + t_{fall})F_{sw} \tag{2}$$

In (2), V_{DC} and I_L represent the block voltage and load current through switch. F_{sw} is the switching frequency. t_{rise} and t_{fall} represent the rise and fall times of switch voltage.

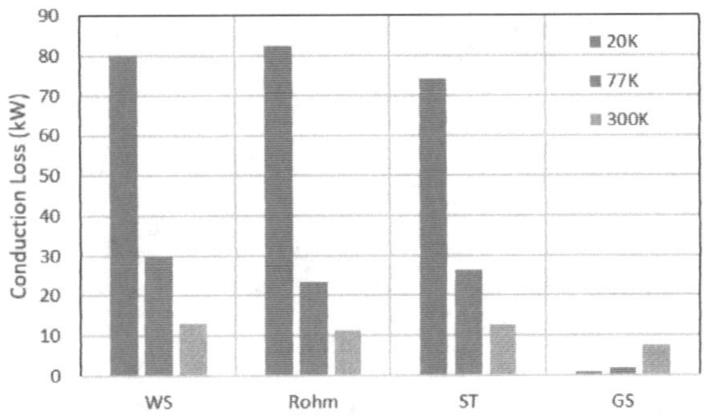

Figure 2. Comparison of conduction losses for SiC and GaN based converters.

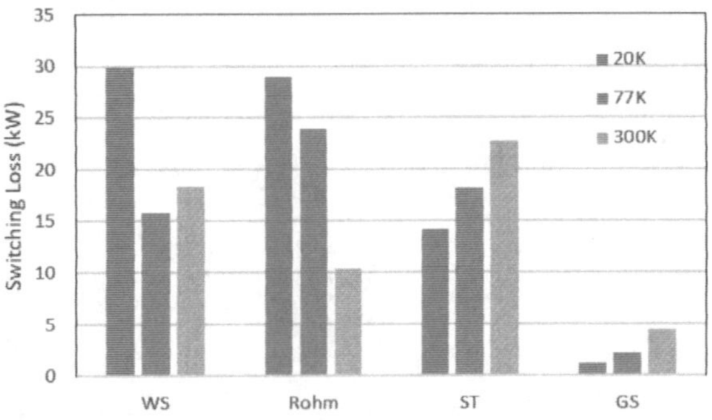

Figure 3. Comparison of switching losses for SiC and GaN based converters.

The conduction loss at different operating temperatures is shown in Figure 2. It can be seen that SiC devices (WS, Rohm and ST) have higher conduction loss at cryogenic temperatures of 20 K and 77 K, as compared to that of room temperature operation at 300 K. This is due to the increased ON state resistance of these devices at low temperatures. GaN devices (GS) have lower conduction loss as compared to SiC devices, for any given temperature. Further, GaN devices perform even better at cryogenic temperatures [7], as can be observed from Figure 2.

Figure 3 shows the switching loss comparison for both SiC and GaN based converters. From the plots shown, it can be inferred that GaN based converters perform better when compared to SiC base converters, at both crygoneic temperatures (20 K, 77 K) and at room temperature (300 K).

The total loss (conduction + switching = Q_{loss}) in the converter for all devices is given in Table 2. SiC devices exhibit poor performance at cryogenic temperatures (Figure 2 & 3). The channel resistance of SiC MOSFETS has negative coefficient and thus the on-state resistance increases rapidly at 20 K. Carrier freeze out is a contributor to this increased channel resistance at lower temperatures [23]. Thus, a large amount of electrons are trapped at lower temperatures and very few electrons are available for conduction. It can be observed that GaN based converter operating at 20 K has 40-50% lower power loss, compared to other devices. Thus, GaN based converters can result in better efficiencies and reduced size at cryogenic temperatures, due to lower conduction as well as switching losses. Table 3 shows the physical properties of the cryogenic liquids and water has been used for evaporation rate calculation from the overall loss.

Table 2. Overall loss (Q_{loss}) comparison

1 MW inverter based on WBG devices	Overall loss (kW)at 20 K	Overall loss (kW) at 77 K	Overall loss (kW) at 300 K
WS	110.1	45.72	31.13
Rohm	111.31	47.2	21.44
ST	88.5	44.28	35.46
GS	2.14	3.91	11.95

Table 3. Physical properties of cryogenic liquids and water

Properties	LH2	LN2	Water
Boiling Temperature at 1 atm. (K)	20.27	77.09	373.15
Enthalpy of vaporization (kJ/kg)	446	199	2257
Liquid Density (kg/m^3)	71	808	997
Specific heat (kJ/kg.K) at normal boiling point	9.78	2.04	4.194

Assuming an adiabatic condition, the rise in temperature (*ΔT*) has been calculated from the following equation [24].

$$\Delta T = \frac{Q}{\rho v C_p} \qquad (3)$$

Q is the overall loss, ρ is the density of the liquid, v is the volume of the liquid and C_p is the specific heat of the liquid. Figure 4 shows the temperature of the SiC and GaN devices if operated for 1 minute at 20 K, 77 K and 300 K. From figure 4, it is observed that the junction temperature (T_{jmax}) of SiC devices will rise above safe operating temperature within 1 minute whereas GaN devices stays below T_{jmax} at 77 K and 20 K operation.

Figure 5 shows the liquid evaporation rate for the SiC and GaN devices if operated for 1 minute at 20 K (liquid hydrogen, LH2), 77 K (liquid nitrogen, LN2) and 300 K (water). From figure 5, it is observed that SiC devices show about 50 times and 17 times higher evaporation rate than GaN devices at 20 K and 77 K respectively. Therefore, GaN based devices are more suitable for cryogenic cooling rather than SiC devices. These liquid evaporation rates have been obtained in steady-state and before temperature equillibrium is attained. To obtain an equilibrium, recondensation of the cryogenic liquid is needed and the time to achieve thermal equilibrium is solely dependent on the output power of the cryocooler. Apart from power devices, polypropylene metallized film capacitors, thin film/metal film resistors, core material made of nano crystalline materials with higher permeability and saturation flux density at low temperatures are preffered for cryogenically cooled devices to reduce thermal loads. The controller circuit will be at room temperature because it will help to reduce additional thermal loads at cryogenic temperatures. Changing the whole circuit to be cryo-compatible will also increase the cost. The next section compares the mass, cost and output power of the cryocoolers which can establish a selection criteria for large scale application like electric aviation.

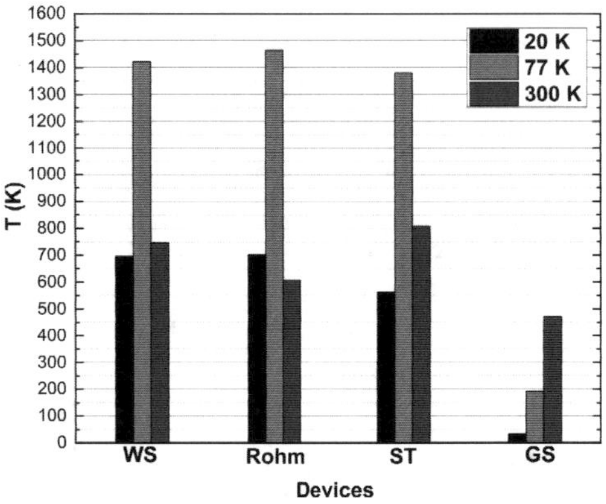

Figure 4. Temperature of the SiC and GaN devices if operated for 1 minute at 20 K, 77 K and 300 K as starting temperature.

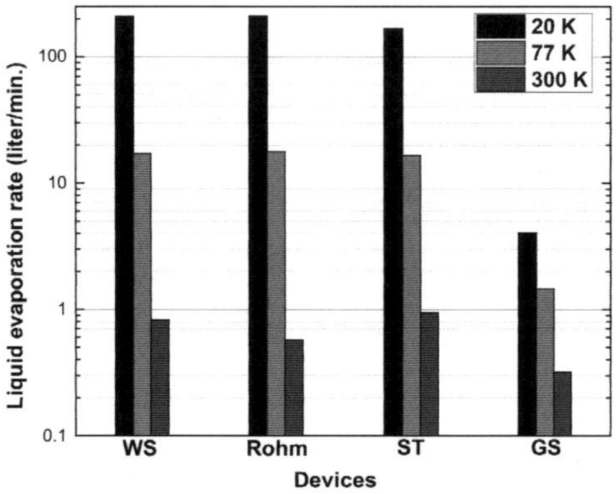

Figure 5. Liquid evaporation rate for the SiC and GaN devices if operated for 1 minute at 20 K (liquid hydrogen, LH2), 77 K (liquid nitrogen, LN2) and 300 K (water).

3. Cryogenic cooling systems for MW-class power electronics

Many cryogenic related applications like superconducting power and energy devices, food preservations, and cryobiology need liquid cryogens to provide the necessary cooling. Such a practice is very common in a laboratory environment, where cryogens are readily available and trained personnel are present to periodically transfer the cryogens. However, for most remote and unmanned operations, periodic cryogen transfer is not desirable. The use of liquid cryogens may be desirable from the standpoint of providing thermal shielding with varying heat loads and for enhancing heat transfer. In such cases the best refrigeration system may be a re-condenser that liquefies a gas such as nitrogen, neon, or helium, which is then circulated passively or actively throughout the superconducting system. The boil-off gas can be re-condensed in a closed loop [24] cryo-cooling (liquefier) system. These closed cryogenic systems are usually more desirable in practical applications like fusion reactors, particle accelerators, MRI, NMR than open systems where the liquid cryogen must be trucked in from a remote liquefier. However, the use of a liquid cryogen, even in a closed system, limits operating temperatures to a region between the cryogen's triple point and the critical point [24, 25].

Cryogenic thermal management is a key technology for superconducting generators and cryogenic power electronics. The cryogenic cooling systems need to be low cost, high reliability, high efficiency, and smaller size. The choice of the cryocooler is based on steady-state cooling, temperature uniformity, control, transient response to dynamic loads, recovery requirements, power requirements in steady-state, availability of cryogens and reliability/safety. Helium and nitrogen, due to their inert nature, are the most widely used cryogenic liquids. Liquid nitrogen has a maximum breakdown strength of 1.6-1.9 MV/cm and dielectric constant of 1.431[26-30]. Liquid hydrogen and liquid oxygens are commonly used for space applications as low-cost cryo-fuels. Liquid hydrogen and Cryo-compressed Hydrogen (CcH_2) are gaining interest due to their low density, low cost, large production, compact storage and can be used as a fuel for net zero emission hybrid electric vehicles [31]. For cryogenic power electronic applications, there are numerous challenges and factors that would affect the cooling requirements. The challenging factors are operating temperature, electric power dissipation (DC/AC), required power density, location of the power electronic system (like deep space, deep sea etc), heat-in-leaks from the outside world (like outdoor and extreme environments in space or sea), geometrical proportions. Applications vary greatly, thus leading to the requirement for many cryocooler types. The primary power requirement of the cryocooler is dictated by the Carnot efficiency [32]. The removal of 1 W at 77 K typically requires 20 W of refrigeration input power, and the removal of 1 W at 20 K typically requires 250 W of refrigeration input power.

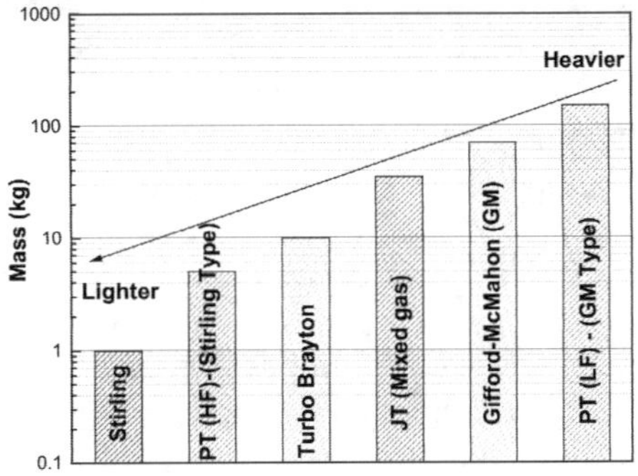

Figure 6. Size and weight comparison for different types of cryocoolers obtained from CILTEC survey 2002 [32, 35, 36].

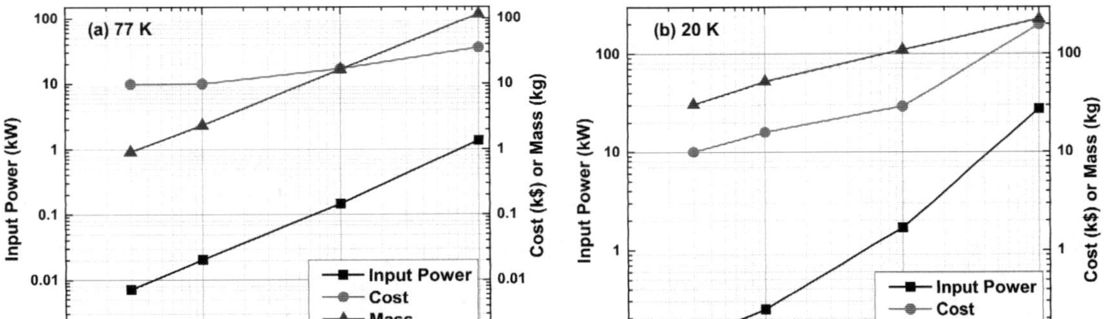

Figure 7. Cooling power versus estimated input power, cooler cost and cooler mass for operating temperature at (a) 77 K and (b) 20 K summarized result from reference [32].

The cooling requirements at low temperatures incur significantly higher operational costs. Although the refrigeration cost may have a small impact on the overall cost, it does affect the selection of a cryogenic plant. Figure 6 shows the size and weight comparison of different types of cryocoolers based on CILTEC 2002 survey as reported in the reference [32]. Stirling cryocoolers are the smallest and lightweight cryocoolers available till now. Many miniatures cryocoolers are commercially available for on-board or airborne applications [23, 33-36]. Brake and Wiegerinck performed a cryocooler survey based on available CILTEC cryocooler data and gathered from publicly available sources and other suppliers [32]. Figure 7 (a) and 7 (b) summarizes the input power requirements, cryocooler cost and cryocooler mass for different cooling powers at 77 K and 20 K operating temperatures. These graphs provide useful estimation of power consumption, cost and mass for a specific operating temperature.

4. Cryogenic power electronics for large-scale HTS applications

High Temperature Superconducting (HTS) power generation and distribution technology is a key technology to achieve high power densities. Figure 8 shows a comparison of power densities for low power conventional electric motors, internal combustion engines (ICE), electric car motors, electric aircraft motors, ship propulsion, actual HTS motors, designed HTS motors for airborne applications and existing commercial-off-the-shelf (COTS) power converters [37-43]. The specific power of cryogenically fuelled rocket engine is about 153 kW/kg [41], whereas the specific power of most conventional electrical motors is limited to about 2 to 4 kW/kg, although specially designed motors can have power densities up to about 6 kW/kg [40]. Ship propulsion has comparatively less specific power 1.3 kW/kg. Some brushless DC motors developed for use on automobiles have power densities as high as 9.0 kW/kg. Some AC motors for electric aircrafts are developed with a high specific power of 11.56 kW/kg [37-40]. High temperature superconducting machines may represent the perfect candidate as they can carry large amounts of current. However, AC loss in superconductors is a challenging factor for any application, and can be minimized by using multifilamentary architecture [44-46]. They therefore enable the development of very high-power density rotating machinery with power densities in excess of 10– 20 kW/kg and torque densities over 35 Nm/kg [42, 43]. Also, the overall power density can be further improved by integrating the cryogenically cooled power converter with the motor/generator. In addition, power density of power converters could be increased by actively removing heat generated by the passive components of the power conversion system. Moreover, superconducting inter-connections between modules will further increase the efficiency of the power converters.

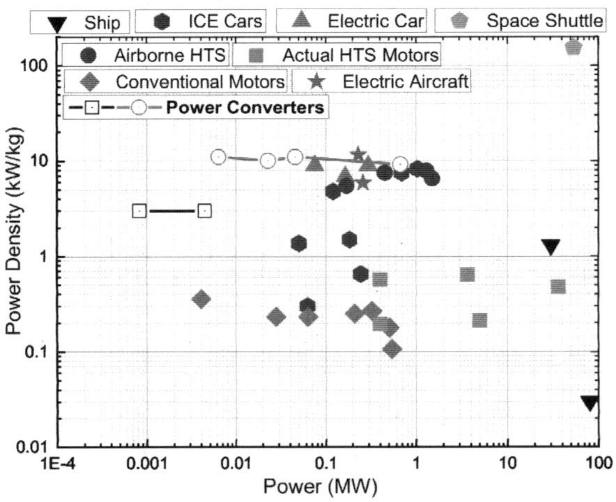

Figure 8. Power density comparison for different types of conventional power applications and HTS applications together with power converters [37-43].

Liquid Hydrogen (LH_2) is arguably the best cryogenic fluid candidate for electric and hybrid electric aircrafts due to three main reasons: (1) LH_2 can be stored in a small place in liquid state and provides a very convenient cooling system at 20 K for superconducting machines and power electronics systems, (2) LH_2 provides an energy density of 39 kW/kg which is about three times that of gasoline and much higher than batteries [37], and (3) LH_2 can be converted into electrical energy through fuel-cells or combusted with very low emissions in high speed turbogenerators. So far, only a limited number of fully cryogenic electric machine designs have been developed and a torque density of 100 Nm/kg is expected. In a review article by Haran *et al*, a complete list of high-power density superconducting rotating machines is provided [47].

5. Conclusions

Several studies have concluded that there is a great potential for cryogenic power conversion in applications such as wind energy, propulsion motors and power generators for on-board ships, future military applications, and aircrafts where size and weight are the primary design considerations. Most of the research results related to the power converters are on the characterization of the operating behavior of the devices, instead of the whole converter/inverter system. Although some research related to converters is presented by researchers, it is mostly at the DC-DC converter level for powering the field windings of a synchronous generator. A few researchers have reported on the design and performance of complete cryogenic converter systems, consisting of both active and passive devices, though much has been written about the superconducting generators. The present conceptual design study shows that the GaN devices-based power converters are smaller in size and overall power loss is much less compared to SiC devices at 20 K. The selection and integration of the right cryogenic system for a given application contributes to the overall performance, efficiency, power density, and cost of the system. Cryogenic power electronics technology is the next step in the evolution of power electronics technology to obtain high power density, high efficiency, and superior performance for various superconducting and non-superconducting applications.

6. References

[1] Rajashekara K and Akin B, 2013 Proc IEEE Int Elect Mach Drive pp 899–904
[2] Reynes H, Maneiro J, Buttay C, Dworakowski P 2017 12th european advanced technology workshop on micropackaging and thermal management (iMaps La Rochelle, France)
[3] Mertens R G et al 2007 Jr Electron Packag vol 129(3) pp 316–323
[4] Arun Kumar C M and Mukesh Kumar P C 2017 Advances in Natural and Applied Sciences vol 11(8) pp 271-279
[5] Iradukunda A-C, Huitink D. R. and Luo F 2020 IEEE Journal of Emerging and Selected Topics in Power Electronics vol 8(1) pp 256-271
[6] Wang F et al 2019 AIAA/IEEE Electric Aircraft Technologies Symposium (EATS Indianapolis, USA) pp 1-9
[7] Gui H et al 2020 IEEE Transactions on Power Electronics vol 35(5) pp 5144-5156
[8] Chakraborty S et al 2019 Energies vol 12 p 1569
[9] Mueller O M and Herd K G 1993 Proceedings of IEEE Power Electronics Specialist Conference (PESC '93 Seattle, WA, USA) pp 772-778
[10] Ray B, Gerber S S, Patterson R L and Myers I T 1995 Proceedings of 1995 IEEE Applied Power Electronics Conference and Exposition (APEC'95 Dallas, TX, USA) vol 2 pp 941-946
[11] Ray B, Gerber S S, Patterson R L and Myers I T 1995 Proceedings of Power Electronics Specialist Conference (PESC '95 Atlanta, GA, USA) vol 1 pp 55-60
[12] Ray B, Gerber S S and Patterson R L 1996 Proceedings of the 31st Intersociety Energy Conversion Engineering Conference (IECEC 96, Washington, DC, USA) vol 1 pp 553-559
[13] Ray B, Gerber S S and Patterson R L and Dickman J E 1996 Proceedings of Applied Power Electronics Conference (APEC '96, San Jose, CA, USA) vol 2 pp 883-888
[14] Perez-Guerrero F F, Ray B and Patterson R L 1997 Proceedings of the Thirty-Second Intersociety Energy Conversion Engineering Conference (IECEC-97 Honolulu, HI, USA) vol 2 pp 1415-1420
[15] Gerber S, Hammoud A, Patterson R L and Elbuluk M 2000 IEEE 31st Annual Power Electronics Specialists Conference. Conference Proceedings (Cat. No.00CH37018 Galway, Ireland) vol 3 pp 1201-1206
[16] Elbuluk M, Gerber S, Hammoud A and Petterson R L 2000 Collection of Technical Papers.

35th Intersociety Energy Conversion Engineering Conference and Exhibit (IECEC Las Vegas, NV, USA) vol 1 pp 1-6
[17] Hennessy M J et al 2006 AIP Conference Proceedings vol 824 pp 367-374
[18] Ward R R et al 2006 Twenty-First Annual IEEE Applied Power Electronics Conference and Exposition (APEC'06 Dallas, TX, USA) p 4
[19] Barth C et al 2017 IEEE Applied Power Electronics Conference and Exposition (APEC Tampa, FL, USA) pp 717-723
[20] Graber L et al 2017 IOP Conf Ser Mater Sci Eng vol 279 p 012011
[21] Gui H et al 2018 IEEE Energy Conversion Congress and Exposition (ECCE 2018) pp 7010-7015
[22] Mayo R E, Bustamante J G and Beechner T L 2017 16th IEEE Intersociety Conference on Thermal and Thermomechanical Phenomena in Electronic Systems (ITherm 2017) pp 530-536
[23] Kim H 2010 Journal of Korean Physical Society 56 (5) pp 1523-1526
[24] Hands B A 1986 Cryogenic Engineering Academic Press
[25] Radebaugh R 2004 Proceedings of the IEEE vol 92 (10) pp 1719-1734
[26] Mathes K N 1967 IEEE Trans. on Electrical Insulation vol 2 (1) pp 24-32
[27] Fujita H et al 1978 Cryogenics vol 18 (4) p 195
[28] Gerhold J 1979 Cryogenics vol 19 (10) p 571
[29] Yoshino K et al 1979 Journal of Electrostatics vol 7 p 103
[30] Naidu S and Kamaraju V 2009 High Voltage Engineering Tata McGraw Hill Education Private Limited p 85
[31] Yang C and Ogden J 2007 Int J Hydrogen Energy vol 32 p 268
[32] ter Brake H J M and Wiegerinck G F M 2002 Cryogenics vol 42 p 705
[33] Chen X P et al 2016 Overview of Micro-Miniature Stirling Cryocoolers for High Temperature Applications International Cryocooler Conference (Cryocoolers 19 Boulder CO) p 115
[34] Raab J and Tward E 2010 Cryogenics vol 50 p 572
[35] Van den Groep W et al 2011 The development of a new generation of miniature long-life linear coolers Cryocoolers 16 p 111
[36] Radebaugh R and Ross Jr. R G 2015 Cryocooler Fundamentals and Space Applications presented in CSA-CEC2015, Tucson, Arizona
[37] Masson P J et al 2007 Supercond. Sci. Technol vol 20 p 748
[38] Nightingale N P (October 1986) 2010 Automotive Stirling Engine – Mod II Design Report NASA Lewis Research Center
[39] Wevers C 2014 The Most Powerful F1 Car Ever – GTspirit
[40] Jewell G 2010 Permanent Magnet Machines and Actuators Symposium on Materials for a Sustainable Future, Birmingham, England, Magnetic Materials Group, University of Birmingham pp 11–18
[41] Blair J C, Ryan R S, and Schutzenhofe L A 2011 Lessons Learned in Engineering NASA technical report NASA/CR—2011–216468
[42] Masson P J and Luongo C A 2005 IEEE Trans on Appl Supercond vol 15 (2) p 2226
[43] Masson P J, Pienkos J E and Luongo C A 2007 IEEE Trans on Appl Supercond vol 17 (2) p 1579
[44] Kanetaka H et al 1995 IEEE Trans on Appl Supercond vol 5 (2) p 737
[45] Kar S, Sarangi S K and Rao V V 2013 IEEE Trans on Appl Supercond vol 23 (3) p 8200404
[46] Yahia A B et al 1995 IEEE Trans on Appl Supercond vol 27 (4) p 5600105
[47] Haran K S et al 2017 Supercond Sci Technol vol 30 p 123002

IOP Conf. Series: Materials Science and Engineering 1241 (2022) 012041 doi:10.1088/1757-899X/1241/1/012041

Electrical characterization of a 1200V GaN HEMT at cryogenic temperatures

Md Maksudul Hossain, Yuqi Wei, Arman-Ur Rashid, Rosten Sweeting, and H. Alan Mantooth

Department of Electrical Engineering, University of Arkansas, Fayetteville, USA

Abstract. The cryogenic operation of power converters integrated with the superconducting electric machines can leverage the lower parasitic, lower core material, and insulation requirement to reduce weight and increase power density in aerospace applications. To that end, Gallium Nitride High Electron Mobility Transistor (GaN HEMT) has recently proven to be the most viable option for the cryogenic converter operation. However, the topology selection and optimization are greatly influenced by the current/voltage rating of the power devices. Although series connected HEMTs can increase the blocking voltage, the dynamic voltage imbalance issue stemming from device parameter fluctuations, the larger number of required isolated gate driver makes the need for single high blocking voltage devices more compelling. This is even more pronounced in GaN due to its higher operating frequency which stems from the lower input capacitance. Although 900V cascode GaN HEMTs are now commercially available, single-chip high voltage HEMTs are attractive due to their zero reverse recovery charge, and lower package bonding parasitics. In this paper, a 1200V enhancement-mode GaN HEMT from GaNPower International has been characterized at cryogenic temperature. A 63% decrease has been demonstrated in the on-resistance, $R_{ds(on)}$ from room temperature to -150°C. The threshold voltage shows a 1.2x increase from room temperature to cryogenic temperature (-178°C). The transconductance (G_{fs}) shows a 1.34x increase for a temperature difference of 175°C. The results translate to a lower conduction loss and a faster switching speed and unfold further possibility of GaN in cryogenic converter applications.

1. Introduction

Owing to the continued need for high efficiency, high power density electrical propulsion system, the research interest in cryogenic electronics is growing in the aircraft industry. A recently proposed fully electric aircraft (FEA) concept by the Cryogenic High-Efficiency Electrical Technologies for Aircraft (CHEETA) considers utilizing LH_2 to use both as a fuel cell and cryogen for the associated power electronics as well as to ensure zero carbon emission. As a key component of the power electronic converter, the power semiconductor devices warrant extensive characterization in a cryogenic environment due to their broad variety of electrical properties with temperature. Unlike power switching operation at high temperatures, silicon carbide (SiC) proves to be an inferior counterpart in cryogenic converter applications. On the contrary, gallium nitride (GaN) switches offer superior characteristics in terms of desired electrical properties for a power converter as found in the literatures [1-2]. Although GaN devices are entitled to better figure-of-merit (FOM), the lack of available high voltage (HV) devices can limit the possibility of its high-power applications. At the time of this writing, commercial offerings of GaN device voltage can be outlined in the range of 600-900V. More recently, a 900V cascode GaN HEMT has become available [3]. Several techniques are described in the literature for achieving high voltage without needing to use high breakdown voltage devices. However, these techniques often involved utilizing MOSFETs/HEMTs in series resulting in introducing more complexity in controller design due to dynamic voltage imbalance issues resulting from die-to-die parametric variations. In addition, this method requires an isolated gate driver circuit for each of the series-connected switches. If one device turns off earlier than others, it may need to block a voltage higher than its rated value which might degrade lifetime and hence reliability [4]. The topology selection and optimization are critically influenced by the current/voltage rating of the power devices. In the recent past, a GaN startup GaNPower International has introduced a 1200V (GPIHV30DDP5L) engineering sample. The device is an N-channel enhancement-mode (E-mode) 1200 V Power GaN HEMTs based on GaN-on-silicon technology. It is housed in TO263-5L package with the source terminal connected to the thermal pad. This paper conducts a comprehensive electrical characterization including dc output curve (I_d-V_{ds}), transfer characteristics, capacitance-voltage (C-V) characteristics, on-resistance, $R_{ds(on)}$ at room temperature and cryogenic temperatures down to -180°C.

2. Need for High Breakdown Voltage GaN

2.1. Series Device Issues

High-voltage, high-current GaN switches are needed for higher voltage and power applications e.g., the aerospace industry. Although series-connected GaN HEMTs can achieve higher voltages, it face the following design challenges [5]: (1) dynamic voltage sharing inequality due to device capacitances variations, mismatch in turn-off delay time due to gate drive signals and device turn-off delay mismatch (2) unequal static voltage sharing due to leakage currents variation among different devices (3) additional add to extra loop inductance due to the connected bus bars between devices increases the device voltage de-rating due to induced voltage experienced by each device during turn-off. Moreover, during switching, the adjacent device from the gate driver switches first which will lead to unequal loss distribution. This delay will require higher the effective dead time and therefore reduce the switching frequency.

2.2. Cascode Structure Issues

Although a 900V cascade GaN HEMT is available now commercially, due to the inherent discrete-die nature of the topology, the interconnect and common source inductance is higher along with added on-resistance, $R_{DS(on)}$ from the low voltage (LV) silicon MOSFET [3].

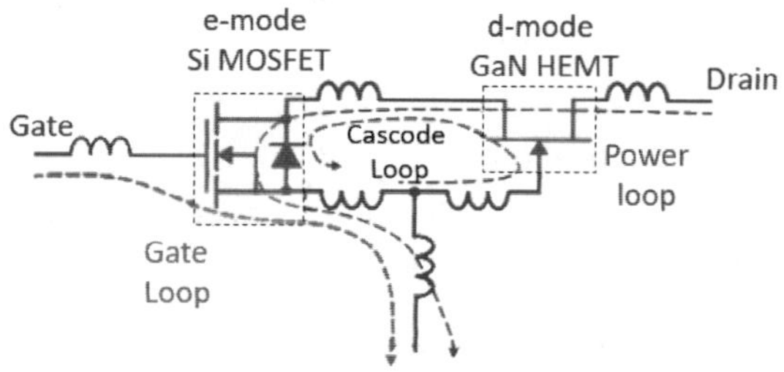

Figure 1. Parasitic components in a GaN cascode structure. [6]

2.3. Topology Selection Flexibility

In cryogenic operation, the topology selection depends on a number of factors. In order to take the advantage of increased current handling capability superconducting (SC) cables and high efficiency SC machines, certain topologies get priority over others. However, the unavailability of devices at the rated current and voltage can limit the selection greatly and might even force to choose devices that are not well suited for low temperature applications. To that end, the need for a single chip HV GaN HEMT is significant.

3. Challenges in Achieving High Voltage for Lateral HEMTs

The available GaN devices are lateral HEMTs i.e., the devices are fabricated on the same plane of the GaN wafer, unlike vertical power MOSFETs. However, the high voltage demand for high power applications requires the gate-drain distance to be longer which in turn translates to more chip area. Although lateral HEMTs can achieve high breakdown voltage (BV) by extending the drift region, the current rating reduces since the effective gate width reduces to keep the same die area. Another fabrication difficulty is growing thick-enough GaN buffer layers on 8-inch wafers. One of several other challenges for achieving high voltage in lateral GaN HEMTs is the high leakage current due to lack of p-n junction in off-state. As a result, the off-state leakage is only limited by gate bias induced barrier which is lowered at higher drain voltages (drain-induced barrier lowering - DIBL) [6], and hence leakage current increases. It is more pronounced in E-mode devices since they are normally off and lack in negative gate voltage to further pinch-off the channel, unlike d-mode devices. Other challenges include the careful management of electric field profiles required between electrodes and hence the use of field plates which in turn slows down the devices by adding to capacitance. Although extensive research for vertical devices [7-8] for HV devices is ongoing, commercial reliable vertical devices are yet to come. Lately, GaNPower International launched its first 1200 V lateral GaN (engineering sample). This paper presents the comprehensive electrical characterization of these devices. This characterization will help evaluate the entrance of lateral GaN devices into MV and HV applications in near future.

4. GaN HEMT Operation Principle

In order to clearly explain the cryogenic device behavior, a simplified lateral GaN HEMT structure is illustrated in Fig. 1. In general, GaN HEMT is a depletion mode device i.e., there is already a channel present in the AlGaN/GaN interface. The piezoelectric polarization induced due to the strained GaN gives rise to a confined 2-dimensional electron gas (2DEG) which increases the unstrained GaN mobility from 1000 cm^2/V-s to 1500-2000 cm^2/V-s and hence the name is high electron mobility transistor (HEMT). However, to ensure fail-safe operation a normally-off device is desired in power electronics applications so that no negative gate bias is needed at the onset of energizing the circuit. However, one of the several available techniques for achieving normally-off E-mode HEMTs includes p-GaN cap layer on top of the AlGaN barrier. The P-type GaN introduces built-in potential which supersedes the polarization induced voltage and essentially depletes the 2DEG. Once the gate voltage reaches the positive threshold voltage, V_{th}, the channel creates a short between drain and source terminal and starts conducting once drain bias is applied. This is on-state behavior, the major contributions of resistances involve 2DEG channel resistance and the drift resistance between gate and drain. As apparent form Fig.1, there is a gate overhang that reshapes the electric field more uniformly to achieve higher breakdown voltage. This gate action controls the 2DEG similar to the actual gate which creates the non-linear behavior of this drift resistance. The inter-electrode parasitic capacitances can also be observed from Fig. 1.

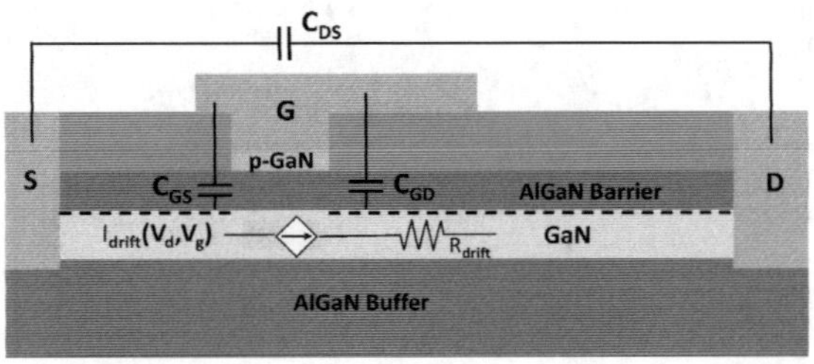

Figure 2. Lateral GaN HEMT device structure

5. Test Setup

The device under test (DUT) has been placed inside a thermal test chamber from Delta-design that can drop to -184°C. The front panel has SMA to BNC electrical feed-through via which the cables connect the power device analyzer and the DUT. The chamber is connected to a thermally insulated LN_2 transfer line and supplied from an LN_2 storage tank. A temperature controller controls the temperature by controlling the solenoid flow valve. An O_2 sensor is placed nearby to monitor the O_2 level in the ambient as a safety precaution. The test setup is demonstrated in Fig. 3. The DUT is a TO-263 package that has been translated to a more convenient TO-247 package via a PCB adapter for characterization in another PCB as shown in Fig. 4(a). As can be seen, the 3-pin header is connected to the SMA connector on the top side. For an accurate measurement of the DUT, a 4-wire method (Kelvin) has been implemented. The working principle is demonstrated in Fig. 5. The 4-wire essentially isolates the force and sense terminal of the DUT so that the cable parasitic voltage drops do not interface with the DUT voltage reading. The current through the DUT is equal to the currents through the force terminal ($I_{DUT} = I_m$) while the sense terminals carry negligible current thanks to the high impedance of the voltmeter leads ($V_{DUT} = V_m$). As evident from Fig. 4(b), both the drain and source terminals have a pair of force and sense terminals as close to the DUT as possible. For gate terminal, usually, one single connection is used since the applied voltage is almost equal to the voltage seen by the DUT gate terminals due to lower current.

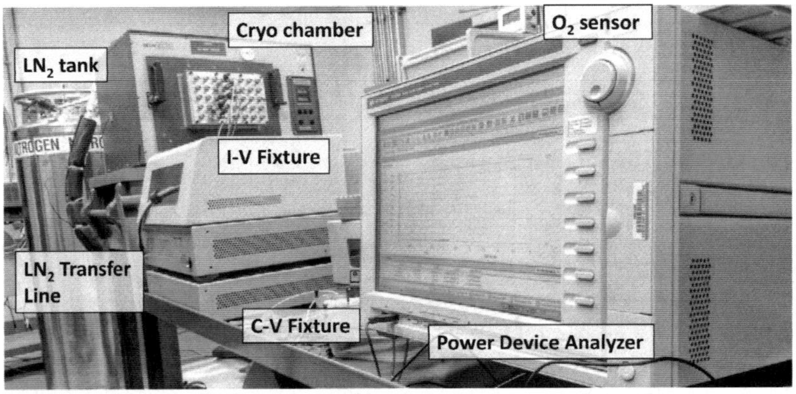

Figure 3. Experimental set up for cryogenic testing

(a) (b)

Figure 4. (a) PCB (bottom side) for Cryogenic characterization (b) PCB (top side) indicating the Kelvin (4 wire) connections.

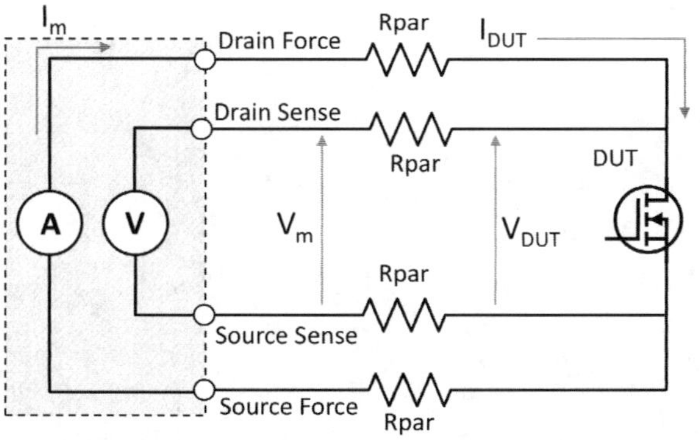

Figure 5. 4-wire measurement (Kelvin connection) working principle.

6. Results and Discussions

In this section, experimental results are presented. The transfer curves at different operating temperatures have been presented in Fig. 6(a). The threshold voltage shift towards the right with a lower temperature down to -180°C. The parasitic interelectrode capacitance is presented in Fig. 6(b). The capacitance is measured with the multi-frequency capacitance measure unit (MFCMU) from Keysight. The different humps in the region between 0-400 V clearly demonstrate the use of multiple field plates as illustrated in Fig. 2 to reshape the interelectrode electrode electric field to achieve higher breakdown without sacrificing reliability. The typical dc output characteristic (I_d-V_{ds}) has been demonstrated in Fig. 7 for both first and third quadrants for two different temperatures. As evident from the figures, the GaN HEMT can conduct in a reverse direction even at $V_{GS} < V_{TH}$ since the electrical drain and source switch their role and the turn-on condition now becomes $V_{GS}+V_{SD} > V_{TH}$. As a result, at a high enough source-drain voltage, the device turns on even if the gate voltage is negative. The on-resistance of both the quadrants decreases as the temperature decreases from 25°C to -180°C. The slope (broken black line) indicates the reduction of $R_{ds(on)}$ with lower temperature which is more clearly depicted in Fig. 8(a). As evident from that figure, GaN HEMTs do not show any carrier freeze-out effect at cryogenic temperatures [9]. This is because the channel carrier originates from piezoelectric effect and not from intentional doping which involves no ionization energy requirement. This reduction in $R_{ds(on)}$ can be explained from the reduced Coulomb scattering and increased 2DEG density of the GaN HEMT. Fig. 8 demonstrates the threshold voltage and transconductance variation with temperature.

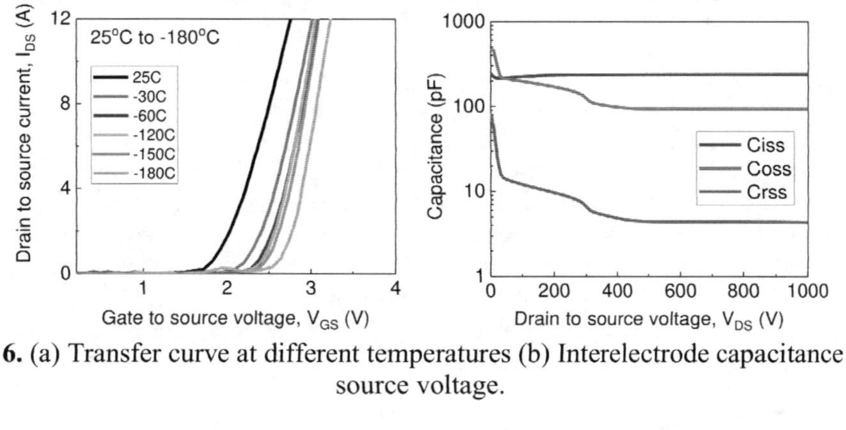

Figure 6. (a) Transfer curve at different temperatures (b) Interelectrode capacitance vs. drain-source voltage.

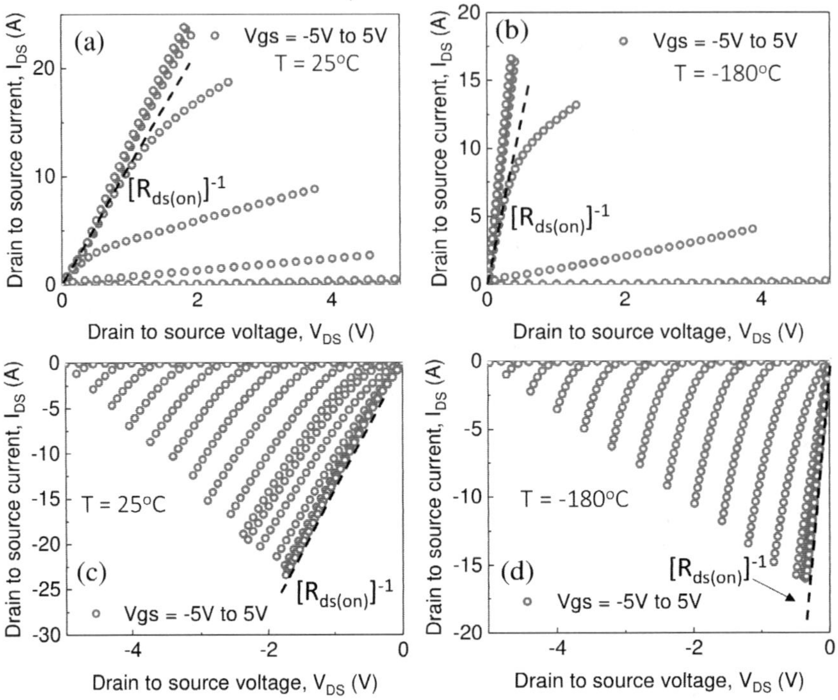

Figure 7. (a) First quadrant dc output curve curves at (a) 25°C, (b) -180°C, third quadrant dc output characteristics at (c) 25°C and (d) -180°C.

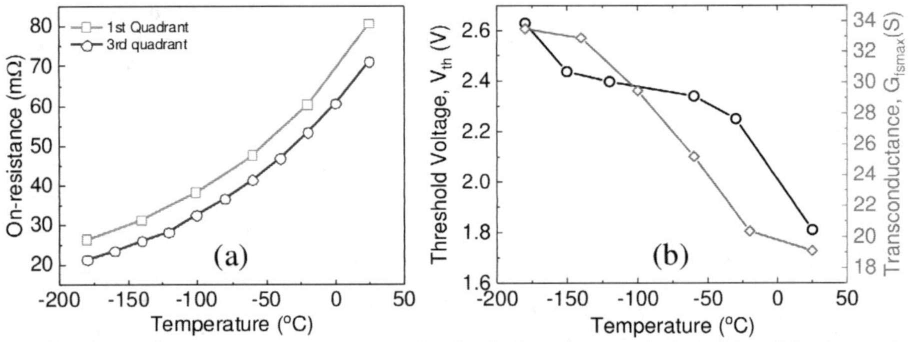

Figure 8. (a) On resistance vs. temperature for both first (orange) and third (blue) quadrant (b) threshold voltage (black) and maximum transconductance (red) vs. temperature.

A 68% decrease has been demonstrated in the first quadrant on-resistance, $R_{ds(on)}$ from room temperature to -180°C while the third quadrant shows a 71% $R_{ds(on)}$ reduction. The threshold voltage shows a 1.46x increase from room temperature to cryogenic temperature (-178°C). The threshold behavior with low temperature depends on the gate structure since the Schottky barrier height and p-GaN junction will behave differently. The behavior here is similar to gate injection transistors (GIT) since the junction potential barrier height increases at a lower temperature which causes the threshold voltage to rise. The maximum transconductance (G_{fs}) shows a 1.34x increase for a temperature difference of 200°C.

The lower $R_{ds(on)}$ at a lower temperature will result in lower conduction losses. The increase in threshold voltage will increase gate driver loss, however, it can be leveraged to reduce false turn-on phenomena. The on-resistance also decreases in the third quadrant, which will further reduce the conduction losses. The transconductance is improved which will lead to improved rise time since it is a function of Miller plateau voltage which decreases as the transconductance decreases [10]. This will reduce the switching loss in turn but increase high dv/dt and di/dt related issues as well. The results translate to a lower conduction loss and a faster switching speed and unfold further possibility of GaN in cryogenic converter applications.

Although the available 1200 V device tested in this paper is a packaged device, usually, the failure related to cryogenic thermal stress originates from packages first. Due to the Coefficient of Thermal Expansion (CTE) mismatch, the wire bonding and die attach adhesives are affected. To isolate the fault, a detailed packaging failure analysis may be required.

The more common 650 V devices come with different device structures e.g., HEMT, cascode, GIT, and also the gate structure might vary e.g., Schottky gate, junction gate, etc. Compared to a 650 V HEMT, the 1200 V cryogenic results are quite similar in terms of on-resistance and threshold voltage. However, depending on the gate structure, the threshold voltage trend can be the opposite. The low-temperature behavior depends on Schottky barrier height, junction barrier potential, and also substrate material being used [11].

The major challenge in characterizing these GaN devices are the oscillation resulting from the larger gain as well as long cable induced stray inductances and capacitive coupling. To combat the oscillation related issues, several measures have been adopted. The drain force and source force connectors are twisted to minimize the inductance by canceling the opposing magnetic flux. Finally, inserting a 50-ohm terminator will help alleviate oscillations [12].

7. Conclusion

This paper presents the cryogenic characterization of a high voltage (1200 V) single-chip GaN HEMT from GaNPower International. The experiments have demonstrated the reliable operation of the device in the entire range of operating temperatures (-180°C) with overall improved performance. The on-resistance improved in both first and third quadrant which translates to lower conduction losses and therefore, higher efficiency. The transconductance increases with reduced temperatures which promises faster switching as cryogenic temperatures. However, the threshold voltage behavior was non-monotonic. Since GaN HEMTs are still not available in higher voltages, the introduction of this 1200 V device will help circuit designers select circuit topologies with higher flexibility. The presented results also verify the possibility of high voltage single-chip lateral GaN HEMTs to enter the medium and high voltage cryogenic converter applications to maximize the efficiency and power density in the extreme ambient.

8. References

[1] H. Mhiesan, M. M. Hossain, A. U. Rashid, Y. Wei and A. Mantooth, "Survey of Cryogenic Power Electronics for Hybrid Electric Aircraft Applications", *2020 IEEE Aerospace Conference*, pp. 1-7, 2020.

[2] H. Gui, R. Chen, J. Niu, Z. Zhang, L. M. Tolbert, F. Wang, et al., "Review of Power Electronics Components at Cryogenic Temperatures", *IEEE Transactions on Power Electronics*.

[3] M. M. Hossain, Y. Wei, A. Rashid, R. Sweeting, A. Mantooth, "Cryogenic Evaluation and modeling of a 900 V Cascode GaN HEMT," *12th International Conference on Electrical and Electromechanical Energy Conversion – ECCE Asia*, 2021, pp. 7-12.

[4] D. A. Grant and J. Gowar, Power MOSFET-Theory and Applications, New York:Wiley, 1989.

[5] K. Vechalapu, S. Bhattacharya and E. Aleoiza, "Performance evaluation of series connected 1700V SiC MOSFET devices," *2015 IEEE 3rd Workshop on Wide Bandgap Power Devices and Applications (WiPDA)*, 2015, pp. 184-191, doi: 10.1109/WiPDA.2015.7369327.

[6] M. Meneghini, G. Meneghesso and E. Zanoni, Power GaN Devices: materials applications and reliability, Springer, 2016.

[7] Y. Zhang *et al.*, "Large-Area 1.2-kV GaN Vertical Power FinFETs With a Record Switching Figure of Merit," in *IEEE Electron Device Letters*, vol. 40, no. 1, pp. 75-78, Jan. 2019, doi: 10.1109/LED.2018.2880306.

IOP Conf. Series: Materials Science and Engineering 1241 (2022) 012041 doi:10.1088/1757-899X/1241/1/012041

[8] D. Ji *et al.*, "Normally OFF Trench CAVET With Active Mg-Doped GaN as Current Blocking Layer," in *IEEE Transactions on Electron Devices*, vol. 64, no. 3, pp. 805-808, March 2017, doi: 10.1109/TED.2016.2632150.
[9] L. Nela, N. Perera, C. Erine and E. Matioli, "Performance of GaN Power Devices for Cryogenic Applications Down to 4.2 K," in *IEEE Transactions on Power Electronics*, vol. 36, no. 7, pp. 7412-7416, July 2021, doi: 10.1109/TPEL.2020.3047466.
[10] "Device Application Note AN608A" [online] Available: https://www.vishay.com/docs/73217/an608a.pdf
[11] J. Kuzmik, R. Javorka, A. Alam, M. Marso, M. Heuken and P. Kordos, "Determination of channel temperature in AlGaN/GaN HEMTs grown on sapphire and silicon substrates using DC characterization method," in *IEEE Transactions on Electron Devices*, vol. 49, no. 8, pp. 1496-1498, Aug. 2002, doi: 10.1109/TED.2002.801430.
[12] "Keysight Technologies B1505A Power Device Analyzer/Curve Tracer" [online] Available: https://www.keysight.com/us/en/assets/9018-05644/installation-guides/9018-05644.pdf

Acknowledgments
This work presented herein was funded in part by the NASA ULI: Development of the Cryogenic Hydrogen-Energy Electric Transport Aircraft (CHEETA) Design Concept under Award Number 80NSSC19M0125.

IOP Conf. Series: Materials Science and Engineering 1241 (2022) 012042 doi:10.1088/1757-899X/1241/1/012042

Cryogenic Performances Comparisons Among Si MOSFET, SiC MOSFET, Cascode GaN, and GaN Devices

Yuqi Wei, Md Maksudul Hossain, and H. Alan Mantooth

University of Arkansas, Fayetteville, Arkansas, 72701, United States

Abstract. Cryogenic power electronics is an emerging technology to achieve high efficiency and high power density. As the key element of the power electronics system, the device performance should be evaluated under cryogenic temperatures. To demonstrate the differences between different materials and technologies, cryogenic temperature performance comparisons among silicon (Si) metal–oxide–semiconductor field-effect transistor (MOSFET), Si insulated-gate bipolar transistor (IGBT), silicon carbide (SiC) MOSFET, Cascode gallium nitride (GaN), and GaN high-electron-mobility transistor (HEMT) from different perspectives are made, which includes on-state resistance, threshold voltage, turn-on and turn-off times, turn-on and turn-off switching losses. The normalized values are adopted and results are drawn in the same pictures to demonstrate the differences. The results demonstrate that traditional Si devices have better performances under low temperatures. The SiC MOSFET is not suitable for cryogenic applications due to its increased on-state resistance and switching loss. Cascode GaN and GaN HEMT are promising candidates for cryogenic applications with reduced conduction loss and switching loss.

1. Introduction

Cryogenic operations are required for the superconducting cables and machines to enable high efficiency. The all-electric aircraft with cryogenic hydrogen system can significantly improve the system efficiency. Power electronic converters are closely connected to the superconducting motors. Traditionally, the converters are designed to operate at room temperature by using an additional heater and designing thermal insulation. Cryogenic power electronics are needed to reduce system cost and weight. Another motivation is semiconductor devices perform better under cryogenic temperatures, improving the system's efficiency when compared to room temperature. [1, 2]. The research in this area can be classified into the following three groups: 1) cryogenic characterizations of semiconductors [3]-[10]; 2) cryogenic characterizations of passive components [11]-[13]; 3) cryogenic design and evaluation of power converters [14]-[18]. In this article, the focus is on the characterizations of semiconductors.

Characterizations and comparisons are made for the 4H-SiC planar and trench gate MOSFETs [3]. The results demonstrate that the on-state resistance and threshold voltage for both planar and trench SiC MOSFETs increase with a decrease of temperature. While the dynamic performance is different, the planar MOSFET has a reduced switching speed under cryogenic temperatures. However, the switching performance of the trench MOSFET is not sensitive to the temperature. The differences can be explained by the weight of three factors in different temperature range [3]: 1) threshold voltage; 2) inversion carrier mobility; and 3) gate drain capacitance. The on-state resistance of a Si MOSFET is investigated under cryogenic temperatures. The results show a reduced on-state resistance under low temperature operation. With further decrease of temperature, the on-state resistance increases [4] due to the carrier freeze out.

Gallium nitride (GaN) based semiconductors show better performance under cryogenic temperature, including a significant reduction of on-state resistance, an improved switching speed, and a reduced switching loss [5, 14]. However, due to the improved switching speed, high *di/dt* and *dv/dt* related issues challenge the design and optimization of these power converters. The *dv/dt* for a commercial 650 V GaN high electron mobility transistor (HEMT) can reach 140 V/ns at 130 K [19].

In this article, commercial semiconductors from different manufacturers with different materials are characterized under cryogenic temperatures. Both static characterizations and dynamic characterizations are evaluated. Comparisons are made by normalizing the key indicators, such as on-state resistance, threshold voltage, turn-on time, turn-off time, turn-on switching loss, and turn-off switching loss. The room temperature (293 K) values for each semiconductor are selected as the base values to derive the normalized values and enable comparisons among different semiconductors. The static characterizations are performed until 93 K, while the dynamic characterizations are conducted until 143 K due to the malfunction of gate driver.

The contributions of this work can be summarized as follows: 1) the cryogenic characterization results for different semiconductor devices are reported; 2) comparisons among different semiconductors are made from different perspectives; 3) discussions on the results are made to guide the semiconductor selection for low temperature applications.

2. Experiment Setup

2.1. Static characterization setup

Figure 1 shows the picture of the static characterization setup. The liquid nitrogen is used to achieve cryogenic temperatures. The power device is connected with the test fixture through the front panel. As discussed in [5], to mitigate the measurement errors caused by the long connection cables from the test fixture to the cryogenic chamber, the Kelvin connection (as illustrated in Fig. 1) is adopted. Another advantage of the designed PCB is that it can enable multiple device characterizations by simply modifying the connections on the front panel, which saves a great amount of time and energy.

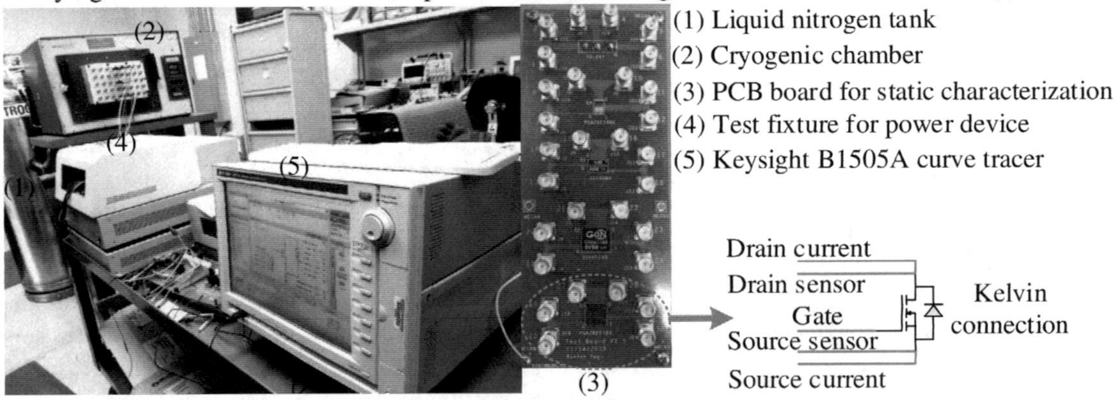

Figure 1. Picture of static characterization setup.

2.2. Dynamic characterization setup

To achieve the dynamic characterizations, a double pulse test (DPT) or clamped inductor load (CIL) test setup is used. Figure 2 shows the dynamic characterization system setup. The gate driver and DPT board are placed inside to evaluate the performance of power devices under cryogenic temperatures. The gate-to-source voltage and drain-to-source voltage are measured by using passive voltage probes from Tektronix P2220 (200 MHz) and Tektronix P5120 (200 MHz), respectively. And the device current is monitored by using current shunt resistor from T&M research products SSDN-015 with very high bandwidth of 1200 MHz. BNC tip adaptors are adopted for the gate-to-source and drain-to-source voltage measurements to reduce the parasitics introduced by the measurement. A differential voltage probe and current probe cannot work properly at low temperatures due to the additional circuits inside the probes. On the other hand, a passive voltage probe and current shunt resistor are appropriate for low temperature measurements.

(1) DC power supply
(2) Pulse generation and data acquisition
(3) Tektronix MS04104 oscilloscope
(4) Auxiliary power supply for gate driver
(5) DSP controller (double pulse)
(6) Fiber transmitter
(7) Load inductor
(8) Cryogenic chamber
(9) DPT test board
(10) Liquid nitrogen tank

Figure 2. Picture of the dynamic characterization setup.

2.3. Summary of the selected semiconductors

In this article, semiconductors of different materials from different manufactures are selected to evaluate their performances under low temperatures. Table I summarizes the selected semiconductors.

Table 1. The selected commercial devices

Part number	Manufacture	Material and structure	Rated voltage/current	Note	Symbol
IXFH26N50P	IXYS	Si MOSFET	500 V/26 A	Static & dynamic	Si MOSFET
FGA5065ADF	ON semiconductor	Si IGBT	650 V/50 A	Static & dynamic	Si IGBT
SCT3120AL	Rohm	SiC trench MOSFET	650 V/21 A	Static & dynamic	SiC MOSFET1
C2M0025120D	CREE	SiC planar MOSFET	1200 V/63 A	Dynamic	SiC MOSFET2
TP90H050WS	Transphorm	Cascode GaN	900 V/34 A	Static & dynamic	Cascode GaN
GS66504B	GaN System	GaN HEMT	650 V/15 A	Static & dynamic	GaN HEMT

3. Results and discussions

3.1. On-state resistance

The normalized on-state resistance is defined as the ratio between the on-state resistance value at a given temperature and the room temperature on-state resistance value. Figure 3 shows the normalized on-state resistance comparison results of different semiconductors, where the base value at room temperature is labelled in the legend. The SiC MOSFET on-state resistance increases dramatically with the decrease of temperature, the on-state resistance value at 93 K is 5.7 times larger than the room temperature value. While all other devices show a decreasing trend with the decrease of temperature. Among them, GaN HEMT and Cascode GaN show the largest on-state resistance reduction under cryogenic temperature, which is around one third of their room temperature resistance values. The on-state resistance drops for the Si MOSFET and Si IGBT can be explained by the increase of carrier mobility. The increased electron mobility under low temperatures is the main cause for the on-state resistance drop in the GaN HEMT. The Cascode GaN FET is composed of a low voltage Si MOSFET and a depletion mode GaN HEMT, which also sees on-state resistance drop under low temperature. On the other hand, there is an increase in interface state density since large amounts of electrons are trapped and few free electrons are available for conduction of the inversion layer of the interface. This is the main cause for the increase of on-state resistance for the SiC MOSFET under low temperature. Thus, from the conduction loss perspective, expect SiC MOSFET, all other semiconductor devices show improved conduction loss performances at low temperature operations.

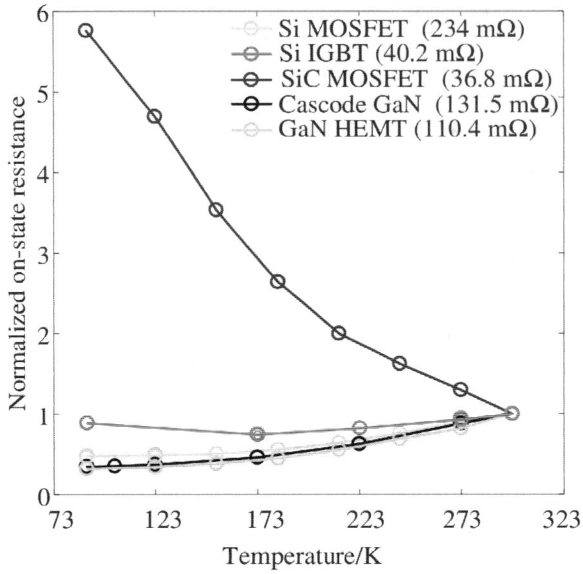

Figure 3. Comparisons of on-state resistance.

3.2. Threshold voltage comparisons

Figure 4 shows the comparison results of the normalized threshold voltage. Aside from the GaN system device GS66504B, all the other power devices witness an increase of the threshold voltage under low temperatures. The threshold voltage increase for the Si MOSFET, SiC MOSFET, Si IGBT, and cascade GaN can be explained by the reduction of intrinsic carrier concentration. The device switching performances are related with the threshold volatge. A large threshold voltage will slow the turn-on process since longer time is required to reach the device threshold voltage. On the other hand, the turn-off process can be accelerated.

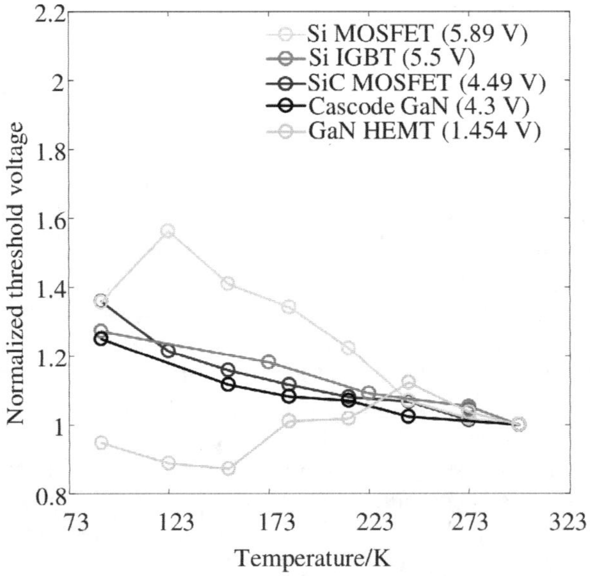

Figure 4. Comparisons of normaized threshold voltage.

IOP Conf. Series: Materials Science and Engineering 1241 (2022) 012042 doi:10.1088/1757-899X/1241/1/012042

3.3. Turn-on time and turn-off time comparisons

Figure 5 shows the turn-on time comparisons among different devices. Figure 6 shows the turn-off time comparisons among different devices. In terms of the turn-on switching time, the SiC MOSFET with planar structure has large turn-on time at low temperature operations, which are mainly caused by the reduction of inversion carrier mobility. The GaN HEMT has the largest reduction on the turn-on switching time due to its increase of electron mobility and reduction of threshold voltage. In terms of the turn-off switching time, the Si IGBT has the largest reduction of turn-off switching time and the long tail current during turn-off process can be alleviated.

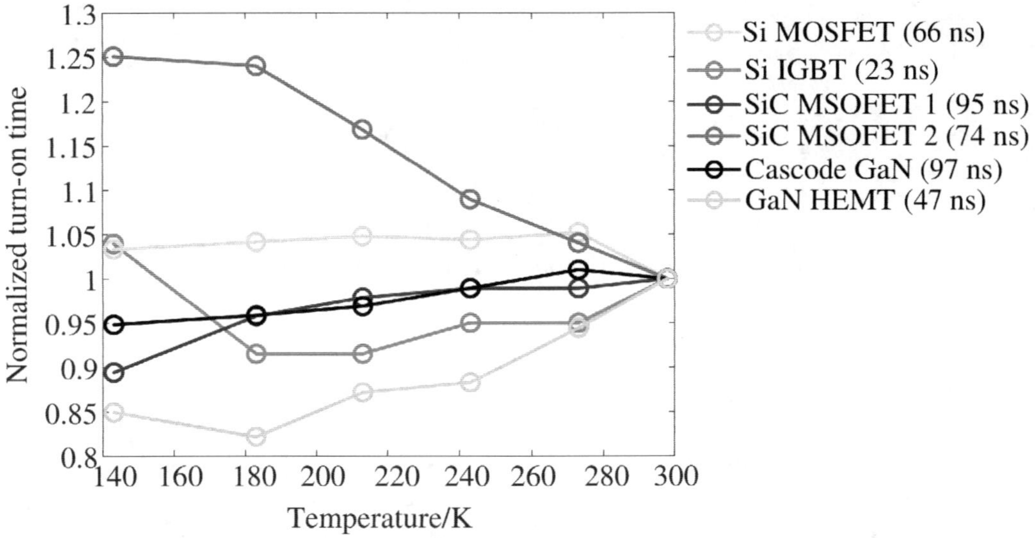

Figure 5. Comparisons of normaized turn-on time.

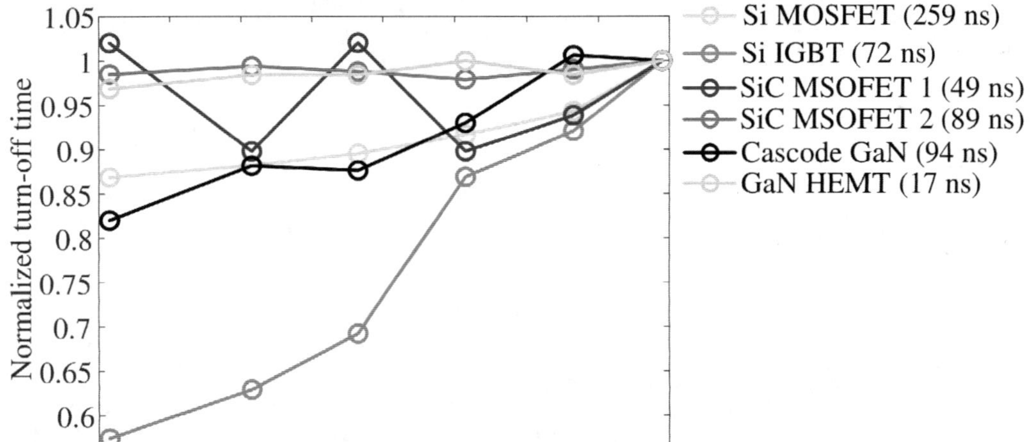

Figure 6. Comparisons of normaized turn-off time.

3.4. Turn-on and turn-off switching loss comparisons

Figure 7 shows the normalized turn-on switching loss comparisons among different devices. Except for the planar SiC MOSFET, the turn-on switching losses for all other devices are decreasing with the decrease of temperature. Among them, the turn-on switching loss reduction for GaN HEMTs are larger than others.

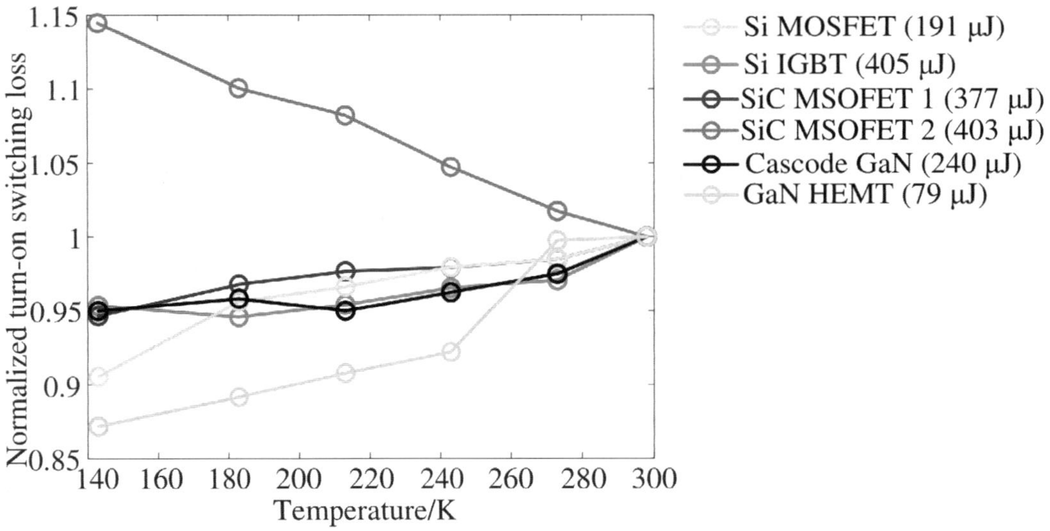

Figure 7. Comparisons of normaized turn-on switching loss.

Figure 8 shows the normalized turn-off switching loss comparisons among different devices. Except planar SiC MOSFET, the turn-off switching loss for all other devices remain almost unchanged or decrease with the decrease of temperature. The Si IGBT has the largest turn-off loss reduction.

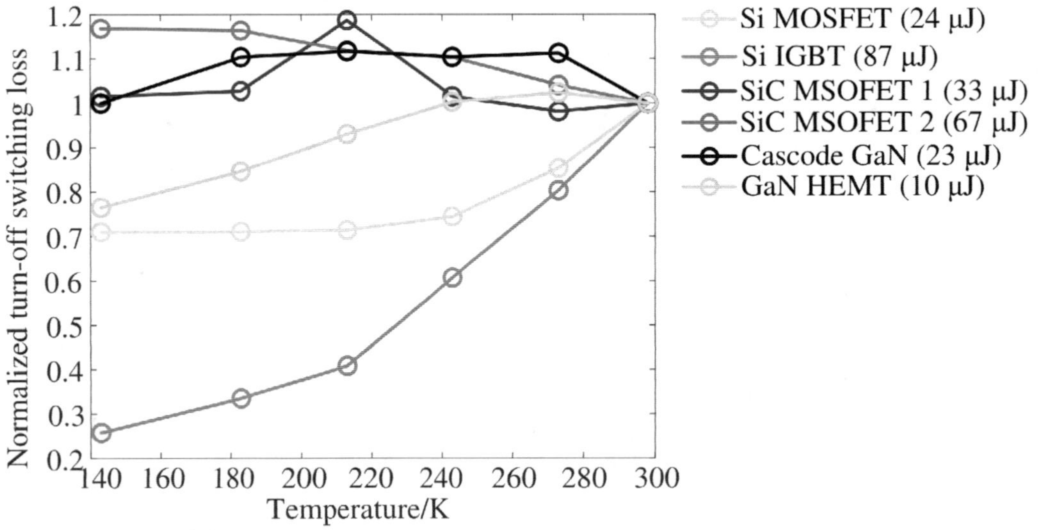

Figure 8. Comparisons of normaized turn-off switching loss.

3.5. Typical experimental waveforms and discussions

Figure 9-Figure 12 show some key experimental waveforms for different semiconductors. From Figure 9 and Figure 10, we can observe that the turn-off processes for the Si MOSFET and Si IGBT have improved significantly at low temperatures. For the planar SiC MOSFET, a "long tail volatge" during the turn-on process is observed, which increases the turn-on switching loss significantly. For the GaN HEMT, the turn-on switching speed at low temperature is improved when compared with room temperature. The turn-on *dv/dt* is increased from 25.93 V/ns to 33.90 V/ns. The explanations can be summarized as follows: 1) for Si MOSFET and Si IGBT, the turn-off processes are shorted due to the increase of threshold voltage and carrier mobility; 2) for planar SiC MOSFET, the turn-on process is degraded because of the increase of threshold voltage and reduction of carrier mobility; 3) for GaN

IOP Conf. Series: Materials Science and Engineering 1241 (2022) 012042 doi:10.1088/1757-899X/1241/1/012042

HEMT, the turn-on process is improved at low temperature due to its reduction of threshold volatge and increase of electron mobility.

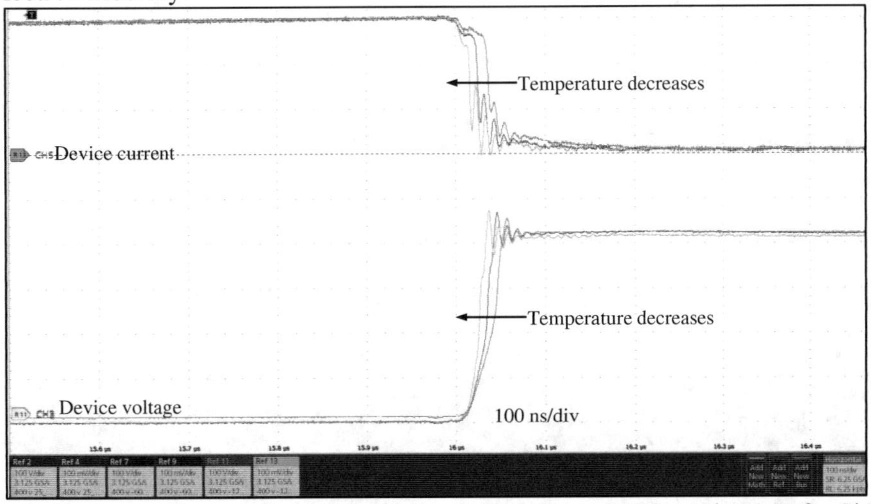

Figure 9. Comparisons of turn-off process at room temperatue, 213 k, and 143 K for Si MOSFET.

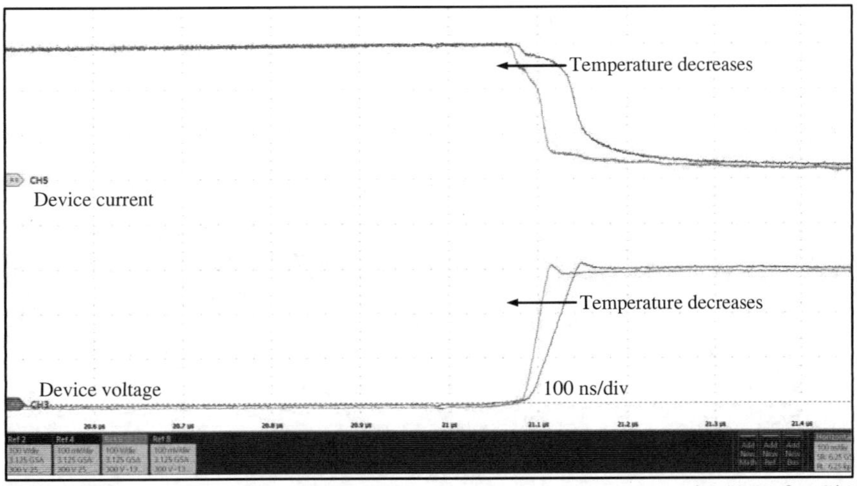

Figure 10. Comparisons of turn-off process at room temperatue and 143 K for Si IGBT.

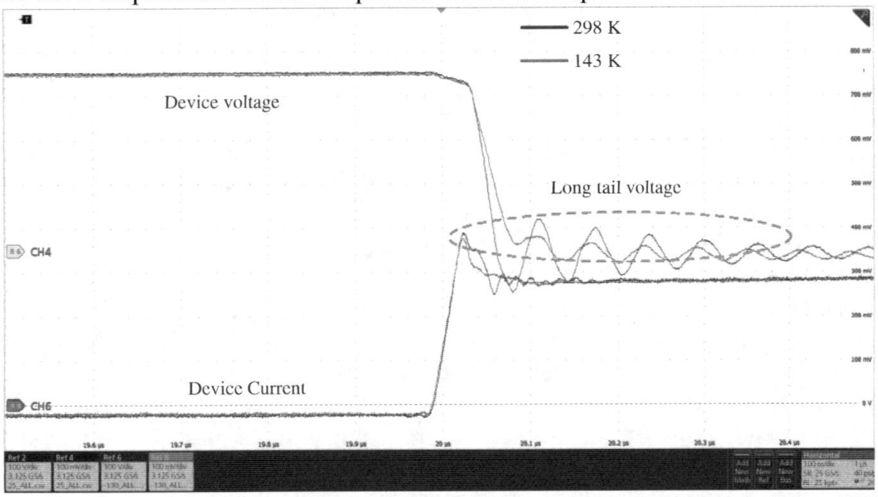

Figure 11. Comparisons of turn-on process at room temperatue and 143 K for planar SiC MOSFET.

IOP Conf. Series: Materials Science and Engineering 1241 (2022) 012042 doi:10.1088/1757-899X/1241/1/012042

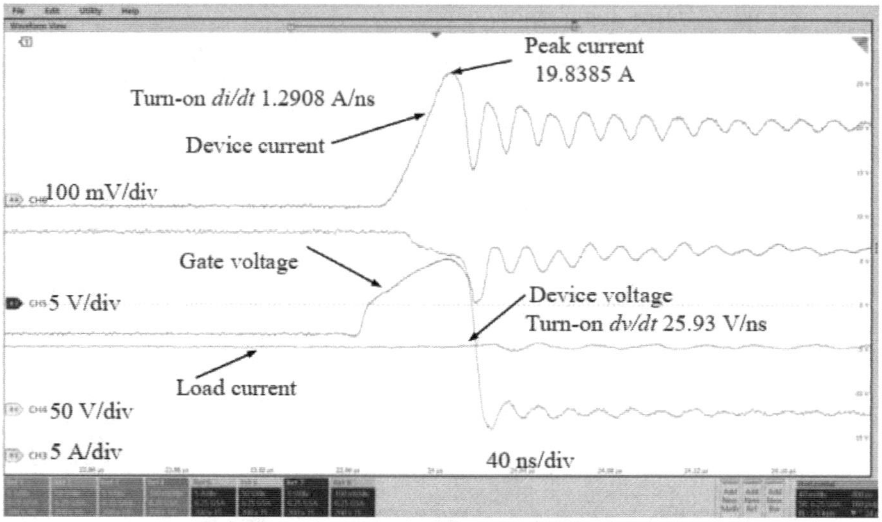

(a) Turn-on process at room temperature

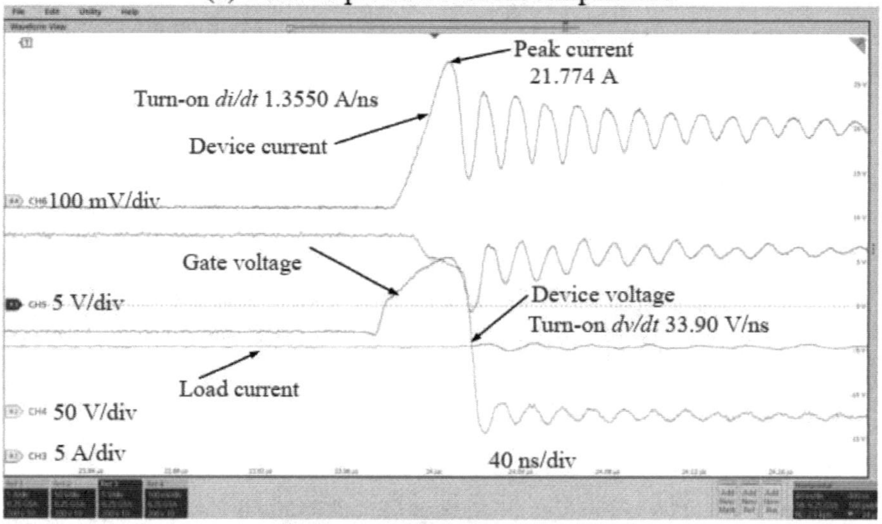

(b) Turn-on process at 143 K

Figure 12. Comparisons of turn-on process at room temperatue and 143 K for planar GaN HEMT.

To provide guidelines for the semiconductor selection at low temperature, the key performances and for different semiconductors are summarized in Table 2. From Table 2, we can conclude that: 1) traditional Si MOSFET and Si IGBT have improved performances at low temperature operations; 2) SiC MOSFET is not advantageous in low temperature operations especially for the planar structure SiC MOSFET; 3) the GaN HEMT is advantageous for cryogenic applications. Both the conduction loss and switching loss can be reduced significantly. In addition, for even lower temperature, like liquid hydrogen and liquid helium temperatures, the GaN HEMT is the only option since no carrier freeze out is observed. The challenges for different semiconductors are also summarized as follows: 1) for traditional Si MOSFET and Si IGBT, the device performance is not comparable with wide band gap devices, which leads to low efficiency operation. Also, the carrier freeze out is observed when the temperature is below 100 K; 2) for SiC MOSFET, although the device performance is much better than traditional Si counterpart, the degraded performance at low temperature makes it not suitable for cryogenic applications. SiC MOSFET can be considered when the system required temperature is not very low. For example, the SiC MOSFET is designed to operate at 257 K in [18]; 3) for GaN HEMT, the reduction of threshold volatge and increased switching speed make it more vulnerable to circuit noises, which reduces the system reliability. Moreover, due to the power limitation of current commercial GaN HEMTs, paralleling operation of several GaN HEMTs is required, which brings more challenge for the application. Overall, the selection of semiconductor for cryogenic application should be made based on the system required operating temperature, cost, power density, and power rating requirements.

Table 2. Comparisons among different evaluated devices

Performance	Si MOSFET	Si IGBT	Trench SiC MOSFET	Planar SiC MOSFET	Cascode GaN	GaN HEMT
On-state resistance	🙂🙂	🙂	☹	☹	🙂🙂🙂	🙂🙂🙂
Threshold voltage	🙂	🙂	🙂	🙂	🙂	☹
Turn-on switching speed	🙂	🙂	😐	☹	🙂	🙂🙂
Turn-off switching speed	🙂🙂	🙂🙂	😐	☹	🙂🙂	🙂
Turn-on switching loss	🙂	🙂	😐	☹	🙂	🙂🙂
Turn-off switching loss	🙂🙂	🙂🙂	😐	☹	🙂🙂	🙂

4. Conclusions

Based on the characterization results, the following conclusions can be drawn: 1) Si MOSFET: Si MOSFET has reduced on-state resistance and switching loss under low temperatures, which would be a suitable candidate for cryogenic power electronics applications; 2) SiC MOSFET: the on-state resistance for the SiC MOSFET increases under low temperatures, which leads to higher conduction loss The switching loss for the evaluated 650 V SiC trench MOSFET is slightly reduced under low temperatures. Overall, SiC MOSFETs are not a promising candidate for cryogenic power electronics applications; 3) Si IGBT: similar to the Si MOSFET, both the on-state resistance and switching loss are reduced under low temperatures; 4) Cascode GaN FET: Cascode GaN devices have better performance under low temperatures from both the conduction loss and switching loss viewpoints, which would be favoured for cryogenic applications. The Cascode GaN FET is better in high switching frequency applications when compared with Si MOSFET due to the small switching loss; 5) GaN HEMT: the GaN HEMT shows excellent performance under low temperatures with reduced conduction loss and switching loss. This should be the focus of cryogenic power electronics.

5. References

[1] K Rajashekara and B Akin, "Cryogenic Power Conversion Systems: The next step in the evolution of power electronics technology," *IEEE Electrific. Mag.*, vol. 1, no. 2, pp. 64–73, Dec. 2013.

[2] H Gui et al., "Review of Power Electronics Components at Cryogenic Temperatures," *IEEE Trans. Power Electron.*, vol. 35, no. 5, pp. 5144-5156, May 2020.

[3] J Qi et al., "Temperature Dependence of Dynamic Performance Characterization of 1.2-kV SiC Power mosfets Compared With Si IGBTs for Wide Temperature Applications," *IEEE Trans. Power Electron.*, vol. 34, no. 9, pp. 9105-9117, Sept. 2019.

[4] K Leong, B Donnellan, A Bryant, and P Mawby, "An investigation into the utilisation of power MOSFETs at cryogenic temperatures to achieve ultra-low power losses," *in Proc. IEEE Energy Convers. Congr. Expo.*, 2010, pp. 2214–2221.

[5] R Ren et al., "Characterization and Failure Analysis of 650-V Enhancement-Mode GaN HEMT for Cryogenically Cooled Power Electronics," *IEEE J. Emerg. Sel. Topics Power Electron.*, vol. 8, no. 1, pp. 66–76, Mar. 2020.

[6] Y Wei, M M Hossain, X Du, R Sweeting and A Mantooth, "Performance Evaluation of 650 V SiC MOSFET Under Low Temperature Operation," *2021 IEEE Energy Conversion Congress and Exposition (ECCE)*, 2021, pp. 5305-5311.

[7] L Nela, N Perera, C Erine and E Matioli, "Performance of GaN Power Devices for Cryogenic Applications Down to 4.2 K," *IEEE Trans. Power Electron.*, vol. 36, no. 7, pp. 7412-7416, Jul. 2021.

[8] Z Zhang et al., "Characterization of Wide Bandgap Semiconductor Devices for Cryogenically-Cooled Power Electronics in Aircraft Applications," *2018 AIAA/IEEE Electric Aircraft Technologies Symposium (EATS)*, Cincinnati, OH, 2018, pp. 1-8.

[9] Y Chen et al., "Experimental Investigations of State-of-the-Art 650-V Class Power MOSFETs for Cryogenic Power Conversion at 77K," *IEEE J. Electron Devices Soc.*, vol. 6, no. 1, pp. 8–18, Oct. 2018.

[10] R Cuerdo et al., "The Kink Effect at Cryogenic Temperatures in Deep Submicron AlGaN/GaN HEMTs," *IEEE Electron Device Lett.*, vol. 30, no. 3, pp. 209–212, Mar. 2009.

[11] S S. Gerber, "Performance of high-frequency high-flux magnetic cores at cryogenic temperatures," *IECEC '02. 2002 37th Intersociety Energy Conversion Engineering Conference*, 2002, 2002, pp. 249-254.

[12] M -J Pan, "Performance of capacitors under DC bias at liquid nitrogen temperature," *Cryog.*, vol. 45, no. 6, pp. 463-467, 2005.

[13] A Hammoud, S Gerber, R L Patterson, and T L MacDonald, "Performance of surface-mount ceramic and solid tantalum capacitor for cryogenic applications," *in Proc. IEEE Conf. Elect. Insul. Dielectr. Phenom.*, 1998, vol. 2, pp. 572-576.

[14] C B Barth et al., "Design, Operation, and Loss Characterization of a 1-kW GaN-Based Three-Level Converter at Cryogenic Temperatures," *IEEE Trans. Power Electron.*, vol. 35, no. 11, pp. 12040-12052, Nov. 2020.

[15] H Gui et al., "Development of High-Power High Switching Frequency Cryogenically Cooled Inverter for Aircraft Applications," *IEEE Trans. Power Electron.*, vol. 35, no. 6, pp. 5670-5682, June 2020.

[16] R Chen et al., "A Cryogenically-Cooled MW Inverter for Electric Aircraft Propulsion," *2020 AIAA/IEEE Electric Aircraft Technologies Symposium (EATS)*, New Orleans, LA, USA, 2020, pp. 1-10.

[17] J Bourne, R Schupbach, B Hollosi, J Di, A Lostetter, and H A. Mantooth, "Ultra-wide temperature (-230°C to 130°C) DC-motor drive with SiGe asynchronous controller," *in Proc. IEEE Aerosp. Conf.*, 2008, pp. 1–15.

[18] R Chen and F F Wang, "SiC and GaN Devices With Cryogenic Cooling," *IEEE Open Journal of Power Electronics*, vol. 2, pp. 315-326, 2021.

Acknowledgments
The information, data, or work presented herein was funded in part by the NASA ULI: Development of the Cryogenic Hydrogen-Energy Electric Transport Aircraft (CHEETA) Design Concept under Award Number 80NSSC19M0125. Testing for this project was conducted at the National Center for Reliable Electric Power Transmission (NCREPT), the University of Arkansas' High-Power Test Facility

IOP Conf. Series: Materials Science and Engineering 1241 (2022) 012043 doi:10.1088/1757-899X/1241/1/012043

Comparative evaluation of different DC-AC converter topologies for cryogenic applications utilizing superconducting materials

M Ul Hassan[1], A I Emon[1], Y Azadeh[1] and F Luo[1]

[1]Stony Brook University, Stony Brook, NY, 11794, United States

E-mail: mustafeez.hassan@stonybrook.edu

Abstract. Modern transportation applications including electric vehicles (EVs), more electric aircraft (MEA) and electric comprise of different power conversion systems. Reduced weight and improved semiconductor efficiency are the major challenges towards their improved overall performance. To overcome them, cryogenic operation of power electronics converters (CPEC), electric motors, cables, and energy storage components has been under research and evaluation. CPEC are being investigated to drive superconducting machines (SCM) because of their lightweight structure and improved efficiency. CPEC can offer improved switching performance, reduced conduction losses and therefore increased efficiency; because of reduced on-resistance and increased carrier mobility of active devices. This article will discuss how can the benefits of lower operating temperature (< 93 K) be utilized in determining the most suitable DC-AC converter configuration. The paper compares three basic converter topologies: (a) two level voltage source inverter (2L-VSI), (b) three level t-type neutral point clamped inverter (3L-TNPC) and (c) two-level current source inverter (2L-CSI). Analysis is provided based on volume of passive components, electromagnetic interference (EMI) spectra and number of active devices, where all components considered are compatible with low temperatures. It can be observed that utilization of cryogenic temperature affects the volume of passive components, and therefore the overall selection of converter.

1. Introduction

Power electronics converters play an integral and enabling role in any power conversion system, especially electric drives, modern transportation systems, and renewable systems. With the ever-increasing demand of improved efficiency, and power density; low temperature or cryogenic power electronics (CPE) has been envisioned to offer an improved overall performance [1]. CPE has been given extensive attention in recent past because of their potential applications in modern transportation, cryogenic instrumentation, wind power generation and super conducting magnetic energy storage systems (SMES). The increased interest towards CPE depends upon the idea of reduced semiconductor losses, increased switching speed, reduced size of passive components and superconducting machines (SCM). These features when combined, offer improved overall system performance, enabling significant reduction in cost and weight of the converter [2].

Considering potential utilization of CPE in future electrification of transportation, NASA has already outlined a roadmap to improve the efficiency beyond 99 % and power density to be 2-3 times the current state of the art [3]. Similarly, DOE/NREL has identified 10-20 MW wind turbine generators which can offer cheaper offshore wind energy by using superconducting technology [4]. These challenges lead to developing and utilizing power electronics to be integrated with systems requiring cryogenic operating conditions e.g., superconducting machines and aircraft-based electronics systems.

Based upon potential applications of CPE, significant research has been conducted in characterization of semiconductor devices including Si-MOSFETs, IGBTs and wideband gap devices (WBG). [5] characterized 650 V SiC devices for low temperature applications whereas [6] characterized a 900 V cascode GaN high electron mobility transistor (HEMT) for cryogenic temperatures. Detailed summary about performance of different switching devices has been reported in [7]. Having known the performance of switching devices, CPEC of different types maybe designed to work for low temperature applications. Several DC-DC power electronics converters have been designed and investigated for their potential applications in cryogenic environments. [8] designed and tested a pulse width modulated (PWM) 175 W buck converter using Si devices and offering efficiency improvement of 1.2 %. Similarly [9] and [10] present the performance of a three-level 60 W buck converter operating at switching frequency of 50 kHz. Boost type DC-DC converters were also designed for operating temperatures around 90 K with moly permalloy powder (MPP) core and high temperature superconducting (HTS) inductor in [11] and [12] respectively. It was concluded that HTS inductor had no significant performance improvement compared to MPP inductor because of its higher AC losses. [13] analyzed the performance difference for a boost converter operating at room temperature (RT) and liquid nitrogen (LN_2 ,77 K) temperature. Performance of a semiconductor switching device was investigated for soft and hard switched configurations. The advantages associated with soft switched converter at LN_2 temperature were more proficient as compared to its RT counterpart, whereas no analysis in terms of using superconducting materials was provided. A 40-kW boost converter was designed in [14] using CoolMOS MOSFETs and HTS inductor at switching frequency of 400 kHz. It was reported to have 85 % reduction in conduction loss down to LN_2 temperature, whereas usage of $GdBa_2Cu_3O_{7-x}$ (GdBCO) inductor enabled reduction of conductor loss from 104 W to mere 9 W.

Unlike DC-DC, DC-AC CPE converters haven't been given much attention and are not reported a lot. In [15], a MW class cryogenically cooled inverter has been designed. Two 500 kW three level active neutral point clamped (3L-ANPC) inverters have been paralleled through interleaved inductors for an output voltage of 600 V. The paper reports hardware results only at room temperature. [16] presented a 1 kW, 3-level GaN based inverter for extreme cold applications. The converter was successfully tested until 133 K only, with input voltage of 150 VDC, and 45 V output. Authors reported a 16 % reduction in losses down to 213 K in comparison to RT.

From the review presented above, all the DC-AC converters employed are primarily voltage source inverters (VSIs) where a DC link storage capacitor is used. On the contrary, no analysis has been conducted for a current source inverter (CSI) for such an application. CSIs historically have been neglected because of non-availability of reverse voltage blocking (RVB) devices, and increased conduction losses. Additionally, DC link storage inductor possess greater challenge in justifying the volume and power density of converter. Considering the CPEC, superconducting magnetic energy storage (SMES) inductor should be thoroughly evaluated in conjunction with a two-level CSI. This paper targets at comparing the volume and efficiency of a 2.5 MW cryogenically cooled inverter for driving a 600 V superconducting machine. The paper considers three basic converter topologies; (a) two level voltage source inverter (2L-VSI), (b) three level t-type neutral point clamped inverter (3L-TNPC) and (c) two-level current source inverter (2L-CSI). Phase leg configuration of all three converter topologies is shown in Fig. 1 (b)-(d) respectively.

2. Evaluation of different DC-AC converter topologies

In this section, the three converter topologies will be evaluated based upon numerous parameters. The mentioned converter is supposed to harvest the energy from a fuel cell, hence operating on a variable DC link voltage; and converts it into alternating current (AC) of 300 Hz. The converter drives a MgB_2 utilized superconducting machine which is highly sensitive to the harmonic content and AC losses.

2.1. Basis for evaluation

To compare different converter topologies, a performance matrix is defined which comprises of contributions from different elements as shown in Fig. 1 (a). A converter will be evaluated based upon semiconductor switch count, input side electromagnetic interference (EMI) spectrum, magnitude and size of capacitors and inductors involved.

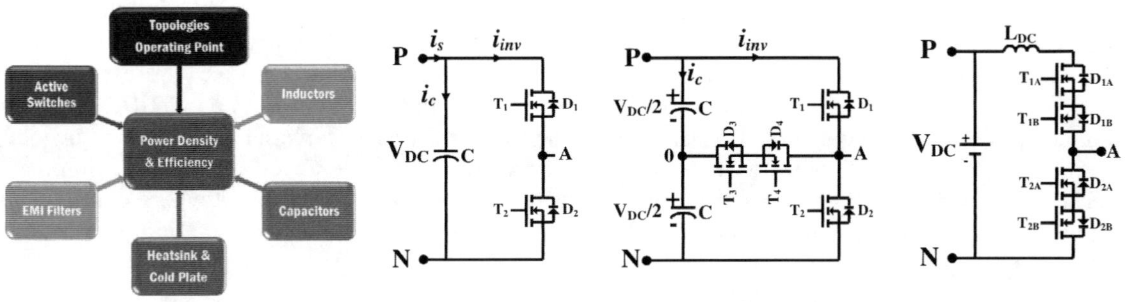

(a) (b) (c) (d)

Figure 1. (a) Important components of a converter **(b)** 2L-VSI **(c)** 3L-VSI **(d)** 2L-CSI

2.2. *Semiconductor switch count*

One of critical aspects in the design of a converter is appropriate device selection. It requires a device to be capable of handling the rated current and blocking the rated voltage. For a fair comparison, all the converter topologies are assumed to have a uniform output of 600 $V_{L\text{-}L}$ and fixed modulation index 'mi'. This results in different DC link voltages for VSIs, and CSI as shown in Fig. 2 (a), as they primarily are buck and boost type topologies respectively. Additionally, only enhancement mode GaN high electron mobility transistors (HEMT) devices available commercially were considered in the analysis. Such devices are selected as they show reduced on state resistance, stable threshold and breakdown voltages, and increased switching speeds [7]. For the comparison, 650 V GaN HEMT was considered in series and anti-series to fulfill the blocking voltages for VSIs and furnish RVB functionality for CSI. Fig. 2 (b) presents the number of devices required for different phase leg configurations where 2L-VSI requires the least number of devices, while 2L-CSI requires the intermediate and fixed number of devices.

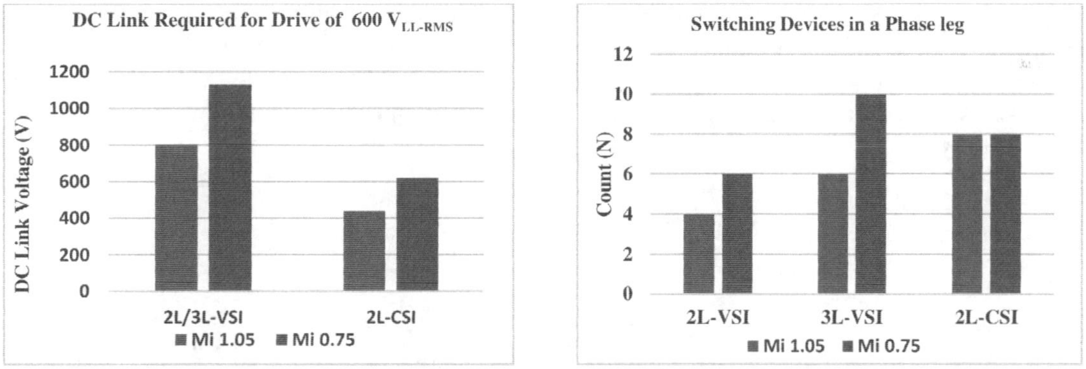

(a) (b)

Figure 2. Mapping of fixed output voltage to **(a)** DC link required **(b)** Device count required

2.3. *Dimensioning and sizing of passive components*

This subsection discusses about the dimensioning of filter components since 30 % of volume of any converter is occupied by them [17], so it is important to have an appropriate estimate of their magnitude, and volume. These filter components primarily comprise of different configurations of inductors and capacitors, where input storage capacitors are required at the DC link for VSIs, and output filter capacitors are required for CSI. Conversely input storage inductors are required at the DC link for CSIs, and LC type output filter is considered for VSIs (to have comparable total harmonic distortion, THD).

2.3.1. Dimensioning of DC link storage. An estimate about input DC link capacitors based upon ripple current can be made from [18], however actual values maybe limited by the current rating [19]. DC link storage capacitor for 2L and 3L VSI can be identified from Fig. 1 (b) and (c) and are estimated in equation (1) and (2). Similarly, DC link storage inductor exists in CSI which can be identified in Fig. 1 (d) and estimated using equation (3) [20]. In the equations, f_{sw} is the converter switching frequency, P_{MAX} is the rated converter power, whereas $V_{LN\text{-}rms}$ represents the converter output voltage. Summary about DC link sizing for all the converter types considered is presented in Table 1 where components with nearest rated voltages are tabulated.

$$C_{2L-VSI} = \frac{P_{MAX}}{(V_{DC}\Delta V_{DC} - 0.5\Delta V_{DC}^2)f_{sw}} \quad (1)$$

$$C_{2L-VSI} = 2C_{3L-VSI} \quad (2)$$

$$L_{DC} = \frac{\sqrt{3}V_{LN-rms}}{4f_{sw}\Delta I_{DC}} \quad (3)$$

Table 1. Summary of DC Link Storage Dimensioning

PROPERTY	MODULATION INDEX	2L-VSI	3L-VSI	2L-CSI
DC LINK CAPACITOR	0.75	1500 uF, 1150 VDC	2 x 3000 uF, 630 VDC	---
	1.05	2700 uF, 850 VDC	2 x 5400 uF, 630 VDC	---
DC LINK INDUCTOR	0.75	---	---	25 uH, 3290 ADC
	1.05	---	---	18 uH, 4600 ADC

2.3.2. Dimensioning of AC link filters. Output filter combination for converter depends upon the type and sensitivity of the load. Although a fully superconducting machine for cryogenic temperature (CT) results in 3.5 times power to weight ratio as compared to other configurations, yet it is very sensitive to AC losses [21]. Therefore, converter output voltage should have lower THD and sinusoidal output voltage. To achieve this, an LC type filter combination is considered for VSIs and can be designed using (4) and (5), whereas already existing $^{dv}/_{dt}$ filter designed using (6) should be sufficient for CSI. In the equations, I_{LN_PK} represents the peak output line current, L_{filter} and C_{filter} are the filtering components, whereas f_{res} is the resonance frequency. Table 2 presents the summary about AC link filter sizing for all the converter types considered.

$$L_{filter} = \frac{V_{DC}}{3\sqrt{2}\pi f_{sw}\Delta I_{max}} \quad (4)$$

$$C_{filter} = \frac{1}{4\pi^2 f_{res}^2 L_{filter}} \quad (5)$$

$$C = \frac{I_{LN\_PK}}{4\Delta V_{LL} f_{sw}} \quad (6)$$

Table 2: Summary of AC Link Filters Dimensioning

PROPERTY	MODULATION INDEX	2L-VSI	3L-VSI	2L-CSI
O/P FILTER CAPACITOR	0.75	3 x 65 uF, 400 VAC	3 x 60 uF, 400 VAC	3 x 430 uF, 400 VAC
	1.05	3 x 70 uF, 400 VAC	3 x 65 uF, 400 VAC	3 x 610 uF, 400 VAC
O/P FILTER INDUCTOR	0.75	6.15 uH, 3402 A pk	4.61 uH, 3402 A pk	---
	1.05	4.35 uH, 3402 A pk	4.27 uH, 3402 A pk	---

2.3.3. Overall volume estimation of DC and AC link filter components. Having determined the magnitudes and values of different filtering components, comparison amongst the topologies is extended to determination of volume occupied by capacitors and inductors. Like the selection of SCM for CT, capacitors considered for the comparison are polypropylene sulphide (PPS) based because of their

negligible variation in capacitance and reduced equivalent series resistance (ESR) at lower temperatures [7]. Capacitors from numerous vendors were selected and the one offering the best combination of volume and cost is being considered in the analysis. Variation of capacitor volume was studied and plotted in Fig. 3 (a) where all three converters were considered under two discrete values of modulation indices. The data used for volume determination was taken from WIMA manufacturer [22].
To determine the volume for inductors, numerous inductor types commonly used for RT applications were considered. From the literature, it can be concluded that ferrite and amorphous materials not only lose their permeability, but their core losses also increase, whereas powdered cores show stable operation with temperature [7]. Considering the bulk amount of power, numerous powdered cores need to be paralleled to justify area product Ap, therefore resulting in huge volume. On the contrary, coreless inductors with superconducting wires/tapes were considered for the analysis and found them to be more suitable for such low temperature and high-power applications. Coreless inductors not only offer zero core losses, but also no limitation on saturation flux density [23]. Numerous superconducting tapes can be utilized for such inductor making applications, but MgB_2 type superconductor was considered with a conservative value of current density to be 200 A/mm^2. Such superconductor was wound in a multilayer cylindrical orientation to furnish the inductors specified in table 1 and 2. The volume computed with such an approach is displayed in Fig. 3 (b), whereas overall volume of whole converter is shown in Fig. 4 (a). It can be seen from there that 2L-CSI occupies least volume in terms of passive components when operated around unity modulation index.

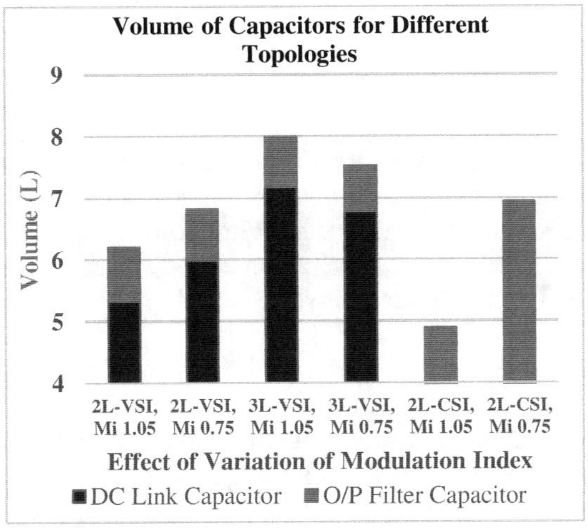

(a)

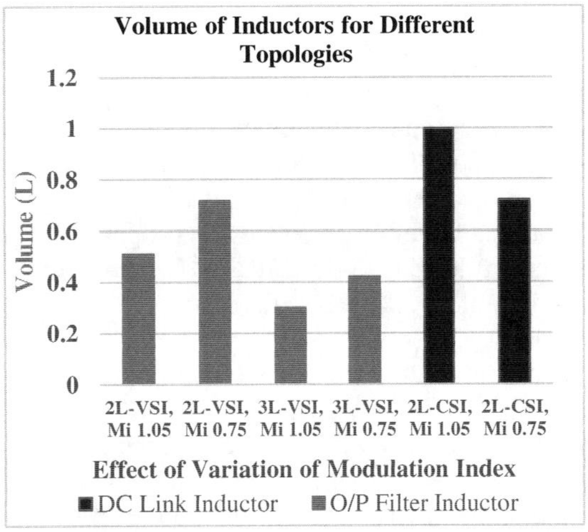

(b)

Figure 3. Effect of topology and modulation index on **(a)** Capacitor volume **(b)** Inductor volume

2.3.4. EMI performance comparison Between 2L-VSI and 2L-CSI. In this subsection, comparison between 2L-VSI and 2L-CSI will be carried out based on conducted electromagnetic interference (EMI). For the analysis, LTspice based simulations were carried out for the two converter topologies. For the simulation models, switch node parasitic capacitances, stray capacitance from DC link to ground, cable parasitic elements and motor model was considered and kept uniform for both the converters as discussed in [24]. Both common mode (CM) and differential mode (DM) EMI spectra was determined and plotted as shown in Fig. 4 (b) – (d).

IOP Conf. Series: Materials Science and Engineering 1241 (2022) 012043 doi:10.1088/1757-899X/1241/1/012043

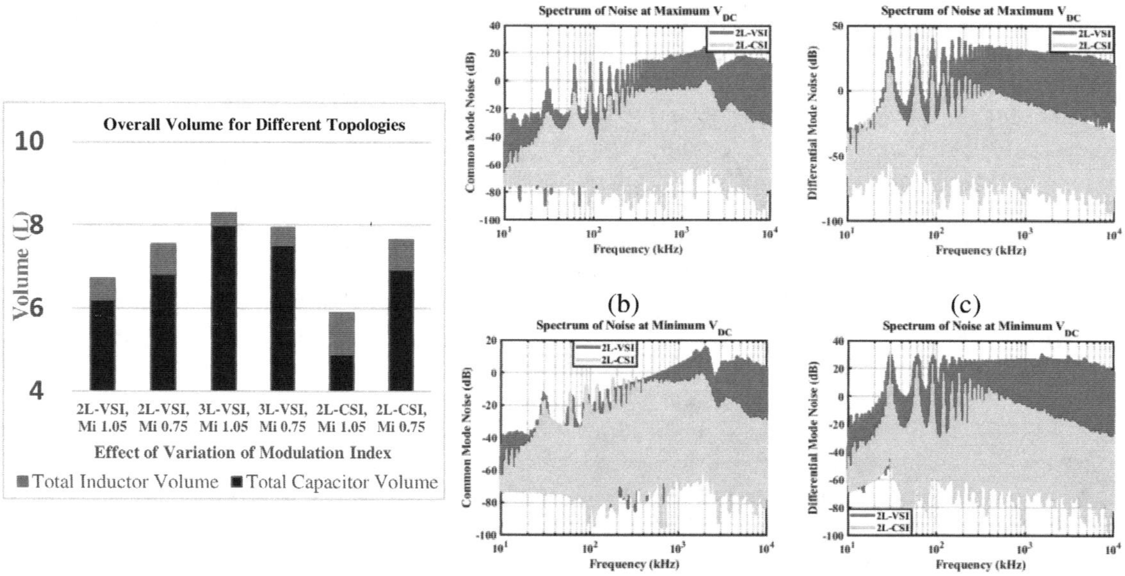

(a) (d) (e)

Figure 4. For a 2.5 MW converter: **(a)** Overall volume **(b)** CM spectrum at Mi=0.75 **(c)** DM spectrum at Mi=1.05 **(d)** CM spectrum at Mi=1.05 **(c)** DM spectrum at Mi=1.05

3. Discussion and conclusion

In this paper, three converter topologies were considered for conversion from DC-AC for cryogenic applications of power electronics converters. Throughout the paper, all the components/materials considered were compatible with cryogenic temperatures as reported in different literature. The analysis begins while considering a 2.5 MW converter which is supposed to drive a superconducting machine working under cryogenic temperatures (< 93 K). To select the optimal converter topology driving such a motor, numerous design parameters like device count variation, buck/boost mode of operation, magnitude and size of passive components and EMI spectra were evaluated and compared. From the analysis, it can be concluded that a 2L-CSI converter needs medium number of cryo compatible switching devices while providing boost mode of operation. This mode of operation is particular to current source inverter and is important when working with variable DC link voltage (e.g., battery bank or fuel cell), where voltage source inverters might need an additional front end boost converter.

Additionally, it was also found that 2L-CSI inverter occupies least volume of passive components for a uniform output power quality. Such power quality is extremely critical in terms of driving superconducting machines which are very sensitive to AC losses. Towards the end of paper, analysis based upon EMI spectra was also presented which clearly shows lesser EMI from 2L-CSI. Such a spectra does not necessarily reflect onto reduced EMI filter required but will also results in better DC side spectra. Considering transmission of this DC side energy from superconducting transmission lines, 2L-CSI can be significantly better as it will reduce the AC losses in such lines.

Based upon the analysis conducted, it can be concluded that 2L-CSI will be more suitable for cryogenic power conversion both in terms of volume, power quality and minimized losses in AC side load and DC side source. Such a configuration of converter, when employed together with superconducting magnetic energy storage can significantly outweigh other counterparts in terms of overall performance.

4. References

[1] Rajashekara K and Akin B 2013 *A review of cryogenic power electronics - status and applications* Int. Electric Machines & Drives Conf. (Chicago) pp 899-904

[2] Elbuluk M and Hammoud *2005 A Power electronics in harsh environments* Fourtieth IAS Annual Meeting. Conf. Record of the Industry Applications pp 1442-48

[3] Felder J A, Brown J V, Kim H and Chu J 2011 Turboelectric distributed propulsion in a hybrid wing body aircraft *in Proc. Int. Symp. Air Breath. Engines Conf.* (Hamoton) pp 1–20

[4] Maples B, Hand M and Musial W 2010 *Comparative Assessment of Direct Drive High Temperature Superconducting Generators in MultiMegawatt Class Wind Turbines* Technical Report NREL/TP- 5000-49086.

[5] Wei Y, Hossain M M, Du X, Sweeting R and Mantooth A 2021 *Performance Evaluation of 650 V SiC MOSFET Under Low Temperature Operation* IEEE Energy Conversion Congress and Exposition (ECCE) (Vancouver)

[6] Hossain M M, Wei Y, Rashid A U, Sweeting R and Mantooth H A 2021 *Cryogenic Evaluation and Modeling of a 900 V Cascode GaN HEMT* IEEE 12th Energy Conversion Congress & Exposition - Asia (ECCE-Asia) (Singapore)

[7] Gui H, Chen R, Niu J, Zhang Z, Tolbert L M, Wang F F, Blalock B J, Costinett D and Choi B B 2020 Review of power electronics components at cryogenic temperatures *IEEE Transactions on Power Electronics* **35** 5144–56

[8] Ray B, Gerber S S, Patterson R L and Myers I T 1995 *Low-temperature operation of a BUCK DC/DC converter* Proc. of IEEE Applied Power Electronics Conf. and Exp. (APEC) (Dallas) pp 941-6

[9] Perez-Guerrero F F, Ray B and Patterson R L 1997 *Low temperature operation of a three-level buck DC-DC converter IECEC* Proc. of the Thirty-Second Intersociety Energy Conversion Engineering Conf. (Honolulu) pp 1415-20

[10] Perez-Guerrero F F, Venkatesan K, Ray B and Patterson R L 1999 *Low temperature performance of a closed loop three level buck converter* Proc. of the IEEE Int. Conf. on Power Electronics and Drive Systems PEDS (Hong Kong) **1** pp 58-62

[11] Ray B and Patterson R L 1995 *Wide temperature operation of a PWM DC-DC converter Conf. Record of the IEEE Industry Applications Conference Thirtieth IAS Annual Meeting* (Orlando) pp 971-976

[12] Ray B, Gerber S S, Patterson R L and Dickman J E 1996 *Low temperature performance of a boost converter with MPP and hts inductors* Proc. of Applied Power Electronics Conf. (APEC) (San Jose) **2** pp 883-8

[13] Li H, Liu D and Luongo C A 2005 Investigation of potential benefits of mosfets HARD-SWITCHING AND soft-switching converters at cryogenic temperature *IEEE Transactions on Appiled Superconductivity* **15** 2376–80

[14] Chen X Y, Jin J X, Tang M G, Feng J, Luo H Y, Li L Y, Xu Q and Zou H L 2016 An efficient boost chopper integrated with cryogenic MOSFETs and HTS inductor *IEEE Transactions on Applied Superconductivity* **26** 1–6

[15] Wang F, Chen R, Gui H, Niu J, Tolbert L, Costinett D, Blalock B, Liu S, Hull J, Williams J, Messer T, Solodovnik E, Paschedag D, Khozikov V, Severns C, Karimi K and Choi B B 2019 MW-Class cryogenically-cooled inverter for electric aircraft applications *AIAA Propulsion and Energy Forum* (Indianapolis) pp 1-9

[16] Barth C, Colmenares J, Foulkes T, Coulson K, Sotelo J, Modeer T, Miljkovic N and Pilawa-Podgurski R C N 2017 Experimental evaluation of a 1 kW, Single-phase, 3-level gallium Nitride inverter in extreme cold environment *IEEE Applied Power Electronics Conf. and Exp. (APEC)* (Tampa) pp 717-23

[17] Karutz P, Round S D, Heldwein M L and Kolar J W 2007 *Ultra compact Three-phase PWM Rectifier* Twenty-Second Annual IEEE Applied Power Electronics Conf. and Exp. (APEC) (Anaheim) pp 816-22

[18] Rixin Lai, Fei Wang, Burgos R, Yunqing Pei, Boroyevich D, Bingsen Wang, Lipo T A, Immanuel V D and Karimi K J 2008 A systematic Topology Evaluation methodology for High-Density THREE-PHASE pwm AC-AC Converters *IEEE Transactions on Power Electronics* **23** 2665–

80
[19] Hava A M, Ayhan U and Aban V V 2012 A DC bus capacitor design method for various inverter applications *IEEE Energy Conversion Cong. and Exp. (ECCE)* (Raleigh) pp 4592-9
[20] Dai H and Jahns T M 2018 Comparative investigation of PWM current-source inverters for future machine drives using high-frequency wide-bandgap power switches *IEEE Applied Power Electronics Conf. and Exp. (APEC)* (San Antonio) pp 2601-8
[21] Corduan M, Boll M, Bause R, Oomen M P, Filipenko M and Noe M 2020 Topology comparison of SUPERCONDUCTING Ac machines for hybrid electric aircraft *IEEE Transactions on Applied Superconductivity* **30** 1–10
[22] WIMA A *The capacitor for high end audio applications* https://www.wima.de/wp-content/uploads/media/e_WIMA_DC_Link_MKP_4.pdf pp 133-59
[23] Park C, Obadolagbonyi O and Graber L 2020 Cryogenic power electronics: Capacitors and inductors *IOP Conf. Series: Materials Science and Engineering* (Hartford) **756**
[24] ul-Hassan M, Emon A I, Peng H, Choksi K and Luo F 2021 *Investigation on Conducted EMI for single and parallel Connected inverters* IEEE 12th Int. Sym. on power electronics for distributed generation systems (PEDG) (Chicago)

Acknowledgments
The authors would like to acknowledge the National Aeronautics and Space Administration (NASA) for lending support for this work, under award number 80NSSC19M0125 as part of the Center for High-Efficiency Electrical Technologies for Aircraft (CHEETA).

LCLS-II-HE Cryomodule Production Status at Fermilab

J Kaluzny*, T Arkan, D Bice, J Blowers, M Checchin, C Grimm, B Hartsell, M Martinello, T Nicol, S Posen, K Premo

Fermilab, Batavia, IL, 60510 USA
*kaluzny@fnal.gov

Abstract. The Linac Coherent Light Source (LCLS) is an X-ray science facility at SLAC National Accelerator Laboratory. The LCLS-II project (an upgrade to LCLS) is in its final stages, and the LCLS-II-HE (High Energy) project is another upgrade to the facility enabling higher energy operation. An electron beam is accelerated using superconducting radio frequency (SRF) cavities built into cryomodules. It is planned to build twenty-five modules total for the LCLS-II-HE project. Fourteen of these modules are planned to be assembled and tested at Fermilab. Procurements for production modules have been awarded, and parts are arriving. The design of the LCLS-II-HE modules is similar to the module design used in LCLS-II. The major cryomodule design change from LCLS-II is the nominal gradient required per cavity, increasing from 16 to 21 MV/m. The additional gradient yields additional heat loads. The first LCLS-II-HE cryomodule, referred to as the verification cryomodule (vCM), was assembled using parts from an LCLS-II module with new SRF cavities. The assembly is complete, and this module is currently being tested at Fermilab. The vCM cavities perform well, exceeding the required gradient and quality factor specifications. The first article module assembly is planned to start later in 2021. The status of the production as well as some of the design and assembly changes will be presented.

1. LCLS-II-HE Overview

SLAC National Accelerator Laboratory (SLAC) is currently finishing the LCLS-II project. The first section of the machine accelerates an electron beam to 4 GeV that will be used to create x-rays [1]. The electron beam is accelerated using superconducting radio frequency (SRF) cavities contained in cryomodules. Fermilab delivered 17 1.3GHz cryomodules for the LCLS-II project and plans to deliver 14 cryomodules for LCLS-II-HE. LCLS-II-HE is an upgrade to LCLS-II that will enable the production of x-rays at higher energies [1]. LCLS-II-HE will increase the electron beam energy from 4 GeV to 8 GeV. For LCLS-II-HE, 25 cryomodules are planned to be assembled, tested and shipped to SLAC, and

Figure 1. LCLS-II-HE verification cryomodule.

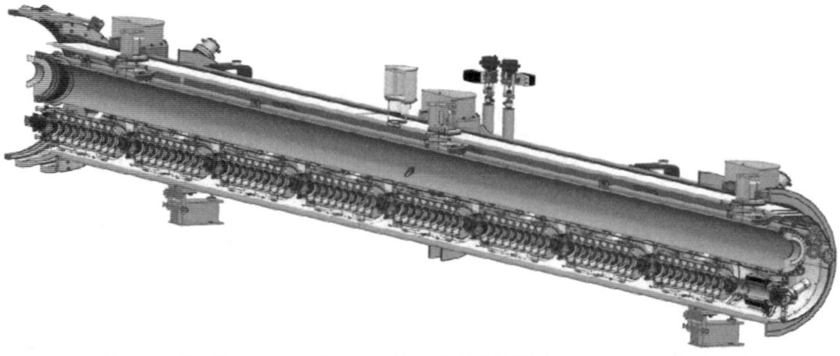

Figure 2. Cross section of LCLS-II-HE cryomodule model.

24 cryomodules are planned to be installed. The plan is to have Fermilab assemble 14 of the modules and Jefferson Lab assemble 11 of the modules.

Fermilab's role in LCLS-II-HE started early in the project with cavity R&D [2] to allow increased cavity performance. This R&D allowed the gradient specification to be increased without changing the mechanical design of the cavity or cryomodule. Fermilab is also the cryomodule designer of record which means that Fermilab maintains the drawing package for the module. The cryomodule component procurements are split between the partner labs (Fermilab, Jefferson Lab and SLAC). Similar to LCLS-II, Fermilab is responsible for procuring the upper cold mass, vacuum vessel, split quadrupole magnet, instrumentation and many other smaller components. Finally, for LCLS-II-HE Fermilab plans to assemble, test and ship to SLAC 14 cryomodules including the verification cryomodule (vCM) seen in Figure 1.

2. Cryomodule Design

The design of the LCLS-II-HE cryomodule, shown in Figure 2, is nearly identical to the LCLS-II cryomodule. The LCLS-II cryomodule was based on the European XFEL module modified for continuous wave operation [3]. Compared to LCLS-II, one major difference is that the cavity gradient specification has increased from 16 MV/m to 21 MV/m. Cavity doping R&D lead to a new recipe that will be used for the LCLS-II-HE project [2]. Another major difference is the extended range tuner. The tuner for LCLS-II-HE, shown in Figure 3, is able to detune the cavity 670kHz allowing for off-frequency operation of the linac [4]. The new tuner design has been tested in a single cavity horizontal test cryostat at Fermilab. The tuner was operated with the cavity at 2K and 4K operating conditions for a total of 625 off-frequency operation cycles.

Figure 3. Extended range tuner and connection between cavities.

Figure 4. MP9 building showing WS2.

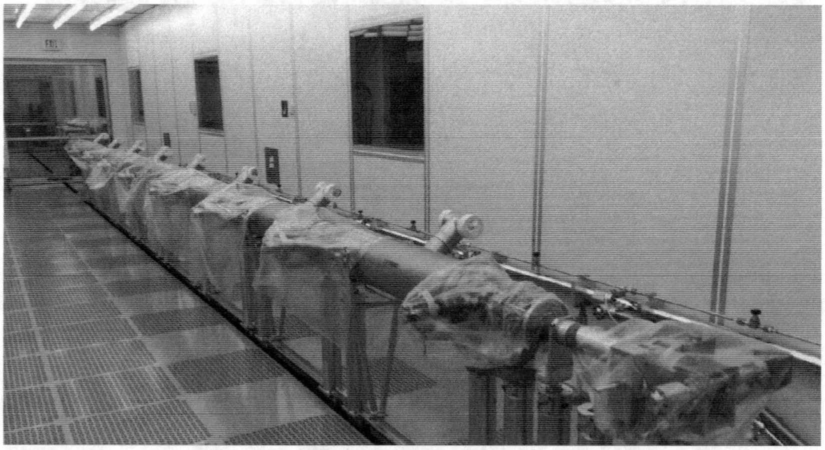

Figure 5. WS1 inside cleanroom at MP9.

For LCLS-II-HE a more formal design change process has been implemented. The process has been used to document changes from the existing LCLS-II design. The design change request process was initiated at Fermilab at the beginning of the project. It is now a SLAC-based process that is followed project wide. The process allows for technical changes to be reviewed and approved prior to implementation. This provides better communication and configuration control of the design.

The project has also held many design reviews for the cryomodule design. Even though the changes made were minor, it was learned from LCLS-II that small changes can have large impacts. Following this lesson from LCLS-II, changes were kept to a minimum. In some cases, "delta" design reviews were held where the reviews were geared toward reviewing the changes rather than the design that has been reviewed and proven for LCLS-II.

A cross section of the LCLS-II-HE cryomodule is shown in Figure 2. There are three helium circuits, 50K, 5K and 2K. The design consists of a 300mm diameter support pipe supported from the top of the vacuum vessel with three support posts. Eight SRF cavities and a quadrupole magnet are hung from this 300mm support pipe with sliding supports that can be seen in Figure 3. The 300mm pipe also acts as the return line for the 2K vapor. The module is designed to connect end to end with the next module in the accelerator. The overall design has been verified with the vCM test at Fermilab showing that the cryomodule design is able to meet the increased total gradient specification required for LCLS-II-HE [5].

3. Cryomodule Assembly

The infrastructure used for LCLS-II and LCLS-II-HE [6,7] cryomodule assembly was developed with the Tesla Technology Collaboration and is similar to infrastructure at Deutsches Elektronen-Synchrotron (DESY) used to assemble modules for FLASH. The assembly is broken into workstations (WS), and at Fermilab, the assembly is performed in two buildings. WS1 and WS2 are performed in the MP9 building shown in Figure 4. WS1 is performed inside the cleanroom shown in Figure 5. During LCLS-II, WS0 was also performed inside the cleanroom and will be mentioned later. WS3-WS6 are performed in the Industrial Center Building shown in Figures 6, 7 and 8. Once the module is assembled, it is transported to another building for testing.

Figure 6. WS3 at Industrial Center Building.

Figure 7. WS4 at Industrial Center Building.

Figure 8. WS5 at Industrial Center Building.

One of the lessons learned from LCLS-II was to document and plan off-normal work. For LCLS-II-HE we have an off-normal work plan that can be completed if a part or assembly does not conform to the standard workflow. If a discrepancy is noted, an off-normal work plan can be drafted to explain the steps required to address the non-conformance and return the part or assembly back into the normal workflow process. The off-normal work plan is approved before the work starts, the outcome of the work is then recorded by the person responsible for it, and the original approver then authorizes that the part or assembly can be returned to the standard workflow. This process allows managers to be informed of the work being performed as well as to document the work that was completed.

Figure 9. NEG/Ion pump used to pump beamline during assembly.

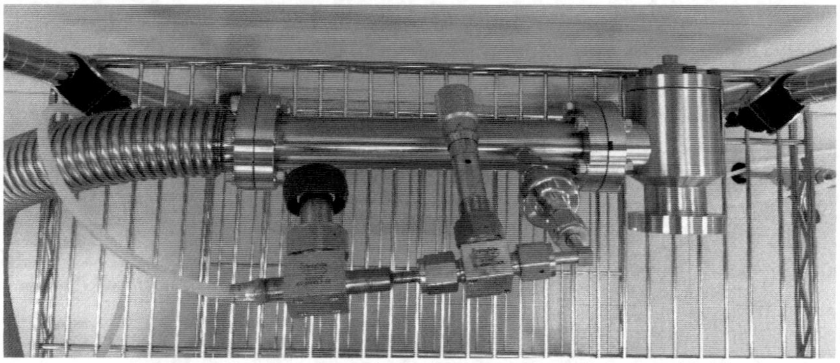

Figure 10. Upgraded cavity/string venting manifold.

A few significant process changes have been implemented since LCLS-II to improve the beamline cleanliness. One is to assemble the cryomodule with the beamline under vacuum. For LCLS-II, the beamline was vented before the module left the cleanroom. To reduce the chance of contamination due to the backfill process, the beamline will be left under vacuum for LCLS-II-HE. In order to complete the module assembly with the beamline under vacuum, some additional steps such as coupler supports are needed to accommodate the vacuum forces. During the assembly, the beamline is pumped with a combination NEG and ion pump shown in Figure 9 [8]. In addition, a Process Failure Modes and Effects Analysis was performed in order to prevent unwanted outcomes related to this change. Another change that has been implemented is to install the power coupler on the cavity after the cavity is on the support post in its string position. For LCLS-II, the cavity was vented, and the coupler was installed at WS0. Using the WS0 workflow, the vented cavity was then valved closed, rolled across the cleanroom and set on the support post for WS1. With the new approach, the vented cavity is not moved except minimal movement along the rail to make the beamline connections. An upgrade to the beamline venting, pump and purge system was completed before the vCM assembly. This included a new assembly that is attached to a cavity or string for venting or purging shown in Figure 10. In Figure 10, the cavity or string would be attached to the valve on the right side of the assembly, and the vent or purge flows through the plastic line on the left.

4. Production Status and Conclusion

The vCM has been assembled and is currently on the test stand at Fermilab. The preparation for the production of cryomodules is progressing well, and a production readiness review was held July 2021. We have a released drawing package for the production module, and the travelers and assembly procedures are being updated based on the vCM assembly experience. All of the major components are on order, and many components have arrived. We are awaiting the arrival of some remaining parts and expect to start assembly of the first article module later in 2021.

The LCLS-II-HE project will build on LCLS-II and take advantage of the lessons learned throughout that project [5]. Fermilab has transitioned from LCLS-II to LCLS-II-HE retaining much of the experienced cryomodule production team. The vCM testing results show that the design is able to meet the increased gradient specification, and Fermilab is capable of producing quality cryomodules for LCLS-II-HE.

LCLS-II-HE will enable the production of higher energy photons in order to give more capability to the LCLS facility. Building on LCLS-II, LCLS-II-HE will add additional cryomodules to the SRF linac to increase the electron beam energy from 4GeV to 8GeV. At Fermilab and Jefferson Lab, the experience of LCLS-II cryomodule production [9,10] will be leveraged to produce cryomodules for LCLS-II-HE. Fermilab plans to assemble, test and deliver 14 cryomodules for the LCLS-II-HE project.

5. References

[1] Ross M C 2019 LCLS-II Status, Issues and Plans 19th Int. Conf. on RF Superconductivity SRF2019, Dresden, Germany doi:10.18429/JACoW-SRF2019-MOFAA1

[2] Gonnella D et al. 2019 The LCLS-II HE High Q and Gradient R&D Program 19th Int. Conf. on RF Superconductivity SRF2019, Dresden, Germany doi:10.18429/JACoW-SRF2019-MOP045

[3] Peterson T et al. 2015 LCLS-II 1.3 GHz Cryomodule Design – Modified Tesla-Style Cryomodule for CW Operation Proceedings of SRF2015, Whistler, BC, Canada THPB119 ISBN 978-3-95450-178-6

[4] Pischalnikov Y et al. 2021 Extended Range SRF Cavity Tuners for LCLS-II HE Project preprint 20th Int. Conf. on RF Superconductivity SRF2021

[5] Arkan T T et al. 2021 LCLS-II Cryomodules Production Experience and Lessons Learned Towards LCLS-II-HE Project preprint 20th Int. Conf. on RF Superconductivity SRF2021

[6] Arkan T et al. 2016 LCLS-II Cryomodules Production at Fermilab Proceedings of LINAC2016, East Lansing, MI, USA TUPLR007 ISBN 978-3-95450-169-4

[7] Arkan T et al. 2015 LCLS-II 1.3 GHz Design Integration for Assembly and Cryomodule Assembly Facility Readiness at Fermilab Proceedings of SRF2015, Whistler, BC, Canada TUPB110 ISBN 978-3-95450-178-6

[8] Ciovati G et al. 2016 Operation of a high-gradient superconducting radio-frequency cavity with a non-evaporable getter pump Nuclear Instruments and Methods in Physics Research Section A, vol. 842, pp. 92-95, https://doi.org/10.1016/j.nima.2016.10.048

[9] Stanek R et al. 2019 Overview of LCLS-II Project Status at Fermilab 19th Int. Conf. on RF Superconductivity SRF2019, Dresden, Germany doi:10.18429/JACoW-SRF2019-MOP092

[10] Arkan T et al. 2019 LCLS-II Cryomodules Production Experience and Lessons Learned at Fermilab 19th Int. Conf. on RF Superconductivity SRF2019, Dresden, Germany doi:10.18429/JACoW-SRF2019-TUP101

Acknowledgments

This manuscript has been authored by Fermi Research Alliance, LLC under Contract No. DE-AC02-07CH11359 with the U.S. Department of Energy, Office of Science, Office of High Energy Physics.

A large millikelvin platform at Fermilab for quantum computing applications

Matthew I Hollister, Ram C Dhuley and Grzegorz L Tatkowski

Fermi National Accelerator Laboratory, P.O. Box 500, Batavia, IL 60510, USA

E-mail: mhollist@fnal.gov

Abstract. The need for larger mK cooling platforms is being driven by the desire to host ever growing numbers of cryogenic qubits in quantum computing platforms. As part of the Superconducting Quantum Materials and Systems Center at Fermilab funded through the Department of Energy under the National Quantum Initiative, we are developing a cryogenic platform capable of reaching millikelvin temperatures in an experimental volume of 2 meters diameter by approximately 1.5 meters in height. The platform is intended to host a three-dimensional qubit architecture based on superconducting radiofrequency accelerator cavity technologies. This paper describes the baseline design of the platform, along with the expected key performance parameters.

1. Introduction

The Superconducting Quantum Materials and Systems Center (SQMS) at Fermilab aims to understand the physics and materials origin of coherence limiting factors in quantum devices. An extremely promising approach to increasing the coherence time of superconducting qubits is to utilize high quality factor superconducting cavities, a technological area that has seen extensive development for use in superconducting radiofrequency (SRF) particle accelerators, combined with two-dimensional devices. In such a configuration, the two-dimensional superconducting circuit will couple to multiple degrees of freedom in the three-dimensional cavity to produce multiple interconnected quantum states that can each function as a discrete qubit. The operation of niobium-based SRF cavities at millikelvin temperatures has previously been demonstrated to exhibit long lifetimes in the quantum regime [1], although there are challenges in cooling large superconducting objects to such low temperatures due to the large heat capacity through the transition temperature and the poor thermal conductivity of the superconductor below the transition. The SRF cavity qubit architecture results in a relatively large object that must be cooled to millikelvin temperature in order to suppress photon noise in the cavity, while the use of multiple such devices drives to a large experimental volume combined with a high cooling capacity in order to cool the cavities to base temperature.

To address the basic cryogenic requirements of the SRF cavity architecture, SQMS plans to construct a large cryogenic platform capable of supporting multiple cavities and/or conventional two-dimensional qubit devices in a large volume cooled to approximately 20 mK with large cooling capacity. The current largest dilution refrigerators available on the commercial market have millikelvin volumes in the 500–1000 mm diameter range and cooling capacities of approximately 40 μW at 20 mK. The SQMS Center targets a platform with a 20 mK volume 2 meters in diameter with a cooling capacity 200–300 μW at 20 mK.

In addition to exploring the coherence time of qubit devices, a second area of interest that is a use case for the large millikelvin platform are "dark photon" searches in which an emitter cavity is used to produce a signal that can be detected in a second receiver cavity that is well-isolated from the emitter (see, for example [2]). The technique has been successfully demonstrated with both the emitter and receiver cavities at 1.4 K in a modified magnet vertical test stand at Fermilab, but in order to increase the sensitivity of the receiver cavity lower temperature operation is required. To this end, the platform is also designed to host an emitter cavity operating at a temperature $\sim$2 K along with several receiver cavities operating at 20 mK.

This paper will discuss the main design features of the millikelvin platform, along with the current status and expected timeline for completion of construction.

2. Challenges of scaling up millikelvin platforms

Large [3] and high-cooling power dilution refrigerators [4] have been previously constructed, but this project calls for a combination of both features. There are two major limiting factors in the construction of larger fridges, which are the surface area of the millikelvin heat exchangers required to overcome the kaptiza boundary resistance of high 3He circulation rates, and the available cooling capacity at the higher temperatures stages of the refrigerators, particularly in the 2–4 K range.

2.1. Overcoming limitations of heat exchangers

The cooling capacity of the dilution process is primarily a function of the rate of circulation of 3He in the circuit. However, the circulation rate cannot be increased arbitrarily since the increased flow rate requires increasing heat transfer between the inflow and outflowing 3He streams in the heat exchangers. Particularly at the lowest temperatures near the mixing chamber, the required area for heat exchange quickly becomes unreasonably large. Rather than attempt to design larger surface area heat exchangers, the approach taken by this project is to use multiple, parallel dilution refrigerators housed in the same cryostat with the common temperature stages tied together by the cold plates. The use of multiple, smaller heat exchanger stacks has the additional advantage that existing commercial design can be utilized, and the exchanger stacks can be added gradually to upgrade the system capacity with time. The platform is designed to mount up to ten dilution stages, each with an independent room temperature pumping and gas handling system.

2.2. Cooling power limitations at intermediate temperature stages

The load at the intermediate stages of the dilution refrigerators come from a combination of conductive loads from supports and wiring, and the load from cooling and condensing the circulating helium. While older designs of dilution refrigerator utilized liquid helium to provide cooling at temperatures around 4 K as a precooling stage and to supply a pumped 4He pot (the "1-K Pot") as a precooling stage, the consumption of liquid helium imposes a practical limit on the heat load that can be absorbed. More modern designs have made use of closed-cycle cryomechanical coolers which, while having the advantage of no cryogen consumption, are extremely limited in their ability to take up heat at temperatures of 4-K. This is addressed in the SQMS platform by using a cooling configuration that is probably more familiar to particle accelerator designs in which a cryoplant is used in refrigeration mode to supply supercritical helium at a temperature of 5 K to the cryostat. The flow is divided, with one channel directly cooling a 5-K cold plate and thermal shield, with the other stream being subcooled in a small heat exchanger and expanded through a Joule-Thomson (J-T) valve to produce a two-phase mixture of liquid and gaseous helium. The pressure of the downstream side of the J-T valve is reduced to approximately 30 mbar (equvialent to a temperature of 2 K) by a room temperature pumping system to provide a powerful condensation stage analogous to the 1-K Pot. This configuration is shown schematically in Fig. 4, with additional discussion in the next section. The cryoplant and pump system are closed-cycle, eliminating the issue of helium consumption. Cooling at the warmer cryogenic stages is provided by an open-cycle liquid nitrogen flow.

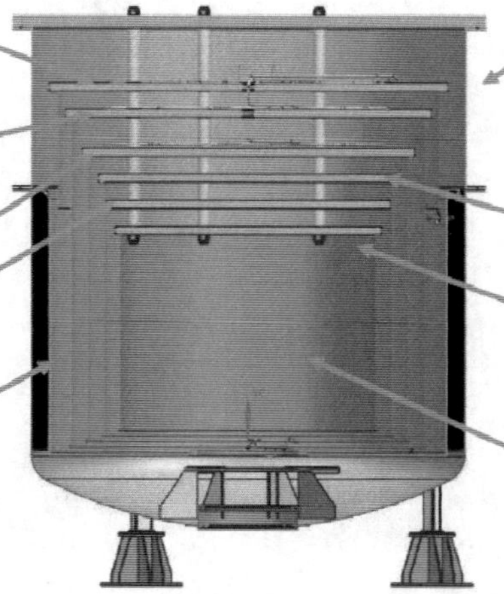

Figure 1. CAD image of the cryostat design shown in cross-section with the major features of the design indicated. The dimensions of the outer vacuum shell are 3.4 meters diameter and 4 meters high. For details and discussion, see text.

3. Design of the SQMS mK platform

The general layout of the millikelvin platform in cross section is shown in Fig. 1. The platform consists of a flat upper head from which the internal assemblies are suspended. This assembly is installed in a stainless steel vacuum vessel which also supports a thermal shield cooled by a liquid nitrogen circuit. The lower vessel and thermal shield were originally used as part of an existing test cryostat, while the flat head and internal assembly are a new construction. The existing cryostat is shown in Fig. 2 with a dished vessel head installed. The structure suspended from the flat head consists of six cold plates at temperatures of 80 K cooled by the liquid nitrogen supply, 5 K and 2 K cooled by the helium flow, and at 1 K, 100 mK and 20 mK cooled by the stages of the dilution refrigerator themselves. The 5 K, 2 K, 1 K (still) plates support shields for thermal radiation control, while the 20 mK plate mounts a shield for control of stray light in the experimental space. The major dimensions and mechanical parameters for the cryostat are summarized in Table 1.

The various thermal stages are suspended from the flat vacuum head using vertical tie rods. To reduce the stress in the rods due to the radial contraction of the large cooled plates, the rods are not rigidly attached but instead connect via hinged joints that allow for radial movement of the rods while the cold plates remain horizontal. The exact configuration of the materials and dimensions of the tie rods is currently being optimized, but will likely consist of a mixture of stainless steel and G-10CR epoxy composite.

For the purposes of maintenance and payload installation, the vacuum vessel head and suspended structures are removed from the vacuum shell using an overhead gantry crane and lowered through the roof of an adjacent clean workroom. The lower thermal shield remains installed in the vacuum shell during this operation.

3.1. Warm cryogenic stages

Cooling at the upper stages of the cryostat cold mass is provided by flow of liquid nitrogen at approximately 80 K and two stages cooled by the flow of supercritical helium. The liquid nitrogen system is fed from a 20000 gallon storage tank outside the building and is supplied to the cryostat via a vacuum jacketed supply line and a custom manifold consisting of pneumatic valving and bayonets. The

Figure 2. Photograph of the existing installation of the vacuum vessel, including the associated work platform and cryogenic infrastructure. The vacuum vessel was formally used as a liquid helium test stand for solenoids for the Mu2e project. Picture credit: Fermilab

Table 1. Dimensions and construction materials of the major cryostat components. For the shield components, "Thickness" refers to the dominant wall thickness of the shell and "Diameter" is the inside diameter of the shield cylinder.

Component	Diameter	Height	Thickness	Material
Vacuum Vessel	3.4 m	3.7 m	10 mm	SS304
Thermal Shield	3.0 m	2.8 m	17 mm	SS304
Helium Shield	2.7 m	2.6 m	5 mm	AL6061
2-K Shield	2.5 m	2.2 m	5 mm	AL6061
Still Shield	2.2 m	2.0 m	5 mm	AL6061
Mixing Chamber Shield	2.0 m	1.5 m	5 mm	OFHC
Vacuum Plate	3.6 m	-	90 mm	SS304
Thermal Shield Plate	3.0 m	-	50 mm	AL6061
Helium Plate	2.8 m	-	50 mm	AL6061
Superfluid Plate	2.5 m	-	50 mm	AL6061
Still Plate	2.3 m	-	25 mm	OFHC
IC Plate	2.2 m	-	25 mm	OFHC
Mixing Chamber Plate	2.0 m	-	25 mm	OFHC

supply line for the nitrogen system, along with the supply and return lines for the helium supply, can be see in the lower left of the cryostat imagine in Fig. 2. Three liquid nitrogen supply bayonets are provided, with one feeding the lower thermal shield, one the Thermal Shield cold plate in the main cold mass, and the third supplying an open-cycle liquid nitrogen dewar that is used to cool the cold traps in the 3He gas system to remove air impurities. Both the stainless steel cooling lines on the cold plate and on the lower thermal shield meander around the structures in order to provide distributed cooling. Copper clamps are attached the cooling line at intervals, and then bolted to the cold plate to provide the thermal connection.

For cooling at helium temperatures, supercritical helium is supplied from a cryogenic plant via a vacuum jacketed transfer line, paired with a corresponding return line. The cryogenic plant, which can be seen in Fig. 3 consists of a compression system and reciprocating expanders that can act either as a helium liquifier or a refrigerator supplying supercritical gas, or a combination of the two modes. [5] For the supercritical supply, the system is capable of providing up to 625 W of cooling at 4.5 K. The cold helium gas is supplied to the cryostat at a pressure of approximately 50 psig and on entering the platform cryostat, is divided into one or more flow paths, as shown schematically in Fig. 3. The primary 5-K flow path is similar to the liquid nitrogen piping system for 80 K cooling on the upper stage, and typically carries approximately 20 % of the flow while the remaining flow is diverted through either the precooling circuit during the initial cool down of the cryostat, or through a subcooling heat exchanger and Joule-Thomson valve in normal operation. The precooling system directs flow to each of the sub-2-K plates in series with the return path connected to the header of the superfluid pumping system discussed below. During normal operation, the precooling system must be evacuated of helium in order to avoid any superfluid flow leaks in the precooling system that would increase the thermal load on the lower temperature stages of the system. To allow control of the cool down process, the supercritical helium entering the cryostat passes through a static mixer with a controlled flow of warm helium gas from the cryoplant suction header provided to regulate the temperature of the gas flow in both the precool circuit and the main 5-K circuit.

Cooling at the 2-K stage of the cryostat is provided by a superfluid helium system. The flow diverted to the system during normal operation is subcooled in a small counterflow heat exchanger with the outflowing cold vapor, and is then expanded through a Joule-Thomson valve to produce a two-phase mixture. The pressure of the mixture is lowered by pumping with a room temperature pumping system consisting of a Kinney vacuum booster with a liquid ring pump. The pump line exiting the cryostat is a 2 inch diameter stainless steel pipe, which transitions to a 4 inch diameter vacuum jacketed pipe outside the cryostat for connection to the pumping system, approximately 200 feet away from the platform vacuum vessel. The heat exchanger and pumping system is sized for a cooling capacity of approximately 50 W. As noted previously, the superfluid cooling system is used to provide an additional thermal intercept for conductive loads in the cryostat, but more importantly provides the condensation stage of the 3He circulating in the dilution circuits. This accounts for only a small fraction of the available cooling power at the superfluid stage (see summary in Table 2), with the remaining capacity available for cooling a transmitter cavity when the platform is used for Dark SRF experiments.

Both the outlet of 2-K pumping system and the return line of the 5-K cooling system connect to low pressure room temperature storage tanks, with a total capacity of 40000 gallons. The collected gas is compressed using a Mycom screw compressor skid, with the high pressure gas being supplied back to the reciprocating engines.

3.2. Sub-1 K cryogenic stages

Cooling of the three coldest stages of the experiment is provided by the dilution unit heat exchanger stacks. As mentioned previously, the platform utilizes commercially-constructed dilution heat exchangers, with capacity for up to ten independent 3He circuits to be installed. It is expected that initially only a subset of the dilution stages will be installed, allowing the platform to operate albeit at reduced cooling capacity, with later expansion to the full system. The specifics of the heat exchanger stacks are being finalized, but based on the currently available commercial dilution units it is likely that

Figure 3. Photograph of the helium cryoplant, including the 2000 liter liquid helium dewar on the left. The plant is rated at a refrigeration capacity of 625 W at 4.5 K or 4.2 g/s of liquid helium.

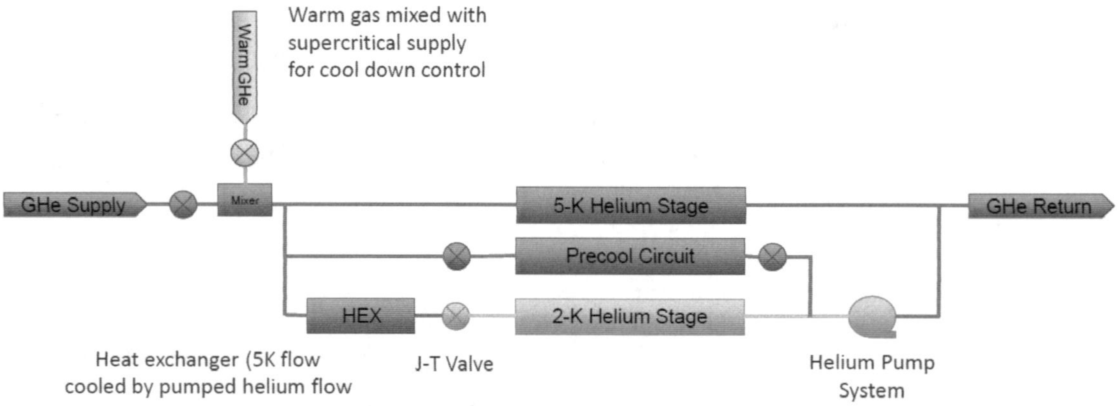

Figure 4. Block diagram schematic of the 5-K and 2-K cooling systems. For detailed discussion, see text.

Figure 5. Example of a pumping system for the superfluid helium system, consisting of a Kinney vacuum booster and liquid ring pump.

each stack will be able to provide 30 μW at 20 mK at the mixing chamber and approximately 10 mW at 800 mK at the still layer, depending on the room temperature pumping system. The mixing chamber of each dilution unit is bolted directly to the 20 mK plate, while thermal connections at the intermediate cold plate and the still layer are made via flexible copper ropes to minimize thermal stresses in the stacks due to contraction of the large cold plates.

The still side of the dilution units are each pumped by an independent gas handling system. Each system has provision for either 1 or 2 large turbo pumps, each turbo system being backed with an Edwards GXS-series dry screw pump. Unlike modern cryogen-free dilution refrigerators that typically require a diaphragm compressor or similar to increase the condensing pressure, the high working outlet pressure of the dry screw pumps combined with the use of a 2 K condensation stage allows this system to operate with the primary pump only. 3He on the return side of the pump system passes through a charcoal cold trap cooled by liquid nitrogen in order to remove any air or other impurities.

On entering the cryostat, the inlet line first passes through a filter packed with metal sinter that is cooled by the 80 K stage of the cyrostat. This filter both removes further impurities from the flow while also acting as the first heat exchanger to cool the incoming gas. The flow then enters a spiral heat exchanger at the 5-K stage of the cryostat consisting of approximately 1 meter of 2 mm diameter cupronickel tube wrapped onto a copper plug of 25 mm diameter that is bolted to the 5-K cold plate. A second, similar spiral heat exchanger is used to the 2-K cold plate in which the 3He is further cooled and condensed, before flowing on to the still heat exchanger via an impedance.

It is estimated that each dilution circuit will be charged with 150 STP liters of 4He and 50 STP liters of 3He. The two helium isotopes are stored in separate room temperature tanks for each dilution circuit to allow for easy tuning of the mixture.

3.3. Thermal modelling

Detailed modelling of the expected thermal loads at each stage of the cryostat was conducted during the design process in order to specify the required capacities for each part of the system. The model

Table 2. Summary of thermal modelling results listed by cryostat stage and major loads.

Load	Thermal Shield	Helium	Superfluid Helium	Still	Intermediate Cold Plate	Mixing Chamber
Nominal Temperature	80 K	5 K	2 K	1 K	100 mK	20 mK
Radiation	23 W	1.8 W	23 μW	0.49 μW	-	-
RF Wiring	7 W	1 W	1.11 W	753 μW	407 μW	2.7 μW
DC Wiring	2.2 W	2.1 W	29 mW	0.31 μW	0.05 μW	-
Supports	33.7 W	4.4 W	9 mW	1 mW	198 μW	2.2 μW
Precool Line Conduction	-	-	370 μW	43 μW	8.5 μW	0.1 μW
Helium-3 Inlet Conduction	1.7 W	0.2 W	0.4 mW	25 μW	-	-
Mixture Cooldown	91.5 W	31.2 W	2.24 W	-	-	-
Still Pumpline Conduction	74.5 W	2.7 W	2 mW	0.49 μW	-	-
2-K Pumpline Conduction	15.8 W	1.8 W	23 μW	-	-	-
Total Load	250 W	65 W	3.4 W	4 mW	614 μW	5 μW
Available Capacity	9kW	200 W	50 W	100 mW	1 mW	300 μW

included thermal radiation, conductive loads, and loads from the circulation of helium mixture in the dilution circuits. The results of the model are summarized in Table 2 in comparison to the available capacity at each stage. To allow for unexpected loads and for possible future expansion, a margin of 60 % or higher has been maintained at each stage. Of particular note is the contribution of the RF readout wiring, which is assumed to consist of a flexible strip line arcitecture using conductive traces on Kapton films, with 1000 readout lines included in the model. If a more conventional semirigid coaxial cable arrangement were used, it would be expected that the contribution of this term would be higher but still easily within the allowable margin of the design. In this platform, the cold HEMT amplifiers are assumed to be at the 2 K stage rather than at the 4 K stage as is commonly seen in other systems. The power dissipation of these cold amplifiers is included in the RF Wiring term at this stage in Table 2.

In addition to the static thermal calculations, the temperature distributions in the large cold plates and thermal shields under power loading was explored using finite-element modelling. These models were used to inform the material selections and dimensions at each stage. An example of these simulation results is shown in Fig. 6 for the mixing chamber cold plate at 10 mK. This is a copper plate with thickness of 50 mm showing a temperature gradient of order 0.5 mK with 50 μW applied uniformly to the outer edge. The use of copper at the lower temperature stages is important due to the discrete point cooling of the dilution units, while at the higher temperature stages with distributed cooling lines carry helium and nitrogen, aluminum alloy can be used with acceptable temperature gradients. Similarly, aluminum alloy shields can used at all stages with the exception of the mixing chamber shield.

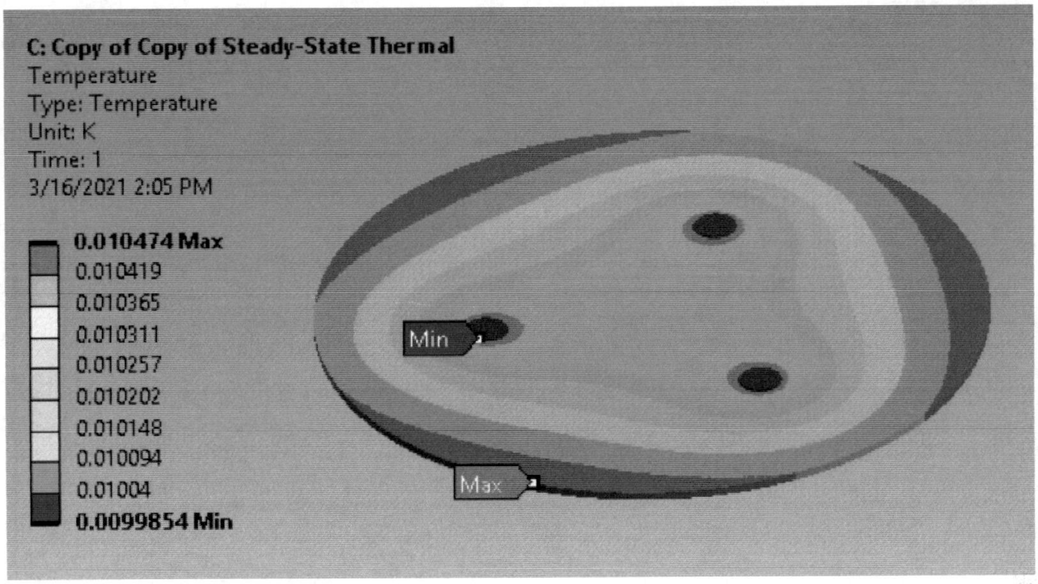

Figure 6. Sample results of FEA analysis for the 20-mK cold plate. The plate is composed of copper with an assumed residual resistivity ratio of 50, 50 mm thick and 2 m diameter. Three points represent the connections between the plate and three mixing chambers. With the mixing chamber contacts fixed at 10 mK and a uniform 50 μW applied to the outer edge of the plate, the maximum temperature is 10.5 mK.

4. Current status and timeline

The detailed final design process is nearing completion as of summer 2021 with major procurements expected to start during 2021. The initial construction phase for the upper cryogenic stages is planned to complete late in 2022, allowing operation of the platform as a large 2-K cryostat for Dark SRF experiment development. Upgrade of the platform to millikelvin operation is planned for 2023 with full operation as a quantum computing platform by the end of 2023.

5. Summary

This paper has discussed the design for a large millikelvin cryogenic platform suitable for use as a quantum computing platform and as a platform of Dark sector photon searches using superconducting radiofrequency cavities. The major design details have been discussed, along with the expected heat loads and thermal performance metrics associated with the system. The platform will address issues associated with scaling up millikelvin cryogenic systems by utilizing existing cryogenic infrastructure for the supply of helium and liquid nitrogen, along with integrating multiple small dilution circuits into a single large cryostat to provide a large, high-cooling power cryostat. The design phase is nearing completion, with procurements expected to begin in 2022 with initial construction and start of operations during 2023.

6. References

[1] Romanenko A, Pilipenk R, Zorzetti S, Frolov D, Awida M, Belomestnykh S, Posen S and Grassellino A 2020 *Phys. Rev. Applied* **13** 034032
[2] Parker S R *et al* 2013 *Phys. Rev.* D **88** 112004
[3] Ligi C *et al* 2016 *J Low Temp. Phys.* **184** 590
[4] Vermeulen G A and Frossati G 1987 *Cryogenics* **27** 139
[5] Rode C H, Andrews R A, Ferry R, Gannon J, Makara J, Martin M, Misek J, Peterson T and Theilacker J 1983 *IEEE Transactions on Nuclear Science* **NS-30** 2892

Acknowledgments
This material is based upon work supported by the U.S. Department of Energy, Office of Science, National Quantum Information Science Research Centers, Superconducting Quantum Materials and Systems Center (SQMS) under the contract No. DE-AC02-07CH11359.

IOP Conf. Series: Materials Science and Engineering 1241 (2022) 012046 doi:10.1088/1757-899X/1241/1/012046

First cryogenic test results of 3D-printed resonators for quantum bits

P R Carriere[1], P Frigola[1], R Agustsson[1], M H Chou[2], A N Cleland[2,3], T Horn[4], M Kelly[3], S V Kutsaev[1], R Povey[2] and A Y Smirnov[1]

[1]RadiaBeam Technologies LLC, Santa Monica, CA, 90404, USA
[2]Institute for Molecular Engineering, University of Chicago, Chicago, IL, 60637, USA
[3]Argonne National Laboratory, Lemont, IL, 60439, USA
[4] North Carolina State University, Raleigh, NC 27607, USA

E-mail: carriere@radiabeam.com

Abstract. Quantum computers (QC) have the potential to efficiently solve problems currently unfeasible on even the fastest generation of classical computers. The building block of a QC is a quantum bit (qubit). Encoding and reading qubit states coupled via high-Q resonant cavity modes is a solution to maintaining qubit states; however there is a need for simple, scalable and robust fabrication techniques capable of realizing high density cavity arrays. RadiaBeam is developing a novel approach utilizing metal additive manufacturing(AM) using both laser and electron beam powder bed fusion. Using a 6GHz quarter wave resonator (QWR), we fabricated several niobium and titanium alloy QWR cavities and characterized their superconducting RF performance. In this letter, we provide the details of the first 3D-printed qubit cavities design and fabrication, and compare their Q-factors, measured at dilution fridge temperatures, against the machined Nb resonators.

1. Introduction

The emerging field of quantum computing (QC) is rapidly growing and has shown immense potential. There is strong evidence that quantum computers possess a capacity exceeding the limits of classical computers and have the potential to efficiently solve problems currently unfeasible with current hardware, drastically advancing the frontiers of high energy physics and science in general [1]. The building block element of a QC is a quantum bit (qubit). Josephson junction circuits [2] are natural candidates for qubit blocks since they operate as non-linear quantum element. However, qubit state degrades rapidly due to the interaction with environmental degrees of freedom and the added control channels [3,4]. To avoid this problem, Josephson circuits can be protected from environmental noise and loss via the encoding qubits states in high-Q resonant cavity modes [5,6,7,8].

Superconducting niobium cavities used in particle accelerators are able to reach quality factors of ~10^{10}, potentially enabling storage times approaching seconds, even at the single-photon level [9]. Traditionally, accelerator superconducting radio frequency (SRF) cavities are fabricated by forming, electron beam welding, chemically etching and heat treating ultra-high purity niobium sheet[[10]]. Generally, this approach is focused on sub-3.9 GHz components, as the frequency dependent BCS surface resistance is comparable to the residual resistance when operating in the superfluid helium bath. However, due to the noise considerations and dilution fridge space limitations, future quantum information science (QIS) systems will likely operate in the frequency range of 6 GHz to 15 GHz and at millikelvin temperatures, enabling a new class of superconducting materials with transition temperatures below 2K. Furthermore, QIS will consist of many individual cavity qubit units coupled together. For these applications, the development of a simple, inexpensive and robust fabrication methods is desired.

In response to this problem, we have successfully fabricated QWRs using two different metal additive manufacturing (AM, a.k.a. 3D-printing) technologies; electron beam powder bed fusion (EB-PBF), and laser powder bed fusion (L-PBF) (See Figure 1).

Figure 1: (Left) Array of niobium QWRs printed using EB-PBF on a titanium build plate; (Center) Ti-6Al-4V (Ti64) L-PBF QWR fitted with SMA coupling ports; (Right) Ti64 SLM QWR undergoing room temperature RF testing.

AM of passive microwave components is of growing interest due the rapid and efficient fabrication of complex geometries. Additive manufacturing of superconducting cavities has been previously demonstrate in titanium (Ti-6V-4Al) and aluminium (Al-12Si), achieving quality factors of $3x10^4$, $4x10^6$ respectively[11,12]. Additively manufactured superconducting niobium resonators have demonstrated quality factors varying from $2x10^5$ to $1x10^9$, with the latter requiring post-print machining [13,14].

Building on these results, we have demonstrated superconducting QWRs using two related AM technologies: EB-PBF and L-PBF. In both powder bed fusion (PBF) processes, the build platform is lowered, fresh powder is recoated over the build platform and the melting process is repeat from 2D slices into a 3D part. In this case, the distinguishing feature between PBF processes is electron beam (EB-PBF) versus laser PBF pencil beam.
Compared to L-PBF, EB-PBF offers a series of unique capabilities including a high vacuum build environment for reduced residual gas contamination, direct beam heating of the part for reduced residual stress and high (6 kW) average beam power. Conversely, L-PBF yields better geometric accuracy and surface finish due to the finer powder, smaller layers thicknesses and smaller spot sizes. Since L-PBF is performed in pressurized inert gas atmosphere, oxygen pickup from powder to printed part can be significantly higher than EB-PBF.

2. Design

Broadly speaking, the Q-factor of the resonator can be defined according to $Q = G/R_s$, where G[Ω] is the cavity geometry factor and Rs[Ω] is the RF surface resistance. In order to maximize the cavity quality factor, and thus the qubit coherence time, it is imperative that the resonator has a high G-factor via shape optimization, as well as low surface resistance, via material selection, fabrication and post-processing. Recently, we have developed a superconducting 3D quarter-wave resonator (QWR) machined from accelerator-grade niobium [15]. This 6 GHz cavity was designed for a G=71Ω, which was chosen as a compromise between machinability and RF performance. For comparison, superconducting elliptical cavities used for accelerators operating in the TM_{010} mode have a corresponding G≈268 Ω, while QWR have a reported G as high as 110Ω [16]. The corresponding 3D cavity design as well as the corresponding electric and magnetic field distributions simulated in COMSOL, are shown in Figure 2.

To compare the performance of the AM QWRs with the machined ones, we replicated their nominal RF design in the printed parts. While useful for comparison, this approach does not fully leverage the advantages of additive manufacturing in terms of component integration or shape optimization. The layer-wise nature of the process results in anisotropy in both the material properties as well as the surface finish. Surface finish is another key difference in as-built AM parts, with the best surface finish achieved on upward facing surfaces and the worst finish associated with downward facing surfaces.

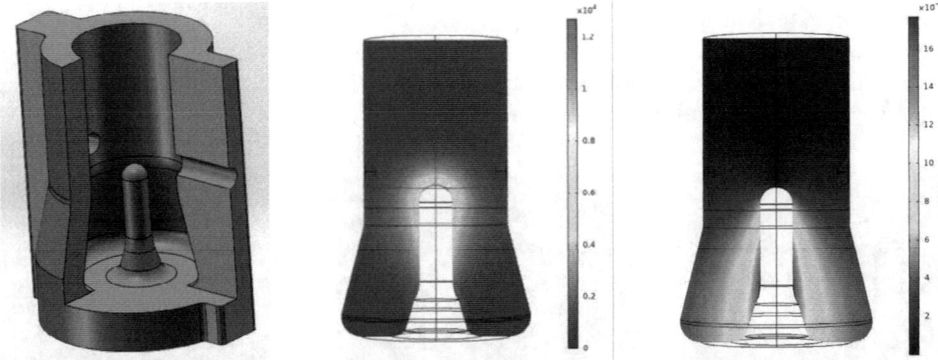

Figure 2 (Right) Sectioned CAD model of QWR, with total part height of 28mm. (Center) Electric field distribution simulated in COMSOL (Right) Magnetic field distribution

3. Fabrication

Within the broader field of additive manufacturing (AM), each material/process pair has achieved a certain technological maturity. We opted to print identical QWR designs from grade 5 titanium (Ti-6wt% Al – 4wt% V), ie Ti64, using the L-PBF process, which is a mature turn-key technology offered from numerous of service bureaus. The transition temperature of Ti64 ranges between T_c = 1.3-6.3K with the variation due to oxidation treatment and annealing state [11]. We also printed QWRs from pure niobium (T_c = 9.26K) using the EB-PBF process, which is relatively undeveloped technology well suited for printing refractory metals such as niobium. Comparison of materials and processes was performed as a first attempt to understand which technology is best suited for further technological development.

Our Ti64 QWR geometries were commercially printed using a Concept laser M2 printer (Castheon Inc, Thousand Oaks, CA). Virgin 15-45μm plasma atomized powder was used with print parameters tuned to reduce the surface roughness of the inner RF surfaces. After printing, the parts were hot isostatically pressed (HIP) using established procedures adapted from aerospace applications (KittyHawk Inc, Garden Grove, CA). The HIP process homogenizes the microstructure, eliminates print-related residual stress and reduces process-induced internal porosity. One Ti64 QWR was sent to a commercial vendor for proprietary Chemical-Mechanical Polishing (REM Surface Engineering, Southington, CT). Approximately 200μm of material was removed from the part, resulting in a surface roughness reduction from Ra = 9.7 to 0.9μm, as measured on the upper bore of the cavity. Both Ti64 cavities underwent a final, 30s etch in Krolls etchant [9% vol HF(48% conc.) 12% vol HNO3(70% conc.), remainder water], resulting in a mass loss of approximately 0.17g. While this number corresponds to a thickness reduction of about 11μm, some powder particles were visually observed in the rinse water, suggesting an overestimation of material removal.

As mentioned, EB-PBF of niobium is a relatively new AM process and was performed in collaboration with North Carolina State University using a custom Arcam S12 printer [17]. The niobium powder was plasma atomized from ASTM B392 type 1, reactor-grade wire. Limited contamination from wire to powder was measured except for an increase of light interstitial elements(C, O, N, H), a feature consistent with the high surface area associated with powder. The Nb powder was sieved to -100/+230 mesh in an inert glove box, yielding a D_{10} = 55μm and D_{90} = 127μm. Some titanium powder cross contamination was observed within the Nb feedstock, likely due to inadequate cleaning of the sieving station.

The Nb parts were printed at a sustained temperature of approximately 1300°C, as measured using a two-color pyrometer focused on the top of build surface. Both Nb cavities were manually polished with an alumina slurry and etched in a buffered chemical etchant [25% vol HF(48%.) 25% vol HNO3(70%.), remainder H3PO4 (95% conc.)] for 120 seconds, with a mass loss of 0.75g and an estimated material removal of 12μm.

Both cavities underwent a small amount of external machining, including milling flat surfaces for contact with the SMA probes and reaming coupling ports for the electric coupling antennas.

4. Resonator Characterization

The accessible surfaces of the QWR's RF surfaces were microscopically inspected for geometric accuracy and surface finish. Images of the centrepin top hemispheres are shown below, comparing the surfaces of various materials and finishing techniques.

Figure 3 Optical microscope images of the 6 GHz QWR center pin. (Left upper) As-machined Nb cavity (Left lower) Machined and etched Nb cavity (Center upper) EB-PBF 3D printed niobium *Cavity 1* (Center lower) EB-PBF Nb *Cavity 2* (Right upper) L-PBF 3D printed Ti64 *Cavity 3* with heavy material removal via abrasive finishing (Right lower) L-PBF 3D printed Ti64 *Cavity 4*

The accelerator grade machined and etched niobium center pins, shown in the first column of Figure 3, exhibit clear machining marks and grain definition in the etched part. Also shown are the centrepins of two niobium EB-PBF printed parts, as wells as the equivalent parts printed in Ti64 using the L-PBF process. From these images, we immediately note that the surface finish of the laser-printed Ti64 parts is superior to the electron beam equivalent, a direct result of smaller spot size and powder size. We also note that the abrasively finished Ti64 *Cavity 3* has a macroscopically smoother finish that the as-printed equivalent, as the finishing process effectively removes adhered powder particles. The differences in niobium *Cavity 1* and *Cavity 2* surface finish is due to the differences in abrasive polishing times applied to the parts, with *Cavity 1* having longer material removal times.

Table 1: Room temperature RF measurements

			Q-factor		
#	Material	Fabrication Method	Expected	Measured	f_0[MHz]
	Copper	Machined	3492	3312	6027.7
	Niobium	Machined, Etched (-89μm)	1190	1069	6045.5
1	Niobium	EB-PBF, abrasive polish and etched (-10um)	1190	425	5991.8
2	Niobium	EB-PBF, abrasive polish and etched (-10um)	1190	487	6113.2
3	Ti64	L-PBF, HIP, tumbled, etched (-10um)	622	345	5962.6
4	Ti64	L-PBF, HIP, etched (-10um)	622	296	5894.4

4.1 Q-factor measurements

Prior to cryogenic testing, the room temperature Q-factors of the cavities were measured, with a summary comparing machined Nb and Cu cavities against the performance of the AM Ti64 and Nb cavities, Table 1.

The room temperature electrical conductivity of each cavity was measured using a 480kHz eddy current probe (Fischer Technology SigmaScope SMP10). From the center frequency and conductivity measurements, we calculate the equivalent room temperature RF surface resistance value (R_s) and assuming that G= 71Ω, estimate the idealized, smooth surface Q factor. While the resonant frequency data shows that the overall cavity shape is accurately rendered by the AM processes, printing and post-processing requires significant improvement.

Following room temperature RF measurements, the cavities were assembled in a magnetically shielded and thermally engineered RF test stand, then fastened to the base of a dilution fridge (Blue Fors BF-LD400) at the University of Chicago. RF tests were performed at a mixing chamber temperature of 20mK, as measured with a calibrated RuO_2 sensor. These tests did not have on-cavity thermometry, therefore precise determination of the superconducting transition temperature was not possible. Subsequent tests with instrumented cavities showed a temperature differential between the mixing chamber and the cavity on the order of 40mK while at base temperature. A schematic of the 4-cavity RF test assembly, as well as an image of the assembly fastened to the dilution fridge, is shown in Figure 4 below.

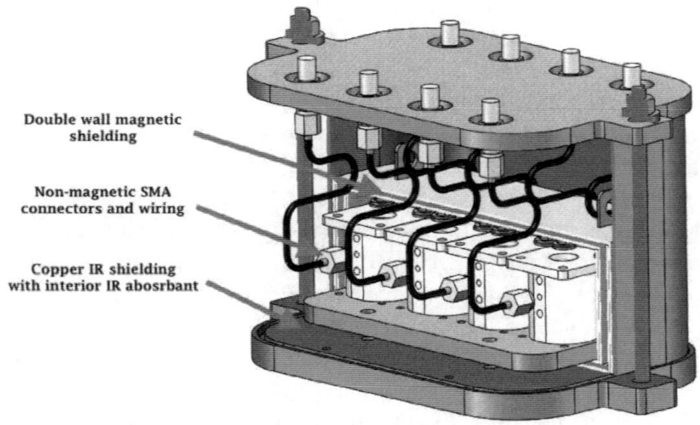

Figure 4 (Left) CAD model of 4-cavity dilution fridge RF test fixture; (Right) Fixture installed on the University of Chicago's dilution refrigerator.

The cryogenic measurements were performed using an RF amplification chain detailed elsewhere[15], which includes -20dB attenuation at 4K and -6dB attenuation at mixing chamber. The complete [S] parameter measurements for all cavities was performed using a vector network analyser with an output power range of -20 to -90dBm. We extracted the loaded quality factors as well as the coupling quality factors under the assumption of equal splitting, as the multiplexed measurement does not allow self-deterministic extraction of the coupling factors [18]. The corresponding resonance frequency, as well as the calculated unloaded quality factor, are summarized in Table 2.

Table 2: Dilution fridge superconducting RF measurements

			Q_0	f_0[MHz]
	Niobium	Machined, Etched (-89µm)	9.2×10^6	6192.9
1	Niobium	EB-PBF, abrasive polish and etched (-10um)	1.6×10^6	6008.7
2	Niobium	EB-PBF, abrasive polish and etched (-10um)	1.2×10^6	6130.4

3	Ti64	L-PBF, HIP, tumbled, etched (-10um)	0.4 x 10^6	5980.5
4	Ti64	L-PBF, HIP, etched (-10um)	2.2 x 10^6	5917.1

In addition to the network analyser measurements, we also performed time-domain ring-down measurements, whereby the cavity is excited at resonance, the drive signal is turned off and the ring-down decay curve is capture on an oscilloscope, providing the loaded Q factor. Our measurements showed good agreement between both time domain and frequency domain measurements. Also, we did not observe meaningful loaded quality factor power dependence over the range of measurements.

5. Analysis

The quality factor of a resonant cavity represents a global measurement of the cavity losses, weighted by the field distribution at specific surfaces. High field regions, such as the center pin, will have an outsize effect on the performance of the cavity, with multiple RF loss channels contributing in parallel. This concept extends to the intrinsic surface anisotropy associated with additive manufacturing, with upward facing surfaces having the best finish and downward facing surfaces the worst. Post-processing is crucial to normalizing the finish and reducing the surface resistance of the entire 3D geometry.

Besides the G factor, normal conducting quality factors are dependent on the conductivity of the bulk material, as well as roughness-enhanced losses. In the case of the precision machined copper and niobium QWRs, we see that measured Q is within 95% and 90% of the calcualted value, respectively. Compared to the RF skin depth, the machined surfaces are smooth with minimal surface area enhancement. Furthermore, the center frequencies are measured to within 1% of the design frequency (6000MHz), with deviations attributable to machining tolerances and etching anisotropy.

The larger center frequency spread of the AM cavities is attributable to the geometric accuracy of the printing process as well as the variability of the surface finishing. In terms of Q, the Nb AM cavities are 35% and 40% of the idealized value. Similarly, the abrasively finished Ti64 AM cavity had an Q of 55% of ideal, while the etched cavity was 47%. Although the absolute Q of the Nb AM cavities is higher than the Ti AM cavities, the comparative value of Nb relative to ideal is lower. This result is to be expected considering the Nb QWR were printed using the EB-PBF process, which is based on coarser powder, thicker layers and larger beam diameters.

In addition to surface roughness, superconducting resonators are sensitive other loss channels, including magnetic flux trapping, two-level losses, contamination, hydride formation, etc. While it is difficult to assign an absolute magnitude to each of these losses without more detailed measurements, the AM Nb cavities had a Q of roughly 15% of the machined part, with the degradation likely a function of both reduced material purity increased surface roughness. Furthermore, the etched Ti64 *Cavity 4* had a higher Q than both AM-Nb cavities. This inversion is surprising, considering the lower superconducting transition temperature of Ti64. Furthermore, it is surprising to note that the Ti64 *Cavity 3*, which underwent heavy abrasive finishing, had the worst performance of all. Our current hypothesis is that these increased losses are attributed to lossy ceramic media embedded in the RF surface. Further destructive analysis of the cavity is required to confirm this result.

While normal conducting measurements provide a rapid method for characterizing the surface finish and geometric accuracy of the printed parts, our results demonstrate that these results cannot be translated to superconducting performance. In fact, the best room temperature AM cavity had the worst superconducting properties; demonstrating that material, printing and post-processing have to all be considered for optimization.

6. Conclusions and future plans

This report represents preliminary RF measurements of additively manufactured quarter wave resonators at millikelvin tempertures. Although both Nb and Ti64 QWRS exhibited Q-factors significantly lower than conventionally machined parts, these initial results are encouraging and suggest the need for further process development. Specifically, AM-specific RF designs, high vacuum heat treatment and electropolishing are all currently under development. Additional thermometry and surface analysis is planned for the future.

These initial results highlight the opportunity for fabricating unmachinable RF geometries with additive manufacturing. For quantum information systems and detectors, seamless arrays of densely packed 3D cavities with customized resonances can be rapidly designed, printed and tested, with minimal engineering and manufacturing effort. Overall, 3D-printing of superconductors has immense potential for future research, and may produce new applications which will be of great benefit in physics and engineering.

7. References

[1] R Raz, A. *Electronic Colloquium on Computational Complexity*, Report No. 107 (2018)
[2] B. D. Josephson, *Phys. Lett.* 1 **(7):** 251–253, 1962
[3] H. Dieter Zeh, *Foundation of Physics*, **vol.** 1, pp. 69-76, (1970).
[4] A Chailloux, *Intl Conf on the and Appl of Cryptology and Information Security*. Springer, Cham, 2017.
[5] P. J. Leek *et al.,*, *Phys Rev Let* **104.10**, 100504 (2010), 0911.4951
[6] M. A. Sillanpaa *et al.*, *Nature* **449**, 438–442, 2007.
[7] J. Gambetta *et al.*, *Phys. Rev. A* **74**, 042318, 2006
[8] A.A. Houck *et al.*, *Phys. Rev. Lett.* **101**, 080502, 2008.
[9] S. Kuhr *et al.*, *Appl. Phys. Lett.* **90**, 164101 (2007).
[10] H Padamsee, *RF superconductivity: science, technology, and applications (2009).* Wiley
[11] E Holland, *et al.*, *Appl. Phys. Lett* **111.20**, 2017
[12] D Creedon *et al.*, *Appl. Phys. Lett* **109.3** (2016)
[13] B McAllister *et al.*, *IEEE Trans on Inst and Meas* **70** (2020): 1-7.
[14] P Frigola *et al Proceedings of SRF2015*, THPB042, Whistler, Canada
[15] S Kutsaev *et al.*, *EPJ Quantum Tech*, **7:7**, 2020.
[16] J Sharping *et al. Jour App Phys* **128.7** (2020): 073906.
[17] C Ledford, *et al. Applied Sciences* **9.19** (2019): 3993.
[18] M Anthony, *et al. Applied Physics Letters* 100.11 (2012): 113510.

Acknowledgements

This work was supported by the U.S. Department of Energy, Office of High Energy Physics, under SBIR grants DE- SC0018753 and DE-SC0019973. The authors would like to thank Dr. Zachary Conway (BNL), Mr. Thomas Reid (ANL), Dr. Alexander Romanenko and Dr. Timergali Khabibouline (FNAL) for their valuable advice.

Cryogenic material properties of additive manufactured 316L stainless steel

K-P Weiss[1], S Hetzler[1], T Kvackaj[2], R Bidulsky[3], M Actis Grande[3] and D Manfredi[4]

[1] Karlsruhe Institute of Technology, Institute for Technical Physics, Hermann von Helmholtz Platz 1, 76344 Eggenstein-Leopoldshafen, Germany

[2] Technical University of Kosice, Faculty of Materials, Metallurgy and Recycling, Department of Plastic Deformation and Simulation Processes, Vysokoskolska 4, 04200, Kosice, Slovakia

[3] Politecnico di Torino, Dept. Applied Science and Technology, Viale T. Michel 5, 15121 Alessandria, Italy

[4] Politecnico di Torino, Dept. Applied Science and Technology, Corso Duca degli Abruzzi 24, 10129 Torino, Italy

E-mail: klaus.weiss@kit.edu

Abstract. Additive manufacturing is recognized as a potential technology to design and create complex geometries as well as a fast track to build prototype components. Different materials are possible to use, depending on the specific requirements of an application. Superconducting applications like magnets or rotating machines are demanding for the structural components. Either high and/or cyclic mechanical loads can be one of the limiting factors in design. In the cryogenic temperature regime austenitic steels are used due to the mechanical performance and the machinability. In this work 316L austenitic steel samples were produced using laser powder bed fusion, also known as Selective Laser Melting (SLM). Followed by different heat treatments to systematically influence to the microstructure evolution. Beside mechanical properties also thermal properties as heat capacity or expansion are investigated. Having the cryogenic application in mind the tests are conducted at room temperature, liquid Nitrogen and liquid Helium temperature. The measured mechanical and thermal properties are compared to industrial cast austenitic steels to qualify the overall performance of the additive manufactured samples.

1. Introduction

While additive printing of polymer or metallic components advances and is already common for some applications the interest to facilitate this technology for cryogenic components is growing [1]. Especially for superconducting magnets as structural former with complex geometries [2] or rotating machines [3] are demanding applications for the structural components. Therefore, thermal and mechanical loads are of high interest. In this work the material of choice is 316L used widely for cryogenic applications, produced by laser powder bed fusion, also known as Selective Laser Melting (SLM).

2. Material and methods

The used powder was obtained commercially (EOS GmbH, Germany) to be close to the chemical composition of bulk material obtained from classical production route like casting and forging (COMTES, Czech). The chemical composition of the powder material was investigated with a Bruker Q4 TASMAN optical emission spectrometer. The results are given in Table 1 together with material coming from classical casting and forging. The composition is in the range of the definition of grade 316L [4]. Further, the powder was examined for example regarding shape, mean size to be used for the SLM process. The powder is of good quality in spherical geometry and a mean diameter of about 32 μm shown in Figure 1.

The fabrication of the specimen were performed using a EOSINT M270, some details of the printing device are summarized in Table 2. The specimen were produced by SLM close to the final geometry to be machined for further experiment.

As the focus is on cryogenic material properties, the aim was to reduce the number of tests to avoid a high effort of costly cryogenic tests. Therefore, it was decided to focus on the z-direction, namely the building direction, as the loading in direction of the layered structure was identified as one of the most crucial. In Figure 2 the printing orientation together with the cylinders for the tensile or fatigue test after removing the not molten powder from the production table are shown.

For the mechanical tests the necessary specimen were machined from cylinders or plates according to the given standards for tensile [5, 6], fatigue [7], fatigue crack growth rate [8], or fracture toughness [9].

All tests were conducted within the cryogenic material test laboratory (CryoMaK) at the Karlsruhe Institute of Technology [10].

Table 1. Chemical composition of powder and cast material in wt.%

316L	C	Si	Mn	Cr	Mo	Ni	P	S	Fe
powder	0.009	0.32	1.68	18.20	2.79	13.98	0.015	<0.005	Bal.
cast & forging	0.068	0.65	2.02	18.32	1.92	14.12	0.009	0.004	Bal.

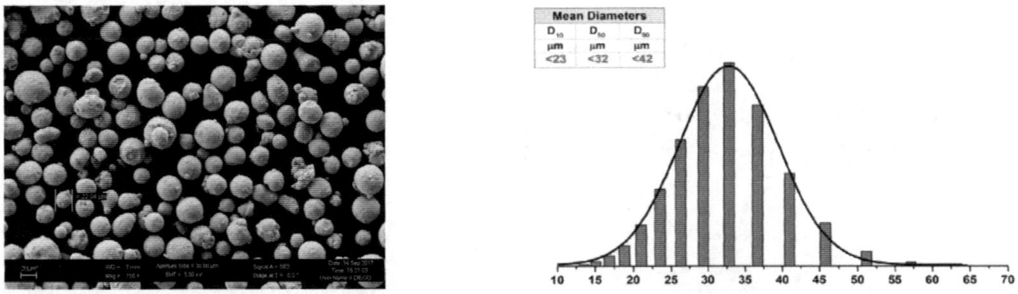

Figure 1. Details of used powder. Left: Electron microscope showing the spherical shape; right: distribution of the diameter.

Table 2. Technical data EOSINT M270 Dual Mode

layer thickness (material-dependent)	20 - 40 µm
laser type	Yb-fibre laser, 200 W
scan speed	50 - 5000 mm/s
laser spot size	100 µm

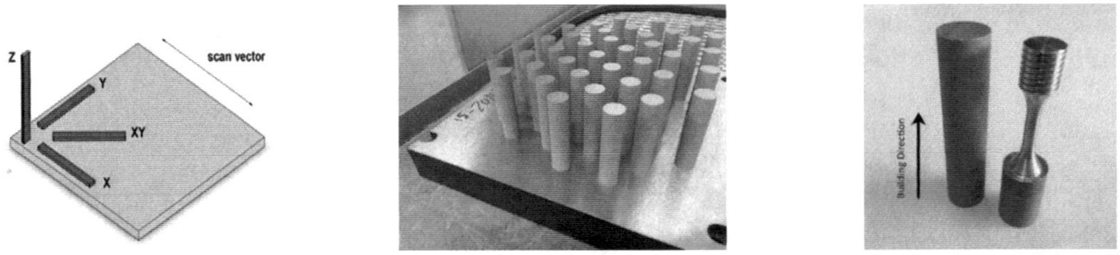

Figure 2. Details of specimen printing. Left: printing directions, focus on z-direction; mid: cylinders after printing; right: example of fatigue specimen machined from cylinder.

To examine the effect of heat treatment four different states were defined, called HT-0, HT-1, HT-2 and HT-3. The separate HT-4 is for the parallel produced 316L bulk material for comparison to influence the microstructure. In Figure 4 the systematic is depicted:

- The logic is to have first the initial state directly coming from the 3D printing without any heat treatment.
- Second, with HT-1 a moderate stress release treatment at 400°C for 3h is performed.
- Further, a full two stage homogenization heat treatment is done. Stage one at 1050°C for 1h followed by stage two at 700°C for 50h.
- Last, a one-step heat treatment at 1050°C for 1h is done, same as for the bulk material.

The microstructure was investigated after heat treatment. The microstructure looking on the printed layers (first row) as well as the lateral view (second and third row) is given in Figure 5.

For HT0 and HT1 the solidification tracks that are formed according to the laser pattern during printing can still be seen in the top view. Moreover, the melt pool boundaries are clearly visible as seen in the pictures from lateral sections while they are completely erased for HT2 and HT3 due to the high temperature impact during the heat treatment. All specimens show a typical elongation of grains along the building direction which is a typical feature of SLM-specimens.

HT0 and HT1 are more or less comparable. This leads to the conclusion that the moderate stress release heat treatment at 400°C does not have a major impact on this level. However, there is a significant difference looking at HT2 and HT3. After this high temperature heat treatment used in both cases of about 1050°C the typical SLM production patterns are lost. There seems to be no major difference after the single high temperature treatment or the two stage.

Moreover, the material properties itself were investigated for the as-built condition (HT0) in terms of different porosity measurements (Archimedes, image analysis with an optical microscope and computer tomography) and surface roughness measurements. Pores are a reported source for

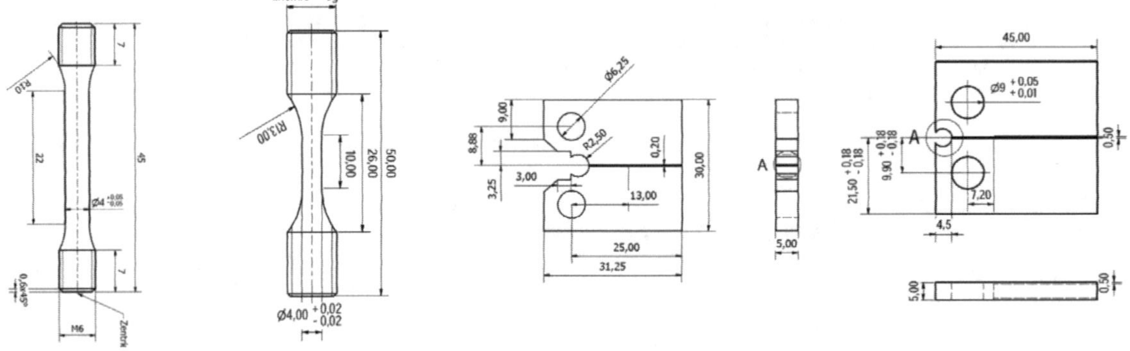

Figure 3. Specimen dimension for tensile, fatigue, fracture toughness, fatigue crack growth rate (from left to right).

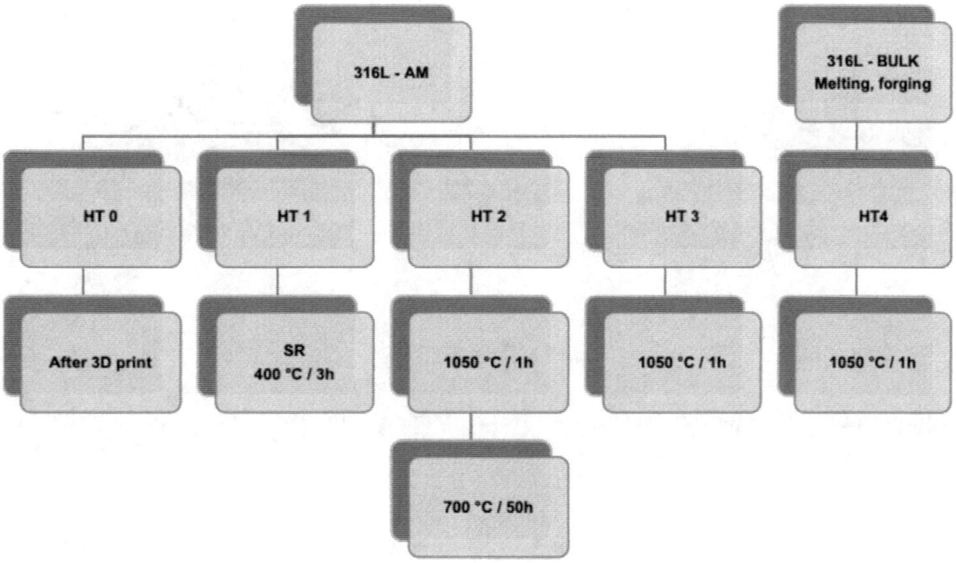

Figure 4. Sspecimen heat treatment systematics for further examination.

premature failure of SLM parts under cyclic loads because they act as localized high stress concentration and crack initiation. Porosity measurements were performed to verify the specification from EOS GmbH (producer of powder and printer) of a porosity below 1%. With a value of 0.5% for the Archimedes measurement and 0.03% for the image analysis, the specification was confirmed.

For the surface roughness measurement, a confocal microscope from NanoFocus was used and the evaluation was executed with a line roughness measurement [11].

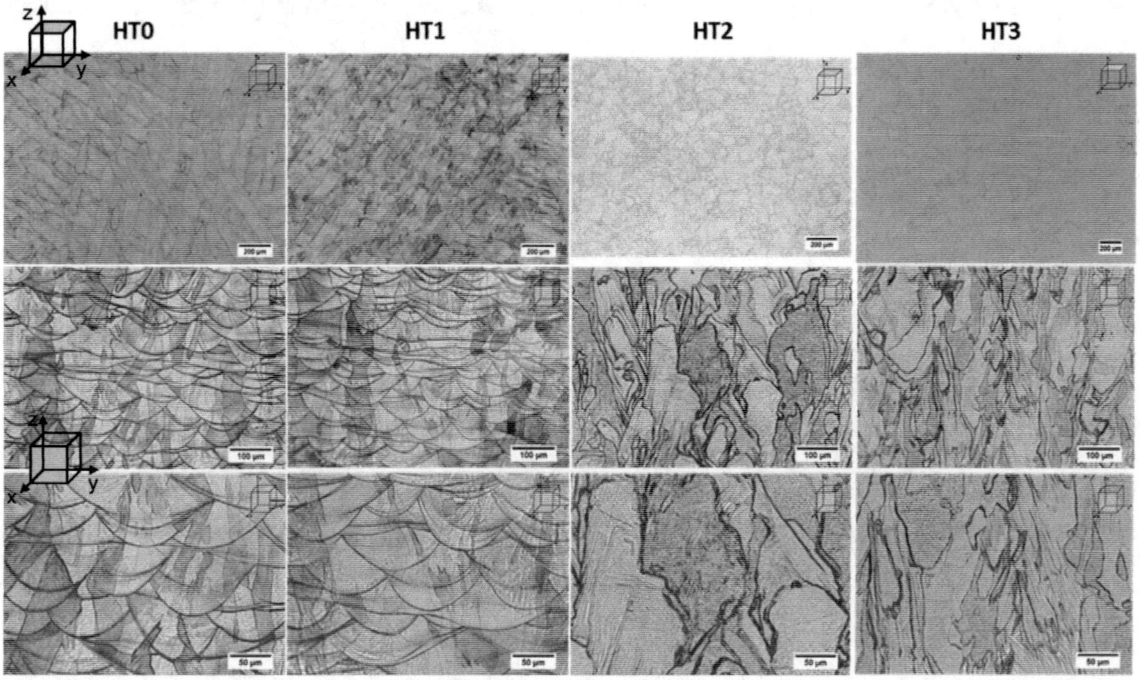

Figure 5. Microstructure overview of HT0, HT1, HT2 and HT3. The observed orientation is indicated by the yellow plane of the inset.

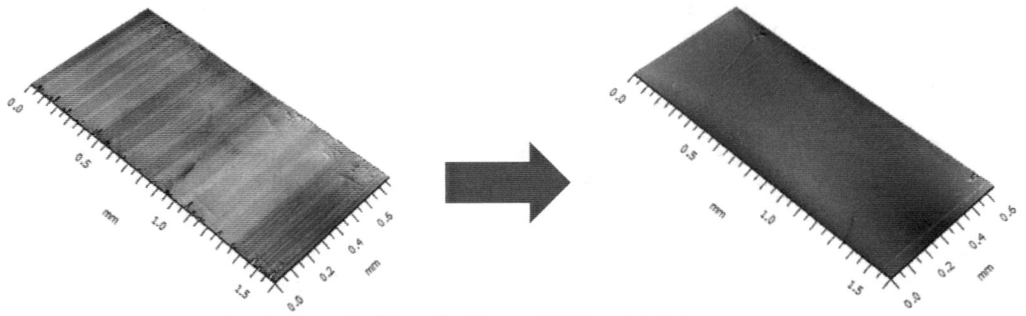

Figure 6. Surface roughness before (left) and after polishing (right).

To avoid premature failure due to surface effects, the specimen were polished to reduce the surface roughness. The roughness was reduced by a factor of approx. ten, reducing the mean maximum peak to valley from 6.5 μm down to 0.7 μm. In Figure 6 a control surface of about 1.6 mm x 0.5 mm is shown, depicting the effect of polishing to obtain a smooth surface.

As well as porosity, rough surfaces are often assumed to act as stress raisers and therefore initiate a crack that leads to failure [12]. The surface of the machined specimen had a mean roughness depth of Rz = 6.53 μm with a mean standard deviation of 1.3 μm. By polishing it was possible to decrease the surface roughness to a mean roughness depth of Rz = 0.7 μm with a mean standard deviation of 0.4 μm.

3. Experimental results

3.1. Thermal properties

In Figure 7 the thermal expansion in z-direction as well as the heat capacity of the printed material is shown measured for all different SLM after heat treatment, measured from about 300K down to 4K [10].

As can be seen the heat treatment has no visible impact on the thermal properties. All specimen results collapse on the same measurement curve. Compared to the standard bulk 316L material [13] there seems to be a small difference in the thermal expansion of about 6%. Looking at the heat capacity the curves are almost identical to the bulk specimen.

Altogether the results indicate, that the production process like casting, SLM or heat treatment is not a dominant factor.

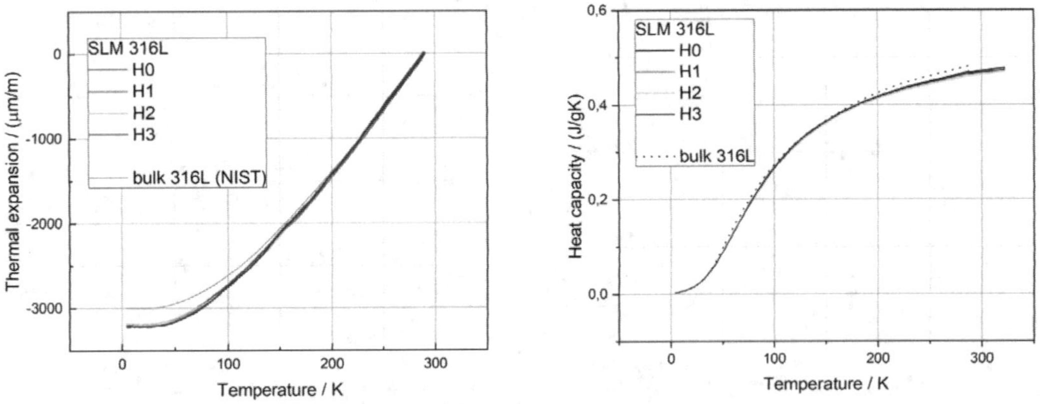

Figure 7. Thermal properties of all SLM material after heat treatment compared to standard 316L for thermal expansion in z-direction (left) and heat capacity (right).

3.2. Tensile tests

Tensile tests have been carried out at ambient 298 K and cryogenic temperature of 77 K and 8 K for all material from HT0 to HT4. The cryogenic test at 77 K was performed in liquid nitrogen. For the 8 K tests an evaporating gas cryostat was used cooled with liquid Helium [10].

The stress-strain curves of all tensile tests are pictured in Figure 3. In addition, a comparison between all heat treatments and traditional manufactured specimens in terms of typical mechanical properties is given in Figure 8. For easier comparison in Figure 9 bar graphs of yield strength, ultimate tensile strength and total elongation are shown.

Looking at the properties of the SLM-built specimens at room temperature, a big decrease in yield strength between HT0, HT1 and HT2, HT3 can be seen. However, HT2 and HT3 exhibit higher strain hardening resulting in a similar ultimate tensile strength than HT0 and HT1. The reason for the poor strain hardening of HT0 and HT1 is still under investigation. Reports from [3], [14], [15] show a similar behavior. Moreover, HT2 and HT3 exhibit a tremendous increase in uniform elongation that is even bigger than for the traditionally manufactured material.

For HT0 and HT1 the mechanical behavior changes significantly at 77 K. Both heat treatments now exhibit strain hardening and the uniform elongation is also in the range of HT2 and HT3 as well as of the traditionally manufactured material.

In summary the mechanical properties of HT2 and HT3 are in the same range as the traditional manufactured material making the SLM-process with an adequate post heat treatment a considerable option for applications under static loads. Disregarding the ductility at room temperature and 8 K, even the yield and ultimate tensile strength of the specimens with no post heat treatment (HT0) are in a similar range to the traditional manufactured material. The typical serrations developing below 40 K are present for all cases.

A detailed discussion of the tensile test results and the microstructure was given in [16].

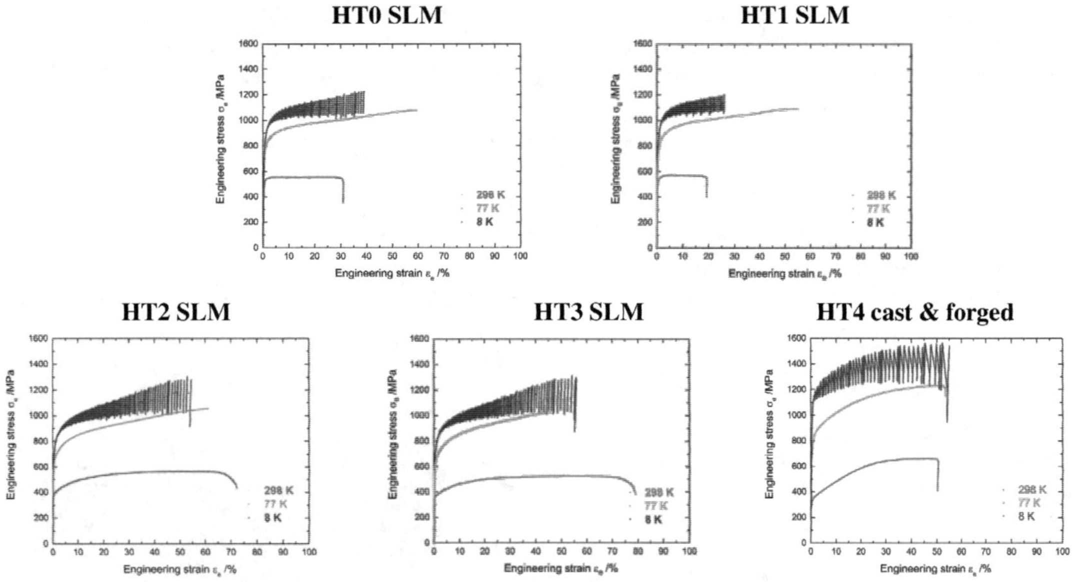

Figure 8. Stress strain curves for all materials HT0 to HT4 measured at room temperature, 77 K and 8 K.

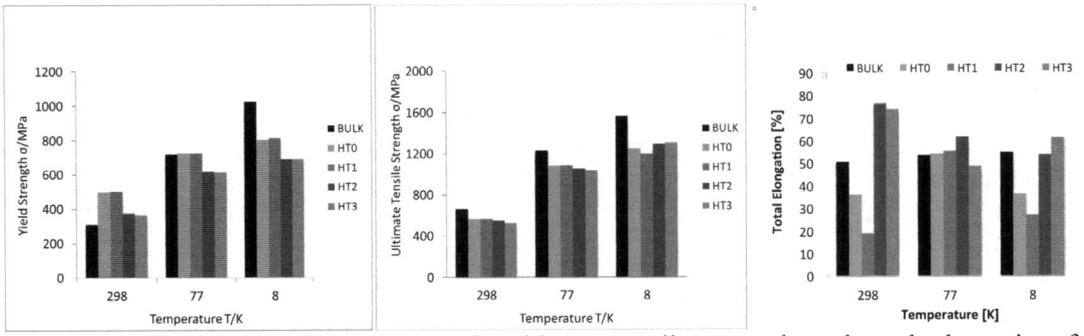

Figure 9. Comparison of yield strength, ultimate tensile strength and total elongation for all specimen.

3.3. Fracture toughness

Further mechanical assessment is done by fracture toughness measurements. Using the J-integral to obtain J_q according to [9] K_{IC} was calculated. To relate the results to the tensile tests the relation between yield strength and fracture toughness K_{IC} is shown in Figure 10 together with the fracture surface of the specimen after test. They show no peculiarities in their appearance. At room temperature high ductility is visible at lower temperatures the typical crack tunnelling is present.

In general, the fracture toughness drops with the test temperature. Having about 300 MPa m^0.5 for the RT tests, between 200 to 240 MPa m^0.5 at 77 K and at 8 K the minimum value is about 160 MPa m^0.5. The results are in line with bulk material measurements [17] considering the error bar.

3.4. Fatigue crack growth rate

Going from quasi static to dynamic behavior under alternating load the fatigue crack growth rate was measured at room temperature and cryogenic temperature at 8 K was measured. The tests were performed under a load ration of 0.1 with constant load. With advancing crack, the differential stress intensity ΔK increased. In Figure 11 the results of all materials are summarized.

The overall crack advance of the SLM specimen da/dn is slightly steeper than the bulk material [17]. However, the slope for all SLM specimen HT0 to HT3 is more or less the same.

Remarkably, at room temperature for the material having the high temperature heat treatment and low yield stress HT2 and the stress release HT3 the critical ΔK_C is reached earlier while the threshold ΔK_{th} seems to be the same for all SLM specimen.

Altogether, it can be concluded that the effect of the SLM production results in a more sensitive fatigue crack growth behavior regardless of heat treatment.

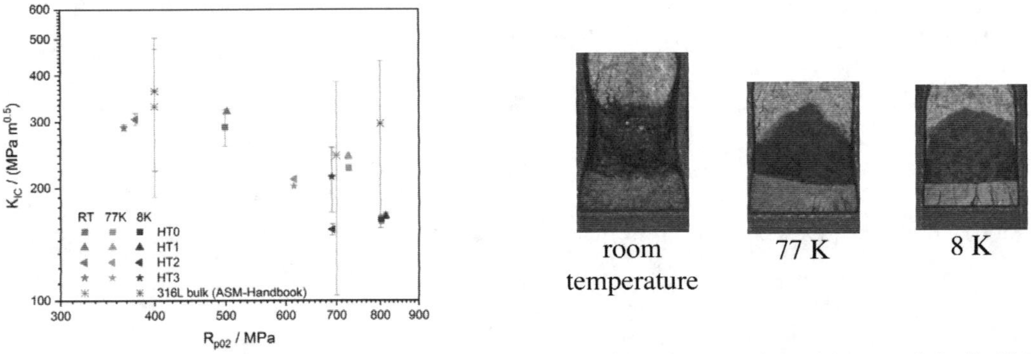

Figure 10. Fracture toughness results versus yield strength for all materials SLM and bulk [17] measured (left). On the right typical crack surface after specimen opening at room temperature, 77 K, and 8 K.

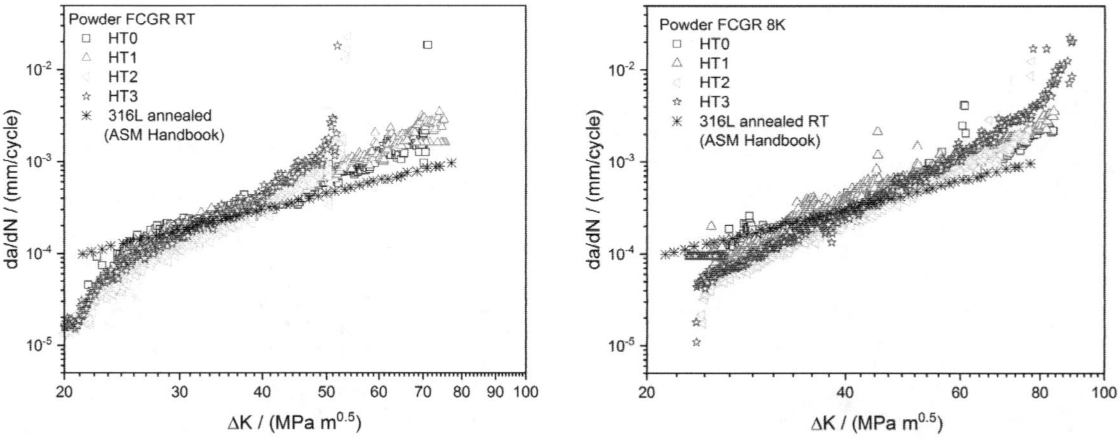

Figure 11. Fatigue crack growth rate measured at room temperature (left) and at 8 K (right) for all materials SLM and bulk material [17].

3.5. *High cycle fatigue test*

Fatigue tests were also performed for all heat treatments. Specimens were loaded with sinusoidal stresses at frequencies between 20 and 30 Hz and a load ratio of $R = 0.1$ and the reached number of cycles N_f at each maximum stress level are given in Figure 12.

At the top left the high cycle tests at room temperature for the SLM HT0 directly from 3D printing is seen. The maximum stress level was chosen incrementally considering yield and ultimate tensile strength from the tensile tests. However, decreasing the maximum stress level below the yield strength no clear evidence of a fatigue life limit could be deduced. All specimen failed. On the contrary, for given maximum stress levels premature failure of specimen occurred. Especially at 420 MPa one specimen reached a cycling number well above 10 million cycles, while another specimen at the same condition failed already shortly after 300 thousand cycles.

The results of the fatigue tests obtained for HT0 at 77 K exhibit more or less the same arbitrary failure behavior. Especially at about 760 to 770 MPa maximum stress level, there were six specimen tested. Remarkably two of them survived 1 million cycles while one specimen failed before 400 thousand cycles and three below 100 thousand cycles. However, for the 77 K test further specimen tested at 760 MPa and below point to a fatigue limit of about 760 MPa.

The described behavior is visible for all measured SLM specimen from HT0 to HT3. Especially for HT2 at 77 K four specimen tested at the same maximum stress level resulted in a surviving specimen but also three premature failures between 100 thousand and 700 thousand cycles.

The calculated uncertainty does not explain this big scatter of results. This points to the conclusion that for the SLM produced specimen the stress level is not the dominating parameter and the fracture surfaces were investigated.

Tests at 8 K were not performed due to the high amount of specimen needed to perform the RT and 77K fatigue tests and, without clarification of the seen effect at RT and 77 K it was decided to avoid the costly 8 K tests at this stage.

3.6. *HCF fracture surface analysis*

In Figure 13 the crack initiation spots of different specimens are pictured. In picture (a) to (c) two specimens are shown, that were tested at the same maximum stress level. While specimen (a) (HT0-L2) reached about 100 thousand cycles before failure, specimen (b)/(c) (HT0-N2) showing both sides of the fracture surface, only reached 30 thousand cycles. In specimen HT0-L2 the striations that are caused by cyclic loading are visible and the crack initiation seems to be induced due to a surface scratch or roughness. In comparison to that, specimen HT0-N2 (b)/(c) shows a macroscopic defect that causes the early failure.

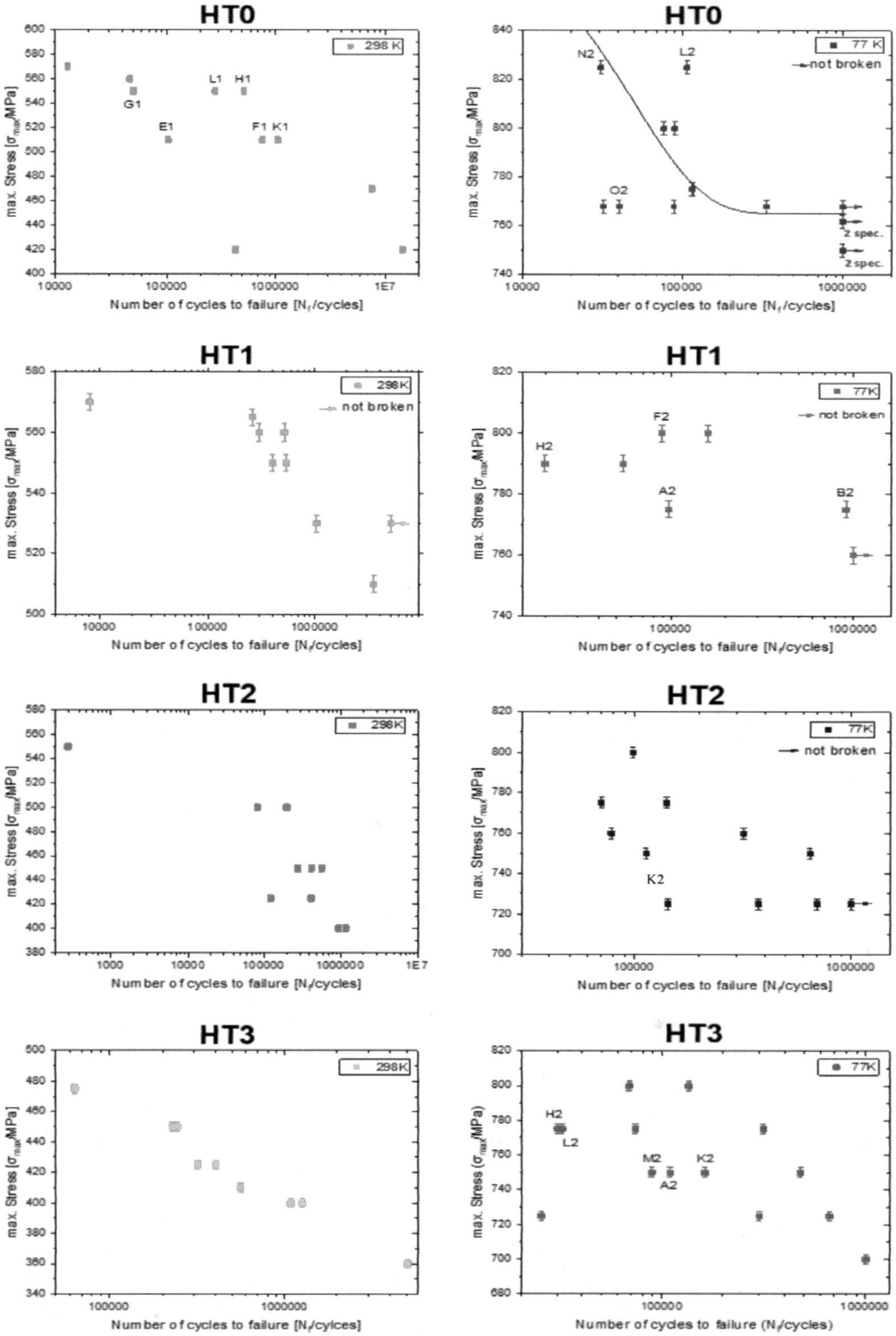

Figure 12. High cycle S-N_f results for all materials HT0, HT1, HT2 and HT3 measured at room temperature (left) and 77 K (right). Named datapoints indicate specimen used for fracture surface investigation.

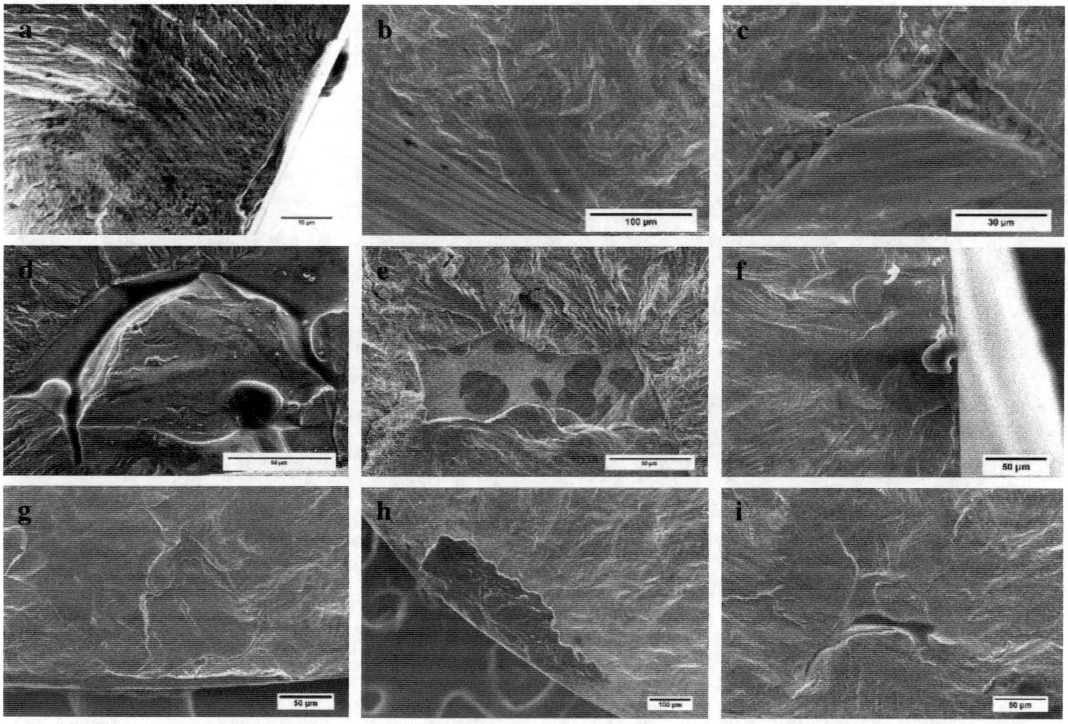

Figure 13. Fracture surfaces of regular (a) and premature failure (b-i). (a) HT0-L2, (b) HT0-N2a, (c) HT0-N2b, (d) HT0-O2, (e) HT0-G1, (f) HT1-F2, (g) HT1-H2a, (h) HT1-H2b, (i) HT2-K2

In picture (b) and (c) small areas with an untypical smooth surface that exhibit no fatigue striations or similar are visible. As expected, similar structures can be seen for specimen that failed after a rather short number of cycles. Such areas are not only found at the direct crack initiation spots but also throughout other areas of some specimens.

These areas appear to come from a lack of fusion between two layers. This could be caused by a localized too low heat input during the SLM-process. While the current layer of powder was molten melt, presumably the heat input was not sufficient to fuse the underlaying layer of solidified material. Such flaws led to a high stress concentration causing early crack initiation and premature failure.

4. Summary and outlook

Although the benefit of additive manufactured 316L is given, great care has to be taken for application. The results coming from the different heat treatment show that the thermal properties are well in line with bulk material from casting and forging. Mechanical tests like tensile, fracture toughness and fatigue crack growth rate can be understood in the light of heat treatment and the influence of the microstructure.

However, high cycle fatigue tests reveal the sensitivity to the quality of the production process. Insufficient fusion of the powder layers lead in this present work to premature failure. This needs to be considered for application. Further possible steps to improve the quality of the material have to be taken into account, like optimization of the SLM process together with a possible quality control (e.g. by CT-scans or ultrasonic testing) to detect such flaws within the material.

5. References

[1] Stautner W, Vanapalli S, Weiss K-P, Chen R, Amm K, Budesheim E, Ricci J, IOP Conference Series: Materials Science and Engineering Volume 278, Issue 1, (2017) 012134

[2] Kirby G et al. IEEE TRANS.APPL.SUPERCOND., VOL. 27, NO. 4, JUNE 2017
[3] Kirchheim A, Zumofen L, Gehrig M "Additive Fertigung mit (316L): Prozess- und Characterisierungsergebnisse" (2014)
[4] Steels A, "Stainless Steel Grade Datasheets," Atlas Steels Tech. Dep., pp. 1–2, 2013.
[5] ASTM E8/E8M-16 "Standard Test Methods for Tension Testing of Metallic Materials"
[6] ASTM E1450-16 "Standard Test Method for Tension Testing of Structural Alloys in Liquid Helium"
[7] ASTM E466-15 "Standard Practice for Conducting Force Controlled Constant Amplitude Axial Fatigue Tests of Metallic Materials"
[8] ASTM E647-15e1 "Standard Test Method for Measurement of Fatigue Crack Growth Rates"
[9] ASTM E1820-20 "Standard Test Method for Measurement of Fracture Toughness"
[10] Sas J, Weiss K-P, Bagrets N, Acta Metallurgica Slovaca 21 (4) (2015), pp. 330-338
[11] DIN EN ISO 4287:2010-07 and DIN EN ISO 4288:1998-04 "Geometrical Product Specifications (GPS) - Surface texture: Profile method"
[12] William D G R, Callister D, "Materialwissenschaften und Werkstofftechnik" 8th ed. Weinheim: Wiley-VCH Verlag GmbH & Co. KGaA, 2013.
[13] Cryogenic material properties: https://trc.nist.gov/cryogenics/materials/materialproperties.htm (2021)
[14] Spierings A B, Starr T L, Wegener K, Rapid Prototyp. J., vol. 19, no. 2, pp. 88–94, 2013.
[15] Lackey A D, "Comparative Study of Mechanical Properties of 316L Stainless Steel Between Traditional Production Methods and Selective Laser Melting",Western Carolina University. ProQuest Dissertations Publishing, 2016. 10103725
[16] Bidulskỳ R, Bidulská J, Gobber F S, Kvačkaj T, Petroušek P, Actis-Grande M, Weiss K-P, Manfredi D, Materials, 2020, 13(15), 3328
[17] ASM Handbook, Volume 19, "Fatigue and Fracture", ISBN: 978-0-87170-385-9 (1996)

IOP Conf. Series: Materials Science and Engineering 1241 (2022) 012048 doi:10.1088/1757-899X/1241/1/012048

Dynamics and Vibrations of Rotors of Electrical Drives with Additively Manufactured Materials at Cryogenic Temperatures

F Steegers[1], C Weidermann[2], M Klöpzig[2] and T Sattel[1]

1 Mechatronics Group, Department of Mechanical Engineering, PF100565, 98684 Ilmenau, Germany
2 Siemens AG, Günther-Scharowky-Str. 1, 91058 Erlangen, Germany

finn@steegers.org

Abstract. The use of additive manufacturing and cryogenic cooling enables engineers to build electric machines with higher power densities than previously seen. Both the new manufacturing technologies and very low temperatures have a significant impact on the mechanical properties of materials. In the present work, a rotordynamics analysis of a permanent magnet synchronous motor concept from the AdHyBau project, funded by the German Federal Ministry for Economic Affairs and Energy, is conducted with an emphasis on the influence of additive manufacturing and cryogenic temperatures on vibrations and natural frequencies.

1. Introduction

The development of lightweight and efficient electric drives is one of the big challenges of the modern transport sector. In the *AdHyBau* project, novel methods of additive manufacturing and fiber-composite-metal hybrid materials are being developed. Together with cryogenic cooling, these new methods will allow for a significantly higher power density in electric drives compared to conventional designs. Lightweight design and cryogenic operating temperatures pose a number of challenges during the design process. Regarding the lightweight mechanical structure, not only does it have to withstand static loads, but it must also exhibit good dynamic behaviour under the influence of several excitations. In the present work, electromagnetic forces on the rotor of a permanent magnet synchronous motor (PMSM) are being investigated as the primary sources of nonsynchronous vibrations in the rotor of an electric drive. To characterize these forces, a finite element model of the investigated machine is created in the simulation environment *COMSOL Multiphysics*. The Maxwell Stress Tensor is used to derive the radial force on the rotor from the magnetic flux density. Then, a rotordynamics model is set up, into which the magnetic forces can be introduced in a simplified form. First, the rotordynamics analysis investigates the position of natural frequencies and the associated mode shapes. Next, the magnetic excitations are considered in the model and resulting deflections are calculated and presented in response plots. Furthermore, the impact of different parameters on natural frequencies and deflections is examined. Those are the operating temperature, a variation of the mechanical structure and the shaft length. This will allow engineers to estimate a range of parameters within which the machine design can vary. Lastly, torsional vibrations of the rotor during run-up from rest are investigated, again considering different operating temperatures as well as different rise times to design the entire system in which the machine operates accordingly.

2. Theoretical Foundations

2.1. Magnetic Forces on the Rotor of the Electric Machine

Corresponding with a cylindrical coordinate system, the magnetic forces on the rotor may be divided into radial, tangential and axial forces. The axial component can be neglected under the assumption of perfect symmetry in that direction. In the present work, the radial forces are being calculated from the magnetic flux density B using the Maxwell Stress Tensor, which is defined as [1]

$$\sigma_{ij} = \epsilon_0 E_i E_j + \frac{B_i B_j}{\mu_0} - \frac{1}{2}(\epsilon_0 E^2 + \frac{B^2}{\mu_0})\delta_{ij} \tag{1}$$

with the magnetic flux density B and the electric field E. In the case at hand, the electric field may be neglected. Further reducing the tensor to the radial component of B in the airgap yields

$$\sigma_r = -\frac{B_r^2}{2\mu_o} \tag{2}$$

Fluctuations of this force may be caused by the interaction of the rotor magnetic field with the stator teeth, resulting in the frequency [2]

$$f_r = 2f_{el}\frac{s}{p} \tag{3}$$

where s is the number of stator teeth and p is the number of pole pairs.

The tangential component of the force on the rotor can be expressed in terms of oscillations of the torque. The torque is calculated as the derivation of the magnetic energy over the rotational angle of the rotor. It is written as [2]

$$T = -\frac{\partial W}{\partial \theta} = -\frac{gL}{2\mu_0}\frac{\partial}{\partial \theta}\int_0^{2\pi} B(\alpha, t)^2 d\alpha \tag{4}$$

In most cases, the greatest cause of torque ripple is the cogging torque which is caused by oscillations of the magnetic reluctance between the permanent magnets and the slots of the stator. Its frequency depends on the rotational speed U, the number of pole pairs p and the number of stator teeth s [3]:

$$f_{tr} = U\left(2\frac{S}{p}\right)p \tag{5}$$

2.2. Fundamentals of Rotordynamics

In the following remarks, the mathematical model of a rotor is first described in its simplest form, being able to perform deviations in the cartesian x and y directions only. Introducing the complex deviation $r = x + jy$, the equation of motion takes the following form [4]:

$$m\ddot{r} + d_{ext}\dot{r} += \sum_i F_i \leftrightarrow \ddot{r} + 2\zeta\omega\dot{r} + \omega^2 r = \frac{1}{m}\sum_i F_i \tag{6}$$

This equation's characteristic polynomial can be solved by the eigenvalues

$$\lambda_{1,2} = -\frac{d_{ext}}{2m} \pm j\sqrt{\frac{k}{m} + \frac{d_{ext}^2}{4m^2}}. \tag{7}$$

The damped eigenfrequencies are expressed by the imaginary term. A critical damping of $d_{crit} = 2\sqrt{km}$ can also be extracted from the equation, although it seldomly becomes relevant in rotors [5]. Usually, the external damping is small, and the natural frequency is close to the that of an undamped system.

Beside the external damping, internal damping may also be present. It results from energy being dissipated by mechanisms of deformation such as micro-slip in press fits and becomes especially relevant in cases of small external damping [6]. After expressing the equation of motion in corotating coordinates and adding the internal damping, the equation takes the form

IOP Conf. Series: Materials Science and Engineering 1241 (2022) 012048 doi:10.1088/1757-899X/1241/1/012048

$$m\ddot{r} + d_{ext}\dot{r} + d_I(\dot{r} - j\Omega r) + kr = \sum_i F_i \tag{8}$$

There is now a skew symmetric share of the stiffness matrix, which can lead to instability for certain values of d_I in proportion to d_{ext}. Stable operations above the natural frequency may now be impossible. The range of stability can be increased by increasing the value of d_{ext}.

Another phenomenon to be considered is the gyroscopic effect, that can gain relevance in rotors with large moments of inertia around the axes that are perpendicular to the axis of rotation. Using a complex rotational deflection $\mathfrak{x}$, the equilibrium of moments can be expressed as [6]

$$J_p\ddot{\mathfrak{x}} - \Omega J_{ax}\dot{\mathfrak{x}} + k_{\mathfrak{x}r}r + k_{r\mathfrak{x}}\mathfrak{x} = \sum_i M_i \tag{9}$$

Again, natural frequencies now depend directly on the rotational speed. The proportion of the two moments of inertia can also be relevant for stability.

Torsional vibration, although outside the reign of classical rotordynamics, may also be a cause for concern in many rotors. In most cases, the torsional motion can be treated as a one-degree-of-freedom system [7]. Its damping is usually very small. The natural frequency can therefore be approximated by the well-known expression $\omega = (k_{tor}/J_p)^{1/2}$ [7].

3. State of the Art

Two publications about the development of electric drives with the consideration of cryogenic temperatures have been released in the past. Filipenko et al. [8] describe an electric motor with superconductive stator coils and cryogenic cooling. The magnetic flux density and the torque ripple are calculated using a finite element analysis. The necessity of qualifying materials for a cryogenic environment is mentioned. Haran et al. [9] describe the increase of the power density by incorporating fiber-composite materials. The increase of the magnetic potential difference allows for the armature mass to be reduced and its structural role to be taken by non-magnetic, more lightweight materials.

4. Methods

Table 1. Properties of the electric machine.

Quantity	Value
rotor outer radius	240 mm
length (yoke and PMs)	38 mm
PM material	NdFeB
rotational speed	2500 rpm
operational torque	1900 Nm

4.1. Electromagnetic Simulation

To calculate the radial forces on the rotor as well as the torque ripple, a two-dimensional representation of the rotor is being set up in the simulation software *COMSOL Multiphysics*. The geometry of the armature and the magnets does not change along the rotational axis. As all electromagnetic properties are symmetrical along that axis, it is feasible to simulate the geometry as two-dimensional. Along the perimeter of the machine, the motor is sector-symmetrical with as many sectors as pole pairs. One of these sectors is depicted in Figure 1. The symmetry condition is defined using the magnetic vector potential **A**. The vector potential at one edge of the sector must be the same as the one on the opposite side to ensure sector symmetry. On the inside and outside of the active parts, no mechanical structure is represented in the electromagnetics model. Instead, it is assumed that the surroundings consist of air with a permeability close to that of the vacuum.

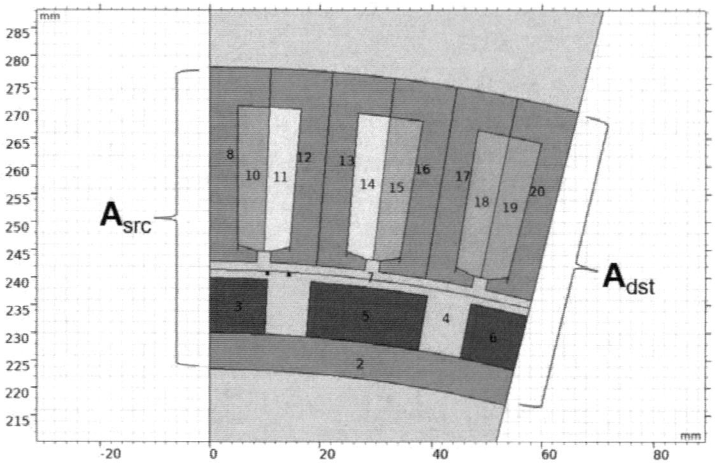

Figure 1. Geometry of the electromagnetic simulation.

For the finite element analysis, the geometry must be divided into discrete elements. The process of generating these elements is referred to as *meshing*. The density of the mesh determines the accuracy of the solution. In the present investigation, field properties in the air gap are of the highest interest. Therefore, the mesh in that area must be particularly fine. To guarantee a sufficient resolution, two boundary layers have been created along the dividing line between the corotating and the static part of the geometry.

Before a time-dependent study of the electromagnetic field during operations can be performed, a stationary study must take place to find the ideal initial rotation of the rotor relative to the stator for maximum torque. After this has been completed, a transient study takes place, calculating the radial force on the rotor using the Maxwell Stress Tensor and the torque fluctuations. Both properties are being investigated globally and, in the case of the radial force, at certain points at the outer diameter of the rotor to extract one-dimensional data that can then be transformed to the frequency domain using the Fast Fourier Transformation.

4.2. Rotordynamic Simulation

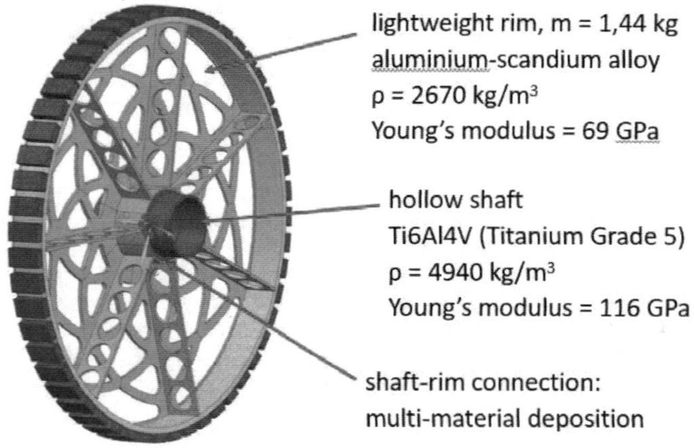

Figure 2. Geometry of the 3D model.

For the rotordynamic calculations, both a 1D beam model and 3D representations of the rotor, one of which is shown in Figure 2, have been created. All connections between different parts of the geometry are considered to be fixed. The shaft and the support structure for the armature may be connected by multi-material-deposition, while the connection between the support structure and the armature may be realized with a press fit in the axial direction. Internal damping, although it may be caused by micro-slip in the press fit, can therefore not be considered in the present investigation. Two different support structures have been investigated. Both are manufactured from a high-performance aluminium-scandium alloy. One of the structures has been designed as an extremely lightweight structure with the benefits of additive manufacturing in mind, while the other one is a simplified version that can easily be achieved with conventional manufacturing methods.

To fully characterize the dynamical behaviour of the rotor, different types of studies must be conducted. First, eigenfrequency analyses are being performed to describe the natural frequencies depending on the rotational speed. If natural frequencies are found to be close to either the rotational speed or one of the excitation frequencies, frequency dependent studies are performed additionally where the deformation response to the excitations can be modelled. All the mentioned studies are conducted for different sets of parameters. The parameters are the operating temperature which affects the modulus of elasticity, the different mechanical structures, and the length of the shaft between the bearings. Several simplifications have been made. External and internal damping have not been considered. This is a reasonable simplification as damping is usually very small in electric drives [5]. Furthermore, the bearings are assumed as perfectly stiff. The magnetic excitations are included in the rotordynamics model in their simplest form without higher harmonics.

5. Results

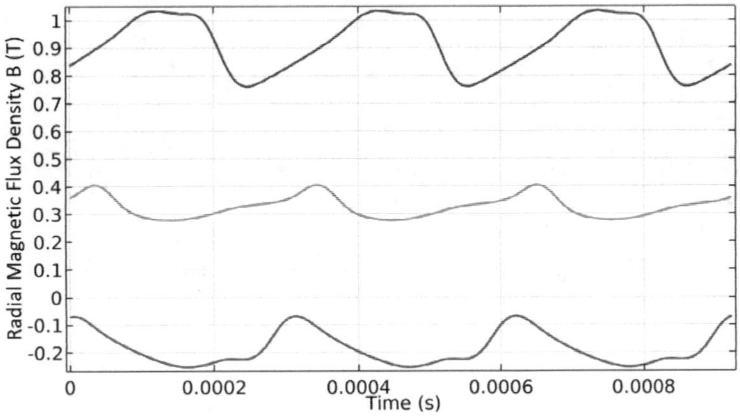

Figure 3. Radial flux density at different points in the air gap.

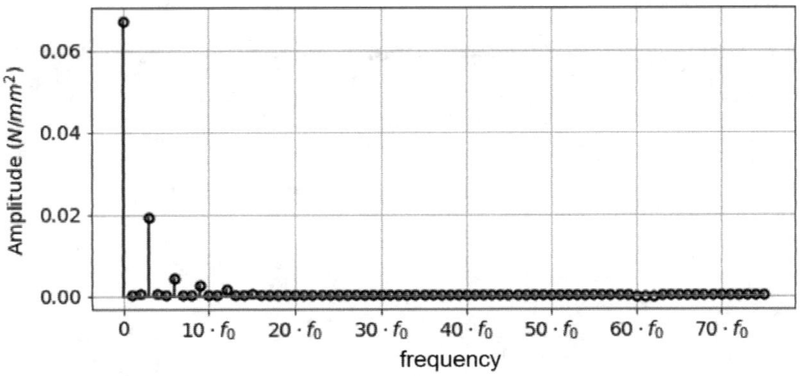

Figure 4. Frequency spectrum of the pressure measured above a PM.

5.1. *Electromagnetic Simulation*

The results of the calculation show that the sign of the radial component of the magnetic flux density in the airgap follows the magnetization of the permanent magnets. The magnetic field of the three permanent magnets per sector accounts for the bigger proportion of the flux density, which is clearly visible in the changes of B in Figure 3, where the top line depicts B in the airgap over the center of a PM. Analyzing B for certain points in the air gap in corotating coordinates shows that although the amount of B at those points depends mostly on the proximity to a permanent magnet, there is also a component that fluctuates while the rotor rotates and is therefore caused by the interaction with the stator geometry. From the values of the radial component of B, the pressure on the rotor can be calculated using the Maxwell Stress Tensor. The Fast Fourier Transformation of the time-dependent data shows the frequency spectrum of the radial forces (Figure 4). The dominating frequency is $f = f_{el} \cdot 3 = 3250$ Hz. Harmonics occur but have a much smaller amount than the fundamental oscillation.

IOP Conf. Series: Materials Science and Engineering 1241 (2022) 012048 doi:10.1088/1757-899X/1241/1/012048

The torque over the time has been evaluated globally for one machine segment. To obtain the overall torque, the data must be multiplied by the number of segments, which is the number of pole pairs. The result of this is shown in Figure 5. It can be used to verify the model by examining the amount of the torque and comparing it to analytical approximations. The results show that the torque fluctuates with $f_{tr} = f_{el} \cdot 6 = 6500$ Hz. This oscillation is most likely being generated by the reluctance between the PMs of the rotor and the teeth of the stator, as described above. The Fast Fourier Transformation shows that harmonics only occur with very small amounts.

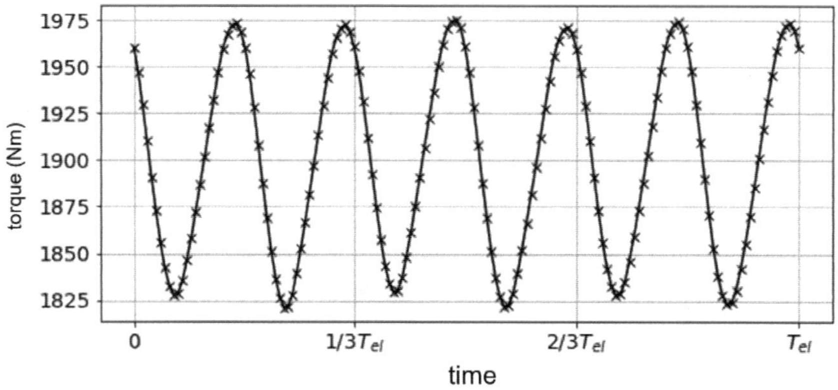

Figure 5. Torque over the time of one electric period.

5.2. Rotordynamic Simulation

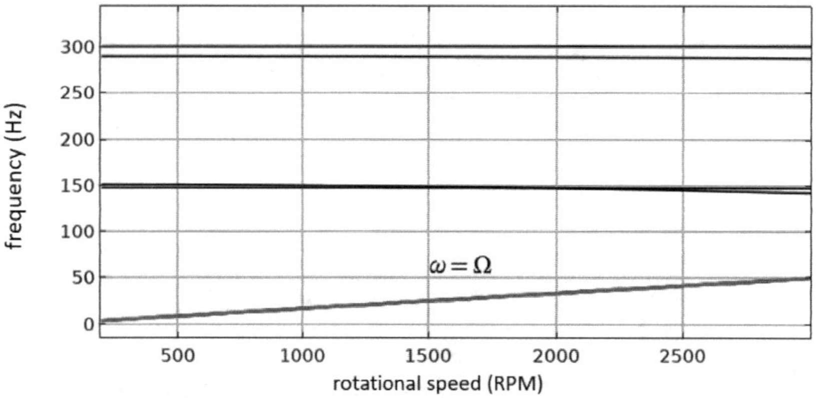

Figure 6. Campbell plot for the lowest frequency range.

The eigenfrequency analysis with the simplified geometry for the lower frequency range shows that the lowest eigenfrequency lies well above the operating rotational speed of 2500 rpm or 43.7 1/s, as can be seen in the Campbell plot in Figure 6. The lowest natural frequencies are those of the torsional and axial mode. The torsional mode could potentially be excited by the torque ripple during run-up, when the torque ripple matches the torsional natural frequency.

Looking at the Campbell plots for the frequencies around the magnetic excitation frequencies (Figures 7 and 8), we find a large number and density of natural frequencies, many of which lie close to the excitations at operating state. When examining the mode shapes, the conclusion can be drawn that the higher natural frequencies almost exclusively concern deformations of small parts of the geometry like membrane-shaped parts of the disk or stiffening fins. Still, response analyses in the frequency domain are necessary to determine if dangerous deformations can occur. The response plots confirm that no significant deflections are to be expected at the excitation frequencies.

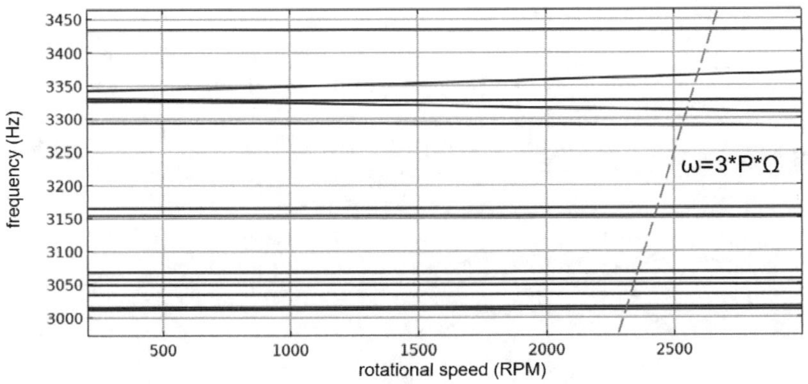

Figure 7. Campbell plot for the frequency range close to the radial excitation.

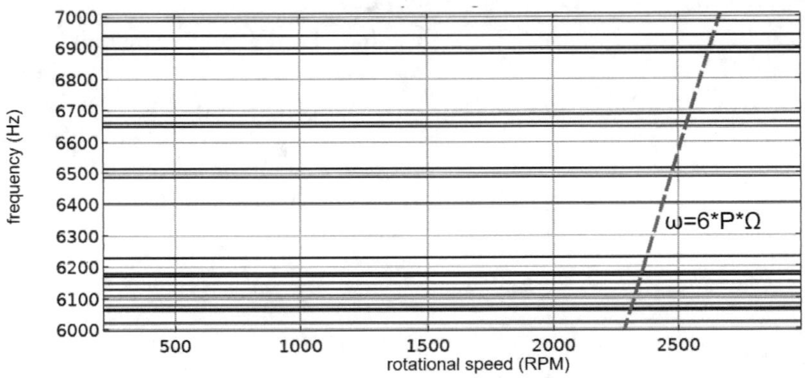

Figure 8. Campbell plot for the frequency range close to the torsional excitation.

The different operating temperatures of 77 K, 300 K and 400 K are considered as an influencing variable to the modulus of elasticity of the structural materials of the shaft and the support structure, titanium and an aluminium-scandium alloy. The data for this is taken from previous studies as well as from [10] and [11].

Table 2. Young's modulus of structural materials for different temperatures.

Temperature	Titanium Ti6Al4V	aluminium-scandium alloy
77 K	132 GPa	76 GPa
300 K	116 GPa	69 GPa
400 K	105 GPa	63.5 GPa

The results show that lowering the temperature leads to an increase of all natural frequencies. This can be seen in Figures 9 and 10 for the example of the torsional eigenfrequency. This result can be explained by the proportionality of the modulus of elasticity to the stiffness, while the mass and the moments of inertia remain constant.

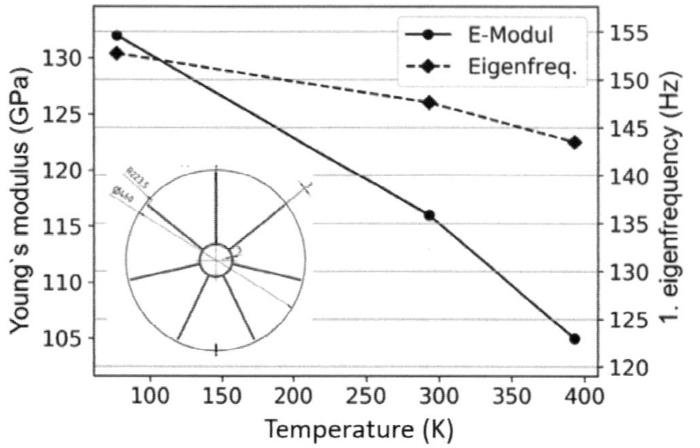

Figure 9. Young's modulus and torsional eigenfrequency for different temperatures (simplified geometry). Radii are 223.5 mm for the stiffeners and 230 mm for the rim. Wall thickness of the stiffeners is 3 mm.

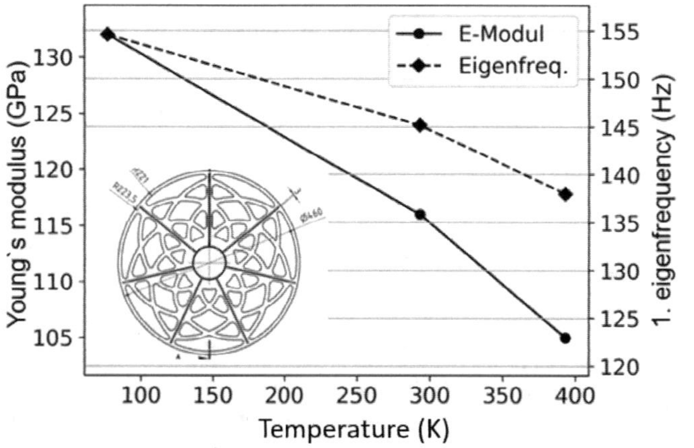

Figure 10. Young's modulus and torsional eigenfrequency for different temperatures (lightweight geometry). Radii are 223.5 mm for the stiffeners and 230 mm for the rim. Wall thickness of the stiffeners is 3 mm.

Evaluating the natural frequencies and deflections for the two different structural designs, the conclusion can be drawn that an additively manufactured lightweight structure can have the same dynamic properties as a much heavier conventional part. The lightweight structure at hand has been designed with a focus on static loads and seeks to have the same behaviour under static loads than a conventional structure. In this case, similar static behaviour appears to imply very similar dynamic behaviour as well.

The shaft length (or distance between bearings) in directly related to the eigenfrequencies as the stiffness of the shaft length is one of the most fundamental variables of its stiffness. We have seen from the abovementioned eigenfrequency analysis that bending modes do not occur close to the rotational speed. For a shaft length of $l = 2000$ mm, the first bending mode matches the frequency of the rotational speed at operating state. The frequency of the torsional mode also decreases with an increase in shaft length due to a reduction of torsional stiffness. As bending modes could not be detected in earlier studies with much smaller shaft length, this can also be used to validate the methods of the analysis itself by proving that bending modes can in fact be detected in the system.

Transient studies of run-up are of interest because only during run-up does the frequency of the torque ripple (cogging torque) momentarily match the natural frequency of the torsional mode. The results show that a significant torsional deflection occurs when the frequency of the cogging torque and the torsional eigenfrequency coincide, which can be recognized in Figure 12. For different operating temperatures, the earlier results are validated as the deflection during run-up becomes smaller for lower temperatures because of the increased stiffness and occurs later due to the increased value of the natural frequencies. Different rise times directly influence the amount of deflection as well.

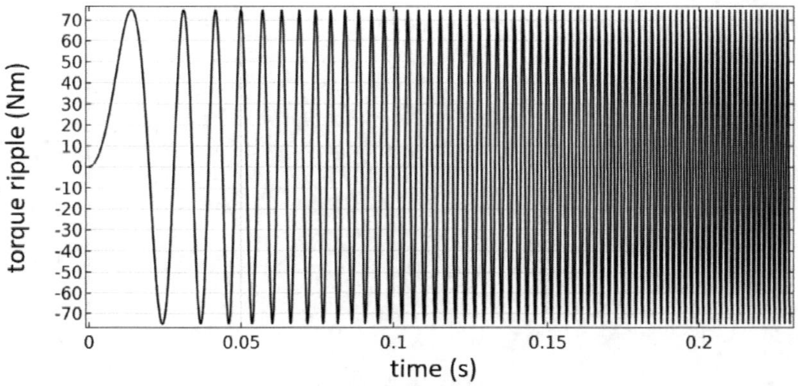

Figure 11. Torque ripple at the beginning of the linear acceleration from rest.

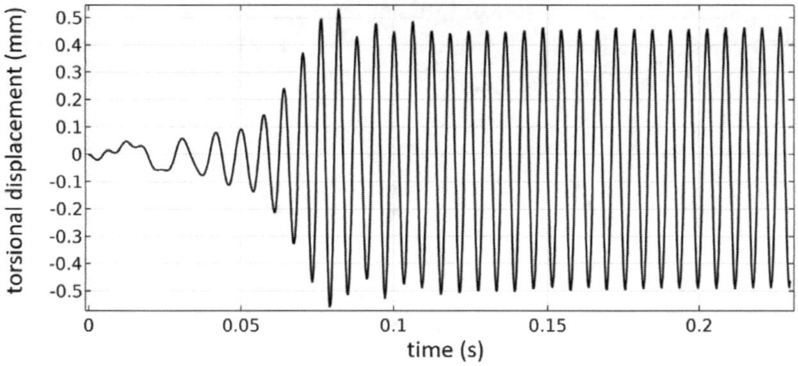

Figure 12. Torsional deflection over time during run-up.

6. Conclusions

The present work has shown that nonsynchronous magnetic forces do not excite significant oscillations in the investigated machine during the operating state. However, during run-up, torsional vibrations may occur due to cogging torque. Cryogenic operating temperatures may lead to stiffer mechanical parts and therefore higher eigenfrequencies, making the system more robust if fatigue is not considered. The evaluation of two different mechanical designs has shown that additively manufactured lightweight parts can have a similar dynamic behaviour to heavier, more conventional mechanical structures. Several tasks remain to be resolved in the future. As soon as the state of the *AdHyBau* project allows for it, a bearing design needs to be conducted and the bearings must be considered in the dynamic evaluation of the rotor. Furthermore, the results of the present research may be validated by incorporating a set of boundary conditions such as external and internal damping. Regarding the vibrations during run-up, a fatigue strength verification needs to be performed to draw the correct conclusions from the present findings.

7. References

[1] Islam M, Islam R and Sebastian T 2014 Noise and vibration characteristics of permanent-magnet synchronous motors using electromagnetic and structural analyses. *IEEE Transactions on Industry Applications* **50** pp 3214–3222
[2] Gieras J, Wang C and Cho Lai J 2006 *Noise of Polyphase Electrc Motors*. (London: Taylor and Francis)
[3] Krishnan R 2010 *Permanent Magnet Synchronous and Brushless DC Motor Drives*. (London: Taylor and Francis)
[4] Matsushita O *et al.* 2017 *Vibrations of Rotating Machinery*. (Tokyo: Springer)
[5] Muszynska A 2005 *Rotordynamics*. (London: Taylor and Francis)
[6] Gasch R, Nordmann R and Pfützner H 2006 *Rotordynamik*. (Berlin, Heidelberg: Springer)

[7] Vance J, Zeidan F and Murphy B 2010 *Machinery Vibration and Rotordynamics*. (Hoboken: John Wiley and Sons)
[8] Filipenko M *et al.* 2020 Concept design of a high power superconducting generator for future hybrid-electric aircraft. *Superconductor Science and Technology*, **33** (054002)
[9] Haran K *et al.* 2017 High power density superconducting rotating machines – development status and technology roadmap. *Superconductor Science and Technology*, **30** (123002)
[10] Su M 2017 Mechanical properties of high strength aluminium alloy at elevated temperatures. *EUROSTEEL*, Copenhagen, Denmark.
[11] McLellan M 1999 *Aerospace Structural Metals Handbook. Titanium Alloys. Types Ti-6Al-4V*. (West Lafayette: Perdue Research Foundation)

Acknowledgments
The corresponding author wishes to thank Prof. Dr. Thomas Sattel at Ilmenau University of Technology for his supervision over the thesis this work is based on, Dr. Christian Weidermann at Siemens AG for his substantial support and Dr. Markus Klöpzig at Siemens AG for the opportunity to be part of his research group.

IOP Conf. Series: Materials Science and Engineering 1241 (2022) 012049 doi:10.1088/1757-899X/1241/1/012049

Using additive manufacturing technologies in high-field accelerator magnet coils

I Novitski[1], J Coghill[1], T Beale[1], J Bergeron[2], R Davidoff[2], B Oshinowo[1], S Raje[2], D Swanson[1], C Wilson[1], AV Zlobin[1]

[1] Fermi National Accelerator Laboratory, Batavia, IL 60510, USA
[2] GE Additive, West Chester, OH 45069, USA

E-mail: novitski@fnal.gov

Abstract. Recent advances with additive manufacturing (AM) technologies using various materials allow them to be considered for the manufacture of precise and complicated metal parts of the magnet coils of high field accelerator magnets from aluminum bronze, titanium alloy, stainless steel, etc. The 3D printing technology is also being used to fabricate prototype models of real parts to test and optimize their geometry. This paper discusses the designs of the complex stress-management coil parts developed at Fermilab, their fabrication using AM technologies, and quality control methods and results.

1. Introduction

Superconducting (SC) dipole and quadrupole magnets are key elements of charged particle accelerators. A strong magnetic field with good field quality in the aperture is required for a stable trajectory of particle beams. SC magnets based on NbTi superconductor with the nominal magnetic fields up to 7-8 T were developed and used in the circular colliders such as Tevatron, HERA, RHIC, LHC [1]. An interest to accelerator magnets with higher field continues. SC magnets with magnetic fields up to 11-12 T based on Nb_3Sn superconductor have been developed during the past two decades and are now being produced for the LHC luminosity upgrade [2]. Modern concepts of a Future Circular Collider (FCC) for High Energy Physics push the operation field of SC magnets even higher to a level of 16-20 T. This level of magnetic field requires using advanced low-temperature and high-temperature superconductors and new magnet technologies.

Most of the present SC accelerator magnets are based on multi strand Rutherford cables, made of round composite wires, and cos-theta coil geometry. Composite round wires, compatible with Rutherford cable technology, are produced by industry based on low-temperature Nb_3Sn and high-temperature Bi2212 class superconductors. Both materials are brittle and, thus, require using Wind-&-React (W&R) technology with multistep high-temperature heat treatment (HT) after coil winding.

Wind-&-React coil technology for both superconductors has been extensively developed over the last two decades. It was found that Nb_3Sn composite wires, which use a Cu matrix, need ~10 days of complex heat treatment at maximum temperatures close to 700°C in an argon atmosphere to form the superconducting phase with optimal structure to achieve the best critical current density. Bi2212 composite wires use a Ag matrix, and to achieve the best critical current density require a multistage HT in a highly corrosive oxygen atmosphere at 50 bar pressure with maximum temperatures up to 900°C for ~2 days. The high-temperature heat treatments call for metallic coil parts and insulating materials which can withstand the high temperatures. After HT all the coils are usually impregnated with epoxy for turn mechanical integrity in the coil and reinforcing their electrical insulation. These impregnated coils work in the temperature range of 300-4 K under cycling pressures up to ~150 MPa.

Coils made of stress/strain sensitive superconductors and operating at high magnetic fields must withstand very high electromagnetic forces at cryogenic temperatures. To provide a stable coil turn position and to keep the strain/stress in the coil within an acceptable range for the superconductor during magnet fabrication and operation, various stress management coil structures have been proposed and are being studied. These structures have complex 3D geometries that are difficult to produce using traditional fabrication tools, and rather tight tolerance requirements. Recent advances with additive manufacturing (AM) technologies using various materials (aluminum bronze, titanium alloy, stainless steel, etc.) allow them to be considered for fabrication of precise and complicated metallic parts used in the magnet coils of high field accelerators. Additionally, AM is also being used for rapid prototyping of parts using inexpensive plastic materials to support design iterations aimed at optimizing part geometry for functionality.

This paper discusses the designs of the complex coil parts developed at Fermilab, their fabrication using AM technologies, quality control methods and the results.

2. Cos-theta coils for high field magnets at Fermilab

Cross-sections of superconducting dipole coils developed at Fermilab based on cos-theta geometry are shown in Figure 1. High currents and strong magnetic fields in coils produce large electromagnetic forces which can degrade or even damage brittle superconductors. These forces significantly increase with the level of magnetic field and magnet aperture and at some level need special treatment. A stress management (SM) concept being developed at Fermilab introduces rigid elements into the cos-theta coil structure to mechanically protect brittle superconductor. In the large-aperture coils shown in Figure 1 c) and d) each coil block is placed in its own compartment and supported separately. As a result, azimuthal and radial components of the electromagnetic force applied to the coil blocks are not accumulated but transmitted to the coil and magnet structure bypassing coil blocks.

The two-layer regular coils (in Figure 1 a, b, and d) have three blocks separated by two wedges in layer 1 and two blocks and one wedge in layer 2. Each layer of the large-aperture SM coils (c and in Figure 1) is split into five blocks, separated by 5 mm wide spacers, with the number of turns approximately following the cos-theta distribution. In addition, the SM coil layers are separated by 5 mm in the radial direction from the inner coil and from each other to provide space for the support structure.

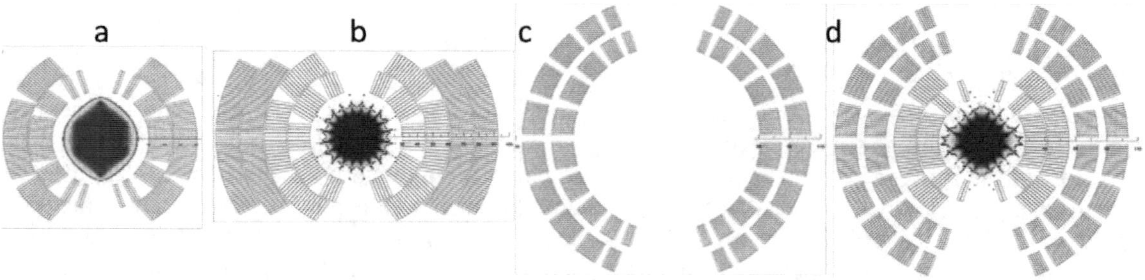

Figure 1. Cross-sections of superconducting dipole coils based on cos-theta geometry developed at Fermilab: a) 60-mm aperture 2-layer coil for 11 T dipole (MBH) for the LHC upgrade [3]; b) 60-mm aperture 4-layer graded coil used in 15 T dipole (MDPCT1) [4]; c) 120-mm aperture 2-layer coil with stress management developed for 11 T dipole [5]; and d) 60-mm 4-layer graded coil with stress management outer layers for 17 T dipole [6]. The cross-section of the inner coil was optimized for best field quality using the ROXIE code [7] and considering the field harmonics produced by the outer coil. The magnetic model included a cylindrical iron yoke with an outer diameter of 600 mm.

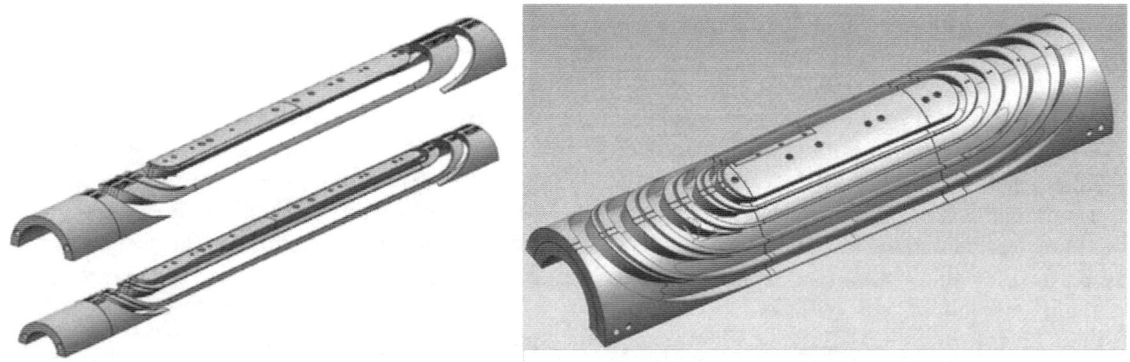

Figure 2. 3D parts of the two-layer cos-theta coils without (left) and with stress management (right). The left and right sides of each figure correspond to the lead and return ends respectively.

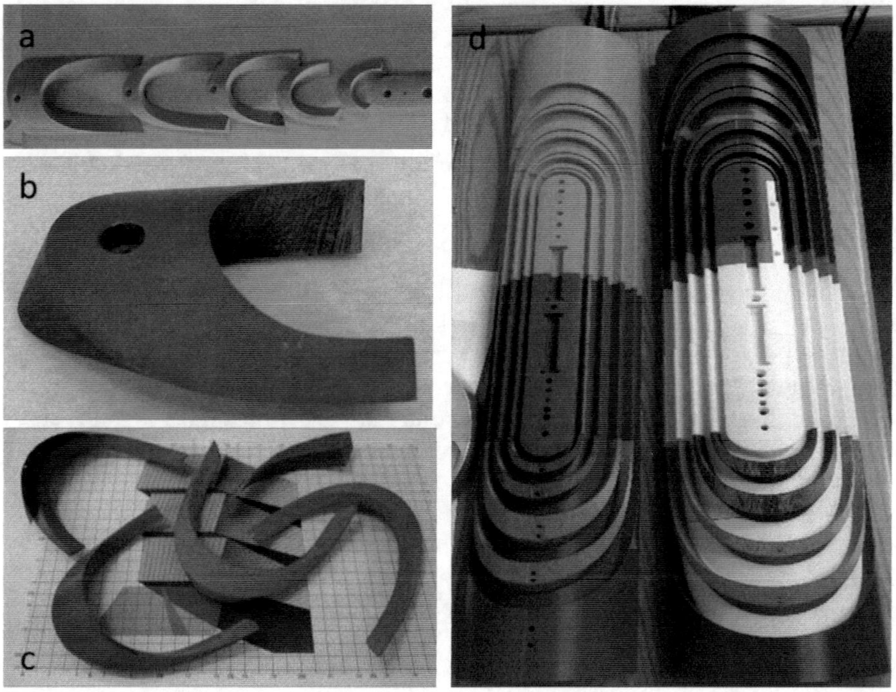

Figure 3. Practice end parts used in various cos-theta dipole coils: a) SLS end spacers; b) waterjet end spacer; c) plastic end spacers; and d) plastic parts for stress management coils.

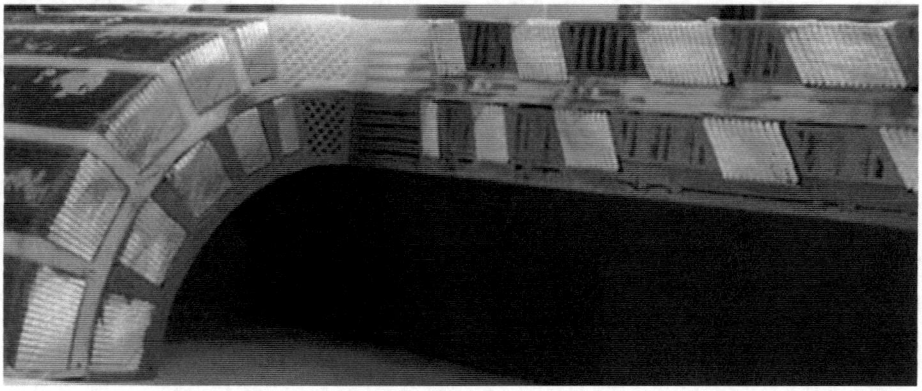

Figure 4. Two-layer SMCT practice coil with axial and transverse cuts of return end.

IOP Conf. Series: Materials Science and Engineering 1241 (2022) 012049 doi:10.1088/1757-899X/1241/1/012049

3. Cos-theta coil part design and fabrication

3.1. End part design and prototyping

The coil ends are 3D regions with complicated cable deformations including bending and twisting. The coil end design has to minimize distortion of the integrated field and reduce peak field enhancement, produce compact and friendly for Rutherford cable winding with minimal stress, use metallic end parts, and limit conductor motion to avoid quench in the SM concept. At Fermilab all the coil ends and end parts are designed using the BEND program [8]. The program creates the conductor surfaces and groups based on the minimization of strain energy accumulated within the cable. It also generates spacers suitable for fabrication without coil gaps. Similar purpose programs were developed at other laboratories, e.g. 3D ROXIE [7] at CERN, analytic codes at CEA-Saclay [9], etc.

3D parts of the 60-mm aperture two-layer inner cos-theta coil without stress management and the 120-mm aperture two-layer outer cos-theta coil with stress management are shown in Figure 2. The left and right sides of each figure correspond to the coil lead and return ends respectively.

Due to the high complexity of end part geometry and the large variation of cable mechanical rigidity, the end parts used in superconducting coils for accelerator magnets need prototyping and qualification in practice windings. The use of AM for parts prototyping increases their design quality and reduces the time and cost of part development. Several AM technologies such as selective laser sintering (SLS), electric discharge machining (EDM), 5-axis waterjet machining, and 3D printing using carbon fiber ABS (Acrylonitrile butadiene styrene) are being used to fabricate complicated coil end parts for the practice windings to finalize part design. Examples of parts produced using the above technologies are shown in Figure 3.

Coil practice winding validates interfaces of end parts with adjacent coil turns. All plastic parts were printed at Fermilab using the Stratasys F170 FDM printer and corresponding CAD models. It allows verifying the part design before procurement, checking important production steps, and testing tooling. In the case of SMCT coils, practice winding also permits learning new procedures of cable winding into slots, checking cable rigidity during winding, and confirms part sizes, cable positioning, room for cable expansion for reaction, etc. A 3D view of practice coil return end with axial and transverse cuts is shown in Figure 4 confirming optimal turn position and minimal gaps inside coil windings.

3.2. Printing of cos-theta coil end spacers

AM is a very attractive technology for producing metallic parts with complex 3D geometries. One type of AM process, used in manufacturing of metal parts, is called Laser-Powder Bed Fusion (L-PBF). This process uses a laser to melt layer-by-layer the powdered material feedstock and to produce a consolidated metal part of a desired shape. A melt pool is generated by the laser, which plays a crucial role in the quality of the final part in terms of surface quality, part geometry, and material properties. The characteristics of that melt pool are developed and controlled by the parameters used by the L-PBF machine, and can include laser power, spot size, speed, offsets, and spacing. Additionally, best practices in Design for Additive Manufacturing (DFAM) are also required to ensure success. One of the key DFAM principles used to manufacture parts with L-PBF is support structure, and it can serve two purposes. It can be used to dissipate heat away from the melt pool to ensure proper cooling rates, and it can act as a physical anchor to withstand thermal gradients created by the melt pool. The proper dissipation of heat away from the melt pool and inclusion of physical anchors into the part build layout ensures conformance to surface quality, mechanical property, and part geometry requirements. Examples of such support structure can be seen on the left photo in Figure 5.

L-PBF processes can use weldable powdered materials, such as 316L, Ti6Al4V as well as nickel, aluminum and copper alloys. The mechanical properties of L-PBF parts are typically better than cast materials and comparable with those of wrought materials [13]. At cryogenic temperatures the printed materials still have tensile characteristics close to bulk material but with compromised ductility [14]. A layering and discretization of the melted zone has a big influence on the build properties. As the melt pool moves in the build plane, the localized heating and cooling of the material results in residual stress buildup inside the part. Therefore, L-PBF parts require a stress relief process to mitigate the residual stresses. Additional heat treatments can be added to the consolidated material to achieve desired material properties. Ultimately, machine parameters, feedstock or powder quality, design for additive manufacturing (DFAM) best practices, and other process controls are crucial to ensure part quality and repeatability.

All of the mentioned above are well-known AM process features, which were used in a conceived design, and were a reason for caution, step-by-step phased use of AM technology. For the entire set of coil parts, a few spacers were first printed and a whole set was printed later for the next coil iteration. The coil end parts were 3D printed from stainless steel 316L and titanium alloy Ti6Al4V and provided by CERN for 11 T (MBH) [3] and 15 T dipole demonstrator (MDPCT1) [4], [15] magnets. The arrangement of coil end spacers during printing and the end spacers printed using titanium alloy Ti6Al4V for inner coils of the 15 T dipole demonstrator are shown in Figure 5. All the magnets worked successfully during cold magnet tests at Fermilab [10], [11]. Moreover, the 15 T inner coils with printed spacers worked extremely well, reaching world record field without end problems.

3.3. SMCT coil part printing

The SMCT parts for coils c) and d) in Figure 1 were printed at GE Additive using L-PBF technology, on a Concept Laser M2 machine. The M2 features a 250×250×350 mm^3 build envelope, a dual laser configuration, and a layer thickness capability of 20 to 80 μm. Available materials included aluminum (AlSi10Mg, AlSi7Mg), cobalt (CoCrMO, CoCrW), steel (316L, 17-4PH, M300), nickel (Ni625, Ni718) and titanium (Ti6Al4V, CP Ti, Ti6242) alloys. The machine provides exceptional surface quality, minimal porosity and mechanical properties equivalent to wrought materials.

Support structures were used in the printing of the SMCT coils in areas where the build angle (vs. horizontal build platform) was less than 45-degrees. Several build and inspect cycles were conducted to understand and compensate for any thermal distortion and feature-specific manufacturing tolerances. Two possible print orientations for the end parts were considered during the build set-up phase, as shown in Figure 6. The orientation on the left shows the part's axis tilted at a 25-degree angle relative to the build plane; the orientation on the right shows the part perpendicular to the build plane. The orientation on the left is self-supporting and the orientation on the right requires support structure, per DFAM best practices. In the self-supporting orientation, the inner diameter (ID) of the part is a slightly downward facing surface, resulting in slightly rougher surface finish (vs. an upward facing surface), fewer parts in a single build, and concerns about anchoring the ends of the semi-circular components to preserve geometric accuracy of the part ID. In order to ensure the best quality surface finish and geometric accuracy on the part ID and maximize the number of parts in a single build, the vertical orientation was chosen for this project. SMCT coil parts printed as two-layer cylinders with removed surface support inside blocks and with technological surface support before removal in narrow inter block channels in LE are shown in Figure 7.

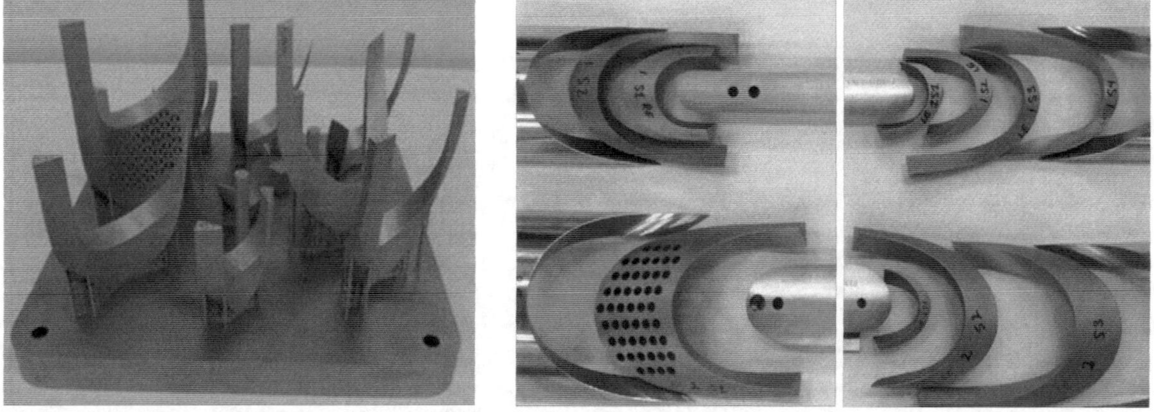

Figure 5. Spacer arrangement during printing (left) and end spacers printed on the SLM 280HL System using Ti6Al4V alloy for inner coils of the 15 T dipole demonstrator [4], [15] (right).

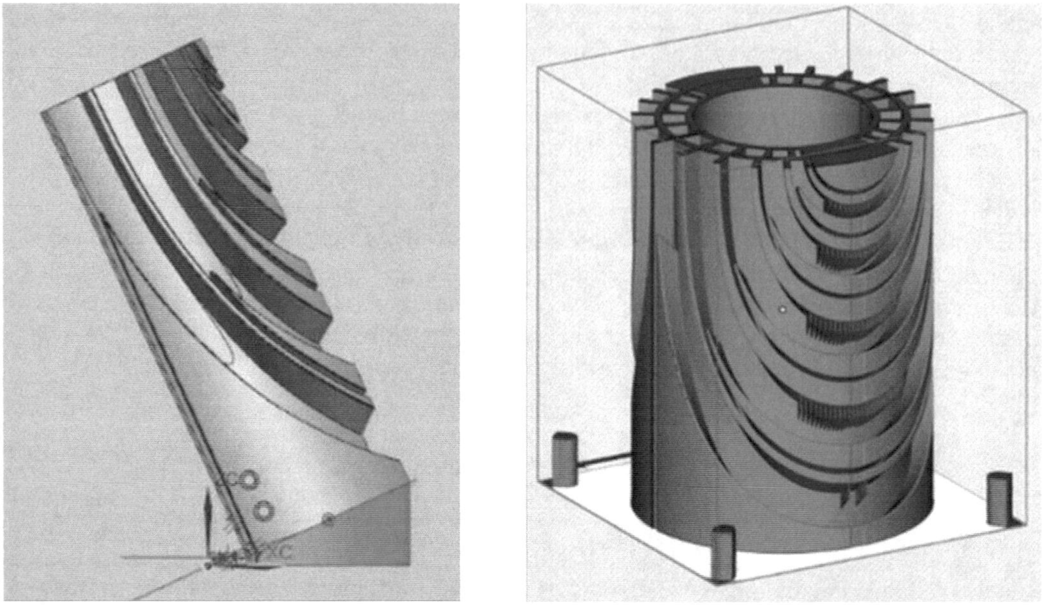

Figure 6. Two orientations of end part printing: 25-degree angle to eliminate support (left) and vertical with surface support shown in blue color (right).

Figure 7. SMCT coil parts printed as two-layer cylinders with removed surface support inside blocks (left) and narrow inter block channels in lead end with surface supports before removal (right).

4. QC inspection of coil parts

Strict tolerances on coil part dimensions require careful size control. It is especially important for evaluating any new fabrication technology of coil parts to be used in high-field superconducting coils. The first quick size control was done using a micrometer with quite encouraging results (see Figure 8). These measurements cannot be considered as completely sufficient since they do not give full information on the accuracy of part geometry. Modern advanced methods of the geometry control used at Fermilab include coordinate measuring machine (CMM) and laser scanning systems.

Examples of CMM and laser scanning measurements of surface deviation from CAD for 3D coil end spacers and SMCT end parts are shown in Figures 9 and 10. Standard CMM machines used at Fermilab allow measuring with a high volumetric accuracy of 2.8 μm. Insufficient support and build rigidity lead to a small amount of part warping. The complex 3D geometry of SMCT coil parts with narrow slots and inclined surfaces made it difficult for the CMM to control their dimensions. These parts were measured by the laser scanning system in the free condition using the ROMER Absolute Arm model 7525SE HP-L-20.8 external scanner.

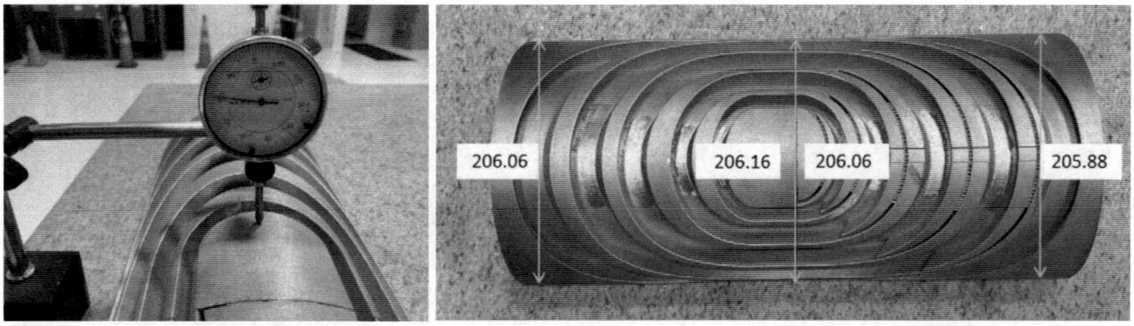

Figure 8. First control of SMCT coil end parts by micrometer on a granite table. The numbers in the right picture are in mm.

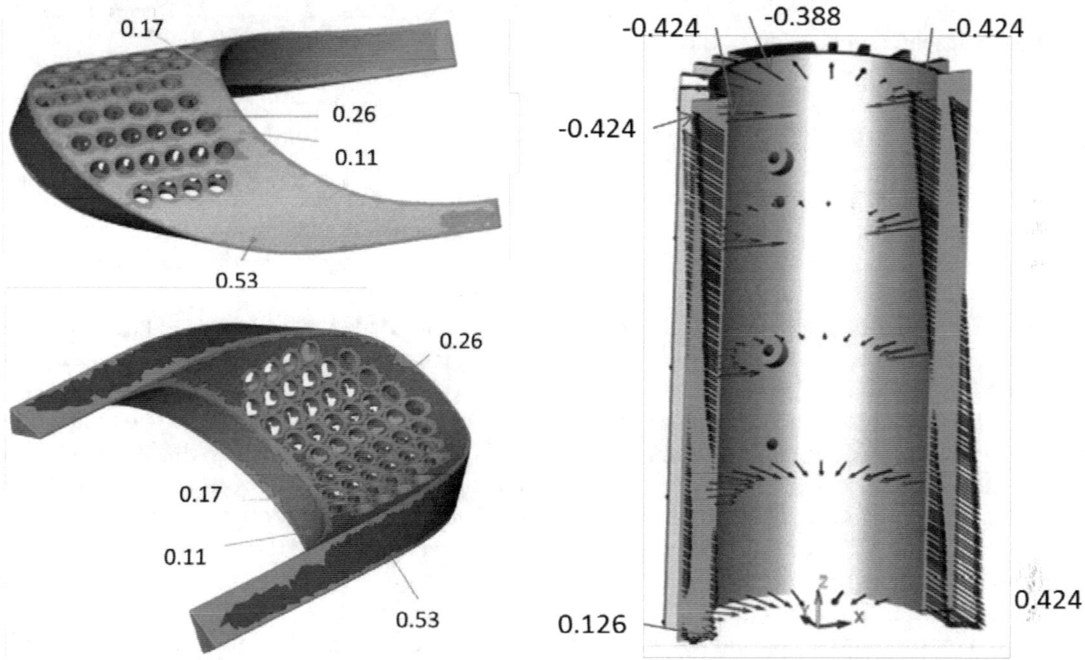

Figure 9. CMM measurements in mm of surface deviation from the CAD for a 3D coil wedge (left) and a straight section part of the SMCT coil (right).

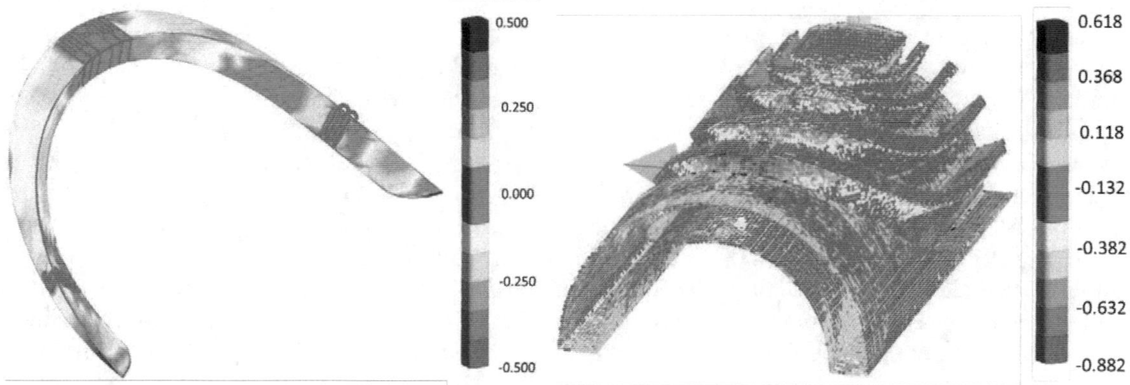

Figure 10. Laser scanning measurements in mm of surface deviation from the CAD for a 3D coil wedge (left) and a return end block of the SMCT coil (right).

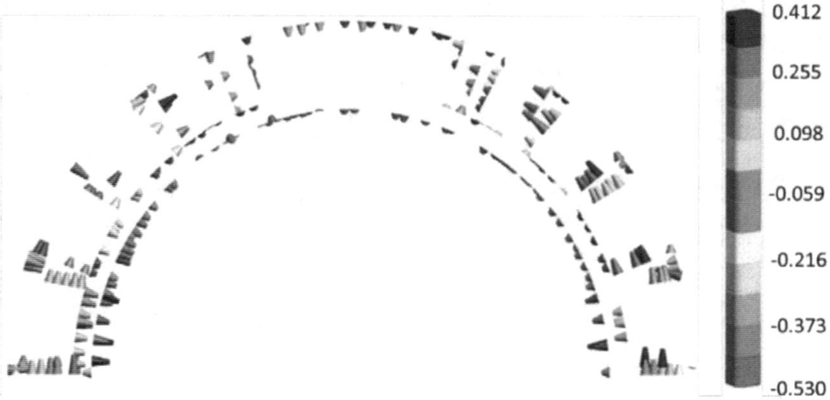

Figure 11. Measured deviations of the SMCT coil cross-section from its CAD model.

The measured part was placed on a steel table and supported by three stands. The stands were high enough that the laser scanner could measure all sides of the part. Measurements on a ceramic sphere were used to verify the accuracy of the Romer Arm external scanner. The scanning was done from one instrument location. The ROMER Absolute Arm with external Laser Scanner was paired with the New River Kinematics Spatial Analyzer™ Software to collect and process the scanned data. The full scan data for the SMCT coil part contain 53,884,621 points. The probing accuracy for the ROMER Absolute Arm with external scanner is 27 μm and scanning accuracy to within 58 μm.

The CAD model for the part was imported into the Spatial Analyzer™ Software and compared to the measured point clouds. A final fit to the part was done using an N-point best-fit to the CAD model with an RMS fit of about 0.2 mm. Next, the 53,884,621 point clouds were filtered to a 2-mm uniform spacing, leaving a subset of the full scan to about 63,058 point clouds. Figures 10 (right) shows a colorized deviation plot produced from this data subset. Figure 11 shows measured deviations of a cross-section of the SMCT coil from its CAD model using the same method. A best fit of cloud data points to the CAD datum gives an ID/OD ratio within ±0.25 mm and 3D surfaces profile to within 0.5 mm. The accuracy of the SMCT coil parts is consistent with the accuracy of 3D wedges. Based on these measurements, the AM part accuracy can be considered as a good result for the first-time printing of medium-large metal objects with complicated 3D geometries and relatively high tolerance requirements. A range of measured surface roughness using a stylus probe Ra is within 5.5–10.5 μm. The red zones in Figure 10 correspond to the areas of surface technological support.

5. Summary

Precise metallic parts of complex 3D geometries are used in superconducting coils of high-field accelerator magnets. It was demonstrated that part prototyping using AM technology and plastic materials is an effective way to finalize part design and reduce their cost during production. This approach is now used on a regular base at Fermilab. Various AM technologies to produce parts for accelerator magnet coils using various materials, such as SS 316L, Ti6Al4V, Inconel, etc., are being developed. These parts have already been used successfully in real superconducting accelerator magnets.

The SMCT coil is a new concept which was proposed at Fermilab for high-field and/or large-aperture accelerator magnets based on low-temperature and high-temperature superconductors with the coil ends of challenging complexity. The first set of SMCT coil parts for Nb_3Sn dipole coils [6] have been produced by GE Additive using DMLM technology. Printed parts have been measured using a laser scanning system with acceptable tolerance results. The SMCT coil winding will start shortly after parts post-processing. Parts for the Bi2212 insert coil [12] to be produced using AM technology are now being designed and optimized using practice plastic parts.

IOP Conf. Series: Materials Science and Engineering 1241 (2022) 012049 doi:10.1088/1757-899X/1241/1/012049

6. References

[1] L. Bottura, S. Gourlay, A. Yamamoto, A.V. Zlobin, "Superconducting Magnets for Particle Accelerators", IEEE Trans. on Nuclear Science, Vol. 63, Issue 2, April 2016.
[2] L. Bottura et al., "Advanced Accelerator Magnets for Upgrading the LHC," IEEE Trans. on Appl. Supercond., Vol. 22, Issue 3, June 2012.
[3] A.V. Zlobin et al, "Development and test of a single-aperture 11T Nb_3Sn demonstrator dipole for LHC upgrades", IEEE Trans. on Appl. Supercond., Vol. 23, Issue 3, June 2013.
[4] I. Novitski et al., "Development of a 15 T Nb_3Sn Accelerator Dipole Demonstrator at Fermilab", IEEE Trans. on Appl. Supercond., Vol. 26, Issue 3, June 2016.
[5] A.V. Zlobin, V.V. Kashikhin, I. Novitski, "Large-aperture high-field Nb_3Sn dipole magnets," Proc. of IPAC2018, WEPML026, 2018, p.2738.
[6] A.V. Zlobin, J. Carmichael, V.V. Kashikhin, I. Novitski, "Conceptual design of a 17 T Nb_3Sn accelerator dipole magnet," Proc. of IPAC2018, WEPML027, 2018, p.2742.
[7] ROXIE code for an electromagnetic simulation and optimization of accelerator magnets, http://cern.ch/roxie.
[8] J.M. Cook, "Strain energy minimization in SSC magnet winding", IEEE Trans. on. Magnetics, Vol. 27, Issue 2, March 1991.
[9] E. Rochepault, et al, "Three-dimensional magnetic optimization of accelerator magnets using an analytic strip model', Journal of Applied Physics 116, 023910 (2014).
[10] A.V. Zlobin et al, "Quench performance of a 1-m long single-aperture 11 T Nb_3Sn dipole model for LHC upgrades", IEEE Trans. on Appl. Supercond., Vol. 24, Issue 3, June 2014.
[11] A.V. Zlobin et al., "Reassembly and Test of High-Field Nb_3Sn Dipole Demonstrator MDPCT1", IEEE Trans. on Appl. Supercond., Vol. 31, Issue 5, 2021.
[12] A.V. Zlobin, I. Novitski and E. Barzi, "Conceptual Design of a HTS Dipole Insert Based on Bi2212 Rutherford Cable," Instruments 2020, 4, 29.
[13] Handbook of Advanced Manufactured Material Properties from TCR Structure Builds at ORNL—FY 2019, ORNL/TM-2019/1328.
[14] Y. Hong et al., "Formation of strain-induced martensite in selective laser melting austenitic stainless steel", Materials Science & Engineering A 740–741 (2019) 420–426.
[15] R. Gerard, "Additive manufacturing at CERN", presented at Additive Manufacturing conference Physics of the two infinities, December 2018.

Acknowledgments

I. Novitski and A.V. Zlobin are thankful to Justine Carmichael (now at PSFC MIT) for the design of 3D coil end parts for the 15 T dipole demonstrator MDPCT1 and Daniel Schoerling (CERN) for the part procurement; Carry Lawless for the SMCT part procurement and to Allen Rusy, James Karambis, Mike Smego, Randy Wyatt, and Craig Bradford (Fermilab) for the technical support of this work.

A review of devices and methods for measuring thermal emissivity at cryogenic temperatures

S Y Zhang[1,2], Y Li*[1] and L F Li[1,2]

[1] Key Laboratory of Cryogenics, Technical Institute of Physics and Chemistry, Chinese Academy of Sciences, Beijing, 100190, China.
[2] University of Chinese Academy of Sciences, Beijing 100049, P. R. China.

Email: liyong1897@mail.ipc.ac.cn

Abstract. Thermal emissivity is an important parameter in characterizing the thermal radiative properties of materials and a critical factor in radiative heat transfer. The radiation of an actual object is different from that of the blackbody: its radiation energy varies irregularly with wavelength and temperature. To investigate the thermal radiative properties, thermal emissivity could be measured at cryogenic temperatures. In this paper, measuring methods and factors affecting the thermal emissivity of metals and infrared coatings at low temperatures were comprehensively reviewed. Two existing methods for measuring thermal emissivity at low temperatures, the calorimetry and radiometry methods, were described. Moreover, this work also discussed several factors such as material properties, surface temperature, and surface conditions. Testing devices for measurement of emissivity at low temperatures based on calorimetry and radiometry were reviewed and analyzed emphatically.

1. Introduction

Deep space is vacuum and cryogenic; therefore, radiation is the only way to transfer heat. To make sure of the temperature and calculate the heat flux, it is crucial to measure the thermal emissivity of metal and infrared coatings at low temperatures in the design of satellite detectors and large-scale cryogenic systems. However, it is a great challenge to accurately measure the low-temperature thermal emissivity of metal and infrared coatings, because the wavelengths of thermal radiation at low temperatures are mainly above 20 μm, and the background radiation can bring great uncertainty. For example, the James Webb Space Telescope (JWST), developed by the National Aeronautics and Space Administration (NASA), in which the detector is designed to operate at 40 K, is one such suitable device. Since the JWST Observatory is too large to test thermal balance indoors, the thermal design relies heavily on simulation analysis. So, accurately measuring the emissivity of materials is of great importance to achieving the goal of tasks [1-5].

The radiation of an actual object is different from that of the blackbody: its radiation energy varies irregularly with wavelength and temperature. If the surface temperature of the object is higher than 0 K, it will emit radiation energy of different wavelengths in different directions in the hemispherical space above the surface and absorb radiation energy from all directions. The material emissivity is defined as the ratio of the total energy emitted at all wavelengths by the actual object into the hemispherical space per unit time and unit area to the total energy radiated by the blackbody at the same temperature [6]. The radiation intensity of actual objects varies with wavelength and spatial direction, so the emissivity can be divided into spectral emissivity, directional emissivity, and average hemispherical emissivity.

In recent years, some researchers have developed vacuum test devices that can be used to measure thermal emissivity at low temperatures based on Planck's and the Stephen-Boltzmann's laws. This paper summarized the measuring methods and factors affecting thermal emissivity at low temperatures. According to the different measuring methods, the low-temperature emissivity test devices were summarized and potential development of this field discussed.

2. Measuring methods and factors affecting thermal emissivity at low temperatures

2.1. *Measuring methods*

Using a spectrophotometer to measure the emissivity, absorptivity, and reflectivity of materials within a specified wavelength and range of incident angle is the common method. This optical method is preferred for determining absorbance and reflectance but is very expensive and time-consuming. Methods for measuring the emissivity of metals and coatings at low temperatures are often divided into: calorimetry and radiometry method [7]. Calorimetry is the earliest method used to measure the thermal emissivity of materials. Its principle is to measure the value of emitted or absorbed heat of the sample and then calculate the thermal emissivity of materials. Calorimetry method can be divided into transient calorimetry and steady-state calorimetry method. In transient calorimetry, the sample is kept at a specified temperature with one side coupled to the external environment through a thermal connection, whereas the thermal connection on the other side is almost negligible. After breaking the thermal equilibrium of the sample, it returns to its initial state. The value of thermal emissivity can be calculated from the relaxation time, sample size, and heat capacity. In steady-state calorimetry, known as static calorimetry, a thermal connection between the sample (or heat receiver) and the heat sink is set up at a constant temperature. The sample under test is the emitter, and a material with a large emissivity is used as the reference. The temperature difference between the two ends of the heat sink is used to calculate the heat flux. The radiation measurement method measures the emitted or reflected electromagnetic radiation to calculate spectral and directional emissivity.

2.2. *Factors affecting emissivity*

The thermal emissivity of metals and coatings depends on the material, its surface temperature, and the condition of the surface. Zhirui et al. [8] summarized the factors affecting the thermal emissivity of infrared ceramics, including crystal structure, ion doping, powder fineness, coating thickness, and surface roughness. The doping method was used to increase the complexity and asymmetry of the spinel and cordierite structures to improve their infrared emissivity. The authors propose that if the particle size is decreased, the coating thickness is reduced or the roughness is increased, the total value of thermal emissivity will be increased.

The temperature and treatment of the surface of coatings also affect the thermal emissivity of the material. Kralik et al. [9] discussed the effects of the temperature and surface treatment, such as chemical and mechanical polishing, roughness, and number of layers, on the hemispherical emissivity. The total values of thermal emissivity increased with temperatures rising from 20 K to 300 K.

2.3. *Uncertainty of thermal emissivity measurement*

There is systematic and measurement uncertainty in emissivity measurement, different testing methods resulting in various uncertainties. In the calorimetric method, the size of the sample, thermometers, and heating power are the main sources of uncertainty. The major sources of uncertainty in testing devices are various thermal leakages from the sample. Thermal losses are as follows:

- thermal conduction from the platform to the sample;
- radiation thermal leakage of the sample;
- thermal conduction from the sample to the sample container walls by residual gas conduction.

In the radiometric method, thermometers, as well as the precision of temperature control of the sample and the blackbody, can affect the accuracy of emissivity measurement [9].

3. Research progress on devices for testing thermal emissivity at low temperatures

The measurement of the thermal emissivity of materials at low temperatures is of great significance in the design of large-scale cryogenic systems and satellite detection systems [10-12]. High emissivity coatings are commonly used as heat-dissipating surface materials in thermal control systems. The hemispherical emissivity of the surface of coatings are usually very high (close to 1) above 100 K but usually drops sharply at low temperatures. Accurate measurement of the relationship between emissivity and temperature is essential for designing low-temperature passive cooling elements. In recent years, researchers have developed different devices to measure the emissivity of materials at low temperatures.

3.1. Testing devices based on calorimetry

Calorimetric measurement can continually monitor thermal emissivity from low temperature to room temperature and has the advantages of simple sample preparation, convenient operation, and real-time monitoring. Table 1 showed the parameters of low-temperature emissivity measurement devices using the calorimetry method.

Table 1. The parameters of low-temperature emissivity measurement devices using the calorimetry method.

No	Author	Temperature/K	Size/mm	Material	Heating	Thermometer
1	D Giulietti	77,195,273,293	170× 9× 0.018	copper foil, 3M black velvet paint	——	microwave dilatometer
2	Duckworth R.C.	25K-35K	Φ200	Silver-plated copper	Heater	Carbon glass resistors
3	Manjunath Hatti	90K-200K	110× 90 ×2.5	Cu, Al, stainless steel, black paint	Stainless steel heater	——
4	Tomas Kralik	20K-300K	Φ40× (1–5)	Al, Cu, DLC , Ti-6Al-4V	Electrical heater	Lake Shore 340
5	J. Tuttle	20K-300K	Φ190	BIRB, stainless steel tubing	Wire-wound heater	——

A study by Giulietti et al. [13-14] showed a calorimetric method for determining the total emissivity of solid materials and coatings at 293 K, 273 K, 195 K, and 77 K. Furthermore, by recording the thermal expansion of the foil instead of directly measuring its temperature, certain systemic uncertainties unique to measuring radiation at low temperatures can be eliminated. Compared with other experimental results and theoretical predictions, the preliminary measurement results at four temperatures of industrial copper foil and black paint were given.

Despite the testing devices achieving thermal emissivity measurement at low temperatures, continuous measurement was not available. A study by Duckworth et al. [15] proposed an experimental device based on calorimetry to measure thermal emissivity of silver-plated copper in the temperature range of 25 K to 35 K, and the parameters of the device were shown in Table 1. The study found that the thermal emissivity of the silver-plated copper surface at 30 K was less than half of that at room temperature, results which were consistent with previous experimental measurements. In addition, the influence of direct water pollution was discussed, and it was found that initial water pollution had the greatest impact on thermal emissivity. Based on steady-state calorimetry, similarly, Hatti et al. [16] fabricated an experimental device for measuring hemispherical emissivity in a vacuum. The device was a cylindrical container filled with liquid nitrogen, with an inner diameter of 215 mm and a height of about 900 mm. The internal vacuum chamber was cooled by concentric liquid nitrogen. They measured the thermal emissivity at temperatures as low as 90 K of various coatings and components, obtaining hemispherical

IOP Conf. Series: Materials Science and Engineering 1241 (2022) 012050 doi:10.1088/1757-899X/1241/1/012050

emissivity for stainless steel, copper, aluminum tape, copper tape, aluminized polyester film, and black coatings.

To measure the hemispherical emissivity for a larger temperature range, Kralik et al. [17-25] designed a testing device based on steady-state calorimetry. They used two parallel disks with very small gaps as the emitter and the absorber, respectively. They developed a device to measure the total hemispherical emissivity and absorptivity of various materials in a vacuum. The technical parameters of the device are shown in Table 1. Emissivity can be calculated by equation (1-3).

$$\varepsilon(T_R)=\frac{1}{\sigma T_R^4}\int_0^{\infty}\varepsilon(T_R,\lambda)e(T_R,\lambda)d\lambda \tag{1}$$

$$\frac{Q_R}{A}=\int_0^{\infty}\frac{e(T_R,\lambda)-e(T_A,\lambda)}{\frac{1}{\varepsilon(T_R,\lambda)}+\frac{1}{\varepsilon(T_A,\lambda)}-1}d\lambda\approx\sigma\varepsilon_{RA}(T_R,T_A)(T_R^4-T_A^4) \tag{2}$$

$$\varepsilon=\varepsilon_{RA}\frac{1}{1-\varepsilon_{RA}(1/\varepsilon_{REF}-1)} \tag{3}$$

where σ stays for Stefan–Boltzmann constant, T_R and T_A are respectively temperatures of radiation and material, $\varepsilon(T, \lambda)$ hemispherical spectral emissivity and $e(T, \lambda)$ is spectral distribution of hemispherical emissive power of a black body. Q_R denotes heat flux transferred between infinite plane parallel surfaces, A is the sample area and ε_{RA} is mutual emissivity of the two surfaces.

Using aluminum as a sample, they conducted a detailed correction analysis of possible interference effects and corrected the radiation leakage caused by the sample structure, the non-gray surface of the test sample, and the lack of reference blackness. They also analyzed the uncertainty of the measuring device. The emissivity uncertainty of aluminum was less than 11% at 30 K and less than 7% when the temperature is above 60 K. They summarized the hemispherical emissivity data of common metals between 20 K and 300 K, as shown in Fig.1(a). The database included 58 materials with different surface treatments, such as mechanical and chemical polishing, fine turning, sandblasting, grinding, etching, and gold deposition. They also compared the hemispherical emissivity of different diamond-like carbon coatings on aluminum and copper substrates from low to high temperatures, as shown in Figure 1(b). The database provided a solid foundation for the calculation of radiant heat flow and thermal design of low-temperature devices.

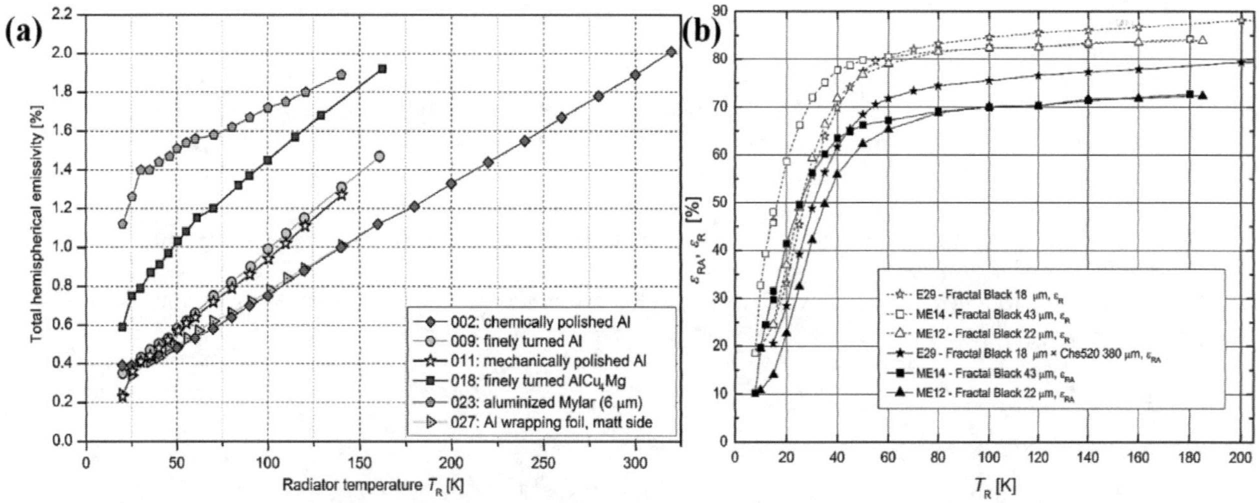

Figure 1. (a) Emissivities of Aluminum, Duralumin and Aluminum foils [17]. (b) The emissivity of Fractal Black™ [18].

The measurement of thermal emissivity at low temperatures usually had the coated sample suspended in a low-temperature chamber, and the heat radiated to the surrounding environment is measured. However, there were several problems with this measuring method: for example, the difference between the wavelength spectrum of the background radiation cavity and the sample, the heat emitted by the sample reflected from the wall of the testing cavity, and non-radiative heat leakage. Moreover, a larger low-temperature chamber was commonly required to reduce the influence of these problems. In the low-temperature chamber, the radiative heat of the sample can be reflected multiple times. Tuttle et al. [26-30] described an accurate and relatively inexpensive method for measurement of thermal emissivity in a small cryostat, as shown in Figure 2. They designed an experimental device for measuring emissivity, and the technical parameters of the device are shown in Table 1. Based on the Stefan-Boltzmann formula, the slope can be used to eliminate systematic uncertainty caused by the calibration deviation of the thermometer, and the uncertainty remains unchanged within a small temperature range. It also eliminated uncertainty caused by any heat leakage of the sample, such as wire conduction due to a constant temperature offset between the sample and the suspension ring. They used this device to measure the thermal emissivity of an aluminum-coated honeycomb and various black coatings between 20 K and 300 K, including a coating, Ball InfraRed Black (BIRB), specifically used in space flight. BIRB has been used in the aerospace field. Experimental results showed that its emissivity was higher than any other qualified coatings used in space below 100 K.

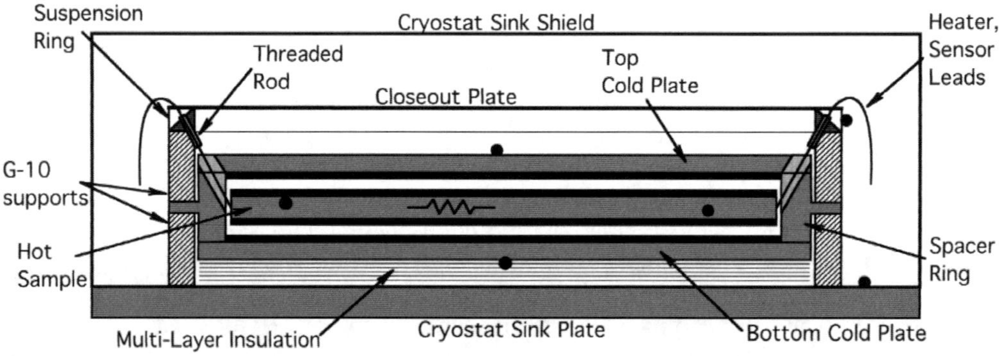

Figure 2. A schematic view of the test apparatus. The BIRB is shown as thick black lines, and thermometers as black dots [26].

3.2. Testing devices based on radiometry

Calorimetric emissivity testing devices are relatively simple and can be used within a wide temperature range with high accuracy. However, this method is applied mainly to measure total hemispherical emissivity. It takes a long time to prepare the sample and to measure emissivity. With the radiometric method, a very sensitive bolometer compares the radiative flux emitted from the surface with the blackbody at the same temperature. This method is suitable for the measurement of directional and spectral emissivity. Table 2 showed the parameters of low-temperature emissivity measuring devices using the radiometry method.

Table 2. Parameters of devices of emissivity measurement at low temperatures by radiometry.

No	Author	Temperature/K	Material	Spectrum/μm	Spectrograph
1	Philippe Hervé	20 K–200 K	coatings	6–800 μm	Thermograph
2	S I Woods	80 K–300 K	SS304L stainless steel coated with silver	Full	ACR
3	Haiyong Gan	100 K–300 K	——	250 nm–26 μm	ACR
4	Jiayu Ma	55 K, 173 K, 213 K, 298 K,	Quartz, stainless steel, and the blackbody	8–14 μm	FTIR

Hervé et al. [31] developed a method for directly measuring the total directional emissivity of various coatings required by satellites. They developed a testing device using a two-stage helium refrigerator as the cooling source, as shown in Fig. 3. The emissivity of different coatings in the temperature range of 20 K–200 K was measured. They analyzed the system uncertainty of the device and the measuring uncertainty of the experiment, and used the iterative difference method and precise adjustment of the cavity wall to greatly reduce the impact of disturbance.

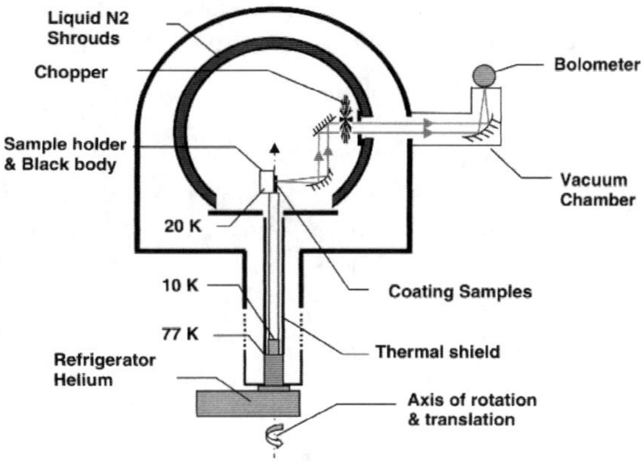

Figure 3. Schematic diagram of the experimental setup [31].

Woods et al. [32-33] developed a technique for measuring the emissivity of high-reflectivity samples at low temperatures based on radiometry. The method carefully quantified any background radiation directly or reflected by the detector. This measuring method used an absolute cryogenic radiometer (ACR) to directly measure the radiant power. The light emitted by the sample was detected by the ACR. The absorption of the ACR in the mid-infrared band was about 99.93% of the radiated power. Absorption was greater than 99% of the radiated power when the wavelength was less than 100 μm and greater than 97% of the radiated power at less than 200 μm. This method can be used to determine the emissivity of large-area, highly reflective samples at temperatures between 80 K and 300 K. The total absolute uncertainty (k = 1) was approximately 0.0002 to 0.002, and the emissivity was approximately 0.0035 to 0.092.

The China National Institute of Metrology has developed an absolute low-temperature radiometer (ACR), which can be used for optical radiation power measurement [34]. The ACR cavity receiver was plated with pure copper, and the inner surface of the ACR cavity receiver was coated with a polymer material mixed with carbon nanotubes. When the optical radiation power was nearly 100 microwatt hours, the relative standard uncertainty of the optical radiation power calibration was 0.042%. When the optical radiation power was nearly 1 microwatt hour, the relative standard uncertainty was 0.22%. The relative standard uncertainty of the ACR system when calibrating a total radiation blackbody radiation source of 100 K was less than 0.5%. After measuring the spectral radiance at a certain wavelength, the spectral emissivity of the radiation source in a wider spectral range can be inferred by measuring the spectral diffuse reflectance at this wavelength.

Ma et al. [35] introduced the experimental system and measurement principle of low-temperature spectral emissivity measurement of thermal control materials in a vacuum. In this device, the Fourier Transform Infrared Spectrometer (FTIR) using step-scan mode can cooperate well with the chopper and the lock-in amplifier, effectively and easily eliminating the influence of background radiation. The sample cooling chamber had an off-axis parabolic gold-plated mirror and a gold-plated plane mirror. The reflectivity of both mirrors was greater than 0.96. A parabolic mirror was used to collect the radiation of the sample and convert it to a parallel beam. Using a plane mirror to change the direction of parallel beam propagation, the detector in the infrared spectrometer can receive radiation. The spectral emissivity of the quartz sample was about 0.9 in the wavelength range of 8–14 μm, with little change in the wavelength range of 8–10 m. The spectral emissivity of the stainless steel sample was between 0.11 and 0.15. The spectral emissivity of the stainless steel decreased with increasing wavelength, and that of the thermal control material had a wavelength range of 8–14 μm between 173 K and 213 K. The comprehensive standard uncertainty of the measurement apparatus was 3.30%.

4. Conclusion

At low temperatures, thermal radiation heat transfer is the main mode of heat conduction. Emissivity is an important parameter for characterizing the thermal radiation of materials and is widely used in the radiative cooling of detectors, optical components, space telescopes, cryogenic cooling equipment, etc [36-38].

Emissivity is greatly affected by temperature and surface conditions, such as surface treatment, roughness, oxidation, etc. Accurate and quick measurement of emissivity at low temperatures is of great significance in aerospace applications, industrial energy conservation, and medical rehabilitation. This paper not only summarized the influencing factors of emissivity but also reviewed testing devices for measurement of emissivity at low temperatures based on calorimetry and radiometry. Furthermore, the measurement principles, structure of apparatus, samples, and parameters of apparatus are summarized. This review makes a major contribution to understanding the measurement technology of emissivity at low temperatures. We believe that through the improvement of experimental devices and testing techniques, the emissivity of materials at lower temperatures can be measured more accurately and quickly.

5. References

[1] Arenberg J W, Adamson J, Harpole G, Macias M, Niedner M B, Bowers C W, Mehalick K I, and Lightsey P A 2014 *Space Telescopes and Instrumentation 2014: Optical, Infrared, and Millimeter Wave* **9143**.
[2] DiPirro M, Tuttle J, Hait T, and Shirron P 2014 *Advances in Cryogenic Engineering* **1573** 110-7.
[3] Franck R A 2014 *Cryogenics* **64** 235-9.
[4] Lightsey P A, Wei Z Y, Skelton D L, Bowers C W, Mehalick K I, Thomson S R, Knollenberg P, and Arenberg J W 2014 *Space Telescopes and Instrumentation 2014: Optical, Infrared, and Millimeter Wave* **9143**.
[5] Shinozaki K, Nishikawa I, Tanaka K, Ando M, and Sugita H 2019 *IOP Conference Series: Mat. Sci. Eng*. **502** 012070.
[6] Fu T, Tan P, and Pang C 2012 *Meas. Sci. Technol.* **23** 025006.
[7] Králík T, Musilová V, Hanzelka P, and Frolec J 2016 *Metrologia* **53** 743-53.
[8] Rui Z Y L, Xin Z. Pengwei J. Xue Z 2020 *Shandong Ceramics* **43** 8-10.
[9] Cárdenas-García D and Monte C 2014 *Int. J. Thermophys* **35** 1299-309.
[10] Hasebe T, Sekimoto Y, Dotani T, Mitsuda K, Shinozaki K, and Yoshida S 2019 *J. Astron. Telesc. Inst.* **5** 9.
[11] Chang C W 2002 *Optical System Contamination: Effects, Measurements, and Control Vii* **4774** 119-28.
[12] Tommasi L, *et al.* 2018 *2018 5th Ieee International Workshop on Metrology for Aerospace* 247-51.
[13] D Giulietti A G, M Lucchesi and R Stampacchia 1979 *J. Phy. D.* **12**.
[14] Lucchesi D G M 1981 *J. Phy. D.* **14 (1981) 877-81**.
[15] Duckworth R C 2006 *AIP Conference Proceedings* **824** 61-8.
[16] Manjunath Hatti S K, Thamme gowda C S 2018 *IJEDR*. **6**.
[17] Frolec J, Králík T, Musilová V, Hanzelka P, Srnka A, and Jelínek J 2019 *Cryogenics* **97** 85-99.
[18] Kralik T and Katsir D 2009 *Proceedings of SPIE - The International Society for Optical Engineering* **7298** 729813.
[19] Hanzelka P, Kralik T, Maskova A, Musilova V, and Vyskocil J 2008 *Cryogenics* **48** 455-7.
[20] Musilova V, Hanzelka P, Kralik T, and Srnka A 2005 *Cryogenics* **45** 529-36.

[21] Musilova V, Kralik T, Hanzelka P, and Srnka A 2007 *Cryogenics* **47** 257-61.
[22] Hanzelka P, Musilova V, and Kralik T 2010 *Cryogenics* **50** 331-5.
[23] Kralik T, Hanzelka P, Musilova V, Srnka A, and Zobac M 2011 *Rev. Sci. Instrum.* **82** 055106.
[24] Kralik T, Hanzelka P, Zobac M, Musilova V, Fort T, and Horak M 2012 *Phys. Rev. Lett.* **109** 224302.
[25] Frolec J, Králík T, and Srnka A 2017 *Int. J. Refrig.* **82** 51-5.
[26] Tuttle J, Canavan E, DiPirro M, Li X, and Knollenberg P 2014 *AIP Conference Proceedings* 590-6.
[27] Tuttle J, Canavan E, DiPirro M, Li X, Franck R, Renbarger M, Copp T, and Hahn M 2014 *Cryogenics* **64** 240-3.
[28] Tuttle J and Canavan E 2015 *IOP Conference Series: Materials Science and Engineering* **102** 012015.
[29] Franck R and Renbarger M 2012 *AIP Conference Proceedings* **1434** 1497-504.
[30] Jahromi A E, Tuttle J G, and Canavan E R 2017 *IOP Conference Series: Mat. Sci. Eng.* **278** 012002.
[31] Philippe Hervé N R, Abdelkader Sadou, David Ramel, Laurent Francou, P. Delouard, Emmanuel Gavila 2008 *HAL* **48** 463-8.
[32] Woods S I, Jung T M, Ly G T, and Yu J 2012 *Metrologia* **49** 737-44.
[33] Woods S I, Jung T M, Sears D R, and Yu J 2014 *Cryogenics* **60** 44-8.
[34] Gan H Y, Xu N, Liu X L, He Y W, Wu H P, Feng G J, Liu W D, and Lin Y D 2019 *AOPC 2019: Optical Sensing and Imaging Technology* **11338** 113382Q.
[35] Jiayu Ma Y Z, Lingnan Wu, Haogeng Li and Lixin Song 2019 *Materials (Basel)* **12** 1141.
[36] Jamotton P, Cucchiaro A, Domken I, Liebecq S, and Delree C 2009 *Vacuum* **83** 1244-7.
[37] Stout K D and Masterson R A 2014 *Modeling, Systems Engineering, and Project Management for Astronomy Vi* **9150**.
[38] Yenni G R, Ambirajan A, Balaji C, and Venkateshan S P 2019 *Heat and Mass Transf.* **55** 1465-76.

Acknowledgments
This work was supported by the CAS Key Laboratory of Cryogenics, Technical Institute of Physics and Chemistry (Grant No. CRYOQN202104).

IOP Conf. Series: Materials Science and Engineering 1241 (2022) 012051 doi:10.1088/1757-899X/1241/1/012051

Design and thermal study of a test device for breakdown characteristics of insulation materials in low temperature vacuum environment

Jun Huang, Qiufu Xu, Guodong Wang*

School of Mechanical Engineering, Hefei University of Technology, Hefei 230009, China

*Corresponding E-mail:gdwang@hfut.edu.cn

Abstract. With the application and development of low-temperature superconductivity technology, the research on electrical properties of insulating materials under different low-temperature conditions has attracted more and more attention. According to the breakdown characteristics of materials in low-temperature environments, a test device for the discharge breakdown characteristics of insulating materials in vacuum low-temperature environments is designed. This device can meet test of the gaseous and solid insulating materials in different low temperature and vacuum conditions. Compared with the traditional low temperature electrical test device, it has the advantages of convenient replacement, short test cycle and multi-functional testing. The finite element simulation software is used to simulate the performance of the cooling device of gas and solid insulating materials, and the effects of liquid helium cooling and refrigerator cooling on the temperature distribution of the device are analyzed and compared. Finally, the flow field analysis of the cooling system defines the internal flow characteristics of liquid helium under liquid helium cooling, which provides theoretical guidance for the design of the low-temperature breakdown test device.

Keywords: low temperature; breakdown characteristics; insulation materials; cooling method.

1. Introduction

With the development of low-temperature superconductivity technology, low-temperature insulating media are more widely used. For different low-temperature conditions and vacuum pressures, the situation of temperature influence on the electrical properties of insulating materials is getting more and more attention. The electrical properties of low-temperature insulating media have become an important factor affecting the performance and reliability of low-temperature superconducting devices[1]. Therefore, it is far-reaching significance to have a discharge device that can efficiently perform dielectric breakdown tests in a wide range of low-temperature and high-vacuum environments for the study of low-temperature insulating materials.

In recent years, scientific researchers have conducted a lot of studies and developed a series of low-temperature dielectric electrical characteristic test devices. G. A. Farrall et al.[2] concluded that the breakdown voltage of vacuum insulation in a uniform electric field at low temperatures is related to the electrode spacing and electrode condition. The breakdown characteristics of vacuum insulation at low temperature and room temperature were compared by Hao Fengnian et al.[3] The spherical electrode attached below the cold bath was cooled by filling liquid nitrogen, and the results showed that the breakdown strength of vacuum insulation was significantly higher at low temperatures for high-voltage electrodes. J.Gerhold[4] found that the low temperature between 4.2-300K had different effects on electrical properties of solid insulating materials such as dielectric constant, dielectric loss and breakdown voltage. E. Husain et al.[5] designed a low-temperature experimental setup with test electrodes immersed in liquid nitrogen to measure and evaluate the AC breakdown voltage of insulating materials under different electrode configurations. The experimental results showed that the breakdown voltage was a function of the electrode geometry and gap length. I. Sauers et al.[6] designed and fabricated a cryogenic vacuum test device using a copper coil connected to a ground electrode and a liquid nitrogen pipe to achieve cooling, so as to study the effect of temperature on the breakdown strength of vacuum insulation and the flashover characteristics along the surface of glass fiber reinforced plastics. However, the controllable temperature range of immersion cooling device is small, and immersion may affect the performance of the insulation material. While the test device by cooling the electrode may lead to a difference between the insulation medium temperature and the electrode temperature. In this paper, a self-designed low-temperature insulation material breakdown characteristics test device is studied. The cooling process and temperature field distribution of the device under different cooling methods are simulated and compared.

2. Design of the test device

Insulation medium at low temperature plays a very critical role in the application of superconductivity technology. Insulation materials of electrical components work under high vacuum and deep low temperature for a long time, which may affect their electrical properties and lead to the generation of insulation medium discharge breakdown phenomenon. As a result, there is a need for a discharge device that can test the breakdown characteristics of insulation media in a low-temperature vacuum environment.

This paper presents a device for testing the discharge breakdown characteristics of insulating media in a low-temperature vacuum environment to meet the test experiment of discharge breakdown characteristics of solid insulating media under different low temperature, vacuum pressure and gas atmosphere conditions, and this device can also realize the test of insulating gas discharge breakdown performance at low temperature after replacing the test sample with a specific insulating gas and replacing the corresponding electrode. Figure 1 shows the schematic diagram of the structure of this device. In ultra-low temperature testing devices, refrigerators are commonly used, and their advantages such as reliability and stability are favored by many experts, so the device also chooses the refrigerator as the cold source to work[7-8].

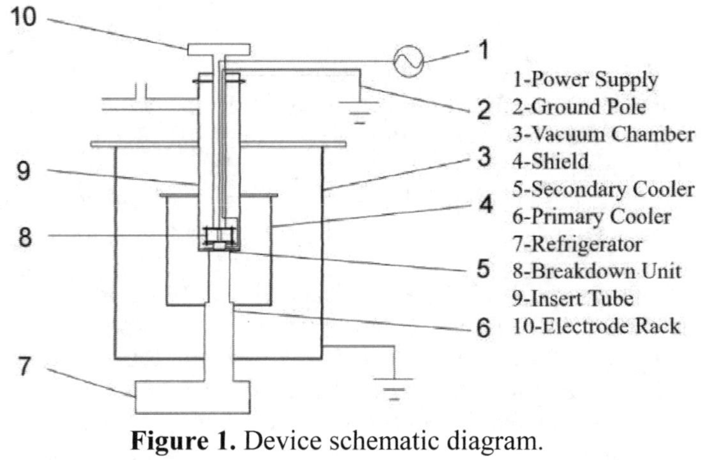

Figure 1. Device schematic diagram.

The vacuum cavity in this device provides a vacuum adiabatic environment for the test, and the refrigerator is used to provide cooling. The primary cold head of refrigerator is connected to the shield, which keeps the shield at a low temperature and plays a role in reducing the radiation heat leakage. The secondary cold head of refrigerator is connected to the copper bottom of the high-voltage electrode holder insertion tube, and the cold capacity is transferred to the test sample through the copper bottom, the insulated cold conductor and the lower electrode in turn, so that the replacement of the test sample does not destroy the vacuum adiabatic chamber vacuum environment. The copper bottom end is equipped with a heater and sensor to achieve the role of temperature control. The insulated cooler is used to transfer the cold and prevent the discharge current from flowing through other devices. The lower end of the high-voltage electrode holder is detachably connected to the high-voltage electrode testing device, which can be easily removed for quick sample replacement. The upper electrode also has a detachable threaded connection, which is used to replace the upper electrode with a different shape for testing. Figure 2 shows the three-dimensional model of the breakdown unit.

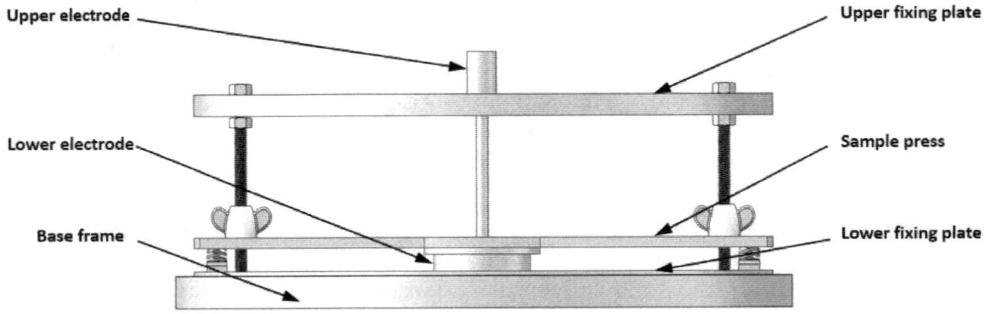

Figure 2. 3D model drawing of breakdown unit.

The upper end of the high voltage electrode holder insertion tube contains side pipe connections for connecting the vacuum pump unit, vacuum gauge and gas source bottle to obtain and monitor the vacuum environment for the high voltage electrode holder insertion tube as well as to introduce specific gases. The gas source bottle interface can be used to access different gas sources to achieve the discharge breakdown characteristics of low temperature solid insulation media under different gas atmospheres for testing and research. After the test sample is replaced with a specific insulating gas, the insulating gas discharge performance test at low temperature can be realized by changing the shape and length of the electrode. Therefore, the device designed in this paper has the advantages of convenient sample replacement, short testing cycle, high testing efficiency and multi-functional testing.

3. Simulation Analysis

3.1 Cooling model and simulation parameter settings

Two forms of cooling using liquid helium or refrigerators as the cooling source are commonly accepted in the design of ultra-low temperature devices. In this section, numerical calculations and simulations of the heat transfer between the two cooling methods are compared by finite element simulation software for this device. According to the literature [9-10], the relevant physical parameters of liquid helium can be obtained. Figure 3 shows the 3D model and its dimensions for the simulation calculation.

The refrigerator cooling model uses the physical field of "solid and fluid heat transfer" for the control solution. The boundary conditions are set as follows: the bottom contact surface temperature is set to 4.2 K, the boundary cold source is set to 1.5 W, and the top surface heat source is naturally connected with the air. The physical process of the liquid helium cooling involves fluid flow, fluid heat transfer, and solid heat transfer. "Solid and fluid heat transfer", "turbulent flow (k-ε)" and "non-isothermal heat transfer multi-physics field" of the multi-physics field coupling module are used for the solution. The boundary conditions are set as follows: the left pipe port is the inlet, and the inlet boundary condition is the mass flow rate; the right pipe port is the outlet, and the boundary condition is the average pressure of fully developed flow, and the reference temperature is 293.15K. One of the difficulties of turbulence calculation is that the flow of fluid near the wall is complicated, so the mesh division at the boundary layer of the wall should be refined to improve the computational convergence and correctness. The inlet mass flow rate is set to 20g/s, 40g/s, 60g/s and 100g/s, and the simulation is compared with a 1.5W refrigerator for comparative analysis.

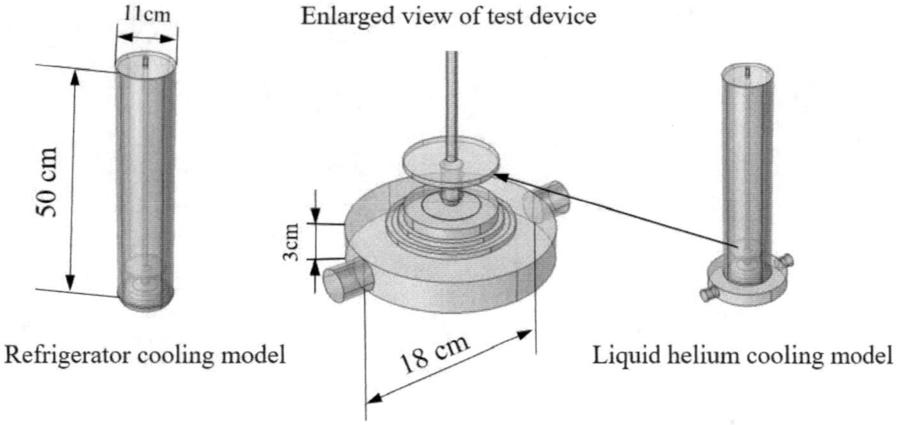

Figure 3: 3D simulation model and its dimensions.

3.2 Analysis of simulation results

3.2.1 Comparison of refrigerator cooling and cooling process under different mass flow.
Both the refrigerator and the liquid helium start to cool down from the device at 293.15 K. Since the sample is closer to the cold source end, the device cools down from the bottom to the top end. Therefore, the cooling rate at the sample end is very fast at the beginning, and the rate of temperature decrease becomes smaller as the cooling time increases. Figure 4 shows the temperature distribution at different cooling moments for a mass flow rate of 20 g/s.

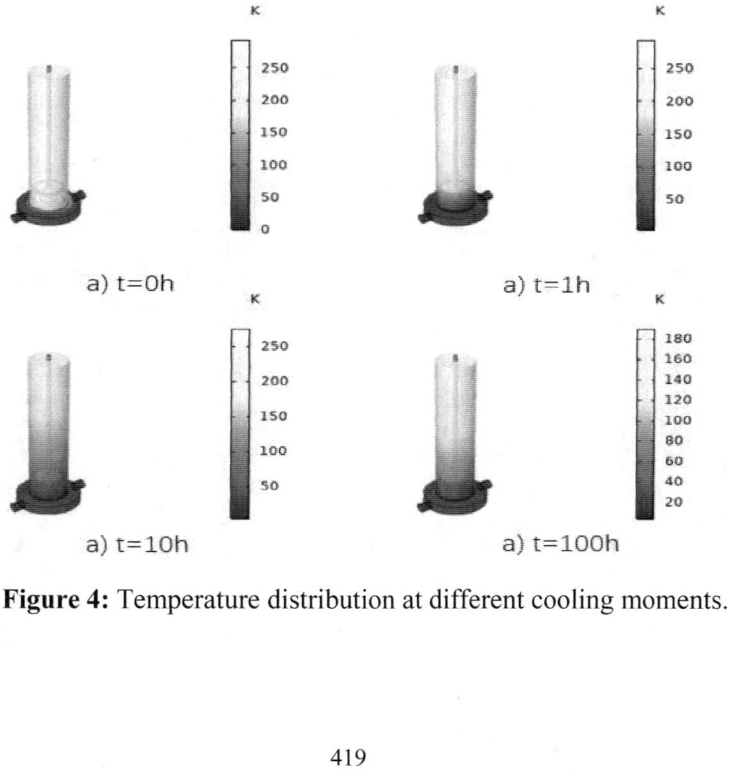

Figure 4: Temperature distribution at different cooling moments.

Figure a) in Figure 5 shows the data curves for the cooling of the refrigerator versus the cooling of the liquid helium at different mass flows. Within one hour, the refrigerator cooling is lower than 20 g/s to 100 g/s at each time period, which indicates that the cooling rate of the refrigerator is highest at the beginning of the cooling phase. Figure b) is a partial enlargement of Figure a). However, after about one hour of cooling, the cooling temperature of liquid helium with a mass flow of 100 g/s starts to be lower than the cooling temperature of the refrigerator. And the higher the mass flow, the faster the cooling rate.

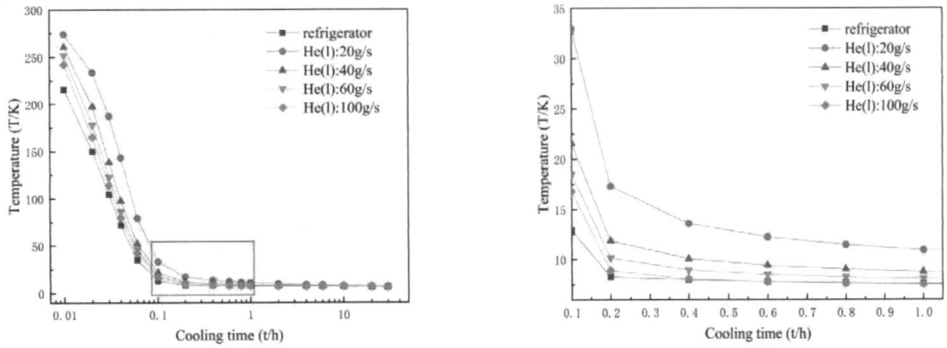

Figure 5: a) The average temperature of the sample varies over time.
b) Local amplification diagram in Figure a).

3.2.2 Comparison of the effect of two cooling forms when cooling is stable.

Figure a) in Figure 6 shows the simulation data of the cooling temperature of the solid breakdown sample at steady state cooling, where the lowest temperature that the refrigerator can lower is similar to the temperature of 60 g/s liquid helium when reaching steady state. It is only when the mass flow exceeds 60 g/s that the temperature of the liquid helium cooling is lower than that of the refrigerator. Figure b) in Figure 6 shows the simulation data of the cooling temperature of the breakdown gas under steady-state cooling. In contrast, when cooling the breakdown gas, the final average temperature of the gas achievable by the refrigerator is lower than that achieved by liquid helium cooling at 20 g/s to 100 g/s. The final temperature that the solid sample to be tested can reach is lower than that of the gas to be tested, because the heat transfer of solids is better than that of gases. The higher the mass flow of liquid helium, the higher the amount of liquid helium consumed, the lower the final temperature and the shorter the cooling time, but the cooling efficiency decreases significantly as the final temperature is approached. Therefore, liquid helium cooling is not a suitable cooling method for this device, and a refrigerator cooling method is more efficient. The simulations show that this device can reach a minimum test temperature of about 5.5 K for solid breakdown samples and below 10 K for gases to be tested.

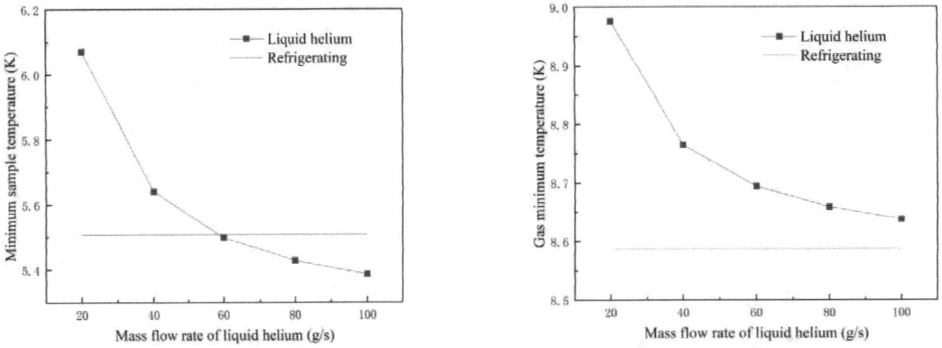

Figure 6: a) Average temperature of the sample to be measured when the cooling is stable.
b) Average temperature of the gas to be measured when the cooling is stable.

3.2.3 Liquid helium tank flow field analysis.

Figure 7 shows the flow field distribution in the liquid helium tank for different mass flows. The liquid helium flows into the liquid helium tank through the inlet and flows out from the outlet. The volume of the liquid helium tank is much larger than that in the pipe, so the flow velocity in the liquid helium tank decreases greatly, and the farther away from the inlet and outlet of the pipe, the smaller the flow velocity is. Due to the circular shape of the liquid helium tank, there is a weak reflux in the liquid helium.

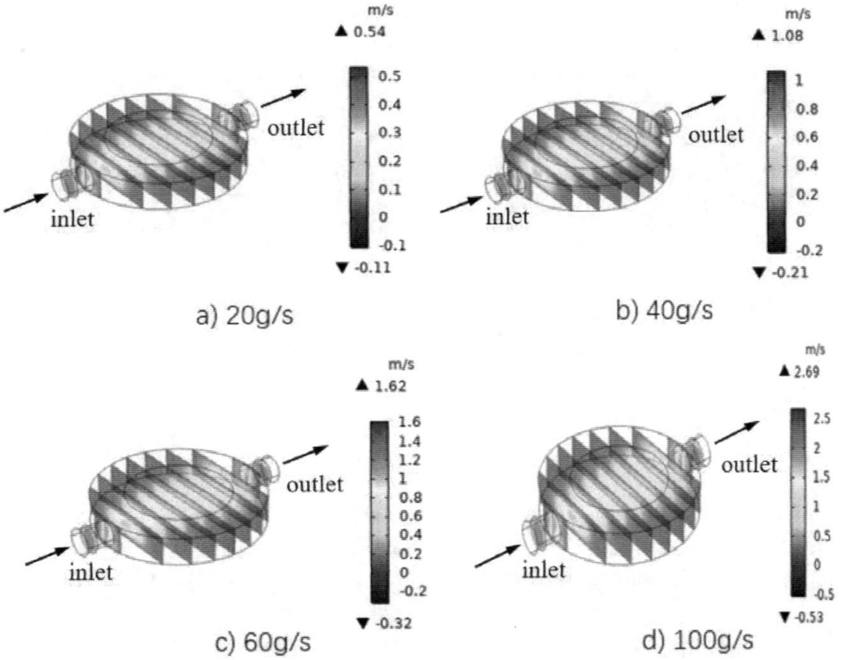

Figure 7: Flow field profiles of liquid helium tank at different mass flow rates.

As shown in Table 1, the average velocity in the liquid helium tank is more than ten times lower than that at the tube port, and the flow rate between the inlet and the outlet is larger, so it may lead to some of the liquid helium flowing out of the tank before complete heat exchange, which is an important reason for the very low cooling efficiency of liquid helium.

Table 1. Flow field data at different mass flow rates.

Inlet mass flow rate(g/s)	20	40	60	100
Inlet flow velocity(m/s)	0.73	1.47	2.2	3.67
Average flow velocity(m/s)	0.060455	0.12096	0.18146	0.30245

4. Conclusion

In this paper, a multi-functional test device that can effectively meet the breakdown characteristics of insulation materials in deep low-temperature vacuum environment is designed. The performance of the device is simulated by finite element simulation software, and the cooling effects under different refrigeration modes are compared. From the aspect of refrigeration efficiency, it is verified that the refrigerator is more suitable to be selected when designing this kind of similar devices. The reasons for the defects of liquid helium refrigeration are also preliminarily analyzed. The expected minimum temperature for the theoretical calculation of the device was obtained. This research method is also applicable and instructive for the design of similar cryogenic devices.

5. References

[1] Gerhold J 2000 *High Voltage in Superconducting Power Equipment — Prospects and Limits* (Boston, MA: Springer)

[2] Farrall G A 1985 Electrical Breakdown in Vacuum *J. IEEE Transactions on Electrical Insulation*, 20(5):815-841.

[3] Fei H and Hu W 1990 Electrical breakdown of vacuum insulation at cryogenic temperature *J. IEEE Transactions on Electrical Insulation*, 25(3):557-562.

[4] Gerhold J 1998 Properties of cryogenic insulants *J. Cryogenics*, 38(11):1063-1081.

[5] Husain E, Mohsin M M, Masood A, Zuberi M U and Alam M S 2002 Dielectric behavior of insulating materials under liquid nitrogen *J. IEEE Transactions on Dielectrics and Electrical Insulation*, 9(6):p.932-938.

[6] Sauers I, James D R, Ellis A R and Pace M O 2002 High voltage studies of dielectric materials for HTS power equipment *J. IEEE Transactions on Dielectrics and Electrical Insulation*, 9(6):922-931.

[7] Liu H M, Xu D, Xu P, Gong L H and Li L F 2011 Development of a low-temperature thermal conductivity measurement device for refrigerators as a cold source *J. Cryogenic Engineering*, 2011(001):1-5.

[8] Dai J B, Fu J, Yu H Q, He C F and Wu Y F 2016 Design and experimental study of low temperature thermal conductivity testing system for thermal insulation materials *J. Cryogenics and Superconductivity*, 044(012), 8-11.

[9] Kent A 1993 *Properties of liquid helium* (London: Palgrave).

[10] Sciver V 2011 Helium cryogenics *J. International Cryogenics Monograph*, 88(11), 4582-6

Acknowledgments

We thank the low-temperature research group of the Institute of Plasma Physics, Hefei Institute of Physical Science, Chinese Academy of Sciences, for its support and help.

Influence of magnetron sputtering process parameters on low-temperature electrical transport characteristics of zirconium oxynitride thin films

X M Sun[1,2], X B Bai[3,4], F C Chen[3,4], R J Huang[1,2*], J Yuan[3,4,5], K Jin[3,4,5], H J Su[1,2], L F Li[1,2,5]

[1] CAS Key Laboratory of Cryogenics, Technical Institute of Physics and Chemistry, Chinese Academy of Science, Beijing 100190, China
[2] College of Materials Science and Opto-electronic Technology, University of the Chinese Academy of Science, Beijing 100049, China
[3] Beijing National Laboratory for Condensed Matter Physics, Institute of Physics, Chinese Academy of Sciences, Beijing 100190, China,
[4] School of Physical Sciences, University of Chinese Academy of Sciences, Beijing 100049, China
[5] Songshan Lake Materials Laboratory, Dongguan, Guangdong 523808, China

E-mail: huangrongjin@mail.ipc.ac.cn

Abstract. In order to determine the influence of process parameters including nitrogen and oxygen flow rate on the structure and electrical transport characteristics of zirconium oxynitride(ZrO_xN_y) thin films, the ZrO_xN_y films were prepared on sapphire substrates by RF reactive magnetron sputtering deposition technology. The crystal orientation and morphology of ZrO_xN_y films prepared at different nitrogen and oxygen flow rate were characterized by XRD and SEM, respectively. The electric transport behavior of ZrO_xN_y films at 300K to 3K was measured by PPMS. The results show that the insulativity of ZrN films is enhanced with the increase of nitrogen flow rate in sputtering atmosphere. With the increase of oxygen flow rate in sputtering atmosphere, the insulativity of ZrO_xN_y film is enhanced.

1. Introduction

In recent years, transition metal oxynitrides have attracted much attention for their high hardness[1, 2], high electrical conductivity, corrosion resistance[3], good thermal stability[4, 5], and diffusion barriers[6-9]. These properties make it have good application prospects in the fields of hard coatings, diffusion barriers and microelectronic devices. In 1975, Izumi et al.[10] prepared zirconium nitride (ZrN) thin films by reactive rf magnetron sputtering, which were used as high-value resistors for integrated circuits. In 1987, Yotsuya et al.[11] developed a ZrN temperature sensing film with high sensitivity, short thermal response time and small magnetoresistance. So far, many transition metal oxynitride films have been used as temperature sensing film in cryogenic thermometers, such as chromium nitride[12] (CrN), Titanium nitride[13] (TiN), hafnium nitride[14] (HfN) and so on.

Impurities will strongly affect the physical properties of ZrN. At present, many studies have used oxygen doping to improve the optical and electrical properties of ZrN[15-20]. Doping a small amount of oxygen into ZrN will result in the formation of ionic metal-oxygen bonds in the matrix of covalent metal-nitrogen bonds. The resistivity of zirconium oxynitride(ZrO_xN_y) is very sensitive to the gas flow during reactive sputtering. The flow of nitrogen and oxygen will affect the structure and composition of the prepared MeN_x (Me = Ti, Zr, Hf) films, thereby changing the characteristics of these films. With the increase of the gas mixture flow rate, the resistivity of the film changes from a few hundred μΩ·cm to 10^{15}μΩ·cm, which leads to a significant decrease in its electrical conductivity[21, 22].

In order to explore the electrical transport characteristics of ZrN and ZrO_xN_y films at low temperatures, this work used reactive magnetron sputtering technology to prepare ZrN and ZrO_xN_y films, and analyzed the crystal structure and electric transport characteristics of ZrN and ZrO_xN_y films grown under different nitrogen pressures and different oxygen flow rates.

2. Experiments

ZrN and ZrO_xN_y films were prepared on sapphire substrates by radio frequency reactive magnetron sputtering. Before sputtering deposition, the sapphire substrates were ultrasonically cleaned in acetone and alcohol for 5 minutes to remove impurities. The background vacuum before deposition was 5×10^{-4} Pa, the total deposition pressure was 2.1 Pa, and the pre-sputtering power and time were 60 W and 10 min, respectively. ZrN films were grown with nitrogen pressure maintained at 0.2, 0.3, 0.6 and 1.0 Pa at 600 °C. Keeping the nitrogen pressure constant at 0.01 Pa, ZrO_xN_y films were prepared at 400°C with oxygen flow rates of 11, 13 and 14sccm, respectively. The deposition parameters of the two films are shown in Table 1.

Table 1. Deposition parameters of ZrN and ZrO_xN_y thin films.

Deposition parameters	ZrN films	ZrO_xN_y films
Base pressure	5×10^{-4} Pa	5×10^{-4} Pa
Substrate to target distance	36mm	36mm
Deposition pressure	2.1 Pa	2.1 Pa
Power	150 W	150 W
Sputtering time	40 min	40 min
Substrate temperature	600 °C	400 °C
N_2 pressure	0.2~1 Pa	0.01 Pa
O_2 flow rate	-	11~14 sccm

The crystal structures of ZrN and ZrO_xN_y films were analyzed by X-ray diffraction (XRD, D8 Focus, Bruker, Germany/ Rigaku SmartLab, Japan) with Cu Kα radiation. Raman spectroscopy (Invia-Qontor, Renishaw, Britain) was used to obtain the structure information at 532 nm. The resistance of ZrN and ZrO_xN_y films in the range of 300K~8K was measured using a physical performance measurement system (PPMS, PPMS-14T (EC-II), Quantum Design, USA).

3. Results and discussion

3.1. *Structure and morphology analysis*

Figure 1(a) is an X-ray θ-2θ scan of ZrN films grown under different nitrogen pressures. It can be found from the figure that the diffraction peaks of the film are mainly 33.95°, 39.48° and 41.73°, which correspond to the (111) and (002) diffraction peaks of the ZrN phase and the (006) diffraction peaks of the Al_2O_3 substrate, respectively. There is no obvious preferred orientation in the ZrN film, and the ZrN film is a polycrystalline film. When the deposited nitrogen pressure increases to 1 Pa, the diffraction peak intensity of the ZrN phase becomes extremely weak, which may be due to the disorder of the crystal structure caused by the excessive nitrogen atoms in the ZrN phase. Figure 1(b) shows the Raman spectra of ZrN thin films grown under different deposition nitrogen pressures. In the 50~900 cm^{-1} spectral range, it is observed that an asymmetric band is located at 493 cm^{-1} and the other band is sited around 172 cm^{-1} in the low frequency at different N_2 pressures. There is a shoulder at 243 cm^{-1} on the right of the 173 cm^{-1} band and 419 cm^{-1} on the left of the 493 cm^{-1} band. The positions of these bands all correspond to the characteristic bands of ZrN phase, which further confirms that the films deposited under different nitrogen pressures belong to the ZrN phase. Moreover, with the increase of nitrogen pressure, the bands at 172 cm^{-1} and 493 cm^{-1} continue to broaden, which is related to the disorder of the crystal structure caused by the disorder of nitrogen atoms, which is also confirmed by XRD.

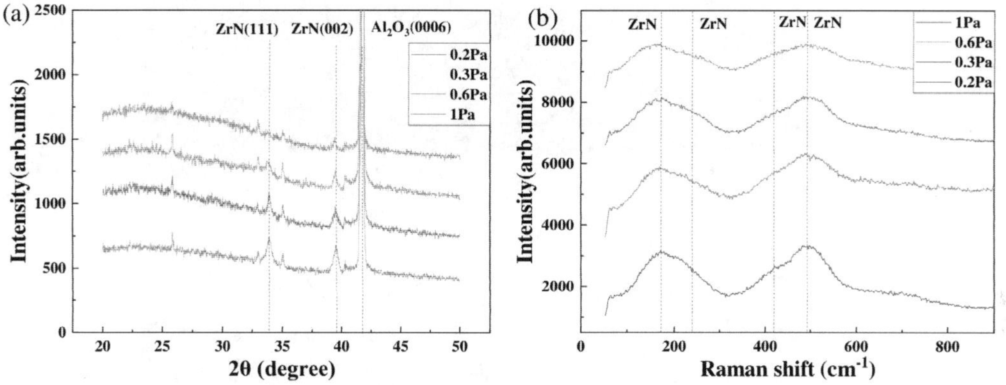

Figure 1. XRD patterns (a) and Raman spectra (b) of ZrN films grown under different nitrogen pressures.

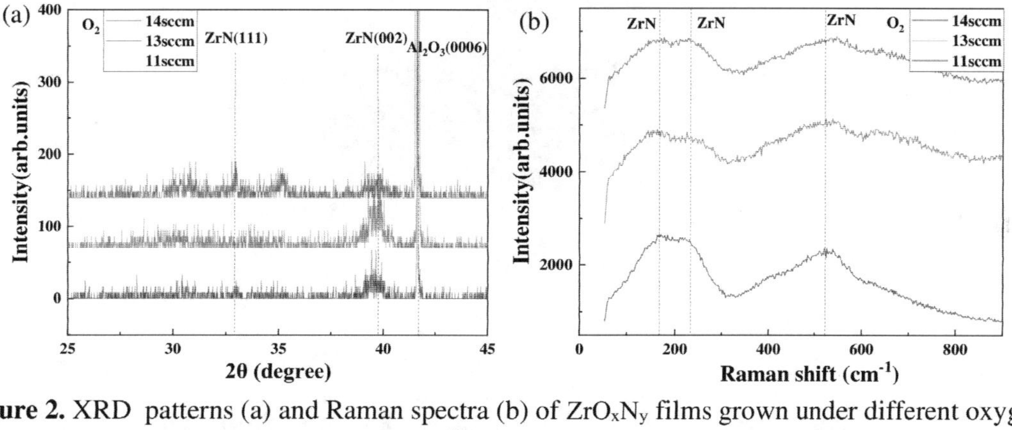

Figure 2. XRD patterns (a) and Raman spectra (b) of ZrO_xN_y films grown under different oxygen flow rates.

Figure 2(a) is an X-ray θ-2θ scan of ZrO_xN_y films grown under different oxygen flow rates. The diffraction peaks of the ZrO_xN_y film are mainly 39.71° and 41.62°, which correspond to the (002) and (006) diffraction peaks of the ZrN phase and the Al_2O_3 substrate, respectively. There is no obvious diffraction peak of zirconia, which indicates that the ZrO_xN_y film still belongs to the ZrN phase. However, the diffraction peak of the ZrN phase is very weak, which may be due to the disorder of the crystal structure of the ZrO_xN_y film caused by the doping of oxygen atoms. Figure 2(b) shows the Raman spectrum of the ZrO_xN_y film. Three asymmetric bands appear at the wavelengths centered at 168, 232, and 522 cm^{-1}. The positions of these bands correspond to the ZrN phase, further confirming that the ZrO_xN_y films grown under different oxygen flow rates belong to the ZrN phase.

Figure 3 is the cross-section morphology of ZrN and ZrO_xN_y films. It can be seen that ZrN and ZrO_xN_y films are uniform and compact with columnar crystal structure. The film thickness of ZrN film varies between 1150 nm and 716 nm, and becomes smaller when the nitrogen pressure is higher. This is because the partial pressure of nitrogen increases while the total pressure remains unchanged, the partial pressure of argon decreases and the amount of ionized argon ions decreases, which eventually leads to low sputtering rate and reduced film thickness. The thickness of ZrO_xN_y film is about 2.6 μm and has little change, indicating that the change of oxygen flow rate in the range of 11-14 sccm had little effect on the deposition rate of ZrO_xN_y film.

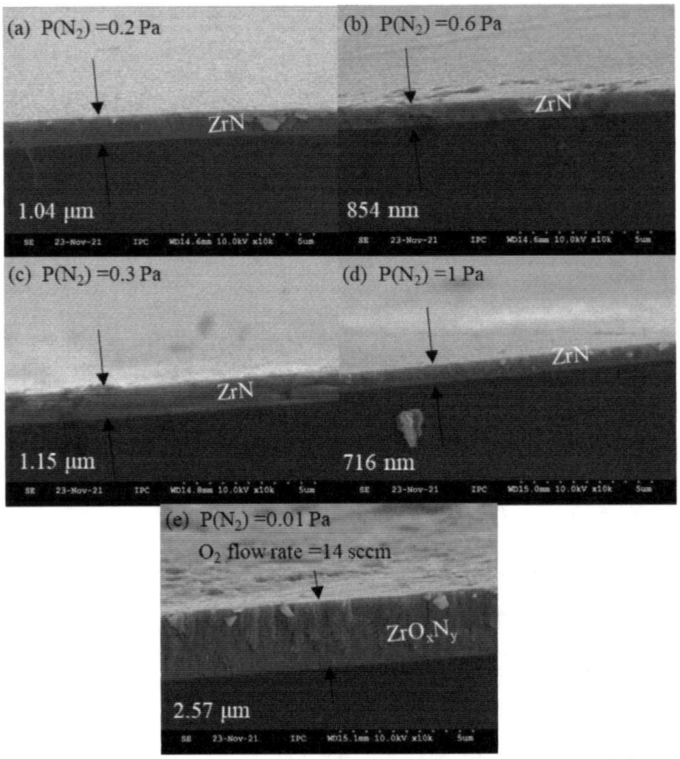

Figure 3. Cross-section morphology of ZrN and ZrO_xN_y films (a)$P(N_2)$=0.2 Pa; (b)$P(N_2)$=0.2 Pa; (c)$P(N_2)$=0.6 Pa; (d)$P(N_2)$=1 Pa; (e) O_2 flow rate =14 sccm.

3.2. Electrical transport behavior

Figure 4(a) shows the relationship between the resistance of the ZrN film grown under different nitrogen pressures and the temperature (300 K~8 K). It can be seen that in the range of 0.2 Pa~1 Pa nitrogen pressure, the resistance of ZrN film decreases with the increase of temperature, which shows obvious semiconductor characteristics. And its absolute resistance increases with the increase of nitrogen pressure. Figure 4(b) shows the normalized resistance of ZrN films grown under different nitrogen pressures as a function of temperature (300 K~8 K). It can be seen from Figure 4(b) that at the same temperature, as the nitrogen pressure increases, the absolute value of the slope of the ZrN film resistance-temperature curve gradually increases. That is, its temperature sensitivity (dR/dT) increases as the nitrogen pressure increases.

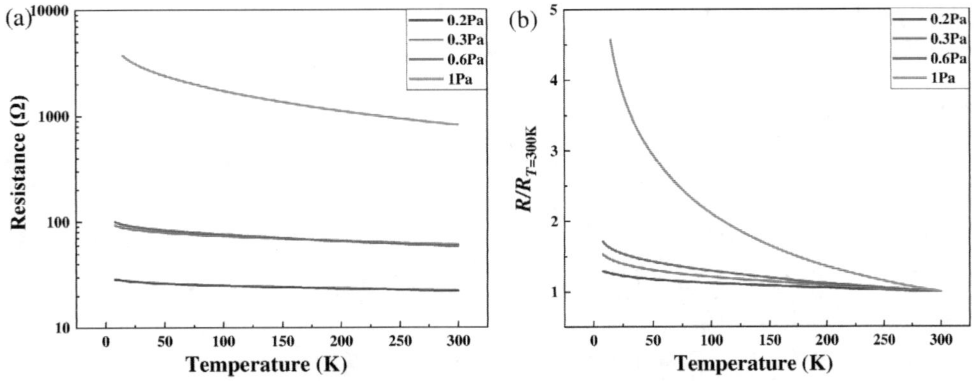

Figure 4. The resistance(a) and the normalized resistance(b) of ZrN films grown under different nitrogen pressures as a function of temperature (300 K~8 K).

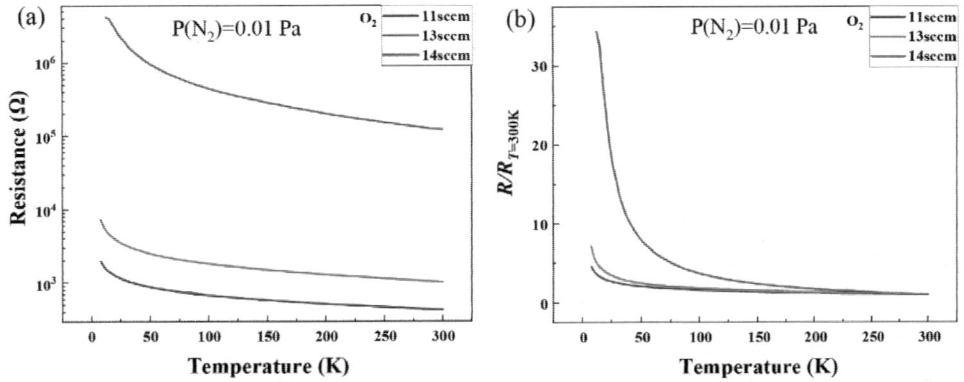

Figure 5. The resistance(a) and the normalized resistance(b) of ZrO_xN_y films grown under different oxygen flow rates as a function of temperature (300 K~8 K).

Figure 5(a) and 5(b) respectively show the resistance and normalized resistance R(T)/R(T=300K) of ZrO_xN_y films grown under different oxygen flow rates with temperature (300K-8K). The electrical transport curve of the ZrO_xN_y thin film exhibits semiconductor conductivity, which is the same as that of the ZrN thin film. With the increase of oxygen flow, the absolute resistance of the ZrO_xN_y film gradually increases, and the better its insulation at room temperature. And from the normalized resistance value of the ZrO_xN_y film, it can be seen that at the same temperature, the temperature sensitivity of the ZrO_xN_y film increases with the increase of the oxygen flow rate. Under the same oxygen flow rate, the temperature sensitivity of the ZrO_xN_y film increases with the decrease of the temperature, which indicates that the ZrO_xN_y film is used as a low-temperature thermometer. The lower the temperature, the greater the sensitivity of the sensing film. Compared with ZrN film, ZrO_xN_y film has stronger insulation and better temperature sensitivity.

4. Conclusion

In this work, reactive magnetron sputtering deposition technology was used to grow ZrN and ZrO_xN_y films under different nitrogen pressures and oxygen flow rates. The crystal structure of the grown films were analyzed by X-ray diffraction and Raman spectroscopy, and the electrical transport characteristics of ZrN and ZrO_xN_y films at 300 K~8 K were tested by PPMS. The results show that the ZrN and ZrO_xN_y films grown under different nitrogen pressures and oxygen flow rates are all ZrN phases. In the nitrogen pressure range of 0.2~1.0 Pa, as the nitrogen pressure increases, the absolute resistance value of the ZrN film increases, and the sensitivity also increases. In the oxygen flow range of 11-14 sccm, with the increase of oxygen flow, the absolute resistance, sensitivity and insulativity of ZrO_xN_y film increase, and the sensitivity is greater than that of ZrN film.

5. References

[1] Ke Y E and Chen Y 2020 *Coatings* **10** 476.
[2] Purandare Y P, Ehiasarian A P, and Hovsepian P E 2011 Journal of Vacuum Science & Technology A: Vacuum, Surfaces, and Films **29**.
[3] Roman D, Bernardi J, Amorim C, Souza F, and Basso R 2011 *Materials Chemistry & Physics* **130** 147-53.
[4] Escobar C, Villarreal M, Caicedo J C, Aperador W, and Prieto P 2013 *Surface & Coatings Technology* **221** 182-90.
[5] Grumski M, Dholabhai P P, and Adams J B 2013 *Acta Materialia* **61** 3799-807.
[6] Huang C L, Lai C H, Tsai P H, Kuo Y L, Lin J C, and Lee C 2014 *Electronic Materials Letters*.
[7] Mayumi, B., Takeyama, Masaru, Sato, Eiji, Aoyagi, Atsushi, and Physics N J J J o A 2010 **49** 05FA06.
[8] Song Z X, Wang J A, Li Y H, Ma F, Xu K W, and Guo S W 2010 *Microelectronic Engineering* **87** 391-93.
[9] Chen C S, Liu C P, Yang H G, and Tsao C Y A 2004 *Journal of Vacuum Science & Technology B* **22** 1075-83.
[10] Izumi, K., Masanobu, D., Ariyoshi, and H. 1975 *Parts Hybrids & Packaging IEEE Transactions on*.
[11] Yotsuya T, Yoshitake M, and Yamamoto J 1987 *Applied Physics Letters* **51** 235-37.
[12] Yotsuya T, Kakehi Y, and Ishida T 2011 *Cryogenics* **51** 546-49.
[13] Lin Z, Zhan G, Wang X, You M, Yang B, Chen X, Zhang W, and Liu J 2018 *Semiconductor Science and Technology* **33** 115002.
[14] Lin Z, Zhan G, You M, Yang B, Chen X, Wang X, Zhang W, and Liu J 2018 *Applied Physics Letters* **113**.
[15] Carvalho P, Vaz F, Rebouta L, Cunha L, Tavares C J, Moura C, Alves E, Cavaleiro A, Goudeau P, Le Bourhis E, Rivière J P, Pierson J F, and Banakh O 2005 *Journal of Applied Physics* **98**.
[16] Venkataraj S, Kappertz O, Jayavel R, and Wuttig M 2002 *Journal of Applied Physics* **92** 2461-66.
[17] Lan K-C, Huang J-H, Ai C-F, and Yu G-P 2011 *Journal of Vacuum Science & Technology A: Vacuum, Surfaces, and Films* **29**.
[18] Carvalho P, Chappé J M, Cunha L, Lanceros-Méndez S, Alpuim P, Vaz F, Alves E, Rousselot C, Espinós J P, and González-Elipe A R 2008 *Journal of Applied Physics* **103**.
[19] Mahmood K, Bashir S, Faizan ul H, Akram M, Hayat A, Rafique M S, and Mahmood A 2019 *Nuclear Instruments and Methods in Physics Research Section B: Beam Interactions with Materials and Atoms* **448** 61-69.
[20] Gharib M, Eshraghi M J, and Bordbari M 2020 *Journal of Materials Science-Materials in Electronics* **31** 21515-27.
[21] Zhan G, Liu J, Guo J, Feng J, Xu B, Yang B, Chen X, and Yang C 2015 *Journal of Materials Science: Materials in Electronics* **26** 9188-94.
[22] Huang J H, Chang K H, and Yu G P 2007 *Surface & Coatings Technology* **201** 6404-13.

Acknowledgments

The authors thank to partly financial support from the Strategic Priority Research Program (B) of Chinese Academy of Sciences (XDB25000000), Key-Area Research and Development Program of Guangdong Province(2020B0101340002). The authors are also grateful to the colleagues for their essential contribution to this work.

IOP Conf. Series: Materials Science and Engineering 1241 (2022) 012053 doi:10.1088/1757-899X/1241/1/012053

The experimental measurement of the thermal emissivity of a black coating from 50 K to 300 K

S Y Zhang[1,2], F Z Shen*[1], H M Liu[1], C J Huang[1] and L F Li[1,2]

[1] Key Laboratory of Cryogenics, Technical Institute of Physics and Chemistry, Chinese Academy of Sciences, Beijing, 100190, China.

[2] University of Chinese Academy of Sciences, Beijing 100049, P. R. China.

Email: shenfuzhi@mail.ipc.ac.cn

Abstract. High-emissivity coatings are commonly applied in satellite detectors and large-scale cryogenic systems. Accurate and quick measurement of emissivity is in high demand for practical applications at cryogenic temperatures. In this paper, a method for the measurement of the total hemispherical emissivity from 50 K to 300 K was presented based on the static calorimetric method. A two-stage GM cryocooler was used as a cooling source in the measuring device without assumption of cryogenic liquids. The thermal emissivity of a black carbon coating was measured by monitoring temperatures timely in the device. The tested results exhibited good accordance with previous reports. This cryogenic measurement device has made it a strong candidate for the radiative heat dissipation at cryogenic temperatures. Design details of the experiment and measurement results will be presented.

1. Introduction

Deep space is vacuum and cryogenic, therefore, radiation is the only way to transfer heat. The total hemispherical emissivity of the material surface, as a symbol determining the thermal radiation capacity of the material, has become an important index to measure the properties of thermal structure materials in aerospace field, and a key parameter for thermal protection design, which can greatly affect the working efficiency and service life of thermal structure components. For example, the thermal shielding system in Tore Supra and ITER provides an opaque shield against thermal load at 80 K, and the thermal load is transferred from the thermal assembly to a superconducting magnetic coil sleeve generating at 4 K. To correctly estimate the power consumption at low temperatures, it is important to know the thermal emissivity of the surface material of large components[1].

It is a great challenge to accurately measure the low-temperature thermal emissivity of metal and infrared coatings, because the wavelengths of thermal radiation at low temperatures are mainly above 20 μm, and the background radiation can bring great uncertainty. Researchers have designed different devices to measure emissivity of materials at low temperatures. Thermal radiative properties of a DLC (diamond-like-carbon) coating were measured and analyzed at cryogenic and room temperature [2]. Tomas et al. compared emissivity and absorptivity of various black coatings on copper and aluminum substrates from cryogenic temperature to room temperature[3, 4]. J. Tuttle et al. measured a coating's emissivity to temperatures at 20 K by using a precise and relatively inexpensive manner [5, 6]. There are some studies focusing on the theory of thermal radiation and theoretical models. A. J. Sievers calculated the thermal radiation from metals and provided a fairly accurate description of the experimentally measured total emissivity of good conductors in terms of the dc resistivity and the radius of the free-electron sphere [7].

In order to calculate the radiation load and infrared background radiation, low-temperature emissivity is usually required. Previous studies of emissivity usually focus on the room temperature or 4.2 K, 77 K and rely heavily on liquid helium and liquid nitrogen. In this paper, we described an emissivity measurement device based on a two stage GM cryocooler. By using this equipment, the total hemispherical emissivity of a carbon coating has been measured and shown from 50 K to 300 K. The total hemispherical emissivity of other materials will be measured and published in the future.

2. Measuring method

Two common methods are used for thermal emissivity measurement: calorimetric method and radiometric method [3]. According to the different state of heat flow, two common types of calorimetric method are used to measure thermal emissivity of a surface: steady-state calorimetry and transient calorimetry . The principle of calorimetry is to calculate thermal emissivity of materials by measuring transferred heat.

Steady-state calorimetric method is applied in this paper for measurement of total hemispherical emissivity from 50 K to 300 K. Thermal emissivity of materials can be measured continuously over the temperature range. The main principle of Steady-state method is to measure the transferred heat between two parallel surfaces. Heat transfers from the radiator (the heated surface) to the absorber (the cold surface). The radiator and the absorber constitute a heat exchange system. The temperatures of the radiator and the absorber are monitored for measurement. The spacing between the radiator and the absorber is very small relative to the diameter of them. If the thermal conductivity of the sample is high enough and the thermal conductivity of residual gas and of thermocouple wires is negligible, mutual thermal emissivity can be calculated by equation (1) based on Stefan-Boltzmann law.

$$\varepsilon = \frac{Q}{A\sigma(T_R^4 - T_A^4)} \tag{1}$$

ε is mutual emissivity, Q is radiative heat flux between the radiator and the absorber, T_R, T_A are temperatures of the radiator and the absorber respectively, A is an area of radiator (as well as absorber), σ is Stefan-Boltzmann constant.

$$\frac{1}{\varepsilon} = \frac{1}{\varepsilon_R} + \frac{1}{\alpha_A} - 1 \tag{2}$$

α_A is the absorptivity of the material on the absorber and ε_R is emissivity of the sample on the radiator. In our experiments, the samples are fixed both on the surface of the radiator and the absorber. We assumed the samples are grey bodies and emissivity is independent of wavelength, so $\varepsilon_R \approx \alpha_A$. Total hemispherical emissivity of sample ε_R can be calculated by equation (2).

3. System description

Based on the steady-state calorimetric method, a measuring system is designed for measurement of the total hemispherical emissivity at low temperatures. The whole measuring system is shown in Figure 1. The process of measurement is controlled and recorded by the computer automatically. The testing process and collection of data are achieved by a program written by Labview. Vacuum is maintained by the cryogenic Pump. A two stage RDK-415D GM cryocooler, as cooling source, is placed backwards on the lower bracket. F-50 Compressor Unit is applied in this system. Digital Temperature Controller LakeShore 332 is used to monitor temperatures of the radiator, the absorber and others. Figure 2 shows the distribution of thermometers.

The apparatus of thermal emissivity measurement is shown in Figure 3. The whole apparatus is cooled by the two stage GM cryocooler. The vacuum shield made of stainless steel 304 is 340 mm in diameter and it is set up on the base to provide a vacuum environment for the system. The radiation sheild is installed on the first stage of the cryocooler to reduce the radiation heat of the testing room.

Testing room consists of the radiation shield and flange which are installed on the second stage of the cryocooler. The main components of testing room are radiation shield, radiator, absorber, thermal resistor, thermal stabilization and thermal switch.

Radiation shield made of copper is cooled by the cryocooler. In order to reduce the temperature different between the upper and the bottom of radiation shield, five heat conducting rods are applied inside the shield and soft red copper braids are used outside the radiation shield.

The mechanical thermal switch is designed to accelerate the cooling of the radiator. It includes displacement device of thermal switch and kinematical seal. They are helpful for forming the vacuum environment. The position of thermal switch can be adjusted with vacuum environment of the testing room intact. One side of the switch is in parallel contact with the radiator. The rod of mechanical thermal switch connected with the radiation shield by soft copper braids which is made of thin copper wires. When the switch is closed before measurement, the cold energy transfers from the radiation shield to the radiator, which can decrease the cooling time of radiator. During the measurement, the mechanical thermal switch is not connected with the radiator.

The radiator and the absorber are 200 mm in diameter and 3 mm in thickness. The coating samples are fixed on the surface of the radiator and the absorber, while thermometers and electric heating pieces are installed in the backside. During the measurement, electric heating pieces can provide the heat to the radiator and then heat flows down from the radiator to the absorber by radiation. To decrease the radiation heat loss from the backside, the radiator and the absorber are covered with Aluminum shield. The radiator is suspended by a hollow epoxy rod which can support the radiation shield and avoid the leakage of heat.

The calibrated thermal resistor is served as a heat flow meter, connecting the absorber with the thermal stabilization. The material of the thermal resistor can be changed for different samples depending on the thermal conductivity The thermal stabilization is composed of the base, insulation layer and copper layer. The base of thermal stabilization is installed on the heat transfer flange, which can provide support and cold energy for the absorber. The insulation layer is installed on the base to provide a buffer layer for the transfer of cold energy and stabilize the temperature of thermal stabilization. Three thermometers are set on the thermal resistor equidistantly for calibration process.

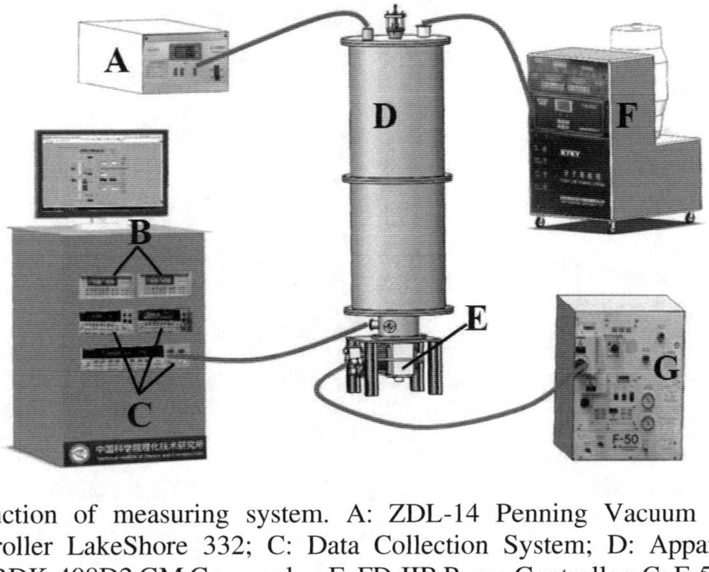

Figure 1. Construction of measuring system. A: ZDL-14 Penning Vacuum Gauge; B: Digital Temperature Controller LakeShore 332; C: Data Collection System; D: Apparatus of emissivity measurement; E: SRDK-408D2 GM Cryocooler; F: FD-IIB Pump Controller; G: F-50 Compressor Unit.

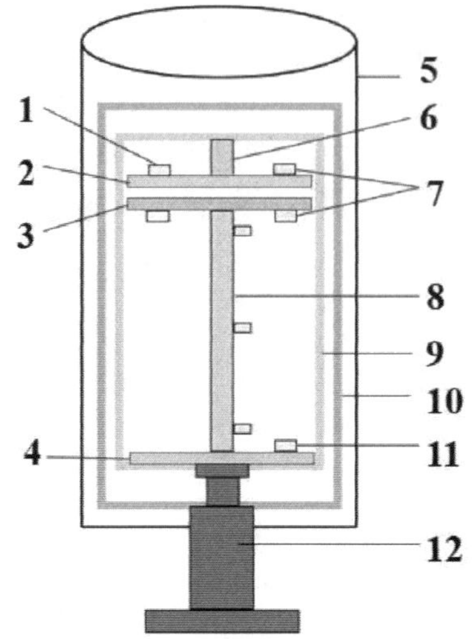

Figure 2. The testing room and distribution of thermometers. 1-Electric heater; 2-Radiator; 3- Absorber; 4-Thermal stabilization; 5-Vacuum shield; 6-Epoxy rod; 7-Thermometers; 8-Thermal resistor; 9-Radiation shield; 10-Shielding case; 11-Thermometers; 12-The cryocooler.

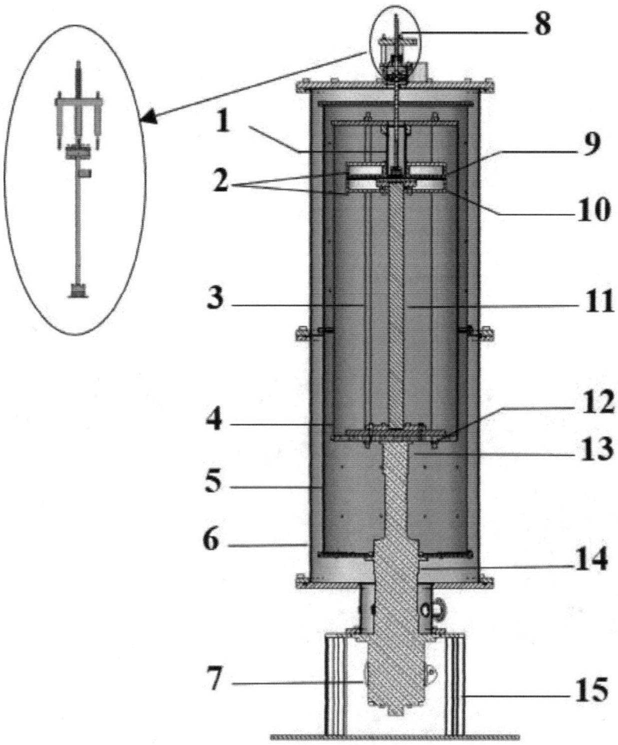

Figure 3 Apparatus of emissivity measurement between two parallel samples. 1-The suspension rod; 2-Aluminum shield; 3- Heat conducting rod; 4-Radiation shield; 5-Shielding case; 6-Vacuum shield; 7-Cryocooler; 8-Thermal switch; 9-Radiator; 10-Absorber; 11-Thermal resistor; 12-Thermal stabilization; 13-The second stage of the cryocooler; 14-The first stage of the cryocooler; 15-Bracket.

4. Experimental results

The carbon coating is from Institute of Electrical Engineering of the Chinese Academy of Sciences. Its ash content is 0.3% and roughness is 0.424 μm. The samples are stuck on two parallel copper plates by Apiezon N. The parameters of the sample were shown in Table 1.

Table 1. Parameters of the sample

Parameters of the sample	
PORE VOLUME	0.7 ml/g
SPECIFIC SURFACE AREA	1600 m^2/g
CAPACITANCE	28 F/g
ASH CONTENT	0.3 %
METAL ANALYSIS	18 ppm
ROUGHNESS (RA)	0.424 μm

After the assembling of apparatus and the fixing of sample, a GM cryocooler is used to cool the whole device in vacuum. When the mechanical thermal switch is closed, the time of cooling in the radiator decreased sharply (Figure 4(a)). The key of measuring method is that the heat generated by electric heating pieces represents the heat transferred between the two parallel surfaces. After calibration process, the radiator is heated by electric heating pieces.

The dependence of thermal emissivity on the temperature of the radiator (T_R) were acquired from our experiments. Fig.4(b) showed experimental values of the carbon coating. We can observe that the total hemispherical emissivity of the sample was 0.7 at room temperature, while emissivity at 50 K was only 0.52. There was a similar dependence on the temperature of the radiator compared with previously published data [2]. In addition, the total hemispherical emissivity the sample leaving for 3 months was near to that of the initial sample.

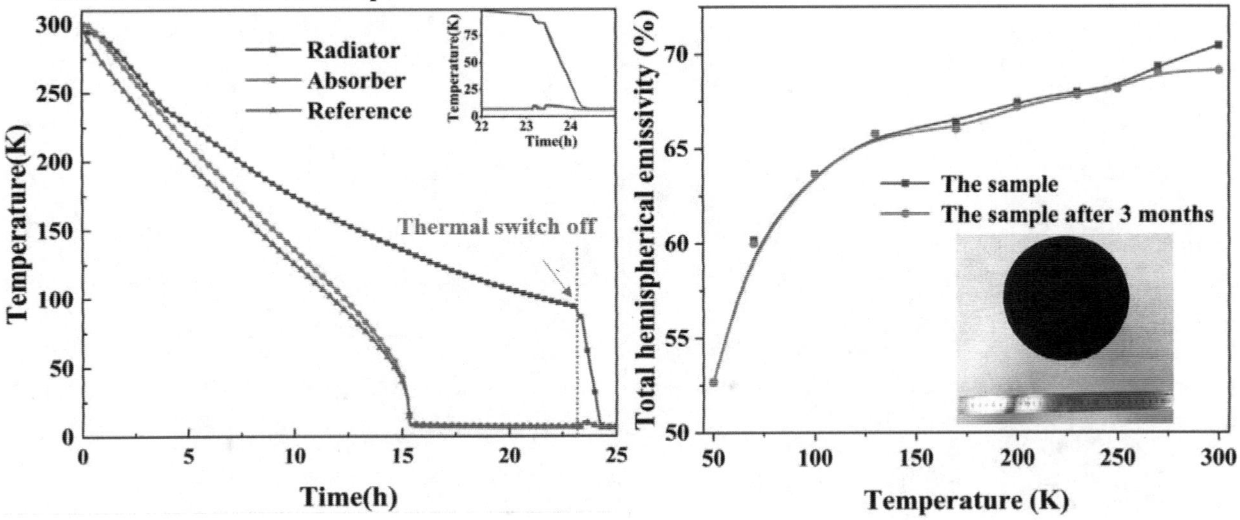

Figure 4. (a)The cooling curves with the thermal switch off before measurement. (b) The total hemispherical emissivity of the sample.

IOP Conf. Series: Materials Science and Engineering 1241 (2022) 012053 doi:10.1088/1757-899X/1241/1/012053

5. Conclusion

We designed a device and measured the total hemispherical emissivity of the carbon coating. The following conclusions can be drawn: when the mechanical thermal switch was closed, the time of cooling in the radiator decreased sharply. We could observe that the total hemispherical emissivity of the sample was 0.70 at room temperature, while the emissivity at 50 K was only 0.52. There was a similar dependence on the temperature of the sample leaving for 3 months compared with the initial sample. The values of emissivity increased with the temperature of the sample rising. The total hemispherical emissivity of other materials will be measured and published in the future.

6. References

[1] Alekseev A, *et al.* 2019 *Fusion Eng. Des.* **146** 144-8.
[2] Hanzelka P, Kralik T, Maskova A, Musilova V, and Vyskocil J 2008 *Cryogenics* **48** 455-7.
[3] Králík T, Musilová V, Hanzelka P, and Frolec J 2016 *Metrologia* **53** 743-53.
[4] Kralik T, Hanzelka P, Zobac M, Musilova V, Fort T, and Horak M 2012 *Phys. Rev. Lett.* **109** 224302.
[5] Tuttle J, Canavan E, DiPirro M, Li X, Franck R, Renbarger M, Copp T, and Hahn M 2014 *Cryogenics* **64** 240-3.
[6] Tuttle J, Jahromi A, Canavan E, and DiPirro M 2016 *Cryogenics* **74** 166-71.
[7] Wang J N, Zhang Y L, Liu Y, Zheng W h, Lee L P, and Sun H B 2015 *Nanoscale* **7** 7101-14.

Acknowledgments

This work was supported by the CAS Key Laboratory of Cryogenics, Technical Institute of Physics and Chemistry (Grant No. CRYOQN202104).

IOP Conf. Series: Materials Science and Engineering 1241 (2022) 012054 doi:10.1088/1757-899X/1241/1/012054

Superconducting-like and magnetic transitions in oxygen-implanted diamond-like and amorphous carbon films, and in highly oriented pyrolytic graphite

N Gheorghiu[1*], C R Ebbing[2], J P Murphy[2], B T Pierce[3], and T J Haugan[3]

[1] UES, Inc., Dayton, OH 45432
[2] University of Dayton Research Institute, Dayton, Ohio 45469
[3] The Air Force Research Laboratory, Wright-Patterson AFB, Ohio 45433

*Email: Nadina.Gheorghiu@yahoo.com

Abstract. In our previously published work, we have reported colossal magnetoresistance, Andreev oscillations, ferromagnetism, and granular superconductivity in oxygen-implanted carbon fibers, graphite foils, and highly oriented pyrolytic graphite (HOPG). In this follow-up research, more results on these oxygen-implanted graphite samples are presented. We show results from transport measurements on oxygen-implanted diamond-like carbon thin coatings, amorphous carbon films, and HOPG. Significantly, a three-order magnitude drop in the electrical resistance of the oxygen-implanted diamond-like carbon films is observed at the 50 K temperature that we have previously reported for the transition to the superconducting state. Below 50 K, the films' resistance oscillates between the high and low resistance states, less when the sample is under a transverse magnetic field. This metastability between the insulating and superconducting-like states possibly reflects the evolution of the amplitude for the superconducting order parameter also known as the longitudinal Higgs mode. Transitions to low resistance state and metastability are also observed for amorphous carbon films. Finally, the HOPG samples' resistance have a thermally activated term that can be understood on the basis of the Langer–Ambegaokar–McCumber–Halperin model applied to narrow SC channels in which thermal fluctuations can cause phase slips. We also find that in oxygen-implanted carbon materials, the electron charge and spin correlations do not compete and their interplay rather facilitates the emergence of high-temperature superconductivity, and thus, additional unexpected effects like Heisenberg spin waves and magneto-structural transitions are observed.

1. Introduction

Following the discovery of high-temperature superconductors/superconductivity (HTSC) like the cuprates, the search for new HTSC materials has been recently directed into ways to avoid the more costly and market restricted rare-earth components or compounds with heavy-metals. One significant, though not always recognized, direction is towards carbon-based or C-content SC materials [1], ranging from several organic, to carbonaceous sulfur hydrides under extremely high pressures [2], and to boron-carbon [3]. The possible role played by sulfur on a-C powders [4] and films [5] has been earlier on investigated.

Carbon (C), hydrogen (H), and oxygen (O) are the building blocks of life. Indeed, the mere existence and evolution of biological structures occurred in the presence of both carbon and water, and life itself is based on hydrogen bonds. In conjugated C-H structures with equal number of C atoms, the delocalized π electrons are like the conduction electrons in a metal. Moreover, the π electrons are like Cooper pairs in SC and their coherence leads to the Josephson tunnel effect [6]. In these materials, there is competition and interplay between magnetic and SC order [7] and electron viscosity can explain the observation of negative differential resistance and SC [8]. Materials having as building blocks C and H atoms are practically small SCs, moreover ferromagnetic (FM), as we have previously found [9,10]. On the other hand, it is well-known the important role played by copper-oxide layers in cuprates. For the C-only materials, if instead of powder mixing rather O plasma is used, charge is implanted into the surface of the target film. The observation of room-temperature SC in water-treated graphite [11] suggested that the C can react with O to benefit HTSC.

SC can occur in the bulk or at the surface of a homogeneous system, also at interfaces of a hybrid system. Graphite can have hybrid Bernal (ABAB…) and rhombohedral (ABCABC…) layer stacking. SC was found in the latter [12]. Solution to the tight-binding Bogoliubov-de Gennes equations revealed that for thin stacks (few layers), surface HTSC survives throughout the bulk due to the proximity effect between ABC/ABA interfaces where the order parameter is enhanced [13].

In this paper, we are following up on previously published work [14] by looking for additional signs of SC, possibly even HTSC, in O-implanted C-based thin films as well as graphite bulk samples.

2. Experimental

The C-based materials used for this study were diamond-like carbon (DLC) thin films, amorphous carbon (a-C) thin films, highly oriented pyrolytic graphite (HOPG, particle size 10 μm) disc-shaped samples (1 mm thick and 3 mm diameter), and graphite foil [15]. The DLC and a-C films were deposited on sapphire (Al_2O_3) substrates using pulsed laser deposition. In order to bring the density of states to levels where possible SC behavior can be observed, the samples were O-ion implanted [16] at two-dimensional (2D) concentrations 7.07×10^{12}, 5.66×10^{15}, and 2.24×10^{16} ions/cm^2, respectively. Details on the ion-implantation efforts can be found in [17], while here we show the data for a-C films and for the graphite foil (Figure 1, top). The implantation energy was 70 keV. For the ion concentrations used, the implantation depth went from ~0.01% to 100% of the either the graphite foil or the a-C samples, from ~0.01% to ~40% for the HOPG samples, and it went all the way though the DLC films (i.e., 100% of the sample). Samples' thickness was: 80 nm for the DLC films, 175 nm for the a-C films, and 1 mm for the graphite foils or the HOPGs. Importantly, the ion implantation changes the hybridization of the C atoms. Thus, as the a-C thin films have a mixture of sp^2 and sp^3 bonds, the ion implantation changes the sp^2/sp^3 ratio such that the bandgap of an a-C is tuned between that of diamond and graphite. In a reverse way, the ion implantation of DLC films results in the sp^3 to sp^2 bond conversion. The four-wire Van Der Pauw square-pattern technique was used for resistivity measurements. The quality of the silver electrical contacts was optically checked using an Olympus BX51 microscope. Raman spectra showed the effect of O-implantation (Figure 1, bottom). Magneto-transport and magnetization measurements were carried out in the 1.9 K - 300 K temperature range and for magnetic fields of induction B up to 9 T using the Quantum Design Physical Properties Measurement System (PPMS) model 6500.

3. Experimental Results

3.1. Superconducting-like transitions and metastability in oxygen-implanted diamond-like C thin films

Temperature-dependent measurements of the electrical resistance R for the DLC films revealed a SC-like transition below $T_c \sim 50$ K (Figure 2). R oscillates between R_{max} and $R_{min} \sim R_{max}/1000$. Below $T_c \sim 50$ K, we observe metastability between the SC and the normal state in this physical system. As known, H.K. Onnes' proof of discovering SC in mercury below $T_c \cong 4.19$ K was the ×10,000 decrease in the electrical resistance. The metastability found here for DLC films reflects oscillations in the amplitude SC order parameter otherwise known as the Higgs mode. We also found that a magnetic field applied normal to the DLC film (i.e., along the c-axis) appears to weaken and possibly destroy the metastability. The weakening effect might be due to the coexistence between singlet SC and antiferromagnetism (AFM) going through a first-order phase transition to the coexistence between triplet SC and FM as the strength of B is increased. This transition was predicted for the occurrence of surface or interface SC in rhombohedral graphite [7]. We have found proof for the AFM-SC (singlet SC) and FM-SC (triplet SC) coexistence below $T_c \sim 50$ K also in hydrogenated C fibers under high magnetic field when comparing the measurements taken during cooling and heating, respectively [9]. On the other hand, the effect played by the magnetic field on cancelling the metastability and lowering of T_c (Figure 2c) relates to the case known for most SC materials where a magnetic field $B \| c$ actually destroys the SC state.

Figure 2 also reveals negative resistance dominating from $T \sim 50$ K to $T \sim 200$ K. Negative R is associated with the viscous flow of electrons, where an electric current running against the applied field (a hole current) is responsible for the observation of a negative nonlocal voltage. The latter may play the

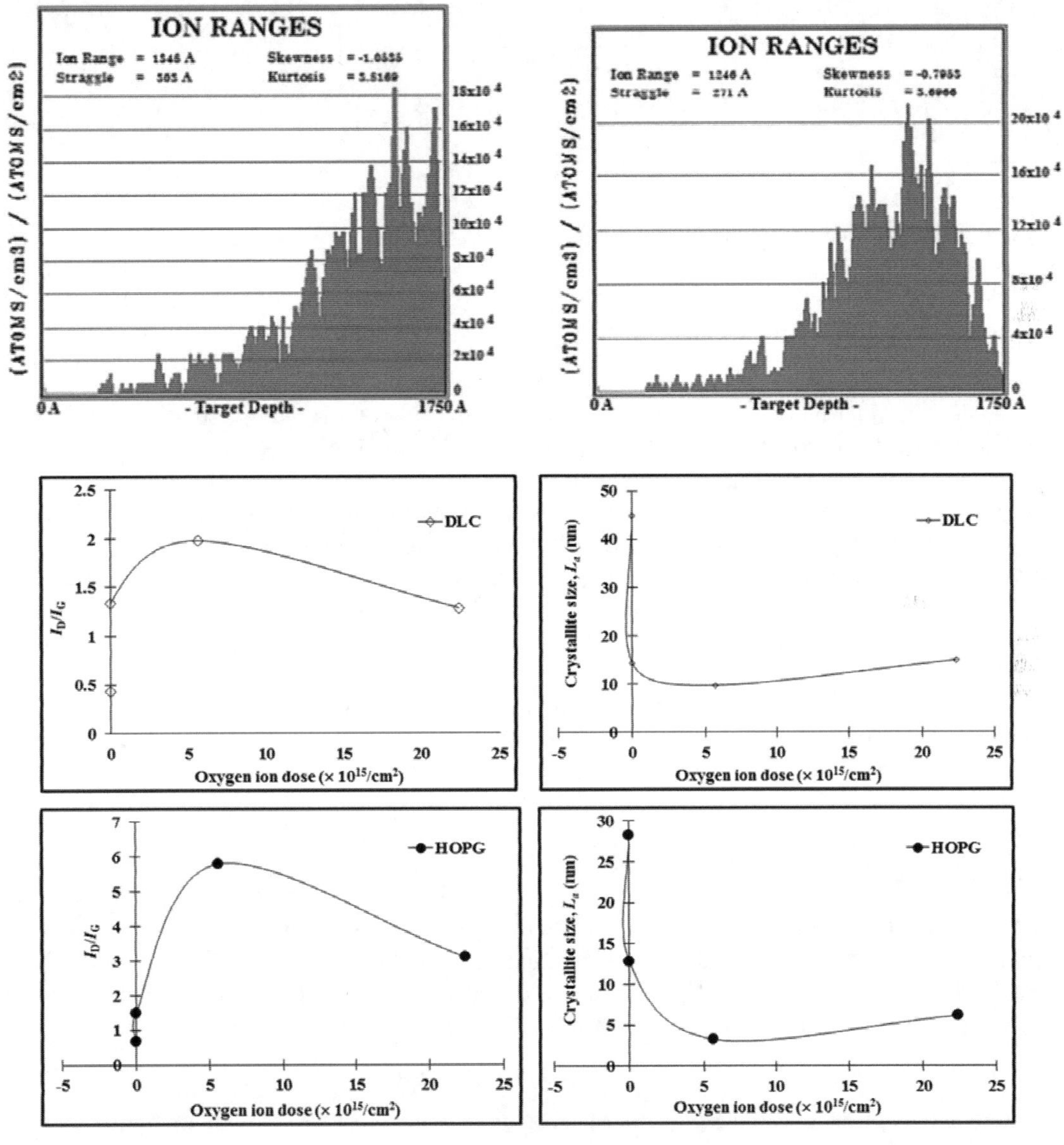

Figure 1. Top: 70 keV ion implantation of O in a-C (top) and graphite (bottom), respectively. Bottom: The effect of O-implantation concentration on the Raman peak I_D/I_G intensity ratio and on the G peak wavevector for the DLC sample (upper plots) and for the HOPG sample (lower plots), respectively.

same role for the viscous regime as zero electrical resistance does for SC [8]. In this case, possible short-range SC fluctuations starting at $T \sim 200$ K eventually grow into long-range SC fluctuations that become critical at $T_c \sim 50$ K. The oxide layer might host both hard and soft SC fluctuations, i.e., the system might be a multigap SC where the larger gap SC phase takes over the smaller gap SC phases. The evolution of the attractor – which is a manifold of equilibrium states – is actually the evolution of the amplitude (Higgs-Anderson mode) and phase for the SC order parameter.

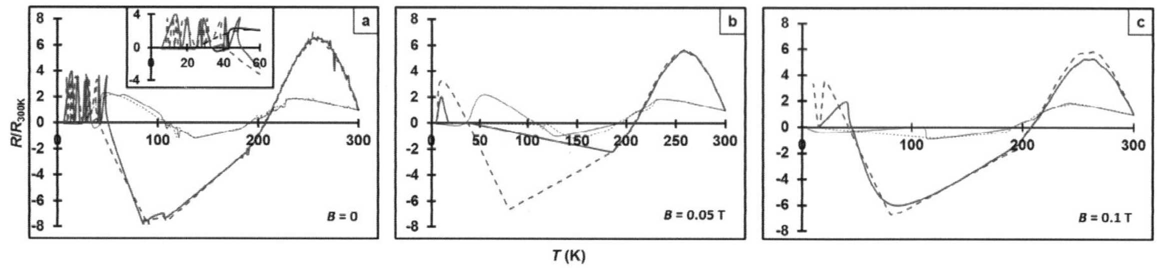

Figure 2. Temperature dependent resistance data (relative to the resistance at 300 K) for DLC films that have been O-implanted at 2.24 x 10^{16} ions/cm^2 (blue) vs. raw samples (black, fine lines), without (a) and with a transverse magnetic field $B_\perp = 0.05$ T (b) and $B_\perp = 0.1$ T (c), respectively. Dash/full line for data acquired during decreasing/increasing the temperature.

Four fluctuation modes can lead to HTSC in flat-band energy systems like the graphitic ones: the total phase (Nambu-Goldstone) mode, the total amplitude (Higgs-Anderson) mode, the relative phase (Leggett) mode, and the relative amplitude mode [18]. At reasonably high temperatures, the Ginzburg-Landau theory predicts that the amplitude mode contribution dominates because of its singularity at $T = T_c$. At low temperatures, the amplitude mode still has a significant contribution.

We also notice that the insulator-metal transition observed at T ~ 250 K is seemingly not affected by B. Notice that we have observed the same insulator-metal transition at T ~ 250 K also in O-implanted carbon fibers [10]. It is known that fullerene C_{60} buckeyball clusters could be formed as a result of O-ion implantation. The metal-insulator transition at T ~ 250 K could be a structural phase transition. In the solid state, the C_{60} molecules assume a simple cubic unit cell that exhibits a phase transition (upon increasing the temperature) to a face-centered-cubic structure upon heating around a transition temperature ~ 250 K [19].

3.2. Superconducting-like transitions in O-implanted amorphous-C thin films

Transition to a possible SC state was also found in amorphous-C (a-C) thin films (Figure 3). The conduction goes from electron to hole carrier type (and vice-versa) in more than one occasion and at different temperatures. Notice also that the sample that has been O-implanted at the largest dose goes to superfluidity (ρ becomes very "negative", i.e., the measured voltage is negative) at $T_c \cong 50$ K. Below T_c, ρ went beyond the measurable range are thus the apparent lack of data. Figure 3 also shows that significant changes occur in all samples at a temperature around 80 K. One might also question if $T_{BG} \sim 80$ K is the Bogoliubov-deGennes temperature for this system. For unconventional SC, electron scattering is due to the interaction with the Bogoliubov excitations or bogolons. In this way, bad conductors in the normal phase (like the ones here) can be good SCs for which an alternative mechanism for high-temperature bogolon-pair mediated SC was proposed [20]. The bogolon-pair-mediated scattering can even dominate over the conventional acoustic phonon channel, over the single-bogolon scattering, and over the scattering on impurities. Thus, T_{BG} is larger than T_c for a phonon-based SC. In a hybrid Fermi-Bose system like graphene and possible the one here, T_c can be larger than T_{BG}.

Just like was the case with DLC films, metastability in $\rho(T)$ is also observed with these a-C films at temperatures below $T \sim 80$ K. This was also observed in indium oxide SC films [21]. The accepted caveat is that in disordered 2D systems, SC is only marginally stable and finite low-T resistivity is always expected [22]. The role of disorder in these amorphous films is, indeed, worth mentioning. Disorder-induced granularity was previously found in amorphous indium oxide SC films [23]. Moreover, moderate disorder actually enhances SC, which can persist up to the critical disorder needed to bring about the Anderson transition. In addition, percolation between grains leads to SC, as we have also found [10]. As for the role of non-magnetic impurities and in contrast to the Anderson theorem, note also that a dirty SC does not always mean weak SC [24].

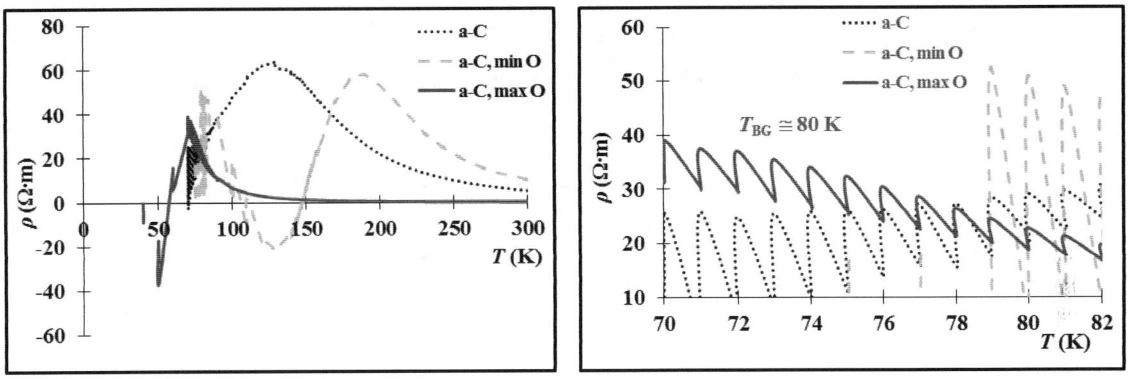

Figure 3. Temperature dependent resistivity data at cooling for a-C films: without ion implantation, O-implanted at 7.07×10^{12} ions/cm^2 (min O), and at 2.24×10^{16} ions/cm^2 (max O), respectively. The zoomed data (to the right) shows that significant changes occur in all samples at $T \sim 80$ K, possible the temperature for the exciton condensation (see text).

3.3. Electric-field driven superconducting-like transition in O-implanted highly HOPG

While the base carriers in HOPG are electrons, the O-ion implantation results in hole doping of the sample. Temperature-dependent resistivity measurements on the O-implanted HOPG samples revealed the strong dependence on the amount of (direct current) input current I. Thus, for $I = 20$ mA, the data is noisy (Figure 4). While ρ values have the order of magnitude expected for HOPG samples, the highest concentration of O-implantation results in lower values for ρ. A larger input current, $I = 5$ mA, results in smoother data and, more significantly, much lower values for ρ (Figure 5). What is being observed has to do with the expected field effect in this metal-oxide-semiconductor hybrid system, in which the silver contact is the metal, the graphite oxide at the top of the sample is the oxide, and the rest of the sample's bulk – graphite – is the semiconductor. When the strength of the input current is increased, the field effect caused by the edge of the silver contact, results in increasingly deeper in the sample charge injection paths.

Yet, perhaps more significant is the analysis of the cooling-heating cycle for $\rho(T)$ using the two-band model for the electrical conduction [25]. We have modified formula (2) for in [25] for the electrical resistance to include contributions from both electron (e) and hole (h) charge carriers, in addition, a logarithmic term log(T) was added. In 2D systems, there are three corrections to the log(T) term: the weak localization of electrons, the electron-electron interaction for a disordered system, and the Kondo effect. The weak localization and the electron-electron interactions are a favourable brewing ground for SC and an increase in the 2D density of charges through material intercalation can enhance the likelihood for the occurrence of quantum phenomena such as 2D SC [26]. Thus, the total electrical resistance is given by: $R(T) = 1/(R_s^{-1} + R_i^{-1})$, where $R_s = a_1 + b_1\log(T) + a_e(T)\exp[E_{g,e}/(2k_BT)] + a_h\exp[E_{g,h}/(2k_BT)]$ is contribution to R that comes from graphite being a semiconductor, while $R_i =$

$R_0 + r_1 T + c_1 \log(T) + R_{2,e}\exp[-E_{a,e}/(k_B T)] + R_{2,h}\exp[-E_{a,h}/(k_B T)]$ is the contribution to R coming from the HOPG's interfaces. For sample thickness larger than ~50 nm, the contribution to R

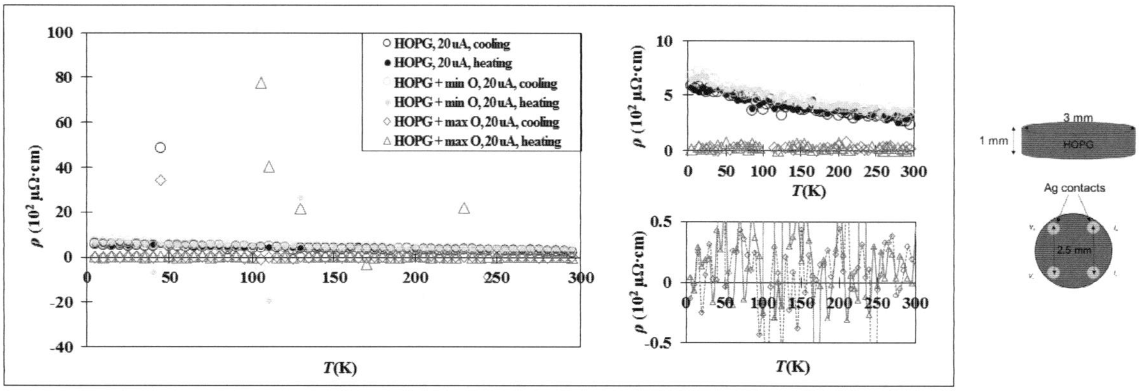

Figure 4. Resistivity data for the HOPG samples. O-implantation concentrations: min O = 7.07 × 10^{12} and max O = 2.24 x 10^{16} ions/cm^2. The input direct current is small, I = 20 mA. Right: sample and its contacts.

coming from the interfaces results in metallic and possible even SC behaviour [25]. The physical description of the most relevant terms is as follows. $a_{e,h}(T)$ depends on the charge mobility through the mean free path and on details of the band structure through the effective mass. We used Gnuplot to fit the resistivity data $\rho(T)$ collected at the cooling and heating of the HOPG O-implanted at the highest concentration (2.24 x 10^{16} ions/cm^2). We used known formula $\rho = R \times A/l$, where $A =$ 2 mm^2 is the cross-sectional area and l = 2 mm is the distance between the current leads. For the energy gap E_g we

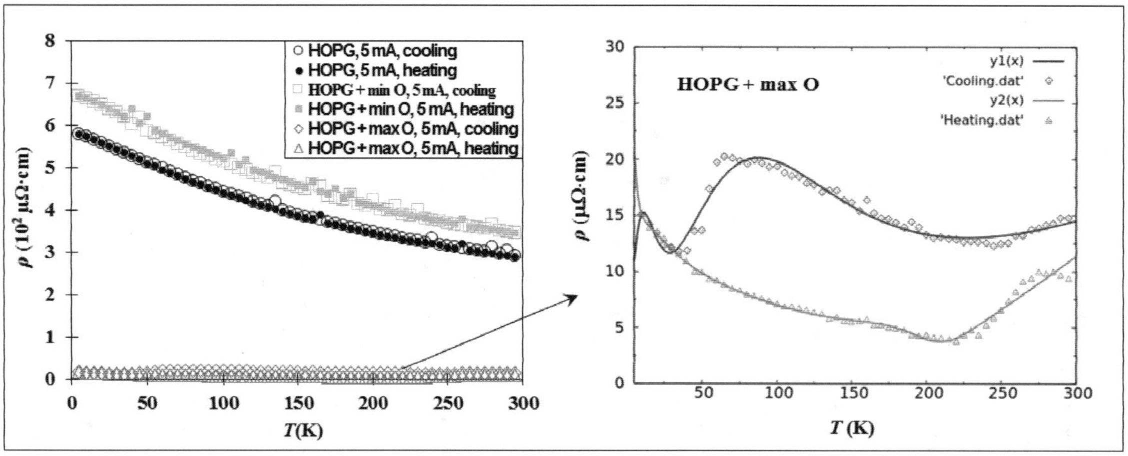

Figure 5. Resistivity data for the HOPG samples. Legend for the O-implantation concentrations is the same as in Figure 4. The input current is large, I = 5 mA. The fit to the data for the maximum concentration of O-implantation (right) is explained in the text.

find: $E_{g,h} \cong 34$ meV and $E_{g,e} \cong -7.8$ meV, respectively. While $E_{g,h}$ is within previously found range, $E_g = (40 \pm 15)$ meV [25], the negative band gap suggests a band overlap (the valence band overlaps slightly the conduction band), i.e., a semimetallic behavior. This was expected, as the thickness of the HOPG sample here is larger than 50 nm. Also obtained from the fit are: the residual resistance $R_0 \cong 56\ \mu\Omega$ and the coefficient for the linear term $r_1 \cong 0.42 \times 10^{-5} \Omega/K$. Also found are the activation energies for the holes $E_{a,h} = k_B T_{a,h} \cong (1.38 \times 10^{-23}\ J/K)(6.0\ K) \cong 0.52$ meV (k_B is the Boltzmann constant) and for electrons $E_{a,e} = k_B T_{a,e} \cong (1.38 \times 10^{-23}\ J/K)(84\ K) \cong 7.3$ meV, respectively. The activation energies, which are positive, have a special meaning in relation to SC. Namely, the thermally activated term E_a is understood on the basis of the Langer–Ambegaokar–McCumber–Halperin (LAMH) model [27,28] that applies to narrow SC channels in which thermal fluctuations can cause phase slips. For clean HOPG, $T_a \cong 40$ K [25], which is close to the average value here, $T_{a,average} \cong 45$ K. A similar dependence was observed in granular Al–Ge [29], suggesting that granular SC might also exist in the O-implanted HOPG. We have also estimated the thermopower $S_{h,e} = E_{g,e(h)}/(|e|T)$, where $e = -1.6 \times 10^{-19}$ C is the electron's charge. The corresponding values are $S_h \cong 113\ \mu V/K$ for holes and $S_e \cong -26\ \mu V/K$ for electrons, respectively. Note that S for pristine graphite is nearly zero. By comparison, the thermopower for a classical gas is $S_{class.\ gas} = k_B/|e| \cong 87\ \mu V/K$. While S is negative for YBCO, it can turn positive by increasing the O deficiency.

3.4. Other effects in HOPG: Spin waves, magneto-structural transitions, and Almeida-Thouless line

We have also found that the temperature-dependent remanent magnetization appears to have the behavior consistent with excitations of spin waves in a 2D Heisenberg model with a weak uniaxial anisotropy [30]. Notice that the activation temperature $T_{a,e} \cong 84$ K found before is close to the temperature where the Heisenberg spin wave has an inflection point at zero remanent magnetization. The latter might be related to a magnetic order phase transition from the AFM to the FM order. In addition, there are two magneto-structural transitions, one at $T \cong 80$ K and another, discontinuous, one at $T \cong 210$ K (Figure 6, left). The discontinuity might be due to the fact that some regions still did not complete the transition from AFM to FM. Zero-field cooled and field-cooled temperature-dependent (direct-current induced) magnetization measurements show that the HOPG is at least a spin glass system (Figure 6, right). Interestingly, $T_c \cong 60$ K (for $H = 0$) extracted from the Almeida-Thouless line is the mean-field T_c for SC correlations in the metallic-H multilayer graphene or in HOPG [31]. The balanced interplay between charge, spin, and lattice orders can favourably result in HTSC.

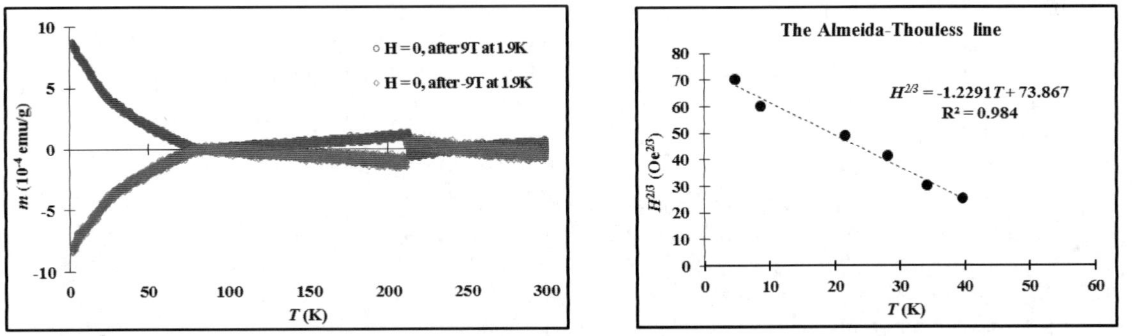

Figure 6. Dynamic and static spin behavior in HOPG. Left: Temperature-dependent remanent magnetization. Right: The Almeida-Thouless line.

4. Conclusions

Carbon, oxygen, and phase transitions are ubiquitous in nature. In this paper, we bring forth new results from transport measurements on O-implanted DLC coatings, a-C films, and HOPG. Significant drop in the electrical resistance for DLC films is observed at ~50 K, which is the temperature previously reported for the transition to the SC state [9]. The metastability between the normal and the SC state observed below 50 K likely reflects the evolution of the amplitude for the SC order parameter, i.e., the longitudinal Higgs mode. Transitions to low-R states and metastability are also observed in the a-C films. Finally, the HOPG samples' resistance contain a thermally activated term that can be understood on the basis of the LAMH model applied to narrow SC channels in which thermal fluctuations can cause phase slips. In O-implanted carbon materials, the electron charge and spin correlations do not compete and together facilitate the emergence of HTSC. Thus, rather unexpected effects like Heisenberg spin waves and magneto-structural transitions are observed.

5. References

[1] Carbon-based superconductors: Towards High-T_c Superconductivity 2015 (Pan Stanford Publishing) edited by J. Haruyama
[2] Snider E *et al.*2020 *Nature* **586** 373
[3] Gheorghiu N, Ebbing CR, and Haugan TJ 2020 *arXiv*:2012.06624
[4] Felner I, Wolf O, and Millo O *2013 J. Supercond. Novel Magn.* **26** 511
[5] Pierce BT, Burke JL, Brunke LB, Bullard TJ, Vier DC, and Haugan TJ 2013 *IEEE Transactions on Applied Superconductivity* **23(3)** 7000205
[6] Kresin V, Litovchenko V, and Panasenko A 1975 *J. Chem. Phys.* **63** 3613
[7] Ojajärvi R, Hyart T, Silaev MA, and Heikkilä TT 2018 *Phys. Rev. B* **98** 054515
[8] Levitov L and Falkovich G 2016 *Nature Phys.* **12** 672
[9] Gheorghiu N, Ebbing CR, and Haugan TJ 2020 *arXiv*:2005.05876
[10] Gheorghiu N, Ebbing CR, and Haugan TJ 2020 *arXiv*:2005.07885
[11] Scheike T, Böhlmann W, Esquinazi P, Barzola-Quiquia J, Ballestar A, and Setzer A 2012 *Adv. Mater.* **24** 5826
[12] Precker CE *et al.* 2016 *New J. Phys.* **18** 113041
[13] Mũnoz WA, Covaci L, and Peeters FM 2013 *Phys. Rev. B* **87** 134509
[14] Gheorghiu N, Ebbing CR, Pierce BT, and Haugan TJ 2020 *IOP Conf. Ser.: Mater. Sci. Eng.* **756** 012022
[15] Graphtek, LLC
[16] Cutting Edge Ions, LLC
[17] Pierce BT 2013 *M.S. Thesis*, University of Cincinnati, Ohio
[18] Kauppila VJ, Hyart T, and Heikkilä TT 2016 *Phys. Rev. B* **93** 024505
[19] Yeshurun Y, Felner I, and Sompolinski H 1987 *Phys. Rev. B* **36(1)** 840
[20] Villegas KHA, Sun M, Kovalev VM, and Savenko IG 2019 *Phys. Rev. Lett.* **123** 095301
[21] Doron A, Levinson T, Gorniaczyk F, Tamir I, and Shahar D 2019 *arXiv*:1911.02818v1
[22] Tamir I *et al.* 2019 *Sci. Adv.* **5** eaau3826
[23] Kowal D and Ovadyabu Z 1994 *Solid State Communic.* **90(12)** 783
[24] Umeda M and Kato M 2019 *J. Phys. Commun.* **3** 105002
[25] Garcia N, Esquinazi P, Barzola-Quiquia J, and Dusari S 2012 *New Journal of Physics* **14** 053015
[26] Kulbachinskii VA 1991 *Synthetic Metals* **41-43** 2697
[27] Langer JS and Ambegaokar V 1967 *Phys. Rev.* **164** 498
[28] McCumber DE and Halperin BI 1970 *Phys. Rev. B* **1** 1054
[29] Shapira Y and Deutscher G 1983 *Phys. Rev. B* **27** 4463
[30] Barzola-Quiquia J, Esquinazi P, Rothermel M, Spemann D, Butz T, and Garcia N 2007 *Phys. Rev. B* **76**, 161403
[31] Garcia N and Esquinazi P 2009 *J. Supercond. Nov. Magn.* **22** 439

Acknowledgments

This work was supported by The Air Force Office of Scientific Research (AFOSR) for the LRIR #14RQ08COR & LRIR #18RQCOR100 and the Aerospace Systems Directorate (AFRL/RQ). T.J. Bullard is recognized for his initial contribution to this research. N. Gheorghiu acknowledges G.Y. Panasyuk for his continuous support and inspiration.

IOP Conf. Series: Materials Science and Engineering 1241 (2022) 012055 doi:10.1088/1757-899X/1241/1/012055

Microwave Characterization of Gamma Ray Irradiated Thin Film Embedded and Non-embedded Nb Resonators

Bhargav Yelamanchili[1], Vaibhav Gupta[1], Simin Zou[1], Robert Christiansen[1], Max C. Cichon[2], John A. Sellers[1], Minseo Park[2], Masoud Mahjouri-Samani[1], David B. Tuckerman[3] and Michael C. Hamilton[1]

[1] Electrical and Computer Engineering Department, Auburn University, AL-36849, USA
[2] Physics Department, Auburn University, AL-36849, USA
[3] Tuckerman & Associates, Inc., Lake Stevens, WA-98258, USA

E-mail: bzy0010@auburn.edu, mch0021@auburn.edu

Abstract. Gamma radiation effects on superconducting microwave transmission line structures, which may find use in future radiation challenging environments, such as satellites or accelerators have been investigated. Two versions of weakly coupled through-type Nb microstrip transmission line resonator material stack-ups were explored. In one version, the Nb signal trace was encapsulated with 20 µm of HD-4110 and in the other version the top Nb was not encapsulated. Exposure of the resonators to gamma radiation was performed for 28 days at room temperature in a sealed vacuum chamber using Co-60 as the gamma radiation source. The quality factors of the resonators were extracted at various cryogenic temperatures below the critical temperature of Nb and resonant frequencies up to 20 GHz. A large dose of gamma radiation used in this work showed a small change in the Nb superconducting properties.

1. Introduction

A crucial challenge when engineering a component for next-generation spacecraft is weight optimization. Reducing weight of the spacecraft allows for more functional equipment to be used in it, resulting in significant cost savings per launch. Given the possibility of taking really powerful equipment into space, we might be able to accomplish processing and compression of complex raw data reducing the data flow between space and ground stations. We believe that thin, flexible and superconducting cables can be an alternative to bulky coaxial cables, reducing the overall weight. For the use of these flexible cables in space, it is important to explore the impact of harsh environments such as high radiation up to 60 Mrad (600kGy) on the conductor and dielectric material. Earlier experiments on the influence of gamma rays on YBCO superconducting properties (critical temperature and critical current) have been reported in [1, 2]. As previously reported, the critical temperature and critical current density of thin superconducting films were not affected by gamma ray exposure up to a cumulative dose of 1.5 Mrad [3].

Various studies have been performed on the mechanical property degradation of polyimide films exposed on the surface of the spacecraft in space radiation environments such as electron, proton and UV rays [4, 5, 6] and extreme temperature conditions [7]. HD-4100 series polyimide products have found their greatest applicability in packaging applications. They have excellent mechanical properties to handle thermal and chemical extremes of post application processing, excellent elongation to prevent cracking, good adhesion and smooth via sidewalls to serves as a dielectric material.

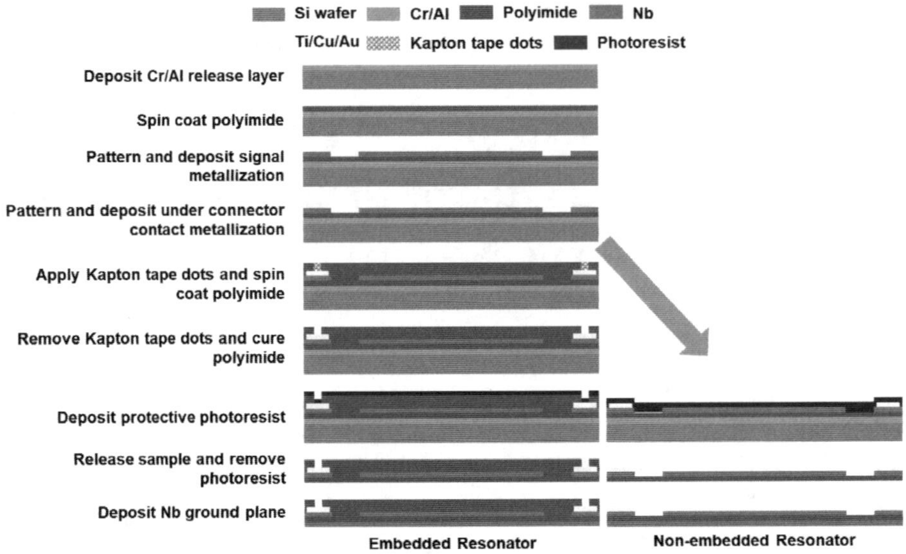

Figure 1. Schematic fabrication flow diagrams for embedded and non-embedded thin film superconducting Nb microstrip transmission line resonators.

Detailed study of the loss mechanisms is difficult due to the minimal loss of superconducting transmission lines of reasonable lengths. Instead, we looked at the quality factors of structurally identical resonators, which provide a sensitive probe of combined dielectric and superconductor losses. In this paper, we investigated gamma radiation effects on thin, flexible superconducting embedded and non-embedded resonators. The dielectric and conductor loss changes after 4 weeks of irradiation have been reported that show promising results to use these cables in high radiation conditions. We describe the design and fabrication procedure in section 2 followed by irradiation and measurement details in sections 3 and 4. Results are discussed in section 5 followed by conclusion in section 6.

2. Design and Fabrication Procedure

Half-wavelength capacitively-coupled microstrip resonators were designed and fabricated comparably to those in our prior work [8, 9, 10, 11]. The width of these thin transmission line resonators were designed for a 50 Ω characteristic impedance. Two versions were investigated in this work, embedded version, where signal layer was enclosed with another layer of polyimide and non-embedded version, where Nb signal layer was not enclosed. Fabrication of these resonators started with a deposition of 25 nm thick Cr and 200 nm thick Al as a sacrificial layer on a couple of oxidized silicon wafers, which act as temporary substrates during fabrication. A photodefinable polyimide from HD Microsystems, HD-4110 was spun on the wafers and cured at 375 °C for one hour to acquire a desired thickness of 20 μm. Wafers were then patterned for signal layer deposition using standard photolithography process and 250 nm thick Nb was sputter deposited followed by a typical lift-off process. Nb was sputter deposited for 30 min with an Ar pressure of 4 mTorr (5.33 mbar) [12]. As a subsequent fabrication process, Ti(50 nm)/Cu(500 nm)/Au(10 nm) deposition was done on the wafers for metallization of connector contacts, followed by lift-off. After this step, the wafer with non-embedded version of samples was put in a dry box until the embedded version wafer had another layer of HD-4110 polyimide spin coated and cured at a lower temperature of 225 °C to protect the superconductivity of embedded Nb, resulting in a 20 μm thick layer after curing. Kapton dots were placed on the connector contact metallization area before spinning the second layer of polyimide on the embedded version samples and were removed before curing. The samples on both the wafers were then protected with a layer of photoresist before releasing in a salt

Figure 2. Experimental irradiation set-up showing (left to right) ^{60}Co source in a deep pool of water when not in use, sample holder with resonator samples and cylindrical chamber into which the sample holder was lowered and held during irradiation.

Table 1. Total absorbed dose for 4 weeks

Material	μ_{en}/ρ (cm$^2/g$)	$D_{material}$ (rad/s)	Total dose (Mrad (kGy))
Air, dry	2.67E-2	23.6	56.4 (564)
Noibium	2.40E-2	21.2	50.7 (507)
Polyimide	2.88E-2	25.5	60.8 (608)

solution with 0.5 V applied to the Cr/Al release layer. After releasing and stripping photoresisit, the samples were inverted and mounted onto a new Si wafer and was followed by a deposition of 250 nm thick Nb ground plane layer. Fig. 1 shows a schematic drawing of the steps involved in fabrication of embedded and non-embedded Nb microstrip transmission line resonators fabricated using HD-4110 polyimide.

3. Irradiation Details

The irradiation of resonator samples was carried out using cobalt-60 as radiation source. The radiation source consisted of eighteen 43.2 cm tall rods. Each rod was encased in stainless-steel and was kept at a radial distance of 10 cm with 20° arc separation between centers of adjoining rods. The samples were under constant exposure of 1.17 Mev and 1.33 Mev gamma photons during the irradiation, commonly employed in total-dose testing for space research [13]. The sample holder with resonator samples mounted radially normal to the gamma source was lowered into a cylindrical chamber and sealed. The radiation source was lifted from its containment pool to surround the resonator samples in the cylindrical vacuum chamber at ambient temperature. During irradiation, a constant vacuum pressure of 1E-6 Torr was maintained in the cylindrical chamber. Sodium sulfide ion chamber dosimetry system was used to measure the dose rate to air at the chamber center, which was 23.6 rad/s. This dose rate was corrected to different materials based on their mass energy-absorption coefficients for 1.25 Mev photons from National Institute of Standards and technology [14]. Absorbed dose or dose rate to any material can be calculated from the absorbed dose to air as:

$$D_{material} = D_{air}\frac{(\mu_{en}/\rho)_{material}}{(\mu_{en}/\rho)_{air}} \tag{1}$$

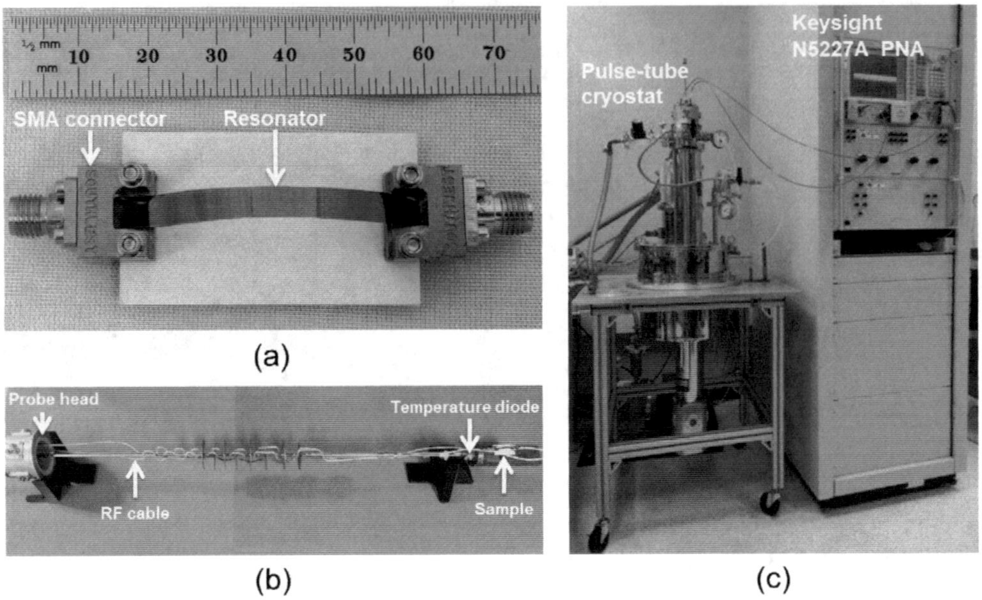

Figure 3. (a) Assembled resonator sample with edge launch SMA connectors and support board, (b) Cryogenic sample probe with resonator sample and RF coaxial cables used for measurements in a (c) pulse tube base closed-cycle cryostat using a Keysight performance network analyzer.

where $D_{material}$ is dose rate to specified material, D_{air} is dose rate to air, $(\mu_{en}/\rho)_{material}$ and $(\mu_{en}/\rho)_{air}$ is mass energy-absorption coefficient at photon energy of interest for specified material and air, respectively. Table 1 gives the total absorbed dose in each material. The absorbed dose values are very high compared to the radiation levels absorbed by electronics on satellites in a geosynchronous orbit in their typical lifetime, which is 3 kGy for 10 years [15]. Fig. 2 shows the gamma radiation source, sample holder with the resonator samples and vacuum chamber. After 4 weeks of irradiation, the radiation source was lowered back into its safe water storage pool, the sample holder was pulled out and the resonator samples were removed.

4. Measurement Details

The resonator samples were baked at 90 °C for 2 hours before each measurement to reduce any humidity effects below detectable levels, which cause shifts in resonance frequency and decrease in measured quality factors at cryogenic temperatures as seen in our previous studies [10]. Southwest Microwave's edge launch SubMiniature version A (SMA) connectors were used to assemble the resonator samples. These heavy SMA connectors may cause damage to the thin flexible cables, so as a prevention and to provide stability, they were mounted on to a support board. A cryogenic sample probe with assembled resonator sample and cryogenic semi rigid RF coaxial cables, which were used to interface the resonator sample through SMA connectors, was lowered into a pulse tube based closed-cycle cryostat system to carry out the measurements at various cryogenic temperatures. The sample temperature was measured using a temperature diode located near the sample on the sample holder. The scattering parameters were measured using a Keysight N5227A performance network analyzer across a substantially wide frequency range around the resonant peaks. Fig. 3 shows the resonator sample with edge launch SMA connectors and support board, cryogenic probe used for measurements inside a pulse-tube based close-cycle cryostat and Keysight PNA. An incident power of -28$\pm$2 dBm at the sample was used in this work

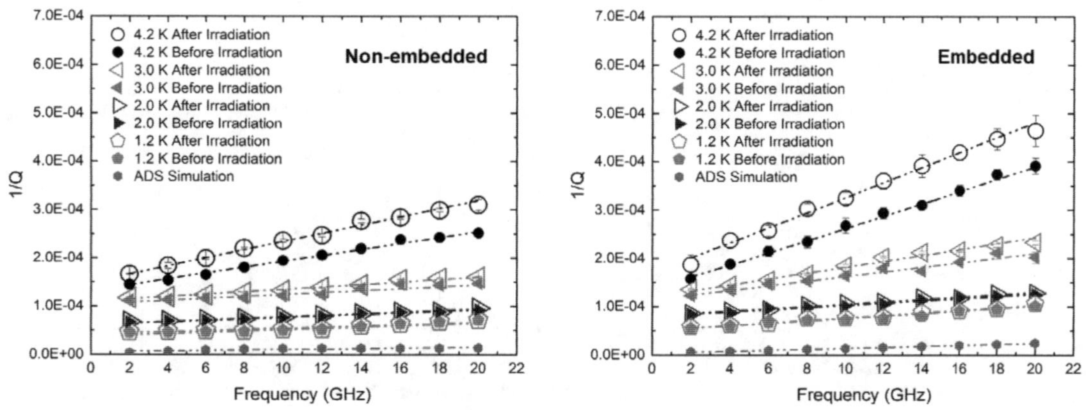

Figure 4. Comparison of $1/Q$ vs. resonant frequency for non embedded and embedded resonators before and after 4 weeks of irradiation. Plots show a slightly larger difference in $1/Q$ values indicating increase in loss at higher temperatures (4.2 K and 3.0 K) compared to difference in values at 2.0 K and 1.2 K after 4 weeks of irradiation.

based on the measurement of similar resonators from our previous work [10], which showed reduction in measured quality factors and non-linearity effects for microwave powers above -25 dBm incident at the sample. After four weeks of irradiation, the resonator samples were reassembled with edge launch SMA connectors within an hour of being removed from the radiation chamber and re-measured using the same measurement approach and set-up.

5. Results and Discussion

The scattering parameters S_{12} and S_{21} were measured up to 20 GHz at various cryogenic temperatures (4.2 K, 3.0 K, 2.0 K, 1.2 K) below critical temperature (T_c). Loaded quality factors (Q_l) of the resonators, which were obtained by fitting the measured S-parameters using a Lorentzian function [16], were compared before and after four weeks of irradiation. Fig. 4 shows $1/Q_l$ versus measured resonant frequencies of the embedded and non-embedded resonators before and after 4 weeks of irradiation. An average of 10 measurements at each frequency were taken and 3σ error bars were added to consider the measurement errors and temperature drifts inside the closed-cycle cryostat. To offer an estimate of the sufficiently low coupling loss at each resonant frequency, ADS simulation data with zero conductor loss and zero dielectric loss have been added. According to BCS theory, there is a reduction of quasi-particles and an exponential increase in superconducting charge carriers as the sample is cooled further below its T_c leading to a reduction in microwave losses. This is consistent with continuous reduction of slope, which corresponds to quasiparticle-induced losses as temperature reduces from 4.2 K to 1.2 K. At 1.2 K, the slope nearly approaches zero indicating that the loss is mostly influenced by the dielectric loss, though not completely. Variance in the plots of embedded and non-embedded resonator samples before irradiation is likely due to the additional embedding fabrication process. The authors believe that solvent and moisture may still remain in lower temperature cured HD-4110, which could explain the variance seen in the embedded samples.

Comparing the plots of extracted Q_l values before and after four weeks of irradiation, it's evident that there is a slight increase in losses at 4.2 K compared to other lower temperatures for both resonator versions indicating no detectable changes in dielectric properties of polyimide at lower cryogenic temperatures. A reasonable approximation of the loss contributors for microstrip resonators can be given as:

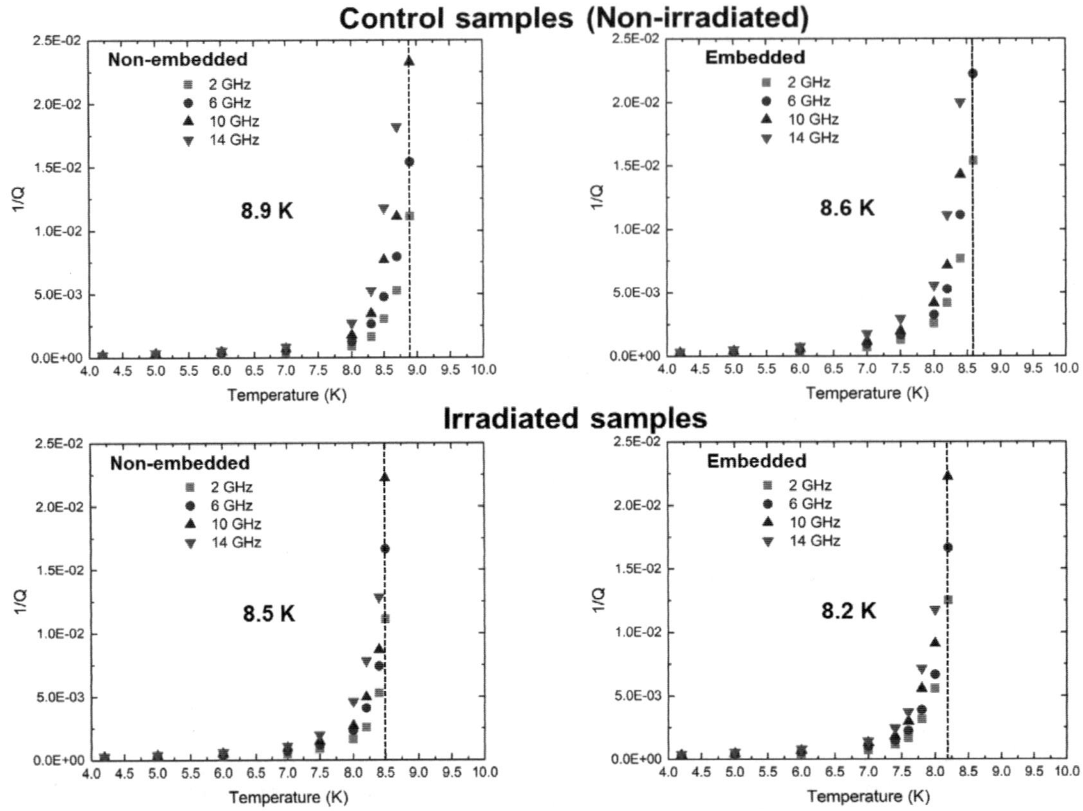

Figure 5. $1/Q$ vs. temperature for non embedded and embedded resonators before and after 4 weeks of irradiation at different frequencies. Plots show decrease in T_c for both resonator versions after 4 weeks of gamma irradiation. The Q_l values decrease close to T_c as a result of increase in normal electron density in the superconductor.

$$\frac{1}{Q_l} = \frac{1}{Q_c} + \frac{1}{Q_d} + \frac{1}{Q_r} + \frac{1}{Q_{coupling}} \tag{2}$$

where Q_c, Q_d, Q_r and $Q_{coupling}$ are the Q-factors associated with conductor loss, dielectric loss, radiation loss and coupling loss, respectively. The radiation loss for these microstrip resonators using equations from [17, 18] was negligible and the coupling loss extracted from the ADS simulations with zero conductor and dielectric loss was also negligible. Thus, equation 2 can be reduced to:

$$\frac{1}{Q_l} = \frac{1}{Q_c} + \frac{1}{Q_d} \tag{3}$$

It is feasible to separate conductor loss and dielectric loss contribution by inspecting the extracted Q_l as a function of temperature and frequency. The Q_l measurement of the resonators was continued above 4.2 K to estimate the T_c of Nb conductor. Control samples, which were fabricated alongside the irradiated resonator samples, were used to estimate the T_c before irradiation. Fig. 5 shows $1/Q_l$ versus temperature for non-embedded and embedded control samples and four-week irradiated samples. At temperatures close to T_c, the normal electron density increases, thus leading to a decrease in the Q_l

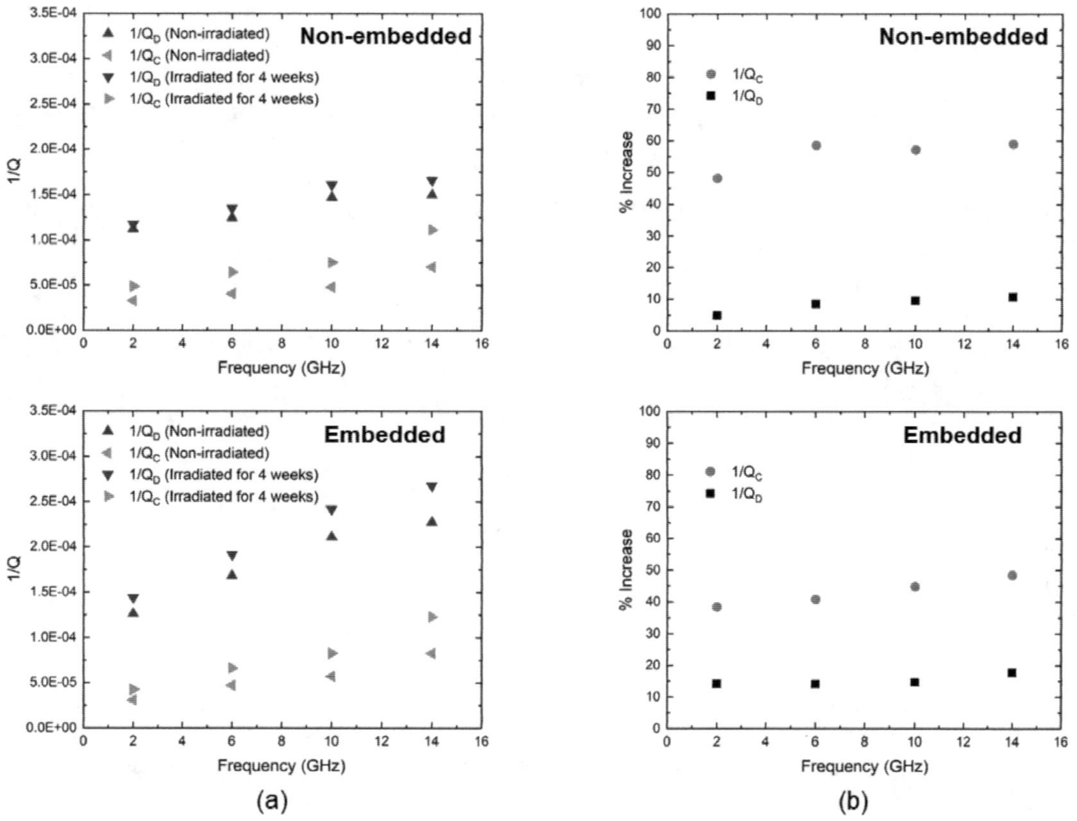

Figure 6. (a) Loss from dielectric ($1/Q_d$) and conductor ($1/Q_c$) vs. frequency for non-embedded (top) and embedded (bottom) resonator samples. (b) Percent increase in dielectric loss and conductor loss after 4 weeks of irradiation. Plots indicate higher relative conductor loss contribution for the difference in the extracted Q_l values at 4.2 K after 4 weeks of gamma irradiation and non-irradiated resonator samples.

values. Measurements were carried out until the Q_l values of the fundamental resonance frequency peak dropped below a reasonably extractable range. It can be seen that the estimated T_c of the embedded sample is less than the non-embedded sample before irradiation. This may be due to the slight degradation on the Nb surface at the conductor and embedded polyimide layer interface, even with a lower temperature curing process. Using the T_c information and characterizing the losses as a function of temperature and frequency under the influence of an external force, we have come up with a way to separate the conductor and dielectric loss in our superconducting resonators, which is part of a pending publication. Fig.6 shows the extracted conductor and dielectric losses and percent increase in the losses for all resonator samples versus frequency at 4.2 K. The plots show a higher increase in conductor loss compared to dielectric loss indicating a slight degradation in thin conductor film for both resonator versions. The results point to a small change in Q_d, which is directly related to dielectric loss factor (tan δ) indicating small changes in the dielectric properties even after subjecting to very high gamma radiation dosages (608 kGy).

6. Conclusion

For large radiation doses, up to 50.7 Mrad (507 kGy) for Nb superconductor and 60.8 Mrad (608 kGy) for polyimide film, show an average of 50% increase in conductor loss and very small increase in the dielectric losses at 4.2 K for both embedded and non-embedded resonator samples. The Q_l value measurements as a function of temperature assisted us to estimate the T_c of various resonator versions. At lower temperatures, undetectable change in the extracted Q_l values was noticed indicating minimal changes in the dielectric properties of the polyimide (HD-4110). Experimental data presented here provide a good encouragement to use these thin, flexible, low mass superconducting cables in radiation challenging environments, such as space.

7. References

[1] Bohandy J, Suter J, Kim B, Moorjani K and Adrian F 1987 *Applied physics letters* **51** 2161–2163
[2] Luo L, Zhang Y, Hu S, Liu W, Zhang G and Hu W 1991 *Physica C: Superconductivity* **178** 11–14
[3] Cooksey J, Brown W, Ang S, Naseem H, Ulrich R and West L 1994 *IEEE transactions on nuclear science* **41** 2521–2524
[4] Russell D A, Connell J W and Fogdall L B 2002 *Journal of spacecraft and rockets* **39** 833–838
[5] Shimamura H and Yamagata I 2009 *Journal of spacecraft and rockets* **46** 15–21
[6] Shen Z, Mu Y, Ding Y, Liu Y and Zhao C 2016 *Selected Papers of the Photoelectronic Technology Committee Conferences held November 2015* vol 9796 (International Society for Optics and Photonics) p 97960U
[7] Plis E A, Engelhart D P, Cooper R, Johnston W R, Ferguson D and Hoffmann R 2019 *Applied Sciences* **9** 1999
[8] Zou S, Cao Y, Gupta V, Yelamanchili B, Sellers J A, Ellis C D, Tuckerman D B and Hamilton M C 2017 *Applied Superconductivity* **27** 1–5
[9] Zou S, Cao Y, Hernandez G, Bai R, Gupta V, Sellers J A, Ellis C D, Tuckerman D B and Hamilton M C 2017 *MRS Advances* **2** 2199–2204
[10] Bai R 2016 *Superconducting resonators on thin film flexible substrates* PhD dissertation Auburn University
[11] Yelamanchili B, Gupta V, Zou S, Christiansen R, Cichon M C, Sellers J A, Park M, Mahjouri-Samani M, Tuckerman D B and Hamilton M C 2021 *IEEE Transactions on Applied Superconductivity* **31** 1–5
[12] Gupta V, Sellers J A, Ellis C D, Yelamanchili B, Zou S, Cao Y, Tuckerman D B and Hamilton M C 2017 *Proc. IMAPS 13th Int. Conf. Exhib. Device Packag.* vol 2017 pp 1–25
[13] Winokur P, Kerris K and Harper L 1983 *IEEE Transactions on Nuclear Science* **30** 4326–4332
[14] X-ray mass attenuation coefficients https://www.nist.gov/pml/x-ray-mass-attenuation-coefficients accessed January 2021
[15] Cressler J D and Mantooth H A 2017 *Extreme environment electronics* (CRC Press)
[16] Petersan P J and Anlage S M 1998 *Journal of applied physics* **84** 3392–3402
[17] Lewin L 1960 *Proceedings of the IEE-Part C: Monographs* **107** 163–170
[18] Sobol H 1972 *IEEE Transactions on Parts, Hybrids, and Packaging* **8** 59–66

Acknowledgment

We gratefully acknowledge funding and technical guidance from Microsoft regarding the flexible superconducting resonators used in this work. We thank the Alabama Micro/Nano Science and Technology Center (AMNSTC) for providing access to fabrication and characterization facilities used in this work.

AUTHOR INDEX

Institute of Physics
Dirac House, Temple Back
Bristol BS1 6BE UK

ISSN: 1757-8981
ISBN 978-1-7138-5809-6

9 781713 858096

CURRAN ASSOCIATES INC.
proceedings.com